Improving Crop Quality and Enhancing Sustainability in Agriculture

Agriculture is facing unprecedented challenges due to climate change, resource depletion, and the growing global population. *Improving Crop Quality and Enhancing Sustainability in Agriculture* presents cutting-edge technologies and practical solutions providing information on sustainable agricultural practices. Edited by Dr. Athar Mahmood, Dr. Muhammad Mansoor Javaid, and Dr. Muhammad Ather Nadeem, the book explores sustainable approaches to improving crop quality while preserving the environment.

This book delves into topics including precision farming, biotechnology, and nanotechnology, and shows how these technologies are transforming agricultural practices. It also highlights organic farming, regenerative agriculture, and eco-friendly pest control methods that offer sustainable alternatives to conventional approaches.

A key focus of the book is the role of healthy soil and nutrient management in improving crop quality. It features information on advanced irrigation techniques, biofertilizers, organic soil amendments, and innovative seed treatments that help crops thrive under challenging conditions. Additionally, the book discusses sustainable fiber production and the repurposing of agricultural waste for bioethanol production, contributing to a more circular agricultural economy.

As the agricultural landscape evolves, *Improving Crop Quality and Enhancing Sustainability in Agriculture* emphasizes the importance of climate-smart farming methods to adapt to climate change and mitigate the impacts of extreme weather conditions such as droughts, heatwaves, and unpredictable rainfall. With contributions from leading scholars and practitioners, this book serves as a vital resource for researchers, agronomists, policymakers, and farmers who are committed to adopting sustainable solutions in their work.

Improving Crop Quality and Enhancing Sustainability in Agriculture

Edited by

Athar Mahmood, Muhammad Mansoor Javaid, and Muhammad Ather Nadeem

CRC Press
Taylor & Francis Group
Boca Raton London New York

CRC Press is an imprint of the
Taylor & Francis Group, an informa business

Designed cover image: Shutterstock

First edition published 2026
by CRC Press
2385 NW Executive Center Drive, Suite 320, Boca Raton FL 33431

and by CRC Press
4 Park Square, Milton Park, Abingdon, Oxon, OX14 4RN

CRC Press is an imprint of Taylor & Francis Group, LLC

ISBN: 978-1-032-79731-1 (hbk)
ISBN: 978-1-032-81222-9 (pbk)
ISBN: 978-1-003-49869-8 (ebk)

DOI: 10.1201/9781003498698

Typeset in Times
by KnowledgeWorks Global Ltd.

Contents

Preface ix
About the Editors xi
List of Contributors xiii

Chapter 1 Introduction 1

Muhammad Mansoor Javaid, Muhammad Ather Nadeem, and Athar Mahmood

Chapter 2 Precision Agriculture for Sustainable Crop Management 20

Asma Zafar, Muhammad Saad Ullah, Aqsa Naseem, Aqsa Hanif, Atman Adiba, Muhammad Mansoor Javaid, Javaria Nargis, and Faiza Fareed

Chapter 3 Biotech Solutions for Crop Production 37

Ayesha Arshad, Abrar Hussain, Javaria Nargis, Zeki Erden, Çağdaş Can Toprak, Hakki Akdeniz, and Jameel Mohammed Al-Khayri

Chapter 4 Organic Farming Opportunities and Challenges 54

Maira Tanveer, Abrar Hussain, Muhammad Ather Nadeem, Aqsa Hanif, Ali Raza, Muhammad Awais, and Romana Iftikhar

Chapter 5 Revolutionizing Agriculture through Drones and Robotics 71

Muaz Ameen, Athar Mahmood, Muhammad Zeeshan Hanif, Muhammad Anjum Zia, Muhammad Junaid Nisar, and Muhammad Awais

Chapter 6 Nanotechnology in Crop Production 90

Filza Ishfaq, Muhammad Mansoor Javaid, Muhammad Ather Nadeem, Muhammad Zubair, Muhammad Anjum Zia, Numra Akbar, and Javaria Nargis

Chapter 7 Cutting-Edge Approaches to Weed Management 105

Noor ul Ain, Muhammad Mansoor Javaid, Muhammad Ather Nadeem, Filza Ishfaq, Muhammad Anjum Zia, Muhammad Zubair, Numra Akbar, and Javaria Nargis

Chapter 8 Use of Modern Approaches to Mitigate Challenges of Resistance 123

Maira Tanveer, Alia Javed, Muhammad Ather Nadeem, Zeki Erden, Çağdaş Can Toprak, Javaria Nargis, and Laiba Sajid

Chapter 9 Revitalizing Irrigation, Soil Moisture Management in Dryland Agriculture 141

Filza Ishfaq, Muhammad Mansoor Javaid, Babar Ali, Rashad Mukhtar Balal, Ikram ul Haq, and Muhammad Awais

Chapter 10 Implications and Potential of Marginal Land 161

Amna Ghafoor, Safura Bibi, Hina Nazir, Muhammad Talha Bin Yousaf, Muhammad Mansoor Javaid, and Muhammad Ather Nadeem

Chapter 11 Impact of Climate Change on Plant Physiology 180

Muhammad Saad Ullah, Umair Gull, Jameel Mohammed Al-Khayri, Umar Raza, and Ali Mardan

Chapter 12 Modernizing Irrigation Management 198

Humma Anwar Rana, Abrar Hussain, Rizwana Kanwal, Zahid Ali, and Basra Manzoor

Chapter 13 Integrated Approaches for Sustainable Forage Production Systems 211

Umair Gull, Sher Afghan, Asif Iqbal, Muhammad Abdullah Saleem, Athar Mahmood, Hakki Akdeniz, and Ayesha Arshad

Chapter 14 Sustainable Fiber Production 228

Usman Tufail, Abrar Hussain, Rabia Zahid, and Ikram ul Haq

Chapter 15 Cutting-Edge Strategies for Environmental Stress: An Overview 245

Abdul Jabbar, Saima Asghar, Athar Mahmood, Muhammad Mazhar Iqbal, Rimsha Kanwal, and Komal Rehman

Chapter 16 Recent Advances in Seed Enhancement 265

Asma Zafar, Umair Gull, Ayesha Arshad, Iqra Anwar, and Muhammad Mansoor Javaid

Chapter 17 Contemporary Nutrient Management 281

Asif Mukhtiar, Wasim Javed, and Ayman El Sabagh

Chapter 18 Cultivating Resilience: Crop Adapting to Elevated CO_2 and Temperature 300

Fouzia Shabbir, Singarayer Florentine, Muhammad Mansoor Javaid, Zeki Erden, Çağdaş Can Toprak, and Hakki Akdeniz

Chapter 19 Farming Systems for Sustainable Resource Utilization 319

Rizwan Maqbool, Nazma Khan, Muhammad Ather Nadeem, Athar Mahmood, and Muhammad Junaid Nisar

Chapter 20 Eco-Friendly Pest Management 336

Ayesha Sadeeqa Muhammad, Muhammad Ahmad Zeshan, Sami Ullah, and Muhammad Anjum Zia

Chapter 21 Bioethanol Production from Agricultural Waste 354

Muhammad Saad Ullah, Sakina Bibi, Athar Mahmood, Muhammad Anjum Zia, Muhammad Awais, and Javaria Nargis

Chapter 22 Nutrients Grow through Organic Soil Enrichment: Revitalizing Soil Health through Organic Nutrient Enrichment 374

Asma Zafar, Ayesha Arshad, Wasim Javed, Muhammad Awais, Javaria Nargis, and Ayman El Sabagh

Chapter 23 Regenerative Agriculture 399

Ghashia Ameen, Muaz Ameen, Adeel Shahid, Athar Mahmood, Muhammad Bilal Goheer, and Muhammad Zubair

Chapter 24 Use of Internet of Things for Modern Agriculture 414

Muaz Ameen, Rizwan Maqbool, Athar Mahmood, and Muhammad Bilal Goheer

Index 437

Chapter 21 Bioethanol Production from Agricultural Waste [illegible]

[illegible]

Chapter 22 [illegible] Soil Health through Organic Manure [illegible]

[illegible]

Chapter 23 Regenerative Agriculture [illegible]

[illegible]

Chapter 24 Use of [illegible] Sustainable Agriculture [illegible]

[illegible]

Index [illegible]

Preface

Agriculture is at a turning point, facing both significant challenges and remarkable opportunities. As climate change intensifies, natural resources become scarcer, and the global population continues to grow, the agricultural sector must adapt to ensure food security while preserving the environment. The need for sustainable agricultural practices has never been greater. This book, *Improving Crop Quality and Enhancing Sustainability in Agriculture*, brings together the latest knowledge, innovative techniques, and cutting-edge technologies aimed at addressing these pressing concerns.

Sustainability in agriculture is about more than just increasing crop yields. It involves protecting natural resources, ensuring economic stability, and supporting farming communities. As the demand for food rises, farmers and researchers must adopt strategies that not only improve productivity but also reduce environmental damage and conserve resources. This book explores a range of modern approaches, including precision farming, biotechnology, nanotechnology, and the Internet of Things (IoT), all of which are transforming the way we cultivate crops. Additionally, organic farming, regenerative agriculture, and eco-friendly pest control are examined as sustainable alternatives to conventional agricultural methods.

Healthy soil and proper nutrient management play a crucial role in improving crop quality. This book highlights modern irrigation techniques, the use of biofertilizers, organic soil amendments, and advanced seed treatments that enhance plant resilience and productivity. It also explores sustainable fiber production and how agricultural waste can be repurposed for bioethanol production, contributing to a more circular economy by reducing waste and making better use of available resources.

One of the biggest challenges modern farmers face is weed and pest resistance. The overuse of chemical pesticides and herbicides has led to resistance problems, making it harder to protect crops effectively. This book presents innovative weed control strategies and eco-friendly pest management techniques that provide long-term, sustainable solutions. Additionally, as climate change continues to disrupt traditional farming practices, adapting to these changes becomes essential. Climate-smart farming methods discussed in this book include strategies to mitigate environmental stressors and protect crops from extreme weather conditions such as droughts, heatwaves, and unpredictable rainfall.

This book is designed to serve a broad audience, including researchers, agronomists, policymakers, extension workers, and farmers who are eager to embrace modern, sustainable agricultural solutions. It combines scientific knowledge with practical applications, making it a useful guide for those looking to implement new methods in their work. Each chapter provides clear explanations, real-world case studies, and insights into future developments, ensuring that readers gain a deeper understanding of sustainable farming practices and their benefits.

The importance of sustainable agriculture cannot be overstated. With the increasing demand for food and the need to protect the planet, adopting environmentally friendly practices is not just a choice but a necessity. *Improving Crop Quality and Enhancing Sustainability in Agriculture* serves as a valuable resource for understanding and implementing sustainable agricultural methods. By encouraging knowledge-sharing and the adoption of modern techniques, this book aims to contribute to the development of a more resilient and productive agricultural system that will benefit both present and future generations.

As agricultural systems continue to evolve, it is vital that we embrace innovative solutions that balance productivity with sustainability. By integrating advanced technologies, adopting climate-smart practices, and fostering collaboration among researchers, policymakers, and farmers, we can develop agricultural systems that are both productive and environmentally responsible. This book provides insights into these evolving trends, offering a roadmap for achieving sustainability in agriculture.

We hope this book serves as a tool to inspire meaningful change. Whether you are a scientist seeking to develop new agricultural technologies, a farmer looking for practical solutions, or a policymaker shaping agricultural policies, the knowledge shared within these pages can help drive positive transformation. By working together, we can ensure a future where food production remains efficient, sustainable, and resilient in the face of global challenges.

We extend our heartfelt gratitude to the contributors whose expertise and dedication have enriched this volume. Their research and insights have played a crucial role in shaping this book, making it a significant contribution to the field of sustainable agriculture. We also acknowledge the invaluable support of institutions, reviewers, and the farming community, whose involvement has been instrumental in bringing this work to fruition.

We encourage readers to explore the ideas presented in this book, apply the knowledge in their agricultural practices, and take part in the collective effort to build a more sustainable agricultural future. Through collaboration and innovation, we can ensure a thriving food system that meets the needs of generations to come while safeguarding our planet's natural resources.

Editors
Athar Mahmood
Muhammad Mansoor Javaid
Muhammad Ather Nadeem

About the Editors

Dr. Athar Mahmood is working as an associate professor in the Department of Agronomy at the University of Agriculture, Faisalabad, Pakistan. He earned his BSc in agriculture (Hons.) and MSc in agronomy (Hons.) from the University of Agriculture, Faisalabad, Pakistan. In 2012, he completed his PhD with a dissertation on "Performance of Sorghum as a Bioenergy Crop for Biogas Production" at Justus Liebig University, Giessen, Germany, under a scholarship sponsored by the Higher Education Commission (HEC), Islamabad, Pakistan.

In 2016, he completed his postdoctoral research at the University of Southern Queensland (USQ), Toowoomba, Australia under the prestigious Endeavour Fellowship Program, sponsored by the Government of Australia. Subsequently, he became a visiting academic at the Institute of Agriculture and the Environment, USQ, Australia.

In 2019, Dr. Athar Mahmood completed a fellowship at the Yantai Institute, China Agricultural University (CAU), China under the Talented Young Scientist Program (TYSP) awarded by the Government of China.

He has published more than 180 peer-reviewed research and review articles, as well as book chapters. He has also served as a reviewer for well-reputed journals published by Springer and Elsevier.

Dr. Muhammad Mansoor Javaid is an associate professor of agronomy at the University of Sargodha, Pakistan. He has been serving in this role since 2021, contributing significantly to teaching, research, and academic leadership. He joined the University of Sargodha in 2012 as an assistant professor and has played an active role in advancing agronomic research and education.

Dr. Javaid earned his PhD in agronomy from the University of Agriculture Faisalabad, Pakistan. He completed his master's and bachelor's degrees at Bahauddin Zakariya University, Multan, Pakistan. His outstanding academic performance was recognized when he was awarded a prestigious gold medal for securing the first position in his master's degree.

His dedication to research and innovation led him to receive the highly competitive Endeavour Award from the Australian government for a postdoctoral fellowship. He successfully completed his postdoctoral research at Federation University, Australia. Additionally, he undertook a six-month fellowship at the University of Florida, Department of Horticultural Sciences, in Gainesville, USA, further expanding his expertise in agricultural sciences.

Dr. Javaid has an extensive research background, having authored more than 135 peer-reviewed research articles and multiple book chapters. His research interests focus on agronomy, weed science, allelopathy, dryland agriculture, and stress physiology. His contributions have been widely recognized, and he continues to play a vital role in advancing knowledge in these fields.

As a dedicated mentor, Dr. Javaid has supervised 34 postgraduate students as either a supervisor or co-supervisor, guiding them in their research and academic growth. His commitment to scientific research is evident in the successful acquisition of research grants for eight projects funded by both national and international agencies.

Dr. Javaid actively participates in national and international conferences, seminars, and workshops, contributing to the global exchange of agricultural knowledge. He has collaborated with renowned institutions such as Lanzhou University, China through the Pak-China Joint Research Center for Dryland Agroecology, where he represented Pakistan in international research projects.

His leadership extends to academic administration, as he serves as a member of the Board of Faculty of Agriculture and the Board of Studies in the Department of Agronomy at the University of Sargodha. His involvement in these roles ensures the development of updated and relevant curricula that align with modern agricultural advancements.

Dr. Javaid has also played a significant role in shaping research strategies, guiding students and fellow researchers toward solutions for critical agricultural challenges. His work has contributed to sustainable farming practices, improved crop resilience, and innovative approaches to weed management, making a lasting impact on the field.

His contributions have been acknowledged by various academic and professional bodies, further highlighting his influence in agronomy. His commitment to fostering scientific collaborations and knowledge dissemination has established him as a highly regarded figure in agricultural sciences.

Dr. Muhammad Ather Nadeem is a distinguished professor of agronomy at the University of Sargodha, Pakistan. He earned his PhD in agronomy from the University of Agriculture, Faisalabad, Pakistan in 2006. To further enhance his expertise, he pursued postdoctoral research in Weed Science at the University of Arkansas, USA through a prestigious Fulbright Fellowship from 2012–2013.

Dr. Nadeem's academic journey is marked by remarkable achievements in both teaching and research. With over 50 MPhil theses and seven PhD dissertations under his supervision, Dr. Nadeem has played a pivotal role in shaping the future of agronomy in Pakistan. His extensive research on weed management, herbicide resistance, and sustainable agricultural practices has earned him recognition in global agricultural research communities.

As a dedicated educator, Dr. Nadeem has contributed significantly to the academic landscape by teaching various undergraduate and graduate-level courses in agronomy at both the University of Sargodha and the University of Agriculture, Faisalabad. His passion for fostering young talent is evident from his supervision of numerous MSc and PhD students, many of whom have gone on to make their own contributions to the field.

Dr. Nadeem's academic excellence has been acknowledged through several awards, including a Research Productivity Award in 2012 and the honor of being named a Fulbright Fellow. His extensive body of work has been published in high-impact journals, and he has actively participated as an editor for various prestigious academic journals. His editorial roles further emphasize his leadership and influence in the scientific community.

In addition to his academic contributions, Dr. Nadeem has taken on several leadership roles within his institution. He served as principal of the College of Agriculture at the University of Sargodha from 2019–2022 and currently holds the position of Dean of the Faculty of Agriculture. As chairman of the Department of Agronomy, he has made significant contributions to the development of the department, further enhancing its reputation both nationally and internationally.

Dr. Nadeem has been a member of several important academic bodies, including Academic Council of the University of Sargodha, and has served on various committees focused on curriculum development, student welfare, and research funding. His leadership has greatly influenced the university's strategic direction in research and academic excellence.

Through his unwavering commitment to agronomy, Dr. Nadeem has advanced agricultural research and education, contributing immensely to the scientific community and to the development of sustainable agricultural practices.

Contributors

Atman Adiba
Regional Agricultural Research Centre of Meknes
National Institute of Agriculture Research
Avenue Ennasr
Rabat, Morocco

Sher Afghan
Department of Botany
University of Agriculture Faisalabad
Faisalabad, Pakistan

Noor ul Ain
Department of Botany
University of Agriculture Faisalabad
Faisalabad, Pakistan

Numra Akbar
Department of Botany
University of Agriculture Faisalabad
Faisalabad, Pakistan

Hakki Akdeniz
Department of Field Crops, Faculty of Agriculture
Iğdır University
Iğdır, Turkey

Babar Ali
Department of Botany
University of Agriculture Faisalabad
Faisalabad, Pakistan

Zahid Ali
Department of Botany
University of Agriculture Faisalabad
Faisalabad, Pakistan

Jameel Mohammed Al-Khayri
Department of Agricultural Biotechnology
College of Agriculture and Food Sciences
King Faisal University
Hofuf, Saudi Arabia

Ghashia Ameen
Department of Chemistry
University of Agriculture Faisalabad
Faisalabad, Pakistan

Muaz Ameen
Department of Botany
University of Agriculture Faisalabad
Faisalabad, Pakistan

Iqra Anwar
Department of Botany
University of Agriculture Faisalabad
Faisalabad, Pakistan

Ayesha Arshad
Department of Botany
University of Agriculture Faisalabad
Faisalabad, Pakistan

Saima Asghar
Department of Botany
University of Agriculture Faisalabad
Faisalabad, Pakistan

Muhammad Awais
College of Agriculture
Bahauddin Zakariya University, Bahadur Sub-Campus
Layyah, Pakistan

Rashad Mukhtar Balal
Department of Horticulture
College of Agriculture
University of Sargodha
Sargodha, Pakistan

Safura Bibi
Department of Botany
Faculty of Sciences
University of Agriculture Faisalabad
Faisalabad, Pakistan

Sakina Bibi
Department of Botany
University of Agriculture Faisalabad
Faisalabad, Pakistan

Zeki Erden
Department of Field Crops, Faculty of Agriculture
Siirt University
Siirt, Turkey

Faiza Fareed
Department of Botany
University of Agriculture Faisalabad
Faisalabad, Pakistan

Singarayer Florentine
The Future Regions Research Centre, Institute of Innovation, Science and Sustainability
Federation University Australia
Ballarat, Victoria, Australia
and
Applied Chemistry and Environmental Science School of Science
STEM College, RMIT University
Melbourne, Victoria, Australia

Amna Ghafoor
Department of Botany
Faculty of Sciences
University of Agriculture Faisalabad
Faisalabad, Pakistan

Muhammad Bilal Goheer
Institute of Animal and Dairy Sciences
Faculty of Animal Husbandry
University of Agriculture Faisalabad
Faisalabad, Pakistan

Umair Gull
Department of Agronomy
University of Agriculture Faisalabad
Faisalabad, Pakistan

Aqsa Hanif
Department of Botany
Government College Women
University of Agriculture Faisalabad
Faisalabad, Pakistan

Muhammad Zeeshan Hanif
Department of Botany
University of Agriculture Faisalabad
Faisalabad, Pakistan

Ikram ul Haq
Department of Plant Breeding and Genetics
College of Agriculture
University of Sargodha
Sargodha, Pakistan

Abrar Hussain
Department of Biosciences
COMSATS University Islamabad, Sahiwal Campus
Islamabad, Pakistan

Romana Iftikhar
Department of Botany
University of Agriculture Faisalabad
Faisalabad, Pakistan

Asif Iqbal
Department of Botany
University of Agriculture Faisalabad
Faisalabad, Pakistan

Muhammad Mazhar Iqbal
Department of Soil and Environmental Sciences
College of Agronomy
University of Sargodha
Sargodha, Pakistan

Filza Ishfaq
Department of Botany
University of Agriculture Faisalabad
Faisalabad, Pakistan

Abdul Jabbar
Department of Botany
University of Agriculture Faisalabad
Faisalabad, Pakistan

Muhammad Mansoor Javaid
Department of Agronomy
College of Agriculture
University of Sargodha
Sargodha, Pakistan

Alia Javed
Department of Botany
University of Agriculture Faisalabad
Faisalabad, Pakistan

Wasim Javed
Water Management Research Centre
University of Agriculture Faisalabad
Faisalabad, Pakistan

Rimsha Kanwal
Department of Botany
University of Agriculture Faisalabad
Faisalabad, Pakistan

Rizwana Kanwal
Department of Botany
University of Agriculture Faisalabad
Faisalabad, Pakistan

Nazma Khan
Department of Botany
University of Agriculture Faisalabad
Faisalabad, Pakistan

Athar Mahmood
Department of Agronomy
University of Agriculture Faisalabad
Faisalabad, Pakistan

Basra Manzoor
Department of Botany
University of Agriculture Faisalabad
Faisalabad, Pakistan

Rizwan Maqbool
Department of Agronomy
University of Agriculture Faisalabad
Faisalabad, Pakistan

Ali Mardan
Department of Botany
Faculty of Sciences
University of Agriculture Faisalabad
Faisalabad, Pakistan

Asif Mukhtiar
Department of Botany
University of Agriculture Faisalabad
Faisalabad, Pakistan

Muhammad Ather Nadeem
Department of Agronomy
College of Agriculture
University of Sargodha
Sargodha, Pakistan

Javaria Nargis
Department of Botany
University of Sargodha
Sargodha, Pakistan

Aqsa Naseem
Department of Botany
University of Agriculture Faisalabad
Faisalabad, Pakistan

Hina Nazir
Department of Botany
Faculty of Sciences
University of Agriculture Faisalabad
Faisalabad, Pakistan

Muhammad Junaid Nisar
Department of Entomology
University of Agriculture Faisalabad
Faisalabad, Pakistan

Humma Anwar Rana
Department of Botany
University of Agriculture Faisalabad
Faisalabad, Pakistan

Ali Raza
Department of Botany
University of Agriculture Faisalabad
Faisalabad, Pakistan

Umar Raza
Department of Botany
Faculty of Sciences
University of Agriculture Faisalabad
Faisalabad, Pakistan

Komal Rehman
Department of Botany
University of Agriculture Faisalabad
Faisalabad, Pakistan

Ayman El Sabagh
Department of Field Crops, Faculty of Agriculture
Siirt University
Siirt, Turkey

Ayesha Sadeeqa Muhammad
Department of Botany
University of Agriculture Faisalabad
Faisalabad, Pakistan

Adeel Shahid
Department of Agriculture
Government College University Lahore
Lahore, Pakistan

Laiba Sajid
Department of Botany
University of Agriculture Faisalabad
Faisalabad, Pakistan

Muhammad Abdullah Saleem
Department of Botany
University of Agriculture Faisalabad
Faisalabad, Pakistan

Fouzia Shabbir
Department of Botany
University of Agriculture Faisalabad
Faisalabad, Pakistan

Maira Tanveer
Department of Botany
University of Agriculture Faisalabad
Faisalabad, Pakistan

Çağdaş Can Toprak
Department of Field Crops
Faculty of Agriculture
Siirt University
Siirt, Turkey

Usman Tufail
Department of Botany
University of Agriculture Faisalabad
Faisalabad, Pakistan

Muhammad Saad Ullah
Department of Botany
Faculty of Sciences
University of Agriculture Faisalabad
Faisalabad, Pakistan

Sami Ullah
College of Agriculture
University of Sargodha
Sargodha, Pakistan

Muhammad Talha Bin Yousaf
Department of Forestry
College of Agriculture
University of Sargodha
Sargodha, Pakistan

Asma Zafar
Department of Botany
University of Agriculture Faisalabad
Faisalabad, Pakistan

Rabia Zahid
Department of Botany
University of Agriculture Faisalabad
Faisalabad, Pakistan

Muhammad Ahmad Zeshan
Department of Plant Pathology
College of Agriculture
University of Sargodha
Sargodha, Pakistan

Muhammad Anjum Zia
Department of Biochemistry
University of Agriculture Faisalabad
Faisalabad, Pakistan

Muhammad Zubair
Department of Chemistry
University of Agriculture Faisalabad
Faisalabad, Pakistan

Muhammad Zubair
College of Plant Protection
Yangzhou University
Jiangsu, China

1 Introduction

Muhammad Mansoor Javaid, Muhammad Ather Nadeem, and Athar Mahmood*

1.1 CURRENT STATUS OF AGRICULTURE

Agriculture sector has always been keystone to human civilization and provides the essential resources for survival and economic growth. Agriculture faces numerous challenges that demand innovative solutions. Global population estimated to reach 9.7 billion by 2050 pose a great pressure on our agricultural systems to produce more food with better quality (Bahar *et al.*, 2020).

Agriculture is crucial for reducing poverty, ensuring food security, and promoting economic growth. By 2050, we need to feed around 10 billion people. Compared to other sectors, agriculture has significant potential to improve the lives of the poorest (Lal, 2016). Globally, agriculture contributes about 4% to gross domestic product (GDP), and in some developing countries, it accounts for more than 25% of GDP (Mohammed, 2016). However, agricultural growth, poverty reduction, and food security are at risk. Figure 1.1 shows the country-wise contribution of agriculture, forestry, and fisheries in GDP. Extreme weather, pests, and conflicts are disrupting food systems, and the goal of ending global hunger by 2030 is offtrack (Chadha *et al.*, 2020; Villarreal, 2022). Conflicts, climate change, and high food prices are increasing food and nutrition insecurity, pushing millions into extreme poverty, and undoing progress. Around 250 million people now face severe food insecurity.

Technological advancements, climate change, and consumer preferences lead to change the global agriculture landscape. This change necessitates new methods of crop production and resources management. High-quality crops not only meet nutritional requirements but also secure the economic viability for stakeholders. Agriculture practices, which are essential for food supply, also contribute to greenhouse gas emission into the atmosphere. It is estimated that 30% of greenhouse gas emission comes from our food system (Tubiello *et al.*, 2021). For example, methane and nitrous oxide gases are produced as a result of livestock rearing, rice cultivation, and use of synthetic fertilizers.

A critical challenge to food security and sustainability is a wastage of food. It is estimated that one-third of global food production is lost or wasted. This waste occurs at supply chain, from production to harvesting, distribution, and consumption. Food loss and waste should be reduced for food security and minimizing the environmental stress (Aziz *et al.*, 2018). Another challenge is poor diets, which cause significant number of deaths. Enough food is not available for population, or they eat wrong types of food, which cause malnutrition. It is estimated that 3 billion people cannot afford healthy diet (Herforth et al., 2020).

1.2 CURRENT CHALLENGES TO AGRICULTURE

To meet the growing demand of food, feed, and fiber, major challenges face the decrease in the capacity of agriculture system. These are interconnected and affect environmental, economic, and social aspects. There is a dire need to mitigate the adverse effects of climate change, resource depletion, and market along with balancing crop productivity. Addressing these issues is crucial for developing agricultural systems that ensure food security. Figure 1.2 illustrates the challenges for agriculture sector.

* Corresponding author: mmansoorjavaid@gmail.com

DOI: 10.1201/9781003498698-1

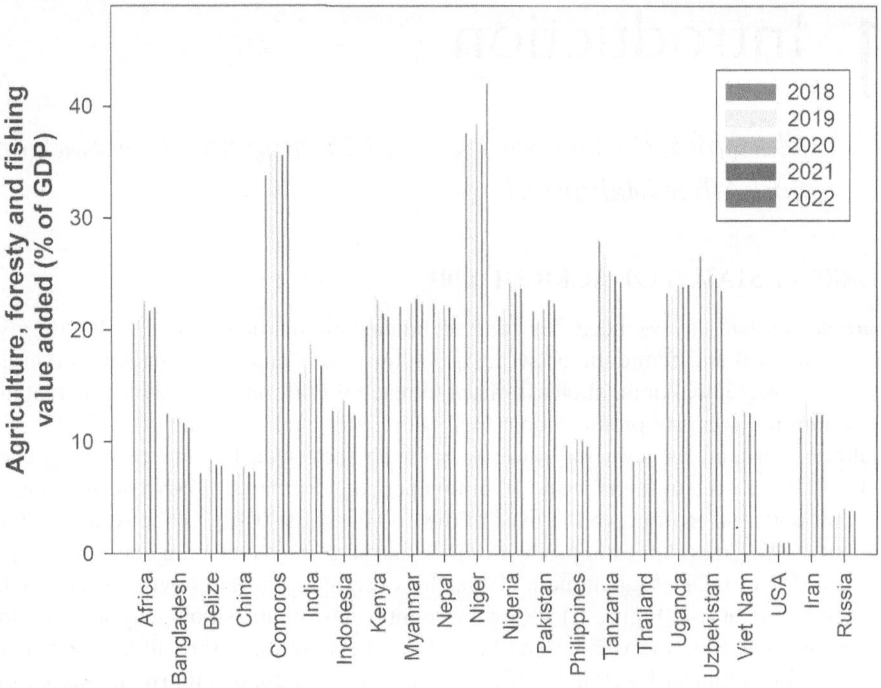

FIGURE 1.1 Contribution of agriculture, forestry, and fisheries in GDP of different countries.

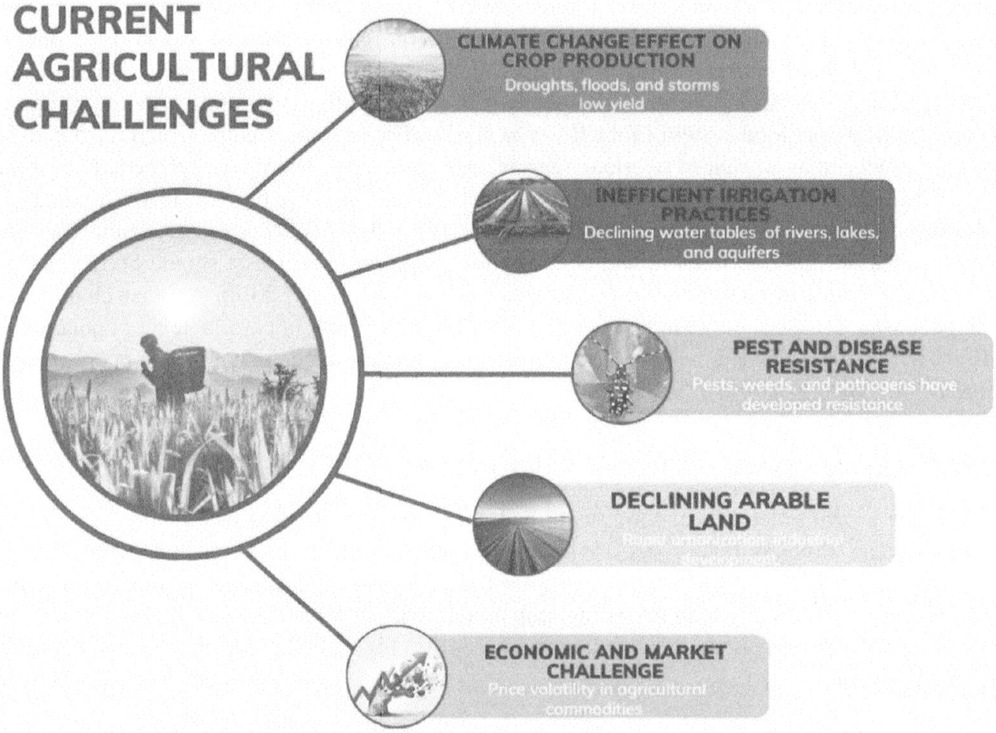

FIGURE 1.2 Current agricultural challenges.

1.2.1 Climate Change and Its Impact on Crop Production

Agriculture has been increasingly challenged by climate change, which affects weather pattern worldwide. Unpredictable and extreme weather events, such as droughts, floods, and storms, damage crops, leaving the farmers to face these challenges. These shifts can destroy crops, lower yields, and disturb planting schedules. The same applies to maize and sorghum grown in Africa – when it is difficult or impossible for farmers from producing countries to grow, which can severely impact food security within those nations with ensuing instability. Another concern of the rise in average global temperatures is heat stress on crops that results from reduced crop growth and production (Chadha *et al.*, 2019). These crops are also less able to withstand pests and diseases.

By disruptions of hydrological cycle, climate change is also destroying the amount of water available in our environment. In some parts there is a scarcity of water availability whereas in other places there is huge water that imperils harvest yields. Between these two extremes, agriculture is harmed. Thus, for instance in California, droughts force farmers to shorten the area of water-intensive crops such as almonds.

Climate change causes soil degradation through soil erosion, depletion, and compaction. All these result in damaging the agricultural production. During extreme weather events, heavy rain and storms exacerbate soil erosion, washing away the topsoil that is richest in nutrients. Long-term drought causes wind erosion (Jafari *et al.*, 2022). This drain of soil fertility makes it more difficult for crops to be grown. In the recent past, prolonged drought worsened the compaction of soil and no deep ploughing left the upper layer impermeable, resulting in hard pan for water absorption and causing problem for root penetration. As a result of these climate change-driven factors, global soil degradation poses serious challenges to food security across the planet.

Decline in soil health affects the entire ecosystem, biodiversity, which disturbs the water cycles. For example, sedimentation in rivers and streams caused by soil erosion affects the aquatic life as well as damages water quality. Loss of soil fertility due to soil erosion contributes to deforestation, which leads to clear more land in search of fertile soil. Soil degradation and nutrient depletion can be tackled by sustainable farming practices that protect and restore soil health. These challenges illustrate how climate change makes it even harder to grow enough food to feed the world's population. With increasingly unpredictable weather, temperatures rise, and water resources being less reliable, farmers must adapt their practices to sustain crop production.

1.2.2 Water Scarcity and Inefficient Irrigation Practices

Fresh water is vital for ecosystem and it is a natural source that is most extracted at the global level. Excessive freshwater consumption is responsible for a scarcity in circulation rate (Mishra, 2023). They are disrupting farmer's ability to grow crops. One key problem relates with the overuse of water resources, particularly where agriculture depends on groundwater or surface water for irrigation. For example, flood irrigation has been one of the oldest and most widely used irrigation techniques. This method causes substantial water loss through runoff and evaporation. This inefficiency not only wastes precious water resources but also makes irrigation less effective. Smart irrigation techniques, such as drip irrigation, pivot irrigation system, and sprinkler systems, are a replacement of traditional irrigation system and they deliver water directly to plants with minimal waste with improved water use efficiency.

During runoff, fields may carry fertilizers, pesticides, and sediments as pollutants (Rashmi *et al.*, 2022). These are leached into the rivers, ponds, or underground waters, resulting in eutrophication and thus damaging aquatic life. Besides, agricultural land runoff pollutes groundwater, which is an important source of drinking water for a lot of communities. That is usually the case in developing countries, where the situation is even worse than anticipated. Water shortage, ineffective irrigation, and pollution act in concert, therefore affecting agriculture production systems (Javaid *et al.*, 2022b).

1.2.3 Pest and Disease Resistance

Nowadays, agriculture is overwhelmed with the excessive use of chemical pesticides and herbicides. Due to the continuous application of these chemicals, it has resulted in pests and diseases resistance. This happens when some pests survive chemicals and then pass on their resistant traits to the next generation, hence leading to strains that become progressively difficult to manage. This increased resistance creates huge problem for farmers (Lucas, 2011; Mehmood *et al.*, 2018). Where other options have to be used, this may require more toxic or higher doses, which is not only expensive but also environmentally hazardous. Control of the resistance is now imperative if agricultural productivity has to be sustained. Alternative control management of pests and diseases must be developed against resistance. Integrated pest management (IPM) is a long-term solution that ensures reduced chemical use by applying the rotation of crops, including resistant crop varieties and biological controls. While IPM may reduce resistance and protect crops, it still demands investment in research, education, and infrastructure, which puts requirements on farmers, especially in the developing nations. For that reason, there is a dire need to manage disease and insect resistance if food security has to be achieved.

1.2.4 Declining Arable Land and the Use of Marginal Lands

The other major threat to the agricultural sector is land degradation. This is mainly because of the rapid growth seen as a result of urbanization and industrialization. Fertile lands are converted into residential areas, factories, and infrastructure with the ballooning cities, thereby reducing the cultivated land (Tang *et al.*, 2022). Specifically, this trend is apprehensive in countries such as China, India, and Pakistan, where agricultural land has been lost due to urban growth that threatens food security (Breisinger *et al.*, 2010; Tanveer *et al.*, 2015).

Land deterioration worsens it by rendering the remaining farmland less productive. Soil degradation through overgrazing, excess application of chemical fertilizers, and deforestation reduces the soil fertility. Climate change is another factor, whereby an increase in temperature and extreme weather conditions like droughts and floods cause damage to land. The global issue is affecting the agricultural productivity, compromising food production (Fonta *et al.*, 2011; Naqve *et al.*, 2021). This puts growing pressure on less fertile, marginal regions. These areas tend to be difficult to cultivate and require more resources like water and fertilizers and are thus more resource expensive and environmentally risky. We not only have to protect the already existing farmlands but also should undertake restoration works of degraded lands by adopting environment-friendly farming practices to achieve the set goals of preserving declining arable land. This is a strategy that shall ensure continued productivity in agriculture, hence meeting future human food security.

1.2.5 Economic and Market Challenges

Economic and market-related challenges are quite stifling the agricultural sustainability and growth. Crop and livestock price volatility, resulting from unpredictable weather, market speculation, and global trade policies, lead to financial uncertainty (Ali *et al.*, 2013; Aina *et al.*, 2024). While the smallholder and subsistence farmers may generally not have access to markets due to poor infrastructure and lack of information on markets, they usually sell their products at low price. The issues can only be addressed by including a holistic approach for the improvement of infrastructure, better market information, and an expansion of the farm sector in financial services. Resilience- and sustainability-enhancing policies and support systems targeted at smallholder farmers in vulnerable regions need to be implemented by governments, financial institutions, and development organizations (Jianing *et al.*, 2022).

1.2.6 Social and Labor Issues

Labor shortages, aging farming population, and rural poverty are critical challenges, faced by agriculture sector. If young people do not join farming in larger numbers, as in Japan or the U.S., there will be a great concern over who will exactly manage farms into the future. This affects the transfer of essential skills and knowledge. Labor shortages, especially during the periods of planting and harvesting, disrupt agricultural productivity. Mechanization may be an answer, but in most cases, this is not possible for small-scale farmers and mainly in developing countries. In a country like the U.S., dependence on seasonal migrant workers has increased labor shortages, further affecting crop yields (Charlton *et al.*, 2021). These are deepened by the rural poverty, reducing access to important services and trapping the communities in low productivity.

1.3 THE NEED FOR IMPROVED CROP QUALITY

With the world population estimated to continue growing rapidly, from 7.3 billion now to around 9.7 billion by 2050, the demands for food, feed, fiber, and fuel will reach dimensions that are presently unquantifiable (Gihring, 2022). A most dramatic increase in agricultural productivity, and hence efficiency, will be needed to feed the increasing billions. In the current scenario, there is an urgent need for the quality of crops to be greatly improved. Improved crop quality is necessary not only to meet the nutritional demands of the increasing population but also for economic sustainability and global market competitiveness.

1.3.1 Factors Driving the Need for Higher Quality Crops

1.3.1.1 Population Growth and Food Security

With a growing world population, more demand was placed upon food production, especially in countries people of which have already been suffering from hunger and malnutrition. Increasing crop yields alone is not enough to meet the challenges. Rather, it is the enhancement of the nutritional qualities that are being produced to combat deficiencies affecting a large number of people worldwide (Godswill *et al.*, 2020). In developing areas where diets are commonly based on a few staple crops, their biofortification assumes significant importance for improving health conditions. Furthermore, nutrient-enrich crops tend to be more resilient to pests, diseases, and environmental stresses. They are able to form the base of both resilient and sustainable food systems that could feed a growing population amid a changing climate.

1.3.1.2 Nutritional Demands

The growing concern for nutrition among health-conscious consumers is driving the demand for food products with enhanced nutritive values, necessitating the development of crops rich in essential nutrients. To reduce the diet-related health issues, there is a need to improve crop quality in terms of increased proteins, vitamins, and minerals (Godswill *et al.*, 2020). By growing nutrient-enrich crops, we will provide better food options that can help prevent diseases, promote healthier eating habits, reduce healthcare costs, and enhance quality of life.

1.3.1.3 Market Standards and Consumer Preferences

Agricultural products, in a globalized economy, have to conform to very strict standards of quality that go beyond yield and embrace nutritional value, taste, texture, shelf life, and even appearance. For producers to compete in the global market, they have to grow high-quality crops that can answer these diversified needs. Changing consumer preferences toward organic and GMO-free crops, in response to concerns over food safety and health, further drive the demand for crops that excel in those traits (Marian, 2014). In addition, this requires crops that have longer shelf lives and less spoilage to lessen food wastage, therefore increasing profitability across the global market. Crop variety breeding for quality will enhance competitiveness and satisfy future market demand.

1.3.1.4 Impact on Food Security and Nutrition

Crop quality is fundamental to providing the food security that guarantees abundant, healthy, and safe food. Focusing on quality will reduce farmers' dependence on costly inputs by offering high-quality crop varieties that resist pests, diseases, and environmental challenges. Furthermore, improved post-harvest qualities in crops, like longer shelf lives and better storage, reduce food loss and provide more food for consumption. Improved nutrition in crops by biofortification leads to avoid a widespread problem of micronutrient deficiencies in many developing countries and contributes to better health.

1.3.1.5 Economic Benefits

High-quality produce opens up new avenues with significant economic gains, starting with better market prices that improve the farmer's income and allow reinvestment in better inputs to enhance productivity. A cycle of causal improvement in the practice of farming is observed. Improved quality further opens doors for value-added products that can provide premium prices and diversify income streams. Moreover, high-quality agricultural products serve to increase the export competitiveness of any country, thereby consequently increasing foreign exchange earnings and encouraging wider system of economic growth through employment opportunities, better infrastructure, and improved living standards.

1.3.1.6 Better Crop Quality through Innovation

Enhanced crop quality results from a blend of traditional knowledge with modern science. Advanced breeding and genetic engineering develop crops that are more nutritious and strong against different stresses, such as drought and pests, thus improving food security. Precision agriculture makes use of global positioning systems, sensors, and data analytics in optimizing resource use to promote better plant health with reduced impacts on the environment. These are sustainable practices like organic and regenerative agriculture that improve crop quality by enhancing soil health and biodiversity. Overall, the approaches ensure higher quality crops, meet growing food demands, and ensure a livable agricultural future.

1.4 THE CONCEPT OF SUSTAINABILITY IN AGRICULTURE

Sustainability in agriculture seeks to balance food production with environmental protection, economic viability, and social well-being. It maintains soil fertility through clean water and vegetation, hence reducing the environmental impact through reduced synthetic chemicals and low greenhouse gas emission. Economically, it ensures fair income for farmers with thriving rural communities, while, socially, it supports fair treatment of workers and equitable access to resources. The goal is to create a farming system that meets current needs while benefiting future generations.

1.4.1 Environmental Dimensions of Sustainable Agriculture

1.4.1.1 Soil Health

The environmental dimension of sustainable agriculture places first priority on soil health. Crop rotation, cover crop, reduced tillage, and organic amendments are examples of sustainable practices that could be used to promote fertility in soils (Crystal-Ornelas *et al.*, 2021). Discuss the importance of these methods in conserving and enhancing natural productive processes within soils. These are responsible for nutrient cycling, water retention, and carbon sequestration, all of which validate long-term agricultural productivity and environmental stability. By keeping soils healthy, sustainable agriculture keeps lands productive while mitigating climate change and conserving natural resources.

1.4.1.2 Water Management

It is important to adopt efficient water management techniques in agriculture at a time when there are growing concerns about water scarcity. Drip irrigation delivers the water right to the root of the plants, and rainwater harvesting collects and stores the rain for later use. Besides, drought-tolerant crop varieties will ensure that such crops do well with minimal water, thus making farming resilient to dry conditions. These practices all combine to realize water efficiency and sustain agricultural systems by ensuring that, at all times, water resources are wisely and effectively utilized.

1.4.1.3 Biodiversity Conservation

Sustainable agriculture conserves biodiversity through the maintenance of natural habitats and adopting practices like sowing crop rotations, multiple cropping, and agroforestry. This diversity underpins ecosystem services so that natural pest control and crop pollination are ensured, enhancing productivity. The enhancement of resilience to environmental stress and reducing reliance on chemical inputs have further implications for farm stability. In the end, biodiversity nurtures long-term health and productivity in both the environment and agricultural systems.

1.4.1.4 Climate Change Mitigation

Agriculture is both a cause and an effect of climate change. Sustainable practices lessen this through optimized fertilizer use, reduced tillage, and better livestock management all of which minimize greenhouse gas emissions and increase the carbon sequestration through cover crops and agroforestry. These also build resilience in the face of climate variability by a diversification of crops and improvement in soil health, together with water-efficient methods. This dual approach enables the long-term protection of the environment while maintaining agricultural productivity under changing climatic conditions, which are directly connected to a more sustainable future.

1.4.2 Economic Dimensions of Sustainable Agriculture

1.4.2.1 Economic Diversification

In sustainable agriculture, profitability, including access to the market, is inseparable from resource efficiency and environmental good practice. No-till farming or reduced tillage, IPM, and organic amendments reduce the input costs through savings on synthetic chemicals and fertilizers, resulting in stable yields and less financial risk (Gomiero, 2019). Moreover, sustainable practices open up access to premium markets that have a liking for eco-friendly products and thus allow farmers to charge premium prices for their produce. This means that by meeting these demands from buyers, farmers meet both their improved financial returns but also contribute to wider economic stability, showing how environmentally friendly farming can affect not just gains at the micro-level in a business but also actually influence macro-economic success and sustainability.

1.4.2.2 Long-term Viability

Investments in sustainability are indispensable to agriculture's long-term viability through the preservation of natural resources and maintenance of productivity (Pretty, 2008). Crop rotation, reduced tillage, and organic farming enhance soil fertility while lowering erosion to achieve constantly high yields. These practices will also reduce dependence on expensive external inputs, such as synthetic fertilizers and pesticides, by means of organic amendments and natural pest control. It, therefore, makes farming more sustainable and profitable in the long run by reducing costs, underpinning stable financial planning, and boosting economic resilience.

1.4.3 Social Dimensions of Sustainable Agriculture

1.4.3.1 Community Development

Community development is supported by sustainable agriculture through local food systems and promotion of family farmers, with the aim of keeping financial benefits local and traditions preserved within rural communities. It reinforces the local economy and jobs in farming, agro-tourism, and food-processing activities by focusing on the production of local food and eliminating long supply chains. It strengthens community resilience and social cohesion by engaging locals in decision-making processes and efforts for cooperation, hence a more successful and supportive rural community.

1.4.3.2 Equitable Access to Resources

For social sustainability in agriculture, equal sharing of resources becomes very important, particularly for small farmers and those on the fringes. Ensuring access to land, education, technology, and finance levels plays the role of increasing productivity and improving income. It reduces inequities, enhances social cohesion, and thus resonates within communities for resilience. Sympathetic communities empower subordinated groups through sustainable agriculture, which facilitates a more equitable and harmonious society that ensures better individual well-being and strengthens the social setting within farming communities.

1.4.3.3 Health and Nutrition

Sustainable agriculture promotes public health by providing safe, nutritious, and diverse food through practices like organic farming and crop diversification. This approach improves dietary quality and reduces health issues by minimizing harmful chemicals and supporting better eating patterns. It also protects farm workers and nearby communities from chemical exposure, ensuring safer food and water. By prioritizing environmental and human health, sustainable agriculture enhances overall well-being and quality of life in agricultural areas.

1.4.4 Practices Promoting Sustainability in Agriculture

1.4.4.1 Organic Farming

Organic farming promotes sustainability by avoiding synthetic inputs and focusing on natural processes. Methods like composting, green manures, and crop rotation enhance soil health, improve water and nutrient retention, and support beneficial microbes. Organic farming also boosts biodiversity and ecological balance, managing pests naturally and maintaining vital ecosystem functions like pollination and nutrient cycling. This holistic approach aligns with sustainable agriculture by fostering environmental health and resilience.

1.4.4.2 Agroecology

Agroecology connects agriculture with ecological principles, focusing on managing natural resources and benefiting from ecosystem services. It makes available a widespread array of practices, inclusive within the field of agroecology, that are manifested as agroforestry, cover cropping, and polycultures that promote improved soil health, water, and nutrient cycling. Agroecology minimizes environmental impacts and enhances agricultural productivity through biodiversity and services like pollination and pest control. This approach will establish a balanced relationship between farming and the environment, enabling long-term sustainability and stable productivity for farmers.

1.4.4.3 Conservation Agriculture

This is a farming procedure that cares for the health and productivity of soil by causing minimal disturbance to the composition, while it is kept covered. Practices such as tillage, mulching, and cover crops minimize erosion, conserve the soil moisture, and improve fertility. Crop rotation helps in preventing the depletion of nutrients, reducing pests and diseases, and enhancing of soil structure.

These approaches are beneficial in ensuring long-term soil health and productivity while enhancing environmental sustainability through such activity supporting the natural processes of land use and ameliorating the effects of conventional farming.

1.4.4.4 Integrated Pest Management (IPM)

It is the most effective approach to pest control. IPM integrates biological, physical, and chemical measures in controlling pests in ways that do not pose any threat to the environment. It uses natural predators, traps, and a mix of chemicals wherever necessary in a way that integrates control measures without posing potential damage to the environment or are used in the eradication of pests. IPM balances the ecosystems, avoids resistance to pesticides, and encourages beneficial organisms; so it increases the sustainability and also the effectiveness in long term for their procedures in any pest management.

1.4.4.5 Agroforestry

Agroforestry is a system of farming that integrates trees and shrubs into overall farming systems in an effort to gain an environmental and economic benefit. There is a benefit to improved biodiversity, soil health via reduction in erosion and increase in nutrient cycling, and management of water through reduction in loss. The added economic value comes from the income generated from timber, fruits, or nuts. The diversification of revenue streams to improve financial stability, and therefore, this practice further enhances sustainability in the farming system by adding improvement to environmental quality, biodiversity, and resilience conferred by it.

1.5 EMERGING SOLUTIONS AND TECHNOLOGIES

The future of agriculture largely depends on how well we are able to devise new innovations and integrate new technologies that can solve the many complexities prevailing in this sector. Innovations like precision agriculture, advanced breeding techniques, and smart farming technologies could make a huge difference in enhanced crop quality, sustainability, and productivity enhancement for better livelihoods. These are the innovations in the optimization of resource utilization, reduction of environmental impact, and enhancement of crop resilience that ensure a more secure and more efficient food system. Recent innovations are summarized in Table 1.1. By adopting state-of-the-art

TABLE 1.1
Recent Innovations in Agriculture

Innovation	Description	References
Vertical farming	Utilizes vertically stacked layers to grow crops in controlled indoor environments, maximizing space and efficiency	Kalantari *et al.* (2018), Beacham *et al.* (2019), Van Gerrewey *et al.* (2021), Akintuyi (2024)
CRISPR technology	Allows precise editing of plant genomes, enabling the development of crop varieties with desired traits	Lee *et al.* (2023), Wang and Doudna (2023), Morshedzadeh *et al.* (2024)
Aquaponics	Integrates aquaculture (fish farming) and hydroponics (soilless plant cultivation), creating a symbiotic ecosystem	Okomoda *et al.* (2023), Samikannu *et al.* (2023); Palm *et al.* (2024)
Precision livestock farming	Utilizes sensors and data analytics to monitor and manage livestock health, welfare, and productivity	Kleen and Guatteo (2023), Qiao *et al.* (2023), Tzanidakis *et al.* (2023)
Blockchain in agriculture	Enables transparent and traceable supply chains, improving food safety and quality assurance	Kumar and Dwivedi (2023), Sajja *et al.* (2023), Ordoñez *et al.* (2024)
Gene editing for disease resistance	Employs gene editing techniques to develop crops resistant to pests and diseases	Dong *et al.* (2023), Saeed *et al.* (2023), Sha *et al.* (2023)
Remote sensing	Uses satellite and drone imagery for crop monitoring, yield prediction, and precision agriculture	Kour *et al.* (2023), Ma *et al.* (2024)

TABLE 1.2
Solutions to Agriculture Challenges

Solution	Description	Reference
Sustainable agriculture practices	Practices such as conservation tillage, crop rotation, and integrated pest management promote soil health, biodiversity, and natural resource conservation, enhancing crop resilience and ecosystem sustainability	Akanmu *et al.* (2023), Mgomezulu *et al.* (2023a), Mgomezulu *et al.* (2023b)
Climate-smart farming techniques	Adoption of practices such as drought-tolerant crop varieties, rainwater harvesting, and agroforestry helps mitigate climate risks, improve water use efficiency, and enhance farm resilience to extreme weather events	Singh (2023), Wakweya (2023), Erekalo *et al.* (2024)
Precision agriculture technologies	Utilization of technologies like GPS-guided machinery, remote sensing, and data analytics enables precise input application, resource management, and decision-making, optimizing yields, minimizing waste, and reducing environmental impact	Ahmad and Sharma (2023), Hundal *et al.* (2023), Kour *et al.* (2023)
Integrated water resource management	Implementation of strategies such as drip irrigation, water recycling, and soil moisture monitoring improves water use efficiency, reduces runoff, and safeguards water quality, ensuring sustainable water management in agriculture	Grison *et al.* (2023), Kalogiannidis *et al.* (2023), Zehtabian *et al.* (2023)
Genetic engineering and biotechnology	Development of genetically modified crops with traits such as pest resistance, drought tolerance, and increased nutrient content enhances crop productivity, reduces chemical inputs, and addresses challenges posed by pests, diseases, and climate stress	Das *et al.* (2023), Eskandar (2023), Kumar *et al.* (2023)
Farmer education and extension services	Provision of training, technical assistance, and access to information on sustainable practices, market trends, and financial management empowers farmers to adopt innovative solutions, improve productivity, and enhance their resilience to challenges	Raji *et al.* (2024), Ullah *et al.* (2024)

solutions, agriculture will be better placed to resolve challenges related to climate change, resource scarcity, and food security, hence setting itself on course toward a more productive and sustainable future for next generations (Table 1.2).

1.5.1 Precision Farming

1.5.1.1 Data-Driven Farming

Data-driven farming involves the use of technologies such as satellite images, drones, and soil sensors to gauge information in quasi-real time on aspects of crop health, soil conditions, and other environmental factors. This information aids farmers in precisely managing resources by using water, fertilizers, and pesticides more effectively in order to increase crop yields while reducing the impact on the environment. For these particular features that enable targeted interventions and provide informed decision-making, data-driven farming enhances efficiency, productivity, and sustainability in agriculture.

1.5.1.2 Variable Rate Technology (VRT)

VRT is one of the major step-through of precision agriculture, which involves applying variable inputs, like fertilizers, pesticides, or water, according to the variation in field conditions. This is possible because of data availed from soil sensors, satellite imaging, and so on, toward efficient resource conservation in agriculture.

1.5.1.3 GPS and GIS Technologies

This gives new meaning to farming. The PS and GIS technologies make it possible to map fields precisely, as well as monitor them. To be brief, GPS helps in the preparation of accurate location data for detailed field maps, while GIS integrates layers of spatial information such as soil types and crop performance. Equipped with this information, farmers are better placed to maximize their efficiencies in field management, reduce inefficiencies, and apply inputs precisely. This, in effect, reduces wastage of resources, lowers operational costs, and improves field productivity, hence making farming more efficient and sustainable.

1.5.1.4 Vertical Farming

This is a new form of agriculture wherein crops are cultivated in vertically stacked layers or vertically inclined surfaces, usually within a controlled indoor environment. It qualifies as the reuse of space and enhances the complete utilization of available city space because it creates expanding inner-city locations and makes use of once underutilized buildings or warehouses, proving to be especially worthwhile where the land extent is small. It is good for vertical farming controls light, temperature, and humidity to create perfect conditions all year round for crop production, increasing yield, and putting available resources into full use. Moreover, producing food near the area where it is required reduces the costs of transportation and environmental degradation from transportation. This kind of sustainable agriculture minimizes water usage and requires no pesticides in yielding healthier produce while simultaneously trying to achieve the food security goal in the midst of densely populated areas.

1.5.2 Biotechnology

1.5.2.1 Genetically Modified Organisms (GMOs)

Genetically modified organisms are key technological advances in agriculture, having been engineered for traits that will help counter a wide array of issues in agriculture. GMOs can be made to repel pests, which would lower the costs of production due to less use of chemical pesticides. They can also be made for drought tolerance, important in maintaining yields in areas under low water availability. Some of the GMOs are also fortified with better nutritional value, hence providing better quality food and bridging the nutritional deficiencies in various populations. These engineered traits mean increased crop yields and can help to guarantee a more secure supply of food. GMOs enable agriculture to be more sustainable because there is reduced dependency on chemical inputs and improved efficiency in the use of resources. It is in the incorporation of traits such as pest resistance and drought tolerance that GMOs can decrease traditional farming's impact on the environment, which includes pesticide runoff and water consumption. The technology increases agricultural productivity but plays a critical role in advancing food security by making crops more resilient to environmental stressors and ensuring stable food supplies for growing populations.

1.5.2.2 CRISPR and Gene Editing

The evolved version, CRISPR technology, is definitely one of the most groundbreaking gene editing tools; it allows for the editing of an organism's DNA with precision. Such a phenomenon, therefore, presents huge potential for designing crop varieties aimed at solving well-targeted agricultural problems. For instance, genes associated with stress responses can be modified using CRISPR to develop crops that can withstand extreme temperatures or drought. It can also increase tolerance against diseases by manipulating genes that give susceptibility to pathogens. The ability to bring such precise changes has huge potential for developing high-yielding and stable crops across varying environmental conditions. Apart from this, resilience CRISPR technology is able to improve nutrient intake via crops, through the creation or boosting of certain specific vitamins and minerals. Such targeted approaches can help remedy nutritional deficiencies and improve general public health. By supporting more efficient development of crops with desired traits than traditional breeding methods, the potential use of CRISPR technology empowers agricultural productivity and sustainability. This also opens up new opportunities for innovation in the development of crops and eventually leads to more adaptable and nutritionally better crops that can respond to the new challenges of global food systems.

1.5.2.3 Biofortification

Biofortification is the latest technology where crops are selectively bred or genetically engineered to increase their levels of some essential vitamins and minerals. Crop breeding is done to provide higher contents of some basic nutrients like iron, zinc, and vitamin A, whose deficiencies have been found to be higher in human bodies. It biofortifies to fight hidden hunger, where populations may consume enough but remain bereft of important micronutrients required for health and development due to the increased nutrient content in staple crops. This technology is particularly valuable in areas with a low intake of a diversified range of foods, where traditional ways of enhancing nutrition may be hard to implement. Biofortified crops provide a practical solution since, from the Equity First International principles, they deliver essential nutrients directly to consumers through frequently consumed foods, increasing general nutritional quality without large increases in food intake. The efforts of biofortification thus go straight to the improvement of public health and support reducing malnutrition and related health problems worldwide.

1.5.3 Drones and Robotics

1.5.3.1 Crop Monitoring and Scouting

Drones equipped with cameras and sensors are changing crop monitoring and scouting by providing real-time insight into crop conditions and field status (Panday *et al.*, 2020). Advanced aerial tools can capture high-resolution images and data about many attributes of plant growth, such as detecting signs pointing toward pest infestation, nutrient deficiency, and diseases. Real-time data gathered by the drone allows much better information and efficient management decisions. For example, when detecting areas with pest problems or issues in plant health, farmers are rapidly able to offer treatments only where necessary and not blanket the entire field. This technology helps in optimizing resource use, reducing input costs, and minimizing environmental impact. On the whole, drone technology improves precision agriculture due to a better accuracy of field assessments and more effective management strategies that would lead into healthier crops with increased productivity.

1.5.3.2 Automated Planting and Harvesting

The use of robotics technology in agriculture, particularly with automated sowing and harvesting processes, has brought down labor costs to a minimum while enhancing operational efficiency (Lowenberg-Deboer *et al.*, 2020). Since such automated planting systems can sow seeds at the right depth and spacing required for optimal growth, the growth would be uniform, and hence yield maximization could be attained. Weeders may identify and destroy weeds (Hanif *et al.*, 2017). The automation mechanics has eased field operations in performing their activities more regularly and precisely. Other than reducing labor demand, robotics technology provides improved crop management through high precision levels that reduce crop destruction and eventually increase the yield quality. For instance, this makes it possible to implement automated harvesting systems that accurately collect mature crops without causing unnecessary disturbances to nearby plants. This precision brings the best in harvesting, minimizes post-harvest losses, and enhances gentle handling of crops. The addition of robotics enables farmers to work more effectively and productively in farming activities, hence opening avenues for both sustainable and scalable farming operations in relation to this.

1.5.3.3 Spraying and Irrigation

Increasingly, drones and robots are used in precision spraying and irrigation. This in turn offers huge efficiency and effectiveness improvements in the spraying and irrigation practices of agriculture. With advanced sensor and imaging capabilities, it becomes possible for a drone to detect areas that may need pesticides or fertilizers and apply such inputs wherever needed (Grift *et al.*, 2008). This reduces the use of chemicals, hence reducing wastages and risks of environmental pollution through runoff into water bodies. Besides, a drone can cover large areas within a short time, hence more effectively and timely compared to the traditional methods used in spraying chemicals. Similarly, robots equipped for precision irrigation are able to deliver water directly into the root zones of plants, thus optimizing their water use and at the same time cutting down on excess irrigation. These are advanced systems that adjust their water supply by making references to real-time information on soil moisture and the needs of plants, thus avoiding excess water use and promoting healthier crops. Right amounts of irrigation and nutrient application according to needs can increase crop productivity while paving ways for sustainable practices through the conservation of resources and lessening impact on the environment. This essentially makes the innovation of drones and robots into spraying or irrigation practices a major step toward advanced technologies for agriculture and cleaner and more efficient farming.

1.5.4 Nanotechnology

1.5.4.1 Nano-fertilizers and Nano-pesticides

Nanotechnology in agriculture develops nano-fertilizers and nano-pesticides that increase efficiency and effectiveness. Nano-fertilizers are ultrafine fertilizers developed to improve nutrient uptake of plants, allowing for enhanced penetration and absorption at the root level (Solanki *et al.*, 2015). This kind of fertilizer can make the nutrients more precisely and controllably available, hence reducing the need for excessive applications, which in turn reduces nutrient loss. This change enhances the plant growing potential and yield while reducing the ecological footprint from fertilization. Similarly, nano-pesticides facilitate targeting of pest management by means of nanoparticles that are able to more effectively deliver active ingredients to specific pests or particular places within plants. Due to the small size of such particles, they more easily penetrate deeper into a plant and are highly selective in their action against so-called target pests, thus reducing undesirable effects on nontarget organisms and the amount of pesticide used. This will help in managing pest populations more effectively while reducing the chances of pesticide resistance and contamination of the environment. The application of nano-fertilizers and nano-pesticides generally stipulates a relevant step forward in agricultural technology by increasing sustainability and resource efficiency in farming.

1.5.4.2 Nano-sensors

With extremely precise and detailed data regarding soil and crop conditions, nano-sensors are changing the face of agricultural monitoring (El-Chaghaby and Rashad, 2024). Changes that occur at a nanoscale in moisture, nutrition, and health levels of both the soil and crops can easily be detected by the nano-sensor. Across large acres, farmers can continuously monitor critical parameters with great accuracy by using nano-sensors ushered into the soil or attached to plants. The real-time data is what allows one to make appropriate and timely changes in management practices, such as scheduling irrigation or nutrient applications based on prevailing conditions. By having the potential to monitor those factors at this level of granularity, it will serve not only to optimize resource use but also enhance crop performance. Nano-sensors are one giant step forward into agricultural technology, offering the possibility of more precise, informed, and efficient management over crops and resources.

1.5.4.3 Better Food Quality

Through advanced means of nano-encapsulation, nanotechnology is playing a very transformational role in enhancing food quality. It involves encasing the sensitive nutrients or bioactive compounds within the nanoscale carriers that safeguard them from degradation during storage and processing. It enables proper storing and efficient delivery of the much-needed nutrients like vitamins and minerals to the body of consumers. This ensures that the food-nutritive value is maintained, and consumers are better able to derive health benefits from it. Besides, nanotechnology can enhance the shelf life of food products by either using nano-materials with antimicrobial properties or creating barriers to environmental factors, because most foods are wasted due to spoilage, and keeping them fresh prevents such spoilage and thus wastage for a longer time. An extended shelf life delivers many consumer-related benefits by bringing better quality food to the consumer but also contributes toward reduced food wastage and, hence, sustainability for the value chain.

1.5.5 Internet of Things (IoT)

1.5.5.1 Smart Farming

Smart farming is making agriculture different by connecting all devices and systems on the farm into an integrated ecosystem, which seems to be empowered through Internet of Things (IoT) technologies (Javaid *et al.*, 2022a). Thereafter, sensors are placed in fields, on machinery, and within the greenhouse environments to collect real-time data regarding the factors of soil moisture, temperature, and crop health. The collected data is sent for analysis on cloud-based platforms to derive actionable insights. These insights enable farmers to make accurate decisions regarding irrigation, fertilizer application, and pest control, thereby optimizing the management of crop growing and resources. Moreover, IoT enables automation between sensors and actuators with farm equipment. For example, irrigation systems can be adjusted over time according to the real-time data from soil moisture, reducing wastage of water while ensuring plants get appropriate amounts of water. Similarly, in greenhouses, lighting, climate, and nutrient supply can be regulated using automated systems to attain better growth and productivity. IoT-enabled smart farming, in general, enhances operational efficiency for sustaining practices based on more precise and data-driven management of agricultural resources.

1.5.5.2 Farm Management Systems

IoT-based farm management systems avail an integrated package of tools that the farmers can use in monitoring and managing every aspect of their operations with improved accuracy and proficiency. Such a system will integrate data originating from sources adjacent to a farm, such as weather forecasts, soil sensors, crop health monitors, and machinery, giving a single platform for real-time observation and analysis. The integration offers the farmer an overview of the whole farm conditions, thus helping to make the correct decisions and well-timed interventions. For instance, weather information combined with soil moisture levels can help optimize irrigation schedules, while crop health monitoring can guide targeted pest control measures. Consequently, these systems raise operational productivity by smoothing and automating activities that are routine in nature for farm management.

1.5.5.3 Supply Chain Traceability

IoT technology greatly enhances supply chain traceability, providing end-to-end traceability of agri-products from the farm down to the consumer (Hu *et al.*, 2021, Abdullah et al., 2022). These sensors, RFID tags, and GPS techniques will allow one to monitor and record each step of the supply line, ensuring products are tracked correctly as they move through different handling and processing stages. Continuous monitoring verifies the origin and journey of food products. The origin is an aspect that holds great importance in ensuring standards concerning food safety, apart from compliance for checking the conformance of the product to the regulatory requirements. Improved traceability through IoT reduces losses by providing real-time monitoring of conditions like temperature and humidity during storage and transportation. Moreover, advanced transparency made possible by IoT systems increases consumer trust and that of other stakeholders in the tracing of food journeys and quality/safety verification. These emerging solutions and technologies represent the future of sustainable, productive agriculture. Through the adoption of best practices and new innovations, we can overcome today's agricultural challenges, enhance the quality of crops, and foster a robust and sustainable food system for future generations.

1.6 OBJECTIVES OF THE BOOK

This book has a wide international spectrum and deals with funny, innovative approaches dealing with contemporary challenges burdening the development of the agriculture sector. In this sense, this is really a serious book showing ways for progressive research, development, and knowledge-shaping that can be challenged to be elaborated and proposed in holistic scientific, technological, organizational, and policy solutions.

1.6.1 Comprehensive Coverage of Advanced Agricultural Practices

The book offers a broad view of this strongly advanced, developed area of agriculture, its technology, and methodologies in shaping modern farming. The book speaks about precision agriculture in optimizing resource use, biotechnology both in genetic modifications and even the new CRISPER, organic farming for sustainability, and drones and robotics for automation. Further, this goes to describe nanotechnology for better nutrient delivery and pest control, and IPM for even effective levels of pest management, ensuring a safe and eco-friendly practice of pest management by realizing new technologies. The book describes these advances, in order to give professionals, researchers, and enthusiasts new ideas about the latest innovations mounted for sustainability, advancement in agriculture.

1.6.2 Sustainable Techniques and Technologies Investigation

It pays specific attention to sustainability from creativity means and technological use that facilitates environmental care, economic success, and social justice in agriculture. For sustainable practices, the book looks at conservation tillage, precision irrigation, and linked water source management because soil health and reaching high water productivity remain a significant attention. The book also includes insight on how to assure, through practices on agroforestry, organic farming, and sustainable pest management, that the ecological balance endures and natural resources are conserved.

1.6.3 Presentation of Practical Modern Agricultural Solutions

The book intimates and addresses certain challenges in modernized agriculture with action points drawn to directly engage farmers, researchers, policymakers, and stakeholders. By reducing the application of acquired theoretical knowledge down to field case studies and best practices, it demonstrates practical strategies to improve crop quality, sustainability, and resource use that are guaranteed. The reason that success stories and examples are included in this book is to incite interest in the readership toward innovative practices; to accomplish this, it gives tools and knowledge to advance agriculture in support of the solution to food and environment issues globally.

1.6.4 Encouragement of Knowledge Sharing and Collaboration

It is a major knowledge-sharing and collaborative platform for agriculture experts, practitioners, and owners of thought leadership. The book sparks interdisciplinary conversations, thereby binding diverse fields closely for it is very instrumental in eliciting innovations. The book highlights successful partnerships and collective action across academia, government, non-profit, and private enterprise aimed at dealing with complicated agricultural challenges.

This approach brings collective diversity resources and expertise to the future realization of enhanced sustainability, productivity, and resilience in agriculture.

1.6.5 Empowerment of Farmers and Agricultural Communities

At its core, the book aims to empower farmers and agricultural communities to thrive in an ever-changing agricultural landscape. By providing accessible, evidence-based information and practical guidance, the book seeks to support farmers in adopting sustainable practices, improving crop quality, enhancing productivity, and ensuring livelihood security. It emphasizes the importance of farmer-centric approaches and participatory decision-making processes in agricultural development.

1.6.6 Inspiration for Future Research and Innovation

The book inspires future research and innovation by highlighting emerging trends, cutting-edge technologies, and knowledge gaps in agriculture. It encourages researchers, practitioners, and policymakers to explore new areas and develop novel solutions to address evolving challenges. By emphasizing the need for continuous adaptation and technological advancement, the book aims to drive the next generation of agricultural innovation, fostering a more sustainable and productive sector for the 21st century.

REFERENCES

Abdullah, Mahmood, A., Bibi, S., Naqve, M., Javaid, M.M., Zia, M.A., Jabbar, A., Ud-Din, W., Attia, K.A., & Khan, N., 2022. Physiological, biochemical, and yield responses of linseed (*Linum usitatissimum* L.) in α-tocopherol-mediated alleviation of salinity stress. *Frontiers in Plant Science*, 13, 867172.

Ahmad, U., & Sharma, L., 2023. A review of best management practices for potato crop using precision agricultural technologies. *Smart Agricultural Technology*, 4, 100220.

Aina, I.V., Ayinde, O.E., Thiam, D.R., & Miranda, M.J., 2024. Climate risk adaptation through livestock insurance: Evidence from a pilot programme in Nigeria. *Climate and Development*, 16, 1–12.

Akanmu, A.O., Akol, A.M., Ndolo, D.O., Kutu, F.R., & Babalola, O.O., 2023. Agroecological techniques: Adoption of safe and sustainable agricultural practices among the smallholder farmers in Africa. *Frontiers in Sustainable Food Systems*, 7, 1143061.

Akintuyi, O.B., 2024. Vertical farming in urban environments: A review of architectural integration and food security. *Open Access Research Journal of Biology and Pharmacy*, 10, 114–126.

Ali, H., Tanveer, A., Nadeem, M., Asghar, H., & Javaid, M., 2013. Germination ecology of *Rhynchosia capitata*: An emerging summer weed in Asia. *Planta Daninha*, 31, 249–257.

Aziz, M., Ashraf, M., & Javaid, M.M., 2018. Enhancement in cotton growth and yield using novel growth promoting substances under water limited conditions. *Pakistan Journal of Botany*, 50, 1691–1701.

Bahar, N.H., Lo, M., Sanjaya, M., Van Vianen, J., Alexander, P., Ickowitz, A., & Sunderland, T., 2020. Meeting the food security challenge for nine billion people in 2050: What impact on forests? *Global Environmental Change*, 62, 102056.

Beacham, A.M., Vickers, L.H., & Monaghan, J.M., 2019. Vertical farming: A summary of approaches to growing skywards. *The Journal of Horticultural Science and Biotechnology*, 94, 277–283.

Breisinger, C., Van Rheenen, T., Ringler, C., Pratt, A.N., Minot, N., Aragon, C., Yu, B., Ecker, O., & Zhu, T., 2010. Food security and economic development in the Middle East and North Africa. *Current State and Future Perspective. IFPRI Discussion Paper,* 985.

Chadha, A., Florentine, S., Chauhan, B.S., Long, B., Jayasundera, M., Javaid, M.M., & Turville, C., 2019. Environmental factors affecting the germination and seedling emergence of two populations of an emerging agricultural weed: Wild lettuce (*Lactuca serriola*). *Crop and Pasture Science*, 70, 709–717.

Chadha, A., Florentine, S., Javaid, M., Welgama, A., & Turville, C., 2020. Influence of elements of climate change on the growth and fecundity of *Datura stramonium. Environmental Science and Pollution Research*, 27, 35859–35869.

Charlton, D., Rutledge, Z., & Taylor, J.E., 2021. Evolving Agricultural Labor Markets. *In Handbook of Agricultural Economics.* Elsevier, 4075–4133.

Crystal-Ornelas, R., Thapa, R., & Tully, K.L., 2021. Soil organic carbon is affected by organic amendments, conservation tillage, and cover cropping in organic farming systems: A meta-analysis. *Agriculture, Ecosystems & Environment*, 312, 107356.

Das, S., Ray, M.K., Panday, D., & Mishra, P.K., 2023. Role of biotechnology in creating sustainable agriculture. *PLOS Sustainability and Transformation*, 2, e0000069.

Dong, B.-R., Jiang, R., Chen, J.-F., Xiao, Y., Lv, Z.-Y., & Chen, W.-S., 2023. Strategic nanoparticle-mediated plant disease resistance. *Critical Reviews in Biotechnology*, 43, 22–37.

El-Chaghaby, G.A., & Rashad, S., 2024. Nanosensors in Agriculture: Applications, Prospects, and Challenges. *Handbook of Nanosensors: Materials and Technological Applications*, 1303–1330.

Erekalo, K.T., Pedersen, S.M., Christensen, T., Denver, S., Gemtou, M., Fountas, S., & Isakhanyan, G., 2024. Review on the contribution of farming practices and technologies towards climate-smart agricultural outcomes in a European context. *Smart Agricultural Technology*, 7, 100413.

Eskandar, K., 2023. Revolutionizing biotechnology and bioengineering: Unleashing the power of innovation. *Journal of Applied Biotechnology & Bioengineering*, 10, 81–88.

Fonta, W., Edame, G., Anam, B.E., & Duru, E., 2011. *Climate Change, Food Security and Agricultural Productivity in Africa: Issues and Policy Directions.*

Gihring, K., 2022. *Life Cycle Sustainability Assessment of Takeaway Food Packaging: A Case Study from the University of Johannesburg.* University of Johannesburg.

Godswill, A.G., Somtochukwu, I.V., Ikechukwu, A.O., & Kate, E.C., 2020. Health benefits of micronutrients (vitamins and minerals) and their associated deficiency diseases: A systematic review. *International Journal of Food Sciences*, 3, 1–32.

Gomiero, T., 2019. Soil and Crop Management to Save Food and Enhance Food Security. In Saving Food, 33–87.

Grift, T., Zhang, Q., Kondo, N., & Ting, K., 2008. A review of automation and robotics for the bio-industry. *Journal of Biomechatronics Engineering*, 1, 37–54.

Grison, C., Koop, S., Eisenreich, S., Hofman, J., Chang, I.-S., Wu, J., Savic, D., & Van Leeuwen, K., 2023. Integrated water resources management in cities in the world: Global challenges. *Water Resources Management*, 37, 2787–2803.

Hanif, Z., Naeem, M., Ali, H.H., Tanveer, A., Javaid, M.M., Peerzada, A.M., & Chauhan, B.S., 2017. Effect of environmental factors on germination of Salsola foetida: Potential species for rehabilitation of degraded rangelands. *Rangeland Ecology & Management*, 70, 638–643.

Herforth, A., Bai, Y., Venkat, A., Mahrt, K., Ebel, A., & Masters, W.A., 2020. *Cost and affordability of healthy diets across and within countries: Background paper for The State of Food Security and Nutrition in the World 2020. FAO Agricultural Development Economics Technical Study No. 9*: Food & Agriculture Org.

Hu, S., Huang, S., Huang, J., & Su, J., 2021. Blockchain and edge computing technology enabling organic agricultural supply chain: A framework solution to trust crisis. *Computers & Industrial Engineering*, 153, 107079.

Hundal, G.S., Laux, C.M., Buckmaster, D., Sutton, M.J., & Langemeier, M., 2023. Exploring barriers to the adoption of Internet of Things-based precision agriculture practices. *Agriculture*, 13, 163.

Jafari, M., Tahmoures, M., Ehteram, M., Ghorbani, M., & Panahi, F., 2022. *Soil Erosion Control in Drylands.* Springer.

Javaid, M., Haleem, A., Singh, R.P., & Suman, R., 2022a. Enhancing smart farming through the applications of Agriculture 4.0 technologies. *International Journal of Intelligent Networks*, 3, 150–164.

Javaid, M.M., Florentine, S.K., Ashraf, M., Mahmood, A., Sattar, A., Wasaya, A., & Li, F.M., 2022b. Photosynthetic activity and water use efficiency of Salvia verbenaca L. under elevated CO_2 and water-deficit conditions. *Journal of Agronomy and Crop Science*, 208, 536–551.

Jianing, G., Yuhong, G., Yijun, G., Rasheed, A., Qian, Z., Zhiming, X., Mahmood, A., Shuheng, Z., Zhuo, Z., & Zhuo, Z., 2022. Improvement of heat stress tolerance in soybean (*Glycine max* L.), by using conventional and molecular tools. *Frontiers in Plant Science*, 13, 993189.

Kalantari, F., Tahir, O.M., Joni, R.A., & Fatemi, E., 2018. Opportunities and challenges in sustainability of vertical farming: A review. *Journal of Landscape Ecology*, 11, 35–60.

Kalogiannidis, S., Kalfas, D., Giannarakis, G., & Paschalidou, M., 2023. Integration of water resources management strategies in land use planning towards environmental conservation. *Sustainability*, 15, 15242.

Kleen, J.L., & Guatteo, R., 2023. Precision livestock farming: What does it contain and what are the perspectives? *Animals*, 13, 779.

Kour, R., Singh, S., Sharma, H.B., Naik, T.S.K., Shehata, N., Pavithra, N., Ali, W., Kapoor, D., Dhanjal, D.S., & Singh, J., 2023. Persistence and remote sensing of agri-food wastes in the environment: Current state and perspectives. *Chemosphere*, 317, 137822.

Kumar, D., & Dwivedi, R.K., 2023. Designing a secure food supply chain system using blockchain in agricultural IoT. *2023 International Conference on Computational Intelligence, Communication Technology and Networking (CICTN)*, IEEE, 26–31.

Kumar, K., Shinde, A., Aeron, V., Verma, A., & Arif, N.S., 2023. Genetic engineering of plants for phytoremediation: Advances and challenges. *Journal of Plant Biochemistry and Biotechnology*, 32, 12–30.

Lal, R., 2016. Feeding 11 billion on 0.5 billion hectare of area under cereal crops. *Food and Energy Security*, 5, 239–251.

Lee, T.-M., Lin, J.-Y., Tsai, T.-H., Yang, R.-Y., & Ng, I.-S., 2023. Clustered regularly interspaced short palindromic repeats (CRISPR) technology and genetic engineering strategies for microalgae towards carbon neutrality: A critical review. *Bioresource Technology*, 368, 128350.

Lowenberg-Deboer, J., Huang, I.Y., Grigoriadis, V., & Blackmore, S., 2020. Economics of robots and automation in field crop production. *Precision Agriculture*, 21, 278–299.

Lucas, J., 2011. Advances in plant disease and pest management. *The Journal of Agricultural Science*, 149, 91–114.

Ma, Y., Chen, S., Ermon, S., & Lobell, D.B., 2024. Transfer learning in environmental remote sensing. *Remote Sensing of Environment*, 301, 113924.

Marian, L., 2014. Organic" and other product attributes: consumer perceptions and buying behaviour regarding organic food products". *PhD, Aarhus University, Denmark.*

Mehmood, A., Naeem, M., Khalid, F., Saeed, Y., Abbas, T., Jabran, K., Sarwar, M.A., Tanveer, A., & Javaid, M.M., 2018. Identification of phytotoxins in different plant parts of *Brassica napus* and their influence on mung bean. *Environmental Science and Pollution Research*, 25, 18071–18080.

Mgomezulu, W.R., Edriss, A.-K., Machira, K., & Pangapanga-Phiri, I., 2023a. Towards sustainability in the adoption of sustainable agricultural practices: Implications on household poverty, food and nutrition security. *Innovation and Green Development*, 2, 100054.

Mgomezulu, W.R., Machira, K., Edriss, A.-K., & Pangapanga-Phiri, I., 2023b. Modelling farmers' adoption decisions of sustainable agricultural practices under varying agro-ecological conditions: A new perspective. *Innovation and Green Development*, 2, 100036.

Mishra, R.K., 2023. Fresh water availability and its global challenge. *British Journal of Multidisciplinary and Advanced Studies*, 4, 1–78.

Mohammed, M., 2016. "Towards a diversified Nigerian economy" the contribution of agriculture to GDP of Nigeria.

Morshedzadeh, F., Ghanei, M., Lotfi, M., Ghasemi, M., Ahmadi, M., Najari-Hanjani, P., Sharif, S., Mozaffari-Jovin, S., Peymani, M., & Abbaszadegan, M.R., 2024. An update on the application of CRISPR technology in clinical practice. *Molecular Biotechnology*, 66, 179–197.

Naqve, M., Wang, X., Shahbaz, M., Fiaz, S., Naqvi, W., Naseer, M., Mahmood, A., & Ali, H., 2021. Foliar spray of alpha-tocopherol modulates antioxidant potential of okra fruit under salt stress. *Plants*, 10, 1382.

Okomoda, V.T., Oladimeji, S.A., Solomon, S.G., Olufeagba, S.O., Ogah, S.I., & Ikhwanuddin, M., 2023. Aquaponics production system: A review of historical perspective, opportunities, and challenges of its adoption. *Food Science & Nutrition*, 11, 1157–1165.

Ordoñez, C.C., Gonzales, G.R., & Corrales, J.C., 2024. Blockchain and agricultural sustainability in South America: A systematic review. *Frontiers in Sustainable Food Systems*, 8, 1347116.

Palm, H.W., Knaus, U., & Kotzen, B., 2024. Aquaponics nomenclature matters: It is about principles and technologies and not as much about coupling. *Reviews in Aquaculture*, 16, 473–490.

Panday, U.S., Pratihast, A.K., Aryal, J., & Kayastha, R.B., 2020. A review on drone-based data solutions for cereal crops. *Drones*, 4, 41.

Pretty, J., 2008. Agricultural sustainability: Concepts, principles and evidence. *Philosophical Transactions of the Royal Society B: Biological Sciences*, 363, 447–465.

Qiao, Y., Guo, Y., & He, D., 2023. Cattle body detection based on YOLOv5-ASFF for precision livestock farming. *Computers and Electronics in Agriculture*, 204, 107579.

Raji, E., Ijomah, T.I., & Eyieyien, O.G., 2024. Improving agricultural practices and productivity through extension services and innovative training programs. *International Journal of Applied Research in Social Sciences*, 6, 1297–1309.

Rashmi, I., Karthika, K., Roy, T., Shinoji, K., Kumawat, A., Kala, S., & Pal, R., 2022. Soil Erosion and Sediments: a Source of Contamination and Impact on Agriculture Productivity. *Agrochemicals in Soil and Environment: Impacts and Remediation*. Springer, 313–345.

Saeed, S., Khan, S.U., Khan, W.U., Abdel-Maksoud, M.A., Mubarak, A.S., Aufy, M., Kiani, F.A., Wahab, A., Shah, M.W., & Saleem, M.H., 2023. Genome editing technology: A new frontier for the treatment and prevention of cardiovascular diseases. *Current Problems in Cardiology*, 48, 101692.

Sajja, G.S., Rane, K.P., Phasinam, K., Kassanuk, T., Okoronkwo, E., & Prabhu, P., 2023. Towards applicability of blockchain in agriculture sector. *Materials Today: Proceedings*, 80, 3705–3708.

Samikannu, R., Koshariya, A.K., Poornima, E., Ramesh, S., Kumar, A., & Boopathi, S., 2023. Sustainable Development in Modern Aquaponics Cultivation Systems Using IoT Technologies. *In Human Agro-Energy Optimization for Business and Industry*. IGI Global, 105–127.

Sha, G., Sun, P., Kong, X., Han, X., Sun, Q., Fouillen, L., Zhao, J., Li, Y., Yang, L., & Wang, Y., 2023. Genome editing of a rice CDP-DAG synthase confers multipathogen resistance. *Nature*, 618, 1017–1023.

Singh, J., 2023. *Advance Techniques in Climate Smart Farming.*

Solanki, P., Bhargava, A., Chhipa, H., Jain, N., & Panwar, J., 2015. Nano-Fertilizers and Their Smart Delivery System. In Nanotechnologies in Food and Agriculture, 81–101. Springer.

Tang, H., Meng, G., Xiang, J., Mahmood, A., Xiang, G., Sanaullah, Liu, Y., & Huang, G., 2022. Toxic effects of antimony in plants: Reasons and remediation possibilities—A review and future prospects. *Frontiers in Plant Science*, 13, 1011945.

Tanveer, A., Javaid, M.M., Irfan, M., Khaliq, A., & Yaseen, M., 2015. Yield losses in chickpea with varying densities of dragon spurge (*Euphorbia dracunculoides*). *Weed Science*, 63, 522–528.

Tubiello, F.N., Rosenzweig, C., Conchedda, G., Karl, K., Gütschow, J., Xueyao, P., Obli-Laryea, G., Wanner, N., Qiu, S.Y., & De Barros, J., 2021. Greenhouse gas emissions from food systems: Building the evidence base. *Environmental Research Letters*, 16, 065007.

Tzanidakis, C., Tzamaloukas, O., Simitzis, P., & Panagakis, P., 2023. Precision livestock farming applications (PLF) for grazing animals. *Agriculture*, 13, 288.

Ullah, A., Zeb, A., Shah, A.A., & Bavorova, M., 2024. From education to rehabilitation: Empowering farming communities through extension services for landscape restoration. *Environment, Development and Sustainability*, 26, 11681–11701.

Van Gerrewey, T., Boon, N., & Geelen, D., 2021. Vertical farming: The only way is up? *Agronomy*, 12, 2.

Villarreal, M., 2022. Can hunger be eradicated by 2030? *SOStenibilità*, 15.

Wakweya, R.B., 2023. Challenges and prospects of adopting climate-smart agricultural practices and technologies: Implications for food security. *Journal of Agriculture and Food Research*, 14, 100698.

Wang, J.Y., & Doudna, J.A., 2023. CRISPR technology: A decade of genome editing is only the beginning. *Science*, 379, eadd8643.

Zehtabian, E., Masoudi, R., Yazdandoost, F., Sedghi-Asl, M., & Loáiciga, H.A., 2023. Investigation of water allocation using integrated water resource management approaches in the Zayandehroud River basin, Iran. *Journal of Cleaner Production*, 395, 136339.

2 Precision Agriculture for Sustainable Crop Management

Asma Zafar, Muhammad Saad Ullah, Aqsa Naseem, Aqsa Hanif, Atman Adiba, Muhammad Mansoor Javaid, Javaria Nargis, and Faiza Fareed*

2.1 INTRODUCTION

According to the European Parliament, precision agriculture (PA) is a farming management concept based on observing, measuring, and responding to inter and intra-field variability in crops, or to aspects of animal rearing (Petrović *et al.*, 2024). A significant concern today is the need to sustain a gradually expanding global population while simultaneously mitigating ecological consequences and conserving natural resources for future generations. The increasing demand for food exerts substantial pressure on the agricultural sector to boost production, often resulting in intensified production methods, with adverse environmental impacts. These methods further increase the competition for the already scarce resources of land and water. Food production and environmental protection interact in complex ways and this interplay is an important issue, too. Hence, finding new ways to establish and run agricultural systems is becoming increasingly essential.

With its beneficial nature and innovative technology approach, PA is the future of the agrarian lifestyle that uses modern technologies to optimize the quantity and quality of crop yield while minimizing the cost and environmental impact (Karunathilake *et al.*, 2023; Zafar *et al.*, 2023). There is traditional farming practices that apply the same inputs across the field, regardless of the variability within it, have led to the degradation of the environment. These practices often lead to excessive use of water, fertilizers, and pesticides that lead to soil degradation, water pollution, reduced biological diversity, and aggravated climate change (Ahmad and Dar, 2020; Hassan *et al.*, 2022). In contrast, precision farming emphasizes the precise management and preservation of individual fields rather than the whole region, employing a disciplined strategy to optimize both economic and environmental benefits (Comi, 2020; Mahmood *et al.*, 2021). PA is an efficient agricultural system that applies technology and management measures to achieve optimal agricultural productivity. Paying attention to best practices in crop planning, tillage, planting, chemical application, harvesting, and post-harvest processing has a substantial impact on productivity, quality of the products, and the extent to which the agricultural activity affects the environment.

Nowadays, PA uses advanced technology, including global satellite systems, remote sensors, and geographic information, to help farmers dynamically observe and assess factors such as soil health, crop diseases, insect infestations, humidity, and temperature levels. Data-driven technology allows farmers to make informed decisions about where and when to plant, fertilize, and harvest, creating a more sustainable and centric agricultural business (Rasheed *et al.*, 2022b; Surendran *et al.*, 2024).

In addition, PA gives farmers the ability to monitor crop conditions, minimize potential damage, and maximize resource use by applying the right amounts of pesticides, fertilizers, and irrigation water (Wasaya *et al.*, 2021). This system would solve the problems of climate change and a growing world population while reducing the use of resources and increasing the crop yield. Sensors are used to measure environmental variables that can modulate the climate to grow every crop, therefore it

* Corresponding author: adiba.atman@gmail.com

 DOI: 10.1201/9781003498698-2

is an ideal approach as a highly effective and efficient agricultural methodology (Hatfield *et al.*, 2020; Rasheed *et al.*, 2022c). This holistic approach allows us to integrate data analysis, decision-making protocols, learning from the past, record-keeping, specialized equipment, and evaluation and revision processes into a single framework. PA is an approach that has also evolved into other disciplines including but not limited to precision horticulture, precision viticulture, and precision animal husbandry.

PA provides significant worth by reducing the ecological impacts of farming, enhancing productivity, raising net profit, promoting resource exploitation proficiency, and reducing the producer's risk to hazard risk of reducing net profit (Steensland and Zeigler, 2021). Moreover, PA reduces the dependency on pesticides, decreases groundwater loss, and offers cost-effective pest and disease management measures, whilst increasing production and protecting the environment (Hanif *et al.*, 2017). It also helps in pest monitoring and forecasting accurately which leads to better crop management. While PA technologies have established benefits and increasing interest aids as the next solution to modern agricultural challenges, planting the seeds for PA technologies adoption can still be difficult, particularly in regions of the world with limited infrastructure and resources. Thus, in order to perform accurate assessments on the adoption of these technologies, a complete analysis covering the costs and the profitability associated with these technologies becomes necessary. It is critical for farmers and policy-makers to understand fully the cost and efficiency of sharp PA uptake. From investment decisions and technology adoption to national food security, our understanding of this issue could have ramifications that reverberate throughout the agricultural supply chain.

PA improves grain production on large-scale farmlands through its combination of tillage, sowing, fertilizing, harvesting, and animal husbandry. The use of variable rate application (VRA) makes an excellent operation, and the greater production of valuable fruits, vegetables, and vineyards with quality also means better profitability. PA also improves the efficiency of irrigation, supports water conservation (e.g. by encouraging appropriate irrigation), and improves the welfare of livestock (Sishodia *et al.*, 2020). Moreover, PA helps farmers comply with high agricultural standards, increase their position as public goods suppliers, and promote sustainability initiatives by measuring their environmental imprint.

Agriculture has been revolutionized, with precision production techniques constantly being needed. The rapid pace of technological advancement has led to an enormous divide between traditional and contemporary techniques, affecting the landscape. The desire for trustworthy, ethically produced products is increasing due to consumer demand and shifting ethical considerations. As a result, farmers are concentrating on growing organic crops and catering to the niche market to increase profits. This specialization includes the production of goods without genetically modified organisms, the cultivation of specialist crops, and the breeding of rare animal species.

A recent systematic review indicates that PA increases agricultural efficiency and decreases environmental impact. The high investment needed at the very beginning is a critical issue with PA Precision agronomic equipment is usually costly, and because agronomic conditions differ, detailed SWOT (strengths, weaknesses, opportunities, and threats) analyses of cost-benefits must be complete (Ara *et al.*, 2021). Larger farms can afford PA technology due to their high initial costs, affecting adoption rates. Countries with large agricultural sectors utilize PA technology more frequently. This review explores the advancements, challenges, and opportunities within PA and smart farming, discussing the role of drones, sensors, and machine learning (ML). It also addresses issues related to data management, technological adoption, and the cost-effectiveness of PA.

2.2 EVOLUTION OF AGRICULTURAL PRACTICES

The term "PA" was first used in 1990, gaining momentum with advancements in information technologies. A pivotal moment occurred in 1983 when U.S. President Ronald Reagan authorized the civilian use of global positioning systems (GPS), marking the onset of the modern era of PA

(Tetik, 2023). The integration of GPS receivers into tractors began in 1991, and by the mid-1990s, the U.S. Department of Defense eliminated selective availability for differential GPS (DGPS), resulting in an improved positioning accuracy of 5–10 m. Additional enhancements were made in the 2000s. Simultaneously, electronic identification (EID) technologies were created to efficiently handle individual animals (Tzanidakis *et al.*, 2023).

Precise farming techniques emerged with the advent of geographic information systems (GIS), GPS, yield monitors, and other data-generating technology in the 1990s that gave rise to accurate agricultural methods. In PA, motorized equipment was initially used solely for executing agricultural tasks while problem identification and decision-making processes remained under human control (Cheng *et al.*, 2023). Technological advancements during the Third Industrial Revolution 3.0 enabled the digitization of PA through the integration of information technology and improving automation capabilities. This digitization marked a transition from traditional agricultural practices reliant on manual tools, animal power, and motorized equipment to a current dependence on digital technology in farming (Giorgio *et al.*, 2024).

Through case studies, data analysis, and stakeholder collaboration, it is clear that variable rate technologies (VRTs), computerized mapping, production monitors, and directing farming systems mainly contribute to PA. In 1988, VRA was first demonstrated in northern Germany and Denmark, soon after the civilian use of GPSs in 1983. Advancements made toward remote sensing technology like satellites, drones, ground sensors, and field staff have facilitated farmers to capture vast field data systematically (Alexopoulos *et al.*, 2023). This enhanced their ability to make accurate crop management assessments significantly. In the early 2000s, yield monitors made it possible for farmers to monitor crop yields in real time, enabling them to manage their land based on specific conditions and changes in the field. PA includes other agricultural production methods in addition to crop cultivation, such as agronomy, animal husbandry, aquaculture, and agroforestry (Misra and Ghosh, 2024). Milestones in the history of PA is shown in Figure 2.1.

2.2.1 Principles of Precision Agriculture

PA operates on several key principles (Figure 2.2). The first demonstration of VRA took place in northern Germany and Denmark in 1988, shortly after the civilian use of GPSs became available in 1983. In the next decade, GPS technology allowed farmers to precisely determine and map the borders of their fields. Innovation in remote sensing technology, including satellites, drones, ground-based sensors, and field personnel, has enabled farmers to collect accurate field data. This approach enhances resource efficiency and reduces environmental impact (Huo *et al.*, 2024; Ullah *et al.*, 2013). Furthermore, it significantly improved their capacity to make accurate assessments of crop management. Integrating data from GPS, sensors, drones, and satellite images is crucial. Decision support systems, utilizing predictive analytics and ML tools, further assist farmers in optimizing operations, maximizing yields, and minimizing costs. PA also allows continuous monitoring and management of farming operations throughout the growth cycle, allowing agricultural technology assessment and adaptation. The primary goal is to attain economic and environmental sustainability by efficiently balancing profitability and environmental stewardship. This is achieved by optimizing resource use and minimizing inputs wherever possible (Yang and Solangi, 2024). PA enables the constant surveillance and regulation of farming activities during the whole cultivation period, enabling the continuing evaluation and modification of agricultural techniques. The adoption of advanced technologies, including Internet of things (IoT), artificial intelligence (AI), and robotics drives innovation and efficiency in agricultural production, fostering more sustainable and productive farming practices overall. PA encompasses a range of advanced technologies, including remote sensing, automated hardware and software, telematics, drones, autonomous vehicles, GPS systems, and robots. These technologies support farmers in three crucial agricultural processes: diagnosis, decision-making, and execution, thus improving overall productivity and sustainability in agriculture (Fathollahi-Fard *et al.*, 2023; Rasheed *et al.*, 2022a).

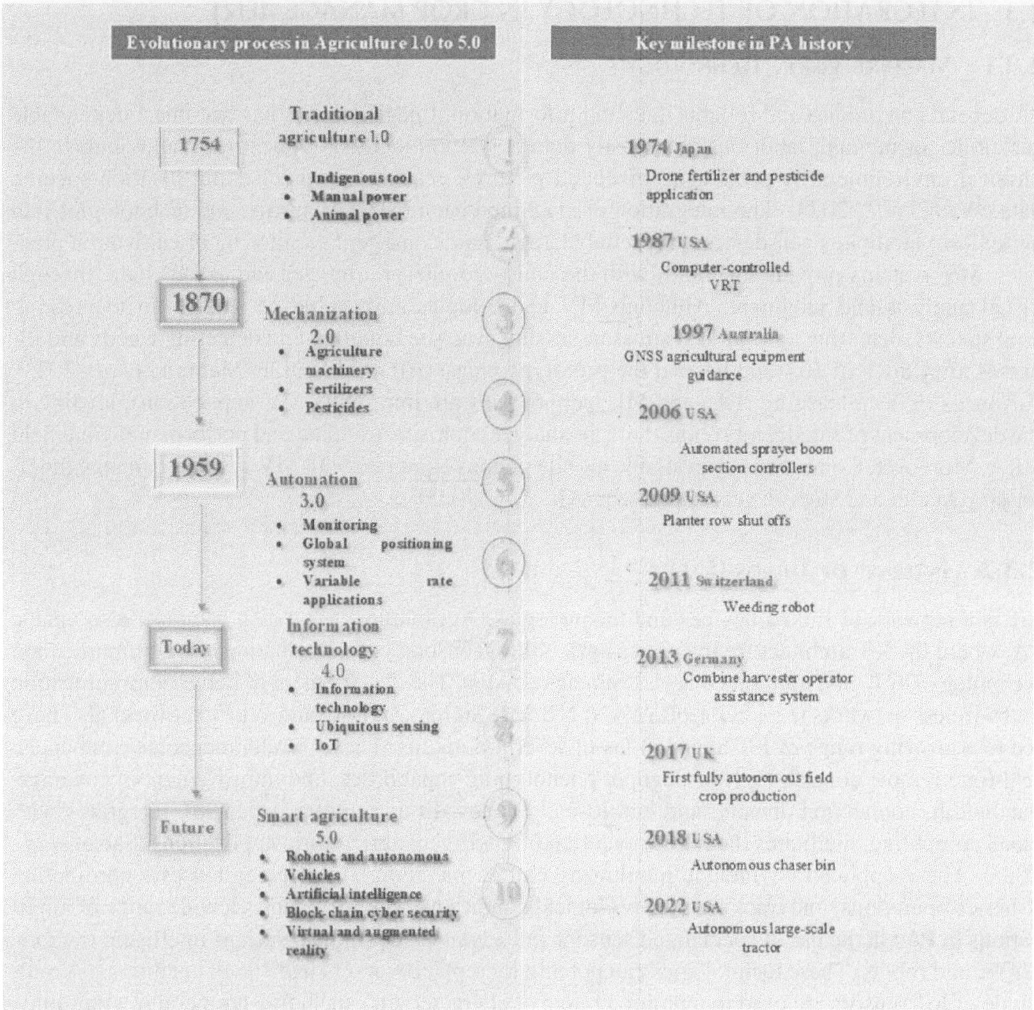

FIGURE 2.1 The milestones in the history of PA.

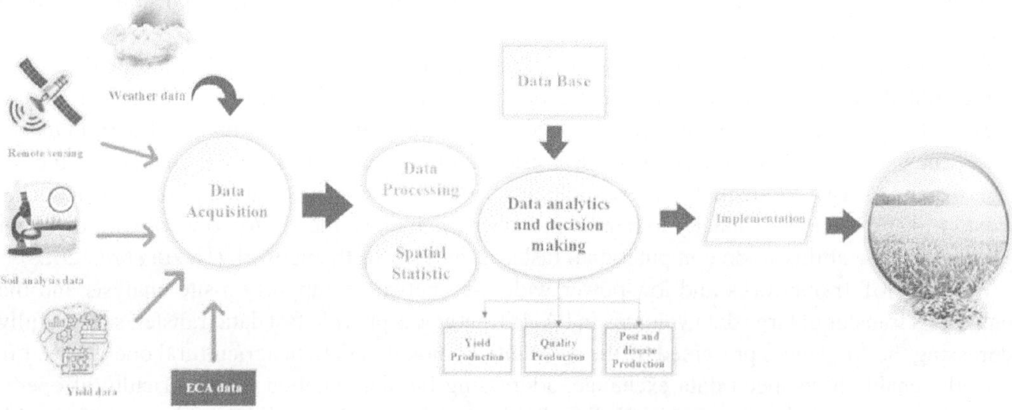

FIGURE 2.2 Key principles of precision agriculture.

2.3 INTEGRATION OF TECHNOLOGY IN CROP MANAGEMENT

2.3.1 Machine Vision Technology

PA depends on precise and reliable data and information. Image analysis has become a dependable substitute for manual, laborious, and costly data collection methods. Machines could analyze the physical environment by processing pixelated pictures, resulting in precise and location-specific data (Wang *et al.*, 2023). The integration of machine vision (MV) or agro-vision technologies into agriculture facilitates non-destructive, reliable, rapid, and consistent monitoring of cultivation practices. MV systems provide machines with the ability to interpret images and extract data, through visual analysis and judgment. Although MV technologies have achieved success in identifying crop species, detecting agricultural stress, assessing crop seed quality, and detecting weeds and illnesses, they are still evolving beyond the prototype phase (Ali *et al.*, 2013b; Mahmud *et al.*, 2023). Advances in deep learning (DL) and ML technologies are improving MV applications, leading to the development of intelligent robots that can analyze multispectral data and perform real-time field VRA. Moreover, Commercially available mobile phones equipped with MV systems can now monitor crop health and stress levels effectively (Ali *et al.*, 2023).

2.3.2 Internet of Things (IoT)

IoT is a network of linked devices and technologies. Agriculture It is a data acquisition to enable PA, where the IoT architecture integrates agricultural sensors with information and communication technology (ICT) and unmanned aerial vehicles (UAVs). The development of better communication and wireless networks (e.g., 5G, LoRaWAN, NB-IoT, Sigfox, ZigBee, and Wi-Fi network) also have led to a growing range of IoT applications in several domains. These developments have enhanced real-time remote control, high-throughput phenotyping capabilities, and improvement in coverage, bandwidth, connection density, and end-to-end latency. In agriculture, when IoT integrates with cloud computing, facilitates the implementation of intelligent agricultural applications (Saban *et al.*, 2023). These applications include monitoring cattle, managing smart greenhouses, coordinating fisheries operations, and tracking meteorological conditions. The IoT enables a wide range of applications in PA via the use of specialized sensors and advanced equipment such as intelligent tractors, UAVs, and robots. These technologies can potentially replace workers in difficult conditions. A multitude of IoT sensors are used to monitor a range of characteristics including temperature, humidity, light intensity, pressure, CO_2 levels, insect infestations, plant health, sunshine intensity, and wind speed. These sensors provide crucial data to cloud-based management systems. Agricultural robots, autonomous platforms, machinery, and weather stations are directly connected to IoT sensors to provide real-time monitoring. UAVs that include (IoT) capabilities provide dependable and accurate 3D monitoring across various geographical regions, delivering high-quality and high-resolution observations (Mohsan *et al.*, 2023).

Wireless IoT frameworks and low-power wide-area networks provide real-time analysis and uninterrupted transfer of substantial volumes of data in PA. However, the use of IoT in agriculture faces challenges such as costs, complex procedures, technological constraints, and data management issues. To address these challenges, it is essential to develop IoT solutions that are both cost-effective and energy-efficient. These technologies should also consider variables such as the delay in data transmission, the capacity to handle increasing demands, the ability to work with different systems, and the ability to do computational tasks at the edge of the network (Hazra *et al.*, 2023).

Wireless IoT frameworks and low-power wide-area networks support on-site analysis and the continuous transfer of large data volumes in PA. IoT networks provide fast data transfer, successfully addressing the challenges presented by the substantial amount of data in agricultural operations. IoT networks enable high-speed data exchange, addressing big data challenges in agricultural operations. Furthermore, advancements in IoT technology extend functionalities such as cattle health and welfare management and weed detection using machines by reducing data package sizes, edge computing also relieves some of the pressure on centralized cloud resources, delivering increased efficiency to real-time data transmission in IoT PA. Internet and communication firms are enhancing the capabilities of cloud services to be used in edge networks, hence improving the efficiency and speed of edge-IoT systems for faster convergence and completion of tasks (Ahmad *et al.*, 2023).

2.3.3 Artificial Intelligence (AI), Machine Learning (ML), and Deep Learning (DL)

AI is crucial for robotics and autonomous systems. Its IoT connectivity enables continuous data feeds, which are essential for analyzing agricultural data and gaining valuable insights. By examining environmental data and historical farming records, AI leverages big data to uncover hidden patterns, aiding in pest identification, disease detection, yield forecasting, and fertilizer planning in agricultural decision support systems (Javaid *et al.*, 2023). AI can minimize food waste, improve production cleanliness, and monitor machinery across agriculture's supply chain, production patterns, soil, crop, water, and disease and insect control. DL and ML are key to AI. ML enhances AI performance by independent learning and explicit feature extraction, whereas DL handles complex problems with layered nonlinear functions. ML algorithms improve detection and efficiency in agriculture's robots, mobile terminals, and smart devices. ML and DL models validate data and increase understanding of dynamic agricultural situations, from plant tissue cultures to breeding (S-Julián *et al.*, 2023).

Farmers with plenty of data still struggle with processing speed and data visualization. Research in big data, IoT, ML, and DL might address these issues by offering accurate forecasts and new agricultural opportunities. Mobile apps have employed AI, ML, DL, and MV technologies, especially when dealing with massive volumes of data, where ML algorithms learn from the data and generate predictions. However, collecting initial ML data from commercial farms can hinder its adoption. Despite this, advancements in IoT sensors and AI-based technologies such as autonomous machines, cloud computing, edge computing, and blockchain are expected to simplify the conversion, storage, and processing of data for creating ML models in PA (Kumar Kasera *et al.*, 2024).

ML techniques like clustering, decision trees, neural networks, ensemble learning, support vector machine, and Bayesian that deal with crop quality, disease and weed detection, species identification, animal welfare, livestock production, water and soil management, and yield prediction. AI and IoT improve soil, crop, disease, and weed control in AI-driven smart farming. Mobile expert systems enable farmers to utilize cell phones for disease diagnosis, species identification, and soil health analysis through applications. AI allows real-time satellite picture analysis to track agricultural growth. AI can adapt to changing agricultural microclimates. However, standardizing heterogeneous solutions remains challenging. Effective collaboration between AI researchers and farmers is essential for understanding agricultural issues and solutions. Addressing accessibility and privacy issues in large datasets will further improve AI's impact on agriculture (Rozenstein *et al.*, 2024).

2.3.4 Guidance Systems

Guidance systems enhance planting, harvesting, and herding by providing farmers with real-time GPS information on equipment and herd-grazing sites. Global Navigation Satellite System (GNSS) has streamlined labor-intensive agricultural activities with more efficient systems like VRA despite few satellites, poor signals, and unpredictable connections. Agriculture 3.0 replaced manual inputs with digitalized machinery. The rising commercialization of agricultural equipment services necessitates good management to minimize overuse or underuse (Moreno and Andújar, 2023). Fully automated steering solutions, enabled by GNSS, save time, labor, and money by optimizing equipment performance. Efficient precision agricultural robots and rovers require high-resolution navigation. Some researchers have combined DL propagation models with GNSS-fused inertial navigation datasets to improve agricultural accuracy. Electric seeders with optical fiber detection have been created and tested. GIS-integrated farm management software automates data collection, analysis, supervision, storage, decision-making, and overall farm management (Ali *et al.*, 2013a; Karydas *et al.*, 2023).

2.3.5 Blockchain Technology

Blockchain is a decentralized, distributed database used primarily in cryptocurrencies to organize information into blocks. The transparency, immutability, and reliability of data within a blockchain boost confidence in the supply chain. Blockchain helps PA by integrating digital technology and addressing smart farming challenges such as insecure data-sharing infrastructure. The "IoT Applied Greenhouse Monitoring System." uses blockchain technology to remotely monitor and manage agricultural equipment (Raj *et al.*, 2024). Lightweight, distributed, decentralized, and transparent security and privacy solutions address decentralization, anonymity, and security in smart farming IoT applications. Blockchain is a fast, safe, and reliable technology for farm monitoring, though it is still evolving. Secure communication in dispersed networks improves scheduling for agricultural activities such as irrigation water management, energy consumption, machine and labor integration, and robot collaboration and autonomous UAV missions. Blockchain is essential for preserving food and resolving unequal and fragmented information caused by supply chain inefficiencies (Yogarajan *et al.*, 2023).

2.3.6 Robotics and Autonomous Systems

Autonomous farming is making extensive use of robots, sensors, drones, and remote sensing to perform tasks such as planting, spraying, harvesting, and weeding, which reduces labor expenditures and increases the efficiency of decision-making. Remote sensing and automation systems (RASs) introduce and assimilate the fundamental modern technology elements into the technological basis of agricultural production processes and patterns. Despite variables and uncertainty, mobile robots combined with sensors, actuators, and ML algorithms can automate operations in farming. RAS applications span various domains including plant manufacturing, 3D food printing, and biodiverse farming (Rowan, 2023).

Enhancements are necessary for the capabilities of RAS to provide precise guidance, autonomous navigation, and accurate identification in ever-changing agricultural environments. These settings include variations in visual appearance, stages of development, atmospheric conditions, and object overlap. Currently, RAS applications encompass various activities such as robot-assisted plant phenotyping, fruit enumeration, harvesting, leaf removal, targeted spraying, and 3D mapping. Navigation systems such as light bars and GNSS are included in autonomous agricultural vehicles, which are used for operations including plowing, sowing, spraying, and harvesting. These systems ensure that the positions of equipment are shown visually to avoid skipping or overlapping, which is essential for implementing VRA (Ahn *et al.*, 2023).

2.3.7 Artificial Satellites, Unmanned Aerial Vehicles (UAVs), and Unmanned Ground Vehicles (UGVs)

Remote sensing utilizes artificial satellites like Landsat (U.S.A.), Sentinel-2 (European), Rapid Eye, GeoEye-1, and WorldView-3 to generate multispectral data. These satellites provide extensive coverage for agricultural monitoring. Recently, affordable drones, UAVs, and airplanes have permitted closer-to-ground and more frequent picture acquisition, improving detail and utility. Unmanned ground vehicles (UGVs) collect high-resolution data for weed identification, targeted pesticide spraying, soil analysis, and crop scouting. Specialized scouting robots like Oz for mechanical weeding, Global Unmanned Spray System (GUSS) for spraying, Row Bot for fertilizing and sowing, and Vine Robots for vineyard management, contribute to achieving agricultural goals. PA relies on satellite, UAV, and UGV imagery for tasks such as vegetation patch identification, weed recognition, insect detection, environmental stress monitoring, and VRT (Javaid *et al.*, 2012; Zualkernan *et al.*, 2023). This data also supports aquaculture, agroforestry, and forestry by providing extensive and repeatable information. Ground survey data paired with ML and DL algorithms transforms satellite, UAV, and UGV pictures into useful information. Hyperspectral and RGB images from UAVs like Sentinel-2 can estimate forest density faster and easier than before. Satellites and drones effectively distinguish vegetative types and tree species, making remote sensing more effective than LiDAR for deforestation monitoring (Sinegalia *et al.*, 2024).

Although there are certain restrictions, the use of drones and remotely piloted aircraft has experienced significant growth. This growth is attributed to their ability to provide accurate and cost-efficient information for PA. This is made possible via the employment of advanced technologies such as hyperspectral sensors and multispectral cameras. These technologies have the potential for monitoring large agricultural fields, cultivated lands, and forested regions (Abdulraheem *et al.*, 2023).

2.3.8 High-Throughput Phenotyping

One of the potential methods is high throughput phenotyping high-throughput phenotyping used both for quantitative and qualitative plant trait measurements technology, behind used is the use of remote sensing, spectral imaging, and robots to gather plant data to accelerate crop breeding. Traits such as growth, yield, disease resistance, and morphology. With insights into how crops perform, these data could empower farmers to make more informed choices on fertilization, irrigation, harvesting, or pest control. It helps breeders invent a variety of crops that perform well in local conditions and give higher output (Potts *et al.*, 2023). ML approaches such as least squares regression (LSR) have been adopted with visible and near-infrared hyperspectral data to estimate biochemical and physicochemical properties, in addition to calculating vegetative indices. In addition to GPS, improved crop rotation strategies, and so on, developing AI-driven, ML-based precision agricultural systems that can be delivered with UGVs, UAVs, hyperspectral images, and high-quality satellite photos that rapidly detect diseases, pests, nutritional deficiencies, and weeds will go a long way. High-throughput phenotyping technology could revolutionize agriculture with precise, real-time, and efficient monitoring of agricultural practices. This breakthrough intends to boost agricultural output, reduce environmental impact, and boost global food security (Potts *et al.*, 2023).

2.4 APPLICATIONS OF PRECISION AGRICULTURE

The field of agricultural robotics and automation aims to enhance efficiency and reduce reliance on manual labor. PA technologies are crucial for enabling greater levels of autonomous operation by providing crucial navigation, perception, and decision-making skills. Figure 2.3 shows the applications of PA. The existing classifications of agricultural robotics and automation encompass:

1. *Guidance systems*: Enable tractors, sprayers, harvesters, and other machinery to autonomously navigate along accurate routes utilizing GNSS location and computerized steering control. They minimize soil compaction by reducing overlaps and allowing for regulated traffic patterns (Torres-Torriti and Burgos, 2023).
2. *Unmanned aerial vehicles*: UAVs with remote sensing payloads, such as RGB, multispectral, hyperspectral, and thermal cameras, are used to collect aerial data for informed management decisions. Applications include irrigation scheduling, disease detection, and weed mapping (Marques *et al.*, 2024).
3. *Autonomous equipment*: Robotic tractors with the ability to carry out tasks such as plowing, sowing, cutting, and transporting are under development and available in limited commercial capacity. These advancements are driven by improvements in computer vision, LiDAR sensors, and AI (Gupta and Gupta, 2024).

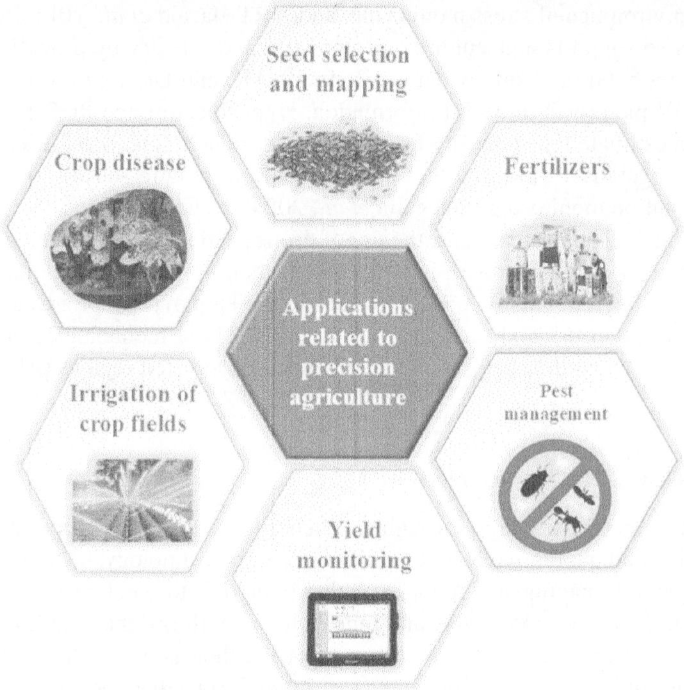

FIGURE 2.3 Applications of precision agriculture.

4. *Fruit-picking robots*: Prototype autonomous robotic harvesters designed for the picking of various crops such as apples, oranges, and strawberries. Utilize designs employ gripper designs, and employ gentle handling techniques to pick crops without causing bruising. Could assist in mitigating workforce shortages (Gupta and Gupta, 2024).
5. *Weed control robots*: Automated weeders use computer vision, mechanical actuation, or electrothermal/chemical treatments to specifically target undesired plants located between rows of crops or individual weeds within agricultural fields (Monteiro and Santos, 2022).
6. *Herbicide reduction*: Agricultural robots and automation are anticipated to play a significant role in the widespread adoption of PA and the promotion of sustainable intensification. The decreasing prices and enhancing capabilities will facilitate higher labor efficiency and reduce environmental impact (Li *et al.*, 2020b).
7. *Weed detection*: Weed infestation spots may be identified for targeted management using DL algorithms that have been trained on field pictures, reducing herbicide use. Tools for prescribing variable rates: Platforms that combine and evaluate several layers of field data to automatically provide application maps for best outcomes. Facilitates the implementation of virtual reality (VR) systems (Ali *et al.*, 2013a; Korkut and Surer, 2023).
8. *Chatbots/Digital advisors*: Deliver customized suggestions and immediate responses to inquiries for farmers by using models and analyzing data collected from the fields. Facilitates widespread access to agricultural knowledge and experience (Steinke *et al.*, 2022).
9. *PA management platforms*: These platforms integrate data collection, analysis, and visualization into a single system designed for informed decisions-making. Advancements in data standardization and integration will improve the capacity of different systems to work together effectively (Boehm *et al.*, 2022).

2.4.1 Soil Selection and Planning

Agriculture depends on soil for crop production. Soil analysis assesses its physical, biological, and chemical properties to guide field management and optimize agricultural practices. The global adoption of innovative technologies is essential to meet food demand, as precision farming maximizes yields from limited arable land. Autumn soil testing based on weather and soil conditions, reveals nutrient levels, fertilizer demands, crop history, soil type, and irrigation needs. Sensor-based soil evaluation and planning systems enable crop selection for soil conditions. Vision-based technology combined with sensors accurately measures seed and plant depth and distance, enabling effective planting. Agribot, which uses GPS, sensors, and autonomous vision systems for seed planting, shows how contemporary technology may optimize crop choices and land utilization. These advances enable farmers to make educated choices, boosting agricultural output and sustainability (Khan *et al.*, 2021).

2.4.2 Irrigation of Crop Fields

Optimizing agricultural productivity and managing global freshwater supplies requires efficient, demand-driven irrigation systems. This means that new improved water management systems need to be introduced as soon as possible. Crop fields rapidly absorb, use, and deplete available water, due to which irrigation scheduling is critical to obtain optimum yield and prevent waste of the already scarce freshwater resources on the surface of the Earth. Only 0.5% of Earth's water is fresh, with most of the rest trapped in polar ice caps or glaciers (Cendón-Flórez and Pippa, 2022). Agriculture accounts for around 70% of this available fresh water, which rises to 85% in underdeveloped countries, emphasizing the dire requirement for improved water management. To overcome the limitations of traditional irrigation, researchers are investigating IoT and wireless sensor network (WSN)-enabled smart irrigation systems. The implementation of WSN-based soil moisture sensors is designed to allow farmers to be notified of irrigation needs to promote more efficient water use. Systems powered by the (IoT), such as crop water stress index (CWSI), allow sensor data to be combined with weather forecasts and satellite images. It allows farmers to make accurate irrigation decisions to provide crops with enough moisture while protecting vital freshwater resources. These technologies ensure proper water consumption for farm field crops, ultimately leading to better agricultural productivity and sustainable environmental practices (Javaid *et al.*, 2022).

2.4.3 Crop Disease

One of the major challenges to the sustainability of many different farming methods around the world is crop diseases, which have plagued agriculture from the time it began. Diseases are caused by many different pathogens, such as fungi, bacteria, viruses, and nematodes. Major challenges in the global agricultural landscape include diseases like leaf spot, apple scab, potato scab, anthracnose, late blight, early blight, and powdery mildew (Sawicka *et al.*, 2020). These disorders result in extensive financial losses of up to billions of dollars. Crop diseases and pests account for 20–40% of global crop losses, as per the Food and Agriculture Organization (FAO). Paying attention to these issues is crucial to avoid much bigger possible losses down the line. Wireless sensors, drones, and other complex IoT devices can be used to monitor agricultural conditions with the help of IoT (Kim *et al.*, 2020; Mehmood *et al.*, 2018). Data is collected about the crops, and it will be evaluated by advanced processors and decision-support systems powered by ML. Advanced techniques have revolutionized our approach to agricultural diseases. This could improve global surveillance and prevention of agricultural diseases (Yadav *et al.*, 2023).

2.4.4 Pest Management and Yield Monitoring

Modern agriculture relies on pest control and monitoring crop yields, which are essential to optimize productivity and sustainability. It is common for crops to be treated with pesticides, herbicides, and insecticides by farmers to protect them from damage. But, excessive use or abuse of these chemicals can upset the natural balance and have an adverse effect on the environment. This is where IoT technologies can track pesticide usage in real-time with environmental monitoring and predictive disease modeling to increase efficiency and prevent negative effects on the environment (Akhbarifar *et al.*, 2023; Chadha *et al.*, 2019). Yield monitoring systems use wireless sensors and associated technology to monitor the development of crops from planting to harvesting. These systems also assist in managing issues related to storage, transportation, and disease control.

2.5 CHALLENGES

PA is an advanced technology that effectively addresses the challenges posed by the growing demand for limited resources. Despite its benefits, farmers and researchers encounter certain hurdles while implementing it (Chadha *et al.*, 2020; Dhillon and Moncur, 2023). Farmers are increasingly interested in IoT, wireless sensor, and actuator business models due to their numerous advantages. Farmers may earn more by collecting agricultural data using IoT. Current IoT platform service providers misuse farmer data by offering limited free and comprehensive services with different subscription levels. This exploitation makes farmers worried about data monitoring (Ayoub Shaikh *et al.*, 2022). Developing smart agriculture solutions demands meticulous cost analysis. Sensors, gateways, and base station infrastructure make IoT costly initially. Maintenance costs also need to be considered. Researchers should save money and boost productivity. Limited awareness of new technology among farmers inhibits IoT adoption in rural areas. Automation and mechanization will lose employment since 60–70% of the world's workforce is in agriculture, especially in Asia, Africa, and Oceania. Large amounts of agronomy workers must be relocated to other sectors, which is challenging in underdeveloped countries like India (Akhter and Sofi, 2022).

Many factors impact IoT system dispersion (Table 2.1). For efficient operation, IoT devices must be portable and linked. The device must connect anywhere, anytime, with little setup. Roaming support and system stability are crucial for accurate results. IoT infrastructure for static devices needs review. Size matters in management zones (Bansal and Kumar, 2020). Maintaining cropland involves assessing the production function to evaluate output and quality influenced by crop and field zones. However, many farmers are ignorant of this change. PA requires farmers to understand production functions (Abdullah *et al.*, 2022; Ahmed *et al.*, 2022). Extra or moderate fertilizer and pesticide use may damage crops. Another key problem is interoperability, the ability of devices, products, services, and systems to communicate (Sinha and Dhanalakshmi, 2022). As original equipment manufacturers (OEMs) develop advanced precision agronomy equipment, devices, and platforms, interoperability is essential. Addressing technical, semantic, syntactic, and

TABLE 2.1
Factors Affecting the Adoption Level of PA Technologies

Factor	Explanation	References
Features of the farm	Farm factors like size, kind, debt, soil texture, and field variability	Zawadzka *et al.* (2021)
Economic factors	Equipment cost, profitability, rental potential	de Oliveira Azevêdo *et al.* (2020)
Features of the technology	Availability, time to learn equipment use, ease of use, technical assistance, system complexity, brand and model compatibility	Almaiah *et al.* (2022)
Technical staff	Quality, quantity, increased labor expenses, availability, and proximity of technical staff	Li *et al.* (2020a)
Multidisciplinary cooperation	Collaboration between engineers, agronomists, and scientists	Sullivan (2023)

organizational interoperability. Instead of smart devices, gateways, or appliances, researchers could build farmer-friendly platforms (Das *et al.*, 2022).

Today, every company and organization use data analytics to boost revenues, including those in smart agriculture. Data analytics presents several obstacles for researchers. Integration is vital, merging data from various sources and formats into one perspective. Due to the different patterns of semi-structured and unstructured data, data integration is difficult yet essential for significant insights. Innovative techniques like edge computing-driven K-Means clustering algorithms are being used to integrate data from different sources effectively (Lazzeri and Robsky, 2024). Knowledge mining is needed to provide solutions for massive IoT data that can be applied to fresh data. Big data complicates data extraction and investigation posing challenges for data mining activities including communication, integration, exploration, extraction, purification, reduction, and transfer (Rani *et al.*, 2022).

ICT-based agricultural gadgets are used for several reasons. Socio-demographic characteristics like age, innovation, income, risk perception, education, and awareness are significant factors. Mobile internet technology is increasingly popular among younger, more inventive farmers (Charania and Li, 2020). Wealthier farmers and those less risk-averse employ ICT applications more. Agronomists and farmers' internet use is heavily influenced by education. Adoption rates are greater on bigger farms with stronger internet connections. Application design, encompassing system development, and user-friendliness, encourages ICT tool adoption (Nguyen *et al.*, 2020). Digital precision agricultural technologies are difficult for farmers to implement. Farmers worry about data abuse and lack of control, which hinders usage. Data flow from cloud platforms to edge computing platforms is slowed by poor infrastructure, slowing processing. Edge computation and scalability are hindered by the urban-rural digital divide, which limits internet and broadband availability (Imran *et al.*, 2024). A software ecosystem (SE) design emphasizes data management, flexible configuration, and organizational cooperation to improve interoperability. However, the link between SE and edge processing is ambiguous, and real-world implementation typically falls short of optimal settings. Precision agricultural technology is hampered by socioeconomic factors including high adoption costs and investment recovery issues (Bosompem, 2021). Technology adoption and socioeconomic issues need incentives and regulations like financial assistance and training. By standardizing data across manufacturers and platforms, governments can help development (Psara *et al.*, 2022).

2.6 FUTURE PERSPECTIVES

The agriculture industry has identified the advantages of using the new digital technologies. The increasing market demand for organic foods leads to a jump from sustainable organic farming to intelligent organic farming. There are several actions that we must take in order to avail ourselves of this emerging opportunity. Education combined with vocational training on the new technological applications can better induce farmers to effectively utilize the new technologies. The need is to connect the expert personnel with the farmer. By investing in energy infrastructure, communication infrastructure, internet connectivity infrastructure, service markets, consultant services, and loan markets, farmers' confidence and operational preparedness for the use of new technologies can be improved. This involves the establishment of requisite physical, economic, legal, and social infrastructures, which can be significantly impacted by policies in governments. Whereas Underground or Underwater transmission capabilities allow wireless power transfer integration enabling the elimination of battery replacement in frequent intervals.

Designing high-quality, high-resolution, and reliable sensors powered by IoT specifically tailored for the agricultural production environment and monitoring plant and animal physiological signs is essential. PA is being revolutionized using MV, smartphone apps, 6G mobile networks, smart hydroponics, and improved breeding methods. These advancements provide useful data, improve decision-making, and foster sustainable agriculture practices. As technology advances, its use in agriculture will increase production, efficiency, and sustainability, assuring a stable food supply in the future. Future developments in precision agricultural technologies can revolutionize the agricultural industry. Education, infrastructure development, sensor technology, communication systems, and new agricultural methods will help us make agriculture more efficient, sustainable, and productive.

2.7 CONCLUSIONS

Through PA, complex technologies, such as AI, ML, and IoT, will change the agricultural environment. The future of agriculture is likely to be fueled by data analytics and automation that will enable efficiency, sustainability, and customization in agricultural processes. Improved relatability and greater inclusivity will mean these technologies will be even more available to a wider range of farming demographics, amplifying advancements no matter where in the world they originate. Nonetheless, it will be critical to tackle ethical and regulatory hurdles in order to make sure the benefits of PA are achieved in a responsible manner. Embracing these innovations and overcoming adjacent barriers will enable the agro sector to enhance production, sustainability, and an equitable global food system.

LIST OF ABBREVIATIONS

AI artificial intelligence
CWSI crop water stress index
DL approaches deep learning approaches
EID technologies electronic identification technologies
FAO Food and Agriculture Organization
GIS geographic information systems
GPS global positioning systems
ICT information and communication technology
IoT Internet of things
LSR least squares regression
ML technologies machine learning technologies
MV systems machine vision systems
OEMs original equipment manufacturers
PA precision agriculture
UAVs unmanned aerial vehicles
VR systems virtual reality systems
VRA variable rate application
VRTs variable rate technologies
WSNs wireless sensor networks

REFERENCES

Abdullah, Mahmood, A., Bibi, S., Naqve, M., Javaid, M.M., Zia, M.A., Jabbar, A., Ud-Din, W., Attia, K.A. & Khan, N., 2022. Physiological, biochemical, and yield responses of linseed (*Linum usitatissimum* L.) in α-tocopherol-mediated alleviation of salinity stress. *Frontiers in Plant Science*, 13, 867172.

Abdulraheem, M.I., Zhang, W., Li, S., Moshayedi, A.J., Farooque, A.A. & Hu, J., 2023. Advancement of remote sensing for soil measurements and applications: A comprehensive review. *Sustainability*, 15, 15444.

Ahmad, S., Shakeel, I., Mehfuz, S. & Ahmad, J., 2023. Deep learning models for cloud, edge, fog, and IoT computing paradigms: Survey, recent advances, and future directions. *Computer Science Review*, 49, 100568.

Ahmad, S.F. & Dar, A.H., 2020. Precision Farming for Resource Use Efficiency. *In* S. Kumar, R.S. Meena & M.K. Jhariya (eds.) *Resources Use Efficiency in Agriculture*. Singapore: Springer Singapore, 109–135.

Ahmed, Z., Gui, D., Qi, Z., Liu, Y., Liu, Y. & Azmat, M., 2022. Agricultural system modeling: Current achievements, innovations, and future roadmap. *Arabian Journal of Geosciences*, 15, 363.

Ahn, J., Park, J., Lee, S.S., Lee, K.-H., Do, H. & Ko, J., 2023. SafeFac: Video-based smart safety monitoring for preventing industrial work accidents. *Expert Systems with Applications*, 215, 119397.

Akhbarifar, S., Javadi, H.H.S., Rahmani, A.M. & Hosseinzadeh, M., 2023. A secure remote health monitoring model for early disease diagnosis in cloud-based IoT environment. *Personal and Ubiquitous Computing*, 27, 697–713.

Akhter, R. & Sofi, S.A., 2022. Precision agriculture using IoT data analytics and machine learning. *Journal of King Saud University – Computer and Information Sciences*, 34, 5602–5618.

Alexopoulos, A., Koutras, K., Ali, S.B., Puccio, S., Carella, A., Ottaviano, R. & Kalogeras, A., 2023. Complementary use of ground-based proximal sensing and airborne/spaceborne remote sensing techniques in precision agriculture: A systematic review. *Agronomy*, 13, 1942.

Ali, A., Hussain, T., Tantashutikun, N., Hussain, N. & Cocetta, G., 2023. Application of smart techniques, internet of things and data mining for resource use efficient and sustainable crop production. *Agriculture*, 13, 397.

Ali, H., Tanveer, A., Nadeem, M., Asghar, H. & Javaid, M., 2013a. Germination ecology of *Rhynchosia capitata*: An emerging summer weed in Asia. *Planta Daninha*, 31, 249–257.

Ali, H., Tanveer, A., Nadeem, M., Javaid, M., Kashif, M. & Chadhar, A., 2013b. Allelopathic effects of *Rhynchosia capitata* on germination and seedling growth of mungbean. *Planta Daninha*, 31, 501–509.

Almaiah, M.A., Alfaisal, R., Salloum, S.A., Hajjej, F., Shishakly, R., Lutfi, A., Alrawad, M., Al Mulhem, A., Alkhdour, T. & Al-Maroof, R.S., 2022. Measuring institutions' adoption of artificial intelligence applications in online learning environments: Integrating the innovation diffusion theory with technology adoption rate. *Electronics*, 11, 3291.

Ara, I., Turner, L., Harrison, M.T., Monjardino, M., Devoil, P. & Rodriguez, D., 2021. Application, adoption and opportunities for improving decision support systems in irrigated agriculture: A review. *Agricultural Water Management*, 257, 107161.

Ayoub Shaikh, T., Rasool, T. & Rasheed Lone, F., 2022. Towards leveraging the role of machine learning and artificial intelligence in precision agriculture and smart farming. *Computers and Electronics in Agriculture*, 198, 107119.

Bansal, S. & Kumar, D., 2020. IoT ecosystem: A survey on devices, gateways, operating systems, middleware and communication. *International Journal of Wireless Information Networks*, 27, 340–364.

Boehm, K.M., Khosravi, P., Vanguri, R., Gao, J. & Shah, S.P., 2022. Harnessing multimodal data integration to advance precision oncology. *Nature Reviews Cancer*, 22, 114–126.

Bosompem, M., 2021. Potential challenges to precision agriculture technologies development in Ghana: scientists' and cocoa extension agents' perspectives. *Precision Agriculture*, 22, 1578–1600.

Cendón-Flórez, Y. & Pippa, R., 2022. Ethics in Water Resource Management: Roles, Frameworks, and Principles. *In* W. Leal Filho, A.M. Azul, L. Brandli, A. Lange Salvia & T. Wall (eds.) *Clean Water and Sanitation*. Cham: Springer International Publishing, 204–213.

Chadha, A., Florentine, S., Chauhan, B.S., Long, B., Jayasundera, M., Javaid, M.M. & Turville, C., 2019. Environmental factors affecting the germination and seedling emergence of two populations of an emerging agricultural weed: Wild lettuce (*Lactuca serriola*). *Crop and Pasture Science*, 70, 709–717.

Chadha, A., Florentine, S., Javaid, M., Welgama, A. & Turville, C., 2020. Influence of elements of climate change on the growth and fecundity of *Datura stramonium*. *Environmental Science and Pollution Research*, 27, 35859–35869.

Charania, I. & Li, X., 2020. Smart farming: Agriculture's shift from a labor intensive to technology native industry. *Internet of Things*, 9, 100142.

Cheng, C., Fu, J., Su, H. & Ren, L., 2023. Recent advancements in agriculture robots: Benefits and challenges. *Machines*, 11, 48.

Comi, M., 2020. The distributed farmer: Rethinking US Midwestern precision agriculture techniques. *Environmental Sociology*, 6, 403–415.

Das, S., Verma, S., Kaushik, T. & Gupta, N., 2022. LoRaWAN IoT based precision smart farming architecture with data visualization. *In 2022 International Conference for Advancement in Technology (ICONAT)*, 1–6.

de Oliveira Azevêdo, R., Rotela Junior, P., Rocha, L.C.S., Chicco, G., Aquila, G. & Peruchi, R.S., 2020. Identification and analysis of impact factors on the economic feasibility of photovoltaic energy investments. *Sustainability*, 12, 7173.

Dhillon, R. & Moncur, Q., 2023. Small-scale farming: A review of challenges and potential opportunities offered by technological advancements. *Sustainability*, 15, 15478.

Fathollahi-Fard, A.M., Tian, G., Ke, H., Fu, Y. & Wong, K.Y., 2023. Efficient multi-objective metaheuristic algorithm for sustainable harvest planning problem. *Computers & Operations Research*, 158, 106304.

Giorgio, A., Penate Lopez, L.P., Bertoni, D., Cavicchioli, D. & Ferrazzi, G., 2024. Enablers to digitalization in agriculture: A case study from Italian field crop farms in the Po River Valley, with insights for policy targeting. *Agriculture*, 14, 1074.

Gupta, N. & Gupta, P.K., 2024. Robotics and Artificial Intelligence (AI) in Agriculture with Major Emphasis on Food Crops. *In* P.M. Priyadarshan, S.M. Jain, S. Penna & J.M. Al-Khayri (eds.) *Digital Agriculture: A Solution for Sustainable Food and Nutritional Security.* Cham: Springer International Publishing, 577–605.

Hanif, Z., Naeem, M., Ali, H.H., Tanveer, A., Javaid, M.M., Peerzada, A.M. & Chauhan, B.S., 2017. Effect of environmental factors on germination of *Salsola foetida*: Potential species for rehabilitation of degraded rangelands. *Rangeland Ecology & Management*, 70, 638–643.

Hassan, M.U., Nawaz, M., Mahmood, A., Shah, A.A., Shah, A.N., Muhammad, F., Batool, M., Rasheed, A., Jaremko, M. & Abdelsalam, N.R., 2022. The role of zinc to mitigate heavy metals toxicity in crops. *Frontiers in Environmental Science*, 10, 990223.

Hatfield, J.L., Antle, J., Garrett, K.A., Izaurralde, R.C., Mader, T., Marshall, E., Nearing, M., Philip Robertson, G. & Ziska, L., 2020. Indicators of climate change in agricultural systems. *Climatic Change*, 163, 1719–1732.

Hazra, A., Rana, P., Adhikari, M. & Amgoth, T., 2023. Fog computing for next-generation Internet of Things: Fundamental, state-of-the-art and research challenges. *Computer Science Review*, 48, 100549.

Huo, D., Malik, A.W., Ravana, S.D., Rahman, A.U. & Ahmedy, I., 2024. Mapping smart farming: Addressing agricultural challenges in data-driven era. *Renewable and Sustainable Energy Reviews*, 189, 113858.

Imran, M.A., Zennaro, M., Popoola, O.R., Chiaraviglio, L., Zhang, H., Manzoni, P., Beek, J.V.D., Stewart, R., Cox, M.A., Mendes, L.L. & Pietrosemoli, E., 2024. Exploring the boundaries of connected systems: Communications for hard-to-reach areas and extreme conditions. *Proceedings of the IEEE*, 112, 912–945.

Javaid, M., Haleem, A., Khan, I.H. & Suman, R., 2023. Understanding the potential applications of artificial intelligence in agriculture sector. *Advanced Agrochem*, 2, 15–30.

Javaid, M., Haleem, A., Singh, R.P. & Suman, R., 2022. Enhancing smart farming through the applications of Agriculture 4.0 technologies. *International Journal of Intelligent Networks*, 3, 150–164.

Javaid, M., Tanveer, A., Ahmad, R., Yaseen, M. & Khaliq, A., 2012. Optimizing activity of herbicides at reduced rate on *Emex spinosa* Campd. with adjuvants. *Planta daninha*, 30, 425–435.

Karunathilake, E.M.B.M., Le, A.T., Heo, S., Chung, Y.S. & Mansoor, S., 2023. The path to smart farming: Innovations and opportunities in precision agriculture. *Agriculture*, 13, 1593.

Karydas, C., Chatziantoniou, M., Stamkopoulos, K., Iatrou, M., Vassiliadis, V. & Mourelatos, S., 2023. Embedding a precision agriculture service into a farm management information system – Ifarma/PreFer. *Smart Agricultural Technology*, 4, 100175.

Khan, N., Ray, R.L., Sargani, G.R., Ihtisham, M., Khayyam, M. & Ismail, S., 2021. Current progress and future prospects of agriculture technology: Gateway to sustainable agriculture. *Sustainability*, 13, 4883.

Kim, W.-S., Lee, W.-S. & Kim, Y.-J., 2020. A review of the applications of the Internet of Things (IoT) for agricultural automation. *Journal of Biosystems Engineering*, 45, 385–400.

Korkut, E.H. & Surer, E., 2023. Visualization in virtual reality: A systematic review. *Virtual Reality*, 27, 1447–1480.

Kumar Kasera, R., Gour, S. & Acharjee, T., 2024. A comprehensive survey on IoT and AI based applications in different pre-harvest, during-harvest and post-harvest activities of smart agriculture. *Computers and Electronics in Agriculture*, 216, 108522.

Lazzeri, F. & Robsky, A., 2024. Searching for the Right Algorithms. *In Machine Learning Governance for Managers.* Cham: Springer International Publishing, 41–71.

Li, J., Shan, Y., Tian, G. & Hao, X., 2020a. Labor cost, government intervention, and corporate innovation: Evidence from China. *Journal of Corporate Finance*, 64, 101668.

Li, Y., Dai, J. & Cui, L., 2020b. The impact of digital technologies on economic and environmental performance in the context of industry 4.0: A moderated mediation model. *International Journal of Production Economics*, 229, 107777.

Mahmood, A., Wang, X., Shahzad, A.N., Fiaz, S., Ali, H., Naqve, M., Javaid, M.M., Mumtaz, S., Naseer, M. & Dong, R., 2021. Perspectives on bioenergy feedstock development in Pakistan: Challenges and opportunities. *Sustainability*, 13, 8438.

Mahmud, M.S., Zahid, A. & Das, A.K., 2023. Sensing and automation technologies for ornamental nursery crop production: Current status and future prospects. *Sensors*, 23, 1818.

Marques, P., Pádua, L., Sousa, J.J. & Fernandes-Silva, A., 2024. Advancements in remote sensing imagery applications for precision management in olive growing: A systematic review. *Remote Sensing*, 16, 1324.

Mehmood, A., Naeem, M., Khalid, F., Saeed, Y., Abbas, T., Jabran, K., Sarwar, M.A., Tanveer, A. & Javaid, M.M., 2018. Identification of phytotoxins in different plant parts of *Brassica napus* and their influence on mung bean. *Environmental Science and Pollution Research*, 25, 18071–18080.

Misra, S. & Ghosh, A., 2024. Chapter 6 – Agriculture Paradigm Shift: A Journey from Traditional to Modern Agriculture. In K. Singh, M.C. Ribeiro & Ö. Calicioglu (eds.) *Biodiversity and Bioeconomy*. Amsterdam, Netherlands: Elsevier, 113–141.

Mohsan, S.H., Othman, N.Q.H., Li, Y., Alsharif, M.H. & Khan, M.A., 2023. Unmanned aerial vehicles (UAVs): Practical aspects, applications, open challenges, security issues, and future trends. *Intelligent Service Robotics*, 16, 109–137.

Monteiro, A. & Santos, S., 2022. Sustainable approach to weed management: The role of precision weed management. *Agronomy*, 12, 118.

Moreno, H. & Andújar, D., 2023. Proximal sensing for geometric characterization of vines: A review of the latest advances. *Computers and Electronics in Agriculture*, 210, 107901.

Nguyen, M., Fujioka, J., Wentlandt, K., Onabajo, N., Wong, I., Bhatia, R.S., Bhattacharyya, O. & Stamenova, V., 2020. Using the technology acceptance model to explore health provider and administrator perceptions of the usefulness and ease of using technology in palliative care. *BMC Palliative Care*, 19, 138.

Petrović, B., Bumbálek, R., Zoubek, T., Kuneš, R., Smutný, L. & Bartoš, P., 2024. Application of precision agriculture technologies in Central Europe – Review. *Journal of Agriculture and Food Research*, 15, 101048.

Potts, J., Jangra, S., Michael, V.N. & Wu, X., 2023. Speed breeding for crop improvement and food security. *Crops*, 3, 276–291.

Psara, K., Papadimitriou, C., Efstratiadi, M., Tsakanikas, S., Papadopoulos, P. & Tobin, P., 2022. European energy regulatory, socioeconomic, and organizational aspects: An analysis of barriers related to data-driven services across electricity sectors. *Energies*, 15, 2197.

Raj, M., N B, H., Gupta, S., Atiquzzaman, M., Rawlley, O. & Goel, L., 2024. Leveraging precision agriculture techniques using UAVs and emerging disruptive technologies. *Energy Nexus*, 14, 100300.

Rani, R., Khurana, M., Kumar, A. & Kumar, N., 2022. Big data dimensionality reduction techniques in IoT: Review, applications and open research challenges. *Cluster Computing*, 25, 4027–4049.

Rasheed, A., Barqawi, A.A., Mahmood, A., Nawaz, M., Shah, A.N., Bay, D.H., Alahdal, M.A., Hassan, M.U. & Qari, S.H., 2022a. CRISPR/Cas9 is a powerful tool for precise genome editing of legume crops: A review. *Molecular Biology Reports*, 49, 5595–5609.

Rasheed, A., Mahmood, A., Maqbool, R., Albaqami, M., Sher, A., Sattar, A., Bakhsh, G., Nawaz, M., Hassan, M.U. & Al-Yahyai, R., 2022b. Key insights to develop drought-resilient soybean: A review. *Journal of King Saud University-Science*, 34, 102089.

Rasheed, A., Raza, A., Jie, H., Mahmood, A., Ma, Y., Zhao, L., Xing, H., Li, L., Hassan, M.U. & Qari, S.H., 2022c. Molecular tools and their applications in developing salt-tolerant soybean (*Glycine max* L.) cultivars. *Bioengineering*, 9, 495.

Rowan, N.J., 2023. The role of digital technologies in supporting and improving fishery and aquaculture across the supply chain – Quo Vadis? *Aquaculture and Fisheries*, 8, 365–374.

Rozenstein, O., Cohen, Y., Alchanatis, V., Behrendt, K., Bonfil, D.J., Eshel, G., Harari, A., Harris, W.E., Klapp, I., Laor, Y., Linker, R., Paz-Kagan, T., Peets, S., Rutter, S.M., Salzer, Y. & Lowenberg-Deboer, J., 2024. Data-driven agriculture and sustainable farming: Friends or foes? *Precision Agriculture*, 25, 520–531.

Saban, M., Bekkour, M., Amdaouch, I., El Gueri, J., Ait Ahmed, B., Chaari, M.Z., Ruiz-Alzola, J., Rosado-Muñoz, A. & Aghzout, O., 2023. A smart agricultural system based on PLC and a cloud computing web application using LoRa and LoRaWan. *Sensors*, 23, 2725.

Sawicka, B., Egbuna, C., Nayak, A.K. & Kala, S., 2020. Chapter 2 – Plant Diseases, Pathogens and Diagnosis. *In* C. Egbuna & B. Sawicka (eds.) *Natural Remedies for Pest, Disease and Weed Control*. Cambridge, Massachusetts, United States: Academic Press, 17–28.

Sinegalia, M.K.S.D., Santoro, G.B. & Molin, P.G., 2024. How have RPAS helped monitor forests and what can we apply in forest restoration monitoring? *Restoration Ecology*, 32, e14061.

Sinha, B.B. & Dhanalakshmi, R., 2022. Recent advancements and challenges of Internet of Things in smart agriculture: A survey. *Future Generation Computer Systems*, 126, 169–184.

Sishodia, R.P., Ray, R.L. & Singh, S.K., 2020. Applications of remote sensing in precision agriculture: A review. *Remote Sensing*, 12, 3136.

S-Julián, R., Lacalle, I., Vaño, R., Boronat, F. & Palau, C.E., 2023. Self-* capabilities of cloud-edge nodes: A research review. *Sensors*, 23, 2931.

Steensland, A. & Zeigler, M., 2021. Productivity in Agriculture for a Sustainable Future. In H. Campos (Ed.), *The innovation revolution in agriculture*. Springer, 33–69.

Steinke, J., Ortiz-Crespo, B., Van Etten, J. & Müller, A., 2022. Participatory design of digital innovation in agricultural research-for-development: Insights from practice. *Agricultural Systems*, 195, 103313.

Sullivan, S., 2023. Ag-tech, agroecology, and the politics of alternative farming futures: The challenges of bringing together diverse agricultural epistemologies. *Agriculture and Human Values*, 40, 913–928.

Surendran, U., Nagakumar, K.C.V. & Samuel, M.P., 2024. Remote Sensing in Precision Agriculture. *In* P.M. Priyadarshan, S.M. Jain, S. Penna & J.M. Al-Khayri (eds.) *Digital Agriculture: A Solution for Sustainable Food and Nutritional Security*. Cham: Springer International Publishing, 201–223.

Tetik, B., 2023. *The Outer Space as a Domain of Competition and Cooperation from the Cold War to Today*. Master's Thesis, Middle East Technical University, Turkey.

Torres-Torriti, M. & Burgos, P.N., 2023. Field Machinery Automated Guidance. *In* Q. Zhang (ed.) *Encyclopedia of Digital Agricultural Technologies*. Cham: Springer International Publishing, 509–526.

Tzanidakis, C., Tzamaloukas, O., Simitzis, P. & Panagakis, P., 2023. Precision livestock farming applications (PLF) for grazing animals. *Agriculture*, 13, 288.

Ullah, H., Mahmood, A., Ijaz, M., Tadesse, B. & Honermeier, B., 2013. Evaluation of anise (*Pimpinella anisum* L.) accessions with regard to morphological characteristics, fruit yield, oil contents and composition. *Journal of Medicinal Plants Research*, 7, 2177–2186.

Wang, L., Guo, Q., Chen, L. & Yan, W., 2023. In-situ experimental and high-fidelity modeling tools to advance understanding of metal additive manufacturing. *International Journal of Machine Tools and Manufacture*, 193, 104077.

Wasaya, A., Abbas, T., Yasir, T.A., Sarwar, N., Aziz, A., Javaid, M.M. & Akram, S., 2021. Mitigating drought stress in sunflower (*Helianthus annuus* L.) through exogenous application of β-aminobutyric acid. *Journal of Soil Science and Plant Nutrition*, 21, 936–948.

Yadav, A., Yadav, K., Ahmad, R. & Abd-Elsalam, K.A., 2023. Emerging frontiers in nanotechnology for precision agriculture: Advancements, hurdles and prospects. *Agrochemicals*, 2, 220–256.

Yang, Z. & Solangi, Y.A., 2024. Analyzing the relationship between natural resource management, environmental protection, and agricultural economics for sustainable development in China. *Journal of Cleaner Production*, 450, 141862.

Yogarajan, L., Masukujjaman, M., Ali, M.H., Khalid, N., Osman, L.H. & Alam, S.S., 2023. Exploring the hype of blockchain adoption in agri-food supply chain: A systematic literature review. *Agriculture*, 13, 1173.

Zafar, S., Afzal, H., Ijaz, A., Mahmood, A., Ayub, A., Nayab, A., Hussain, S., Maqsood, U.-H., Sabir, M.A. & Zulfiqar, U., 2023. Cotton and drought stress: An updated overview for improving stress tolerance. *South African Journal of Botany*, 161, 258–268.

Zawadzka, D., Strzelecka, A. & Szafraniec-Siluta, E., 2021. Debt as a source of financial energy of the farm—What causes the use of external capital in financing agricultural activity? A model approach. *Energies*, 14, 4124.

Zualkernan, I., Abuhani, D.A., Hussain, M.H., Khan, J. & Elmohandes, M., 2023. Machine learning for precision agriculture using imagery from unmanned aerial vehicles (UAVs): A survey. *Drones*, 7, 382.

3 Biotech Solutions for Crop Production

Ayesha Arshad, Abrar Hussain, Javaria Nargis, Zeki Erden, Çağdaş Can Toprak, Hakki Akdeniz, and Jameel Mohammed Al-Khayri*

3.1 INTRODUCTION

Crop production is a component of the agricultural sector that includes crop development through field cultivation, vegetable and fruit production, and other operations (Sharma *et al.*, 2024). Agricultural firms have significant production capacity considering the challenging production circumstances arising from excessive costs of production materials, limited appeal of rural regions and difficulties in obtaining financial matters (Mpala and Simatele, 2024). Additional chemical ingredients, such as chemical pesticides and fertilizers, may successfully substitute for the functions often performed by prolonged crop rotations. Although extended rotations, such as grasses or grass-leys, are not as common as they used to be, they nevertheless have importance in conventional or mixed-farming methods. The decision to plant a crop for each season is mostly driven through economic factors compelling farmers to optimize their income by cultivating the same crop consecutively on the same plot of land (Saiz-Rubio and Rovira-Más, 2020).

Biotechnology is the use of naturally occurring processes to develop new goods, procedures, and organisms in order to improve human well-being and the development of the modern world. Biotechnology, also known as biotech, has been present since the inception of human civilization, including activities such as the cultivation of plants and animals, as well as the identification of fermentation processes (Kordi *et al.*, 2022). Plant biotechnology includes a range of methodologies used to modify plants in order to meet particular requirements or benefit on favourable conditions (Patra *et al.*, 2021). As the global population continues to increase, the need for food is also rising. Biotechnology has the potential to improve the farming systems (Das *et al.*, 2023). The scientists have engineered plants which are able to resist both biotic and abiotic stresses.

We have used numerous techniques, including tissue culture, mutagenesis, and transformation (Figure 3.1). Advanced functional genomics research improves our knowledge of the plant genome and allows us to change it with greater efficiency. RNA interference, state-of-the-art DNA sequencing, and nanotechnology have become effective approaches for enhancing agricultural productivity in preparation for future demands. Mutational breeding is an effective technique for developing plant varieties with desired features that are advantageous for both food crops and landscaping. Presently, about 2000 distinct crop varieties that undergo induced mutations are produced and commercialized for economic purposes (Noor Ul *et al.*, 2024).

Biotechnology has revolutionized agricultural production by replacing traditional techniques with more precise ones. Scientists can use biotechnological technologies, such as gene editing technology, to boost the production of crops. These techniques have resulted in crops that generate an abundance of food, are more resistant to insects, pests, and illnesses, and have contributed to global food security. Transgenic crop production may affect biodiversity, soil quality, and water resources;

* Corresponding author: ayeshaarshadd6677@gmail.com

DOI: 10.1201/9781003498698-3

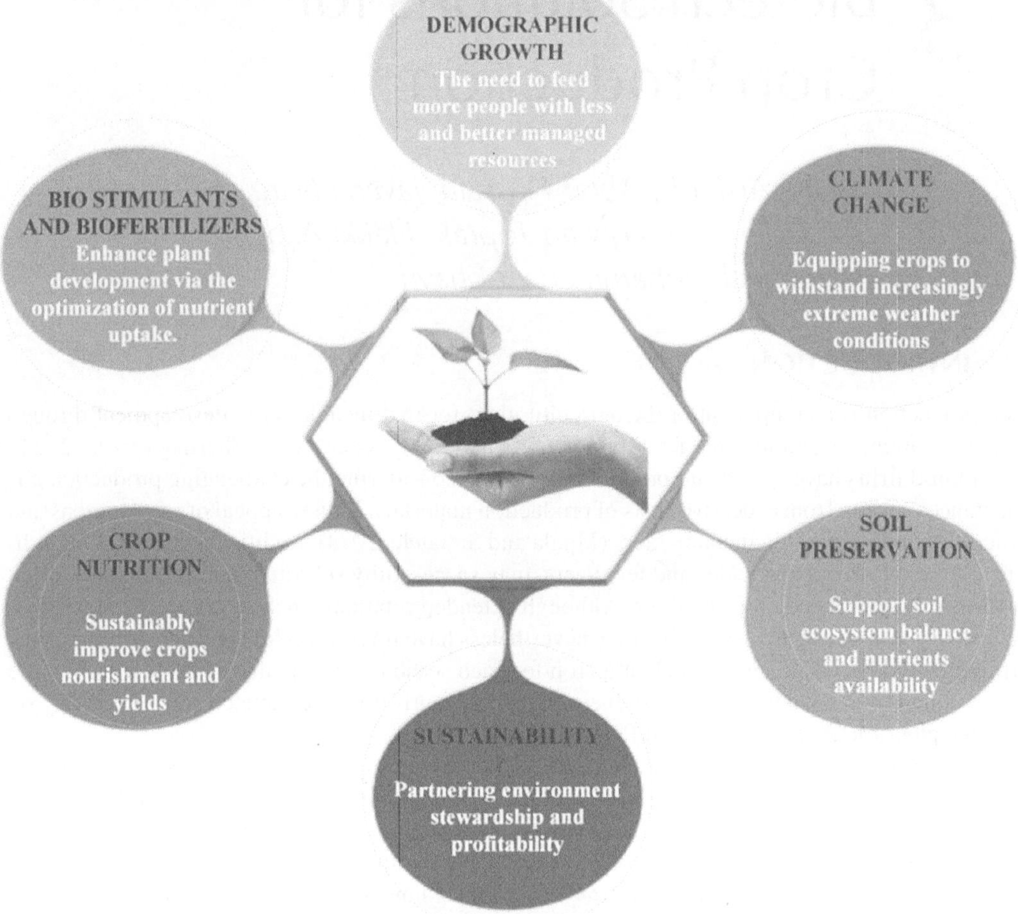

FIGURE 3.1 Schematic diagram representing as bio solutions for enhancing crop production.

thus, they must be considered. Biotech solutions may affect agricultural processes, market dynamics, and food prices, which might have major socioeconomic effects. Researchers, leaders, and consumers must work together to responsibly and sustainably apply biotech solutions in agriculture. Biotechnology holds good promise for improving crop yields and enhancing agricultural sustainability in response to changing global needs and limitations.

3.2 BIOTECH SOLUTIONS REPLACING TRADITIONAL METHODS

Traditional agriculture is a fundamental method of food production and farming that depends extensively on local knowledge, land use, traditional tools, natural resources, organic fertilizers, and farmers' religious and cultural beliefs. Traditional agriculture, supplying food for almost 50% of the global population, encompasses both recognized and unrecognized methods of protecting crops (Pramanik *et al.*, 2023). However, in the face of changing climate, modern biotechnological applications are increasingly being integrated to enhance the resilience of traditional farming systems by improving stress tolerance, crop productivity, and resource-use efficiency. Applications of biotechnology by changing the climate is presented in Table 3.1.

TABLE 3.1
Applications of Biotechnology by Changing the Climate

Climate Change	Biotechnology	Application	Reference
Reduced fuel usage	Developing resistance to herbicides in plants to minimize the need for treatment	Genetically modified (GM) soybeans and genetically modified canola	Rahman *et al.* (2023)
	Developing resistance to insects in agriculture will minimize the need for pesticides	*B. thuringiensis* maize, cotton fibres, and the eggplants	Tudi *et al.* (2021)
Adaptation to both living (biotic) and non-living (abiotic) challenges	Creating resistance to drought, salt water, and temperature in technology	Genetically modified rice and tomato	Ali *et al.* (2020)
Carbon sequestering	Biotechnology allows farming without tilling, renewable energy, and nitrogen-efficient genetically modified crops	Resistant to herbicides Genetically modified soybeans and canola Genetically modified nutrition crop Highly effective Genetically modified canola	Turnbull *et al.* (2021)

3.2.1 Tillage Method

Tillage is the primary agricultural activity involving the mechanical manipulation of soil, which involves actions, such as digging, stirring, and preparing the seedbed. The use of agricultural mechanization is necessary in order to increase productivity and fulfil the growing demand for food among growing populations (Pawlak and Kołodziejczak, 2020). Tillage operations modify the physiochemical properties of soil to inhibit weeds while creating a suitable seedbed for crop plants (Ali *et al.*, 2013a; Tanveer *et al.*, 2013). Additionally, tillage practices integrate crop residue into the soil, loosen the soil, and enhance chemical responses. As a result, the soil's physiochemical condition improves, leading to higher growth and yields of crops (Wasaya *et al.*, 2019).

3.2.2 Traditional Cropping Method

Traditional cropping methods mostly depend on inputs, such as fertilizers and insecticides, to enhance crop yields. Nevertheless, an abundance of inputs leads to a rise in production costs and results in a greater release of greenhouse gases into the atmosphere, which has a detrimental effect on the environment. The main goal of contemporary agriculture is to discover sustainable methods to enhance crop production while minimizing or eliminating negative environmental impacts. Crop rotations increase the process of converting atmospheric CO_2 into plant biomass, leading to the storage of additional carbon into the soil (Chadha *et al.*, 2020; Meena *et al.*, 2020).

3.2.3 Irrigation

Irrigation enhances economic profitability and has the potential to increase output for up to 400%. Conversely, irrigation might result in undesirable effects on the environment. Approximately 33% of the world's irrigated regions are losing productivity due to poor irrigation practices, which cause water logging and salinity. However, modern irrigation techniques are more water-efficient than traditional ways (Gebrehiwot, 2018).

3.2.4 Sustainable Agricultural Method

To sustainably feed the world's rapidly rising population of several billion people, it is critical to improve the production of nutritious food while decreasing the environmental footprint. However, this task will be exceptionally difficult due to the impact of global environmental change (Poore and Nemecek, 2018). Plant breeders should prioritize features that have the highest capacity to enhance crop productivity. While agricultural education institutions have been steadily incorporating environmental objectives, the evaluation of the possibility of various agricultural systems is still mostly conducted using cost-benefit analysis. Due to the exclusion of ecological, social, and cultural factors, the present investigation has significant limitations in its capacity to compare the long-term viability of various agricultural systems. In order to successfully overcome such challenges, it is essential to introduce original and sustainable agricultural methods, including precision farming, agroecology, organic farming, and the use of modern technology, such as hydroponics, vertical farming, and genetic engineering (Balal *et al.*, 2017; Daniar *et al.*, 2024). These strategies may aid in reducing environmental effects, bolstering resistance to climate change, enhancing the efficiency of resource use, and improving food security (Chadha *et al.*, 2019). Plant diseases are causing significant reductions in staple crop production worldwide, which inhibits the critical goal of achieving a 60% increase in order to fulfil demand for food. This effort is further complicated by climatic changes. The primary consequences are the decrease in both the quantity and quality of food. Crop diseases undermine food safety by introducing pesticides and/or poisons (Misiou and Koutsoumanis, 2022).

3.2.5 Recombinant DNA Technology

The use of recombinant DNA technology has greatly enhanced traditional methods of crop development and has considerable potential in aiding plant breeders in meeting the projected increase in food demand throughout the 21st century. The last 20 years have seen significant advancements in the modification of genes from various and unpredictable sources (Mehmood *et al.*, 2018; Lenaerts *et al.*, 2019). Afterwards, these genes are inserted into microorganisms and agricultural plants to provide them with the ability to fight insect pests and illnesses, as well as withstand herbicides, dehydration, soil salinity, and aluminium toxicity (Abdullah et al., 2022). This process also improves the quality of crops after they are harvested, makes them absorb nutrients better and have more nutritional value, speeds up photosynthesis, creates sugars and starch, makes biological pesticides work better, helps us learn more about gene function and metabolic pathways, and makes it possible to make drugs and vaccines in plants (D'amelia *et al.*, 2021). Biotechnology is critical in maintaining agricultural productivity and expanding the use of scientific approaches to improve agriculture in a sustainable manner. In the previous few decades, agricultural biotechnologies have achieved significant advances through an extensive use of unique, fast, and effective technologies (Prasad *et al.*, 2017; Benedek *et al.*, 2023).

3.3 BIOTECHNOLOGICAL TOOLS FOR CROP IMPROVEMENT

Plants have a vital role in the extensive ecology of our planet. Primary food producers are essential for maintaining a balanced environment by generating oxygen, which in turn supports stable and sustained climates (Chattopadhyay and Chattopadhyay, 2020). Given the ongoing growth of the population and the growing damage to the environment, it is essential to undertake an extensive makeover of our agricultural systems. Plants have many applications beyond their usage as food, including their use as energy sources, construction materials, or in the production of cosmetics. Ultimately, they produce many compounds that are appropriate for pharmaceutical use. The increasing awareness of the significance of plants has occurred with advancements in plant molecular biology. These specific traits make them ideal for gene editing and genetic research generally (Mao *et al.*, 2019).

The majority of nations have implemented biosafety frameworks via dedicated laws that regulate genetically engineered, or genetic modification (GM), agricultural plants created using DNA that is rDNA technology (Turnbull *et al.*, 2021). Typically, these biosafety contexts depend on essential principles that ensure the safety of food and feed and evaluate the risks to the environment linked to crops produced using modern biotechnology. The FAO/WHO and the OECD are international organizations that have created these guidelines. Researchers have developed a diverse array of novel genetically modified technologies (nGMs) with the purpose of modifying organisms, including plants, for study or agriculture, in addition to conventional GM technologies. These fresh genetic alterations are sometimes mentioned as "new methods" or "novel breeding techniques" (Gentile and La Malfa, 2019). Site-directed nucleases (SDNs), such as CRISPR-directed nucleases, transcription activator-like effector nucleases (TALENs), zinc finger nucleases (ZFNs), or mega nucleases, can accomplish genome editing. Application of advanced techniques likeTrans and CRISPR/Cas9 is presented in Table 3.2. We can also use SDN-based methodologies for simultaneous genome editing and base editing. Genome editing and base editing are techniques used to modify genetic material, whereas changing transcriptional regulation involves modifying the gene expression process (Biswas *et al.*, 2021). RNA-directed DNA methylation is a technique used to modify the epigenetic regulation of gene expression. We have developed haploid induction and fast breeding techniques to facilitate the implementation of extensive breeding plants (Sharma *et al.*, 2022).

TABLE 3.2
Application of Trans and CRISPR/Cas9 in Woody Fruit Species

Plant Species	Name of Gene	Trait	Applications	References
Malus domestica	Phytoene desaturase (PDS)	Produce mutation in the PDS genome	Overall, 31.1% of the reproduced plants exhibited both clear and partial albino phenotypes, and the DNA sequencing confirmed that there were changes in both alleles of the apple PDS gene	Ren *et al.* (2023)
Vitis vinifera L.	L-idonate dehydrogenase gene (ldnDH)	Targeted gene mutation and genome editing	The mutation's frequency in the recombinant mass of cells and plant regeneration-producing sgRNA1/Cas9 is 100%	Wang *et al.* (2021)
V. vinifera L. and *M. domestica* protoplasts	Grape gene MLO-7 and the apple genes DIPM-1, 2, and 4	Grape and apple powdery mildew resistance and fire blight and disease resistance	Efficient and precise gene modification in grape MLO-7 and apple DIPM1, DIPM2, and DIPM4 protoplasts.	Fizikova *et al.*, (2021)
Citrus sinensis Osbeck	EBE $_{PthA4}$ of the CsLOB1 promoter	EBE $_{PthA4}$ effector mutation causes citrus canker resistance	Enhancing resistance to high-a yeast infection in citrus by altering the EBE $_{PthA4}$ effector	Jia and Lin (2022)
C. sinensis	CsPDS	Generate mutations in the CsPDS gene	A change in the CsPDS locus has been found at the intended site in those treated sweet orange leaves	Huang *et al.* (2020)
Prunus avium	PNRSV	Non-transgenic scions tolerate PNRSV	The transfer of hpRNA-derived siRNAs, which conferred resistance to PNRSV in the non-transgenic sweet cherry seedling when it was grafted onto the transgenic roots	Sidorova *et al.* (2021)

3.3.1 Marker-assisted Breeding

Plant breeding, along with advancements in agricultural technologies like agrochemicals, has significantly improved crop yields for more than a century. Plant breeders must constantly adapt to many changes. Agricultural techniques evolve, necessitating the development of genotypes with certain agronomic traits (Varshney *et al.*, 2018; Mustafa *et al.*, 2022). Secondly, the habitats and the species inside them are ever-evolving. Fungal and insect pests constantly adapt and surpass the resilience of host plants. The expansion of agriculture into new regions brings about altered growing conditions, necessitating the development of innovative crop breeds that not only flourish in these environments but also fulfil the changing demands of consumers (Kaiser *et al.*, 2020).

A plant breeder must find ways to improve the selection process and speed up breeding to meet the demands of evolving markets for crop varieties with marker-assisted breeding (MAB). Molecule MAB is the incorporation of DNA markers in effective breeding as well as selection techniques to enhance plant development (Hasan *et al.*, 2021). It offers substantial benefits in comparison to traditional breeding techniques. MAB experienced an important increase in recognition and implementation across several crop species during the 1990s. The discussion focuses on the fundamental procedures, theoretical concepts, and practical aspects of these techniques from the perception of a plant breeder (He *et al.*, 2017; Rasheed *et al.*, 2022b).

Recombinant DNA technology has greatly improved traditional crop development and has great potential for helping plant breeders manage the increasing demand for food predicted for the 21st century (Mahmood *et al.*, 2015; Saini *et al.*, 2020). Significant advancements have been achieved in the last 20 years in transferring genes from various sources into microorganisms and crop plants. The benefits represent modified diseases and pest resistance, raised tolerance to pesticides, drought, soil salinity, and metal toxicity; raised post-harvest quality; increased nutrient uptake and nutritional value; elevated photosynthetic rate; augmented sugar and starch production; improved efficiency of biological controls; deeper comprehension of gene function and metabolic pathways; and the ability to produce medicinal products and vaccines in crop plants (Aziz *et al.*, 2018; Giri *et al.*, 2023).

Modern plant breeding offers various techniques with potential advantages for crop improvement. These include polyploidy, wide crossing, haploidy, mutagenesis, and in vitro methodologies. Mutant breeding is now out of vogue and questionable; however, in vitro techniques used in a judicious manner allow breeders to avoid saying something speculative about the implications of genetic engineering (Van Harten, 2002; Zafar *et al.*, 2023).

3.3.2 Genome Editing Technologies

The CRISPR/Cas9 system enables precise DNA modifications using the Cas9 nuclease associated with a single-guide RNA (sgRNA). The sgRNA binds unambiguously to the specified DNA target sequence, directing the Cas9 endonuclease to automatically detect and cleave the corresponding DNA target (Filippova et al., 2019). The process of cutting produces the formation of double-strand breaks (DSBs), which activate cellular repair systems and result in mutations at or near the DSB sites. The CRISPR/Cas9 technology has significantly transformed the genome editing process since its inception, leading to its extensive use in altering the genetic makeup of many animal species. Often utilized for genome editing, the CRISPR/Cas9 system is highly efficient, precise, and simpler than alternative methods like ZFNs or TALENs (Akram *et al.*, 2023).

As many key food crops hit their maximum production limit, researchers must find methods to advance their progress further. To do so, we implement a highly predicting method called *strigolactone biosynthesis* (SB) (Chimmili *et al.*, 2022). This approach has been studied in a number of crop species and holds particular promise for pre-breeding and breeding processes performed in controlled environments. Breeding programmes can incorporate (SB) through several genomics-based approaches, such as marker-assisted selection (MAS), genomic selection (GS), pollen-based selection (PBS), and overexpression/knockdown transgenic and genome editing. Translational GM accuracy and efficiency (Basso *et al.*, 2020).

CRISPR/Cas9 technology holds great promise for enhancing plant growth and resilience. It has exhibited the ability to alter the genes of plants to bestow desired properties, such as resistance to diseases, drought, salinity, and cold. Nevertheless, the use of CRISPR/Cas9 has the potential to introduce a variety of issues unique to its application in plant breeding, issues that require careful consideration and study. This technique is yet to reach its full potential for boosting crop response to biological and environmental stressors in agricultural systems (Ali *et al.*, 2013b; Kumar *et al.*, 2023). In addition, scientists are investigating other potential means of chemical delivery, using technologies based on proteins or RNA that eliminate the need for transgenes to be integrated into the plant's genomic structure. This decision is out of concern for genetically modified organisms (GMOs). As genome-editing techniques continue to advance (Figure 3.2), we expect that CRISPR/

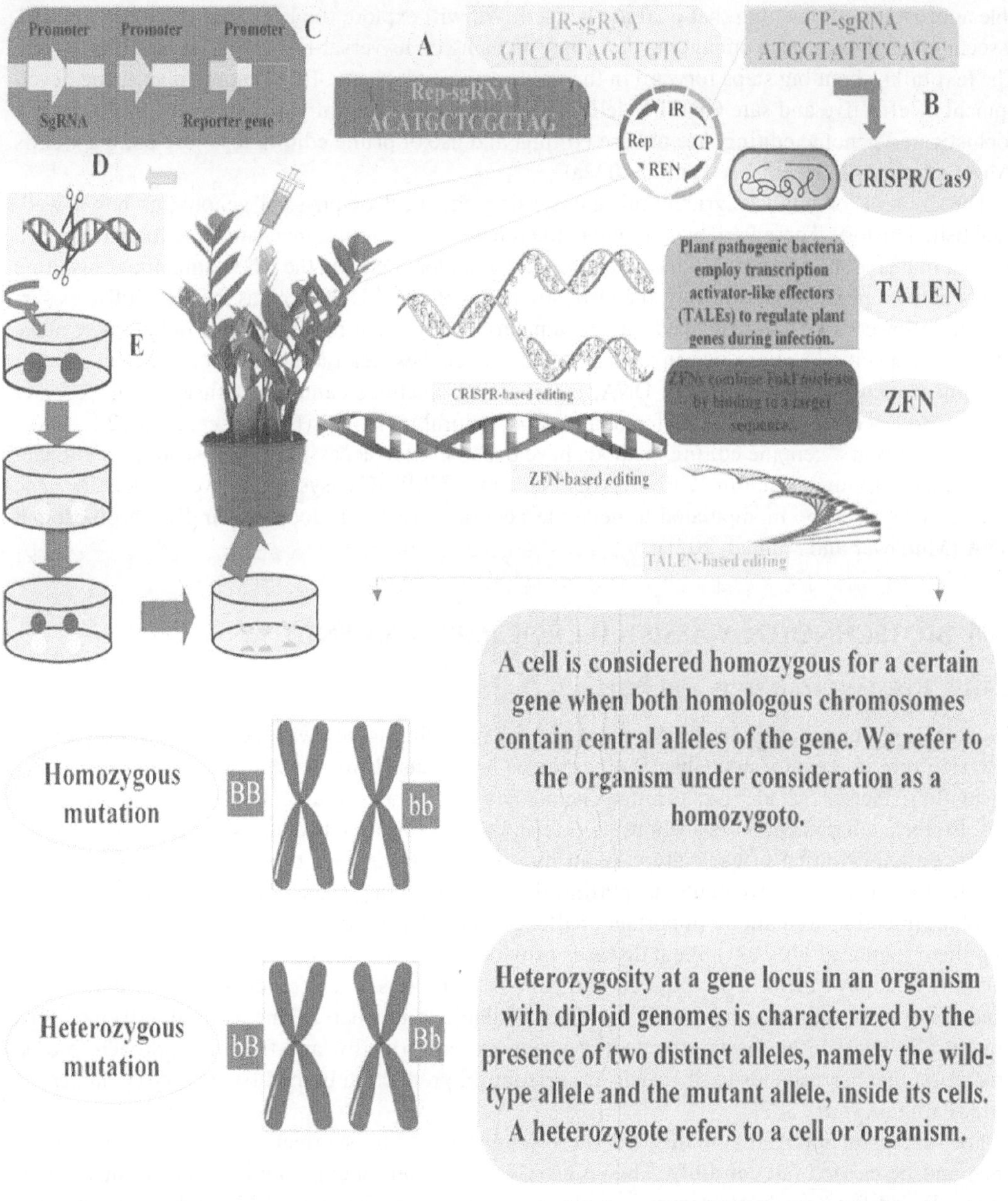

FIGURE 3.2 Applying genome editing techniques for improving resistance to diseases in plants.

Cas9 will remain an essential tool to improve crops, boost resistance to pathogens, and enhance tolerance to challenging environments (Mahmood *et al.*, 2013; Mutezo *et al.*, 2021).

3.3.3 Genome Editing Tools

The technique allows for precise GMs in a variety of animals and tissues. The Cas9 protein is an RNA-guided endonuclease, which creates precise double-stranded breaks at a specific target sequence recognized by a shorter RNA sequence, and is successfully used in both animal and plant species (Gupta *et al.*, 2019). Genetically altered crops provide a potential and adaptable solution for agricultural sustainability in order to meet the food demands of a rising population in a changing environment, much as traditional or mutant breeding approaches do (Ciren *et al.*, 2024).

Advances in genome editing technology make global food security achievable, but sustainable approaches to reaching that goal are needed. We will explore the latest advances of CRISPR-associated technologies in enhancing the crops, focusing on its versatility, potential, and challenges. The text talks about big steps forward in the field, such as transgene-free genome plants, the development of effective and safe CRISPR delivery systems for editing of plant genomes, multiplexing, sophisticated genome editing, use of base editing, and use of prime editing for more complex GMs (Mushtaq *et al.*, 2021; Rasheed *et al.*, 2022a).

The creation of mutant agricultural plants using chemical or physical genotoxins has a well-established history. These variables stimulate the inherent DNA repair mechanism to fix damage in a manner that is prone to errors. Radiation causes many random DSBs in the DNA, sometimes resulting in a beneficial phenotype. Recently, the utilization of synthetic SDNs, such as the CRISPR/Cas system, has allowed scientists to take advantage of natural DNA repair mechanisms to deliberately make changes in specific locations within plant genomes. Since these alterations do not always involve the permanent incorporation of foreign DNA, the resulting creatures cannot be considered genetically engineered since they cannot be differentiated from natural variations (Pugsley *et al.*, 2022).

In recent years, genome editing methods have become more accessible and essential for enhancing productivity in agriculture. ZFNs, TALENs, and CRISPR/Cas systems serve as tools for gene editing. They may be manipulated to generate accurate breaks in double-stranded chromosomal DNA (Munawar and Ahmad, 2021).

3.4 BIOTECHNOLOGY-ASSISTED CROP IMPROVEMENT

3.4.1 Resistant to Insects and Pests

Biotechnology has proved an effective method for resolving issues with pest and disease control in agriculture in recent years. Chemical pesticides have caused worries due to their environmental effect; these factors include risks to human health and an increase in susceptibility in the desired species. Biotech solutions provide accurate, eco-friendly, and sustainable options (Singh *et al.*, 2020).

Since the beginning of agriculture, plant diseases and pests have been a problem for farmers. Despite the world's achievements in splitting the atom, reaching the moon, and global connectivity, plant infections remain an important challenge to food availability despite our best attempts to stop them (Sattar *et al.*, 2021). Plant diseases provide a serious issue in agriculture despite our best attempts to control them. Utilizing GM and genome editing is a very efficient and durable way of controlling plant diseases, enhancing the tools available to breeders. Genetically modifying crops to naturally resist insect pests provides a convenient, eco-friendly, and friendly approach to crop protection that corresponds to the needs of agricultural production in the 21st century (Ullah *et al.*, 2013; Pathania *et al.*, 2021).

For successful application in the field, these solutions need to be effective, not harm plant agronomy, and be carried out carefully. They must also avoid introducing any possible sensitivities or toxins. World food production must grow by a minimum of 60% before 2050 while maintaining the current land use. Everyone has to react immediately and consider the new possibilities that are provided by Biotech to control plant diseases (Dias *et al.*, 2021).

3.4.2 Transgenic Plants

Transgenic plants are created by agrobacteria and other techniques of direct DNA transfer. Several transgenes providing resistance to insects, diseases, and herbicides have been introduced into agricultural plants from various plant and bacterial sources (Javaid *et al.*, 2012; Gosal and Wani, 2018). Most transgenic plants retain expressed genes permanently in their offspring without causing harm to the acquiring plant. Transgenic plants adapted to environmental factors consistently exhibit elevated levels of insect resistance. Transgenic crops cover 44.2 million hectares worldwide. Over the last 15 years, GMs have been carried out on over 100 plant species (Kumar *et al.*, 2020).

Transgenic technology should be included in an integrated approach for environmentally safe and sustainable pest control. Considerations, including intellectual property, safety concerns, and public views, must be taken into account before releasing transgenic organisms. Gene manipulation becomes an efficient method for genetically engineering plants, with an ability to regulate gene expression in higher plants. Previous studies introduced various genes to express modified *Bacillus thuringiensis* (Bt) toxins. In controlled laboratory and field conditions, excellent progress has been made in controlling Bt-sensitive pests. Consequently, the first Bt transgenic crops available on the market were constructed (Fisher *et al.*, 2017).

3.4.3 Resistant to Diseases

Genetic engineering involves changing the genetic makeup of living things, usually plants and animals, so they are more resilient to diseases. This way has become particularly widespread in agriculture, specifically in GMs of crops to increase their tolerance against infections, pests, and environmental pressures. But modern agriculture has successfully provided enough food to keep pace with the burgeoning population of the world, which the United Nations estimates will reach at least 9.8 billion people by 2050, up from 7.3 billion in 2015. The difficulty can be made worse by crop losses resulting from illnesses. Bacterial and fungal infections result in a reduction in agricultural production by around 15%, whereas viruses lead to a fall in yields by 3%. Applying genetic engineering to promote disease resistance in crops, which increased agricultural productivity, reduced reliance on chemicals, and improved the food supply. Nevertheless, contemporary technology also gives rise to ethical, environmental, and societal concerns that need careful examination and resolution (Kamila and Jasrotia, 2023).

Plant disease resistance genes are crucial for effective control of crop diseases. Uncovering and describing plant genes might be difficult, but identifying the substances that detect diseases is comparatively easy. Pathogens secrete tiny proteins known as effectors, which can have a significant impact on organism's ability to fight itself against a biological attack. Effectors play an important role in directing resistance gene selection and generating defensive responses. They are critical components of integrated disease control systems used to prevent harm to crops (Prasad *et al.*, 2019).

3.4.4 High Yield

Given the constant rise in human population, it is critical to give priority to improving the agricultural sector. To achieve this, we can use gene modification to enhance important traits that influence crop productivity, such as photosynthesis, biomass distribution between shoots and roots, inflorescence structure, regulation of stomatal conductance, nutrient use efficiency and utilization, interactions with beneficial microorganisms, ability to withstand environmental stresses like drought, salinity, flooding, extreme temperatures, and resistance to pests and diseases (Ullah *et al.*, 2021; Wasaya *et al.*, 2021).

Such solutions have to be effective to use, not damaging to plant agronomy, and need to be performed with caution to be applied in the field successfully. They also have to avoid any potential sensitivities or toxins from introducing. This requires at least a 60% increase in food production in the world before 2050 while keeping the land use intact. We all must act and assess the new opportunities that Biotech give us for plant disease management (Tylecote, 2019).

3.5 ENVIRONMENTAL IMPACTS OF BIOTECH CROP PRODUCTION

Biotechnology is a very powerful tool for improving the environment. It can be used for developing solutions that can effectively tackle climate change, food production, pollution, waste, restoration of ecosystems, and conservation of biodiversity. It seeks to have a potentially huge environmental impact (Munaweera et al., 2022). Biotechnology has enhanced agricultural productivity by integrating characteristics such as disease resistance and enhanced drought tolerance into crops. Previous studies have demonstrated that it is possible to introduce genes neighbouring loci linked to disease resistance from unrelated animal species into plants of agricultural importance. This study examines the global economic impact of GM technology on agricultural profits and environmental effects, specifically focusing on pesticide use and greenhouse gas emissions in nations that have been growing GM crops since 1996. The study found significant economic benefits in the agricultural industry, totalling $5 billion in 2005 and $27 billion over ten years. From an agronomic standpoint, there have been significant advancements in agricultural production per unit of land, mostly due to reduced losses coming from improved insect and weed control (Yadav et al., 2021). Synthetic pesticides are lethal to numerous living organisms, including insects, weeds, rodents, and much more. At the same time, they are destroying the quality of the soil. Pesticides have a major adverse impact on non-target organisms through the disruption of natural ecosystems (Waring *et al.*, 2018).

3.5.1 Socioeconomic Implications of Biotech Solutions

Socioeconomic assessment of IPM is important to assess the value of public spending on IPM research and extension. There is an urgent need to assess what the environmental and human health benefits of IPM implementation achieve. The detrimental consequences of over-reliance on synthetic pesticides are widely acknowledged.

Green bio-refineries (GBRs) provide both economic and environmental advantages by transforming the use of land from growing cereals to cultivating grass, producing protein nutrition for animals from grasses, and creating valuable byproducts. The potential for GBRs depends on many factors, such as the productivity of the land, sustainability, distance feedstock must be transported, and the local economy. Biotech crops, their accessibility and affordability have been widely discussed and analysed with rationales proposed and support for the widespread application (Corona *et al.*, 2018).

3.5.2 Biotech Crop Adoption and Development

With enhanced accessibility, farmers have embraced biotech crops in various parts of the world, especially, Brazil, the United States of America, India, and Argentina. Molecular breeding has the potential to accelerate crop development and allow for the utilization of a broader range of genetic sources. Progress with fruits and vegetables has faltered due to the long development timelines for molecular breeding, regulatory costs, and political barriers faced by biotech crops in numerous countries. Biotechnology research and development are primarily driven by the private sector, but there are substantial market failures in leveraging this expertise towards assisting poor producers and consumers in developing countries. The public sector, both national and international, must take on a significant role in overcoming this gap. In order to do this, it needs to strengthen its capacity to form creative collaborations with the commercial sector to obtain vital research tools and technology (Mahmood *et al.*, 2021; Ricroch *et al.*, 2022).

3.5.3 Biotechnology for Developing Countries

Biotechnology research is dominated by the private sector; however, there are major market failures when it comes to using this expertise to help underprivileged farmers and consumers in developing nations. To close this gap, the public sector—both domestic and foreign—must assume a significant role. While other biotechnology programmes are building the ability to adapt to other projects' tools and procedures, many nations still have research systems that are essentially incapable of doing molecular biology research. Each type of programme offers different challenges and chances to get access to new technologies (Alston and Pardey, 2021).

3.6 FUTURE DIRECTIONS AND CHALLENGES

Utilizing biological systems or currently existing living things, biotechnology produces, develops, or produces crops. This chapter evaluates the current state of biotechnology and examines its probable future paths. Today, biotechnology has significant role in business and agriculture. In medicine, vaccines are the most effective means of preventing infectious illnesses. However, their production still relies on biological systems. Antibodies and RNA/DNA probes have been very beneficial in detecting and treating infections. Using biotechnology, crops with improved nutrition profiles, higher yields, and reduced pesticide treatment requirements are being produced in agriculture. Biotechnology is being used in industry to produce chemicals, process food, process metallic ore, and lower pollution and energy use. The importance of progress in biotechnology research was highlighted during the discussions at the FAO International Symposium in February 2016. The highlight was the futuristic aspect of showing various goods resulting from research in agricultural biotechnology. The side-shows discussed the current worldwide advantages of agricultural biotechnology and the possibility of using them on a large scale by implementing careful policy strategies and refining farmers' methods to optimize the results of research in agricultural biotechnology (Wagh *et al.*, 2024).

In the last century, progress has been made in the part of plant tissue culture (PTC) via the adoption of new technologies, concepts, and their application. Scientists have advocated for the artificial cultivation of plants from the start to reduce dependency on natural resources. The rise in demand for plants with high yields and disease resistance led to the large-scale manufacture of plants using explants, resulting in the creation of thousands of plants. Genetic resources of plants Plant growth regulators are a crucial part of agro biodiversity, supporting the needs for food, fodder, fibre, and fuel. Furthermore, the promotion of plant genetic resources, environmental conservation, and the recovery of local and unique resources are expediting the development of new industrial products and scientific breakthroughs while also assuring food security. The threat to biodiversity is growing due to the continuous degradation of worldwide agricultural resources. Production needs in agricultural systems are displacing indigenous varieties, indigenous peoples, and other genetic resources (Pathirana and Carimi, 2022).

The public discussion around agricultural biotechnology revolves around its stated goal of promoting industrialization in agriculture. The critics and supporters have participated in an implied moral discussion about the path of technological and scientific advancement using the language of "risk". Critics have questioned the biotechnology research and development agenda for linking social and agricultural issues to genetic shortcomings while also continuing the risks associated with extensive monoculture. Facilitating international collaboration to create standardized regulatory protocols and ethical principles for the sustainable advancement and use of biotechnological innovations in agriculture. Biotech technologies show promise for tackling agricultural production difficulties but need strong regulatory frameworks, ethical concerns, and continuous research to assure safe, fair, and sustainable deployment (Khan *et al.*, 2021).

3.7 CONCLUSION

Crop production is vital, and biotechnology can help improve its characteristics. Traditional approaches have limitations; hence, biotechnology solutions are being developed to overcome these constraints and improve crop traits. Biotechnology-derived products should undergo continual improvement and refinement throughout the development, evaluation, and release stages, taking into consideration the most current discoveries and interactions. GM techniques, marker-assisted breeding, and genome editing are critical biotechnological tools that play a significant role in this process. This chapter examines the utilization of DNA markers breeding programmes, with a particular focus on novice breeders with little understanding of DNA markers. We now understand DNA markers better and can see the genetic diversity in agricultural plants thanks to progress in molecular plant breeding, plant genetics, genomics selection, and genome editing. This has led to big improvements in plant breeding methods. Biotechnology plays a significant role in agricultural production issues by offering sophisticated solutions that enhance crop productivity, resistance to stress, and nutritional value, while minimizing environmental impact. Biotechnology plays a leading role in tackling agricultural production problems by providing novel technologies that improve productivity, resilience, and sustainability in agriculture. We expect significant and transformative progress in industrial biotechnology, despite its unpredictable history. The advancements in wet- and dry-lab technologies are providing the opportunities for the development of cell factories and associated products. The rapid growth of obtaining important requirements for commercial investments has increased the attractiveness of the biotech industry to those who invest. The chapter emphasizes the pivotal role of biotechnology in revolutionizing agricultural output, addressing challenges, and advancing sustainable agriculture. There is a need for continuous examination, collaboration, and meticulous execution of biotechnological remedies to tackle global food security concerns and improve the long-term viability of agriculture.

LIST OF ABBREVIATIONS

Bt	*Bacillus thuringiensis*
DSBs	double-strand breaks
GBRs	green bio-refineries
GM	genetic modification
GMOs	genetically modified organisms
GS	genomic selection
MAB	marker-assisted breeding
nGMs	novel genetically modified technologies
PBS	pollen-based selection
PTC	plant tissue culture
SB	*Strigolactone biosynthesis*
SDNs	site-directed nucleases
sgRNA	single-guide RNA
TALENs	transcription activator-like effector nucleases
ZFNs	zinc finger nucleases

REFERENCES

Abdullah, Mahmood, A., Bibi, S., Naqve, M., Javaid, M.M., Zia, M.A., Jabbar, A., Ud-Din, W., Attia, K.A., & Khan, N., 2022. Physiological, biochemical, and yield responses of linseed (*Linum usitatissimum* L.) in α-tocopherol-mediated alleviation of salinity stress. *Frontiers in Plant Science*, 13, 867172.

Akram, F., Sahreen, S., Aamir, F., Haq, I.U., Malik, K., Imtiaz, M., Naseem, W., Nasir, N., & Waheed, H.M., 2023. An insight into modern targeted genome-editing technologies with a special focus on CRISPR/Cas9 and its applications. *Molecular Biotechnology*, 65, 227–242.

Ali, H., Tanveer, A., Nadeem, M., Asghar, H., & Javaid, M., 2013a. Germination ecology of *Rhynchosia capitata*: An emerging summer weed in Asia. *Planta Daninha*, 31, 249–257.

Ali, H., Tanveer, A., Nadeem, M., Javaid, M., Kashif, M., & Chadhar, A., 2013b. Allelopathic effects of *Rhynchosia capitata* on germination and seedling growth of mungbean. *Planta Daninha*, 31, 501–509.

Ali, Q., Shabaan, M., Ashraf, S., Rehman, A., & Asghar, H.N., 2020. Genetically Modified Rice Stacked with Antioxidants for Nutrient Enhancement and Stress Tolerance. *In* A. Roychoudhury (ed.) *Rice Research for Quality Improvement: Genomics and Genetic Engineering: Volume 2: Nutrient Biofortification and Herbicide and Biotic Stress Resistance in Rice.* Singapore: Springer Singapore, 433–467.

Alston, J.M., & Pardey, P.G., 2021. Chapter 75 - The Economics of Agricultural Innovation. *In* C.B. Barrett & D.R. Just (eds.) *Handbook of Agricultural Economics.* Amsterdam, Netherlands: Elsevier, 3895–3980.

Aziz, M., Ashraf, M., & Javaid, M.M., 2018. Enhancement in cotton growth and yield using novel growth promoting substances under water limited conditions. *Pakistan Journal of Botany*, 50, 1691–1701.

Balal, R.M., Shahid, M.A., Javaid, M.M., Iqbal, Z., Liu, G.D., Zotarelli, L., & Khan, N., 2017. Chitosan alleviates phytotoxicity caused by boron through augmented polyamine metabolism and antioxidant activities and reduced boron concentration in *Cucumis sativus* L. *Acta Physiologiae Plantarum*, 39, 1–15.

Basso, M.F., Arraes, F.B.M., Grossi-De-Sa, M., Moreira, V.J.V., Alves-Ferreira, M., & Grossi-De-Sa, M.F., 2020. Insights into genetic and molecular elements for transgenic crop development. *Frontiers in Plant Science*, 11.

Benedek, A., Rokicki, T., & Szeberényi, A., 2023. Bibliometric evaluation of energy efficiency in agriculture. *Energies*, 16, 5942.

Biswas, S., Zhang, D., & Shi, J., 2021. CRISPR/Cas systems: Opportunities and challenges for crop breeding. *Plant Cell Reports*, 40, 979–998.

Chadha, A., Florentine, S., Chauhan, B.S., Long, B., Jayasundera, M., Javaid, M.M., & Turville, C., 2019. Environmental factors affecting the germination and seedling emergence of two populations of an emerging agricultural weed: Wild lettuce (*Lactuca serriola*). *Crop and Pasture Science*, 70, 709–717.

Chadha, A., Florentine, S., Javaid, M., Welgama, A., & Turville, C., 2020. Influence of elements of climate change on the growth and fecundity of *Datura stramonium*. *Environmental Science and Pollution Research*, 27, 35859–35869.

Chattopadhyay, S., & Chattopadhyay, D., 2020. Coal and Other Mining Operations: Role of Sustainability. *In* R. Malhotra (ed.) *Fossil Energy.* New York: Springer New York, 333–356.

Chimmili, S.R., Kanneboina, S., Hanjagi, P.S., P.S, B., Sakhare, A.S., P, S., Kumar, S., & Kota, S., 2022. Integrating Advanced Molecular, Genomic, and Speed Breeding Methods for Genetic Improvement of Stress Tolerance in Rice. *In* M. Gowdra Mallikarjuna, S.C. Nayaka & T. Kaul (eds.) *Next-Generation Plant Breeding Approaches for Stress Resilience in Cereal Crops.* Singapore: Springer Nature Singapore, 263–283.

Ciren, D., Zebell, S., & Lippman, Z.B., 2024. Extreme restructuring of cis-regulatory regions controlling a deeply conserved plant stem cell regulator. *PLOS Genetics*, 20, e1011174.

Corona, A., Ambye-Jensen, M., Vega, G.C., Hauschild, M.Z., & Birkved, M., 2018. Techno-environmental assessment of the green biorefinery concept: Combining process simulation and life cycle assessment at an early design stage. *Science of the Total Environment*, 635, 100–111.

D'amelia, V., Docimo, T., Crocoll, C., & Rigano, M.M., 2021. Specialized metabolites and valuable molecules in crop and medicinal plants: The evolution of their use and strategies for their production. *Genes*, 12, 936.

Daniar, S., Karami, M., & Khosravi Sarvenoee, A., 2024. A curve fitting–based directional protection of transmission lines using genetic programing. *Journal of Sensors*, 2024, 1107906.

Das, S., Ray, M.K., Panday, D., & Mishra, P.K., 2023. Role of biotechnology in creating sustainable agriculture. *PLOS Sustainability and Transformation*, 2, e0000069.

Dias, T.a.D.C., Lora, E.E.S., Maya, D.M.Y., & Olmo, O.a.D., 2021. Global potential assessment of available land for bioenergy projects in 2050 within food security limits. *Land Use Policy*, 105, 105346.

Filippova, J., Matveeva, A., Zhuravlev, E., & Stepanov, G., 2019. Guide RNA modification as a way to improve CRISPR/Cas9-based genome-editing systems. *Biochimie*, 167, 49–60.

Fisher, K.E., Mason, C.E., Flexner, J.L., Hough-Goldstein, J., & Mcdonald, J.H., 2017. Survivorship of Z-pheromone race European corn borer (Lepidoptera: Crambidae) on a range of host plants varying in defensive chemistry. *Journal of Economic Entomology*, 110, 978–985.

Fizikova, A., Tikhonova, N., Ukhatova, Y., Ivanov, R., & Khlestkina, E., 2021. Applications of CRISPR/Cas9 system in vegetatively propagated fruit and berry crops. *Agronomy*, 11, 1849.

Gebrehiwot, K.A., 2018. A review on waterlogging, salinization and drainage in Ethiopian irrigated agriculture. *Sustainable Water Resources Management*, 4, 55–62.

Gentile, A., & La Malfa, S., 2019. New Breeding Techniques for Sustainable Agriculture. *In* M. Farooq & M. Pisante (eds.) *Innovations in Sustainable Agriculture.* Cham: Springer International Publishing, 411–437.

Giri, V.P., Shukla, P., Tripathi, A., Verma, P., Kumar, N., Pandey, S., Dimkpa, C.O., & Mishra, A., 2023. A review of sustainable use of biogenic nanoscale agro-materials to enhance stress tolerance and nutritional value of plants. *Plants*, 12, 815.

Gosal, S.S., & Wani, S.H., 2018. Plant Genetic Transformation and Transgenic Crops: Methods and Applications. *In* S.S. Gosal & S.H. Wani (eds.) *Biotechnologies of Crop Improvement, Volume 2: Transgenic Approaches.* Cham: Springer International Publishing, 1–23.

Gupta, D., Bhattacharjee, O., Mandal, D., Sen, M.K., Dey, D., Dasgupta, A., Kazi, T.A., Gupta, R., Sinharoy, S., Acharya, K., Chattopadhyay, D., Ravichandiran, V., Roy, S., & Ghosh, D., 2019. CRISPR-Cas9 system: A new-fangled dawn in gene editing. *Life Sciences*, 232, 116636.

Hasan, N., Choudhary, S., Naaz, N., Sharma, N., & Laskar, R.A., 2021. Recent advancements in molecular marker-assisted selection and applications in plant breeding programmes. *Journal of Genetic Engineering and Biotechnology*, 19, 128.

He, J., Meng, S., Zhao, T., Xing, G., Yang, S., Li, Y., Guan, R., Lu, J., Wang, Y., Xia, Q., Yang, B., & Gai, J., 2017. An innovative procedure of genome-wide association analysis fits studies on germplasm population and plant breeding. *Theoretical and Applied Genetics*, 130, 2327–2343.

Huang, X., Wang, Y., Xu, J., & Wang, N., 2020. Development of multiplex genome editing toolkits for citrus with high efficacy in biallelic and homozygous mutations. *Plant Molecular Biology*, 104, 297–307.

Javaid, M., Tanveer, A., Ahmad, R., Yaseen, M., & Khaliq, A., 2012. Optimizing activity of herbicides at reduced rate on *Emex spinosa* Campd. with adjuvants. *Planta Daninha*, 30, 425–435.

Jia, Z., & Lin, B., 2022. CEEEA2.0 model: A dynamic CGE model for energy-environment-economy analysis with available data and code. *Energy Economics*, 112, 106117.

Kaiser, N., Douches, D., Dhingra, A., Glenn, K.C., Herzig, P.R., Stowe, E.C., & Swarup, S., 2020. The role of conventional plant breeding in ensuring safe levels of naturally occurring toxins in food crops. *Trends in Food Science & Technology*, 100, 51–66.

Kamila, M.K., & Jasrotia, S.S., 2023. Ethics and marketing responsibility: A bibliometric analysis and literature review. *Asia Pacific Management Review*, 28, 567–583.

Khan, N., Ray, R.L., Kassem, H.S., Hussain, S., Zhang, S., Khayyam, M., Ihtisham, M., & Asongu, S.A., 2021. Potential role of technology innovation in transformation of sustainable food systems: A review. *Agriculture*, 11, 984.

Kordi, M., Salami, R., Bolouri, P., Delangiz, N., Asgari Lajayer, B., & Van Hullebusch, E.D., 2022. White biotechnology and the production of bio-products. *Systems Microbiology and Biomanufacturing*, 2, 413–429.

Kumar, K., Gambhir, G., Dass, A., Tripathi, A.K., Singh, A., Jha, A.K., Yadava, P., Choudhary, M., & Rakshit, S., 2020. Genetically modified crops: Current status and future prospects. *Planta*, 251, 91.

Kumar, M., Prusty, M.R., Pandey, M.K., Singh, P.K., Bohra, A., Guo, B., & Varshney, R.K., 2023. Application of CRISPR/Cas9-mediated gene editing for abiotic stress management in crop plants. *Frontiers in Plant Science*, 14.

Lenaerts, B., Collard, B.C.Y., & Demont, M., 2019. Review: Improving global food security through accelerated plant breeding. *Plant Science*, 287, 110207.

Mahmood, A., Hussain, A., Shahzad, A.N., & Honermeier, B., 2015. Biomass and biogas yielding potential of sorghum as affected by planting density, sowing time and cultivar. *Pakistan Journal of Botany*, 47, 2401–2408.

Mahmood, A., Ullah, H., Shahzad, A., Ali, H., Ahmad, S., Zia-Ul-Haq, M., Honermeier, B., & Hasanuzzaman, M., 2013. Dry matter yield and chemical composition of sorghum cultivars with varying planting density and sowing date. *Sains Malaysiana*, 42, 1529–1538.

Mahmood, A., Wang, X., Shahzad, A.N., Fiaz, S., Ali, H., Naqve, M., Javaid, M.M., Mumtaz, S., Naseer, M., & Dong, R., 2021. Perspectives on bioenergy feedstock development in Pakistan: Challenges and opportunities. *Sustainability*, 13, 8438.

Mao, Y., Botella, J.R., Liu, Y., & Zhu, J.-K., 2019. Gene editing in plants: Progress and challenges. *National Science Review*, 6, 421–437.

Meena, R.S., Kumar, S., & Yadav, G.S., 2020. Soil Carbon Sequestration in Crop Production. *In* R.S. Meena (ed.) *Nutrient Dynamics for Sustainable Crop Production.* Singapore: Springer Singapore, 1–39.

Mehmood, A., Naeem, M., Khalid, F., Saeed, Y., Abbas, T., Jabran, K., Sarwar, M.A., Tanveer, A., & Javaid, M.M., 2018. Identification of phytotoxins in different plant parts of *Brassica napus* and their influence on mung bean. *Environmental Science and Pollution Research*, 25, 18071–18080.

Misiou, O., & Koutsoumanis, K., 2022. Climate change and its implications for food safety and spoilage. *Trends in Food Science & Technology*, 126, 142–152.

Mpala, T.A., & Simatele, M.D., 2024. Climate-smart agricultural practices among rural farmers in Masvingo district of Zimbabwe: Perspectives on the mitigation strategies to drought and water scarcity for improved crop production. *Frontiers in Sustainable Food Systems*, 7.

Munawar, N., & Ahmad, A., 2021. CRISPR/Cas System: An Introduction. *In* A. Ahmad, S.H. Khan & Z. Khan (eds.) *CRISPR Crops: The Future of Food Security*. Singapore: Springer Singapore, 1–35.

Munaweera, T.I.K., Jayawardana, N.U., Rajaratnam, R., & Dissanayake, N., 2022. Modern plant biotechnology as a strategy in addressing climate change and attaining food security. *Agriculture & Food Security*, 11, 26.

Mushtaq, M., Ahmad Dar, A., Skalicky, M., Tyagi, A., Bhagat, N., Basu, U., Bhat, B.A., Zaid, A., Ali, S., Dar, T.-U.-H., Rai, G.K., Wani, S.H., Habib-Ur-Rahman, M., Hejnak, V., Vachova, P., Brestic, M., Çığ, A., Çığ, F., Erman, M., & El Sabagh, A., 2021. CRISPR-based genome editing tools: Insights into technological breakthroughs and future challenges. *Genes*, 12, 797.

Mustafa, A., Athar, F., Khan, I., Chattha, M.U., Nawaz, M., Shah, A.N., Mahmood, A., Batool, M., Aslam, M.T., & Jaremko, M., 2022. Improving crop productivity and nitrogen use efficiency using sulfur and zinc-coated urea: A review. *Frontiers in Plant Science*, 13, 942384.

Mutezo, W., Sedibe, M.M., Mofokeng, A., Shargie, N., & Soko, T., 2021. The application of CRISPR/Cas9 technology in the management of genetic and nongenetic plant traits. *International Journal of Agronomy*, 2021, 9993784.

Noor Ul, A., Arif, A., Jamil, A., Munir, A., Ahmad, A., & Munawar, N., 2024. Chapter 26 – Regulatory, Ethical, and Social Aspects of CRISPR Crops. *In* K.A. Abd-Elsalam, A. Ahmad & B. Zhang (eds.) *CRISPRized Horticulture Crops*. Cambridge, Massachusetts, United States: Academic Press, 439–470.

Pathania, A., Singh, L., & Sharma, P.N., 2021. Host Plant Resistance: An Eco-Friendly Approach for Crop Disease Management. *In* M. Kaushal & R. Prasad (eds.) *Microbial Biotechnology in Crop Protection*. Singapore: Springer Singapore, 395–449.

Pathirana, R., & Carimi, F., 2022. Management and utilization of plant genetic resources for a sustainable agriculture. *Plants*, 11, 2038.

Patra, P., Das, M., Kundu, P., & Ghosh, A., 2021. Recent advances in systems and synthetic biology approaches for developing novel cell-factories in non-conventional yeasts. *Biotechnology Advances*, 47, 107695.

Pawlak, K., & Kołodziejczak, M., 2020. The role of agriculture in ensuring food security in developing countries: Considerations in the context of the problem of sustainable food production. *Sustainability*, 12, 5488.

Poore, J., & Nemecek, T., 2018. Reducing food's environmental impacts through producers and consumers. *Science*, 360, 987–992.

Pramanik, A., Sinha, A., Chaubey, K.K., Hariharan, S., Dayal, D., Bachheti, R.K., Bachheti, A., & Chandel, A.K., 2023. Second-generation bio-fuels: Strategies for employing degraded land for climate change mitigation meeting United nation-sustainable development goals. *Sustainability*, 15, 7578.

Prasad, P., Savadi, S., Bhardwaj, S.C., Gangwar, O.P., & Kumar, S., 2019. Rust pathogen effectors: Perspectives in resistance breeding. *Planta*, 250, 1–22.

Prasad, R., Bhattacharyya, A., & Nguyen, Q.D., 2017. Nanotechnology in sustainable agriculture: Recent developments, challenges, and perspectives. *Frontiers in Microbiology*, 8.

Pugsley, K., Scherer, S.W., Bellgrove, M.A., & Hawi, Z., 2022. Environmental exposures associated with elevated risk for autism spectrum disorder may augment the burden of deleterious de novo mutations among probands. *Molecular Psychiatry*, 27, 710–730.

Rahman, S.U., Mccoy, E., Raza, G., Ali, Z., Mansoor, S., & Amin, I., 2023. Improvement of soybean; A way forward transition from genetic engineering to new plant breeding technologies. *Molecular Biotechnology*, 65, 162–180.

Rasheed, A., Barqawi, A.A., Mahmood, A., Nawaz, M., Shah, A.N., Bay, D.H., Alahdal, M.A., Hassan, M.U., & Qari, S.H., 2022a. CRISPR/Cas9 is a powerful tool for precise genome editing of legume crops: A review. *Molecular Biology Reports*, 49, 5595–5609.

Rasheed, A., Mahmood, A., Maqbool, R., Albaqami, M., Sher, A., Sattar, A., Bakhsh, G., Nawaz, M., Hassan, M.U., & Al-Yahyai, R., 2022b. Key insights to develop drought-resilient soybean: A review. *Journal of King Saud University-Science*, 34, 102089.

Ren, J., Li, W., Guo, Z., Ma, Z., Wan, D., Lu, S., Guo, L., Gou, H., Chen, B., & Mao, J., 2023. Whole-genome resequencing and transcriptome analyses of four generation mutants to reveal spur-type and skin-color related genes in apple (*Malus domestica* Borkh. Cv. Red delicious). *BMC Plant Biology*, 23, 607.

Ricroch, A.E., Martin-Laffon, J., Rault, B., Pallares, V.C., & Kuntz, M., 2022. Next biotechnological plants for addressing global challenges: The contribution of transgenesis and new breeding techniques. *New Biotechnology*, 66, 25–35.

Saini, P., Saini, P., Kaur, J.J., Francies, R.M., Gani, M., Rajendra, A.A., Negi, N., Jagtap, A., Kadam, A., Singh, C., & Chauhan, S.S., 2020. Molecular Approaches for Harvesting Natural Diversity for Crop Improvement. *In* R.K. Salgotra & S.M. Zargar (eds.) *Rediscovery of Genetic and Genomic Resources for Future Food Security*. Singapore: Springer Singapore, 67–169.

Saiz-Rubio, V., & Rovira-Más, F., 2020. From smart farming towards agriculture 5.0: A review on crop data management. *Agronomy*, 10, 207.

Sattar, M.N., Iqbal, Z., Al-Khayri, J.M., & Jain, S.M., 2021. Induced genetic variations in fruit trees using new breeding tools: Food security and climate resilience. *Plants*, 10, 1347.

Sharma, D., Gahtyari, N.C., Sharma, P., Khulbe, R.K., Pal, R.S., & Kant, L., 2022. Doubled Haploidy: An Accelerated Breeding Tool for Stress Resilience Breeding in Cereals. *In* M. Gowdra Mallikarjuna, S.C. Nayaka & T. Kaul (eds.) *Next-Generation Plant Breeding Approaches for Stress Resilience in Cereal Crops*. Singapore: Springer Nature Singapore, 199–240.

Sharma, K., Kumar, R., Kumar, A., Balabantaray, S., & Arora, M., 2024. A digital ecosystem for sustainable fruit supply chain in Uttarakhand: A comprehensive review. *Environment, Development and Sustainability*, 26, 13217–13252.

Sidorova, T., Miroshnichenko, D., Kirov, I., Pushin, A., & Dolgov, S., 2021. Effect of grafting on viral resistance of non-transgenic plum scion combined with transgenic PPV-resistant rootstock. *Frontiers in Plant Science*, 12.

Singh, A., Gautam, P.K., Verma, A., Singh, V., Shivapriya, P.M., Shivalkar, S., Sahoo, A.K., & Samanta, S.K., 2020. Green synthesis of metallic nanoparticles as effective alternatives to treat antibiotics resistant bacterial infections: A review. *Biotechnology Reports*, 25, e00427.

Tanveer, A., Khaliq, A., Javaid, M., Chaudhry, M., & Awan, I., 2013. Implications of weeds of genus euphorbia for crop production: A review. *Planta Daninha*, 31, 723–731.

Tudi, M., Daniel Ruan, H., Wang, L., Lyu, J., Sadler, R., Connell, D., Chu, C., & Phung, D.T., 2021. Agriculture development, pesticide application and its impact on the environment. *International Journal of Environmental Research and Public Health*, 18, 1112.

Turnbull, C., Lillemo, M., & Hvoslef-Eide, T.a.K., 2021. Global regulation of genetically modified crops amid the gene edited crop boom – A review. *Frontiers in Plant Science*, 12.

Tylecote, A., 2019. Biotechnology as a new techno-economic paradigm that will help drive the world economy and mitigate climate change. *Research Policy*, 48, 858–868.

Ullah, A., Bano, A., & Khan, N., 2021. Climate change and salinity effects on crops and chemical communication between plants and plant growth-promoting microorganisms under stress. *Frontiers in Sustainable Food Systems*, 5.

Ullah, H., Mahmood, A., Ijaz, M., Tadesse, B., & Honermeier, B., 2013. Evaluation of anise (*Pimpinella anisum* L.) accessions with regard to morphological characteristics, fruit yield, oil contents and composition. *Journal of Medicinal Plants Research*, 7, 2177–2186.

Van Harten, A.M., 2002. Mutation Breeding of Vegetatively Propagated Ornamentals. *In* A. Vainstein (ed.) *Breeding For Ornamentals: Classical and Molecular Approaches*. Dordrecht: Springer Netherlands, 105–127.

Varshney, R.K., Thudi, M., Pandey, M.K., Tardieu, F., Ojiewo, C., Vadez, V., Whitbread, A.M., Siddique, K.H.M., Nguyen, H.T., Carberry, P.S., & Bergvinson, D., 2018. Accelerating genetic gains in legumes for the development of prosperous smallholder agriculture: Integrating genomics, phenotyping, systems modelling and agronomy. *Journal of Experimental Botany*, 69, 3293–3312.

Wagh, M.S., Sowjanya, S., Nath, P.C., Chakraborty, A., Amrit, R., Mishra, B., Mishra, A.K., & Mohanta, Y.K., 2024. Valorisation of agro-industrial wastes: Circular bioeconomy and biorefinery process – A sustainable symphony. *Process Safety and Environmental Protection*, 183, 708–725.

Wang, X., Tu, M., Wang, Y., Yin, W., Zhang, Y., Wu, H., Gu, Y., Li, Z., Xi, Z., & Wang, X., 2021. Whole-genome sequencing reveals rare off-target mutations in CRISPR/Cas9-edited grapevine. *Horticulture Research*, 8.

Waring, R.H., Harris, R.M., & Mitchell, S.C., 2018. Plastic contamination of the food chain: A threat to human health? *Maturitas*, 115, 64–68.

Wasaya, A., Abbas, T., Yasir, T.A., Sarwar, N., Aziz, A., Javaid, M.M., & Akram, S., 2021. Mitigating drought stress in sunflower (*Helianthus annuus* L.) through exogenous application of β-aminobutyric acid. *Journal of Soil Science and Plant Nutrition*, 21, 936–948.

Wasaya, A., Yasir, T.A., Ijaz, M., & Ahmad, S., 2019. Tillage Effects on Agronomic Crop Production. *In* M. Hasanuzzaman (ed.) *Agronomic Crops: Volume 2: Management Practices*. Singapore: Springer Singapore, 73–99.

Yadav, P., Jaiswal, D.K., & Sinha, R.K., 2021. 7 – Climate Change: Impact on Agricultural Production and Sustainable Mitigation. *In* S. Singh, P. Singh, S. Rangabhashiyam & K.K. Srivastava (eds.) *Global Climate Change*. Amsterdam, Netherlands: Elsevier, 151–174.

Zafar, S., Afzal, H., Ijaz, A., Mahmood, A., Ayub, A., Nayab, A., Hussain, S., Maqsood, U.-H., Sabir, M.A., & Zulfiqar, U., 2023. Cotton and drought stress: An updated overview for improving stress tolerance. *South African Journal of Botany*, 161, 258–268.

4 Organic Farming Opportunities and Challenges

Maira Tanveer, Abrar Hussain, Muhammad Ather Nadeem, Aqsa Hanif, Ali Raza, Muhammad Awais, and Romana Iftikhar*

4.1 INTRODUCTION

An organic agricultural system that relies on crop rotation, green manure, composting processes, and biological pest control. Similar to alternative farming, organic farming (OF) does not depend on chemicals to provide nutrients for the crops, but employs organic materials instead (Durán-Lara *et al.*, 2020). Clover, for instance, has a significant amount of nitrogen, which farmers use to enhance soil quality. Animal waste and compost are also used to improve soil fertility. Furthermore, these fertilizers aid in soil conservation rather than causing its degradation quickly (Mckenna *et al.*, 2018). Crop rotation is used by organic farmers to build soil quality and prevent monoculture. Chemical pesticides harm the population of natural predators of pests, such as birds or frogs, by either killing them or reducing their strength. Furthermore, they can eliminate the insects that regulate a large population of pests. By providing water and nutrients, OF facilitates the establishment of new habitats for wasps, bugs, beetles, and flies. The use of specialized machinery manages weeds. Hay, straw, and wood chips are used as mulch to suppress weed growth between the rows of plants (Costa *et al.*, 2023). A wide range of agricultural goods may be cultivated using organic methods. Meat, dairy products, and eggs are derived from animals that are fed organic and have access to outdoor grazing. Various crops and animals are cultivated on the farm and systematically cycled throughout different seasons, including designated times of inactivity (Mie *et al.*, 2017; Rasheed *et al.*, 2022c). Implementing a mixed farming technique effectively disrupts the recurring patterns of pests and diseases while also enhancing soil fertility. Cattle, such as cows, are often confined to meadows and fields. Organic techniques are used to grow vegetables and fruit, whereas genetically modified (GM) crops and components are prohibited (Giampieri *et al.*, 2022).

Based on FAO data from 2014, it is projected that around 805 million people suffer from starvation worldwide, accounting for one out of every nine people. Sub-Saharan Africa has a prevalence of 25% of this statistic, which means one out of every four people is affected. In general, 98% of people experiencing hunger live in developing countries. In particular, 37 million individuals in Latin America, 227 million in Africa, and 526 million in Asia are running out of food. While there has been a significant decrease in these figures, particularly in Latin America, compared to previous years, there is still a considerable distance in the journey toward eliminating hunger (Ullah *et al.*, 2013; Mc Carthy *et al.*, 2018). With the global population and consumption rising, there will be an increased need for food, feed, and fuel. With increasing global population and consumption, there will be a greater demand for food, feed, and fuel. Furthermore, emerging nations' nutrition patterns are shifting, putting more strain on natural resources as dairy and meat consumption increases. By 2050, the world's population will be around 9 billion, resulting in a 1.1% increase in annual demand for agricultural goods (Hemathilake and Gunathilake, 2022).

* Corresponding author: abrar.hussain@cuisahiwal.edu.pk

DOI: 10.1201/9781003498698-4

A primary complaint of organic agriculture is its reduced productivity, which poses a challenge as the need for food production rises significantly to fulfill the needs of an expanding and more affluent worldwide population. Organic agriculture is often criticized for being inefficient, particularly in terms of land use. Due to the increasing global demand for food, it is recognized that the existing agricultural area must be improved. As a result, the expansion of organic agriculture will lead to further encroachment on pristine ecosystems (Hanif *et al.*, 2017; Rasheed *et al.*, 2022a). Some experts believe that the concept of "ecological intensification" in agricultural production systems represents the most efficient strategy for meeting future food demand while also reducing environmental constraints in a sustainable manner. This involves using and managing biodiversity's ecosystem services rather than relying on human-made inputs (Hayat *et al.*, 2010; Raj *et al.*, 2021).

According to FAO data, agriculture consumes 11% of the planet's land area and 70% of its freshwater resources. The global distribution of arable land is characterized by inequality, with high-income nations possessing more than 50% of the cultivated land per individual. In contrast, low-income countries have the opposite trend. In some areas of the globe, such as Africa, the indigenous agricultural technique consists primarily of slash-and-burn agriculture with a couple-year hibernation period. Despite this, farmers are reducing the time they allow their lands to fallow as the population increases (Pereira *et al.*, 2018; Abdullah *et al.*, 2022). To meet demand, most small-scale farmers plant annually, which causes severe soil erosion and nutrient degradation. As a result, to replicate the process, these farmers are compelled to disregard their properties and acquire new land. According to the FAO, the amount of arable land per person worldwide has declined from 0.38 hectares in 1970 to 0.2 hectares in 2013. It is expected to reach about 0.15 hectares by 2050. Various studies show that the quantity of cultivable land and water resources is seriously decreasing in developing countries (Jouzi *et al.*, 2017; Naqve *et al.*, 2021). Small-scale producers have many opportunities and challenges while adopting OF.

This chapter aims to analyze the major pros and cons of organic agriculture for small-holders in developing countries. This chapter will begin by looking at the health and nutritional benefits of organic foods and the environmental, economic, and social benefits of OF. In this chapter, we will focus on the primary challenges of OF, which include low productivity and poor nutrient control, market and certification concerns, as well as research and educational barriers. Therefore, we will evaluate what degree policymakers have to invest in OF to be used as an instrument to improve the living quality of farmers below those farms.

4.2 PRINCIPLES OF ORGANIC FARMING

Understanding the underlying principles of organic agriculture is crucial to understanding the rationale for OF. These principles contain the essential objectives and precautions that are considered vital for producing high-quality food, fiber, and other items ecologically sustainably (Figure 4.1). OF relies on the dynamic interaction of the soil, plants, animals, human beings, ecosystem, and environment (Van Geel *et al.*, 2017; Jianing *et al.*, 2022).

4.2.1 Soil Health in Organic Farming

Over an extended period, the scientific community has dedicated significant attention to the subject of soil quality or health. The term "soil quality" refers to the ability of the soil to sustain biological activity, preserve environmental conditions, and promote the well-being of plants and animals in an ecological system. It is very improbable that the same chemical qualities of the soil would be desired in all locations or management methods (Toor *et al.*, 2021; Rasheed *et al.*, 2022b). The integration of physical and chemical factors and biological indicators is used to assess the quality of the soil rather than a predefined standard (Table 4.1).

The interpretation of these indicators considers the ecosystem in which the soil is located, regardless of whether the soil is natural or exposed to management processes. The dynamics of

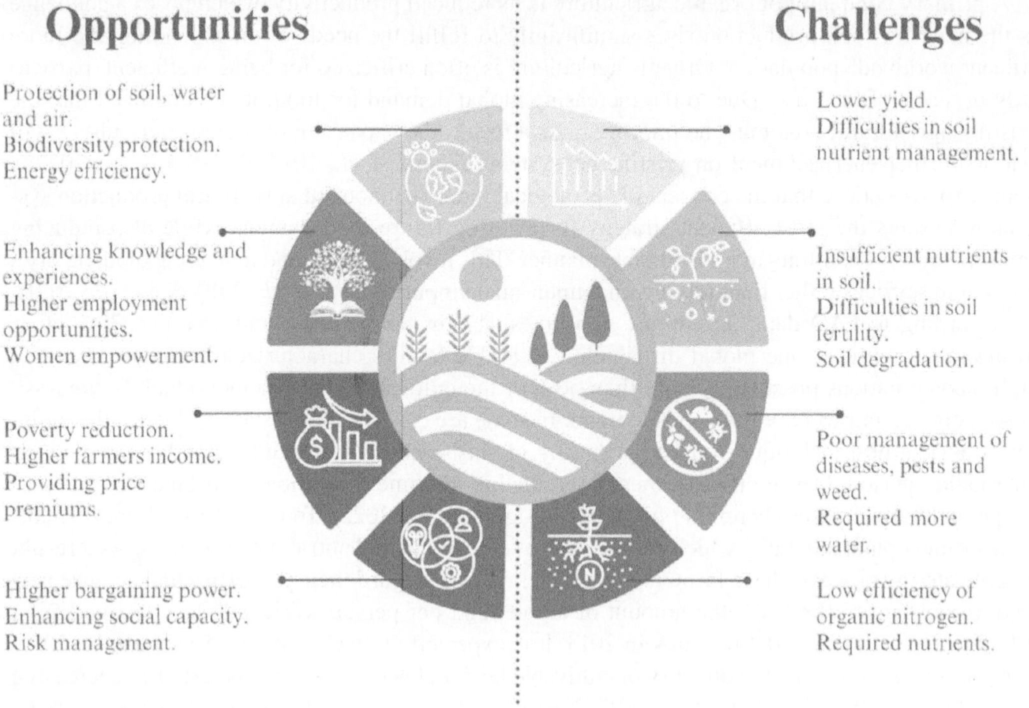

FIGURE 4.1 Main principles and outcomes of organic farming.

TABLE 4.1
Impacts of Organic Farming Practices on Soil Health

Organic Components	Soil Properties	Impacts on Soil Properties	References
Farm-yard manure, vermicompost, household waste, green manure, and sewage sludge	Physical	Enhancing soil structure, porosity, moisture retention capacity in the soil	Ahmad *et al.* (2021)
	Chemical	Provides several macro- and micronutrients to the plants	Goswami *et al.* (2017)
	Biological	The soil micro-organisms are the living portion of the soil organic matter	Das *et al.* (2021)
Mulching	Physical	Resulted in better and less run off-of water in field	Yang *et al.* (2020)
	Chemical	The mulching materials decompose and enrich the soil with organic matter and other nutrients	Kader *et al.* (2017)
	Biological	Mulching also increases the density, diversity, and activity of the macro-fauna in soil	Noor Ul *et al.* (2022)
Crop rotation	Physical	Various root systems of several crops with crop rotation associated with the Bayesian network were used for architectural formation, which affects the physical structure of the soil	Bacq-Labreuil *et al.* (2018)
	Chemical	Soil pH, available phosphate, exchangeable K^+, and Ca^{2+} in soil significantly increased under rotation	Kwiatkowski and Harasim (2020)
	Biological	Crop rotation reduces soil-born pathogen damage and enhances some of soil chemical properties and soil microbial biomass	Wen *et al.* (2020)

soil carbon (C) and organic matter (OM) are closely interconnected regarding the soil's quality. These dynamics, in turn, directly impact the soil's physical, chemical, and biological functions. The incorporation of OM inside soil a crucial role in the stabilization of aggregates, erosion prevention, water retention capacity, and slow release of nutrients (Levi *et al.*, 2021). Soil organic matter (SOM) depletion is not only highly susceptible to management strategies, but it is also associated with a decrease in soil quality. It is now well understood that there is a link between soil quality over time and environmental quality. Furthermore, many individuals recognize the need to protect the soil as a valuable resource for the future. (Ramesh *et al.*, 2019; Tang *et al.*, 2022). Approximating "soil health" with "soil quality" is standard practice. However, most scientists tend to favor soil quality, while farmers tend to favor soil health. According to the opinions of some individuals, the word "soil health" should be reserved for the biological aspects of soil quality, particularly the prevention of diseases. Usually, the term "soil tilth" pertains to both the physical characteristics of the soil and its management. Nevertheless, the characterization of soil tilth closely resembles that of soil quality (Brock *et al.*, 2021).

It has been demonstrated that management techniques such as diversified and extensive crop rotations, cover crops, and animal manures have a noteworthy effect on the health of plants and soil. Since the mid-20th century, numerous lasting investigations have been conducted on agricultural systems, with some of these studies still ongoing. To a greater extent than conventional farms in the surrounding area, organic farms that are well-maintained often have higher soil quality (Baldwin-Kordick *et al.*, 2022). Recent reviews and meta-analyses have shown that OF methods usually improve soil organic carbon, a vibrant indication of soil quality. This enhancement occurs even when there is no net input of nutrients into the soil.

4.2.2 Nutrient Cycling in Organic Farming

Soil fertility is maintained in organic systems by implementing crop rotations that include organic amendments (e.g., composted or raw animal manure), incorporating soil-building green manures, and utilizing biochemical methods that change nutrients into forms that plants can use. Plants can absorb organic forms of nitrogen, especially in certain circumstances; this may be significant in agricultural systems that depend on the mineralization of OM for their nitrogen source (Rodríguez-Espinosa *et al.*, 2023). On the other hand, the most effective management of organic nitrogen requires the prompt discharge of inorganic nitrogen from OM and the subsequent absorption of this nitrogen by plants. To maintain soil quality, it is necessary to mineralize carbon and nitrogen in SOM for short-term crop assimilation and to retain carbon and nitrogen in SOM for conservation purposes. The second aspect has significant implications for carbon and nitrogen budgets on a regional and global scale, as well as for water quality and carbon storage in soil (Ashraf *et al.*, 2013; Ukalska-Jaruga *et al.*, 2020).

On the other hand, additional scientific research needs to be done on techniques to maximize biological turnover in OF. Using manures and composts in significant quantities over a long period of time may increase soil quality matter and SOM while preserving or even boosting yields in comparison to conventional farming techniques (Table 4.2).

However, excessive reliance on manures can exacerbate excess nutrient accumulation and eventual losses, possibly damaging the ecosystem (Tian *et al.*, 2017). On organic farms, there is often a reduction in losses, particularly when considering the crop yield per unit area. However, in cases when there is an excess accumulation of nutrients, the quantity of nitrogen (N) and phosphorus (P) that gets eliminated from organic systems may be more substantial compared to conventional farms (Ahmad *et al.*, 2010; Möller, 2018). Such excessive nutrient accumulation is most likely in organic systems that rely heavily on external supplements to boost yields. Under cold-temperate climates. Bargaz *et al.* (2018) believe that it may be especially difficult to balance the supply of nitrogen with the demand of plants. For OF, there is room for more research to increase nitrogen use efficiency by better-reconciling soil supply and plant demand.

TABLE 4.2
Strategies to Increase Yields in Organic Arable Crops

Intervention Area	Crop	Strategies to Increase Yield	References
Plant nutrients	All crops (except nitrogen for legume crops)	• Optimal use of legumes in rotations • Effective use of manures • Enhance recycling and harvesting nutrients from society	Benincasa *et al.* (2017)
Soil fertility	Soybean and Wheat (especially stockless farms)	• More diverse crop rotation • Intercropping • New technologies for reduced tillage • Strengthened collaboration between agents working on livestock farms with those working on stockless farms	Zikeli and Gruber (2017)
Crop-weed competition	All crops (especially stockless systems without perennial leys)	• False seedbed technique application • Precision farming and robots • Crop rotation design and management	Merfield (2023)
Diseases Control	Potatoes and legumes	• Use of tolerant or resistant crop varieties Crop rotation design and management • Replace the copper that is currently used. • Use of certified and dressed seeds	Olle and Williams (2021)
Pests Control	Oilseed rape and potatoes	• Crop rotation design • Physical/biological methods such as nets, traps, and repellents • Selective pest control products with low negative side effects	Ratnadass *et al.* (2023)

4.2.3 Green Manure and Crop Rotation in Organic Farming

Good organic farms use manures and composts that are appropriate for the needs of their crops. These sources often supply phosphorus, potassium, and other macro- and micronutrients. These farms also depend on legume cover crops and forages for their nitrogen supply, and they also grow winter catch crops in order to reduce leaching losses of nutrients. Crop rotations, which may include legumes and other soil-improving crops, are commonly employed. Integrating forage legumes and green manures into long rotations can improve soil physical characteristics, decrease erosion and the risk of nitrogen leaching, and increase SOM. It is possible that enhanced rotations might reduce the need for external inputs by as much as 85%. Green manure may be used as cover crops before termination, but it can also be used to promote soil fertility, prevent soil erosion, absorb soil carbon, boost soil water penetration and storage, and inhibit weeds. Weed control and soil quality improvement are two of the main functions cover crops perform in OF systems (Fageria, 2012; Ashraf *et al.*, 2016). Research has shown that a mixture of barley (*Hordeum vulgare* L.), rye (*Secale cereale* L.), and wheat (*Triticum* spp.) with hairy vetch (*Vicia villosa* Roth) and red clover (*Trifolium incarnatum* L.) have been proven to be effective under organic crop production systems. This is due to the ability of these cover crops to develop rapidly, competitively, and simply to remove by mechanical options (Singh *et al.*, 2021).

4.2.4 Plant Disease Management in Organic Farming

In OF, proper pest and disease management depends on the maintenance of soil fertility with balanced crop rotation. These rotations include nitrogen-fixing crops and winter cover, intercrops, manure and compost inputs, and reduced soil tillage. Crop rotations are modified to maximize nutrient availability and to reduce the number of weeds, diseases, and pests. Improving ecological diversity can also leverage enhanced biological control of pests and diseases. This is achieved, among other things, by practicing intercropping and introducing flowering plants, shrubs, trees, and wild grasses (Rusinamhodzi, 2020).

This crop protection method emphasizes managing the environment to help plants resist potential attacks as opposed to aggressively eliminating pathogens. For example, many organic farmers employ cultural plant protection practices. Thus, if artificial fertilizers are be replaced by organic inputs and manure, the resulting system shall be microbial-dominated and have variable micronutrient availability. This will have major implications for plant resilience and the balance of pathogens and beneficial microbes in the soil. The current study will examine four essential techniques for disease management. (1) The implementation of preplant measures to prevent the introduction of pathogens; (2) The reduction of pathogen entry by minimizing the initial inoculum; (3) The control of pathogen establishment by reducing the host and environment's suitability for infection and reproduction; (4) The management of pathogens through the utilization of curative methods to limit further dissemination (Shaw *et al.*, 2021).

4.3 OPPORTUNITIES IN ORGANIC FARMING

4.3.1 Environmental Consciousness

Research shows that the rural impoverished are particularly vulnerable to environmental degradation because they live in fragile ecosystems and rely heavily on natural resources for their livelihoods. The financing limitations they face, in an environment where environmental degradation severely limits income, can further lead to the depletion of natural resources and thus the endless cycle of poverty and ecological degradation. IFOAM states that OF is governed by cosmic principles rotating around natural ecological processes and cycles (Figure 4.2) (Ahmed *et al.*, 2020; Javaid *et al.*, 2022). OF is a more comprehensive agricultural method that highly prioritizes the long-lasting sustainability of the ecosystem and concentrates on generating food in an environmentally considerate manner. Environmental benefits of OF include the development of soil, water, biodiversity, indoor air holding, and energy benefits. OF is said to do well in terms of environmental impact: through metrics, especially area-normalized metrics. OF often has positive environmental outcomes relative to the amount of land used, but this may not be feasible for the quantity of product produced. Organic systems exhibit a decrease in nitrogen emissions per unit of land area, while simultaneously experiencing an increase in nitrogen leaching per unit of product. Although there was a drop in energy use, there was an increase in land requirements, resulting in a higher potential for eutrophication and acidification per unit of product (Panday *et al.*, 2024).

The quality and quantity of the arable lands that belong to small-holder farmers are declining because many of them live on degraded land and engage in agricultural practices that are not sustainable. Within the framework of the OF system, soil plays an essential part in production and can enhance soil quality. Restoring degraded lands and mitigating further deterioration in vulnerable regions, such as sub-Saharan Africa, are both outcomes of the soil management practices that are used in OF. Agroforestry, soil bunds, contour cultivation, terraces, growing cover crops, mulching, and a lesser amount of tilling of the ground are some of the strategies that are used in the OF system to conserve the soil. According to these studies, the quantity of organic soil matter that is present in OF systems is noticeably larger than that which is present in traditional systems. Through the diversification of soil-food webs, OM improves the nitrogen cycle within the soil, which in turn protects water resources (Huang *et al.*, 2022).

A significant advantage is offered by the OF system in comparison to traditional systems in terms of the amount of energy that is used. In comparison to conventional farms, organic corn production showed a 31% reduction in fossil energy inputs, and soybean production demonstrated a 17%

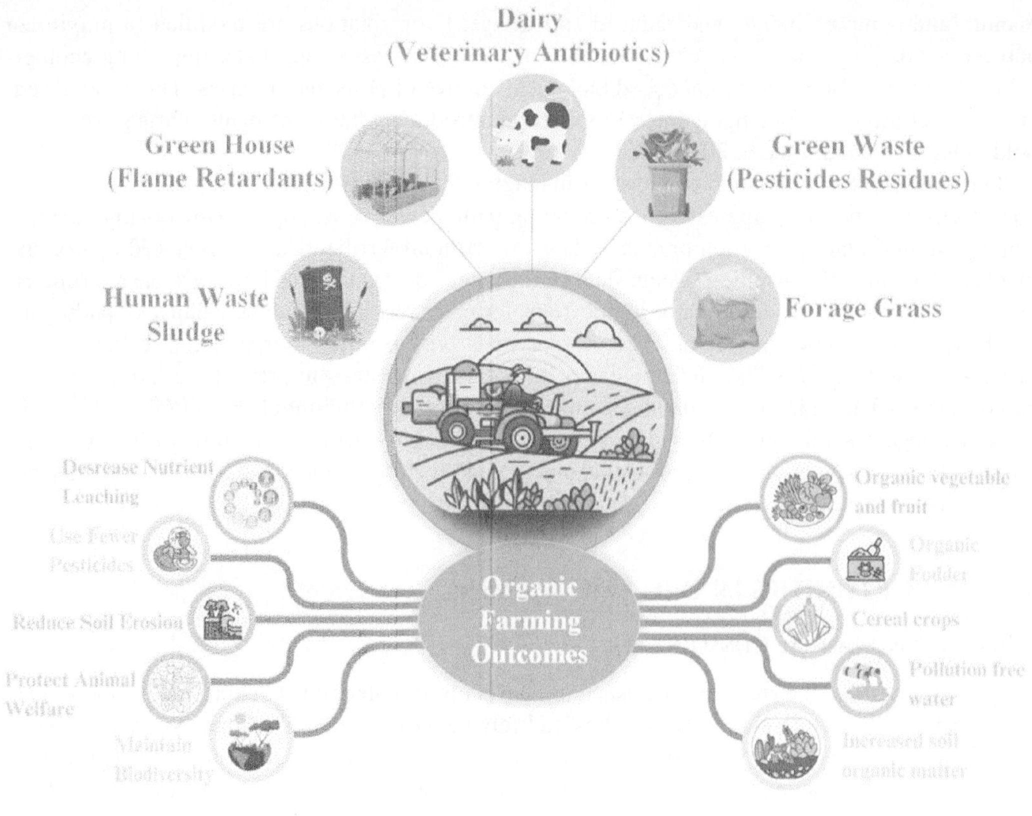

FIGURE 4.2 Opportunities and challenges faced by organic farming.

decrease in fossil energy inputs (Abbas *et al.*, 2021). A further study conducted on OF in Central Europe showed a reduction in energy use and fertilizer ranging from 34 to 53%. Because high fuel costs have lately led to a rise in food prices, it is becoming more reasonable to acknowledge the pressing need to transition to agricultural techniques that are more environmentally friendly in general and that of OF in particular. A decrease in the cost of energy results in a reduction in inputs that small-scale farmers have to spend in inputs (Paris *et al.*, 2022).

Moreover, given that the labor force is the most important asset owned by small-holders, they are more adaptable to changing environmental circumstances inside the OF system. As a result, they can modify their product patterns and practices with greater ease. Adaptation techniques that are founded in multi-cultures are advocated by OF. These measures can not only mitigate the risk of crop loss but also strengthen the resilience to adverse weather. Also, farmers have the capacity to create many kinds of well-suited and even resistant to harsh conditions crops by using the native knowledge they possess. Furthermore, organic farming (OF) can be retained, which helps in reducing greenhouse gas emissions including carbon and nitrogen oxides. It can also improve how much carbon the soil absorbs (Gunathilaka *et al.*, 2018; Chadha *et al.*, 2020). In general, OF has the potential to increase the climate resilience of ecosystems using both mitigation and adaptation strategies. Both are known as "climate change adaptation." However, research has demonstrated the impact of OF on the environment to be more dramatic, per product unit, than that of its competitor (Raza *et al.*, 2019). For this reason, they argue that integrated systems that employ the best practices of both conventional and OF could produce higher yields while reducing their negative effects on the environment.

4.3.2 Nutritional and Health Benefits

Through efficient use of local resources and a skill set, OF can meet the nutritional requirement and provide affordable access to food to low-income and resource-poor farmers, resulting in growing different varieties of crops at an economical cost. Small-scale farmers in arid locations may enhance their food output by using OF methods. A report conducted by UNEP-UNCTAD in 2008 examined 114 organic or near-organic applications across 24 African states, revealing a significant increase of 128% in average output. Most farmers in regions like Africa operate on a small scale and depend on minimum chemical inputs for cultivating crops. Hence, adopting OF practices is a viable option for individuals to augment their crop productivity and improve food accessibility (Bottrell and Schoenly, 2018). OF encourages multi-culture which helps in decreasing the probability of crop failure and food insecurity. This is due to food shortages in rural areas are often the result of crop failures in monoculture systems. Because OF is a multi-culture, the dietary variety of subsistence farmers also rises, which is another essential problem that should be taken into consideration.

This is due to the fact OF creates nutritional deficiencies. Studies also prove that the process of gradually switching to OF can increase yield and lead to improved food supply for the small-holder farmers. It could also increase the revenue for smaller farmers, which improves purchasing power (Suresh Reddy, 2017). Studies show that organic foods have fewer chemical residues than non-organic foods in terms of food safety and quality. Additionally, organic food has less nitrate. By going organic, farmers eliminate synthetic inputs in farms and therefore the risk of being exposed to chemical pesticides. According to research, 99% of pesticide-linked closures globally happen in developing nations where levels of illiteracy and poverty are high in rural population groups. The farmers there are typically poor and not adequately educated about the safety protocols for applying chemical pesticides (Atandi *et al.*, 2017; Chadha *et al.*, 2019). The French Agency for Food Safety (AFSSA) conducted a review investigation which indicated that organic plant products contained higher levels of dry matter, minerals such as iron and magnesium and healthful micronutrient antioxidants, indicating superior nutritional quality. They go on to note that organic animal products contain higher levels of polyunsaturated fatty acids than conventional animal products. Moreover, a recent meta-analysis analysis encompassing 343 studies has demonstrated significant dissimilarities in nutritional composition between organic and conventional food products. The current study found that organic foods have a higher antioxidant concentration, while, at the same time, showing a Lower level of harmful heavy metals including cadmium and pesticide residues. This discovered organic foods are of higher quality than non-organic foods. At present there is little appetite among consumers for organic products, which will require more public and government support. Organic foods offer proven health and nutritional benefits (Thakur, 2017).

4.4 ECONOMIC BENEFITS OF ORGANIC FARMING

The global market for organic food has seen significant growth, increasing from 15.2 billion USD in 1999 to 72 billion USD in 2013. As a result, organic companies have become one of the fastest-growing categories of the food industry. Most organic markets are in the European Union and the United States, while emerging nations have very small organic markets (Figure 4.3).

In addition, OF is recognized to be a sustainable means for alleviating poverty, particularly for subsistence farmers and farmers in countries in the developing world that have limited access to resources (Mehmood *et al.*, 2018; Maclaren *et al.*, 2020). Crowder and Reganold (2015) conducted a global meta-analysis of the economic competitiveness of OF systems across five continents. Even though OF has lower yields, the research showed it is far more economically profitable (22–35% more) than conventional agriculture. The profitability of OF can be attributed to the greater rates imposed for organic products, as shown by their research. A recent study comparing conventional and OF practices in India also reported high profitability, with a 9.2% loss in crop production amounting to a 22% gain in farmers' net profit due to a 20–40% price premium and an 11.7%

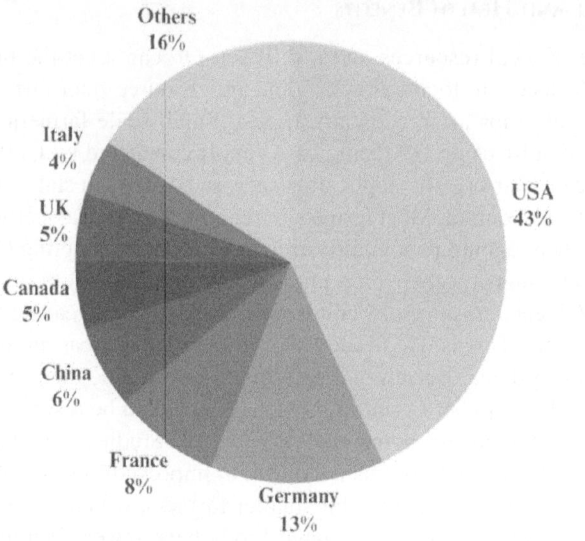

FIGURE 4.3 Organic food marketing on a global scale.

reduction in production costs. OF in under-developed countries results in more profit due to higher crop yields, lower expenses, and the ability to charge higher prices for organic goods.

Important economic benefits of OF for small-scale producers include access to organic markets and a reduction in external inputs, which enables them to offer their products at higher prices. Organic goods have price premiums ranging from 10 to 300%. Farmers are predicted to get 44–50% of these price premiums, which might potentially help in reducing poverty in developing nations. An organic agricultural system possesses the capacity to decrease production expenses through the substitution of chemical inputs with locally attainable organic inputs. On the other hand, OF is a labor-intensive method of producing food coupled with the reality that small-scale producers' family members often spend their time working on subsistence farms may result in even lower production costs. The management of risks is yet another significant component that has to be addressed (Meemken and Qaim, 2018). Small-holder farmers often have a limited ability to deal with risk since they do not have access to risk reduction solutions such as crop insurance at their disposal. On the other hand, OF has a wonderful potential to favorably alter the risks that small-scale farmers face by diversifying their goods via agroforestry, intercropping, and rotation to assist them in lowering the chance of their biggest crops failing. Reducing input costs will mitigate the vulnerability of small-scale farmers to crop failure affected by climate change. Therefore, OF, a technique with less danger, is a suitable choice for farmers with reduced risk levels (Tanveer *et al.*, 2015; Zerssa *et al.*, 2021).

4.5 SOCIAL BENEFITS OF ORGANIC FARMING

According to a recent study published in the journal Organic Agriculture (2015), the global count of organic food producers reached 2 million in the year 2013. Furthermore, more than 80% of these farmers, which amounted to 1.7 million, and around 25% of organic lands, comprising a total area of 11.7 million hectares, are situated within developing countries. The majority of off-grid OF producers in Africa are small-holder farmers (2–3 hectares) who are focused on exporting their products and get help mostly from non-governmental organizations (NGOs) rather than from the government. For instance, over 90% of the organic farmers in Uganda are small-scale farmers. With the maximum number of organic producers and the largest cultivated area, Uganda is the originator of organic agriculture in Africa. Small-scale farmers also practice these systems of agriculture which are not sustainable either, and therefore, it leads to further degradation of the landscape. The aforementioned circumstances may be attributed to adverse socio-economic conditions, with a lack of elements such as market accessibility, suitable technology, financing availability, natural resource availability, and unstable land tenure (Ali *et al.*, 2013; Muller *et al.*, 2017). The integration of OF practices, with a focus on local and indigenous knowledge, has promise for enhancing social capacity and simultaneously improving natural resources in an ecological system. OF increases social capital by supporting local or regional social organizations and NGOs and introducing new norms, as well as regulating resources by small farmers. OF supports farmers' associations, which provide several advantages for small-scale farmers. Benefits include increased negotiating power, improved access to financing and markets, opportunities for knowledge sharing, reduced certification costs, and enhanced contributions to policy institutions. Because OF is labor-intensive, it may provide additional employment in rural areas and enable farmers to access improved education and healthcare facilities because of the greater revenues generated by OF. Small-scale tea producers in Kenya managed to cover educational and medical costs by using OF methods (Deka and Goswami, 2021).

Regarding the social advantages of OF, there are apprehensions about its effects on female farmers. Engaging in OF may provide the chance to boost revenue by cultivating low-input cash crops, but it may also lead to an increased burden, resulting in the delegation of more tasks to their daughters. Moreover, the potential enhancement of household nutrition via increased income from working abroad is dependent upon women having enough bargaining power and the ability to share in family decision-making. World Watch Institute (2006) states that transitioning to OF leads to a more reliable rise in output, particularly in distant regions, which may help support impoverished farmers in such areas (Rahman *et al.*, 2022).

4.6 CHALLENGES IN ORGANIC FARMING

4.6.1 Low Production Rate

Some experts suggest that a widespread transition to OF might lead to a 40% decrease in global agricultural yields, resulting in crop failures that could potentially impact the food supply for about 2.5 billion people. They argue that this transformation might result in a significant worldwide food shortage. In about 1900, agricultural methods were similar to those with few external resources, capable of supporting a population of approximately three billion (Gomiero, 2021). The primary causes for low yield in OF systems are inadequate soil nutrients, limited soil enrichment choices, and ineffective control of diseases, pests, and weeds (Javaid *et al.*, 2012). Furthermore, some academics believe that the primary reason for poor agricultural productivity in underdeveloped nations is the limited availability of proper chemical fertilizers and inadequate crop and water protection technology. To enhance agricultural crop yields, a new agricultural production system must tackle these three difficulties. Due to its prohibition of artificial fertilizers and pesticides, OF is not a feasible solution for this agricultural issue.

Although reduced yield in OF is a topic of debate rather than a widespread occurrence, there is a significant amount of literature on the subject. We did not include research that just focused on the yield disparity in developed nations. An analysis of 362 published studies shows that OF yields are around 80% of conventional farming yields. In this field investigation, researchers found increased output gaps due to challenges in soil nutrient management. Lesur-Dumoulin *et al.* (2017) state that scientific research indicates that OF methods globally produce 25 to 50% less than conventional systems. They also contend that the quantity of accessible animal dung is very essential in this context. Naab *et al.* (2017) assert that the yield in OF is 30–50% less than that of conventional and conservation agriculture. Recent research comparing two farming methods found that with better management techniques, organic yields are typically 19.2% lower than conventional systems.

While many food policy experts and scientists believe that OF overall can produce enough food to feed the world, the basic problem is that organic farms produce less than conventional farms. Yield does not decisively solve "food security or insecurity" as it is influenced by a range of social, political, and economic factors. The reduced yield in OF is a controversial issue. It has been proved through various research. Certain studies claim OF methods outperform traditional systems in total output, while others do the opposite. In terms of yield, the yield difference between organic and conventional farming can vary greatly depending on the geography and specific crops being grown (Knapp and Van Der Heijden, 2018). In highly developed countries, organic yields are on average 15% lower than those of conventional systems. But OF yields in developing countries are still 16% greater than in conventional systems, and 116% for less developed countries. The yield gap is not the same across areas, as already mentioned. Incremental increases in production when small-holder coffee growers transitioned from conventional to OF were recorded from 15% in Mexico to 67% in Guatemala. However, in Costa Rica, organic yields were 22% lower than conventional production the growing demand for organic food (Pearce *et al.*, 2015).

4.6.2 Nutrient Losses

A significant issue that contributes to eutrophication is the escape of nitrogen (N) and phosphorus (P) from agricultural systems into channels. This is especially problematic in coastal regions. Additionally, agriculture is the primary source of airborne nitrogen trioxide emissions, mostly because of the management of manure. There is a significant possibility that increased inputs of fertilizers, particularly nitrogen, might result in increased yields in OF. On the other hand, there is a larger risk of nutrient losses associated with higher N inputs, which is something that must be carefully considered. The danger is highest when the amount of nitrogen that is released from organic fertilizers does not correspond to the amount that is taken up by the crop, or when the rates of nitrogen fertilization begin to approach or surpass the "economic optimal level," which is determined by the yield response to nitrogen mineral fertilization. Beyond the optimum range, the yield response is reversed, and nitrogen leaching losses increase significantly (Lazcano *et al.*, 2016). It has been shown via simulations that it is possible to boost yields by increasing the amount of manure or other organic fertilizer inputs, without having a detrimental impact on nitrogen leaching. Two examples include better management of animal dung in the form of covering manure storage facilities and more precise spreading to reduce the loss of nitrogen monoxide. NH_3 emissions may be reduced, the efficiency with which nitrogen is used can be increased, and yield levels can be elevated by distributing bands on growing crops and incorporating manure directly in the soils (Ershadi *et al.*, 2020).

One of the most important sources of nitrogen in organic systems is the biological fixation of nitrogen by legumes, both annual and perennial. An increase in the number of legumes in the crop rotation may result in an increase in the risk of nitrogen losses. This is because it is difficult to coordinate the time of nitrogen release with specific crop needs. Incorporating nitrogen-rich crop residues in the fall before planting winter cereals, for instance, increases the risk of leaching. This is because the amount of nitrogen that is mineralized in the autumn is often higher than the amount of nitrogen that is taken up by the crop. Using management methods that are appropriate might help decrease such risks (Barbieri *et al.*, 2023).

This already includes restrictions on the availability of nitrogen and other nutrients in forms that are permissible for organic production. In order to achieve the aim of increasing organic yields and organic area, the recycling of nutrients within society, as well as the production of "new" nutrients, will need to increase. The extent to which we can recycle more effectively and schedule nitrogen fertilization more accurately is limited by the fact that organic regulations do not permit the widespread use of biogas digestate (Mahmood *et al.*, 2015). That implies that it has the potential to enhance nitrogen use efficiency and yields, but also to minimize nitrogen losses. The result of this is that it has the ability to enhance the efficiency of nitrogen utilization and production levels, as well as minimize nitrogen losses (Chojnacka and Moustakas, 2024). As a means of supplying "new nitrogen" that might be delivered with a high degree of accuracy, there are others who believe that a limited usage of mineral nitrogen fertilizers that are created by renewable energy may be an intriguing alternative to take into consideration. The OF method is based on the fundamental premise of feeding the soil rather than the plant, which is not permitted under organic laws. This presents a challenge to the fundamental principle. Furthermore, it contributes to the development of an OF model that focuses on the replacement of inputs rather than the redesign of the system. In the context of precise nitrogen fertilizer, the use of diluted urine might be considered a less contentious alternative. This would include the separation of human wastes from their sources (Migliorini and Wezel, 2017). A number of proposals are now being considered to permit the use of phosphorus fertilizers, such as struvite, which are produced from human excrement. There is no way around the need to develop fresh techniques for nutrient provision if organic yields are to grow and organic agriculture is to massively expand.

4.6.3 Certification and Market Competition

Overall, there are two distinct systems of OF that may be found. Non-certified production is mostly aimed at local markets in poor nations, whereas certified manufacturing fetches a premium price and is primarily marketed to organic markets in rich countries. Noting that certified items from poor nations are mostly geared toward export is a key point to keep in mind. Certification is expensive since it requires infrastructures for monitoring and recording producers; as a result, many farmers who run on a small scale and have limited resources are unable to purchase certification. Furthermore, it is important to note that certification offers absolutely no benefits to farmers who depend on subsistence farming or to those who reside in areas where there is no organic market that can be relied upon. However, certified items might often be more expensive than non-certified ones (Oya *et al.*, 2018). Research conducted over a decade on 327 coffee cultivators in Nicaragua compared the economic outcomes of organic, fair trade, and conventional coffee production. Despite the higher prices received by certified coffee producers at the farm gate, organic producers saw a decrease in wealth compared to conventional producers owing to reduced productivity. Organic and organic-fair trade certified coffee were priced 8% and 11% more than regular coffee, respectively. The premium of around 0.2 US$/kg was insufficient to offset the expenses associated with additional labor and land requirements. Organic producers need to hire workers because their relatives did not meet the demand for labor completely. The labor cost of OF is usually 7–13% above that of conventional systems. OF profitability depends on price premiums for organic products, which typically vary from 29 to 32% (Adamtey *et al.*, 2016). Research on small-scale coffee producers in Uganda found that certified farmers had greater living standards than their traditional counterparts. On the other hand, organic certification did not have a statistically significant and favorable effect on the farmers' means of subsistence. On the other hand, a fair-trade certification raises the quality of living of the family by 30% and lessens the vulnerability of the farmers. A variety of causes may be responsible for this phenomenon. Price promises are provided to farmers who participate in fair trade, and they are given greater leeway in terms of how their goods are marketed (Makita, 2016). Furthermore, farmers who participate in fair trade sell their goods after they have been milled, while organic farmers sell their coffee in its raw state which is intended for export.

Market access is another crucial problem that must be resolved. Access to marketplaces is approximately 43% for people in rural areas of developing countries within two hours by motorized transport. So, in parts of sub-Saharan Africa, it accounts for only 25% of the population. Changes in food marketing chains brought about by economic evolution and urbanization in parts of the developing world like Latin America, parts of Southeast Asia and parts of China However, those very requirements lead to great pressure on small-scale farmers that strive to meet the requirements imposed by supermarket chains with regards to quality, quantity, traceback, timeliness, and flexibility. Due to limited resources and education, small-scale farmers struggle to compete with wealthier farmers. Strict regulations and high customer expectations in wealthy nations hinder the access of farmers from developing countries to the export market (Amin and Song, 2023).

4.7 CONCLUSION

Global agricultural expansion is essential to eliminate hunger and poverty and foster development. In regions such as sub-Saharan Africa, the rural population is mostly comprised of small-scale farmers, and the economy relies significantly on agriculture. It is essential to enable small-scale farmers to formulate policies in this area. Policymakers globally are considering many strategies to enhance the well-being of small-holder farmers in emerging nations, each with its own advantages and disadvantages. This chapter has examined the environmental, economic, social, and nutritional advantages of OF, which is receiving more attention. Moreover, we discussed the main challenges that OF faces and whether OF can bolster food security in developing countries. Thus, when considering all of the challenges and opportunities and although OF may represent a serious challenge for small-scale farmers, it could be regarded as a potential contributor to improving farmer livelihoods within an integrated approach utilizing the advantages of a range of production systems, it is possible that OF could be adopted as a component of the solution. It is possible to consider OF to be a useful development strategy for the purpose of alleviating poverty and empowering small-scale farmers in nations that are yet in terms of development. Increasing yields, decreasing costs, and offering premium prices are the three primary techniques that may be used by OF to enhance the standard of living of small-scale livestock producers. The effectiveness of each mechanism is determined by the baseline agricultural system and the level of market integration. Various studies indicate that transitioning from resource-limited and subsistence farming to OF in poor nations may lead to higher crop yields. Poor farmers may improve their crop production by using OF techniques that are primarily rooted in agroecological principles. Another kind of farmer is those that use external inputs. Since synthetic inputs are prohibited in OF, producers may lower their production expenses by transitioning to OF. Additionally, they may also take advantage of organic certification and sell their products following their transition period.

LIST OF ABBREVIATIONS

AFSSA French Agency for Food Safety
FAO food and agriculture organization
GM genetically modified
IFOAM international federation of organic agriculture movements
NGOs non-governmental organization
OF organic farming
OM organic matter
SOM soil organic matter

REFERENCES

Abbas, A., Zhao, C., Ullah, W., Ahmad, R., Waseem, M. & Zhu, J., 2021. Towards sustainable farm production system: A case study of corn farming. *Sustainability*, 13, 9243.

Abdullah, Mahmood, A., Bibi, S., Naqve, M., Javaid, M.M., Zia, M.A., Jabbar, A., Ud-Din, W., Attia, K.A. & Khan, N., 2022. Physiological, biochemical, and yield responses of linseed (*Linum usitatissimum* L.) in α-tocopherol-mediated alleviation of salinity stress. *Frontiers in Plant Science*, 13, 867172.

Adamtey, N., Musyoka, M.W., Zundel, C., Cobo, J.G., Karanja, E., Fiaboe, K.K.M., Muriuki, A., Mucheru-Muna, M., Vanlauwe, B., Berset, E., Messmer, M.M., Gattinger, A., Bhullar, G.S., Cadisch, G., Fliessbach, A., Mäder, P., Niggli, U. & Foster, D., 2016. Productivity, profitability and partial nutrient balance in maize-based conventional and organic farming systems in Kenya. *Agriculture, Ecosystems & Environment*, 235, 61–79.

Ahmad, K., Ejaz, A., Khan, Z.I., Gondal, S., Fardous, A., Hussain, A., Sher, M., Valeem, E.E. & Ullah, S., 2010. Evaluation of dynamics of iron and manganese from pasture to buffaloes: A case study at rural livestock farms. *Pakistan Journal of Botany*, 42, 3415–3421.

Ahmad, S., Ghaffar, A., Rahman, M.H.U., Hussain, I., Iqbal, R., Haider, G., Khan, M.A., Ikram, R.M., Hussnain, H. & Bashir, M.S., 2021. Effect of application of biochar, poultry and farmyard manures in combination with synthetic fertilizers on soil fertility and cotton productivity under arid environment. *Communications in Soil Science and Plant Analysis*, 52, 2018–2031.

Ahmed, N., Thompson, S. & Turchini, G.M., 2020. Organic aquaculture productivity, environmental sustainability, and food security: Insights from organic agriculture. *Food Security*, 12, 1253–1267.

Ali, H., Tanveer, A., Nadeem, M., Asghar, H. & Javaid, M., 2013. Germination ecology of *Rhynchosia capitata*: An emerging summer weed in Asia. *Planta Daninha*, 31, 249–257.

Amin, N. & Song, H., 2023. The role of renewable, non-renewable energy consumption, trade, economic growth, and urbanization in achieving carbon neutrality: A comparative study for South and East Asian countries. *Environmental Science and Pollution Research*, 30, 12798–12812.

Ashraf, E., Hayat, Z., Khan, M., Atif, F. & Haider, M., 2016. Impact of dairy farm management training workshop on the knowledge level of participants. *International Journal of Agriculture and Applied Sciences (Pakistan)*, 4.

Ashraf, E., Shah, F., Luqman, M., Younis, M., Aziz, I. & Farooq, U., 2013. Use of untreated wastewater for vegetable farming: A threat to food safety. *International Journal of Applied Agricultural Sciences*, 5.

Atandi, J.G., Haukeland, S., Kariuki, G.M., Coyne, D.L., Karanja, E.N., Musyoka, M.W., Fiaboe, K.K.M., Bautze, D. & Adamtey, N., 2017. Organic farming provides improved management of plant parasitic nematodes in maize and bean cropping systems. *Agriculture, Ecosystems & Environment*, 247, 265–272.

Bacq-Labreuil, A., Crawford, J., Mooney, S.J., Neal, A.L., Akkari, E., Mcauliffe, C., Zhang, X., Redmile-Gordon, M. & Ritz, K., 2018. Effects of cropping systems upon the three-dimensional architecture of soil systems are modulated by texture. *Geoderma*, 332, 73–83.

Baldwin-Kordick, R., De, M., Lopez, M.D., Liebman, M., Lauter, N., Marino, J. & Mcdaniel, M.D., 2022. Comprehensive impacts of diversified cropping on soil health and sustainability. *Agroecology and Sustainable Food Systems*, 46, 331–363.

Barbieri, P., Starck, T., Voisin, A.-S. & Nesme, T., 2023. Biological nitrogen fixation of legumes crops under organic farming as driven by cropping management: A review. *Agricultural Systems*, 205, 103579.

Bargaz, A., Lyamlouli, K., Chtouki, M., Zeroual, Y. & Dhiba, D., 2018. Soil microbial resources for improving fertilizers efficiency in an integrated plant nutrient management system. *Frontiers in Microbiology*, 9.

Benincasa, P., Tosti, G., Guiducci, M., Farneselli, M. & Tei, F., 2017. Crop Rotation as a System Approach for Soil Fertility Management in Vegetables. *In* F. Tei, S. Nicola & P. Benincasa (eds.) *Advances in Research on Fertilization Management of Vegetable Crops*. Cham: Springer International Publishing, 115–148.

Bottrell, D.G. & Schoenly, K.G., 2018. Integrated pest management for resource-limited farmers: Challenges for achieving ecological, social and economic sustainability. *The Journal of Agricultural Science*, 156, 408–426.

Brock, C., Jackson-Smith, D., Culman, S., Doohan, D. & Herms, C., 2021. Soil balancing within organic farming: Negotiating meanings and boundaries in an alternative agricultural community of practice. *Agriculture and Human Values*, 38, 449–465.

Chadha, A., Florentine, S., Chauhan, B.S., Long, B., Jayasundera, M., Javaid, M.M. & Turville, C., 2019. Environmental factors affecting the germination and seedling emergence of two populations of an emerging agricultural weed: Wild lettuce (*Lactuca serriola*). *Crop and Pasture Science*, 70, 709–717.

Chadha, A., Florentine, S., Javaid, M., Welgama, A. & Turville, C., 2020. Influence of elements of climate change on the growth and fecundity of *Datura stramonium*. *Environmental Science and Pollution Research*, 27, 35859–35869.

Chojnacka, K. & Moustakas, K., 2024. Anaerobic digestate management for carbon neutrality and fertilizer use: A review of current practices and future opportunities. *Biomass and Bioenergy*, 180, 106991.

Costa, C.A., Guiné, R.P.F., Costa, D.V.T.A., Correia, H.E. & Nave, A., 2023. Chapter 3 – Pest Control in Organic Farming. *In* Sarathchandran, U. M.R, S. Thomas & D.K. Meena (eds.) *Organic Farming* (Second Edition). Woodhead Publishing, 111–179.

Crowder, D.W. & Reganold, J.P., 2015. Financial competitiveness of organic agriculture on a global scale. *Proceedings of the National Academy of Sciences*, 112, 7611–7616.

Das, S., Pradhan, A.K., Beura, K., Das, R. & Singh, R., 2021. Improvement of Soil Quality by Solid Waste Recycling: A Global Perspective. *In* A. Rakshit, S.K. Singh, P.C. Abhilash & A. Biswas (eds.) *Soil Science: Fundamentals to Recent Advances*. Singapore: Springer Singapore, 637–667.

Deka, N. & Goswami, K., 2021. Economic sustainability of organic cultivation of Assam tea produced by small-scale growers. *Sustainable Production and Consumption*, 26, 111–125.

Durán-Lara, E.F., Valderrama, A. & Marican, A., 2020. Natural organic compounds for application in organic farming. *Agriculture*, 10, 41.

Ershadi, S.Z., Dias, G., Heidari, M.D. & Pelletier, N., 2020. Improving nitrogen use efficiency in crop-livestock systems: A review of mitigation technologies and management strategies, and their potential applicability for egg supply chains. *Journal of Cleaner Production*, 265, 121671.

Fageria, N.K., 2012. Role of soil organic matter in maintaining sustainability of cropping systems. *Communications in Soil Science and Plant Analysis*, 43, 2063–2113.

Giampieri, F., Mazzoni, L., Cianciosi, D., Alvarez-Suarez, J.M., Regolo, L., Sánchez-González, C., Capocasa, F., Xiao, J., Mezzetti, B. & Battino, M., 2022. Organic vs conventional plant-based foods: A review. *Food Chemistry*, 383, 132352.

Gomiero, T., 2021. Chapter 2 – Organic Agriculture: Impact on the Environment and Food Quality. *In* C.M. Galanakis (ed.) *Environmental Impact of Agro-Food Industry and Food Consumption*. Academic Press, USA, 31–58.

Goswami, L., Nath, A., Sutradhar, S., Bhattacharya, S.S., Kalamdhad, A., Vellingiri, K. & Kim, K.-H., 2017. Application of drum compost and vermicompost to improve soil health, growth, and yield parameters for tomato and cabbage plants. *Journal of Environmental Management*, 200, 243–252.

Gunathilaka, R.P.D., Smart, J.C.R. & Fleming, C.M., 2018. Adaptation to climate change in perennial cropping systems: Options, barriers and policy implications. *Environmental Science & Policy*, 82, 108–116.

Hanif, Z., Naeem, M., Ali, H.H., Tanveer, A., Javaid, M.M., Peerzada, A.M. & Chauhan, B.S., 2017. Effect of environmental factors on germination of *Salsola foetida*: Potential species for rehabilitation of degraded rangelands. *Rangeland Ecology & Management*, 70, 638–643.

Hayat, Z., Pasha, T., Khattak, F., Jabbar, M., Nasir, Z. & Samiullah, S., 2010. Consumer's perception and willingness to buy nutrient enriched designer eggs in Pakistan. *Education*, 5, 0–067.

Hemathilake, D.M.K.S. & Gunathilake, D.M.C.C., 2022. Chapter 31 – Agricultural Productivity and Food Supply to Meet Increased Demands. *In* R. Bhat (ed.) *Future Foods*. Academic Press, USA, 539–553.

Huang, B., Yuan, Z., Zheng, M., Liao, Y., Nguyen, K.L., Nguyen, T.H., Sombatpanit, S. & Li, D., 2022. Soil and water conservation techniques in tropical and subtropical Asia: A review. *Sustainability*, 14, 5035.

Javaid, M., Tanveer, A., Ahmad, R., Yaseen, M. & Khaliq, A., 2012. Optimizing activity of herbicides at reduced rate on *Emex spinosa* Campd. with adjuvants. *Planta Daninha*, 30, 425–435.

Javaid, M.M., Florentine, S.K., Ashraf, M., Mahmood, A., Sattar, A., Wasaya, A. & Li, F.M., 2022. Photosynthetic activity and water use efficiency of *Salvia verbenaca* L. under elevated CO_2 and water-deficit conditions. *Journal of Agronomy and Crop Science*, 208, 536–551.

Jianing, G., Yuhong, G., Yijun, G., Rasheed, A., Qian, Z., Zhiming, X., Mahmood, A., Shuheng, Z., Zhuo, Z. & Zhuo, Z., 2022. Improvement of heat stress tolerance in soybean (*Glycine max* L), by using conventional and molecular tools. *Frontiers in Plant Science*, 13, 993189.

Jouzi, Z., Azadi, H., Taheri, F., Zarafshani, K., Gebrehiwot, K., Van Passel, S. & Lebailly, P., 2017. Organic farming and small-scale farmers: Main opportunities and challenges. *Ecological Economics*, 132, 144–154.

Kader, M.A., Senge, M., Mojid, M.A. & Ito, K., 2017. Recent advances in mulching materials and methods for modifying soil environment. *Soil and Tillage Research*, 168, 155–166.

Knapp, S. & Van Der Heijden, M.G.A., 2018. A global meta-analysis of yield stability in organic and conservation agriculture. *Nature Communications*, 9, 3632.

Kwiatkowski, C.A. & Harasim, E., 2020. Chemical properties of soil in four-field crop rotations under organic and conventional farming systems. *Agronomy*, 10, 1045.

Lazcano, C., Tsang, A., Doane, T.A., Pettygrove, G.S., Horwath, W.R. & Burger, M., 2016. Soil nitrous oxide emissions in forage systems fertilized with liquid dairy manure and inorganic fertilizers. *Agriculture, Ecosystems & Environment*, 225, 160–172.

Lesur-Dumoulin, C., Malézieux, E., Ben-Ari, T., Langlais, C. & Makowski, D., 2017. Lower average yields but similar yield variability in organic versus conventional horticulture. A meta-analysis. *Agronomy for Sustainable Development*, 37, 45.

Levi, N., Hillel, N., Zaady, E., Rotem, G., Ziv, Y., Karnieli, A. & Paz-Kagan, T., 2021. Soil quality index for assessing phosphate mining restoration in a hyper-arid environment. *Ecological Indicators*, 125, 107571.

Maclaren, C., Storkey, J., Menegat, A., Metcalfe, H. & Dehnen-Schmutz, K., 2020. An ecological future for weed science to sustain crop production and the environment. A review. *Agronomy for Sustainable Development*, 40, 24.

Mahmood, A., Hussain, A., Shahzad, A. & Honermeier, B., 2015. Biomass and biogas yielding potential of sorghum as affected by planting density, sowing time and cultivar. 47, 2401–2408.

Makita, R., 2016. A role of fair trade certification for environmental sustainability. *Journal of Agricultural and Environmental Ethics*, 29, 185–201.

Mc Carthy, U., Uysal, I., Badia-Melis, R., Mercier, S., O'donnell, C. & Ktenioudaki, A., 2018. Global food security – Issues, challenges and technological solutions. *Trends in Food Science & Technology*, 77, 11–20.

Mckenna, P., Cannon, N., Conway, J. & Dooley, J., 2018. The use of red clover (*Trifolium pratense*) in soil fertility-building: A review. *Field Crops Research*, 221, 38–49.

Meemken, E.-M. & Qaim, M., 2018. Organic agriculture, food security, and the environment. *Annual Review of Resource Economics*, 10, 39–63.

Mehmood, A., Naeem, M., Khalid, F., Saeed, Y., Abbas, T., Jabran, K., Sarwar, M.A., Tanveer, A. & Javaid, M.M., 2018. Identification of phytotoxins in different plant parts of *Brassica napus* and their influence on mung bean. *Environmental Science and Pollution Research*, 25, 18071–18080.

Merfield, C.N., 2023. Chapter 2 – Integrated Weed Management in Organic Farming. *In* Sarathchandran, U. M.R, S. Thomas & D.K. Meena (eds.) *Organic Farming (Second* Edition*)*. Woodhead Publishing, Cambridge UK, 31–109.

Mie, A., Andersen, H.R., Gunnarsson, S., Kahl, J., Kesse-Guyot, E., Rembiałkowska, E., Quaglio, G. & Grandjean, P., 2017. Human health implications of organic food and organic agriculture: A comprehensive review. *Environmental Health*, 16, 111.

Migliorini, P. & Wezel, A., 2017. Converging and diverging principles and practices of organic agriculture regulations and agroecology. A review. *Agronomy for Sustainable Development*, 37, 63.

Möller, K., 2018. Soil fertility status and nutrient input–output flows of specialised organic cropping systems: A review. *Nutrient Cycling in Agroecosystems*, 112, 147–164.

Muller, A., Schader, C., El-Hage Scialabba, N., Brüggemann, J., Isensee, A., Erb, K.-H., Smith, P., Klocke, P., Leiber, F., Stolze, M. & Niggli, U., 2017. Strategies for feeding the world more sustainably with organic agriculture. *Nature Communications*, 8, 1290.

Naab, J.B., Mahama, G.Y., Yahaya, I. & Prasad, P.V.V., 2017. Conservation agriculture improves soil quality, crop yield, and incomes of smallholder farmers in north western Ghana. *Frontiers in Plant Science*, 8.

Naqve, M., Wang, X., Shahbaz, M., Fiaz, S., Naqvi, W., Naseer, M., Mahmood, A. & Ali, H., 2021. Foliar spray of alpha-tocopherol modulates antioxidant potential of okra fruit under salt stress. *Plants*, 10, 1382.

Noor Ul, A., Aslam, A. & Haider, F.U., 2022. Effects of Mulching on Soil Biota and Biological Indicators of Soil Quality. *In* K. Akhtar, M. Arif, M. Riaz & H. Wang (eds.) *Mulching in Agroecosystems: Plants, Soil & Environment*. Singapore: Springer Nature Singapore, 15–40.

Olle, M. & Williams, I.H., 2021. Organic Cultivation of Vegetables. *In* E. Lichtfouse (ed.) *Sustainable Agriculture Reviews 52*. Cham: Springer International Publishing, 1–19.

Oya, C., Schaefer, F. & Skalidou, D., 2018. The effectiveness of agricultural certification in developing countries: A systematic review. *World Development*, 112, 282–312.

Panday, D., Bhusal, N., Das, S. & Ghalehgolabbehbahani, A., 2024. Rooted in nature: The rise, challenges, and potential of organic farming and fertilizers in agroecosystems. *Sustainability*, 16, 1530.

Paris, B., Vandorou, F., Balafoutis, A.T., Vaiopoulos, K., Kyriakarakos, G., Manolakos, D. & Papadakis, G., 2022. Energy use in open-field agriculture in the EU: A critical review recommending energy efficiency measures and renewable energy sources adoption. *Renewable and Sustainable Energy Reviews*, 158, 112098.

Pearce, B.D., Smith, L.G. & Williams, A.G., 2015. The energy efficiency of organic agriculture: A review. *Renewable Agriculture and Food Systems*, 30, 280–301.

Pereira, L., Wynberg, R. & Reis, Y., 2018. Agroecology: The future of sustainable farming? *Environment: Science and Policy for Sustainable Development*, 60, 4–17.

Rahman, M.M., Khan, I., Field, D.L., Techato, K. & Alameh, K., 2022. Powering agriculture: Present status, future potential, and challenges of renewable energy applications. *Renewable Energy*, 188, 731–749.

Raj, A., Jhariya, M.K., Khan, N., Banerjee, A. & Meena, R.S., 2021. Ecological Intensification for Sustainable Development. *In* M.K. Jhariya, R.S. Meena & A. Banerjee (eds.) *Ecological Intensification of Natural Resources for Sustainable Agriculture*. Singapore: Springer Singapore, 137–170.

Ramesh, T., Bolan, N.S., Kirkham, M.B., Wijesekara, H., Kanchikerimath, M., Srinivasa Rao, C., Sandeep, S., Rinklebe, J., Ok, Y.S., Choudhury, B.U., Wang, H., Tang, C., Wang, X., Song, Z. & Freeman Ii, O.W., 2019. Chapter One – Soil Organic Carbon Dynamics: Impact of Land Use Changes and Management Practices: A Review. *In* D.L. Sparks (ed.) *Advances in Agronomy*. Academic Press, USA, 1–107.

Rasheed, A., Barqawi, A.A., Mahmood, A., Nawaz, M., Shah, A.N., Bay, D.H., Alahdal, M.A., Hassan, M.U. & Qari, S.H., 2022a. CRISPR/Cas9 is a powerful tool for precise genome editing of legume crops: A review. *Molecular Biology Reports*, 49, 5595–5609.

Rasheed, A., Li, H., Nawaz, M., Mahmood, A., Hassan, M.U., Shah, A.N., Hussain, F., Azmat, S., Gillani, S.F.A. & Majeed, Y., 2022b. Molecular tools, potential frontiers for enhancing salinity tolerance in rice: A critical review and future prospective. *Frontiers in Plant Science*, 13, 966749.

Rasheed, A., Raza, A., Jie, H., Mahmood, A., Ma, Y., Zhao, L., Xing, H., Li, L., Hassan, M.U. & Qari, S.H., 2022c. Molecular tools and their applications in developing salt-tolerant soybean (*Glycine max* L.) cultivars. *Bioengineering*, 9, 495.

Ratnadass, A., Deberdt, P., Martin, T., Sester, M. & Deguine, J.-P., 2023. Impacts of crop protection practices on human infectious diseases. *One Health, Human, Animal, and Environment Triad*, 287–308.

Raza, A., Razzaq, A., Mehmood, S.S., Zou, X., Zhang, X., Lv, Y. & Xu, J., 2019. Impact of climate change on crops adaptation and strategies to tackle its outcome: A review. *Plants*, 8, 34.

Rodríguez-Espinosa, T., Papamichael, I., Voukkali, I., Gimeno, A.P., Candel, M.B.A., Navarro-Pedreño, J., Zorpas, A.A. & Lucas, I.G., 2023. Nitrogen management in farming systems under the use of agricultural wastes and circular economy. *Science of the Total Environment*, 876, 162666.

Rusinamhodzi, L., 2020. Managing Crop Rotations in No-till Farming Systems. *In* Y.P. Dang, R.C. Dalal & N.W. Menzies (eds.) *No-Till Farming Systems for Sustainable Agriculture: Challenges and Opportunities*. Cham: Springer International Publishing, 21–31.

Shaw, B., Nagy, C. & Fountain, M.T., 2021. Organic control strategies for use in IPM of invertebrate pests in apple and pear orchards. *Insects*, 12, 1106.

Singh, J., Wang, T., Kumar, S., Xu, Z., Sexton, P., Davis, J. & Bly, A., 2021. Crop yield and economics of cropping systems involving different rotations, tillage, and cover crops. *Journal of Soil and Water Conservation*, 76, 340–348.

Suresh Reddy, B., 2017. Prospects of Organic Farming. *In* M.S. Khan & M.S. Rahman (eds.) *Pesticide Residue in Foods: Sources, Management, and Control*. Cham: Springer International Publishing, 167–194.

Tang, H., Meng, G., Xiang, J., Mahmood, A., Xiang, G., Sanaullah, Liu, Y. & Huang, G., 2022. Toxic effects of antimony in plants: Reasons and remediation possibilities—A review and future prospects. *Frontiers in Plant Science*, 13, 1011945.

Tanveer, A., Javaid, M.M., Irfan, M., Khaliq, A. & Yaseen, M., 2015. Yield losses in chickpea with varying densities of dragon spurge (Euphorbia dracunculoides). *Weed Science*, 63, 522–528.

Thakur, N., 2017. Organic Farming, Food Quality, and Human Health: A Trisection of Sustainability and a Move from Pesticides to Eco-friendly Biofertilizers. *In* V. Kumar, M. Kumar, S. Sharma & R. Prasad (eds.) *Probiotics in Agroecosystem*. Singapore: Springer Singapore, 491–515.

Tian, J., Lou, Y., Gao, Y., Fang, H., Liu, S., Xu, M., Blagodatskaya, E. & Kuzyakov, Y., 2017. Response of soil organic matter fractions and composition of microbial community to long-term organic and mineral fertilization. *Biology and Fertility of Soils*, 53, 523–532.

Toor, G.S., Yang, Y.-Y., Das, S., Dorsey, S. & Felton, G., 2021. Chapter Four – Soil Health in Agricultural Ecosystems: Current Status and Future Perspectives. *In* D.L. Sparks (ed.) *Advances in Agronomy*. Academic Press, USA, 157–201.

Ukalska-Jaruga, A., Siebielec, G., Siebielec, S. & Pecio, M., 2020. The impact of exogenous organic matter on wheat growth and mineral nitrogen availability in soil. *Agronomy*, 10, 1314.

Ullah, H., Mahmood, A., Ijaz, M., Tadesse, B. & Honermeier, B., 2013. Evaluation of anise (*Pimpinella anisum* L.) accessions with regard to morphological characteristics, fruit yield, oil contents and composition. *Journal of Medicinal Plants Research*, 7, 2177–2186.

Van Geel, M., Verbruggen, E., De Beenhouwer, M., Van Rennes, G., Lievens, B. & Honnay, O., 2017. High soil phosphorus levels overrule the potential benefits of organic farming on arbuscular mycorrhizal diversity in northern vineyards. *Agriculture, Ecosystems & Environment*, 248, 144–152.

Wen, Y.-C., Li, H.-Y., Lin, Z.-A., Zhao, B.-Q., Sun, Z.-B., Yuan, L., Xu, J.-K. & Li, Y.-Q., 2020. Long-term fertilization alters soil properties and fungal community composition in fluvo-aquic soil of the North China Plain. *Scientific Reports*, 10, 7198.

Yang, F.K., He, B., Zhang, L., Zhang, G. & Gao, Y., 2020. An approach to improve soil quality: A case study of straw incorporation with a decomposer under full film-mulched ridge-furrow tillage on the semiarid Loess Plateau, China. *Journal of Soil Science and Plant Nutrition*, 20, 125–138.

Zerssa, G., Feyssa, D., Kim, D.-G. & Eichler-Löbermann, B., 2021. Challenges of smallholder farming in Ethiopia and opportunities by adopting climate-smart agriculture. *Agriculture*, 11, 192.

Zikeli, S. & Gruber, S., 2017. Reduced tillage and no-till in organic farming systems, Germany—Status quo, potentials and challenges. *Agriculture*, 7, 35.

5 Revolutionizing Agriculture through Drones and Robotics

Muaz Ameen, Athar Mahmood, Muhammad Zeeshan Hanif, Muhammad Anjum Zia[*]*, Muhammad Junaid Nisar, and Muhammad Awais*

5.1 INTRODUCTION

Nowadays, farming faces a number of financial difficulties associated with cost- and productivity-effectiveness as well as a growing labor shortage, which is primarily brought on by the population loss in rural areas. Global concerns such as population growth, urbanization, environment degradation, increasing demand for animal proteins due to migration and dietary changes, and climatic changes further complicate the situation (Castillejo *et al.*, 2020; Ameen *et al.*, 2023b). As the primary global source of food, agriculture must address rising demand, food safety concerns, and the requirement for environmental sustainability to mitigate health risks, trade disruptions, and financial costs. Effective management of pathogen detection, identification, and quantification besides other factors influencing the well-being of crops and livestock are also critical (Table 5.1). With the global population projected to reach 9.7 billion by 2050, the demand for water and food will grow significantly (King *et al.*, 2017; Ameen *et al.*, 2023a). Smart farming encompasses the use of advanced technology, such as information communication technologies, to improve the efficiency of agricultural practices reducing farmer populations, insufficient agricultural land, and environmental concerns highlighting the necessity for innovative and persistent agricultural approaches. To tackle these difficulties, the integration of innovative technologies like smart farming and precision agriculture has garnered significant interest (Karunathilake *et al.*, 2023).

The use of drones and robots represents a significant breakthrough in modern agricultural methods, transforming farm management and resource optimization. Precision agriculture is a farming approach that emphasizes managing particular areas utilizing advanced technology such as artificial intelligence (machine learning (ML) and deep learning), Internet of Things (IoT), computer technologies, Wireless Sensor Networks (WSN), blockchain, and big data (Sivakumar *et al.*, 2021). Remote sensing technologies, such as satellites, manned aircraft, and unmanned aerial vehicles (UAVs), are crucial for enhancing smart and precise agriculture. The ability to recognize and track pests, weeds, and diseases is just one of the numerous benefits of using drones, which can also take high-resolution photos in various conditions, are inexpensive, simple to install and maintain, and evaluate data swiftly (Mohsan *et al.*, 2022; Jabbar *et al.*, 2023). Moreover, drones use environmental data to provide precise watering as well as weed, insect, and disease management.

This chapter examines the considerable impact of automated systems and aerial robots on enhancing agricultural productivity, sustainability, and efficiency. Being an essential component of global food security and economic growth, agriculture continually seeks innovative solutions to challenges like labor shortages, environmental concerns, and increasing crop yield. Agricultural robots, including autonomous harvesters, robotic weed eaters, and fruit harvesting machines perform various tasks. Drones equipped with modern technologies, such as remote sensing and artificial intelligence, offer remarkable capabilities for efficient crop monitoring, soil conditions analysis, and pest management. The multidisciplinary connection of drones, robotics, and

* Corresponding author: anjum.zia@uaf.edu.pk

DOI: 10.1201/9781003498698-5

TABLE 5.1
Conventional versus Drones and Robotics Agriculture

Points	Conventional Agriculture	Drones and Robotics Agriculture	References
Global agricultural drone market size	–	The market for drones used in agriculture was expected to increase at a rate of 29.4% between 2024 and 2032, and it was predicted to reach several billion dollars by 2024.	Kapustina *et al.* (2021)
Number of farms using drones	A significant number of farms globally utilize conventional agricultural techniques, with minimal reliance on sophisticated technology.	There is a growing rate of acceptance, as evidenced by the survey results showing that over 1000 farmers are either utilizing or planning to use drones in their operations.	Skevas and Kalaitzandonakes (2020)
Cost savings with drone technology	Drones and other technological innovations have cost-saving features that conventional agriculture might not be able to take advantage effectively.	Farmers and customers have benefited from greater cost savings throughout operations brought about by the adoption of agricultural drone technology.	Kendall *et al.* (2022)
Increased crop yields	Potential yield gains may be limited by the use of traditional agricultural practices that do not make use of improved monitoring or precision techniques.	Through enhanced monitoring and precision agriculture techniques, drones and robotics have led to higher crop yields and increased productivity.	Karunathilake *et al.* (2023)
Environmental benefits	Conventional agricultural methods, such as overuse of fertilizers and pesticides, soil erosion, and water pollution, can have a detrimental impact on the environment.	Environmental sustainability has been enhanced by the use of drones and robotics in agriculture, which has led to lower use of pesticides and fertilizers, less soil erosion, and better water management techniques.	Balyan *et al.* (2024)
Jobs created in the drone industry	Traditional agriculture utilizes laborers in the fields, throughout the process of farming, harvesting, and other associated tasks. However, it may not generate as many specialized employment opportunities as the drone industry.	Economic growth has been stimulated by the drone sector, which has produced jobs in manufacturing, maintenance, data analysis, and software development.	Beltozar-Clemente *et al.* (2023)
Investment in agricultural drone technology	While machines and equipment may be included in technology investments for conventional agriculture, they may not be as large as those made in the drone industry.	Based on substantial investment and market potential, the worldwide agriculture drone market is expected to increase from $4.98 billion in 2023 to $18.22 billion by 2030.	Singh (2023)
Technological advancements	Traditional agriculture uses little in the way of technology and is primarily dependent on manual labor and simple machines.	New developments in robotics and drones for agriculture have increased productivity and efficiency. These developments include autonomous navigation systems, AI-powered analytics, and better sensors.	Maddikunta *et al.* (2021)
Challenges and opportunities	Market instability, land degradation, and climate change are some of the problems conventional agriculture faces. Opportunities can be found in integrating technology and using sustainable practices.	Even though there are many possibilities for innovation and expansion in this industry, there are obstacles to general acceptance such as upfront expenditures, data privacy issues, and legal barriers.	Maddikunta *et al.* (2021)

TABLE 5.1 (*Continued*)
Conventional versus Drones and Robotics Agriculture

Points	Conventional Agriculture	Drones and Robotics Agriculture	References
Future outlook	It's possible that selective technological utilization and a greater adoption of sustainable practices will characterize conventional agriculture in the future.	Agriculture is predicted to undergo a revolution with the integration of robotics and drones, as this would increase productivity, lower expenses, and allow for sustainable practices. AI integration and greater automation are emerging trends.	Maddikunta *et al.* (2021)

agriculture is examined in this chapter, with an emphasis on the most recent advancements, challenges, and potential future directions. These technologies reduce operating expenses and optimize labor-intensive operations.

5.2 OVERVIEW OF AGRICULTURE CHALLENGES

5.2.1 Precision Farming

One of the primary challenges facing the agricultural sector today is precision agriculture. The demands of a growing global population are becoming too great for conventional agricultural methods to meet. The most challenging aspect is developing cost-effective and efficient ways to achieve the required level of accuracy (Mehmood *et al.*, 2018; Weiss *et al.*, 2020). The diversity of soil types, variable weather patterns, and distinct crop needs necessitates a more precise and targeted approach. Precision farming aims to optimize the use of resources like herbicides, fertilizers, and water by tailoring their application to the specific needs of various fields.

5.2.2 Crop Monitoring

The agricultural sector faces a critical challenge in proper and efficient crop monitoring. Conventional methods of manual monitoring are time-consuming, laborious, and often lack the necessary accuracy. Identifying diseases, nutritional deficiencies, and insect infestations at an early stage is essential to prevent reductions in crop yield (Franzen *et al.*, 2021, Abdullah et al., 2022). Enhancing irrigation and fertilizing methods is another benefit of crop health monitoring. As agricultural operations grow in complexity, solutions that provide comprehensive, up-to-date crop status data are needed. Utilizing land effectively is crucial for balancing the increasing food demand and the protection of natural ecosystems, as energy-intensive agricultural practices contribute to environmental degradation (Chadha *et al.*, 2020; Babu *et al.*, 2023). The aim is to develop sustainable techniques that preserve agricultural quality and yield while optimizing resource utilization.

5.2.3 Climate Variability

The growing number and complexity of agricultural techniques result from climate change. Farmers face challenges due to the unpredictable nature of weather conditions, disastrous events, and changing geographical regions with favorable conditions. These issues disrupt their ability to rely on past weather patterns for making informed decisions (Gull *et al.*, 2023). To cope with these changes flexible and adaptable solutions are necessary to effectively respond to evolving environmental circumstances.

5.3 AGRICULTURAL DRONES

A drone, technically speaking, refers to an aircraft that operates without a human operator on board. These aircraft weigh between 2 and 20 kg, and they are often referred to as unmanned aircraft systems (UASes) or UAVs (Elmeseiry *et al.*, 2021). A drone is a kind of flying robot that can be operated remotely or fly independently using pre-programmed flight paths controlled by software in its internal systems. Drones function with built-in sensors and GPS to navigate and perform tasks (Figure 5.1). The Hewitt-Sperry Automatic Airplane, sometimes referred to as the "flying bomb," had its first flight on September 12, 1917, during World War I. This flight served as a demonstration of the unmanned aircraft idea (Kim *et al.*, 2019). In 1921, a method that is often referred to as crop dusting was carried out by the United States Army Signal Corps and the United States Department of Agriculture research station in Ohio. The aircraft was used to spray insecticides on crops (Kumar *et al.*, 2022). Drones have a diverse array of uses, such as farming, photography, and military applications, but their main purposes are navigation and flying. Drones are equipped with a power source, such as a battery or fuel, that enables them to achieve flight. Drone frames are often constructed using lightweight composite materials to reduce weight and enhance flexibility during flight (Ibrahim *et al.*, 2023). Drones need a controller operated by a remote operator to initiate, control, and land the drone. Controllers use Wi-Fi and radio waves for communication with the drone.

There are notable distinctions between drones and satellites. For instance, drones have the capability to capture images with a resolution of a few centimeters per pixel, while commercial satellites possess a resolution of 50 centimeters per pixel (Ruwaimana *et al.*, 2018). Due to their ability to operate below cloud cover, drones can record real-time photos with enhanced clarity and accuracy, whereas satellite photography is adversely affected by the presence of clouds.

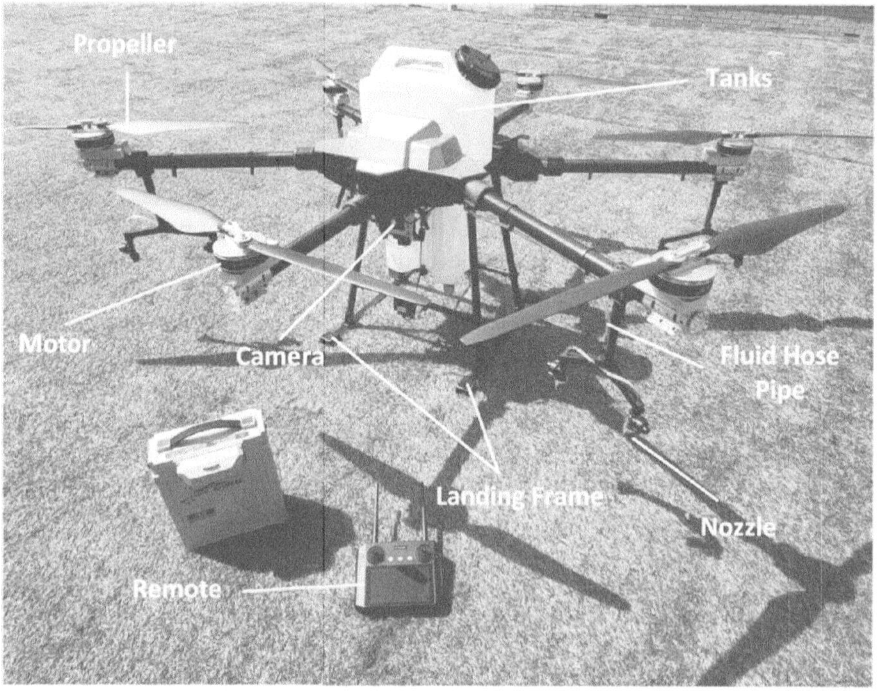

FIGURE 5.1 Basic components of a drone.

5.4 APPLICATION OF DRONES IN AGRICULTURE

Intensive agriculture has several detrimental effects on the environment. It contributes substantial and ecologically harmful quantities of nitrogen and phosphorus to land-based ecosystems. Moreover, the excessive use of fertilizers may pose pollution hazards, while inadequate use of fertilizers to replenish nitrogen and phosphorus lost due to heavy cropping can result in soil deterioration and a decline in fertility (Tyagi *et al.*, 2022). It is essential to consider the health risks associated with the utilization of pesticides in agriculture in addition to the environmental challenges. Moreover, soil degradation has been prominent throughout Europe, posing a significant threat to the fertility and productivity of various soils. Furthermore, the contamination of waterways and aquatic environments, along with the subsequent deterioration of water-dependent ecosystems, is increasing as agricultural chemicals infiltrate adjacent water sources (Ferreira *et al.*, 2022).

Agricultural chemicals pose a potential danger to agricultural workers, families and residents near crops, vineyards, and farming sites (Aziz *et al.*, 2018). Pesticides are absorbed by crops and natural resources such as water and soil, ultimately becoming hidden components in the food chain. This poses a growing threat to both animals and humans, resulting in significant adverse effects on public health (Curl *et al.*, 2020; Dereumeaux *et al.*, 2020). Autonomous precision farming may help mitigate these consequences by selectively applying chemicals, such as fertilizer and insecticides, just to the areas where they are required, rather than spreading them over a wide region (Figure 5.2). Recently, drones have been employed in agriculture to survey large regions and provide precise

FIGURE 5.2 Application of drones in agriculture sector.

watering and fertilizing. The use of a drone equipped with an infrared camera has the potential to identify regions requiring significant irrigation or experiencing the development of foliage diseases (Abbas *et al.*, 2023). This technology helps agronomists conserve time, water, and agrochemical chemicals. Using novel agricultural techniques could lead to higher crop and better quality simultaneously (Ali *et al.*, 2013b).

5.4.1 Soil and Field Analysis

Drones have emerged as a breakthrough tool for conducting soil and field experiments in agriculture, radically changing traditional farming methods. Equipped with modern sensors and image technology, drones enable farmers to monitor soil health, evaluate field conditions, and enhance crop management tactics with exceptional precision. The rapid capture of high-resolution images and data by these drones allows for accurate assessment of nutrient deficiencies, moisture content, and soil composition (Inoue, 2020). By utilizing drones for soil and field studies, farmers can apply data-driven strategies to maximize productivity, conserve resources, and reduce environmental impact. Additionally, extensive field mapping made it possible to use drones for aerial examinations to help farmers quickly identify problem areas like insect outbreaks or crop diseases. Furthermore, the timely and detailed information obtained through drone monitoring supports sustainable agricultural practices (Liang and Shah, 2023). This involves optimizing resource allocation and adopting customized crop management strategies tailored to the specific characteristics of the soil and field.

Drones equipped with remote sensing cameras use the electromagnetic spectrum to gather data from the ground for soil and field analysis. Unmanned aerial systems collect unprocessed data and use algorithms to convert it into valuable information. Various elements exhibit distinct wavelengths, enabling their differentiation (Aasen *et al.*, 2018). Consequently, they may be used in many agricultural contexts to monitor the following parameters:

1. Crop health deterioration caused by pests, nutritional shortages, and color alteration resulting from pest infestation.
2. Vegetation catalogs include measurements of leaf area, treatment efficiency, phenology, and yield.
3. The development, height, leaf area index (LAI), and density of the plant.
4. Stand number, planter failures, plant size, plant inspection, compromised field, and field statistics.
5. Water requirements, water requirements based on climatic conditions, and water-stressed areas of the orchard or field that require irrigation.
6. Soil analysis, nutrient availability, and nutrient concentration within the plant in support of plant nutrient management. This data assists agricultural professionals in optimizing their sowing, crop, and soil management strategies.

5.4.2 Seed Sowing from Air

Annually, around 15 billion trees are cut down to facilitate horizontal agricultural expansion, mining operations, and particularly for the expansion of urban areas (Mandal *et al.*, 2023). Nevertheless, drone-seeders can rapidly accelerate reforestation efforts. These drones use a pneumatic firing mechanism to propel seed pods farther into the ground, particularly in challenging landscapes like the rocky mountains or mangrove forests. A pair of aerial drones can sow up to 40,000 seeds in a single day. In just over 10 minutes, a drone can plant the same amount that an ordinary person can plant. It has the capability to achieve an emergence rate of 90% and reduce planting expenses by 85% (Mohan *et al.*, 2021).

Utilizing drones to plant seeds from the air is a revolutionary advancement in contemporary agriculture. Seed dispersion systems integrated with drones provide a multitude of advantages and opportunities for improving agricultural farming:

1. Precise seed planting promotes sustainable agriculture practices by minimizing inefficiencies and reducing the negative ecological effects of excessive seed consumption.
2. Drones offer flexibility in seed dispersal, enabling personalization based on soil conditions, crop type, and land topography. Because they need to handle various farm sizes, they may additionally adapt their operations.
3. Drones make it possible to plant seeds in remote or unreachable areas, boosting agricultural potential in areas that were previously challenging to develop or access.
4. To improve agricultural productivity and resource utilization, drones offer targeted seed planting services that improve seed dispersal across areas.
5. Drones efficiently reduce the time and labor required for manual seeding by employing automation in the seeding process. This allows farmers to effectively cover more areas in less time.

5.4.3 Spraying Operation in Fields

Drones can transport reservoirs of fertilizers, insecticides, herbicides, and plant growth regulators (PGRs) of suitable sizes to enable faster spraying. Applying chemical substances to eradicate pests and undesirable vegetation such as weeds is crucial for crops' well-being (Hanif *et al.*, 2017; Das *et al.*, 2024). Manual spraying operations may be challenging due to the height of the crops. To mitigate this, smart farms use drones for spraying, minimizing human exposure to hazardous chemicals. The spraying capacity of drones is significantly higher than that of conventional equipment, allowing them to cover a one-hectare field in less than 40 minutes up to five times faster. This method also reduces pesticide use by 30% (Jalu *et al.*, 2024).

Agriculture requires production and protection materials to achieve high productivity. Considering various crops and land conditions, drones can precisely apply chemicals like insecticides and fertilizers. Agricultural fertilizers and chemicals are often necessary to eradicate insects and promote crop development. Integrating UAVs into sprayer systems is a promising approach for vector control and pest management over large agricultural area, offering targeted and accurate treatment. The number of pesticides sprayed may need to be adjusted based on the agricultural conditions or the intensity of insect-pest invasion (Javaid *et al.*, 2012; Li *et al.*, 2021). Due to its compact size, a quadcopter is suitable for both indoor and outdoor agriculture farming. Unmanned drones, such as quadcopters, can self-spray pesticides and fertilizer. UAVs with powerful lifting capabilities are essential for widespread spraying areas (Lambey and Prasad, 2021). Through a Bluetooth connection, an Android smartphone can communicate with the quadcopter in real time, enhancing agricultural productivity, and addressing field-related issues. The installation of a Pulse Width Modulation (PWM) controller in the spraying system mounted on the UAV improves the pesticide application efficiency (Maddikunta *et al.*, 2021). An Aerial Automated Pesticide Sprayer (AAPS) was developed by combining a blimp with a quadcopter. This system is designed for spraying pesticides at lower altitudes using GPS coordinates (Suman *et al.*, 2023). To address the issue of cost-effectiveness and user flexibility, the "Freyr" pesticide spraying drone was created. This drone can be operated using an Android application. Laboratory and field evaluations assess the pressure, discharge rate, droplet size and density, and spray uniformity of a hexacopter-mounted sprayer. An electrostatic sprayer, based on electrostatic spray technology, was developed to minimize pesticides waste (Mustafi *et al.*, 2021).

According to the World Health Organization (WHO), pesticide exposure results in over 1 million cases and more than 100,000 deaths annually, especially in developing countries (Boedeker *et al.*, 2020). Insecticides can affect the human nervous system and induce physiological abnormalities.

Remote-controlled UAVs are used to apply pesticides and fertilizer reducing human exposure to these chemicals (Hassan *et al.*, 2022). Crop dusting: Drones equipped with tanks of fertilizers and pesticides are capable of spraying crops with far greater accuracy compared to a tractor. This method effectively reduces both costs and the risk of workers' exposure to pesticides. Drone-based pesticide application is versatile, particularly in labor-scarce areas. It helps reduce environmental pollution by spraying from lower altitudes and improves pest management in both small and large through precise and targeted applications. A group of scientists conducted a study on the effects of various UAV spraying parameters (UAV N-3) at varying heights and insecticide concentrations on wheat canopy to specifically prevent powdery mildew in Asian nations (Wang *et al.*, 2022). As a result, the application of chemicals becomes more efficient, leading to a decrease in their negative effects on the environment, namely in terms of soil and water contamination (Wasaya *et al.*, 2021). Therefore, it may contribute to the development of sustainable agriculture.

Drones administer chemicals at a higher velocity compared to other techniques. Additionally, it may lead to a reduction in the quantity of chemicals used, resulting in cost savings. The use of fertilizers, such as phosphate (P), nitrogen (N), potassium (K), and micronutrients like sulfur, magnesium, and zinc, is a significant economic factor in agriculture. Fertilizer is distributed using either on-ground machinery, such as sprayers operated by tractors or pressurized irrigation systems, or by manned airplanes (Sinha and Tandon, 2020). Producers with extensive land units often prefer the latter option. Typically, a uniform application rate is applied across all fields during fertilizer spraying due to the limitations posed by fluctuating wind speed and direction, as well as the aircraft's altitude, which hinder more accurate application (Chávez *et al.*, 2020). The UAVs can estimate the nutritional status of crops, assisting producers or agronomic consultants in determining appropriate application rates for the entire field. Research suggests that scientific UAVs should be equipped with specialized camera sensors (thermal or optical) for monitoring purposes. These sensors can also be enhanced with specialized optical filters (hyperspectral/Red Edge) (Silva *et al.*, 2023). The use of accelerometer and gyroscope sensors in spraying systems helps minimize both time and human labor.

5.4.4 Crop Health and Irrigation Monitoring

Crop health can be tracked over time and the efficacy of corrective measures can be evaluated using drones equipped with sensors that can scan crops with visible and near-infrared (NIR) light. This technology can be trained to recognize specific characteristics such as Normalized Difference Vegetation Index (NDVI), nutritional deficiencies, and water stress (Virnodkar *et al.*, 2020). A greater understanding of crops and novel ways to maximize yields while reducing damage can be gained by farmers using data acquired from modern sensors, which can be presented in two or three dimensions. This data can also be utilized by farmers or government organizations to enhance crop protection or support future projections (Smith *et al.*, 2018). Drones with remote sensing and thermal cameras can segment fields according to varying moisture levels, thereby optimizing irrigation strategies. Equipped with photogrammetric and navigation systems, drones used by the Food and Agriculture Organization (FAO) in the Philippines can achieve a ground resolution as precise as 3 centimeters (Singh and Gupta, 2024).

Throughout the growing season, aerial drones facilitate timely and focused interventions by monitoring crop conditions. Various sensors detect visible, NIR, and thermal infrared radiation, enabling the computation of several multispectral indices based on reflection patterns at different wavelengths (Lingli *et al.*, 2017). Indicators such as water stress, insect pest infestation, nutritional stress, and diseases are used to assess crops conditions. Drones equipped with these sensors can identify diseases or weaknesses before visible symptoms appear, thus serving as a rapid disease diagnosis method. By immediately taking the necessary measures in response to stress levels, drones might be utilized as an early warning system (Fascista, 2022). The UAV can cover extensive areas of land in a single flight; measuring plant canopy reflectance with thermal and multispectral cameras. These cameras take one picture per second, storing the data in memory and transmitting it via telemetry to a ground station. The NDVI (GNDVI/NDVI) serves as a geographic indicator to assess the multispectral camera data (Mogili and Deepak, 2018).

Drones can monitor agricultural conditions throughout the growing season, enabling timely and targeted interventions that may prevent crop loss. This technology eliminates the need for farmers to visually inspect crops, allowing for remote observation and tracking of horticultural and other crops, even in geographically challenging areas such as hilly regions. In addition, they have the ability to effectively monitor the height of crops and trees, a task that would be difficult for farmers to perform manually.

5.4.5 Water Stress Monitoring

Characterizing water stress on crops is a challenging endeavor due to the intricate interplay between drought impacts and other influencing variables (Hassan *et al.*, 2020). Thermal imaging variables often rely on modest temperature fluctuations to identify stresses and other events. Different genotypes of a crop may exhibit distinct canopy temperatures under identical conditions due to intrinsic variations in stomatal conductance and transpiration rates. Scientists used a range of sensors and models to detect instances of water stress (Sellami *et al.*, 2022; Vidican *et al.*, 2023). Multispectral or hyperspectral pictures, along with the photochemical reflectance index (PRI) (Falcioni *et al.*, 2023; Marques *et al.*, 2023), allow for the detection of reflectance changes highly responsive to variations in carotenoid pigments in leaves. Cheung et al. (2021) used thermal infrared images to determine the difference between canopy and air temperatures (Tc – Ta). Some studies, directly measure canopy temperature. The use of thermal infrared images and the crop water stress index (CWSI), as described by (Romero-Trigueros *et al.*, 2019; Jamshidi *et al.*, 2021), relies on calculating the disparity between air and canopy temperatures, adjusted by the vapor pressure deficit (VPD). This is because water stress reduces gas movement via the stomata and heat dissipation in plants, causing a noticeable rise in the plant's surface temperature. RGB photos are often used in conjunction with multispectral or thermal images to calculate hybrid variables like the chlorophyll fluorescence and Water Deficit Index (WDI) (Barbedo, 2019).

5.4.6 Diseases and Weeds Monitoring

Crop diseases, categorized as fungal, bacterial, or viral, can have severe consequences. Drones equipped with infrared cameras provide a distinct visual representation of plant health by penetrating the plant canopy. Image-based methods are valuable for identifying and categorizing plant diseases when human evaluation is inappropriate, untrustworthy, or inaccessible. This is particularly true for UAVs, which capture a wider range of data. RGB and multispectral photos are the primary means of obtaining information, with hyperspectral and thermal images also used to identify signs of water stress caused by specific (Ortega *et al.*, 2020).

Weeds, undesirable plants that compete for resources like water and space, can significantly reduce crop yields. In India, herbicide use is the predominant method for managing weeds (Ali *et al.*, 2013a). Traditional agriculture involves applying herbicides uniformly across fields, even in weed-free areas, leading to herbicide-resistant weeds and negative impacts on crop development (Khan *et al.*, 2019). Hyperspectral pictures help differentiate spectral characteristics of weeds with varying glyphosate resistance, while RGB sensors categorize different weed species (Tanveer *et al.*, 2015).

Scientists use drones with hyperspectral sensors to measure chlorophyll content and leaf density of the plant canopy, essential for creating precise weed cover maps. Drones can collect field-wide data to generate accurate maps indicating where herbicides are needed. Agro-drones for weedicide spraying effectively control weeds pre- and post-emergence, suitable for all field conditions, including muddy, weedy, and insect-infested areas, and in both sunny and drizzly weather (Yadav *et al.*,

2024). Drones usage for weedicide spraying is efficient, maximizing herbicide effectiveness, and is easy to operate, transport, and maintain, promoting good health through remote operation.

5.4.7 Nutrient Status and Deficiency Monitoring

Plants need optimal quantities of nutrients to flourish and deliver a robust harvest (Waheed *et al.*, 2019). Adequate nitrogen levels are essential for healthy vegetation and foliage growth, while sufficient phosphorous is necessary for strong root and stem development (Mustafa *et al.*, 2022). Additionally, appropriate potassium levels are crucial for enhancing disease resistance and improving crop quality. If the soil is deficient in any of these nutrients, the plant will experience stress and have difficulty thriving. NDVI Index mosaics provide the opportunity to precisely pinpoint sections of crops that are experiencing stress or difficulties, allowing for targeted intervention in these specific locations. The UAVs can use NIR/multispectral photography to detect these management zones well before they become evident to the human eye (Bhatt and Maclean, 2023). This early detection allows for intervention before crop growth and productivity are significantly impacted. Currently, the prevailing method for assessing nutritional status involves visual inspection using plant color guides, which cannot provide precise quantitative evaluations.

Conducting precise assessments requires leaf analyses in a controlled environment. This process is time-consuming and demands proper procedures for accurate interpretation of the findings. Despite this, the forecasts may not always be precise (Mahmood *et al.*, 2015; Mehrabi and Sepaskhah, 2022). Therefore, significant resources have been devoted to developing novel techniques for identifying and assessing nutritional issues in plants. Nitrogen is extensively researched due to its strong correlation with biomass and crop productivity. Potassium and sodium have also garnered considerable attention. While multispectral pictures are commonly used to extract significant features and indices, RGB and hyperspectral images are also widely utilized. Research has also explored data fusion, which involves integrating two or three types of sensors, including multispectral, RGB, and temperature sensors. Most studies extract vegetation indices (VI) from photographs and use a regression model (often linear) to correlate them with nutrient content. In addition to the more often used variables, other kinds of variables have been used to input data into the regression models (Prado Osco *et al.*, 2019).

5.4.8 Additional Benefits of Drones in Agriculture

Previously, satellite or aerial imaging was used to provide a comprehensive overview of the farm and detect potential issues. However, these photos were not only costly but also lacked the accuracy that drones can provide. It offers both live video and time-based animation that display the crop progress in real time. Here are some notable advantages:

1. This solution requires fewer human resources and is outsourced. Consequently, there is less dependence on departmental personnel.
2. The outcomes might be acquired within a brief duration (about 3–4 weeks), facilitating prompt processing.
3. A drone can fly in any weather conditions. While drones can be capable of withstand water, image quality may be compromised in wet weather.
4. Drones optimize the utilization of resources (such as seeds, fertilizers, and water) for farmers, enhance their ability to promptly address risks (such as weeds, pests, and fungi), reduce the time spent on crop scouting (by verifying treatments and actions taken), enable real-time adjustments in variable-rate prescriptions, and estimate crop yield from a specific field.
5. Drones can surveil various crops in any geographical setting. Due to its recent development, the market and use of this agricultural technology are projected to see substantial growth in the coming years.

6. The method has also shown its usefulness in obtaining a full understanding of plant emergence and population, which may assist in making replanting choices.
7. Drone data's exceptional level of detail can be utilized to evaluate crop fertility, enabling agricultural experts to apply fertilizer with greater accuracy, minimize wastage, and strategize irrigation systems.
8. Considering the extensive expanse of land that requires surveying, drones prove to be more efficient, enabling users to swiftly capture high-resolution imagery compared to alternative methods.

5.5 LIMITATIONS OF DRONES IN AGRICULTURE

Agricultural drones possess the accompanying drawbacks or disadvantages:

1. Significant initial investment.
2. A conventional farmer may lack the ability to analyze drone photos. Proficiency in fundamental knowledge and abilities is required to operate agricultural drones, including picture editing software and expertise.
3. One of the significant environmental consequences is the endangerment of animals, especially birds.
4. UAVs, known as drones, have the potential to cause accidents if they lose control while flying. Sharing the same airspace as commercial aircraft, they may disrupt the flight path of manned planes. Severe weather conditions pose challenges to flying drones. Severe precipitation, dense fog, or strong gusts of wind might all impede the devices from taking flight or capturing the necessary area.
5. Rain can damage the electrical components of a drone. Image capturing needs substantial sunshine.
6. Government authorization is required to use drones.
7. Due to the dynamic nature of drone flying rules, the increasing number of limitations on airspace may lead to potential financial or legal consequences.
8. The battery life of a drone is a constraint, reducing the duration of its flight.

5.6 AGRICULTURAL ROBOTS

Agricultural robotics involves the use of automation technology across various biological systems, including agriculture, forestry, greenhouses, horticulture, and fisheries. It aims to replace traditional practices with more efficient methods. Agricultural robots are specifically designed and used for agricultural tasks. Currently, the primary use of robots in agriculture is for harvesting crops. However, there is potential for the development of robots and drones designed for weed management. Implementing automation in agriculture fosters several developments in the industry, enabling farmers to save both time and money (Oliveira *et al.*, 2021; Rasheed *et al.*, 2022a). Notable progress has been made in the field of robotics, leading to the creation of mobile robots, flying robots, forester robots, and dimeter robots specifically designed for agricultural use. Even in underdeveloped nations, farmers are increasingly adopting robots for tasks such as tending to crops, harvesting fruit, and maintaining livestock.

5.6.1 TYPES OF ROBOTS IN AGRICULTURE

1. *Farming robots*: Multifunctional robots on farms perform a wide variety of tasks, including maintenance, transportation, and observation.
2. *Fertilizing robots*: Autonomous robots are employed to distribute fertilizers to crops, providing precise nutrient management tailored to specific areas within the field.

3. *Grading robots*: These automated machines evaluate agricultural goods based on dimensions, configuration, and quality to guarantee consistent quality standards.
4. *Harvesting robots*: Designed for accurate and efficient crop-gathering, these robots save labor costs and increase yield.
5. *Planting and seeding robots*: These robots automate the planting process, ensuring precise seed placement and optimal spacing to improve crop productivity.
6. *Weeding robots*: Engineered to identify and eradicate weeds in agricultural zones, thereby reducing the necessity for pesticides and manual labor.

5.6.2 Role of Robots in Modern Agriculture

Agricultural robots have been created to replace the fieldwork traditionally done by migrant laborers. Robots can execute the repeated movements involved in crop tending with high efficiency and accuracy. Nevertheless, this doesn't mean that people will no longer be required in the agricultural sector. It simply implies that future agricultural laborers will need more sophisticated abilities to collaborate with robots (Sparrow and Howard, 2021). The current agricultural labor force will transition into managing networks of robotic farming devices and autonomous vehicles. There will be a need for the creation of devices that regulate the operations of equipment, such as remotely powering it on and off over the Internet and controlling it via smartphone applications (Figure 5.3). In addition, the robots will need routine maintenance carried out by human operators.

Humans will continue to be responsible for agricultural research and making difficult decisions. Individuals will be required to construct, market, and maintain the equipment. Due to new laws in several regions, farmers can now include other enterprises on their farms to generate income. These projects, which include wineries and breweries, will need a larger workforce. UAVs will continue to gather substantial quantities of up-to-the-minute data on agricultural lands. The continuous expansion of this data, combined with ML, will further enhance our understanding of agriculture (Rasheed *et al.*, 2022b; Hu *et al.*, 2023). Activities such as crop irrigation can provide the sector with more precise maps of land erosion, crop vitality, and water drainage. As a result, humans will be making more well-informed agricultural decisions than ever before.

FIGURE 5.3 Application of robots in agriculture sector.

There are many advantages of using robots in agriculture:

1. Robots have the capability to decrease a farm's pesticide use by as much as 80%. In the future, robots may take on additional or alternative duties. Robots may provide employment opportunities for those involved in robot manufacturing and maintenance.
2. Robots are immune to disease and exhaustion and do not need breaks or time off. They can function with precise tolerances, ensuring maximum field capacity. The machines can reliably detect better-grade items, with fewer faults and operate at greater rates.
3. Robots are widely used in several agricultural applications, including the Merlin Robot Milker, Ro-sphere, Harvest Automation, Orange Harvester, lettuce bot, and the weeder. Horticulture is another area where this technology is applied. Harvest Automation Inc. has developed RV 100 for horticultural use.
4. Machines can be made lighter and more affordable by eliminating the driver's seat, controls, and cab. Robots have many applications in the agricultural industry. The robots can navigate effortlessly among the trees, rocks, ponds, and other obstacles.
5. The RV100 is used for managing and arranging potted plants, including adjusting spacing, gathering plants, and consolidating them. It provides exceptional placement precision, with the added advantage of autonomous operation both indoors and outdoors. Additionally, it offers decreased manufacturing expenses.
6. Autonomous robots can be relied upon to carry out many agricultural tasks, including spraying, mechanical weed control, and fruit harvesting. This technology enables farmers to minimize environmental harm, enhance accuracy and efficiency, and adopt innovative approaches to managing individual plants.
7. Robotics are increasingly expanding into several fields, providing practical solutions and a return on investment by potentially replacing human operators. They perform hazardous jobs that pose risks to workers' health, ensuring worker safety.
8. Rice farmers cultivate rice in the fields, automated milkers extract milk from cows in barns or milking facilities, and fruit harvesters collect fruit from orchards. Robots are trained according to their specific jobs. The robotic milker and the fruit picker have several functions, however the rice planter does not.
9. Fruit harvesting robots, autonomous tractors/sprayers, and sheep shearing robots have been developed to substitute human work. Prior to beginning the work, many factors must be considered, including the dimensions and hue of the fruit to be harvested.
10. Robots may also be used for other horticultural activities, like trimming, weeding, spraying, and monitoring. Livestock robots may be used for various purposes, including automated milking, washing, and castrating.
11. The concept of using robotic technology in agriculture is somewhat novel. The agricultural sector offers vast prospects for enhancing production via the use of robots, which are increasingly being deployed on farms in different forms.
12. Robots in the agriculture sector provide superior quality fresh food, reduced production expenses, and less reliance on physical labor. Drones can mechanize labor-intensive jobs, such as spraying weed or bracken, in situations where it is hazardous for operators to employ tractors or other human vehicles.
13. Robots may protect human workers from the detrimental consequences of manual chemical handling and high-pressure spraying systems. Additionally, they have the potential to decrease a farm's pesticide use by up to 80%.
14. The deployment of unmanned options includes operations such as distributing heavy chemicals, dispensing pharmaceuticals, and spreading fertilizers. Agricultural robots use a combination of advanced sensors, cameras, software, and technology.

5.7 DISADVANTAGES OF AGRICULTURAL ROBOTS

1. The production or purchase of robots incurs significant expenses. Regular maintenance is necessary to ensure their continued functioning.
2. The farmers may face unemployment. The robots can alter the cultural and emotional aspects of agriculture, while the energy requirements come with a high price tag.
3. The non-farming community may find the robots more acceptable. Jobs in the agricultural sector need intellect and agility, and robots may replace human operators.
4. Advantages and cons of automatic or robotic milking.
5. Robotic uses in the agriculture business. Types, applications, and significance of autonomous agricultural robots.

5.8 CONCLUSIONS AND FUTURE PERSPECTIVES

The integration of drones equipped with modern sensors is expected to greatly enhance precision agriculture. These drones will play a crucial role in closely monitoring crops and identifying concerns such as nutrient deficits and insect infestations at early stages, providing unprecedented levels of detail. The combination of technology and robots will enhance productivity and reduce labor costs by allowing precise activities like sowing, watering, and harvesting to be performed autonomously. Making well-informed decisions will lead to increased production and profitability in agricultural activities. By using the data gathered via these technologies, artificial intelligence algorithms will enable farmers to make data-driven decisions, providing valuable information on improving irrigation, predicting crop yields, and determining the best periods for planting and harvesting. Furthermore, the use of drones and robots in agriculture will promote sustainability by enabling precise administration of fertilizers and pesticides, reducing environmental harm, and encouraging eco-friendly methods such as weed eradication without herbicides.

The increasing prominence of automation in agriculture will result in the emergence of drones and robots that can function independently, optimizing efficiency and production. By using cutting-edge imaging technology, farmers will be able to promptly identify crop diseases and stressors, allowing them to take proactive actions to protect their crops and guarantee food security. Moreover, the adaptation of farming techniques using data analysis would maximize the use of resources and improve crop productivity, leading to a more effective and environmentally friendly agricultural system. This will enable farmers to have enhanced control and supervision of operations from a distant location, while also allowing for more flexibility. Due to the advancement and reduced cost of these technologies, their worldwide implementation is unavoidable, which will help small-scale farmers in developing countries and contribute to tackling food security issues while improving rural lives. The widespread access to agricultural technology highlights its capacity to transform farming methods on a global scale.

The modern era is progressing toward intelligent technology, but agronomy, which is an essential ecological component, is also experiencing its proportionate advances. Mechatronics, as a multidisciplinary engineering field, aims to assist the agricultural industry by using robots and automation (Ur Rehman *et al.*, 2022). Relevant literature indicates a growing trend in the use and exploration of electromechanical systems that have the potential to aid farmers in several areas of agronomy. Various systems have been developed to automate labor-intensive tasks in agriculture, including seed implantation, weed cutting or spraying, and fruit and produce collection. These systems use precision actuators that are controlled by advanced control systems enhancing the accuracy, efficiency, and profitability of agricultural processes. UAVs and ground robots have been suggested and possess their own advantages and disadvantages. Ground robots are more practical but challenging to operate in humid and harsh terrains, whilst UAVs have superior range but limited energy and capabilities.

However, research in this area indicates the development of intelligent systems for gathering and accessing online information using the IoT and advanced monitoring systems that use cutting-edge

ML and clustering techniques (Lv and Xie, 2022). The availability of greenhouse and outdoor temperature monitoring, soil humidity tracking, nutrition monitoring, and the ability to access data remotely and in real time has made it easier for farmers to acquire essential information. While the suggested ML techniques for identifying and separating diseases, organs, and weeds are quite precise, they need further development to ensure their resilience and adaptability to various terrains and plant species.

Mechatronics may provide many benefits to the sector via the following methods:

1. The use of automation and intelligent technology in irrigation would enhance irrigation efficiency and reduce water wastage.
2. The use of automation and intelligent technologies in the food business also assists in identifying the optimal chemicals and pesticides to be employed for agricultural reasons.
3. Automatic and advanced tools facilitate the identification and control of weeds. The multifunctional robots will assist in agricultural operations and processes, while also accelerating the completion of tasks, maintaining product quality, and minimizing the need for human labor.
4. Thanks to the hybrid agricultural systems, we can provide data on humidity, temperature, and water level.
5. By effectively utilizing automation and IoT in the field of agriculture, we can optimize the agricultural process and make informed decisions on the selection of weeds, crops, and pesticides.
6. Alternatively, automation and IoT technologies provide solutions that enhance the predictability of agricultural operations.
7. These technologies will aid in reducing the need for human labor and increasing output.

LIST OF ABBREVIATIONSC

AAPS aerial automated pesticide sprayer
CWSI crop water stress index
FAO food and agriculture organization
IoT Internet of Things
LAI leaf area index
ML machine learning
NDVI normalized difference vegetation index
NIR near-infrared
PGRs plant growth regulators
PRI photochemical reflectance index
UAVs unmanned aerial vehicles
UASes unmanned aircraft systems
VI vegetation indices
VPD vapor pressure deficit
WHO World Health Organization
WSN wireless sensor networks
WDI water deficit index

REFERENCES

Aasen, H., Honkavaara, E., Lucieer, A., & Zarco-Tejada, P.J., 2018. Quantitative remote sensing at ultra-high resolution with UAV spectroscopy: A review of sensor technology, measurement procedures, and data correction workflows. *Remote Sensing*, 10, 1091.

Abbas, A., Zhang, Z., Zheng, H., Alami, M.M., Alrefaei, A.F., Abbas, Q., Naqvi, S.a.H., Rao, M.J., Mosa, W.F.A., Abbas, Q., Hussain, A., Hassan, M.Z., & Zhou, L., 2023. Drones in plant disease assessment, efficient monitoring, and detection: A way forward to smart agriculture. *Agronomy*, 13, 1524.

Abdullah, Mahmood, A., Bibi, S., Naqve, M., Javaid, M.M., Zia, M.A., Jabbar, A., Ud-Din, W., Attia, K.A., & Khan, N., 2022. Physiological, biochemical, and yield responses of linseed (*Linum usitatissimum* L.) in α-tocopherol-mediated alleviation of salinity stress. *Frontiers in Plant Science*, 13, 867172.

Ali, H., Tanveer, A., Nadeem, M., Asghar, H., & Javaid, M., 2013a. Germination ecology of *Rhynchosia capitata*: An emerging summer weed in Asia. *Planta Daninha*, 31, 249–257.

Ali, H., Tanveer, A., Nadeem, M., Javaid, M., Kashif, M., & Chadhar, A., 2013b. Allelopathic effects of *Rhynchosia capitata* on germination and seedling growth of mungbean. *Planta Daninha*, 31, 501–509.

Ameen, M., Mahmood, A., Ahmad, M., Mansoor Javaid, M., Nadeem, M.A., Asif, M., Balal, R.M., & Khan, B.A., 2023a. Impacts of Climate Change on Fruit Physiology and Quality. *In Climate-Resilient Agriculture, Vol 1: Crop Responses and Agroecological Perspectives*. Cham: Springer international publishing, 93–124.

Ameen, M., Zafar, A., Javaid, M.M., Zia, M.A., Mahmood, A., Naqve, M., & Bibi, S., 2023b. Climate-Resilient Technology for Maize Production. *In* M. Hasanuzzaman (ed.) *Climate-Resilient Agriculture, Vol 2: Agro-Biotechnological Advancement for Crop Production*. Cham: Switzerland Springer International Publishing, 157–188.

Aziz, M., Ashraf, M., & Javaid, M.M., 2018. Enhancement in cotton growth and yield using novel growth promoting substances under water limited conditions. *Pakistan Journal of Botany*, 50, 1691–1701.

Babu, S., Das, A., Singh, R., Mohapatra, K.P., Kumar, S., Rathore, S.S., Yadav, S.K., Yadav, P., Ansari, M.A., Panwar, A.S., Wani, O.A., Singh, M., Ravishankar, N., Layek, J., Chandra, P., & Singh, V.K., 2023. Designing an energy efficient, economically feasible, and environmentally robust integrated farming system model for sustainable food production in the Indian Himalayas. *Sustainable Food Technology*, 1, 126–142.

Balyan, S., Jangir, H., Tripathi, S.N., Tripathi, A., Jhang, T., & Pandey, P., 2024. Seeding a sustainable future: Navigating the digital horizon of smart agriculture. *Sustainability*, 16, 475.

Barbedo, J.G.A., 2019. A review on the use of unmanned aerial vehicles and imaging sensors for monitoring and assessing plant stresses. *Drones*, 3, 40.

Beltozar-Clemente, S., Iparraguirre-Villanueva, O., Pucuhuayla-Revatta, F., Sierra-Liñan, F., Zapata-Paulini, J., & Cabanillas-Carbonell, M., 2023. Contributions of the 5G network with respect to decent work and economic growth (sustainable development goal 8): A systematic review of the literature. *Sustainability*, 15, 15776.

Bhatt, P., & Maclean, A.L., 2023. Comparison of high-resolution NAIP and unmanned aerial vehicle (UAV) imagery for natural vegetation communities classification using machine learning approaches. *GIScience & Remote Sensing*, 60, 2177448.

Boedeker, W., Watts, M., Clausing, P., & Marquez, E., 2020. The global distribution of acute unintentional pesticide poisoning: Estimations based on a systematic review. *BMC Public Health*, 20, 1875.

Castillejo, P., Johansen, G., Cürüklü, B., Bilbao-Arechabala, S., Fresco, R., Martínez-Rodríguez, B., Pomante, L., Rusu, C., Martínez-Ortega, J.-F., Centofanti, C., Hakojärvi, M., Santic, M., & Häggman, J., 2020. Aggregate farming in the cloud: The AFarCloud ECSEL project. *Microprocessors and Microsystems*, 78, 103218.

Chadha, A., Florentine, S., Javaid, M., Welgama, A., & Turville, C., 2020. Influence of elements of climate change on the growth and fecundity of *Datura stramonium*. *Environmental Science and Pollution Research*, 27, 35859–35869.

Chávez, J.L., Torres-Rua, A.F., Woldt, W.E., Zhang, H., Robertson, C.C., Marek, G.W., Wang, D., Heeren, D.M., Taghvaeian, S., & Neale, C.M.U., 2020. A decade of unmanned aerial systems in irrigated agriculture in the Western U.S. *Applied Engineering in Agriculture*, 36, 423–436.

Cheung, P.K., Jim, C.Y., & Hung, P.L., 2021. Preliminary study on the temperature relationship at remotely-sensed tree canopy and below-canopy air and ground surface. *Building and Environment*, 204, 108169.

Curl, C.L., Spivak, M., Phinney, R., & Montrose, L., 2020. Synthetic pesticides and health in vulnerable populations: Agricultural workers. *Current Environmental Health Reports*, 7, 13–29.

Das, A., Kadawla, K., Nath, H., Chakraborty, S., Ali, H., Singh, S., & Dubey, V.K., 2024. Drone-Based Intelligent Spraying of Pesticides: Current Challenges and Its Future Prospects. *In* S.S. Chouhan, U.P. Singh & S. Jain (eds.) *Applications of Computer Vision and Drone Technology in Agriculture 4.0*. Singapore: Springer Nature Singapore, 199–223.

Dereumeaux, C., Fillol, C., Quenel, P., & Denys, S., 2020. Pesticide exposures for residents living close to agricultural lands: A review. *Environment International*, 134, 105210.

Elmeseiry, N., Alshaer, N., & Ismail, T., 2021. A detailed survey and future directions of unmanned aerial vehicles (UAVs) with potential applications. *Aerospace*, 8, 363.

Falcioni, R., Antunes, W.C., Demattê, J.a.M., & Nanni, M.R., 2023. Biophysical, biochemical, and photochemical analyses using reflectance hyperspectroscopy and chlorophyll a fluorescence kinetics in variegated leaves. *Biology*, 12, 704.

Fascista, A., 2022. Toward integrated large-scale environmental monitoring using WSN/UAV/crowdsensing: A review of applications, signal processing, and future perspectives. *Sensors*, 22, 1824.

Ferreira, C.S.S., Seifollahi-Aghmiuni, S., Destouni, G., Ghajarnia, N., & Kalantari, Z., 2022. Soil degradation in the European Mediterranean region: Processes, status and consequences. *Science of the Total Environment*, 805, 150106.

Franzen, D.W., Miao, Y., Kitchen, N.R., Schepers, J.S., & Scharf, P.C., 2021. Sensing for Health, Vigour and Disease Detection in Row and Grain Crops. *In* R. Kerry & A. Escolà (eds.) *Sensing Approaches for Precision Agriculture*. Cham: Springer International Publishing, 159–193.

Gull, T., Mahmood, A., Shaheen, C., Javaid, M.M., Zia, M.A., Naqve, M., Bibi, S., Nadeem, M.A., Ameen, M., Nargis, J., & Khan, S.R., 2023. Climate Change and Nutrient Use Efficiency of Plants. *In* M. Hasanuzzaman (ed.) *Climate-Resilient Agriculture, Vol 1: Crop Responses and Agroecological Perspectives*. Cham: Springer International Publishing, 291–312.

Hanif, Z., Naeem, M., Ali, H.H., Tanveer, A., Javaid, M.M., Peerzada, A.M., & Chauhan, B.S., 2017. Effect of environmental factors on germination of *Salsola foetida*: Potential species for rehabilitation of degraded rangelands. *Rangeland Ecology & Management*, 70, 638–643.

Hassan, M.U., Chattha, M.U., Barbanti, L., Mahmood, A., Chattha, M.B., Khan, I., Mirza, S., Aziz, S.A., Nawaz, M., & Aamer, M., 2020. Cultivar and seeding time role in sorghum to optimize biomass and methane yield under warm dry climate. *Industrial Crops and Products*, 145, 111983.

Hassan, M.U., Nawaz, M., Mahmood, A., Shah, A.A., Shah, A.N., Muhammad, F., Batool, M., Rasheed, A., Jaremko, M., & Abdelsalam, N.R., 2022. The role of zinc to mitigate heavy metals toxicity in crops. *Frontiers in Environmental Science*, 10, 990223.

Hu, L., Luo, X., & He, J., 2023. Unmanned Farm. *In* Q. Zhang (ed.) *Encyclopedia of Digital Agricultural Technologies*. Cham: Springer International Publishing, 1501–1505.

Ibrahim, M.R., Azman, M.F., Ariffin, A.H., Mansur, M.N., Mustapa, M.S., & Irfan, A.R., 2023. Overview of Unmanned Aerial Vehicle (UAV) Parts Material in Recent Application. *In* A.H. Ariffin, N.A. Latif, M.F.B. Mahmod & Z.B. Mohamad (eds.) *Structural Integrity and Monitoring for Composite Materials*. Singapore: Springer Nature Singapore, 179–189.

Inoue, Y., 2020. Satellite- and drone-based remote sensing of crops and soils for smart farming – A review. *Soil Science and Plant Nutrition*, 66, 798–810.

Jabbar, A., Ibrahim, H., Mahmood, A., Javaid, M.M., Wasaya, A., Nadeem, M.A., Ismael, M., Awais, M., & Khan, S.R., 2023. Modern Agronomic Measurement for Climate-Resilient Agriculture. *In* M. Hasanuzzaman (ed.) *Climate-Resilient Agriculture, Vol 2: Agro-Biotechnological Advancement for Crop Production*. Cham: Springer International Publishing, 81–105.

Jalu, M., Ambaliya, P., & Modhvadiya, Y., 2024. Comparative performance of the developed mini tractor operated sprayer cum weeder with the various types of existing spraying and weeding methods. *International Journal of Environment, Agriculture and Biotechnology*, 9, 8–19.

Jamshidi, S., Zand-Parsa, S., & Niyogi, D., 2021. Assessing crop water stress index of Citrus using in-situ measurements, Landsat, and Sentinel-2 data. *International Journal of Remote Sensing*, 42, 1893–1916.

Javaid, M., Tanveer, A., Ahmad, R., Yaseen, M., & Khaliq, A., 2012. Optimizing activity of herbicides at reduced rate on *Emex spinosa* Campd. with adjuvants. *Planta Daninha*, 30, 425–435.

Kapustina, L., Izakova, N., Makovkina, E., & Khmelkov, M., 2021. The global drone market: Main development trends. *SHS Web of Conferences*, 129, 11004.

Karunathilake, E.M.B.M., Le, A.T., Heo, S., Chung, Y.S., & Mansoor, S., 2023. The path to smart farming: Innovations and opportunities in precision agriculture. *Agriculture*, 13, 1593.

Kendall, H., Clark, B., Li, W., Jin, S., Jones, G.D., Chen, J., Taylor, J., Li, Z., & Frewer, L.J., 2022. Precision agriculture technology adoption: A qualitative study of small-scale commercial "family farms" located in the North China Plain. *Precision Agriculture*, 23, 319–351.

Khan, M.A., Ahmad, S., & Raza, A., 2019. Integrated Weed Management for Agronomic Crops. *In* M. Hasanuzzaman (ed.) *Agronomic Crops: Volume 2: Management Practices*. Singapore: Springer Singapore, 257–281.

Kim, J., Kim, S., Ju, C., & Son, H.I., 2019. Unmanned aerial vehicles in agriculture: A review of perspective of platform, control, and applications. *IEEE Access*, 7, 105100–105115.

King, T., Cole, M., Farber, J.M., Eisenbrand, G., Zabaras, D., Fox, E.M., & Hill, J.P., 2017. Food safety for food security: Relationship between global megatrends and developments in food safety. *Trends in Food Science & Technology*, 68, 160–175.

Kumar, A., Ahamad, S., Kumar, M., Bihari, C., Singh, S., Pandey, V., Alam, K., Singh, S., & Gautam, P., 2022. An innovative drones technology in agriculture: A review. *The Pharma Innovation Journal*, 11, 279–286.

Lambey, V., & Prasad, A.D., 2021. A review on air quality measurement using an unmanned aerial vehicle. *Water, Air, & Soil Pollution*, 232, 109.

Li, X., Giles, D.K., Andaloro, J.T., Long, R., Lang, E.B., Watson, L.J., & Qandah, I., 2021. Comparison of UAV and fixed-wing aerial application for alfalfa insect pest control: Evaluating efficacy, residues, and spray quality. *Pest Management Science*, 77, 4980–4992.

Liang, C., & Shah, T., 2023. IoT in agriculture: The future of precision monitoring and data-driven farming. *Eigenpub Review of Science and Technology*, 7, 85–104.

Lingli, Z., Juha, S., Jingbin, L., Juha, H., Harri, K., & Henrik, H., 2017. A Review: Remote Sensing Sensors. *In* B.R. Rustam, H. Sabina & H.Z. Mahfuza (eds.) *Multi-Purposeful Application of Geospatial Data*. Rijeka: IntechOpen, 1–24.

Lv, Z., & Xie, S., 2022. Artificial intelligence in the digital twins: State of the art, challenges, and future research topics [version 2; peer Review: 2 approved]. *Digital Twin*, 1.

Maddikunta, P.K.R., Hakak, S., Alazab, M., Bhattacharya, S., Gadekallu, T.R., Khan, W.Z., & Pham, Q.V., 2021. Unmanned aerial vehicles in smart agriculture: Applications, requirements, and challenges. *IEEE Sensors Journal*, 21, 17608–17619.

Mahmood, A., Hussain, A., Shahzad, A.N., & Honermeier, B., 2015. Biomass and biogas yielding potential of sorghum as affected by planting density, sowing time and cultivar. *Pakistan Journal of Botany*, 47, 2401–2408.

Mandal, M., Popek, R., Przybysz, A., Roy, A., Das, S., & Sarkar, A., 2023. Breathing fresh air in the City: Implementing avenue trees as a sustainable solution to reduce particulate pollution in urban agglomerations. *Plants*, 12, 1545.

Marques, P., Pádua, L., Sousa, J.J., & Fernandes-Silva, A., 2023. Assessing the water status and leaf pigment content of olive trees: Evaluating the potential and feasibility of unmanned aerial vehicle multispectral and thermal data for estimation purposes. *Remote Sensing*, 15, 4777.

Mehmood, A., Naeem, M., Khalid, F., Saeed, Y., Abbas, T., Jabran, K., Sarwar, M.A., Tanveer, A., & Javaid, M.M., 2018. Identification of phytotoxins in different plant parts of *Brassica napus* and their influence on mung bean. *Environmental Science and Pollution Research*, 25, 18071–18080.

Mehrabi, F., & Sepaskhah, A.R., 2022. Leaf nitrogen, based on SPAD chlorophyll Reading can determine agronomic parameters of winter wheat. *International Journal of Plant Production*, 16, 77–91.

Mogili, U.M.R., & Deepak, B.B.V.L., 2018. Review on application of drone systems in precision agriculture. *Procedia Computer Science*, 133, 502–509.

Mohan, M., Richardson, G., Gopan, G., Aghai, M.M., Bajaj, S., Galgamuwa, G.a.P., Vastaranta, M., Arachchige, P.S.P., Amorós, L., Corte, A.P.D., De-Miguel, S., Leite, R.V., Kganyago, M., Broadbent, E.N., Doaemo, W., Shorab, M.a.B., & Cardil, A., 2021. UAV-supported forest regeneration: Current trends, challenges and implications. *Remote Sensing*, 13, 2596.

Mohsan, S.a.H., Khan, M.A., Noor, F., Ullah, I., & Alsharif, M.H., 2022. Towards the unmanned aerial vehicles (UAVs): A comprehensive review. *Drones*, 6, 147.

Mustafa, A., Athar, F., Khan, I., Chattha, M.U., Nawaz, M., Shah, A.N., Mahmood, A., Batool, M., Aslam, M.T., & Jaremko, M., 2022. Improving crop productivity and nitrogen use efficiency using sulfur and zinc-coated urea: A review. *Frontiers in Plant Science*, 13, 942384.

Mustafi, S., Ghosh, P., Roy, K., Dan, S., Mukherjee, K., & Nath mandal, S., 2021. Drones for Intelligent Agricultural Management. *In* P. Krause & F. Xhafa (eds.) *IoT-Based Intelligent Modelling for Environmental and Ecological Engineering: IoT Next Generation EcoAgro Systems*. Cham: Springer International Publishing, 81–100.

Oliveira, L.F.P., Moreira, A.P., & Silva, M.F., 2021. Advances in agriculture robotics: A state-of-the-art review and challenges ahead. *Robotics*, 10, 52.

Ortega, S., Halicek, M., Fabelo, H., Callico, G.M., & Fei, B., 2020. Hyperspectral and multispectral imaging in digital and computational pathology: A systematic review [Invited]. *Biomedical Optics Express*, 11, 3195–3233.

Prado Osco, L., Marques Ramos, A.P., Roberto Pereira, D., Akemi Saito Moriya, É., Nobuhiro Imai, N., Takashi Matsubara, E., Estrabis, N., De Souza, M., Marcato Junior, J., Gonçalves, W.N., Li, J., Liesenberg, V., & Eduardo Creste, J., 2019. Predicting canopy nitrogen content in citrus-trees using random forest algorithm associated to spectral vegetation indices from UAV-imagery. *Remote Sensing*, 11, 2925.

Rasheed, A., Barqawi, A.A., Mahmood, A., Nawaz, M., Shah, A.N., Bay, D.H., Alahdal, M.A., Hassan, M.U., & Qari, S.H., 2022a. CRISPR/Cas9 is a powerful tool for precise genome editing of legume crops: A review. *Molecular Biology Reports*, 49, 5595–5609.

Rasheed, A., Raza, A., Jie, H., Mahmood, A., Ma, Y., Zhao, L., Xing, H., Li, L., Hassan, M.U., & Qari, S.H., 2022b. Molecular tools and their applications in developing salt-tolerant soybean (*Glycine max* L.) cultivars. *Bioengineering*, 9, 495.

Romero-Trigueros, C., Bayona Gambín, J.M., Nortes Tortosa, P.A., Alarcón Cabañero, J.J., & Nicolás Nicolás, E., 2019. Determination of crop water stress index by infrared thermometry in grapefruit trees irrigated with saline reclaimed water combined with deficit irrigation. *Remote Sensing*, 11, 757.

Ruwaimana, M., Satyanarayana, B., Otero, V., M. Muslim, A., Syafiq A, M., Ibrahim, S., Raymaekers, D., Koedam, N., & Dahdouh-Guebas, F., 2018. The advantages of using drones over space-borne imagery in the mapping of mangrove forests. *PLoS One*, 13, e0200288.

Sellami, M.H., Albrizio, R., Čolović, M., Hamze, M., Cantore, V., Todorovic, M., Piscitelli, L., & Stellacci, A.M., 2022. Selection of hyperspectral vegetation indices for monitoring yield and physiological response in sweet maize under different water and nitrogen availability. *Agronomy*, 12, 489.

Silva, L., Conceição, L.A., Lidon, F.C., & Maçãs, B., 2023. Remote monitoring of crop nitrogen nutrition to adjust crop models: A review. *Agriculture*, 13, 835.

Singh, A.K., & Gupta, S., 2024. Agriculture and the Use of Unmanned Aerial Vehicles (UAVs). *Advances in Aerial Sensing and Imaging*. 87–108.

Singh, P., 2023. *Drones in Indian Agriculture: Trends, Challenges, and Policy Implications*. India: Indian Chamber of Food and Agriculture.

Sinha, D., & Tandon, P.K., 2020. An Overview of Nitrogen, Phosphorus and Potassium: Key Players of Nutrition Process in Plants. *In* K. Mishra, P.K. Tandon & S. Srivastava (eds.) *Sustainable Solutions for Elemental Deficiency and Excess in Crop Plants*. Singapore: Springer Singapore, 85–117.

Sivakumar, R., Prabadevi, B., Velvizhi, G., Muthuraja, S., Kathiravan, S., Biswajita, M., & Madhumathi, A., 2021. Internet of Things and Machine Learning Applications for Smart Precision Agriculture. *In* S. Ishwar, G. Zhen & M. Carmine (eds.) *IoT Applications Computing*. Rijeka: IntechOpen, 1–30.

Skevas, T., & Kalaitzandonakes, N., 2020. Farmer awareness, perceptions and adoption of unmanned aerial vehicles: Evidence from Missouri. *International Food and Agribusiness Management Review*, 23, 469–485.

Smith, L.N., Zhang, W., Hansen, M.F., Hales, I.J., & Smith, M.L., 2018. Innovative 3D and 2D machine vision methods for analysis of plants and crops in the field. *Computers in Industry*, 97, 122–131.

Sparrow, R., & Howard, M., 2021. Robots in agriculture: Prospects, impacts, ethics, and policy. *Precision Agriculture*, 22, 818–833.

Suman, D., Ajit Kumar, S., Bhabani Prasad, M., Debashis, P., & Kiranmoy, P., 2023. Perspective Chapter: Digital Inclusion of the Farming Sector Using Drone Technology. *In* V. Ramana (ed.) *Human-Robot Interaction*. Rijeka: IntechOpen, 1–16.

Tanveer, A., Javaid, M.M., Irfan, M., Khaliq, A., & Yaseen, M., 2015. Yield losses in chickpea with varying densities of dragon spurge (*Euphorbia dracunculoides*). *Weed Science*, 63, 522–528.

Tyagi, J., Ahmad, S., & Malik, M., 2022. Nitrogenous fertilizers: Impact on environment sustainability, mitigation strategies, and challenges. *International Journal of Environmental Science and Technology*, 19, 11649–11672.

Ur Rehman, U., Iqbal, T., Hussain, S., Cheema, M.J.M., Iqbal, F., & Basit, A., 2022. Drone and robotics roadmap for agriculture crops in Pakistan: A review. *Environmental Sciences Proceedings*, 23, 39.

Vidican, R., Mălinaş, A., Ranta, O., Moldovan, C., Marian, O., Gheţe, A., Ghişe, C.R., Popovici, F., & Cătunescu, G.M., 2023. Using remote sensing vegetation indices for the discrimination and monitoring of agricultural crops: A critical review. *Agronomy*, 13, 3040.

Virnodkar, S.S., Pachghare, V.K., Patil, V.C., & Jha, S.K., 2020. Remote sensing and machine learning for crop water stress determination in various crops: A critical review. *Precision Agriculture*, 21, 1121–1155.

Waheed, H., Javaid, M.M., Shahid, A., Ali, H.H., Nargis, J., & Mehmood, A., 2019. Impact of foliar-applied Hoagland's nutrient solution on growth and yield of mash bean (*Vigna mungo* L.) under different growth stages. *Journal of Plant Nutrition*, 42, 1133–1141.

Wang, L., Huang, X., Li, W., Yan, K., Han, Y., Zhang, Y., Pawlowski, L., & Lan, Y., 2022. Progress in agricultural unmanned aerial vehicles (UAVs) applied in China and prospects for Poland. *Agriculture*, 12, 397.

Wasaya, A., Abbas, T., Yasir, T.A., Sarwar, N., Aziz, A., Javaid, M.M., & Akram, S., 2021. Mitigating drought stress in sunflower (*Helianthus annuus* L.) through exogenous application of β-aminobutyric acid. *Journal of Soil Science and Plant Nutrition*, 21, 936–948.

Weiss, M., Jacob, F., & Duveiller, G., 2020. Remote sensing for agricultural applications: A meta-review. *Remote Sensing of Environment*, 236, 111402.

Yadav, M., Vashisht, B.B., Vullaganti, N., Jalota, S.K., Yadav, S.L., Singh, G., Kumar, A., & Kothiyal, S., 2024. IoT-Enabled Unmanned Aerial Vehicle: An Emerging Trend in Precision Farming. *In* K. Pandey, N.L. Kushwaha, C.B. Pande & K.G. Singh (eds.) *Artificial Intelligence and Smart Agriculture: Technology and Applications*. Singapore: Springer Nature Singapore, 271–292.

6 Nanotechnology in Crop Production

Filza Ishfaq, Muhammad Mansoor Javaid, Muhammad Ather Nadeem, Muhammad Zubair, Muhammad Anjum Zia, Numra Akbar, and Javaria Nargis*

6.1 INTRODUCTION

The rapidly developing area of nanotechnology connects the fields of biology, chemistry, and physics by working with structures as small as one billionth of a meter (Khan *et al.*, 2022b). It includes the design, characterization, construction, and use of systems and structures by nanoscale material manipulation (Khan *et al.*, 2022b, 2023c). This interdisciplinary field has made significant strides in biomedical research, particularly in clinical applications such as imaging, diagnostics, therapies, drug administration, and tissue engineering. Nanoparticles (NPs) are utilized in photothermal therapy and optical scattering for imaging and diagnostics (Khan *et al.*, 2019a). The use of cell-specific compounds, such as peptides and antibodies, linked to NPs enhances targeting by binding to specific receptors on the membranes of diseased cells (Fróna *et al.*, 2019; Khan *et al.*, 2023b).

The Food and Agriculture Organization (FAO) estimates that by 2050, there will be 10 billion people on the planet, necessitating a 50% increase in food production. Developing nations will feel this need the most acutely (Fróna *et al.*, 2019). At the same time, another 2 billion cannot but go hungry, bringing the global total to 22 from today's 19 billion (Rajput *et al.*, 2021b).

According to recent studies (Goyal *et al.*, 2023; Khan *et al.*, 2023a), it's observed that nanotechnology has the ability to completely transform the agricultural sector by increasing the effectiveness of agricultural inputs and offering solutions to agricultural and environmental issues that would boost food production and security. These have prompted a noticeable interest in using nanotechnology in agriculture. Each usage of nanomaterials (NMs) in agriculture is critically analyzed, including biotic and abiotic soil remediation, nanopesticides, nanofertilizers, and nanobiosensors (Bayda *et al.*, 2019). NMs have significant effects on crops and soil health, especially on the morphological and physiological reactions of plants, soil biota, soil organic matter (SOM), and nutrient release in target soils (Pandey, 2018; Khan *et al.*, 2022a). Khan *et al.* (2019b) discuss the mechanisms related to absorption and transport of NMs in plants.

6.2 EVOLUTION OF NANOTECHNOLOGY IN AGRICULTURE

One of the most recent developments in this field is NP engineering. Initially, Professor Norio Taniguchi used the term "nanotechnology" in 1974 (Tyagi *et al.*, 2022). In the agricultural and food processing machinery industries, productivity has been secured through significant technological advances and innovation to address the challenges to food security and sustainable production posed by rising world population. These continuing advancements in agriculture are vital if, with both natural resources and technical means, the rapidly expanding population is to be provided with its growing food requirements (Arora *et al.*, 2018).

Nanotechnology presents practical solutions to many agricultural challenges. NPs are a focus of significant interest because they bridge the gap between atomic or molecular structures and bulk

* Corresponding author: anjum.zia@uaf.edu.pk

 DOI: 10.1201/9781003498698-6

materials (Verma *et al.*, 2020; Rasheed *et al.*, 2022). Although fertilizer is necessary to increase agricultural yield, excessive fertilizer use irreversibly alters the chemical ecology of the soil, reducing the amount of land available for crop production (Kabir *et al.*, 2018; Wang *et al.*, 2024).

Sustainable agriculture can protect the ecosystem and prevent the extinction of several species by using less agrochemicals (Wang *et al.*, 2021). Specifically, NMs improve crop yield by creating more efficient agricultural inputs, which allows for site-targeted and regulated nutrient delivery and reduces the need for agri-inputs (Afzal *et al.*, 2020). Moreover, the primary challenge in producing agricultural output is to facilitate plant acclimatization to more severe climate change parameters like high temperature, low water availability, salinity, alkalinity, and heavy metal pollution without compromising the fragile existing ecosystems (Zain *et al.*, 2023; Ameen *et al.*, 2024).

6.3 FUNDAMENTALS OF NANOTECHNOLOGY IN AGRICULTURE

The concept of nanotechnology has been developing for decades; however, the purpose of NPs in improving cultivation is a recent scientific achievement (Acharya and Pal, 2020). Recent developments in NM manufacturing, featuring an abundance of sizes and shapes, have been found useful in agriculture, food processing, medicine, and environmental studies (Khan *et al.*, 2021; Mukhtiar *et al.*, 2023).

Recently, wireless networking and downsizing of sensors for monitoring, analyzing, and regulating agricultural activities have led to precision agriculture (Ali *et al.*, 2019). Site-specific crop management, which includes both pre- and post-production activities according to the capabilities of each site, can be applied to all crop types from field crops to horticultural crops (Figure 6.1) (Hassanisaadi *et al.*, 2022). Tissue engineering and engineered NMs have further transformed this by enabling the targeted delivery of CRISPR (clustered regularly interspaced short palindromic

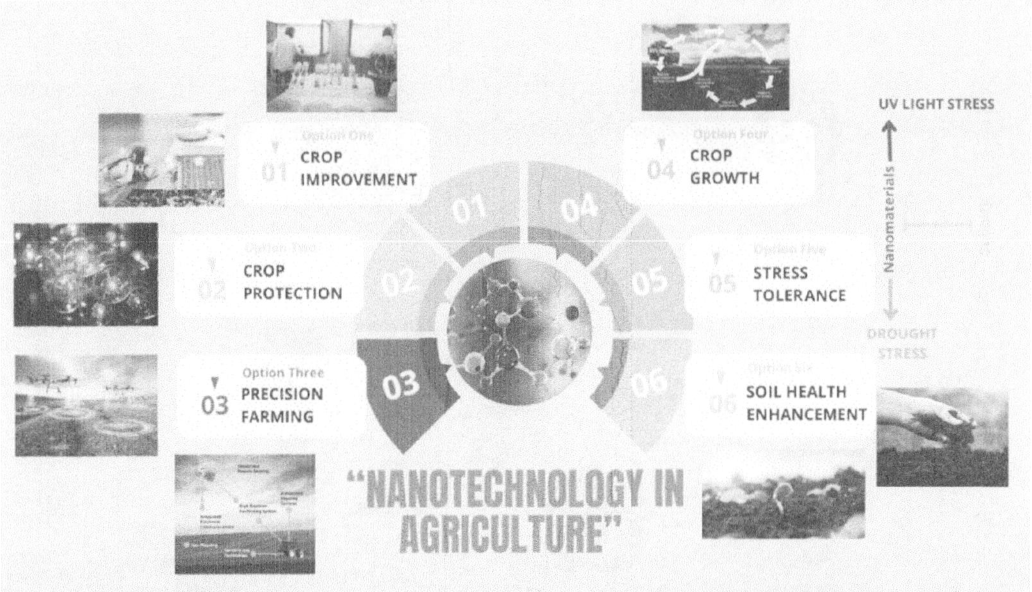

FIGURE 6.1 Application of nanotechnology in agriculture. Uses of nanotechnology in the field of agriculture. Crop productivity, yield, and growth are all enhanced by controlled-release nanofertilizers. For crop improvement, a target delivery strategy based on nanotechnology (gene transfer) is employed. Crop protection may be effectively achieved by using nanopesticides. Precision farming is substantially enhanced by the use of electronic controls and nanosensors. Additionally, plants may be made more resilient to stress by using NPs in the soil. Readers seeking further information are directed to the text.

repeats)/CAS (CRISPR-associated protein) mRNA and sgRNA for crop genetic modification (GM) (Rai-Kalal and Jajoo, 2021).

Nanotechnology also offers creative remedies to the growing variety of environmental concerns. Development of nanosensors is an example with potential to monitor stressful environments and increase the resistance of crops to infections (Maroufpoor *et al.*, 2019). Consequently, future developments in nanotechnology are a source of great promise in terms of wider social and economic impact that focuses on addressing potential issues and enabling collaborative capacity building to ensure long-term sustainability in agriculture (Lang *et al.*, 2021).

6.3.1 Nanotechnology in Regulating Seed Germination and Plant Growth

In order to increase crop output and quality, nanotechnology is now a commonly utilized technique in a variety of agricultural production processes, including seed germination and plant growth (Shang *et al.*, 2019). According to recent studies (Chattha *et al.*, 2017; An *et al.*, 2022), NPs, including SiO_2, ZnO, TiO_2, and Au, have beneficial impacts on seed germination in agricultural crops, including soybean, barley, maize, tomato, wheat, rice, and pearl millet (Figure 6.2).

Antioxidant enzyme levels and rate of oxygen and water uptake are closely linked to seed germination (Demirer *et al.*, 2021). For instance, Au NPs promote the germination rate of pearl millet compared to untreated plants. Similarly, seed germination is higher in wheat seeds that are treated with ZnO NPs compared to control samples (Rai-Kalal and Jajoo, 2021). Exogenous TiO_2 NP treatment allows the seeds to better absorb water and oxygen, thereby reducing germination time. TiO_2 NPs serve as additives that enhance the germination of seeds (Almajidi *et al.*, 2023).

In another study(Maroufpoor *et al.*, 2019), TiO_2 NPs were found to enhance seed growth and significantly reduce the average germination time in wheatgrass. Furthermore, nonmetal NPs, such as multi-walled carbon nanotubes (MWCNTs), not only enhance seed water absorption but also promote seed germination in various crops (Malakar *et al.*, 2021).

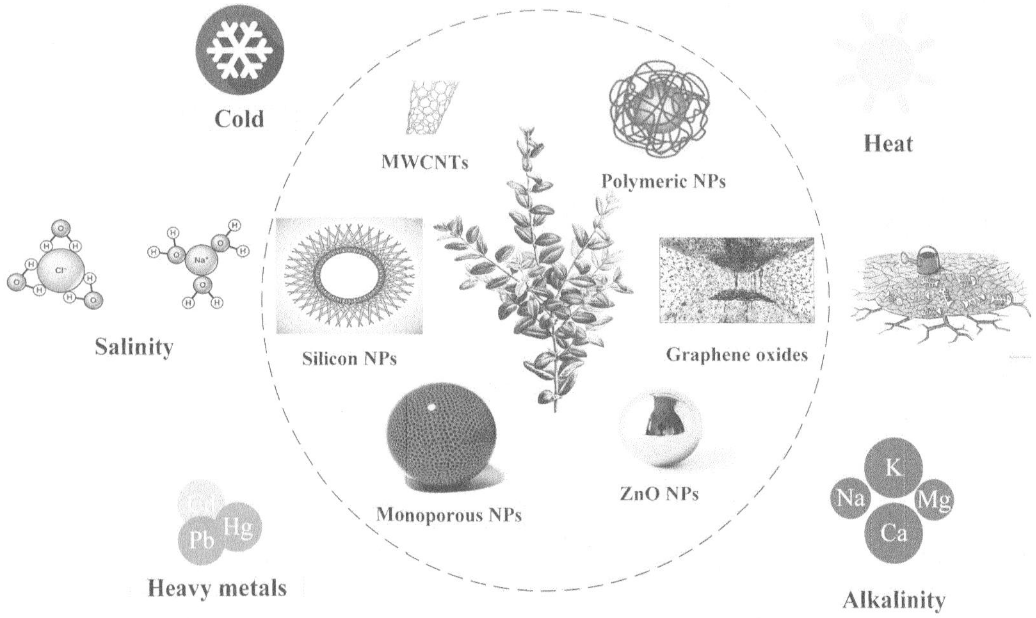

FIGURE 6.2 Overview of applications of NPs in stress tolerance. NPs, including MWCNTs, mesoporous NPs, silicon NPs, polymeric NPs, graphene oxides, and ZnO NPs, have been successfully applied to help crops adapt to different stresses.

6.3.2 Nanostructures for Soil Health and Genetics

Nanotechnology is one of the most significant technological advancements of the 21st century, capable of improving agricultural processes and enabling sustainable growth by improving conservation and management measures and reducing input loss. Delivery of agrochemicals, as well as organic molecules, such as genetic material or oligonucleotides, into the cells of the potential hosts in plants is an important aspect of efficient precision microbiome farming and continual efficiency of agricultural production (Aamer *et al.*, 2021; Lang *et al.*, 2021).

The recent developments in nanotechnology have proved to be promising as new herbicides, insecticides, and slow- or controlled-release fertilizers are becoming available, increasing their use in sustainable agricultural practices (Shang *et al.*, 2019). Because of their huge surface area, ease of attachment, and rapid mass transfer, NPs provide significant benefits for the administration of pesticides. As a result NPs enhance the efficacy of agrochemicals, improve their stability, protect them from deterioration, and reduce environmental impact (Rofeal *et al.*, 2022).

In addition to its utilization in agriculture, biotechnology and nanotechnology integration offers new tools for molecular transporters that can be used to modify genes and create entirely new species (Liu *et al.*, 2021; Aslam *et al.*, 2023). In nanobiotechnologies, for example, foreign DNA and compounds that aid in modifying target genes are carried by NPs, nanocapsules, and nanofibers. However, viral gene-delivery vectors confront several challenges, including a limited range of hosts, constraints on the size of genetic material, difficulties with cell membrane transport, and issues related to nuclear trafficking (Dobrescu *et al.*, 2023).

Silicon dioxide NPs have been created through genetic engineering and they transfer DNA sequences to the target species (Khaeso *et al.*, 2018). Additionally, novel insect-resistant crop types are developed using the NP-assisted delivery approach. Recently created siRNA delivery vehicles encapsulated in chitosan NPs provide novel prospects for controlling agricultural insect pests. This occurs through the effective RNA binding capacity of chitosan and the cell membrane penetration of chitosan (Mustafa *et al.*, 2023).

6.3.3 Nanomaterials for Soil Remediation

NMs can interact with organic and mineral fractions, natural colloids, and other components in the soil, resulting in their partitioning among different components of the soil system, including solid and aqueous phases (Pariona *et al.*, 2019; Khan *et al.*, 2023d). Mechanisms of mobility, stability, and toxicity of NMs in soil are shown in Figure 6.3. Upon contact with soil, physical, chemical, or biological changes may occur in the NPs (Kalemba and Synowiec, 2019).

Aggregate formation is the key process that occurs spontaneously according to soil engineering principles when NMs are introduced into soil. Therefore, aggregation reduces the surface area for NM interaction, which affects their reactivity. With increasing aggregate size, the mobility of NMs in a porous medium decreases, affecting their behavior and responsiveness. NMs' interaction with different particles in the environment is called heteroaggregation, while homoaggregation describes the interaction between NMs themselves (Dhiman *et al.*, 2022). Heteroaggregation is favored when concentration of external particles is more than NMs, particularly when the pH of the environment is in between the point of zero charge (PZC) of natural colloid and NMs, whereas heteroaggregation is most often dominated by homoaggregation (Vanti *et al.*, 2020). This is likely attributed to the oppositely surface charges between natural colloids and NMs, which reduces the electrostatic repulsion in this pH range (Ramezani *et al.*, 2019; Jianing *et al.*, 2022).

Soil colloids have been shown to have such transport capacity, and using electron microscopy and analyzing split-slice images of direct and semi-direct deposition, the study showed that aggregates of firmly bound nano-ZnO particles were clearly deposited to soil minerals (Pramanik *et al.*, 2020). Consequently, in natural environments, heteroaggregation is more probable to be in charge of NM destiny. Clay and iron minerals are considered significant sinks for NMs in soil particles and nutrients (Shinde *et al.*, 2022).

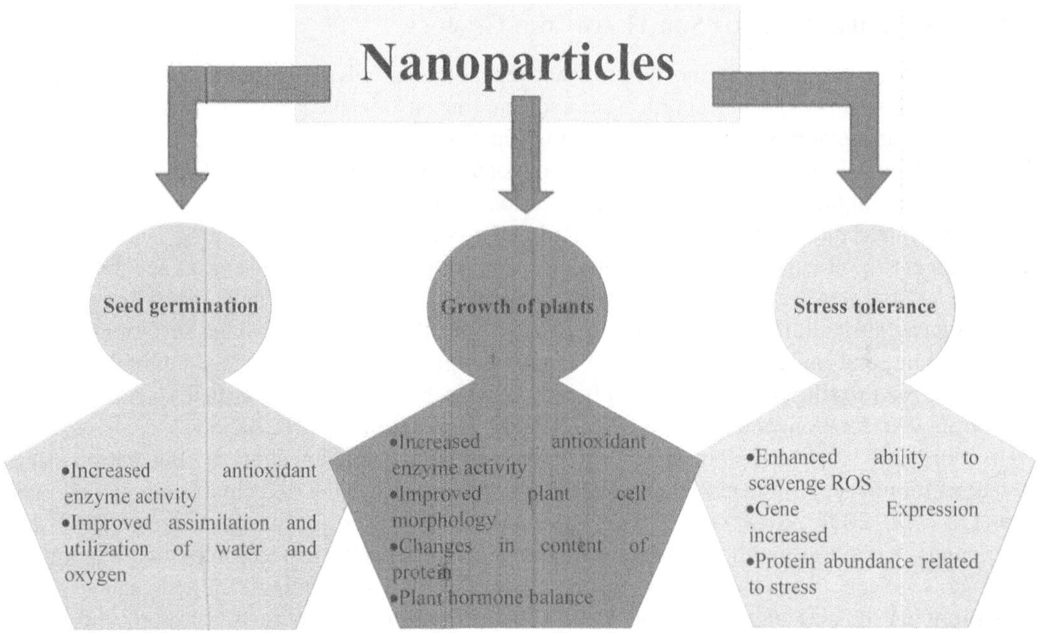

FIGURE 6.3 The mechanism of NP-induced enhancement in seed germination, plant growth, and stress tolerance. (Molecules | Free Full-Text | The Applications of Nanotechnology in Crop Production (mdpi.com).)

The surfaces of NMs may become coated by the adsorption of SOM, which would decrease their active surface area and perhaps lessen their effects. For instance, after humic substances (HS) were adsorbed at concentrations as low as 0.05 mg L^{-1}, the toxicity of nC60 was removed. Depending on its composition, SOM may have conflicting impacts on the stability and mobility of NMs. On the one hand, SOM can bridge flocculation and speed up NM aggregation. However, boosting their electrostatic stability might improve NM stability (Elmer and White, 2018) (Table 6.1).

6.3.4 Role of Nanotechnology in Crop Protection

Nowadays, pesticides, medicines, and fertilizers are administered to soil, plant, or animal surfaces by means of sprays, drenches, or injection systems (Rasheed *et al.*, 2021; Kohls *et al.*, 2022). The premise of these nanoscale devices is that they would be able to identify and treat an infection, nutrient deficiency, or other condition before visible signs could be detected at the macro-scale (Saliev, 2019). This proactive approach could significantly enhance early intervention and precision in agricultural and medical applications. Moreover, integrating nanoscale technologies can lead to more targeted and efficient delivery systems, reducing overall chemical use and minimizing environmental impact.

NMs are utilized in agriculture sector in various ways like the development of biosensors and enzyme encapsulation technologies. The foundation of nanotechnology lies in the precise manipulation and positioning of molecules to create innovative frameworks and devices. In agriculture, NPs are increasingly used to address various challenges, such as improving the effectiveness of treatments and targeting specific issues more accurately (Farjadian *et al.*, 2019). Size, shape, sedimentation, drug encapsulation efficiency, targeted pattern of release, dispersion in the plant, circulation, and cost are some of the main properties of NPs that should be kept in mind (Castillo *et al.*, 2019). This reduces the risk of anomalies and improves plant protection through targeted plant-pathogen interactions (Ramesh *et al.*, 2022).

TABLE 6.1
Effects of NPs on Crop Protection

Nanoparticle Types	Concentration	Active Compound	Pesticide Type	Target Pest/Disease	References
Chitosan	20 mg/L	*Aspergillus niger*	Insecticide	*Holotrichia* sp. (white grubs)	Philibert *et al.* (2017)
Zinc oxide (ZnO)	200 mg/L	–	Antiviral pesticide	Tobacco mosaic virus (TMV)	Liu *et al.* (2021)
Carbon nanotubes (CNTs)	500 mg/L	–	Fungicide	Various fungi such as *Fusarium oxysporum*, *Colletotrichum gloeosporioides*, and *Dematophora necatrix*	Wang *et al.* (2022)
Nanogel	12 mg/mL	Methyl eugenol	Insecticide	*Bactrocera dorsalis* (fruit flies)	Bhagat *et al.* (2013)
Chitosan	10 µg/L	Hexaconazole	Fungicide	*Ganoderma boninense* (basal stem rot)	Mustafa *et al.* (2023)
Copper (Cu)	0.5 mg/mL	–	Fungicide	Various fungi, including *Fusarium solani* and *Fusarium oxysporum*	Pariona *et al.* (2019)
Nanogel	15 mL/L	Savory essential oil	Herbicide	Weeds	Gogoi *et al.* (2025)
Chitosan	10 mg/L	Spinosad/permethrin	Insecticide	*Drosophila melanogaster* (fruit flies)	Xiao *et al.* (2024)
Aluminum oxide (Al_2O_3)	2 g/kg	–	Insecticide	*Sitophilus oryzae* L. (rice weevils)	Abo-Arab *et al.* (2014)
Solid lipid	0.3 kg/ha	Atrazine and simazine	Herbicide	*Raphanus raphanistrum* (wild radish)	De Albuquerque *et al.* (2020)
Silver (Ag)	15 mM	Leaf extracts	Bactericide	*Xanthomonas axonopodis* pv. punicae (bacterial blight)	Chinnathambi *et al.* (2023)

6.3.5 Mesoporous Silica Nanoparticles

NPs can serve as "magic bullets" in agriculture to target herbicides, chemicals, or genes to precise parts of the plant (Clemente and Huntsman, 2019). They carry genes, which are introduced into the plant tissues after chemical treatment (Munir *et al.*, 2023). This method improves treatment time by targeting active ingredients directly at their destination. Mesoporous silica nanoparticles (MSNs) also help to mitigate the overall amount of chemicals released into the environment, making agriculture more sustainable.

6.4 NANOSENSORS

Due to excellent sensitivity and fast response time, nanoscale materials are widely utilized in electrochemical biosensors. In agri-food sectors, bioanalytical nanosensors are used to discriminate and identify trace levels of pollutants, including bacteria, viruses, toxins, and biohazard products (Ali *et al.*, 2020).

Complex systems composed of an oil phase, a surface active agent, and water that are optically isotropic and kinetically stable colloidal solutions in which the droplet size is between 20 and 200 nm are commonly termed "nanoemulsions". The ability of these nanoemulsions to trap functional ingredients inside their droplets also minimizes chemical degradation, enhancing the stability of the embedded compounds (Dhakate *et al.*, 2022).

6.5 SILVER NANOPARTICLES

NPs are becoming increasingly prevalent in consumer products, including powders and dispersions that are supposed to be antimicrobials. Nanosilver finds extensive application in agriculture. Numerous investigations have been conducted on plant responses to chemically reduced nanosilver. A silver nanomolecular solution method decreases the prevalence of root infections. A colloidal nanosilver solution can significantly enhance the development and well-being of a variety of plants (Kanipandian and Thirumurugan, 2014; Torabian *et al.*, 2017). These applications highlight the potential of nanosilver to enhance agricultural productivity and plant resilience. Ongoing research continues to explore the optimal formulations and applications of nanosilver to maximize its benefits while minimizing potential risks.

6.6 NANOTECHNOLOGY AND CROP RESILIENCE

As the global population grows and natural resources become scarcer owing to changes in the climate, food safety is becoming an increasingly pressing concern. Progressive climate change pertains to alterations in the fundamental parameters of the climate over extended periods, encompassing temperature fluctuations, water scarcity, low temperatures, salt, alkalinity, and heavy metal contamination (Chaud *et al.*, 2021).

Research on NM engineering shows that nanofertilizers increase agricultural yields in challenging conditions. Nano-SiO_2 has been shown to promote seed germination, increase plant biomass, and chlorophyll, and promote proline accumulation in NaCl-stressed tomato and squash plants (Khan *et al.*, 2017). Additionally, the result of salinity stress tolerance that applied nanoparticle deposit of iron sulfate ($FeSO_4$) through foliar spray to sunflower cultivars confirmed a positive response to salinity stress tolerance (Ali *et al.*, 2021). Foliar spray of nano-based $FeSO_4$ enhances leaf area, plant biomass, gas exchange parameters, and fluorescence parameters (Yusefi-Tanha *et al.*, 2020).

AgNPs synthesized with the extract of cotton plant (*Gossypium hirsutum*) show potent antibacterial activity (Shiraz *et al.*, 2024). Biotic factors (e.g. pests and diseases) greatly influence crop productivity. Farmers have long relied on preventative applications of pesticides to avoid crop losses, compromising human and environmental health by doing so. In recent studies (Gupta *et al.*, 2018), agricultural NMs have been tested to reduce the incidence of diseases and pests, minimize crop failures, and reduce ecological risks.

Plant growth and stress tolerance induced by NMs are still mostly unexplored. The enhanced activity of the enzymatic system may, at least partially, account for the significant impacts of NMs on crop development under adverse circumstances (Van Aken, 2015). The use of nano-SiO_2 or nano-ZnO increases the uptake of nutrients and water, and the development of amino acids and enzymes that protect against damage such as catalase, peroxidase, nitrate reductase, glutathione reductase, and superoxide dismutase (Mohamed and Hussien, 2016).

6.6.1 Negative Influence of Nanomaterials on Plants

While NMs can have several positive impacts, their application may also lead to phytotoxic effects, as shown in Figure 6.3. Moreover, the NM size, composition, dosage, exposure time, and environmental factors all affect the toxicity and beneficial effects. For instance, it was shown that dissolved Cu^{2+} and smaller CuO NPs with a diameter of 25 nm are more phytotoxic to soybeans than bigger CuO NPs (diameters of 50 and 250 nm) (Song *et al.*, 2020).

The physiological attributes of plants affected by NP toxicity include germination and development. The NPs can shorten plant elongation, reduce seed germination, and sometimes even kill plants (Ghasempour *et al.*, 2019). Negative effects could include altered sub-cellular metabolism, altered photosynthetic rate, chromosomal abnormalities, oxidative damage to biological membranes, disruption of plant water transport and status, plant growth hormone reduction, and modification to gene transcription profiles (Berni *et al.*, 2019). Plant development is impacted by changes in plant gene expression and related biological processes because plant cells interact with NPs.

It has been reported that the genomic DNA was harmed by the use of TiO_2 NPs (Rajput *et al.*, 2021a). The NP affected the gene-up- and gene-down-regulation pathway and demonstrated the high toxicity of the hairy structure in higher plants, as shown by transcriptome analysis (Ashraf *et al.*, 2021). Moreover, LR1 and RTCS expression was increased and RTH1 and RTH3 expression was decreased in maize exposed to single-walled carbon nanotubes.

With respect to gene expression, NPs increased the expression of several genes, including stress response and water channel genes. However, high-concentration quota applications can be damaging. The DNA structure of the soybeans was also impaired after an administration of doses of 2000 and 4000 mg L^{-1} of CeO_2 NPs (Meena *et al.*, 2017). Maize suffers damage at 10 mM and tobacco DNA at 2 mM when treated with TiO_2 NPs (Sajid and Płotka-Wasylka, 2020). Oxidative damage occurs when plants are exposed to metallic NPs, inducing the generation of reactive oxygen species (ROS) and the activation of their antioxidant response system (Ao *et al.*, 2022).

Antioxidant defense also involves enzymatic antioxidants such as ascorbate peroxidase, catalase, superoxide dismutase, guaiacol peroxidase, and glutathione reductase, and non-enzymatic antioxidants such as glutathione, ascorbate, thiols, and phenolics (Soni *et al.*, 2022). Superoxide dismutase also plays a crucial role in converting superoxide anion to hydrogen peroxide, while catalase and guaiacol peroxidase are involved in the reduction of ROS and the subsequent reduction of peroxy radicals (Wani *et al.*, 2023). ROS generation can associate with photocytotoxicity of TiO_2 NPs; nevertheless, these ROS radicals can also be signal molecules to stimulate the antioxidant defense system and detoxify the free radicals by the plant (Sekhon, 2014).

6.7 FUTURE OF NANOTECHNOLOGY IN PLANT IMPROVEMENT

Sustainable agriculture, food security, and nutrition are the three crucial objectives for the century. Nanotechnology can play an important role in enhancing the availability of nutrients for plants and minimizing the losses in soil (Ventura-Aguilar *et al.*, 2023). Developmental processes also combine with ecological dynamics in the form of plant growth and consumption, food digestion, etc. The application of nanotechnology in soil modification can dramatically influence crop production, as well as their nutrient status, and even potentially promote the nutritional quality of the harvested produce and organisms (Ansari, 2023). Nanochemicals exhibit better properties and unique applications owing to larger surface area and quantum dimensions (Singh *et al.*, 2019).

Moreover, plants may take up the nutrients they need and leave the rest in the soil. It is unclear whether the toxic compounds are absorbed or become non-toxic; either way the toxic agent would remain in the habitat (Petersen *et al.*, 2015). In order to eliminate all the misconceptions and implement nanotechnology in green chemistry and sustainable farming, there is a shift in NM synthesis from toxic chemicals to biological agents (Prajitha *et al.*, 2019).

Some of the potential benefits this unique transformation are offered to agricultural ecosystems, which include restoring depleted soil nutrients, administering green nanochemicals for combatting weeds and pests, providing macro- and micronutrients to ensure healthy plant growth, improving nutrient retention, and fortifying food crops (Harish *et al.*, 2023). In addition to crop enhancement, another agricultural application of this innovative technology is in crop and food packaging, where it is used to create nanobarcodes with gold and silver stripes (Ansari, 2023).

Graphene oxide has been used in biosensing to improve aflatoxins in food item detection. Hemoglobin is immobilized by gold nanorods coated in silica, serving as a biosensor. The use of NMs in biosensing is still relatively new, and in agriculture, the activities are more advantageous in the food or before and after crop sowing (Cerón García, 2021). As a result of additional efforts to expand the use of nanocomposites, gelatin–gold–single-walled nanotubes with a biocompatible surface have become more hydrophilic. These nanotubes could be used in detecting, but they might additionally enhance nutrient uptake by plant cells and promote growth (Kóczán and Pásztory, 2024).

6.8 BIOSAFETY OF NANOMATERIALS IN SUSTAINABLE AGRICULTURE

NM safety evaluation is a challenging task as there are several variables that affect the materials, including the chemicals or salts used, the processing technique, biological substrates, size and structure, and interactions in the application media (Buonvino *et al.*, 2023). During toxicity studies, aggregation can cause initially stable NPs to become unstable, which may alter the required dosages and affect the accuracy of safety assessments (Obrador *et al.*, 2022).

The safety of NMs might vary according to the synthesis technique used. There have been reports of hazardous properties for NPs produced chemically, physically, and biologically. Because the chemical agents utilized in their production release slowly, chemically produced NPs are more hazardous. More biocompatible and comparatively safer for the ecosystem and living things are the biologically designed NPs (Ahmed *et al.*, 2023).

NMs are characterized by variable properties, as various techniques and reaction conditions are used in their synthesis, with sizes and morphologies that range from nanoscale to microscale (Dong *et al.*, 2023). Consequently, these NMs have proved to cause a multitude of diverse effects. Toxicity and behavior of such materials are influenced by the production materials and their particle sizes, respectively (Kaphle *et al.*, 2018). At the nanoscale, materials seem harmless, but they can lead to toxicity when interacting with other media in unwanted catalytic integration reactions. One recent development is the shift from chemical to biological production of NMs, which enables this new technology to be safe and sustainable (Munir *et al.*, 2023).

When NPs are utilized in material fabrication, individual bionanomaterials react differently since they are made up of potentially hazardous proteins (Nikzamir *et al.*, 2021). In order to measure the degree of toxicity associated with these NMs, it is crucial to keep researching how they behave in living things as well as the compounds that result from their breakdown. Environmental factors significantly influence the behavior and effects of NMs. For example, fullerene (C_{60}) NPs have been found to be hazardous in aquatic environments but non-toxic to soil microorganisms (Shukla *et al.*, 2022).

Zinc oxide NPs are a good source of zinc micronutrients and can serve as a fertilizer feedstock without being too toxic. On the other hand, using zinc oxide NPs in acidic soil has more lethal effects than in calcareous soils. For example, they worked with bean and tomato plants and found that if the level of zinc in the calcareous soil increased, plants' photosynthetic pigmentation increased (Hamzah Saleem *et al.*, 2022). Soil pH significantly influences the toxicity of zinc oxide NPs (Usman *et al.*, 2023).

It is evident that NMs increase plant nutrient uptake which leads to crop productivity (Harte *et al.*, 2023). We are unable to totally eliminate the risks associated with nanotechnologies, despite the fact that there have been successful breakthroughs in this area of application. As a means of achieving unmatched progress, contemporary research has embraced this technology and, as such, ought to devote more resources to assessing the materials' safety in order to enhance their characterization, processing, and general application standards (Francis, 2024).

6.9 CONCLUSIONS

Nanotechnology has been utilized in agriculture to enhance agricultural systems and increase the quantity and quality of crops. Due to the upsurge in world food consumption brought about by the rapid world population growth, the use of engineering NMs and their applications in sustainable agriculture has become a new paradigm shift in world agriculture. However, sustainable agriculture is also concerned about the best approach to maintain the artificial environment, and simultaneously, NMs can provide a perfect way for the better administration of input and maintain the environment in an advanced phase of plant growth. NPs hold the promise of a new green revolution with minimal hazards in agriculture and they have the potential to bring it. Even as researchers worry about the potential dangers of some NPs, nanotechnology continues to be one of the most promising directions for agricultural advancement. To be more commercially viable and feasible in agriculture, the plant-mediated biological methods of synthesis, where raw materials could be left-over vegetables, plant extracts, flowers, plant barks, roots, fruit peels, and leather cuttings, should be investigated and expanded. However, for the practical use of NMs, we must move from natural science to real-world knowledge. Seamless monitoring of these NMs during the nursery stage of crop production for their effective bearing of advantages could also directly help in tackling some of the challenges involved in this process.

LIST OF ABBREVIATIONS

CAS CRISPR-associated protein
CRISPR clustered regularly interspaced short palindromic repeats
GM genetic modification
HS humic substances
MWCNTs multi-walled carbon nanotubes
NMs nanomaterials
NPs nanoparticles
PZC point of zero charge
ROS reactive oxygen species
SOM soil organic matter

REFERENCES

Cerón García, S.P., 2021. Solution-processed electrodes for printed electronics.

Aamer, M., Bilal Chattha, M., Mahmood, A., Naqve, M., Hassan, M.U., Shaaban, M., Rasul, F., Batool, M., Rasheed, A., & Tang, H., 2021. Rice residue-based biochar mitigates N_2O emission from acid red soil. *Agronomy*, 11, 2462.

Abo-Arab, R., Amal, M.H., & Hashem, A., 2014. Comparative bioactivity of Aluminum Oxide (Al_2O_3), Titanium Dioxide (TiO_2) nanoparticles and malathion on *Sitophilus oryzae* L. and *Sitophilus zeamais* (Motsch.).

Acharya, A., & Pal, P.K., 2020. Agriculture nanotechnology: Translating research outcome to field applications by influencing environmental sustainability. *NanoImpact*, 19, 100232.

Afzal, S., Sirohi, P., & Singh, N.K., 2020. A review of CRISPR associated genome engineering: Application, advances and future prospects of genome targeting tool for crop improvement. *Biotechnology Letters*, 42, 1611–1632.

Ahmed, A., He, P., He, P., Wu, Y., He, Y., & Munir, S., 2023. Environmental effect of agriculture-related manufactured nano-objects on soil microbial communities. *Environment International*, 173, 107819.

Ali, M., Hayat, S., Ahmad, H., Ghani, M.I., Amin, B., Atif, M.J., & Cheng, Z., 2019. Priming of *Solanum melongena* L. seeds enhances germination, alters antioxidant enzymes, modulates ROS, and improves early seedling growth: Indicating aqueous garlic extract as seed-priming bio-stimulant for eggplant production. *Applied Sciences*, 9, 2203.

Ali, Q., Ali, S., El-Esawi, M.A., Rizwan, M., Azeem, M., Hussain, A.I., Perveen, R., El-Sheikh, M.A., Alyemeni, M.N., & Wijaya, L., 2020. Foliar spray of Fe-Asp confers better drought tolerance in sunflower as compared with $FeSO_4$: Yield traits, osmotic adjustment, and antioxidative defense mechanisms. *Biomolecules*, 10, 1217.

Ali, S., Mehmood, A., & Khan, N., 2021. Uptake, translocation, and consequences of nanomaterials on plant growth and stress adaptation. *Journal of Nanomaterials*, 2021, 6677616.

Almajidi, Y.Q., Gupta, J., Sheri, F.S., Zabibah, R.S., Faisal, A., Ruzibayev, A., Adil, M., Saadh, M.J., Jawad, M.J., & Alsaikhan, F., 2023. Advances in chitosan-based hydrogels for pharmaceutical and biomedical applications: A comprehensive review. *International Journal of Biological Macromolecules*, 127278.

Ameen, M., Mahmood, A., Sahkoor, A., Zia, M.A., & Ullah, M.S., 2024. The role of endophytes to combat abiotic stress in plants. *Plant Stress*, 100435.

An, C., Sun, C., Li, N., Huang, B., Jiang, J., Shen, Y., Wang, C., Zhao, X., Cui, B., & Wang, C., 2022. Nanomaterials and nanotechnology for the delivery of agrochemicals: Strategies towards sustainable agriculture. *Journal of Nanobiotechnology*, 20, 11.

Ansari, M.A., 2023. Nanotechnology in food and plant science: Challenges and future prospects. *Plants*, 12, 2565.

Ao, M., Chen, X., Deng, T., Sun, S., Tang, Y., Morel, J.L., Qiu, R., & Wang, S., 2022. Chromium biogeochemical behaviour in soil-plant systems and remediation strategies: A critical review. *Journal of Hazardous Materials*, 424, 127233.

Arora, N.K., Fatima, T., Mishra, I., Verma, M., Mishra, J., & Mishra, V., 2018. Environmental sustainability: Challenges and viable solutions. *Environmental Sustainability*, 1, 309–340.

Ashraf, S.A., Siddiqui, A.J., Abd Elmoneim, O.E., Khan, M.I., Patel, M., Alreshidi, M., Moin, A., Singh, R., Snoussi, M., & Adnan, M., 2021. Innovations in nanoscience for the sustainable development of food and agriculture with implications on health and environment. *Science of the Total Environment*, 768, 144990.

Aslam, A., Mahmood, A., Ur-Rehman, H., Li, C., Liang, X., Shao, J., Negm, S., Moustafa, M., Aamer, M., & Hassan, M.U., 2023. Plant adaptation to flooding stress under changing climate conditions: Ongoing breakthroughs and future challenges. *Plants*, 12, 3824.

Bayda, S., Adeel, M., Tuccinardi, T., Cordani, M., & Rizzolio, F., 2019. The history of nanoscience and nanotechnology: From chemical–physical applications to nanomedicine. *Molecules*, 25, 112.

Berni, R., Luyckx, M., Xu, X., Legay, S., Sergeant, K., Hausman, J.-F., Lutts, S., Cai, G., & Guerriero, G., 2019. Reactive oxygen species and heavy metal stress in plants: Impact on the cell wall and secondary metabolism. *Environmental and Experimental Botany*, 161, 98–106.

Bhagat, D., Samanta, S.K., & Bhattacharya, S., 2013. Efficient management of fruit pests by pheromone nanogels. *Scientific Reports*, 3, 1–8.

Buonvino, S., Ciocci, M., Nanni, F., Cacciotti, I., & Melino, S., 2023. New vegetable-waste biomaterials by *Lupin albus* L. as cellular scaffolds for applications in biomedicine and food. *Biomaterials*, 293, 121984.

Castillo, R.R., Lozano, D., González, B., Manzano, M., Izquierdo-Barba, I., & Vallet-Regí, M., 2019. Advances in mesoporous silica nanoparticles for targeted stimuli-responsive drug delivery: An update. *Expert Opinion on Drug Delivery*, 16, 415–439.

Chattha, M.B., Khan, I., Mahmood, A., Chattha, M.U., Anjum, S.A., Ashraf, U., Zain, M., & Bilal, U., 2017. Seed priming with thiourea improves the performance of late sown wheat. *Journal of Agricultural Research*, 55, 29–39.

Chaud, M., Souto, E.B., Zielinska, A., Severino, P., Batain, F., Oliveira-Junior, J., & Alves, T., 2021. Nanopesticides in agriculture: Benefits and challenge in agricultural productivity, toxicological risks to human health and environment. *Toxics*, 9, 131.

Chinnathambi, A., Alharbi, S.A., Joshi, D., Saranya, V., Jhanani, G., On-Uma, R., Jutamas, K., & Anupong, W., 2023. Synthesis of AgNPs from leaf extract of *Naringi crenulata* and evaluation of its antibacterial activity against multidrug resistant bacteria. *Environmental Research*, 216, 114455.

Clemente, J.S., & Huntsman, P., 2019. Potential climate change effects on the geochemical stability of waste and mobility of elements in receiving environments for Canadian metal mines south of 60° N. *Environmental Reviews*, 27, 478–518.

De Albuquerque, F.P., De Oliveira, J.L., Moschini-Carlos, V., & Fraceto, L.F., 2020. An overview of the potential impacts of atrazine in aquatic environments: Perspectives for tailored solutions based on nanotechnology. *Science of the Total Environment*, 700, 134868.

Demirer, G.S., Silva, T.N., Jackson, C.T., Thomas, J.B., W. Ehrhardt, D., Rhee, S.Y., Mortimer, J.C., & Landry, M.P., 2021. Nanotechnology to advance CRISPR–Cas genetic engineering of plants. *Nature Nanotechnology*, 16, 243–250.

Dhakate, P., Kandhol, N., Raturi, G., Ray, P., Bhardwaj, A., Srivastava, A., Kaushal, L., Singh, A., Pandey, S., & Chauhan, D.K., 2022. Silicon nanoforms in crop improvement and stress management. *Chemosphere*, 305, 135165.

Dhiman, A., Sharma, A.K., & Agrawal, G., 2022. Polymer based engineered materials for sustainable agriculture. *ACS Agricultural Science & Technology*, 2, 693–711.

Dobrescu, C.M., Dorobăț, L.M., & Angela, M., 2023. 10 Health, Safety and Environmental Management and Risk Mitigation of Nanomaterials. *Nanoformulations for Sustainable Agriculture and Environmental Risk Mitigation*, 177.

Dong, B.-R., Jiang, R., Chen, J.-F., Xiao, Y., Lv, Z.-Y., & Chen, W.-S., 2023. Strategic nanoparticle-mediated plant disease resistance. *Critical Reviews in Biotechnology*, 43, 22–37.

Elmer, W., & White, J.C., 2018. The future of nanotechnology in plant pathology. *Annual Review of Phytopathology*, 56, 111–133.

Farjadian, F., Roointan, A., Mohammadi-Samani, S., & Hosseini, M., 2019. Mesoporous silica nanoparticles: Synthesis, pharmaceutical applications, biodistribution, and biosafety assessment. *Chemical Engineering Journal*, 359, 684–705.

Francis, L.F., 2024. *Materials Processing: A Unified Approach to Processing of Metals, Ceramics, and Polymers*. Elsevier.

Fróna, D., Szenderák, J., & Harangi-Rákos, M., 2019. The challenge of feeding the world. *Sustainability*, 11, 5816.

Ghasempour, M., Iranbakhsh, A., Ebadi, M., & Oraghi Ardebili, Z., 2019. Multi-walled carbon nanotubes improved growth, anatomy, physiology, secondary metabolism, and callus performance in *Catharanthus roseus*: An in vitro study. *3 Biotech*, 9, 404.

Gogoi, R., Chanderwal, A., Deka, P., Das, J., Barooah, M., & Bora, S.S., 2025. Role of Green Nanomaterials for Sustainable Agriculture. *Sustainable Green Nanomaterials*. Apple Academic Press, 121–133.

Goyal, V., Rani, D., Ritika, Mehrotra, S., Deng, C., & Wang, Y., 2023. Unlocking the potential of nano-enabled precision agriculture for efficient and sustainable farming. *Plants*, 12, 3744.

Gupta, S.D., Agarwal, A., & Pradhan, S., 2018. Phytostimulatory effect of silver nanoparticles (AgNPs) on rice seedling growth: An insight from antioxidative enzyme activities and gene expression patterns. *Ecotoxicology and Environmental Safety*, 161, 624–633.

Hamzah Saleem, M., Usman, K., Rizwan, M., Al Jabri, H., & Alsafran, M., 2022. Functions and strategies for enhancing zinc availability in plants for sustainable agriculture. *Frontiers in Plant Science*, 13, 1033092.

Harish, V., Ansari, M., Tewari, D., Yadav, A.B., Sharma, N., Bawarig, S., García-Betancourt, M.-L., Karatutlu, A., Bechelany, M., & Barhoum, A., 2023. Cutting-edge advances in tailoring size, shape, and functionality of nanoparticles and nanostructures: A review. *Journal of the Taiwan Institute of Chemical Engineers*, 149, 105010.

Harte, J., Holdren, C., Schneider, R., & Shirley, C., 2023. *Toxics A to Z: A Guide to Everyday Pollution Hazards*. University of California Press.

Hassanisaadi, M., Barani, M., Rahdar, A., Heidary, M., Thysiadou, A., & Kyzas, G.Z., 2022. Role of agrochemical-based nanomaterials in plants: Biotic and abiotic stress with germination improvement of seeds. *Plant Growth Regulation*, 97, 375–418.

Jianing, G., Yuhong, G., Yijun, G., Rasheed, A., Qian, Z., Zhiming, X., Mahmood, A., Shuheng, Z., Zhuo, Z., & Zhuo, Z., 2022. Improvement of heat stress tolerance in soybean (*Glycine max* L), by using conventional and molecular tools. *Frontiers in Plant Science*, 13, 993189.

Kabir, E., Kumar, V., Kim, K.-H., Yip, A.C., & Sohn, J., 2018. Environmental impacts of nanomaterials. *Journal of Environmental Management*, 225, 261–271.

Kalemba, D., & Synowiec, A., 2019. Agrobiological interactions of essential oils of two menthol mints: *Mentha piperita* and *Mentha arvensis*. *Molecules*, 25, 59.

Kanipandian, N., & Thirumurugan, R., 2014. A feasible approach to phyto-mediated synthesis of silver nanoparticles using industrial crop *Gossypium hirsutum* (cotton) extract as stabilizing agent and assessment of its in vitro biomedical potential. *Industrial Crops and Products*, 55, 1–10.

Kaphle, A., Navya, P., Umapathi, A., & Daima, H.K., 2018. Nanomaterials for agriculture, food and environment: Applications, toxicity and regulation. *Environmental Chemistry Letters*, 16, 43–58.

Khaeso, K., Andongma, A.A., Akami, M., Souliyanonh, B., Zhu, J., Krutmuang, P., & Niu, C.-Y., 2018. Assessing the effects of gut bacteria manipulation on the development of the oriental fruit fly, *Bactrocera dorsalis* (Diptera; Tephritidae). *Symbiosis*, 74, 97–105.

Khan, A., Khan, A.L., Muneer, S., Kim, Y.-H., Al-Rawahi, A., & Al-Harrasi, A., 2019a. Silicon and salinity: Crosstalk in crop-mediated stress tolerance mechanisms. *Frontiers in Plant Science*, 10, 1429.

Khan, B.A., Nadeem, M.A., Iqbal, M., Yaqoob, N., Javaid, M.M., Maqbool, R., Elnaggar, N., & Oraby, H., 2023a. Chitosan nanoparticles loaded with mesosulfuron methyl and mesosulfuron methyl+ florasulam+ MCPA isooctyl to manage weeds of wheat (*Triticum aestivum* L.). *Green Processing and Synthesis*, 12, 20228152.

Khan, B.A., Nadeem, M.A., Javaid, M.M., Maqbool, R., Ikram, M., & Oraby, H., 2022a. Chemical synthesis, characterization, and dose optimization of chitosan-based nanoparticles of clodinofop propargyl and fenoxaprop-p-ethyl for management of *Phalaris minor* (little seed canary grass): First report. *Green Processing and Synthesis*, 11, 1118–1127.

Khan, B.A., Nadeem, M.A., Najeeb Alawadi, H.F., Javaid, M.M., Mahmood, A., Qamar, R., Iqbal, M., Mumtaz, A., Maqbool, R., & Oraby, H., 2023b. Erratum to "Synthesis, characterization, and evaluation of nanoparticles of clodinofop propargyl and fenoxaprop-P-ethyl on weed control, growth, and yield of wheat (*Triticum aestivum* L.). *Green Processing and Synthesis*, 12, 20230301.

Khan, B.A., Nadeem, M.A., Najeed Alawadi, H.F., Javaid, M.M., Mahmood, A., Qamar, R., Iqbal, M., Mumtaz, A., Maqbool, R., & Oraby, H., 2023c. Synthesis, characterization, and evaluation of nanoparticles of clodinofop propargyl and fenoxaprop-P-ethyl on weed control, growth, and yield of wheat (*Triticum aestivum* L.). *Green Processing and Synthesis*, 12, 20230105.

Khan, I., Ali, S.M., Chattha, M.U., Barbanti, L., Calone, R., Mahmood, A., Albishi, T.S., Hassan, M.U., & Qari, S.H., 2023d. Neem and castor oil–coated urea mitigates salinity effects in wheat by improving physiological responses and plant homeostasis. *Journal of Soil Science and Plant Nutrition*, 23, 3915–3931.

Khan, I., Saeed, K., & Khan, I., 2019b. Nanoparticles: Properties, applications and toxicities. *Arabian Journal of Chemistry*, 12, 908–931.

Khan, M., Khan, M.S.A., Borah, K.K., Goswami, Y., Hakeem, K.R., & Chakrabartty, I., 2021. The potential exposure and hazards of metal-based nanoparticles on plants and environment, with special emphasis on ZnO NPs, TiO_2 NPs, and AgNPs: A review. *Environmental Advances*, 6, 100128.

Khan, M.N., Mobin, M., Abbas, Z.K., Almutairi, K.A., & Siddiqui, Z.H., 2017. Role of nanomaterials in plants under challenging environments. *Plant Physiology and Biochemistry*, 110, 194–209.

Khan, Y., Sadia, H., Ali Shah, S.Z., Khan, M.N., Shah, A.A., Ullah, N., Ullah, M.F., Bibi, H., Bafakeeh, O.T., & Khedher, N.B., 2022b. Classification, synthetic, and characterization approaches to nanoparticles, and their applications in various fields of nanotechnology: A review. *Catalysts*, 12, 1386.

Kóczán, Z., & Pásztory, Z., 2024. Overview of natural fiber-based packaging materials. *Journal of Natural Fibers*, 21, 2301364.

Kohls, A., Maurer Ditty, M., Dehghandehnavi, F., & Zheng, S.-Y., 2022. Vertically aligned carbon nanotubes as a unique material for biomedical applications. *ACS Applied Materials & Interfaces*, 14, 6287–6306.

Lang, C., Mission, E.G., Fuaad, A.a.-H.A., & Shaalan, M., 2021. Nanoparticle tools to improve and advance precision practices in the Agrifoods Sector towards sustainability—A review. *Journal of Cleaner Production*, 293, 126063.

Liu, C., Zhou, H., & Zhou, J., 2021. The applications of nanotechnology in crop production. *Molecules*, 26, 7070.

Malakar, A., Kanel, S.R., Ray, C., Snow, D.D., & Nadagouda, M.N., 2021. Nanomaterials in the environment, human exposure pathway, and health effects: A review. *Science of the Total Environment*, 759, 143470.

Maroufpoor, N., Mousavi, M., Hatami, M., Rasoulnia, A., & Lajayer, B.A., 2019. Mechanisms involved in stimulatory and toxicity effects of nanomaterials on seed germination and early seedling growth. *Advances in Phytonanotechnology*, 1, 153–181.

Meena, V.S., Meena, S.K., Verma, J.P., Kumar, A., Aeron, A., Mishra, P.K., Bisht, J.K., Pattanayak, A., Naveed, M., & Dotaniya, M., 2017. Plant beneficial rhizospheric microorganism (PBRM) strategies to improve nutrients use efficiency: A review. *Ecological Engineering*, 107, 8–32.

Mohamed, H.R., & Hussien, N.A., 2016. Genotoxicity studies of titanium dioxide nanoparticles (TiO_2NPs) in the brain of mice. *Scientifica*, 2016, 6710840.

Mukhtiar, A., Lihong, W., Mahmood, A., Ameen, M., Zia, M.A., Arshad, T., Naqve, M., Wahab, H.A., Rasheed, A., & Asghar, S., 2023. The role of endophytes and rhizobacteria to combat drought stress in wheat. *Notulae Botanicae Horti Agrobotanici Cluj-Napoca*, 51, 13453–13453.

Munir, N., Gulzar, W., Abideen, Z., Hancock, J.T., El-Keblawy, A., & Radicetti, E., 2023. Nanotechnology improves disease resistance in plants for food security: Applications and challenges. *Biocatalysis and Agricultural Biotechnology*, 51, 102781.

Mustafa, I.F., Hussein, M.Z., Idris, A.S., Hilmi, N.H.Z., & Fakurazi, S., 2023. The effect of surfactant on fungitoxicity of dazomet-micelle nanodelivery system against *Ganoderma boninense*. *Chemical Papers*, 77, 4007–4018.

Nikzamir, M., Akbarzadeh, A., & Panahi, Y., 2021. An overview on nanoparticles used in biomedicine and their cytotoxicity. *Journal of Drug Delivery Science and Technology*, 61, 102316.

Obrador, A., González, D., Almendros, P., García-Gómez, C., & Fernández, M.D., 2022. Assessment of phytotoxicity and behavior of 1-year-aged Zn in soil from ZnO nanoparticles, bulk ZnO, and Zn sulfate in different soil-plant cropping systems: From biofortification to toxicity. *Journal of Soil Science and Plant Nutrition*, 22, 150–164.

Pandey, G., 2018. Challenges and future prospects of agri-nanotechnology for sustainable agriculture in India. *Environmental Technology & Innovation*, 11, 299–307.

Pariona, N., Mtz-Enriquez, A.I., Sánchez-Rangel, D., Carrión, G., Paraguay-Delgado, F., & Rosas-Saito, G., 2019. Green-synthesized copper nanoparticles as a potential antifungal against plant pathogens. *RSC Advances*, 9, 18835–18843.

Petersen, E.J., Diamond, S.A., Kennedy, A.J., Goss, G.G., Ho, K., Lead, J., Hanna, S.K., Hartmann, N.B., Hund-Rinke, K., & Mader, B., 2015. Adapting OECD aquatic toxicity tests for use with manufactured nanomaterials: Key issues and consensus recommendations. *Environmental Science & Technology*, 49, 9532–9547.

Philibert, T., Lee, B.H., & Fabien, N., 2017. Current status and new perspectives on chitin and chitosan as functional biopolymers. *Applied Biochemistry and Biotechnology*, 181, 1314–1337.

Prajitha, N., Athira, S., & Mohanan, P., 2019. Bio-interactions and risks of engineered nanoparticles. *Environmental Research*, 172, 98–108.

Pramanik, P.K.D., Solanki, A., Debnath, A., Nayyar, A., El-Sappagh, S., & Kwak, K.-S., 2020. Advancing modern healthcare with nanotechnology, nanobiosensors, and internet of nano things: Taxonomies, applications, architecture, and challenges. *IEEE Access*, 8, 65230–65266.

Rai-Kalal, P., & Jajoo, A., 2021. Priming with zinc oxide nanoparticles improve germination and photosynthetic performance in wheat. *Plant Physiology and Biochemistry*, 160, 341–351.

Rajput, V.D., Harish, Singh, R.K., Verma, K.K., Sharma, L., Quiroz-Figueroa, F.R., Meena, M., Gour, V.S., Minkina, T., & Sushkova, S., 2021a. Recent developments in enzymatic antioxidant defence mechanism in plants with special reference to abiotic stress. *Biology*, 10, 267.

Rajput, V.D., Singh, A., Singh, V.K., Minkina, T.M., & Sushkova, S., 2021b. Impact of Nanoparticles on Soil Resource. *In Nanomaterials for Soil Remediation*. Elsevier, 65–85.

Ramesh, M., Janani, R., Deepa, C., & Rajeshkumar, L., 2022. Nanotechnology-enabled biosensors: A review of fundamentals, design principles, materials, and applications. *Biosensors*, 13, 40.

Ramezani, M., Ramezani, F., & Gerami, M., 2019. Nanoparticles in pest incidences and plant disease control. *Nanotechnology for agriculture: crop production & protection*, 233–272.

Rasheed, A., Li, H., Nawaz, M., Mahmood, A., Hassan, M.U., Shah, A.N., Hussain, F., Azmat, S., Gillani, S.F.A., & Majeed, Y., 2022. Molecular tools, potential frontiers for enhancing salinity tolerance in rice: A critical review and future prospective. *Frontiers in Plant Science*, 13, 966749.

Rasheed, A., Seleiman, M.F., Nawaz, M., Mahmood, A., Anwar, M.R., Ayub, M.A., Aamer, M., El-Esawi, M.A., El-Harty, E.H., & Batool, M., 2021. Agronomic and genetic approaches for enhancing tolerance to heat stress in rice: A review. *Notulae Botanicae Horti Agrobotanici Cluj-Napoca*, 49, 12501–12501.

Rofeal, M., Abdelmalek, F., & Steinbüchel, A., 2022. Naturally-sourced antibacterial polymeric nanomaterials with special reference to modified polymer variants. *International Journal of Molecular Sciences*, 23, 4101.

Sajid, M., & Płotka-Wasylka, J., 2020. Nanoparticles: Synthesis, characteristics, and applications in analytical and other sciences. *Microchemical Journal*, 154, 104623.

Saliev, T., 2019. The advances in biomedical applications of carbon nanotubes. *C - Journal of Carbon Research*, 5, 29.

Sekhon, B.S., 2014. Nanotechnology in agri-food production: An overview. *Nanotechnology, Science and Applications*, 20, 31–53.

Shang, Y., Hasan, M.K., Ahammed, G.J., Li, M., Yin, H., & Zhou, J., 2019. Applications of nanotechnology in plant growth and crop protection: A review. *Molecules*, 24, 2558.

Shinde, M.U., Patwekar, M., Patwekar, F., Bajaber, M.A., Medikeri, A., Mohammad, F.S., Mukim, M., Soni, S., Mallick, J., & Jawaid, T., 2022. Nanomaterials: A potential hope for life sciences from bench to bedside. *Journal of Nanomaterials*, 2022, 5968131.

Shiraz, M., Imtiaz, H., Azam, A., & Hayat, S., 2024. Phytogenic nanoparticles: Synthesis, characterization, and their roles in physiology and biochemistry of plants. *Biometals*, 37, 23–70.

Shukla, S., Sharma, M., Yadav, S., Raghupathy, A., Shukla, K., Varma, A., & Mishra, A., 2022. Synthesis and applications of nanoparticles: State of the art and future perspective. *Nanoscience & Nanotechnology-Asia*, 12, 2–15.

Singh, A.V., Laux, P., Luch, A., Sudrik, C., Wiehr, S., Wild, A.-M., Santomauro, G., Bill, J., & Sitti, M., 2019. Review of emerging concepts in nanotoxicology: Opportunities and challenges for safer nanomaterial design. *Toxicology Mechanisms and Methods*, 29, 378–387.

Song, L.-Y., Wu, Y.-Z., Pei, X.-X., Li, R., Chen, H.-T., & Sun, X.-Z., 2020. Pulmonary toxicity and RNA sequencing analyses of mouse in response to exposure to cellulose nanofibrils. *Inhalation Toxicology*, 32, 388–401.

Soni, R.A., Rizwan, M.A., & Singh, S., 2022. Opportunities and potential of green chemistry in nanotechnology. *Nanotechnology for Environmental Engineering*, 7, 661–673.

Torabian, S., Zahedi, M., & Khoshgoftar, A.H., 2017. Effects of foliar spray of nano-particles of $FeSO_4$ on the growth and ion content of sunflower under saline condition. *Journal of Plant Nutrition*, 40, 615–623.

Tyagi, J., Ahmad, S., & Malik, M., 2022. Nitrogenous fertilizers: Impact on environment sustainability, mitigation strategies, and challenges. *International Journal of Environmental Science and Technology*, 19, 11649–11672.

Usman, M., Zia-Ur-Rehman, M., Rizwan, M., Abbas, T., Ayub, M.A., Naeem, A., Alharby, H.F., Alabdallah, N.M., Alharbi, B.M., & Qamar, M.J., 2023. Effect of soil texture and zinc oxide nanoparticles on growth and accumulation of cadmium by wheat: A life cycle study. *Environmental Research*, 216, 114397.

Van Aken, B., 2015. Gene expression changes in plants and microorganisms exposed to nanomaterials. *Current Opinion in Biotechnology*, 33, 206–219.

Vanti, G.L., Kurjogi, M., Basavesha, K., Teradal, N.L., Masaphy, S., & Nargund, V.B., 2020. Synthesis and antibacterial activity of *Solanum torvum* mediated silver nanoparticle against *Xanthomonas axonopodis* pv. punicae and *Ralstonia solanacearum*. *Journal of Biotechnology*, 309, 20–28.

Ventura-Aguilar, R.I., Bautista-Baños, S., Mendoza-Acevedo, S., & Bosquez-Molina, E., 2023. Nanomaterials for designing biosensors to detect fungi and bacteria related to food safety of agricultural products. *Postharvest Biology and Technology*, 195, 112116.

Verma, K.K., Song, X.-P., Li, D.-M., Singh, M., Rajput, V.D., Malviya, M.K., Minkina, T., Singh, R.K., Singh, P., & Li, Y.-R., 2020. Interactive role of silicon and plant–rhizobacteria mitigating abiotic stresses: A new approach for sustainable agriculture and climate change. *Plants*, 9, 1055.

Wang, T., Xu, X., Wang, C., Li, Z., & Li, D., 2021. From smart farming towards unmanned farms: A new mode of agricultural production. *Agriculture*, 11, 145.

Wang, X., Wang, J., Zou, Y., Bie, Y., Mahmood, A., Zhang, L., Liao, L., Song, Z., Liu, G., & Zhang, C., 2024. Urea fertilization increased CO_2 and CH_4 emissions by enhancing C-cycling genes in semi-arid grasslands. *Journal of Environmental Management*, 356, 120718.

Wang, Y., Tian, J., Wang, Z., Li, C., & Li, X., 2022. Crop-safe pyraclostrobin-loaded multiwalled carbon nanotube delivery systems: Higher fungicidal activity and lower acute toxicity. *ACS Agricultural Science & Technology*, 2, 534–545.

Wani, M.Y., Ganie, N., Dar, K., Dar, S., Khan, A.H., Khan, N.A., Zahmatkesh, S., Manzar, M.S., & Banerjee, R., 2023. Nanotechnology future in food using carbohydrate macromolecules: A state-of-the-art review. *International Journal of Biological Macromolecules*, 239, 124350.

Xiao, Z., Ji, H., Yue, L., Chen, F., Yan, X.-P., Wang, Z., & Rasmann, S., 2024. Nano-chitosan boosts sesame plant anti-herbivore defenses and seed nutritional metabolites. *Environmental Science: Nano*, 11, 797–811.

Yusefi-Tanha, E., Fallah, S., Rostamnejadi, A., & Pokhrel, L.R., 2020. Root system architecture, copper uptake and tissue distribution in soybean (*Glycine max* (L.) Merr.) grown in copper oxide nanoparticle (CuONP)-amended soil and implications for human nutrition. *Plants*, 9, 1326.

Zain, M., Ma, H., Nuruzzaman, M., Chaudhary, S., Nadeem, M., Shakoor, N., Azeem, I., Duan, A., Sun, C., & Ahamad, T., 2023. Nanotechnology based precision agriculture for alleviating biotic and abiotic stress in plants. *Plant Stress*, 10, 100239.

7 Cutting-Edge Approaches to Weed Management

Noor ul Ain, Muhammad Mansoor Javaid, Muhammad Ather Nadeem, Filza Ishfaq, Muhammad Anjum Zia, Muhammad Zubair, Numra Akbar, and Javaria Nargis*

7.1 INTRODUCTION

Weed's biology, as outlined by "Baker's Rules," significantly influences the success of each species as a weed (Lau and Funk, 2023). This might be attributed to the seeds' behavior, the competitive abilities of the plants, or their seed output. Biology interacts with crops or management of land, such as wheat, apples, or amenity grassland. Weed seeds' behavior significantly influences a species' propensity to grow into a weed. Mall and Singh (2013) recorded the quantity of seeds in the soil among the first observations. He collected dirt from a pond, put it in a cup, and observed the seedlings that sprouted over a period of six months. He acquired 537 seeds from 210 g of mud with a dry weight. Weed seed population often exceeds the standard sowing rates of different crops by a significant margin. Barley seeding occurs at a rate of around 3 million seeds per hectare; by comparison, poppy species (*Papaver* spp.) have been found to have 279 million seeds per hectare; brome species (*Bromus* spp.) have 24 million; and blackgrass species (*Alopecurus myosuroides*) have 55 million seeds per hectare. The estimated total weed seed population in arable soils typically ranges from 1000 to 10,000 seeds/m^2. Weed seed concentrations in the soil are practically significant because they need ongoing weed management over many years. This is due to the delayed germination of weed seeds and the limited impact of herbicides on dormant weed seeds (Tanveer *et al.*, 2013; Florentine *et al.*, 2016). The weed seed bank is crucial for the survival of the species of plant and the emergence of freshly emerged seedlings in future harvests. The quantity of seeds generated and their persistence in the soil are crucial factors (Mahmood *et al.*, 2013; Long *et al.*, 2015). Short-lived seeds like barren brome (*Bromus diandrus*) will mainly affect crops in the next year, whereas long-lasting seeds like poppy might potentially infest fresh crops for many years. Seed dispersion plays a crucial role, since seeds with characteristics that aid in long-distance transit are more likely to spread to new locations in comparison with seeds that only fall near the parent plant. Agricultural practices, such as plowing, may impact seed behavior, generally leading to higher seed retention opposed to surface cultivation. We focus on creating integrated weed management (IWM) systems in the Australian cotton sector (Embrapa *et al.*, 2013). Implementing IWM is crucial for achieving sustainable weed control in these systems over the long run. IWM aims to avoid future occurrences of resistance to herbicides. IWM involves using a number of weed control approaches in conjunction with each other, not alone, to avoid herbicide resistance. IWM will eventually result in decreased dependence on herbicides for weed control, minimized development of herbicide resistance and species displacement, and reduced environmental effects of herbicides. The article on IWM has been placed to the front of WEED Pak, right after the weed identification and information guide, so that users may easily access it, given the significance of IWM to the agricultural system (Tong, 2024) comprehending the biology and ecology of weeds is crucial for creating a sustainable weed control strategy. There is still a need for more studies to get essential information on several

* Corresponding author: anjum.zia@uaf.edu.pk

DOI: 10.1201/9781003498698-7

significant species. Various methods have been used in farmers' fields to reduce the impact of weeds in efficient and cost-effective ways. Herbicide-resistant (HR) weed growth is linked to elements such as selection pressure, genetic diversity of the weed, patterns of transmission, gene flow, herbicide characteristics, agroecosystem parameters, and more (Qasem, 2013). Therefore, a single management technique may not always be feasible for controlling and avoiding the expansion of HR weeds, rather than an IWM technique is necessary (Ali *et al.*, 2013a, b; Chadha *et al.*, 2019).

7.2 STRATEGIES FOR CONTROLLING RESISTANT WEEDS USING INTEGRATED WEED MANAGEMENT

1. Consistently survey fields for weeds and promptly address fluctuations in weed populations to inhibit weed proliferation.
2. Select herbicides according to the specific kinds of weeds present and apply them carefully.
3. Crop rotation is a successful strategy for disrupting weed cycle, and certain kinds of crops are better at managing specific weed issues than others.
4. Avoid applying the same herbicide or another herbicide with the identical mechanism of action consecutively for at least two years in the same field.
5. Consider using a tank combination or applying herbicides sequentially with different mechanisms of action.
6. Maintain cleanliness of tillage and harvesting equipments to prevent the spread of weeds across fields.

Weed control is a significant concern in agriculture. Weeds compete with crops, limiting production. If weeds are not removed, they leave seeds in the soil, causing future issues (Tanveer *et al.*, 2015). One mechanical weed management practice is called "weed harrowing" (Rueda-Ayala *et al.*, 2010). This method involves both weeding inside rows and distressing between rows. Controlled sowing and harrowing may be combined using exact geolocation. Another alternative is to control the life cycles of weeds. Weed inflorescences, often positioned above crops, may be harvested early using a header. This method, now used in organic agriculture, aids in diminishing the weed seed reservoir and may be modified for organic farming via the development of new equipment suitable for various crops. Australian engineers have advanced by creating a combine that includes a mechanism for recovering weed seeds, known as harvest weed seed control (Walsh *et al.*, 2019).

7.3 EVOLUTION OF WEED MANAGEMENT STRATEGIES

Weeds in agriculture, chosen by human farming practices, are a relatively new occurrence in terms of ecology and evolution. Agricultural weeds are the ecological and evolutionary reaction of the indigenous and imported plant life in an area to the advantages and difficulties brought about by human agricultural practices (Chadha *et al.*, 2020; Javaid *et al.*, 2022). The field of evolutionary ecology should play a key role in shaping ideas and methods in practical crop-weed control. Despite some acknowledgment of the significance of weed evolution in weed management, the lack of widespread incorporation of the concepts and techniques from evolutionary environment into specific crop-weed research has significantly impacted and will continue to harm weed management. Chauhan *et al.* (2017) propose that crop protection corporations, crop breeders, farmers, and weed scientists are involved in a "weed management arms race," like the "evolutionary arms race" seen between diseases and their hosts.

Weeds' connections with other trophic levels and soil health are crucial for providing ecological benefits. The benefits provided may include providing food, shelter, and habitat for predators of crop pests and insect pollinators, providing plant cover during non-cropped instances of the rotation to prevent soil erosion, and improving soil quality and function. Weed variety in use may enhance crop output by decreasing losses from insect pests and preserving or improving soil health (Ali *et al.*, 2013a, b). Trophic interactions may help control weed populations by including processes such as weed seed ingestion and microbial breakdown of seed that is viable in the soil seedbank. Conservation agriculture (CA) is an agricultural management technique that involves minimal soil disturbance, maintaining residue for soil cover, and rotating primary crops. One way to reduce weed pressure in certain crops is to change the period of crop planting. Planting wheat earlier in North India provides the crop with a competitive edge against *Phalaris minor*, a harmful grassy weed species. During the early part of the season, the ecological and climatic circumstances are not conducive for the seed germination of *P. minor* (Pons and Fenner, 2000). Implementing no-till and early wheat planting in North India was beneficial for farmers by reducing issues caused by *P. minor*.

7.4 INTEGRATE WEED MANAGEMENT: A HISTORIC APPROACH

Current agricultural goals include achieving a crop output that is consistent in quantity, quality, and timing, all while maintaining environmental sustainability. In 1991, the United Nations Framework Convention on Environment and Development chose Integrated Pest Management (IPM) as the primary method for sustainable agriculture to decrease the use of pesticides and promote eco-friendly pest control measures (Trivedi and Ahuja, 2011). Since 2009, the IPM, which also includes IWM, has been required in the European Union following Directive 2009/128/EC about sustainable pesticide usage. IWM is crucial for weed control in modern agricultural systems in industrialized countries, particularly in the European Union, but is not often used in underdeveloped countries. Weeds are often known as strictu sensu and are intimately associated with agricultural practices (McNeill, 1976). The idea of cannabis is subjective and not universal.

Various definitions of weed have been offered by the scientific community from diverse perspectives such as agronomic, biological, and ecological. The current commonly accepted definitions are those established by the European Weed Investigation Society in 1986 ("any plant or vegetation, eliminating fungi, that hinders the goals or needs of people") and by the Weed The field of science The community of America in 1989 ("a plant growing in a location where it is not wanted") (Byrd, 1990) (Figure 7.1).

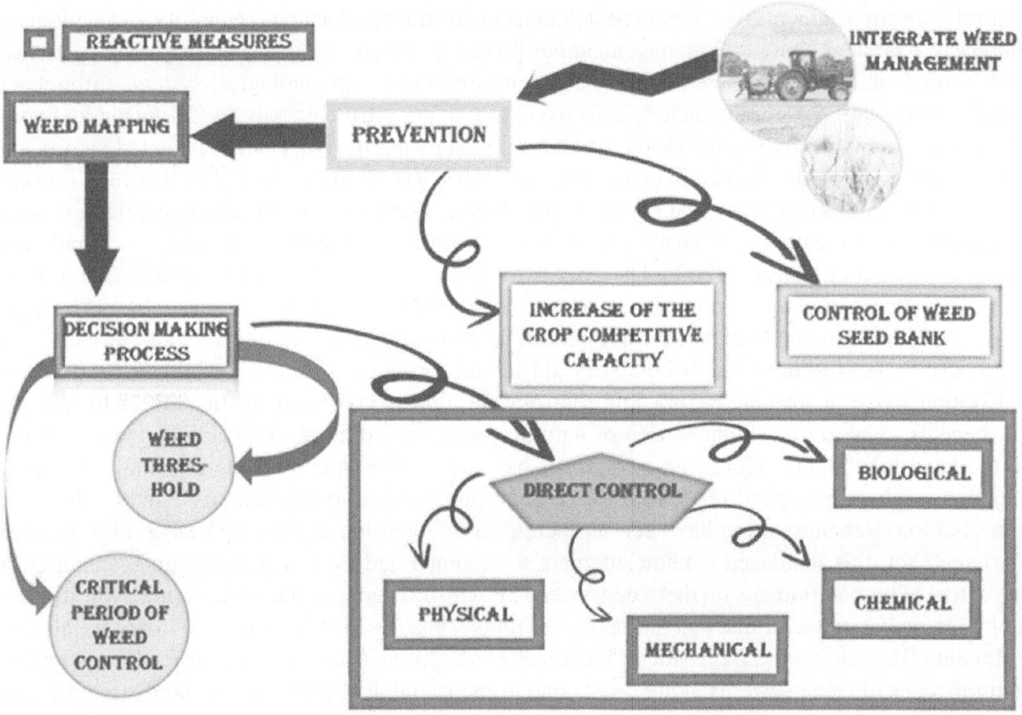

FIGURE 7.1 Integrate weed management, proactive strategies, and reactive measure.

Enhanced weed control is necessary, but it has been hindered by low pricing for agricultural goods. Poor financial outcomes in raisin-grape cultivation have caused many vineyards to be abandoned or led to lower weed control efforts in the south-Central Valley of California. This phenomenon might be the cause of the recent surge of horseweed (*Erigeron canadensis*) in the south-Central Valley (Shrestha *et al.*, 2004). While IWM, as well as other non-chemical methods, may be effective, herbicides are the most cost-effective choice. As previously stated, herbicide expenses are becoming more unaffordable, particularly when combined with other economic challenges encountered by growers. Justifying substantial investments in weed control is challenging (Pannell *et al.*, 2016). Significant resources are being allocated to the research and creation of HR crops, especially in North America. HR crops might tip the scales back in favor of reductionist methods, even though several ecological and socioeconomic variables are pushing a trend toward increased holism in weed control (Ali *et al.*, 2013a, b; Hanif *et al.*, 2017). Some question whether weed research is repeating the same error in the development and advocacy of HR cultivars as seen with herbicides. The unforeseen ecological impacts of genetically modified crops should be carefully examined. Weed community management needs enduring techniques, unlike managing typical insect pests and illnesses. A progressed IWM strategy should impact weed population dynamics at multiple phases of their life cycle (Riemens *et al.*, 2022).

a. Preventing weed development from seeds, rhizomes, or tubers.
b. Minimizing the negative effects of formed invasive plants on the crop, and
c. Decreasing the regrowth of the seed or by vegetative means bud bank.

7.5 BIOLOGICAL CONTROL: HARNESSING NATURE'S SOLUTIONS

Classical biological control's effectiveness stems from using nature's defenses against pests to provide a sustainable and enduring solution. Introducing native predators that target a particular pest reduces the chance of harming beneficial creatures and non-target species. Classical biological control is an environmentally friendly pest management strategy due to its reliability and ecological emphasis. Classical biological management has played a crucial role in managing many pests that have caused significant damage to crop for agriculture, forests, and biological systems throughout time (Batra, 1982). Examples include effectively managing citrus blackfly in California and controlling a cassava mealybug in Africa. The accomplishments have not only prevented significant economic losses but also helped in conserving biodiversity and environmental well-being (Tanveer *et al.*, 2013). Traditional biological control is a technique that involves introducing natural enemies like predators, parasitoids, or viruses from the pest's original habitat to regulate its population (Van Driesche and Hoddle, 2009). The goal is to create a sustainable equilibrium between pests and their natural enemies, decreasing reliance on chemical pesticides and lowering harm to crops, forests, and ecosystems. Mangroves are essential for providing a variety of ecological services (Ali *et al.*, 2013a, b; Husain *et al.*, 2020). They aid in soil formation, use photosynthesis for primary cultivation, assist in nutrient cycling, and control water flow in coastal areas. In addition to ecological benefits, they are a crucial source of natural goods, including proteins, tannins, and timbers to the local inhabitants. Moreover, mangroves have a key role in controlling essential ecosystem functions such as biological regulation, nutrient cycling, air quality management, and biodiversity conservation. Recently, there has been an increasing recognition of the importance of mangrove wetlands. Yet, this increased acknowledgment is accompanied by a concerning understanding of the extent of human impacts on these ecosystems. Phyto-filtration, also known as rhizo-filtration, is a phytoremediation technique that harnesses the roots' capacity to cleanse water of toxic metals and pollutants (Hassan *et al.*, 2022; Mocek-Płóciniak *et al.*, 2023). Mangroves are crucial for acting as natural water filtering systems along coasts due to their complex root systems and distinctive foliage. Water passing through the intricate system of mangrove roots gets rid of suspended particles and contaminants, leading to enhanced water clarity. Mangroves have a unique capacity to reduce nutrient runoff from nearby land regions, therefore decreasing the flow of excessive nutrients like nitrogen and phosphorous into adjacent waterways. Mangroves are important plant groupings that thrive in the transitional zone between land and coast, making them effective at limiting the transfer of harmful heavy metals and chemicals between grassland and coastal ecosystems (Srivastava and Prasad, 2019). Utilizing mangroves for phytoremediation has great potential as a sustainable method for removing pollutants. Genetic engineering, chemical substances, and microbes may boost their potential. It has been extensively studied and used in many systems, including rice, soybean, potatoes, lettuce, blueberries, and soybeans, to employ blooming plants to increase the efficiency of natural enemies (Lee *et al.*, 2012). Recently, there has been a move from focusing on unique insect-plant interactions to adopting a trait-specific strategy to develop more generalizable guidelines, reducing the need to test many potential plants for each new pest-crop system. Multiple research projects have observed the disparity between the potential and actual impact of biological control, analyzed the reasons for this shortfall, and suggested strategies to enhance its use (Vanloqueren and Baret, 2017). Identify five scenarios where natural habitat may not be able to sustain conservation biological control.

1. Pest populations lack effective natural predators in the area.
2. The natural environment produces greater numbers of bugs than natural predators.
3. Crops offer more resources for natural predators than the natural environment.
4. The natural environment fails to provide enough suitable conditions to support sufficient predator populations for pest control.
5. Farming practices hinder the establishment of natural predators and biological pest management provided by the natural environment.

Various variables may impact the significance of organic habitat for biological control, and these considerations must be considered when developing strategies to preserve or improve the natural environment for biocontrol purposes (Baker *et al.*, 2020).

7.6 CHEMICAL INNOVATIONS IN WEED CONTROL

Chemicals that are used to destroy or impede the development of plants are known as herbicides. The use of herbicides has been steadily rising since 1944. Currently, several novel chemical compounds are now accessible for weed management (Javaid *et al.*, 2012; Umetsu and Shirai, 2020). Currently, herbicides can effectively address all types of weed issues. They often provide the most practical, efficient, and cheap method for controlling weeds. Kumar *et al.* (2022) reported that in Indian agriculture, the most widely used herbicides were butachlor (6032 tons) and glyphosate (6003 tons), next to paraquat (2068 tons), pretilachlor (2418 tons), and pendimethalin (1444 tons). India currently has a yearly total capacity of herbicide manufacture of around 6000 tons. Many herbicides in India are mostly used for cultivation in plantations. Low-dose and high-effectiveness herbicides are being used more often, replacing traditional high-dose herbicide like 2,4-D and isoproturon (de Castro Marcato *et al.*, 2017). Although fewer herbicides are being used, more land is being used for herbicidal weed control. Butachlor in rice and isoproturon in wheat are being mainly replaced with low-dose herbicides such as pyrazosulfuron in rice and sulfosulfuron in wheat. Butachlor use has steadily declined since 2005. Pretilachlor has gained greater popularity than butachlor in the rice market. Over the last five years, pretilachlor has maintained a consistent output level of around 1900 tons per year (Manisankar *et al.*, 2022).

7.7 RECENT TRENDS IN CHEMICAL WEED MANAGEMENT

7.7.1 Herbigation

Herbigation is the simultaneous application of herbicide to the soil with irrigation water. Initiated in 1960, this strategy quickly achieved widespread favor. Tap and sprinkler watering methods have expanded rapidly in recent years and have been used in many regions. Using herbicides with irrigation water is usually successful in preventing weeds in just a part of the area reached by the water. Utilizing irrigation systems for herbicide application is a new advancement in weed prevention technology. Herbigation with a trickle irrigation technology may be more widely accepted in dry settings (Fernández-Aparicio *et al.*, 2016).

7.7.2 Nano-Herbicides

Nano-herbicides are herbicide formulations that use nanomaterials to deliver chemicals effectively to a specific spot. Overuse and improper application of herbicides that contain chemicals have led to widespread issues of food and pollution to the environment. Nanomaterial-based formulations may enhance the effectiveness of the herbicide, increase solubility, and decrease toxicity when compared to traditional herbicides (Rasheed *et al.*, 2022). Nanomaterials are used to enhance the bioavailability of herbicides, leading to more effective weed control. Nano-herbicides improved the capacity of agricultural formulations to moisten and disperse. Nano-emulsions, nano-encapsules, nanocontainers, and nanocages are types of nano-herbicide formulations (Sharma *et al.*, 2023).

7.7.3 Herbicide Mixture or Herbicide Combinations

Combining herbicides has many benefits, including reducing dose, preventing one herbicide in the mixture from being detoxified, and providing broad-spectrum weed control. There are two types of herbicide mixes: tank compounds, which are prepared using specific herbicides and rates right before application, and formulated herbicide mixtures, which are created by firms during production. Many herbicide mixes are created by numerous manufacturers with various trade names to target distinct weed situations. Combining herbicides from various chemical families may enhance the management of certain weed populations. For instance, 2,4-D is used in conjunction with dicamba to manage broadleaf weeds. The herbicide mixtures have the capacity to manage many kinds of weeds simultaneously (Joseph, 2014; Barbieri *et al.*, 2022).

7.7.4 New Generation Spot Sprayer Technology

The agricultural sector faces difficulties and strain because of the overuse of pesticides. Repeated use of the same herbicides leads to the development of herbicide resistance. The ineffectiveness is mostly due to the inadequate accuracy of traditional sprayers. New cutting-edge technology is now accessible for intelligent implementation of remedies, resulting in a 90% reduction in herbicide expenses via targeted weed control. Manufacturers have only slightly utilized spot spraying technology as now. The weed catcher is an excellent demonstration of this technology, using computer algorithms to identify green plants and trigger spray nozzles to deliver a nonselective herbicide to the target (Bajwa *et al.*, 2015).

7.7.5 Precision Weed Management: Technology at the Forefront

Precision Agriculture (PA) is a management approach that considers variations in time and space to enhance the sustainability of agricultural output (Bongiovanni and Lowenberg-DeBoer, 2004).

Precision agriculture is often used for tasks such as identifying weeds, monitoring crop and soil health, managing sites for activities like tillage, sowing, mechanical weeding, and fertilizer distribution, estimating crop output, and detecting and collecting fruits and vegetables (Rehman *et al.*, 2023). An innovative weeding technology that minimizes the usage of herbicides is required promptly. Weed robots are being developed using real-time image recognition to quickly identify and manage weeds (Sujaritha *et al.*, 2017).

Site-specific weed management (SSWM) typically involves four processes:

1. Utilize various tools for data collecting.
2. Detect weeds using accurate sensors to get real-time information, including weed location, size, and kind, or create a weed map.
3. Select appropriate techniques and chemicals for weeding based on the provided information.
4. Assess the impact of weeding for future enhancements.

Optic sensors are the predominant technology utilized for weed identification during data collecting. Several sensor types, such as machine vision, visibility and near infrared (Vis-NIR) spectroscopy, multi-/hyper-spectral pictures, and distance sensing approaches, have been evaluated. These sensors may be categorized as either airborne remote sensing or ground-based approaches. The captured photos are analyzed offline to create weed maps for future SSWM tasks. Remote sensing methods are beneficial for map-based SSWM because of their suitability for broad regions, albeit they are not instantaneous and provide poorer spatial resolution than ground-based methods. Ground-based methods gather and quickly analyze weed data to facilitate real-time SSWM activities (Rai *et al.*, 2023). Compound signaling is a technology that may be used for detecting crops and weeds. Precision agriculture utilizes field-specific data to determine the most effective inputs required for productive farming. Growers using precision agricultural technologies have access to more information and often dedicate more time to strategizing crop management and methods to improve yields and earnings. Site-specific IPM research and application focus on the ability to target regions of the field for pest management and to optimize crop health by adjusting inputs within those areas rather than throughout the whole field (Vreysen *et al.*, 2007). These provide significant promise for decreasing expenses and environmental footprint. Site-specific control of weeds is a rapidly evolving field. Site-specific agriculture offers a more accurate method for sampling and controlling farms. Integrating soil, crop, pest, disease, and environmental data into a single program allows for more efficient crop management, leading to better economic outcomes and minimizing environmental harm. Watkins *et al.*, (2012) observed that detailed processes are often unneeded, and that broader descriptions may be suitable when using sampling to gather information for decision-making. Emphasizing the threshold of pests that need management action is more important than refining sample methodologies. Finding the position and severity of pests in the field is important information for data sampling in pest control. Instead of sampling randomly over the field, it could be more effective to focus on certain regions where population changes are anticipated, creating boundaries for constructing maps. The size and distribution of pest clusters are more significant than the average value. The weed type is significantly affected by the preceding crops, and the location of cereal in the succession of crops possesses a major role in determining the weed community type and size that develops in a particular year. The significance of crop variety, including perennials, annuals, cool-season annuals, and warm-season annuals, is crucial in avoiding the prevalence of a single weed species (Derksen *et al.*, 2002). Autonomous robots in precise farming provide an achievable option for environmentally friendly agriculture and reducing chemical treatments. Autonomous robots with autonomous weed detecting systems may enhance the efficiency of precision farming by adjusting herbicide spraying based on the extent of weed infestation. Unmanned aerial vehicles (UAVs) are a very successful technology used in precision agriculture. Unmanned vehicle systems are movable platforms, either aerial or terrestrial, that provide many benefits for carrying out and overseeing agricultural tasks. UAVs are particularly beneficial since they enable SSWM (Velusamy *et al.*, 2021).

SSWM is an enhanced method for managing weeds that is very effective and ecologically friendly. It allows for accurate and ongoing tracking and mapping of infestations of weeds. SSWM agrees to customize weed control measures for each unique agricultural scenario. Utilizing UAVs equipped with sophisticated cameras and sensors capable of identifying certain weeds, together with global positioning system (GPS) technologies that provide geographical data for field mapping, may allow accurate monitoring of vast regions within a short timeframe (Quamar *et al.*, 2023). Improved planning of weed management in sustainable smallholder weed management can enhance the effectiveness of mechanical methods and decrease herbicide usage. This can lead to significant agroecological and economic benefits such as lower production costs, decreased weed resistance, enhanced biodiversity, and reduced environmental impacts. Utilizing UAVs for weed management may enhance the environmental sustainability of the eventual agricultural production systems required to meet the needs of a growing global population (Monteiro and Santos, 2022).

7.7.6 Genetic Engineering and Weed Resistance

Herbicide resistance was a rare issue for over 30 years. Some believed that the infrequent occurrence of new HR instances meant that herbicide resistance in weeds was unlikely to proliferate and was not a significant concern for the future. Undeniably, HR weeds have proliferated and become a significant economic challenge for farmers globally. Some argue that herbicide resistance, if not controlled, might jeopardize the viability of certain agricultural production methods (Green and Owen, 2011). Weed experts acknowledged the significance of herbicide combinations as a crucial strategy for managing resistant weeds well in advance of resistance becoming a significant issue. Subsequently, the use of several herbicides in combinations has been shown through practical experience to decrease the chances of resistant weed development in real-world settings. Weed management is a challenging issue that involves intricate interactions between natural and human systems (Buhler, 2002). Finding straightforward methods for managing invasive and agricultural weeds is becoming more challenging. A serious problem that is quickly arising is linked to the widespread development of herbicide resistance. Another pressing concern is to invade alien weeds, which may be challenging to manage due to their tendency to infiltrate delicate, rough, and isolated regions where traditional control methods may be impractical, unfeasible, or prohibitively costly. Gene drives may be used to manage weeds in two distinct methods (Kumam *et al.*, 2023): (i) population control or (ii) population awareness.

In the first case, suppression drives could effectively manage exotic alien weeds in both natural and agricultural landscapes by targeting important features like fertility, establishment, and persistence. In the second case, sensitizing drives might be used in the second scenario to make a target demographic more receptive to a certain management solution.

For instance, weeds that have developed a resistance to herbicides may have their herbicide susceptibility restored by employing an endonuclease-based motive to disrupt the inability to respond alleles at herbicides target sites and replace them with wild-type susceptibility alleles. This would restore the effectiveness of previous chemistries for controlling important species. Weeds are the primary pest complex affecting humans. The resistance to herbicides in crop weeds has increased expenses for producers; however, it should not be solely attributed to the recent development of genetically modified herbicide tolerant crops (Hassan *et al.*, 2021). Weed resistance to herbicides has long posed a danger to worldwide food supply. The emergence of glyphosate resistance has increased the risk to global agricultural production and highlighted the need of using IWM strategies. Growers' actions to reduce HR weeds in their fields will be expensive and need to be weighed against the advantages, whether short-term or long-term. Herbicides are applied to over 90% of crop acres in the USA each year and account for 65% of the pesticide costs incurred by farmers

(Gianessi and Reigner, 2007). Without herbicides, agricultural output in the USA would decrease by at least 20%. The development of HR weed biotypes due to herbicide usage leads to lower productivity, increased herbicide costs, and significant growth in the weed seedbank. Challenging aspects include altering producer perceptions and behaviors to implement required changes in weed control programs. How this can be achieved on a large enough scale to tackle the emerging challenges posed by developed HR weed strains in the USA is yet to be determined.

7.7.7 Emerging Technologies in Weed Detection and Monitoring

Nano-sensors, nanomaterials such as nanoparticles, and innovative smart delivery systems are gaining more interest for potential use for agriculture (Ashraf *et al.*, 2021; Khan *et al.*, 2023). These innovations and supplies have the potential to enhance the use and spread of herbicides, as well as to address obstacles that hinder the practical deployment of smart sensors and smart monitoring systems. Remote sensing offers a practical option to significantly reduce the need for laborious ground-based weed scouting. Remote sensing technologies have been increasingly used in farming operations over the past two decades. They have been applied in mapping of soils and fertility assessments, irrigation scheduling, evaluating crop growth status, and monitoring abiotic and biotic stress factors. Remote sensing offers accurate and fast data gathering, aiding in the execution of short- and long-term crop management programs (Sishodia *et al.*, 2020). Integrating remote sensing into a comprehensive weed control system helps optimize herbicide use and decrease the likelihood of herbicide resistance development in weeds. Aerial technologies provide distinct benefits compared to ground-based sensing in some scenarios. Satellite remote sensing is preferable to ground-based sensors for collecting large-scale spatial data, especially at local and continental levels, and is often used for monitoring crops and estimating yields. Yet, satellite imaging is hindered by cloud cover and usually has limited spatial, spectral, and radiometric resolutions. Satellite-based sensing is inadequate for precise weed control at the level of individual fields. Precision agriculture involves using sensors and geospatial methods to detect field differences and address them via specific tactics, as outlined by Sishodia *et al.* (2020). Enhancements in sensor abilities have increased the precision of identifying and categorizing weeds. Sensors provide object identification, and when paired with adaptive algorithms, they can initiate a precise mechanical reaction. The first crucial stage in achieving automated weeding is weed detection. Tractor-mounted equipment for real-time weed identification and targeted spraying of weed in agricultural areas has the potential to reduce economic and environmental expenses while ensuring effective weed management. Early research examined the effectiveness and dependability of using various light spectra and basic image processing methods (such as color thresholds, differentiation, and filtering) to distinguish crops from weeds. Another study utilizes fuzzy logic and associated methods to transform visual data into instructions for operating pesticide sprayers. Recent progress in sensors, computer power, and algorithms has led to significant advancements in weed identification in recent years. There are autonomous weeding robots on the market that use robots and computer vision to identify and eliminate weeds. Weed identification using digital picture processors and artificial retinas is a highly researched and often used method. Digital photographs may be differentiated between weeds and crops based on spectral information, biological morphology, visual materials, geographical contexts, and patterns (Su, 2020). Transplanted lettuce plants are often larger than weeds due to their longer growth period, making size a useful characteristic for weed detection. These traits are usually identified via observations and experience, then incorporated into traditional image processing approaches through programming and design. Another common method to address the issue of weed detection is by machine learning (ML). This method uses statistical models to automatically extract characteristics from sample photos. Some advanced models not only identify weeds but also have the capability to pinpoint several weeds in a single image. Image processing is a technology frequently used for weed identification. The usual processes involved include preliminary processing, the process of segmentation, feature extraction, and classification.

TABLE 7.1
Some Tools/Techniques for Weed Management

Tools/Implement	Purpose
CRIDA Manual weeder	For weeding in uplands crops
TNAU improved dryland weeder	For weeding in row crops for removing shallow rooted weeds. It has been designed ergonomically for easy operation. Useful in dryland and garden land crops and is ideal at a soil moisture content of 8–10%
Self-propelled power weeder	For weeding and intercultural operation in upland crops viz., groundnut, soybean, maize, and pigeon pea
Power rotary weeder	For mechanical control of weeds in crops such as sugarcane, tapioca, cotton and orchards. Useful for weeding between rows of crops like tapioca, cotton, sugarcane, maize, tomato and pulses whose rows spacing is more than 45 cm
Tractor drawn, weeding cum, earthing up equipment	Tractor Operated Multi Row Rotary Weeder
Tractor operated multi row rotary weeder	For weeding and intercultural operations in between row crops like sugarcane, cotton, maize, etc.
Cono weeder	For weeding between rows of paddy crop
Two row finger type	For weeding in paddy row crops
Paddy rotary weeder	
Battery-operated portable wetland weeder	For weeding in SRI field

Weeds in Australia cause a yield loss of 2.76 million tons of grains from various crops such as barley, wheat, oats, canola, sorghum, and pulses. Weeds generate an estimated yearly global economic loss of over $100 billion U.S. dollars, although worldwide annual pesticide sales totaling about $25 billion. Herbicides are the second highest-selling kind of insecticides in Europe. In 2018, they represented 35% of total pesticide sales, surpassing insecticides and acaricides (Lykogianni *et al.*, 2021). By 2050, the global population is projected to double, exceeding 9.15 billion people. The anticipated rise in food demand is unlikely to be satisfied by the existing production system. Climate change will provide an extra challenge to the human supply of food soon. Weed control is expected to be one of the most challenging procedures impacting agricultural output. The drawbacks associated with mechanical and pharmacological weed control methods may hinder their effectiveness for future weed control. Mechanical techniques are not very effective, and herbicides have a significant ecological impact. IWM is a method that reduces the limitations related to mechanical and chemical weed management (Swanton and Murphy, 1996). Table 7.1 demonstrates the emerging technologies in weed detection and monitoring.

7.7.8 Data-driven Approaches to Weed Management

Data-driven technologies have already transformed several areas of the economy. The potential advantages of a revolution in agriculture are fueling increased interest in using Information and Communication Technology (ICT) in the agricultural sector. Data collecting, modeling, analysis, and sharing are now fundamental problems, presenting opportunities for innovation and serving as a growing sector for commercial advancement (Allahyari and Chizari, 2010). Extensive data is gathered using several types of sensors such as proximate, aerial, or satellite-based detectors, in situ sensors like soil moisture sensors, on-farm weather stations, and examined farm equipment. According to Chen and Zhang (2014), this meets the criteria for Big Data, which includes large volume, high velocity, and diverse data types that need specialized technology and analytical techniques for extracting value. Farmers may use three sorts of data-intensive technology to tackle the difficulties stated above:

1. Farm Management Information Systems (FMIS) are a structured system designed to gather, process, store, and distribute data in a format necessary for a farm's activities.
2. Precision agriculture focuses on managing spatial and temporal variations to enhance profits and minimize environmental harm.
3. Agricultural automation and robotics involve implementing robotics, automatic control, and artificial intelligence (AI) methods across all stages of agricultural production.

Advanced tools and systems are being created to enhance throughput, which refers to the quantity of plants/genotypes that may be phenotyped in a single experiment or during a year. Nevertheless, expenses sometimes rise, making it difficult for researchers or organizations with minimal funds to attain these accomplishments. Static or plant-to-sensor facilities, whether inside or field-based, have restricted capacity, allowing only a certain number of projects to be included in the pipeline. Consequently, there is an increasing interest in cost-effective sensor systems and platforms with mobility that may be moved to the plants instead of moving the plants to the platform. The user has various demands in productivity, cost-effectiveness, accuracy, temporal and spatial resolution, correctness, user-friendliness, level of automation, and variety of data output (Mangla *et al.*, 2022). Therefore, a variety of sensors are necessary to meet a wide range of specific and goal-oriented requirements. The industry offers a limited number of commercially accessible sensors that do not meet all the various and specialized demands of researchers and end users. Recent advancements in phenotyping techniques have enabled the extensive collection of datasets from various sensors such as hyperspectral cameras, across different scales (field and controlled conditions), as well as levels of organization (canopy and leaf level). These datasets are valuable resources that provide detailed information on various characteristics that describe plant function and structure. The main challenge is managing the large volume of data produced by characterization facilities and devices. While many informatics solutions for high-throughput techniques have concentrated on creating specialized image analysis tools, there has been limited focus on developing information systems to manage, combine, and analyze the vast amount of sensor-derived data, which is complex due to its diverse origins. ML is one of the rapidly expanding fields within AI. ML is a growing discipline that provides solutions for handling the issues posed by Big Data. ML is a set of modeling approaches or algorithms that may be discovered from information as well as make decisions without human involvement. ML methods are beneficial when dealing with substantial datasets that are relevant to the desired output variables. ML offers a flexible and modular approach for analyzing data in the context of Big Data issues. New technological trends such as AI allow farmers to use data-driven methods to gather and evaluate large amounts of data (Linaza *et al.*, 2021). This helps them gain insights into the current condition of their fields in real-time, enhancing farm yield and reducing risks of weeds, pests, and diseases. The advancement of smart sensors, instruments, and machinery is becoming increasingly important in the fields, which are influenced by various factors like environmental conditions, soil properties, water supply, harvesting methods, plant diseases, weeds, and pests. In the future, the combination of automated data gathering, AI algorithms, and decision support systems will provide sophisticated capabilities for PA (Shaikh *et al.*, 2022). In addition, robotic devices on land and in the air will play a significant role in implementing precision agriculture and digitalization by performing tasks like harvesting, pest management, and data collecting. The current acceptance rate of AI innovations in the agriculture industry is low primarily because of issues such as a lack of understanding of its practical applications. Several research have focused on identifying the primary AI technologies utilized in physical activity and documenting their implementations. Mohyuddin *et al.* (2024) provided an in-depth assessment of research focused on the use of ML in systems for farming. The main areas of application were classified as crop management, water management, and soil management. It is indisputable that agriculture will gain advantages from ML technology, enabling farm management systems to transform into real-time decision-support mechanisms via the use of AI algorithms. Cravero *et al.* (2022), Oikonomidis *et al.* (2023), and Van Klompenburg *et al.* (2020) conducted a systematic literature review (SLR)

FIGURE 7.2 Typical image processing procedures for weed detection.

on the ML techniques and characteristics used in research predicting agricultural production. The findings indicate that no definitive conclusion can be made about the optimal model, but they demonstrate that some ML models, including a random forest approach, neural networks, regression using linear regression, and gradient-boost tree, are more often used than others. (Hassan *et al.*, 2020, 2022) conducted an extensive review on the use of deep learning (DL) methods for identifying plant leaf stress. The studied strategies were categorized into vegetables, fruits, and other crops based on stress type, dataset size, training/test size, and the extensive network used. Research has revealed many gaps, including the need to detect and classify all degrees of stress intensity to identify infectious illnesses and to generalize algorithms for other crops. AI and robots are becoming important in the agriculture industry, either assisting or replacing manual labor across a wide range of applications. Robotics has been crucial in agricultural productivity by autonomously carrying out various tasks, including weeding, irrigating, and controlling individual plants. Liu *et al.* (2023) have presented an in-depth overview of advanced robotic grippers, gripping techniques, sensor-based control approaches, and their use in agricultural robotics. Agricultural grabbing demands are increasing because of the delicate nature of food and agricultural goods, which are often fragile, prone to bruising, and slippery. Drones are being used in agriculture for tasks such as monitoring crop health, checking irrigation systems, and identifying weeds. Black *et al.* (2023) outlined many features of self-governing agricultural vehicles created at Hokkaido University. A comparison and discussion were conducted on the development techniques and attributes of each autonomous aerial vehicle. Case studies have been offered as actual examples of autonomous vehicles design and development processes, showcasing typical issues encountered throughout development. Further the diagram illustrates the data-driven approaches to weed management (Figure 7.2).

7.8 FUTURE PERSPECTIVES IN WEED MANAGEMENT

Recent evaluations of the potential future of weed control provide comprehensive overviews of the existing strengths, deficiencies, and requirements of weed science. Several writers have suggested that enhancing the formulation of queries for study to uncover fundamental concepts of weed ecological and evolution, enhancing the use of combined weed control strategies (Kumar *et al.*, 2013), enhancing the integration of fundamental knowledge with practical weed science (Swanton *et al.*, 2008), and capitalizing on new research possibilities, we mostly agree with the assessments and suggestions provided. However, we provide an alternative viewpoint on how weed management may evolve in the far future. Approximately 400,000 secondary metabolites exist in plants, although only a few numbers have been studied for their biochemical or herbicidal effects. The biologic or herbicidal capabilities of the other chemicals have not been investigated yet. The research examines the current situation and future possibilities of managing significant crop weeds to promote sustainable agricultural output. Substantial progress has been made in weed control via research, leading to enhanced approaches that have enabled farmers to boost crop output. However, the threat of weeds is growing in both cultivated and uncultivated areas.

Future studies should concentrate more on the following:

1. To effectively manage weeds and minimize their negative effects on crop yield, it is essential to have a comprehensive grasp of the weed's life cycle, growth patterns, vulnerable development phases, and reproductive capabilities. Therefore, there is a need to focus more on fundamental studies related to weed ecology and biology.
2. Traditional universities with a solid basis in fundamental sciences should be encouraged to conduct basic weed science research to study the environmental factors and biology of weeds for effective weed management. Regular evaluation of weed dynamics is essential to properly analyze new weeds and control them promptly.
3. *CA and permanent weed control*: CA has gained significance in promoting sustainable crop production. Conservation tillage strategies are likely to cause a transition from yearly to everlasting weed dominance in weed populations owing to reduced soil disturbance. Perennial weeds are harder to control. Perennial weed control solutions in CA systems need both basic and practical research to advance.
4. HR weed monitoring and prevention are essential owing to the rising herbicide usage in India, driven by manpower shortages and cost concerns. HR weeds are more likely to arise with the increased usage of herbicides in India. HR weeds have become a significant issue in worldwide agriculture in recent decades, and their control is crucial for promoting sustainable farming practices. In India, it is crucial to implement preventative steps to delay the emergence and spread of HR weeds. Weed science research should focus on developing resistance management practices and monitoring HR weeds to ensure profitable and sustainable agricultural systems.
5. Allelopathy and biocontrol are being researched for practical applications in India. Indian weed science research should focus on creating allelopathic crop types for IWM, and developing, synthesizing, formulating, and registering commercial allelochemical pesticide products. Research on enhancing practical techniques for biological weed management will be increased systematically.
6. Monitoring and developing efficient strategies to avoid herbicide residue buildup in soil, water, and the food chain is crucial owing to the rising herbicide usage in India. Globalization may lead to an increase in the threat of invasive weeds if proper control strategies are not implemented.

7. To effectively manage herbicide tolerant crops, it is important to practice careful stewardship by rotating crops and herbicides that have various ways of action, using verified seeds, and avoiding planting herbicide tolerant crops in the same field for multiple seasons to prevent the emergence of resistant weeds. Considering recent incidents in the USA involving dicamba application and 2,4-D herbicide tolerant crops, as well as Malaysia's issues with herbicide tolerant rice leading to an increase in herbicide tolerant weedy rice, India should take a cautious approach, like the government's current stance, when developing and utilizing herbicide tolerant crop varieties. The technology can be used successfully only if all stewardship techniques and measures are adopted by all parties involved, particularly the farmers. Teaching farmers how to utilize advanced technology safely and correctly is essential for them to start using it.
8. International cooperation is required. The expansion of global relations is reducing the limits of weed distribution, leading to the emergence of various weeds as global species via diverse dissemination techniques, posing a threat worldwide. Collaborative international efforts among Indian Weed Scientists will aid in developing management techniques to reduce the proliferation and effect of weeds.
9. International institutions such as CGIAR institutions like international rice research institute (IRRI), international crop research institute for the Semi-Arid tropics (ICRISAT), and international maize and wheat improvement center (CIMMYT) should take a leading role in conducting strategic studies on weeds and weed management. This can be achieved through collaboration with Indian research institutions and agricultural universities in India. Herbicides will be essential for weed control in the future. Increasing herbicide usage, along with inappropriate application and insufficient knowledge about the correct use of existing herbicides on farms, will lead to a higher prevalence of HR weeds in India. Therefore, the herbicide weed science research in India requires industry and weed scientists to take a more active part in training farmers and extension personnel on the prudent and sustainable use of pesticides. There is a need to prioritize the development of interdisciplinary techniques within IWM systems by using optimum mixtures of physical, chemical, biological, and ecological methodologies. Weed scientists in India should prioritize developing and promoting IWM solutions to control weeds successfully and economically in an environmentally friendly way for farmers.

LIST OF ABBREVIATIONS

AI artificial intelligence
CA conservation agriculture
DL deep learning
FMIS Farm Management Information Systems
GPS global positioning system
HR herbicide-resistant
ICT Information and Communication Technology
IPM Integrated Pest Management
IWM integrated weed management
ML machine learning
SLR systematic literature review
SSWM site-specific weed management
UAVs unmanned aerial vehicles
Vis-NIR visibility and near infrared

REFERENCES

Ali, H., Abid, S., Ahmad, S., Sarwar, N., Arooj, M., Mahmood, A., & Shahzad, A., 2013a. Impact of integrated weed management on flat-sown cotton (*Gossypium hirsutum* L.).

Ali, H., Abid, S.A., Ahmad, S., Sarwar, N., Arooj, M., Mahmood, A., & Shahzad, A.N., 2013b. Integrated weed management in cotton cultivated in the alternate-furrow planting system. *Journal of Food, Agriculture and Environment*, *11*(3&4), 1664–1669.

Ali, H., Tanveer, A., Nadeem, M., Asghar, H., & Javaid, M., 2013c. Germination ecology of *Rhynchosia capitata*: An emerging summer weed in Asia. *Planta Daninha*, 31, 249–257.

Ali, H., Tanveer, A., Nadeem, M., Javaid, M., Kashif, M., & Chadhar, A., 2013d. Allelopathic effects of *Rhynchosia capitata* on germination and seedling growth of mungbean. *Planta Daninha*, 31, 501–509.

Allahyari, M., & Chizari, M., 2010. Potentials of new information and communication technologies (ICTS) in agriculture sector. *Nong Ye Ke Xue Yu Ji Shu*, *4*(4), 115.

Ashraf, S.A., Siddiqui, A.J., Abd Elmoneim, O.E., Khan, M.I., Patel, M., Alreshidi, M., Moin, A., Singh, R., Snoussi, M., & Adnan, M., 2021. Innovations in nanoscience for the sustainable development of food and agriculture with implications on health and environment. *Science of the Total Environment*, 768, 144990.

Bajwa, A.A., Mahajan, G., & Chauhan, B.S., 2015. Nonconventional weed management strategies for modern agriculture. *Weed Science*, *63*(4), 723–747.

Baker, B.P., Green, T.A., & Loker, A.J., 2020. Biological control and integrated pest management in organic and conventional systems. *Biological Control*, 140, 104095.

Barbieri, G.F., Young, B.G., Dayan, F.E., Streibig, J.C., Takano, H.K., Merotto, A. Jr, & Avila, L.A., 2022. Herbicide mixtures: Interactions and modeling. *Advances in Weed Science*, *40*(spe 1), e020220051.

Batra, S.W., 1982. Biological control in agroecosystems. *Science*, *215*(4529), 134–139.

Black, J.A., Nakanishi, H., & Kobayashi, Y.H., 2023. Governance and the provision of roads and mobility in five Japanese 'societies'. *Case Studies on Transport Policy*, 12, 101000.

Bongiovanni, R., & Lowenberg-DeBoer, J., 2004. Precision agriculture and sustainability. *Precision Agriculture*, 5, 359–387.

Buhler, D.D., 2002. 50th anniversary—Invited article: Challenges and opportunities for integrated weed management. *Weed Science*, *50*(3), 273–280.

Byrd, J., 1990. Report of the 1989 cotton weed loss committee. *Report of the 1989 cotton weed loss committee.*, 365–366.

Chadha, A., Florentine, S., Chauhan, B.S., Long, B., Jayasundera, M., Javaid, M.M., & Turville, C., 2019. Environmental factors affecting the germination and seedling emergence of two populations of an emerging agricultural weed: Wild lettuce (*Lactuca serriola*). *Crop and Pasture Science*, *70*(8), 709–717.

Chadha, A., Florentine, S., Javaid, M., Welgama, A., & Turville, C., 2020. Influence of elements of climate change on the growth and fecundity of *Datura stramonium*. *Environmental Science and Pollution Research*, 27, 35859–35869.

Chauhan, B.S., Matloob, A., Mahajan, G., Aslam, F., Florentine, S.K., & Jha, P., 2017. Emerging challenges and opportunities for education and research in weed science. *Frontiers in Plant Science*, 8, 1537.

Chen, C.P., & Zhang, C.-Y., 2014. Data-intensive applications, challenges, techniques and technologies: A survey on Big Data. *Information Sciences*, 275, 314–347.

Cravero, A., Pardo, S., Sepúlveda, S., & Muñoz, L., 2022. Challenges to use machine learning in agricultural big data: A systematic literature review. *Agronomy*, *12*(3), 748.

de Castro Marcato, A.C., de Souza, C.P., & Fontanetti, C.S., 2017. Herbicide 2,4-D: A review of toxicity on non-target organisms. *Water, Air, & Soil Pollution*, 228, 1–12.

Derksen, D.A., Anderson, R.L., Blackshaw, R.E., & Maxwell, B., 2002. Weed dynamics and management strategies for cropping systems in the northern Great Plains. *Agronomy Journal*, *94*(2), 174–185.

Embrapa, P.A.V.B., Roman, M.E., & Fonseca, M.L.A., 2013. Report of the round table for biotechnology in cotton.

Fernández-Aparicio, M., Reboud, X., & Gibot-Leclerc, S., 2016. Broomrape weeds. Underground mechanisms of parasitism and associated strategies for their control: A review. *Frontiers in Plant Science*, 7, 135.

Florentine, S.K., Weller, S., Graz, P.F., Westbrooke, M., Florentine, A., Javaid, M., Fernando, N., Chauhan, B.S., & Dowling, K., 2016. Influence of selected environmental factors on seed germination and seedling survival of the arid zone invasive species tobacco bush (*Nicotiana glauca* R. Graham). *The Rangeland Journal*, *38*(4), 417–425.

Gianessi, L.P., & Reigner, N.P., 2007. The value of herbicides in US crop production. *Weed Technology*, *21*(2), 559–566.

Green, J.M., & Owen, M.D., 2011. Herbicide-resistant crops: Utilities and limitations for herbicide-resistant weed management. *Journal of Agricultural and Food Chemistry*, *59*(11), 5819–5829.

Hanif, Z., Naeem, M., Ali, H.H., Tanveer, A., Javaid, M.M., Peerzada, A.M., & Chauhan, B.S., 2017. Effect of environmental factors on germination of *Salsola foetida*: Potential species for rehabilitation of degraded rangelands. *Rangeland Ecology & Management*, *70*(5), 638–643.

Hassan, M.U., Chattha, M.U., Barbanti, L., Mahmood, A., Chattha, M.B., Khan, I., Mirza, S., Aziz, S.A., Nawaz, M., & Aamer, M., 2020. Cultivar and seeding time role in sorghum to optimize biomass and methane yield under warm dry climate. *Industrial Crops and Products*, 145, 111983.

Hassan, M.U., Chattha, M.U., Khan, I., Chattha, M.B., Barbanti, L., Aamer, M., Iqbal, M.M., Nawaz, M., Mahmood, A., Ali, A., & Aslam, M.T., 2021. Heat stress in cultivated plants: Nature, impact, mechanisms, and mitigation strategies—A review. *Plant Biosystems – An International Journal Dealing with all Aspects of Plant Biology, 155*(2), 211–234. https://doi.org/10.1080/11263504.2020.1727987

Hassan, M.U., Mahmood, A., Awan, M.I., Maqbool, R., Aamer, M., Alhaithloul, H.A.S., Huang, G., Skalicky, M., Brestic, M., Pandey, S., El Sabagh, A., & Qari, S.H., 2022. Melatonin-induced protection against plant abiotic stress: Mechanisms and prospects [Review]. *Frontiers in Plant Science, 13*. https://doi.org/10.3389/fpls.2022.902694

Hassan, M.U., Nawaz, M., Mahmood, A., Shah, A.A., Shah, A.N., Muhammad, F., Batool, M., Rasheed, A., Jaremko, M., & Abdelsalam, N.R., 2022. The role of zinc to mitigate heavy metals toxicity in crops. *Frontiers in Environmental Science*, 10, 990223.

Husain, P., Al Idrus, A., & Ihsan, M.S., 2020. The ecosystem services of mangroves for sustainable coastal area and marine fauna in Lombok, Indonesia: A review. *Jurnal Inovasi Pendidikan dan Sains, 1*(1), 1–7.

Javaid, M., Tanveer, A., Ahmad, R., Yaseen, M., & Khaliq, A., 2012. Optimizing activity of herbicides at reduced rate on *Emex spinosa* Campd. with adjuvants. *Planta Daninha*, 30, 425–435.

Javaid, M.M., Mahmood, A., Nadeem, M.A., Sarwar, N., Safdar, M.E., Ahmad, M., Hasanuzzaman, M., & Ahmad, S., 2022. Weed Management Strategies in Direct Seeded Rice. *In* N. Sarwar, R. Atique ur, S. Ahmad, & M. Hasanuzzaman (eds.) *Modern Techniques of Rice Crop Production* (pp. 327–347). Springer Singapore. https://doi.org/10.1007/978-981-16-4955-4_19

Joseph, D., 2014. Evaluation of 2,4-D and Dicamba Based Herbicide Programs for Weed Control in Tolerant Soybean.

Khan, B.A., Nadeem, M.A., Alawadi, H.F.N., Javaid, M.M., Mahmood, A., Qamar, R., Iqbal, M., Mumtaz, A., Maqbool, R., Oraby, H., & Elnaggar, N., 2023. Synthesis, characterization, and evaluation of nanoparticles of clodinofop propargyl and fenoxaprop-P-ethyl on weed control, growth, and yield of wheat (*Triticum aestivum* L.). *Green Processing and Synthesis, 12*(1). https://doi.org/10.1007/978-3-031-22269-6_5

Kumam, Y., Trick, H.N., Vara Prasad, P., & Jugulam, M., 2023. Transformative approaches for sustainable weed management: The power of Gene drive and CRISPR-Cas9. *Genes, 14*(12), 2176.

Kumar, V., Singh, S., Chhokar, R.S., Malik, R.K., Brainard, D.C., & Ladha, J.K., 2013. Weed management strategies to reduce herbicide use in zero-till rice–wheat cropping systems of the Indo-Gangetic Plains. *Weed Technology, 27*(1), 241–254.

Kumar, V., Swain, H.S., Roy, S., Das, B.K., Upadhyay, A., Ramteke, M.H., Kumar, V., Kole, R.K., & Banerjee, H., 2022. Integrated biomarker approach strongly explaining in vivo sub-lethal acute toxicity of butachlor on *Labeo rohita. Comparative Biochemistry and Physiology Part C: Toxicology & Pharmacology*, 261, 109427.

Lau, J.A., & Funk, J.L., 2023. How ecological and evolutionary theory expanded the 'ideal weed' concept. *Oecologia, 203*(3), 251–266.

Lee, J.W., Na, D., Park, J.M., Lee, J., Choi, S., & Lee, S.Y., 2012. Systems metabolic engineering of microorganisms for natural and non-natural chemicals. *Nature Chemical Biology, 8*(6), 536–546.

Linaza, M.T., Posada, J., Bund, J., Eisert, P., Quartulli, M., Döllner, J., Pagani, A., G. Olaizola, I., Barriguinha, A., & Moysiadis, T., 2021. Data-driven artificial intelligence applications for sustainable precision agriculture. *Agronomy, 11*(6), 1227.

Liu, Y., Hou, J., Li, C., & Wang, X., 2023. Intelligent soft robotic grippers for agricultural and food product handling: A brief review with a focus on design and control. *Advanced Intelligent Systems, 5*(12), 2300233.

Long, R.L., Gorecki, M.J., Renton, M., Scott, J.K., Colville, L., Goggin, D.E., Commander, L.E., Westcott, D.A., Cherry, H., & Finch-Savage, W.E., 2015. The ecophysiology of seed persistence: A mechanistic view of the journey to germination or demise. *Biological Reviews, 90*(1), 31–59.

Lykogianni, M., Bempelou, E., Karamaouna, F., & Aliferis, K.A., 2021. Do pesticides promote or hinder sustainability in agriculture? The challenge of sustainable use of pesticides in modern agriculture. *Science of the Total Environment*, 795, 148625.

Mohyuddin, G., M.A. Khan, A. Haseeb, S. Mahpara, M. Waseem and A.M. Saleh. 2024. Evaluation of Machine Learning approaches for precision farming in Smart Agriculture System-A comprehensive Review. IEEE Access *46*(9) 31–57.

Mahmood, A., Ullah, H., Shahzad, A., Ali, H., Ahmad, S., Zia-Ul-Haq, M., Honermeier, B., & Hasanuzzaman, M., 2013. Dry matter yield and chemical composition of sorghum cultivars with varying planting density and sowing date. *Sains Malaysiana*, *42*(10), 1529–1538.
Mall, U., & Singh, G.S., 2013. Soil Seed Bank Dynamics: History and Ecological Significance in Sustainability of Different Ecosystems. *In Environment and Sustainable Development* (pp. 31–46). Berlin: Springer.
Mangla, M., Shinde, S.K., Mehta, V., Sharma, N., & Mohanty, S.N., 2022. *Handbook of Research on Machine Learning: Foundations and Applications*. CRC Press.
Manisankar, G., Ghosh, P., Malik, G.C., & Banerjee, M., 2022. Recent trends in chemical weed management: A review. *The Pharma Innovation*, *11*(4), 745–753.
McNeill, J., 1976. The taxonomy and evolution of weeds. *Weed Research*, *16*(6), 399–413.
Mocek-Płóciniak, A., Mencel, J., Zakrzewski, W., & Roszkowski, S., 2023. Phytoremediation as an effective remedy for removing trace elements from ecosystems. *Plants*, *12*(8), 1653.
Monteiro, A., & Santos, S., 2022. Sustainable approach to weed management: The role of precision weed management. *Agronomy*, *12*(1), 118.
Oikonomidis, A., Catal, C., & Kassahun, A., 202). Deep learning for crop yield prediction: A systematic literature review. *New Zealand Journal of Crop and Horticultural Science*, *51*(1), 1–26.
Pannell, D.J., Tillie, P., Rodríguez-Cerezo, E., Ervin, D., & Frisvold, G.B., 2016. *Herbicide resistance: Economic and environmental challenges*.
Pons, T.L., & Fenner, M., 2000. Seed responses to light. *Seeds: The Ecology of Regeneration in Plant Communities*, 2, 237–260.
Qasem, J.R., 2013. Herbicide Resistant Weeds: The Technology and Weed Management. *In Herbicides–Current Research and Case Studies in Use*. UK: InTech, 445–471.
Quamar, M.M., Al-Ramadan, B., Khan, K., Shafiullah, M., & El Ferik, S., 2023. Advancements and applications of drone-integrated geographic information system technology—A review. *Remote Sensing*, *15*(20), 5039.
Rai, N., Zhang, Y., Ram, B.G., Schumacher, L., Yellavajjala, R.K., Bajwa, S., & Sun, X., 2023. Applications of deep learning in precision weed management: A review. *Computers and Electronics in Agriculture*, 206, 107698.
Rasheed, A., Li, H., Tahir, M.M., Mahmood, A., Nawaz, M., Shah, A.N., Aslam, M.T., Negm, S., Moustafa, M., Hassan, M.U., & Wu, Z., 2022. The role of nanoparticles in plant biochemical, physiological, and molecular responses under drought stress: A review [review]. *Frontiers in Plant Science*, *13*. https://doi.org/10.3389/fpls.2022.976179
Rehman, H., Mahmood, A., Ishfaq, F., Javaid, M.M., Haider, A., Asif, M., Nadeem, M.A., & Nargis, J., 2023. Sustainable Crop Management for Drylands. *In* M. Hasanuzzaman (ed.) *Climate-Resilient Agriculture, Vol 1: Crop Responses and Agroecological Perspectives* (pp. 435–464). Springer International Publishing. https://doi.org/10.1007/978-3-031-37424-1_20
Riemens, M., Sønderskov, M., Moonen, A.-C., Storkey, J., & Kudsk, P., 2022. An integrated weed management framework: A pan-European perspective. *European Journal of Agronomy*, 133, 126443.
Rueda-Ayala, V., Rasmussen, J., & Gerhards, R., 2010. Mechanical weed control. *Precision Crop Protection-the Challenge and Use of Heterogeneity*, 279–294.
Shaikh, T.A., Rasool, T., & Lone, F.R., 2022. Towards leveraging the role of machine learning and artificial intelligence in precision agriculture and smart farming. *Computers and Electronics in Agriculture*, 198, 107119.
Sharma, N., Choudhary, B., & Puri, N. 2023 Nanoherbicides: Formulations and Applications. *In Nanopesticides, Nanoherbicides, and Nanofertilizers* (pp. 83–93). Florida: CRC Press.
Shrestha, A., Clements, D.R., & Upadhyaya, M.K., 2004. Weed management in agroecosystems: Towards a holistic approach. *Recent Research Developments in Crop Science*, *1*(Part II), 451–477.
Sishodia, R.P., Ray, R.L., & Singh, S.K., 2020. Applications of remote sensing in precision agriculture: A review. *Remote Sensing*, *12*(19), 3136.
Srivastava, J., & Prasad, V., 2019. Evolution and paleobiogeography of mangroves. *Marine Ecology*, *40*(6), e12571.
Su, W.-H., 2020. Advanced machine learning in point spectroscopy, RGB-and hyperspectral-imaging for automatic discriminations of crops and weeds: A review. *Smart Cities*, *3*(3), 767–792.
Sujaritha, M., Annadurai, S., Satheeshkumar, J., Sharan, S.K., & Mahesh, L., 2017. Weed detecting robot in sugarcane fields using fuzzy real time classifier. *Computers and Electronics in Agriculture*, 134, 160–171.
Swanton, C.J., Mahoney, K.J., Chandler, K., & Gulden, R.H., 2008. Integrated weed management: Knowledge-based weed management systems. *Weed Science*, *56*(1), 168–172.

Swanton, C.J., & Murphy, S.D., 1996. Weed science beyond The weeds: The role of integrated weed management (IWM) in agroecosystem health. *Weed Science*, *44*(2), 437–445.
Tong, P.S. 2024. A case study on weed management awareness in Kampar: Farmers' perceptions, weed flora composition, and effect of fertilizers on weed growth, *UTAR*, *57*(7), 533–547.
Tanveer, A., Javaid, M.M., Irfan, M., Khaliq, A., & Yaseen, M., 2015. Yield losses in chickpea with varying densities of dragon spurge (*Euphorbia dracunculoides*). *Weed Science*, *63*(2), 522–528.
Tanveer, A., Khaliq, A., Javaid, M., Chaudhry, M., & Awan, I., 2013. Implications of weeds of genus euphorbia for crop production: A review. *Planta Daninha*, 31, 723–731.
Tanveer, A., Mumtaz, K., Javaid, M., Chaudhry, M., Balal, R., & Khaliq, A., 2013. Effect of ecological factors on germination of horse purslane (*Trianthema portulacastrum*). *Planta Daninha*, 31, 587–597.
Trivedi, T., & Ahuja, D., 2011. Integrated pest management: Approaches and implementation. *Indian Journal of Agricultural Sciences*, *81*(11), 981–993.
Umetsu, N., & Shirai, Y., 2020. Development of novel pesticides in the 21st century. *Journal of Pesticide Science*, *45*(2), 54–74.
Van Driesche, R., & Hoddle, M., 2009. *Control of Pests and Weeds by Natural Enemies: An Introduction to Biological Control*. New Jersey: John Wiley & Sons.
Van Klompenburg, T., Kassahun, A., & Catal, C., 2020. Crop yield prediction using machine learning: A systematic literature review. *Computers and Electronics in Agriculture*, 177, 105709.
Vanloqueren, G., & Baret, P.V., 2017. How Agricultural Research Systems Shape a Technological Regime That Develops Genetic Engineering but Locks Out Agroecological Innovations 1. *In Food Sovereignty, Agroecology and Biocultural Diversity* (pp. 57–92). Oxfordshire: Routledge.
Velusamy, P., Rajendran, S., Mahendran, R.K., Naseer, S., Shafiq, M., & Choi, J.-G., 2021. Unmanned aerial vehicles (UAV) in precision agriculture: Applications and challenges. *Energies*, *15*(1), 217.
Vreysen, M., Robinson, A., Hendrichs, J., & Kenmore, P., 2007. Area-wide integrated pest management (AW-IPM): Principles, practice and prospects. *Area-wide control of insect pests: from research to field implementation* (pp. 3–33). Springer
Walsh, M.J., Broster, J., Chauhan, B., Rebetzke, G.J., & Pratley, J., 2019. Weed Control in Cropping Systems–Past Lessons and Future Opportunities. *In Australian Agriculture in 2020: From Conservation to Automation* (pp. 153–172). Australian Society of Agronomy.
Watkins, R., Meiers, M.W., & Visser, Y., 2012. *A Guide to Assessing Needs: Essential Tools for Collecting Information, Making Decisions, and Achieving Development Results*. World Bank Publications.

8 Use of Modern Approaches to Mitigate Challenges of Resistance

Maira Tanveer, Alia Javed, Muhammad Ather Nadeem, Zeki Erden, Çağdaş Can Toprak*, Javaria Nargis, and Laiba Sajid*

8.1 INTRODUCTION

Agricultural resistance is an ongoing and changing issue that farmers globally face, which presents substantial risks to crop yield, environmental health, and food safety (Savary *et al.*, 2017). The use of agrochemicals like fertilizers and pesticides has become an important component of modern agricultural systems in the last century to bring about a significant increase in crop yields. It is estimated that 35%–45% of crop production is lost due to improper protection of crops against the pest attack of untimely insects, weeds, and diseases. Global data regarding the use of pesticides, including insecticides, acaricides, fungicides, rodenticides, herbicides, and antibiotics used for the protection of crops, were around 2.72 million metric tons in 2012 (Khan *et al.*, 2019). About 45% of pesticides in the world are consumed in Europe, followed by the United States (25%) and the rest of the world (30%), including India. Consumption of herbicides (about 47.5%) followed by insecticides (29.5%), fungicides (17.5%), and only 5.5% are the share of other classes of pesticides among pesticide classes. In addition to pest control in agriculture, pesticides are used to control disease vectors in humans and livestock and to prevent or control organisms affecting other human activities. India has an estimated market loss due to pest attack by US $43 million in different regions. Cotton is the crop with the highest pesticide consumption, followed by paddy, wheat, and sugarcane. In contrast to the world market, where herbicides lead the way for crop protection, insecticides account for nearly 60% (Yadav *et al.*, 2015; Balal *et al.*, 2017). Fungicides and herbicides are the fastest growing segments comprising 18% and 16%, respectively, of the total market for crop protection chemicals. The use of pesticides in India appears to be the least in the world (0.6 kg/ha) compared to 5–7 kg/ha in the UK (0.7–1 times) and almost 20–30 times (13 kg/ha) in China (Rajan *et al.*, 2023).

Sustainable crop protection (SCP) is a crucial aspect of food security in the present, as well as for the future. However, this continuous and widespread use of the pesticides in the last few decades has brought serious resistance to pest species, leading to pest resurgence as well. The effectiveness of crop protection products declines when pests develop resistance to a specific pesticide, thus limiting pest management choices (Sparks *et al.*, 2021). Moreover, pesticide resistance development necessitates enhanced chemical application for crop protection, increasing adverse effects on human health and the environment because of the residual toxicity of pesticides (Chadha *et al.*, 2019). Similar to the case of the efficacy of antibiotics to control human diseases around the globe, emerging infectious resistant bacterial strains, other microbial strains, can also make the control of crop pests be under threat by the evolution of resistant strains, such as against insecticides, fungicides, and herbicides (Paul and Mandal, 2019).

Pest resistance to pesticides is an acute global concern that seriously restricts the application of pesticides in pest management. For over 50 years, the development of resistance to pest

* Corresponding authors: cagdascan.toprak@siirt.edu.tr; zeki.erden@siirt.edu.tr

DOI: 10.1201/9781003498698-8

management tools has been a key driver of pest management decisions. They were first reported almost 100 years ago. Afterward, a rapid increase in occurrences of insecticide resistance followed with the synthetic organic insecticides (dichlorodiphenyltrichloroethanes (DDTs)) introduction in the 1940s because of their widespread use. The amount of products and number of species with insecticide resistance increased over time. The first reports about weed resistance to herbicides and similar mechanisms among some of the organisms to fungicides appeared in the 1960s and 1970s. But authors found that the rates of insecticide resistance were higher than herbicide and fungicide resistance (Popivanov *et al.*, 2015).

8.2 TYPES OF RESISTANCE IN AGRICULTURE

Agricultural resistance refers to pest and disease resistance; however, many adaptive mechanisms are used that pests, diseases, and plants use to overcome the inhibitory effects of pesticides. In agriculture, resistance is considered a major challenge to the cultivation and production of crops globally. Disease resistance is the ability of pathogens, such as bacteria, fungus, virus, and nematodes, to evolve mechanism to overcome the effect of biocides used for their control and also to exhibit physiological and biochemical effects on plants (Figure 8.1) (Rajwade *et al.*, 2020).

Pathogens could increase their capability for detoxification via change in target site, efflux pumps, or involving enzymes. This ability could result in overcoming their control method, thus increasing the disease burden. The issue of insect resistance to chemical pesticides is a high-priority concern involving the ability of arthropods, like parasites and pests, to overcome chemical inhibition by either of a few processes. These include metabolic detoxification, target-site mutations, or

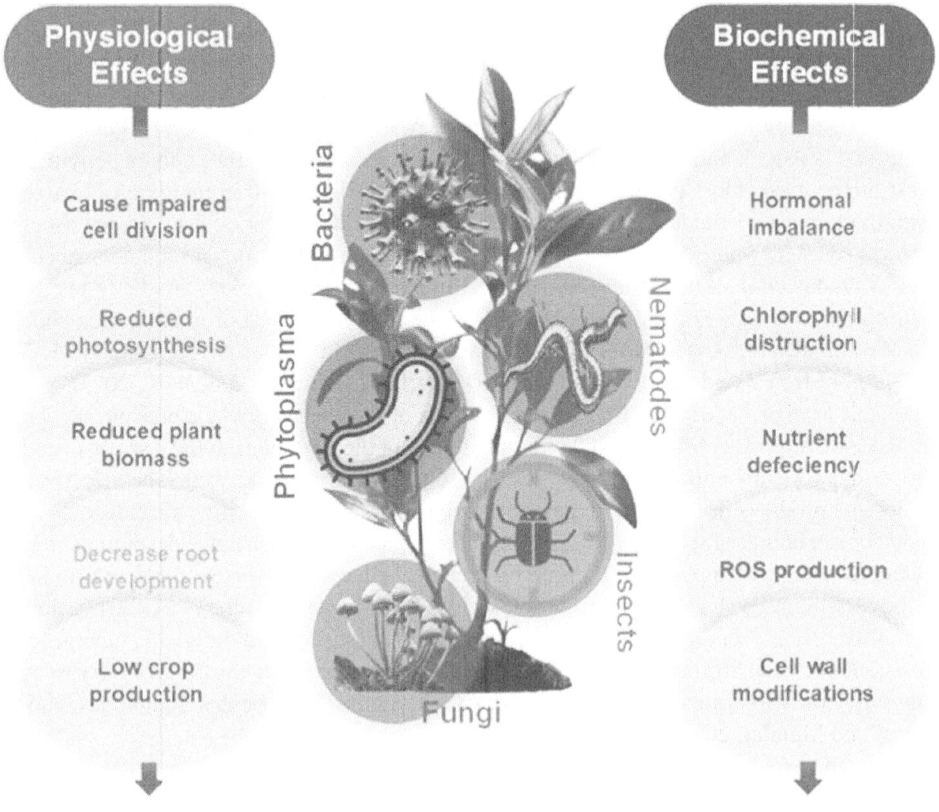

FIGURE 8.1 Physiological and biochemical effects of different disease-causing agents on plants.

behavioral adaptations that decrease the effectiveness of standard pharmacologic control methods (Parankusam *et al.*, 2018).

Herbicide resistance in agriculture is a key source of resistance where weeds evolve to tolerate or resist the herbicides, diminishing the effectiveness of the chemical agents to control weed populations. This can occur via target-site alterations in weed species, metabolic degradation of herbicides, or a more controlled uptake of herbicides. This has resulted in increased weed infestation and the potential decreased yield of agricultural crops (Beckie *et al.*, 2019). The evolution of resistance to the non-chemical control measures, like cultural or mechanical methods via adaptive modifications of pests, diseases, or weeds that allow them to overcome those control methods, has added more difficulty in the management (Nawaz *et al.*, 2019). A better understanding of the wide range of resistance routes in agriculture is essential to create responsive and sustainable solutions that can effectively alleviate the consequences of resistance on the success, reliability, and sustainability of pest, disease, and weed management, and the well-being of agriculture in general around the world. This can only be achieved by a combination of many methods of pest/weed/disease control; those most important include, but are not limited to, biological control methods, cultural methods, chemical control methods, and genetic engineering (GE) methods. They should be tailored for particular pressures from pests, diseases, and weeds and impact minimization on natural selection and fostering an ecologically balanced ecosystem within agricultural territories. Furthermore, overcoming resistance requires the collaboration and sharing of knowledge and experience between farmers and academics and the interaction and support from policymakers and industry. The lateral is equally important, as economically sustainable and comparatively fair solutions are the only credible resistance solutions (Lamichhane *et al.*, 2016). Around pursuing the above point, agriculture in a changing world could address one more option, concerning the resistance, enhance productivity of the crop, sustainable programs of environmental stewardship, and global food security with a challenging proactive integrated resistance management solution (Ali *et al.*, 2023). Various resistances in each category and how they affect different crops can be seen in Table 8.1.

TABLE 8.1
Various Types of Resistance and Their Effects on Different Crops

Type of Resistance	Crops Affected	Effects on Crops	References
Pesticide resistance	Corn, cotton, soybean, rice	Development of resistance in pests to chemical pesticides due to repeated and/or improper use	Nandula (2019)
Disease resistance	Wheat (rusts), potato (late blight), tomato (bacterial wilt)	Resistance in crops to specific diseases caused by pathogens like fungi, bacteria, or viruses	Saharan *et al.* (2016)
Herbicide resistance	Corn, soybean, cotton, canola	Development of resistance in weeds to herbicides due to repeated and/or improper use	Jhala *et al.* (2021)
Insect resistance	*Bt* cotton (against bollworm), *Bt* corn (against corn borer)	Resistance in crops to specific insect pests due to genetic modifications or natural traits	Sheikh *et al.* (2017)
Biotic stress resistance	Soybean (soybean cyst nematode), wheat (aphids)	Resistance in crops to various biotic stresses like nematodes, mites, or aphids	Choudhary *et al.* (2021)
Abiotic stress resistance	Rice (submergence tolerance), maize (drought tolerance)	Resistance in crops to abiotic stresses like drought, salinity, or extreme temperatures	Nachimuthu *et al.* (2017)

8.3 FACTORS CONTRIBUTING TO RESISTANCE DEVELOPMENT

Resistance development is a complex issue in agriculture, affected by multiple intertwined and interacted factors inside the agroecosystems. Genetic variability among populations of pests, pathogens, or weeds necessarily exists; it is the basis for the development and spread of resistance. The genetic variability of populations leads to the presence of organisms with traits that confer resistance to control measures usually based on the application of chemical pesticides, herbicides, or fungicides. Such successive application of these measures leads to selective pressure on the populations by the survival and reproduction of resistant individuals (Vats, 2015). Furthermore, weak surveillance and control apply relative mild selection pressure definitive of the virulent form, since all resistant individuals are allowed to develop unchecked which can amplify the resistance. For instance, increased high dependency on one mode of action, and failure to alternate control options, may enhance the intensity of selection and the rate of spread of the resistance (Clapp, 2021). Misapplication of chemicals, such as underdosed or overdosed, or if application guidelines are not followed, might also encourage resistance, by selecting for resistant organisms and decreasing the effectiveness of control measures. The resistance dynamic is also influenced by environmental factors in agricultural landscapes. Resistance development and dissemination (Pacilly *et al.*, 2018): temperature, humidity, or cropping practices may affect the rate at which resistance develops and disseminates. Planting the same crop repetitively promotes conditions conducive for the exponential proliferation of resistance. In this regard, uniform control measures exert uniform high selection pressure. In addition, the viability of some resistance is determined by social and economic issues. As the availability of resources, levels of knowledge, or quality of education can impact resistance management, this can also influence the evolution of resistance (Wuest *et al.*, 2021). Inadequate access to alternative control measures and a lack of knowledge may hinder farmers' adoption of control measures. Replacing nitrogen fertilizers with alternative practices may be effective to a point, but a more holistic response to resistance stemming from research, education, policy, and stakeholder engagement is required. This approach aims to develop and apply proactive strategies for resistance management. It must place emphasis on keeping the effectiveness of current control strategies, decreasing the selection pressure, and emphasizing sustainable agricultural practices that decrease the chances of resistance emergence. Identifying the factors that promote resistance development and addressing these involves complex and interacting factors that must be understood and accounted for so that stakeholders can work together to foster plant health, increase productivity, and safeguard the sustainability of agricultural systems (Schröder *et al.*, 2019).

8.4 MODERN APPROACHES TO RESISTANCE MANAGEMENT

Integrated systems for resistance management include integrated pest management (IPM) strategies, GE, precise and diversified cropping systems, crop rotations, biological control, and sustainable agriculture practices. To control this increasing threat due to resistance, new mechanisms and natural enemies are being investigated that lead to developing resistant crops, maximizing the strategic utilization of resources (Baker *et al.*, 2020). Conversely, if the three methods are applied equally, they can be used to manage pests, diseases, and weeds effectively while avoiding the damage of the environment in line with the sustainability of the agriculture system over the long run.

8.4.1 Integrated Pest Management (IPM)

In the 1970s, in response to rising awareness of the negative side effects of pesticide use, the concept of "Integrated Pest Management" was first introduced in agriculture. The strategy builds on the integration of artistic methods and pest biology for managing insect pests in crops. Such a complex technology will require a comprehensive experiential-based learning approach, such as farmer field schools, to help farmers appreciate how to use and what it means for IPM. Development of threshold theory and harmonic control tactics for pest management has been researched for about 50 years in certain parts of Europe, Canada, and the United States (Deguine *et al.*, 2021). Most of the cultural, mechanical, genetic, chemical, and biological approaches are included in the methods and practices of IPM to keep the number of harmful crop insects below the level where the amount of damage results in economic loss (Romeh, 2019). The spotted *Alfalfa aphid* and the *Alfalfa caterpillar* were the first pests for which early "supervised" and "integrated" management schemes were created (Pretty and Bharucha, 2015). IPM has since been described as the predominant crop protection paradigm, but adoption rates are still modest. Due to people moving to new locations within or across nations and the globalization of the food industry, more pest species have been introduced into areas where they did not previously exist. IPM's gradual expansion in popularity has led to increased recognition of it as an ecologically sustainable strategy that can also ensure the output and quality of agricultural goods. IPM, then, may be defined as the adoption of a superior pest management method that reduces agricultural yields while minimizing chemical toxicity to the beneficial organisms that are essential to the success of the actual management practice (Tamanreet and Mandeep, 2020). Currently, countries in Africa and Latin America are using integrated production and pest management together and concurrently. The central tenets of IPM are (a) encouraging healthy soil and crops, (b) fending off natural predators, (c) frequently monitoring crop health and pest activity, and (d) imparting technological expertise to the farming society. IPM strategies expand on these needs to combine the ability to control the behavior of numerous insects that are detrimental to agricultural production with enhanced environmental conservation due to reduced chemical pesticides (Torres and Bueno, 2018).

8.4.1.1 Principles of IPM

The principles of IPM are rooted in ecological, economic, and social considerations, aiming to minimize reliance on chemical pesticides while maximizing the effectiveness, sustainability, and economic viability of pest control practices. IPM revolves around the prevention of pest development and about identifying and maintaining unfavorable conditions for those entities to thrive. These conditions are maintained through practices such as crop rotation, use of pest-resistant varieties, and healthy soils and ecosystems (Karlsson Green *et al.*, 2020). Scouting and monitoring are staples of IPM practice, allowing farmers to assess the size of the pest problem and take appropriate measures to address it. If intervention becomes necessary, IPM ensures that various non-chemical and less toxic measures are adapted—cultural practices, biological control, and mechanical and physical measures are taken to reduce effects on human health, non-target organisms, and the environment (Brown *et al.*, 2022). Chemicals are always the last option to use and apply since the selection process is based on the substance's toxicological properties and the risk of resistance from pests. Lastly, the other aspect of IPM is learning, modification, and stakeholder consultation (Begum *et al.*, 2017). Farmers, scientists, extension officers, and others frequently share information and update protocols to fit the trends in the environment. Because of these factors, the IPM is the most reliable way to deal with pest issues.

8.4.1.2 Implementation Strategies of IPM

The implementation of IPM relies on several essential strategies that, on the practical level, should be tailored to the needs of each agricultural system. Most importantly, effective monitoring and identification of pests should be done using usual weather scouting to ascertain the presence of species and find the cause of action. We may incorporate cultural control methods into the system, such as crop rotation, resistant varieties, and rational irrigation and fertilization to generate disease or undesirable circumstances for the pest species (Panth *et al.*, 2020). Moreover, biological control is a cornerstone of IPM that involves the augmentation of these natural enemies (predators, parasitoids, and pathogens) in large enough populations to suppress the pest population. Finally, in the last case, pesticides may provide some help when selected based on IPM, where there are certain thresholds considered, e.g., pesticides would not result in detrimental impact on the environment and/or where application methods would not be off-target. To this end, education is crucial, a role fulfilled by extension services, training workshops, and farmer networks (Jacquet *et al.*, 2022). Research and innovation complement this by enabling us to maximize the impact of what we know already, alongside the development of new tools and strategies that will help us rethink IPM and push sustainable agriculture forward. *Interventions*: the different strategies involved in IPM are integrated and pragmatic system of approaches that conserve not only crop health but also farmer health, environmental health, and long-term sustainability of agriculture. Thus, successful implementation of IPM can only be achieved through a comprehensive and systemic approach integrating scientific knowledge with traditional experience and the involvement of all relevant stakeholders to designing resilient and sustainable agricultural systems that acknowledge the issue of pests, while empowering their destructive capacity where it is useful in human life and mitigating their damaging effect in a safe manner. Thus, IPM will assist farmers in reducing the use of pesticides and supporting biodiversity throughout sustainable agriculture (van Lenteren and Nicot, 2020).

8.4.2 Genetic Engineering and Biotechnology

On a molecular level, transfer of certain genes, which express the above factors into crop plants, confers the above resistance traits through genetic manipulation. Therefore, GE and biotechnology are utilized to reroute the plants for durability to address certain fungal toxins, insects, and environmental stresses (Ahanger *et al.*, 2017; Hanif *et al.*, 2017). In addition, resistance traits can be viewed as those conferring insect, disease, herbicide, and abiotic stress tolerances (Abdullah *et al.*, 2022). A widely implemented example of GE to impart resistance to crops is the introgression of genes for resistance to insect pests into genetically modified crops expressing insecticidal proteins derived from *Bacillus thuringiensis*. In this case, the above technology is environmentally friendly and reduces the application of chemical insecticides, which cause severe effects on the environment via non-target impacts (Mahmood *et al.*, 2013; Koul, 2020). Similarly, the use of GE has ensured the above resistance traits through the transference of resistance genes from wild relatives or other organisms. Moreover, abiotic resistance traits such as diseases and late blight resistance have been conferred to enhance resistance against specific diseases. This is achievable through transformation technology to transfect the desired genes or clustered regularly interspaced short palindromic repeats and CRISPR-associated protein 9 (CRISPR-Cas9) technology, which may require foreign DNA insertion to transfect the gene of interest (van Esse *et al.*, 2020; Rasheed *et al.*, 2022a). Despite the above advantages, the introduction of genetically engineered crops may still face criticism. GE critics argue that genetically engineered crops may lead to unexpected ecological impacts. These include the evolution of pest resistance and contamination to wild relatives. Overcoming the above challenges requires comprehensive risk assessment, open regulatory processes, and public participation to enhance the sustainable and safe practice of GE. Some of the principles of agroecology and biodiversity strategies might also provide some collateral benefits when integrated into GE. Overall, this integration of GE will facilitate the production of resistant crops to improve sustainability in agriculture and fulfill the increasing nutritional demand (Ali *et al.*, 2022; Rasheed *et al.*, 2022b).

8.4.2.1 Development of Resistant Crop Varieties

Using GE techniques, scientists have been able to add specific genes into crop plants, including resistance to herbicides, pests, or diseases, without losing the agronomically desirable traits. Perhaps the most successful genetically engineered resistance is the insertion of *B. thuringiensis* genes into crops (e.g., corn, cotton), which are lethal to certain insect pests. It has also helped in bio-safe and effective pest management, as the Bt crops did not require any synthetic insecticides (Anderson *et al.*, 2019). Likewise, GE has also given rise to varieties resistant to viral, bacterial, or fungal pathogens, thereby preventing crop loss due to diseases. Moreover, herbicide-resistant crop varieties have also been developed to enhance crop tolerance, thus enabling the killing of weeds only sparing crops. This has been applied in revolutionizing weed control, such as the introduction of no-till farming, which involves direct planting of crop seeds in weeds (Ali *et al.*, 2013). Thus, GE could alleviate the concern of reducing crop productivity and losses that have deprived the world of billions of people (Bigini *et al.*, 2021). In addition to resistance to pests and disease, enhancing resistance to abiotic stress, such as heat drought, salinity, and extreme temperatures, was considered using GE (Wasaya *et al.*, 2021; Tanveer *et al.*, 2023). Importing genes associated with drought tolerance into crop plants would make it possible to exploit crops even under adverse conditions, thus reducing is essential to increase agricultural productivity (Tanveer *et al.*, 2012; Fahad *et al.*, 2017). Although the GE of crops is faced with controversies, such as environmental risks, human and animal health impacts, and socio-economic impacts, it is guided by a regulatory framework that monitors testing, approval, and release of genetically modified crops. However, public awareness through informatics awareness programs will be the primary success of GE for the world to realize the full potential of crop biotechnology. In summary, crop biotechnology remains the future of increasing crop production, reducing ecological footprints under sustainable agricultural practice (Ullah *et al.*, 2013; Trivedi *et al.*, 2021).

8.4.2.2 Gene Editing Technique

Novel techniques such as gene editing have emerged as a state-of-the-art approach to combat resistance challenges in agriculture. Gene editing platforms like CRISPR-Cas9 are revolutionary as they allow the precise alteration of the genetic architecture of species, providing unparalleled opportunities to create crop varieties with enhanced resistance traits (Bisht *et al.*, 2019). In the case of pest resistance, gene editing may be employed to insert or alter the genes responsible for pest resistance mechanisms such as toxin synthesis or recognition proteins into crop plants. Gene editing can utilize an already existing gene encoding the defense protein or the signaling pathways involved in plant immunity to put disease resistance on enhancing the host's prohibition of infection by the pathogen. Furthermore, gene editing demonstrates significant potential in solving the increasing cases of herbicides' resistance by some weeds. The GE of crops will contain new herbicide tolerance traits, preferably the editing of herbicide molecules affecting the crop sites, hence preventing resistance (Kushalappa *et al.*, 2016).

On the other hand, gene editing is similar to genuine conventional breeding methods in which it permits quick and precise modifications and adjustments as opposed to the years taken to develop resistant species. And also, gene editing is likely to alter resistance management too, since this technology permits resistance traits expression against fresh dangers or improving existing ones against new pest, disease, or weed populations (Tanveer *et al.*, 2015; Nerkar *et al.*, 2022). However, the utilization of gene editing in agriculture has moral, regulatory, and acquiescence consequences, but the most crucial is the safety of consuming genetically modified organisms (GMOs) and other unexpected environmental consequences. Hence, utilizing this technique responsibly in resistance management requires a good balance of these, as well as clear jurisdiction and control to guarantee that this powerful tool is used in agriculture responsibly. Agriculture may then move toward more competent, productive, and sustainable systems that can effectively manage resistance and fulfill the evolving demands of worldwide food security while using gene editing and other advanced biotechnology tactics strategically (Parray *et al.*, 2019).

8.5 CROP ROTATION AND DIVERSIFICATION

The use of crop rotation and diversification are among the vital approaches to tackle resistance in agriculture, offering benefits in controlling pests sustainably, health of the soil and entailing resilience of the agroecosystem. Crop rotation is farming where crops are rotated through the same area during a certain period, while crop diversification is the growing of various crops with the same farming system. They break the pest life cycle interrupting its overflow, increasing competitive natural enemies, and providing high potential impact on resistance development and spread (Shah *et al.*, 2021). Crop rotation slows down the transition movements of pests across the farm that peculiar crop specificity; hence, they prevent the succession generation and control the use of chemical pesticides. Moreover, concerns in nutrient cycling, erosion prevention, and weed control are reduced, making a strong resilient and productive agroecosystem (Babar *et al.*, 2009; Bass and Nauen, 2023). Diversification is done on the farming system level though creates heterogeneity to battle pests and diseases. The benefits in crop diversification are the lowering of risks associated with total loss of crops in monocropping; nutrient utilization of leguminous crops happens to be very important with nitrogen fixation, as the soil fertility is improved, hence reducing or replacing commercial and nitrogen-synthesized fertilizers (Waheed *et al.*, 2019; Thomine *et al.*, 2022). Furthermore, genetic properties in adopted pests provide genetic resource use and increase diversity resistance in pest populations. However, the planting time and planting compatibility, market demand, and other changes from the ordinary way of planting need to be planned specifically. Weed management is also crucial when diversifying crops as these cropping systems come with cover cropping, tillage, and organic mulching that regulate weeds and prevent a weed shift. Finally, education is fundamental for implementing crop rotation and diversification through extension, including financial incentives from the government's policies to encourage the adoption of highly productive practices (Cordeau, 2022).

8.5.1 Benefits in Mitigating Resistance

Crop rotation and diversification provide several benefits when it comes to resistance management and sustainability in agriculture. Crop rotation is the systematic shifting of different crops through a particular order over a while, while diversification is cultivating several crops in one field or farm. Both options help to disrupt the pests and disease life cycles by limiting the availability of host plants and making the conditions hostile for pests and pathogens (Peltonen-Sainio *et al.*, 2017; Mehmood *et al.*, 2018). Rotating crops such as maize and soybeans, which share the same life cycle, or similar root and growth characteristics, continue the food availability and habitat conditions conducive for pests to breed; this helps the pests to continue multiplying in the soil, making the infestation inevitable (Yu *et al.*, 2022). In addition, diversification also creates a healthy system where a variety of crops are coming up, hence making a more favorable condition for natural enemies that control pest infestation. On the other hand, when different types of crops are planted in the same field, it provides a healthy balance between the two species of insects, which regulate each other. Other than pests and disease control, crop rotation and diversification also help to manage water and improve soil conditions. For instance, crop rotation maintains soil moisture by selecting the crop to be grown at a particular time. Furthermore, diversification helps to add some nutrients and controls soil-borne diseases (He *et al.*, 2019). Leguminous crops such as peas or beans sequester some atmospheric nitrogen, thereby adding some necessary nutrients to the soil and minimizing the use of chemical fertilizers (Meena and Lal, 2018; Mustafa *et al.*, 2022).

8.5.2 Examples of Successful Rotation Practices

Crop rotation is a cornerstone agricultural practice that involves organized changing of crops cultivated in a specific field within a certain season or years framework. For several centuries, crop rotation has shown its positive results in improving soil quality, pest and disease control, and overall crop productivity. One of the many successful examples of crop rotation practices is that of the medieval three-field system. It refers to dividing the field into three parts: one of them sown with winter crops such as wheat or rye, the second with spring crops like barley or oats, the third remaining fallow or sown with legumes such as peas or beans (Wang *et al.*, 2020). Particularly, the field left uncultivated or planted with legumes allowed for predicting soil nitrogen through the fixation of atmospheric nitrogen by legumes. In addition, the spread over three fields prevented pest and disease spread and weed proliferation (Scavo *et al.*, 2022).

In modern agriculture, the corn-soybean rotation widely practiced in the United States serves a prime example. Corn and soybeans are grown in rotation for maximum yield, each contributing to soil fertility renewal with its cycle. In addition, soybeans' ability to fix nitrogen and high nutrient demand by corn provides stable yields of both crops and continues to replenish the soil (Grassini *et al.*, 2015). Crop rotation happens to be among the most discussed topics in a scientific dialogue about crop circles. Diversified crop rotation inclusion in organic systems is possible, where the primary goal is to have the soil covered, eliminate the need for using herbicides for weed control and preserve soil (Sharma *et al.*, 2021). Cover crops such as clover or vetch increase the soil structure and maintain biodiversity. Last but not least, one should remember about successful examples of crop rotation in the form of intercropping, a cultural method, including growing several crops on one field. Intercropping dramatically influences resource requirements and resources uptake by crops. As a result of reducing disease spread and blight development, the cultivated crop's health is ensured, and the soil is maintained stable. Cultivation of crops clearly highlights the applicability of crop rotation to certain conditions and benefits (Garbelini *et al.*, 2022).

8.5.3 Biopesticides and Their Applications

Biopesticides are a rapidly growing category of agricultural products derived from natural sources such as microorganisms, plants, and biochemicals. They are considered a sustainable alternative to synthetic chemical pesticides since they are environmentally friendly. Biopesticides are a diverse group of products with varying modes of action and target pests categorized into microbial pesticides, plant-derived pesticides, and biochemical pesticides (Prabha *et al.*, 2016). Various biopesticides and their applications are involved in different plants for managing resistance (Table 8.2).

Microbial pesticides contain appropriate microorganisms such as bacteria, fungi, viruses, and protozoa targeting only specific pests, pathogens, or weeds. The control mechanisms of these microorganisms may include pathogen competition or parasite production through releasing toxins to kill the target organisms (Sarwar, 2015).

For instance, *B. thuringiensis* is a bacterium that controls insect pests while *Beauveria bassiana* and *Metarhizium anisopliae* are fungal organisms applied to kill insect pests and soil-borne pathogens. Unlike the living organisms contained in microbial pesticides, plant-based are derived from botanical extracts and oils or secondary metabolites produced by plants (Iqbal *et al.*, 2021). Plant-based pesticides can be insecticidal, fungicidal, or herbicidal and can be used to eliminate a variety of pest species and diseases. Biochemical pesticides comprising natural components like pheromones, insect growth regulators, and enzymes are used to alter the behavior, development, or physiology of pests. For instance, pheromones are sexual attractants for insects, which are used by scientists to interrupt mating behavior and kill pests (Mandal, 2019). Compared to artificial chemicals, biopesticides have several advantages, including their lower environmental risk, negligible exposure to early levels and residues on food and water, and compatibility with IPM. Biopesticides are also less hazardous to human health and kill non-target pests like pollinators and natural enemies. Furthermore, there are fewer restrictions on the production and distribution of new biopesticides, which shortens the time to market. Nonetheless, there are numerous factors knot that restrict the effective commercialization and implementation of biopesticides within the agricultural sector. The first one is the reduced efficacy and control of biopesticides in various field situations (Sachdev and Singh, 2016). Under various situations, including temperature, moisture, and microbial relationships, biopesticides' efficacy can be limited due to various factors. Furthermore, biopesticides result in less residual activity after application than the compared synthetic chemicals. As a result, they may need to be utilized extra times, resulting in

TABLE 8.2
Various Biopesticides and Their Applications for Managing Resistance in Different Plants

Crops	Biopesticide	Active Ingredient	Target Pests/ Diseases	Mode of Action	References
wheat, corn, tomato	*Trichoderma* spp.	Fungal species	Soil-borne pathogens (e.g., Fusarium, Rhizoctonia)	Competes with and parasitizes plant pathogens, enhancing plant defense mechanisms	Mukesh *et al.* (2015)
Soybean, tomato, potato	*Beauveria bassiana*	Fungal spores	Various insects (e.g., beetles, aphids)	Infects and kills target pests by penetrating their exoskeleton and causing fungal growth inside them	Corrêa *et al.* (2020)
Corn, cotton	*Bacillus thuringiensis* (Bt)	Bt toxin	Lepidopteran pests (e.g., caterpillars)	Produces toxins that target specific insect pests, disrupting their digestive system	Abbas (2018)
cucumber, pepper, rice	*Pseudomonas fluorescens*	Bacterial strain	Soil-borne diseases (e.g., Pythium, Fusarium)	Produces antibiotics and enzymes that inhibit pathogen growth and enhance plant defenses	Verma *et al.* (2019)
Cotton, tomato, eggplant	Neem (Azadirachtin)	Azadirachtin	Various insects (e.g., beetles, caterpillars)	Disrupts insect growth and development, repels pests, and has antifeedant properties	Agbo *et al.* (2019)
Potato, apple, citrus	Spinosad	Spinosyns	Lepidopteran pests, thrips	Acts on the insect's nervous system, causing paralysis and death	Cubillos-Salamanca *et al.* (2020)
Lettuce, cabbage, beans	*Bacillus subtilis*	Lipopeptides	Various bacterial and fungal diseases	Produces antifungal compounds and induces systemic resistance in plants	Lastochkina *et al.* (2019)
Strawberry, grape, citrus	Essential oils (e.g., thyme, clove)	Natural oils	Various insects, mites	Disrupts insect nervous system, repels pests, and has antifeedant properties	Singh *et al.* (2015)

increased costs to customers (Lengai and Muthomi, 2018). The low production and distribution of certain biopesticides make them unaffordable for smallholder farmers in several developing countries. Similarly, the stringent registration process prevents the fast creation and distribution of novel biopesticides. To resolve the limitations of biopesticides, stakeholders should solve the problem through collaboration and fund joint research and development to address the limitations (Wiedemann *et al.*, 2022).

8.5.4 Organic Farming Methods

Organic farming methods are mostly holistic in nature to agriculture, and thus they are an adequate way of addressing resistance issues common to the conventional systems. The fundamental aspects of organic agriculture of the building of healthy soils, biodiversity, and reduction in environmental consequences of agriculture minimize the generation and spread of resistance in agricultural pests, diseases, and noxious plants. The most substantial organically based tool in the battle against resistance is crop rotation (Jouzi *et al.*, 2017). Rotating crops over many seasons interrupts the life cycles of pests and diseases in the soil and diminishes the buildup of the pests and diseases while reducing chemical input. In addition, crop diversification poses virtually friendly environments and helps increase beneficial insects. In addition, cultural approaches associated with organic farming, including interplanting common and widely spaced planting systems, can generally increase resistance. This approach actually tends toward ecological diversity and predator-prey relationship that exists naturally. This can discourage pests and promote stability that can resist every outbreak (van Bruggen *et al.*, 2016).

Soil health is especially critical in organic farming, where living soils provide the basis for plant nutrition, growth, and natural defense against pests and diseases. Organic farmers naturally focus on soil fertility approaches, such as the use of compost, green manures, and crop residues, as sources of medicinal nutrition and microorganisms (Aulakh *et al.*, 2022). Healthy soil implies that nutrient levels are appropriate to support a plant's immune system. In particular, plants become more resilient to pest and disease pressure when they are well-fed by healthy soil. Moreover, stable soil structure and appropriate moisture levels, achieved using agricultural practices like reduced tillage, mulching, and so on, promote good growing conditions for such beneficial microorganisms without causing disruption to soil life. Over the long term, activities that focus on soil health are crucial for pest and disease control as they contribute to a diverse and well-balanced microbiome in the soil that can better kill pathogens through competitive methods and support plant health (Iorizzo *et al.*, 2022). Also, organic farming restricts the use of inputs, mainly synthetic pesticides and fertilizers (Figure 8.2). Unlike inorganic farming where synthetic pesticides are used, organic standards prohibit synthetic pesticides use and only allow natural and organic means for pest and disease management (Muneret *et al.*, 2018). The farmers use natural pesticides and mostly botanical extracts, microbial pesticides, and insect traps, among others, which are specific, leaving the beneficial organisms and non-target species. But through chemical use, the farmer creates pressure to be exerted on the pest, making resistance development less likely.

Third, it focuses on promoting biodiversity both above and below ground. Planting crop rotation and polyculture body and planting broad-based crops with agroforestry create habitation for different other organisms. That is why organic farmers conserve natural habitats too (Deangelis and Domeignoz Horta, 2020).

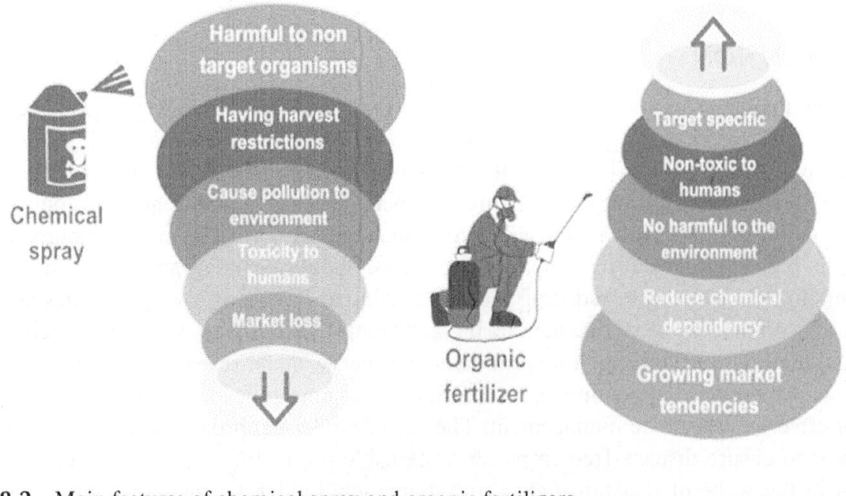

FIGURE 8.2 Main features of chemical spray and organic fertilizers.

8.6 CHALLENGES AND FUTURE DIRECTIONS

Use of modern approaches to mitigate challenges of resistance in agriculture faces a myriad of both existing obstacles and future uncertainties. One of the primary challenges is presented by the evolution of resistance, which is currently overcoming the development of mechanisms for effective control. The rapid adaptation of pests, diseases, and weeds to conventional management practices demands even more innovation to beat the developing mechanisms of resistance (Liebman *et al.*, 2016). Additionally, the use of chemical pesticides and herbicides has resulted in pesticide pollution, soil degradation, and collateral damage to target and non-target organisms. It is also unrealistic to reverse the evolution of pests and diseases. As such, the future directions in resistance management must seek sustainable alternatives that cause minimal environmental damage while remaining effective in pest management. Furthermore, with the globalization of trade and travel, resistant pests and pathogens can quickly spread across regions, challenging pest management at the local level and requiring international cooperation (Barzman *et al.*, 2015). Additionally, socio-economic dynamics, including financial pressures among farmers, influence the uptake of modern resistance management approaches. High costs of inputs, change in market preferences, and limited resources influence the uptake and sustainability of modern resistance management approaches (Peerzada *et al.*, 2019). Future resistance management efforts should align with the existing socio-economic issues in agriculture and offer supportive mechanisms for farmers to efficiently take up the approaches. Another component in the future resistance management efforts is modern technologies that have the potential to revolutionize the fight against resistance. Technologies such as precision agriculture, remote sensing, and digital tools for farm management open up new frontiers for real-time data-based management systems that boost the efficiency of resistance management measures (Huang *et al.*, 2018). Moreover, GE technologies, such as biotechnology and gene-editing techniques, offer hope in creating crops that are resistant to pests, diseases, and herbicides (Alemu, 2020). However, the regulatory environment, ethical issues, and market and consumer acceptance of GMOs pose significant risks to the development and application of biotechnological resistance solutions. Additionally, policy incentives and rewards play a crucial role in driving sustainable practices in resistance management and also in creating conducive environments for the uptake of modern approaches. Ultimately, solving the current challenges in the fight against resistance in agriculture requires a concerted effort from all stakeholders to provide lasting solutions to farm health and productivity. With the spirit of innovation, teamwork, and coexisting with nature, the agricultural sector can overcome resistance challenges and continue to amplify sustainability and resilience in the coming generations (Calicioglu *et al.*, 2019).

8.7 CONCLUSION

In conclusion, the above analysis reveals that the adoption and use of modern approaches to resistance mitigation in agriculture play an essential role in ensuring the sustainability, productivity, and resilience of the global food systems. It is evident that resistance showcasing, whether as pests, diseases, or weeds, has a pronounced nature and wide coverage, thus requiring comprehensive, innovative, and diversified strategies to enhance its management. Various modern approaches, such as IPM, GE, precision agriculture impact, and others, should be integrated since they empower stakeholders to fight resistance, reducing the oversaw impact. Besides, the approaches are sustainable since they emphasize proactive monitoring and control agent diversification, reducing reliance on chemical methods. Accordingly, it is fundamental that farmers, researchers, policymakers, and industries work closely in supporting research, enhancing knowledge and promoting practical applications for effective resistance management. The use of modern approaches already informs farmers how best to ensure disease-free crops for sustainable production and food security for future generation in the wake of resistance changing dynamics. Evidently, modern approaches provide platforms for ensuring viable healthy crops, and this way, one cannot succeed if not investing in the development and use of modern approaches to resistance mitigation in agriculture.

LIST OF ABBREVIATIONS

CRISPR-Cas9	clustered regularly interspaced short palindromic repeats and CRISPR-associated protein 9
DDTs	dichlorodiphenyltrichloroethanes
GE	genetic engineering
GMOs	genetically modified organisms
IPM	integrated pest management
SCP	sustainable crop protection

REFERENCES

Abbas, M.S.T., 2018. Genetically engineered (modified) crops (*Bacillus thuringiensis* crops) and the world controversy on their safety. *Egyptian Journal of Biological Pest Control*, 28, 1–12.

Abdullah, Mahmood, A., Bibi, S., Naqve, M., Javaid, M.M., Zia, M.A., Jabbar, A., Ud-Din, W., Attia, K.A. & Khan, N., 2022. Physiological, biochemical, and yield responses of linseed (*Linum usitatissimum* L.) in α-tocopherol-mediated alleviation of salinity stress. *Frontiers in Plant Science*, 13, 867172.

Agbo, B., Nta, A. & Ajaba, M., 2019. Bio-pesticidal properties of neem (*Azadirachta indica*). *Advances and Trends in Agriculture Sciences*, 1, 17–26.

Ahanger, M.A., Akram, N.A., Ashraf, M., Alyemeni, M.N., Wijaya, L. & Ahmad, P., 2017. Plant responses to environmental stresses—From Gene to biotechnology. *AoB PLANTS*, 9.

Alemu, M., 2020. Trend of biotechnology applications in pest management: A review. *International Journal of Applied Sciences and Biotechnology*, 8, 108–131.

Ali, H., Tanveer, A., Nadeem, M., Asghar, H. & Javaid, M., 2013. Germination ecology of *Rhynchosia capitata*: An emerging summer weed in Asia. *Planta Daninha*, 31, 249–257.

Ali, M.A., Abdellah, I.M. & Eletmany, M.R., 2023. Towards sustainable management of insect pests: Protecting food security through ecological intensification. *International Journal of Chemical and Biochemical Sciences*, 24, 386–394.

Ali, Q., Yu, C., Hussain, A., Ali, M., Ahmar, S., Sohail, M.A., Riaz, M., Ashraf, M.F., Abdalmegeed, D., Wang, X., Imran, M., Manghwar, H. & Zhou, L., 2022. Genome engineering technology for durable disease resistance: Recent progress and future outlooks for sustainable agriculture. *Frontiers in Plant Science*, 13.

Anderson, J.A., Ellsworth, P.C., Faria, J.C., Head, G.P., Owen, M.D.K., Pilcher, C.D., Shelton, A.M. & Meissle, M., 2019. Genetically engineered crops: Importance of diversified integrated pest management for agricultural sustainability. *Frontiers in Bioengineering and Biotechnology*, 7.

Aulakh, C.S., Sharma, S., Thakur, M. & Kaur, P., 2022. A review of the influences of organic farming on soil quality, crop productivity and produce quality. *Journal of Plant Nutrition*, 45, 1884–1905.

Babar, B.H., Tanveer, A., Tahir, M., Aziz, A., Ahmad, A.U.H., Nadeem, M.A. & Javaid, M.M., 2009. Allelopathic potential of wild onion (*Asphodelus tenuifolius*) on the germination and seedling growth of chickpea (*Cicer arietinum*). *Weed Biology and Management*, 9, 146–151.

Baker, B.P., Green, T.A. & Loker, A.J., 2020. Biological control and integrated pest management in organic and conventional systems. *Biological Control*, 140, 104095.

Balal, R.M., Shahid, M.A., Javaid, M.M., Iqbal, Z., Liu, G.D., Zotarelli, L. & Khan, N., 2017. Chitosan alleviates phytotoxicity caused by boron through augmented polyamine metabolism and antioxidant activities and reduced boron concentration in *Cucumis sativus* L. *Acta Physiologiae Plantarum*, 39, 1–15.

Barzman, M., Lamichhane, J.R., Booij, K., Boonekamp, P., Desneux, N., Huber, L., Kudsk, P., Langrell, S.R., Ratnadass, A. & Ricci, P., 2015. Research and development priorities in the face of climate change and rapidly evolving pests. *Sustainable Agriculture Reviews*, 17, 1–27.

Bass, C. & Nauen, R., 2023. The molecular mechanisms of insecticide resistance in aphid crop pests. *Insect Biochemistry and Molecular Biology*, 156, 103937.

Beckie, H.J., Ashworth, M.B. & Flower, K.C., 2019. Herbicide resistance management: Recent developments and trends. *Plants*, 8, 161.

Begum, A., Alam, S.N. & Jalal Uddin, M., 2017. Management of Pesticides: Purposes, Uses, and Concerns. *In* M.S. Khan & M.S. Rahman (eds.) *Pesticide Residue in Foods: Sources, Management, and Control.* Cham: Springer International Publishing, 53–86.

Bigini, V., Camerlengo, F., Botticella, E., Sestili, F. & Savatin, D.V., 2021. Biotechnological resources to increase disease-resistance by improving plant immunity: A sustainable approach to save cereal crop production. *Plants*, 10, 1146.

Bisht, D.S., Bhatia, V. & Bhattacharya, R., 2019. Improving plant-resistance to insect-pests and pathogens: The new opportunities through targeted genome editing. *Seminars in Cell & Developmental Biology*, 96, 65–76.

Brown, M.E., Mugo, S., Petersen, S. & Klauser, D., 2022. Designing a pest and disease outbreak warning system for farmers, agronomists and agricultural input distributors in East Africa. *Insects*, 13(3), 232.

Calicioglu, O., Flammini, A., Bracco, S., Bellù, L. & Sims, R., 2019. The future challenges of food and agriculture: An integrated analysis of trends and solutions. *Sustainability*, 11, 222.

Chadha, A., Florentine, S., Chauhan, B.S., Long, B., Jayasundera, M., Javaid, M.M. & Turville, C., 2019. Environmental factors affecting the germination and seedling emergence of two populations of an emerging agricultural weed: Wild lettuce (*Lactuca serriola*). *Crop and Pasture Science*, 70, 709–717.

Choudhary, V., Kumar, K.K., Sahu, B., Muthappa, S.K., Dixit, A. & Das, A., 2021. Agronomic innovations in biotic stress management and its combined effect with abiotic stresses in crop production. *Indian Journal of Agronomy*, 66, 237–257.

Clapp, J., 2021. Explaining growing glyphosate use: The political economy of herbicide-dependent agriculture. *Global Environmental Change*, 67, 102239.

Cordeau, S., 2022. Conservation agriculture and agroecological weed management. *Agronomy*, 12, 867.

Corrêa, B., Da Silveira Duarte, V., Silva, D.M., Mascarin, G.M. & Júnior, I.D., 2020. Comparative analysis of blastospore production and virulence of *Beauveria bassiana* and *Cordyceps fumosorosea* against soybean pests. *BioControl*, 65, 323–337.

Cubillos-Salamanca, Y.P., Rodríguez-Maciel, J.C., Pineda-Guillermo, S., Silva-Rojas, H.V., Berzosa, J., Tejeda-Reyes, M.A. & Rebollar-Alviter, Á, 2020. Identification of thrips species and resistance of *Frankliniella occidentalis* (Thysanoptera: Thripidae) to malathion, spinosad, and bifenthrin in blackberry crops. *Florida Entomologist*, 102, 738–746.

Deangelis, K. & Domeignoz Horta, L., 2020. *Integrative Experience: Soil Microbes and the Sustainability of Organic Agriculture*, *6*, 301–331.

Deguine, J.-P., Aubertot, J.-N., Flor, R.J., Lescourret, F., Wyckhuys, K.A.G. & Ratnadass, A., 2021. Integrated pest management: Good intentions, hard realities. A review. *Agronomy for Sustainable Development*, 41, 38.

Fahad, S., Bajwa, A.A., Nazir, U., Anjum, S.A., Farooq, A., Zohaib, A., Sadia, S., Nasim, W., Adkins, S., Saud, S., Ihsan, M.Z., Alharby, H., Wu, C., Wang, D. & Huang, J., 2017. Crop production under drought and heat stress: Plant responses and management options. *Frontiers in Plant Science*, 8.

Garbelini, L.G., Debiasi, H., Junior, A.A.B., Franchini, J.C., Coelho, A.E. & Telles, T.S., 2022. Diversified crop rotations increase the yield and economic efficiency of grain production systems. *European Journal of Agronomy*, 137, 126528.

Grassini, P., Specht, J.E., Tollenaar, M., Ciampitti, I. & Cassman, K.G., 2015. Chapter 2 – High-Yield Maize–Soybean Cropping Systems in the US Corn Belt. *In* V.O. Sadras & D.F. Calderini (eds.) *Crop Physiology (Second* Edition*)*. San Diego: Academic Press, 17–41.

Hanif, Z., Naeem, M., Ali, H.H., Tanveer, A., Javaid, M.M., Peerzada, A.M. & Chauhan, B.S., 2017. Effect of environmental factors on germination of *Salsola foetida*: Potential species for rehabilitation of degraded rangelands. *Rangeland Ecology & Management*, 70, 638–643.

He, H.-M., Liu, L.-N., Munir, S., Bashir, N.H., Wang, Y., Yang, J. & Li, C.-Y., 2019. Crop diversity and pest management in sustainable agriculture. *Journal of Integrative Agriculture*, 18, 1945–1952.

Huang, Y., Chen, Z.-X., Tao, Y., Huang, X.-Z. & Gu, X.-F., 2018. Agricultural remote sensing big data: Management and applications. *Journal of Integrative Agriculture*, 17, 1915–1931.

Iorizzo, M., Letizia, F., Ganassi, S., Testa, B., Petrarca, S., Albanese, G., Di Criscio, D. & De Cristofaro, A., 2022. Functional properties and antimicrobial activity from lactic acid bacteria as resources to improve the health and welfare of honey bees. *Insects*, 13, 308.

Iqbal, M., Gogi, M.D., Atta, B., Nisar, M.J., Arif, M.J. & Javed, N., 2021. Assessment of pathogenicity of *Beauveria bassiana, Metarhizium anisopliae, Verticillium lecanii* and *Bacillus thuringiensis* var. kurstaki against *Bactrocera cucurbitae* Coquillett (Diptera: Tephritidae) via diet-bioassay technique under controlled conditions. *International Journal of Tropical Insect Science*, 41, 1129–1145.

Jacquet, F., Jeuffroy, M.-H., Jouan, J., Le Cadre, E., Litrico, I., Malausa, T., Reboud, X. & Huyghe, C., 2022. Pesticide-free agriculture as a new paradigm for research. *Agronomy for Sustainable Development*, 42, 8.

Jhala, A.J., Beckie, H.J., Peters, T.J., Culpepper, A.S. & Norsworthy, J.K., 2021. Interference and management of herbicide-resistant crop volunteers. *Weed Science*, 69, 257–273.

Jouzi, Z., Azadi, H., Taheri, F., Zarafshani, K., Gebrehiwot, K., Van Passel, S. & Lebailly, P., 2017. Organic farming and small-scale farmers: Main opportunities and challenges. *Ecological Economics*, 132, 144–154.

Karlsson Green, K., Stenberg, J.A. & Lankinen, Å, 2020. Making sense of integrated pest management (IPM) in the light of evolution. *Evolutionary Applications*, 13, 1791–1805.

Khan, S.M., Ali, S., Nawaz, A., Bukhari, S.H., Ejaz, S. & Ahmad, S., 2019. Integrated Pest and Disease Management for Better Agronomic Crop Production. *In* M. Hasanuzzaman (ed.) *Agronomic Crops: Volume 2: Management Practices*. Singapore: Springer Singapore, 385–428.

Koul, B., 2020. Genetically Modified (GM) Crops Harbouring *Bacillus thuringiensis* (BT) Gene(S) to Combat Biotic Stress Caused by Insect Pests. *In* M. Kumar, V. Kumar & R. Prasad (eds.) *Phyto-Microbiome in Stress Regulation*. Singapore: Springer, 21–61.

Kushalappa, A.C., Yogendra, K.N., Sarkar, K., Kage, U. & Karre, S., 2016. Gene discovery and genome editing to develop cisgenic crops with improved resistance against pathogen infection. *Canadian Journal of Plant Pathology*, 38, 279–295.

Lamichhane, J.R., Aubertot, J.-N., Begg, G., Birch, A.N.E., Boonekamp, P., Dachbrodt-Saaydeh, S., Hansen, J.G., Hovmøller, M.S., Jensen, J.E., Jørgensen, L.N., Kiss, J., Kudsk, P., Moonen, A.-C., Rasplus, J.-Y., Sattin, M., Streito, J.-C. & Messéan, A., 2016. Networking of integrated pest management: A powerful approach to address common challenges in agriculture. *Crop Protection*, 89, 139–151.

Lastochkina, O., Seifikalhor, M., Aliniaeifard, S., Baymiev, A., Pusenkova, L., Garipova, S., Kulabuhova, D. & Maksimov, I., 2019. Bacillus spp.: Efficient biotic strategy to control postharvest diseases of fruits and vegetables. *Plants*, 8, 97.

Lengai, G.M. & Muthomi, J.W., 2018. Biopesticides and their role in sustainable agricultural production. *Journal of Biosciences and Medicines*, 6, 7–41.

Liebman, M., Baraibar, B., Buckley, Y., Childs, D., Christensen, S., Cousens, R., Eizenberg, H., Heijting, S., Loddo, D., Merotto, A., Jr, Renton, M. & Riemens, M., 2016. Ecologically sustainable weed management: How do we get from proof-of-concept to adoption? *Ecological Applications*, 26, 1352–1369.

Mahmood, A., Ullah, H., Shahzad, A., Ali, H., Ahmad, S., Zia-Ul-Haq, M., Honermeier, B. & Hasanuzzaman, M., 2013. Dry matter yield and chemical composition of sorghum cultivars with varying planting density and sowing date. *Sains Malaysiana*, 42, 1529–1538.

Mandal, S., 2019. Impact of pest control chemicals on biological activity of biocontrol agents. *In The souvenir of National Seminar on "Agro-Chemical Inputs and Its Extension Approaches towards Food-Security and Bio-Safety: Prospects and Challenges (AEFS-2019)"*, 80–84.

Meena, R.S. & Lal, R., 2018. Legumes and Sustainable Use of Soils. *In* R.S. Meena, A. Das, G.S. Yadav & R. Lal (eds.) *Legumes for Soil Health and Sustainable Management*. Singapore: Springer Singapore, 1–31.

Mehmood, A., Naeem, M., Khalid, F., Saeed, Y., Abbas, T., Jabran, K., Sarwar, M.A., Tanveer, A. & Javaid, M.M., 2018. Identification of phytotoxins in different plant parts of *Brassica napus* and their influence on mung bean. *Environmental Science and Pollution Research*, 25, 18071–18080.

Mukesh, S., Mohammad, S., Sonika, P., Vipul, K., Anuradha, S., Shubha, T. & Srivastava, Y., 2015. Trichoderma: A scientific approach against soil borne pathogens. *African Journal of Microbiology Research*, 9, 2377–2384.

Muneret, L., Mitchell, M., Seufert, V., Aviron, S., Djoudi, E.A., Pétillon, J., Plantegenest, M., Thiéry, D. & Rusch, A., 2018. Evidence that organic farming promotes pest control. *Nature Sustainability*, 1, 361–368.

Mustafa, A., Athar, F., Khan, I., Chattha, M.U., Nawaz, M., Shah, A.N., Mahmood, A., Batool, M., Aslam, M.T. & Jaremko, M., 2022. Improving crop productivity and nitrogen use efficiency using sulfur and zinc-coated urea: A review. *Frontiers in Plant Science*, 13, 942384.

Nachimuthu, V.V., Sabariappan, R., Muthurajan, R. & Kumar, A., 2017. Breeding Rice Varieties for Abiotic Stress Tolerance: Challenges and Opportunities. *In Abiotic Stress Management for Resilient Agriculture* (Springer), 5, 339–361.

Nandula, V.K., 2019. Herbicide resistance traits in maize and soybean: Current status and future outlook. *Plants*, 8, 337.

Nawaz, A., Sufyan, M., Gogi, M.D. & Javed, M.W., 2019. Sustainable Management of Insect-Pests. *In* M. Farooq & M. Pisante (eds.) *Innovations in Sustainable Agriculture*. Cham: Springer International Publishing, 287–335.

Nerkar, G., Devarumath, S., Purankar, M., Kumar, A., Valarmathi, R., Devarumath, R. & Appunu, C., 2022. Advances in crop breeding through precision genome editing. *Frontiers in Genetics*, 13.

Pacilly, F.C.A., Hofstede, G.J., Lammerts Van Bueren, E.T., Kessel, G.J.T. & Groot, J.C.J., 2018. Simulating crop-disease interactions in agricultural landscapes to analyse the effectiveness of host resistance in disease control: The case of potato late blight. *Ecological Modelling*, 378, 1–12.

Panth, M., Hassler, S.C. & Baysal-Gurel, F., 2020. Methods for management of soilborne diseases in crop production. *Agriculture*, 10, 16.

Parankusam, S., Katamreddy, S., Bommineni, P.R., Bhatnagar-Mathur, P. & Sharma, K.K., 2018. Insights into Insect Resistance in Pulse Crops: Problems and Preventions. *In* S.H. Wani & M. Jain (eds.) *Pulse Improvement: Physiological, Molecular and Genetic Perspectives*. Cham: Springer International Publishing, 137–173.

Parray, J.A., Yaseen Mir, M. & Shameem, N., 2019. Plant Biotechnology: Tool for Sustainable Agriculture. *In* J.A. Parray, M. Yaseen Mir & N. Shameem (eds.) *Sustainable Agriculture: Biotechniques in Plant Biology*. Singapore: Springer Singapore, 1–50.

Paul, D. & Mandal, S.M., 2019. Microbial Adaptation and Resistance to Pesticides. *In* S.M. Mandal & D. Paul (eds.) *Bacterial Adaptation to Co-Resistance*. Singapore: Springer Singapore, 233–249.

Peerzada, A.M., Bukhari, S.H., Dawood, M., Nawaz, A., Ahmad, S. & Adkins, S., 2019. Weed Management for Healthy Crop Production. *Agronomic Crops: Volume 2: Management Practices*, 2, 225–256.

Peltonen-Sainio, P., Jauhiainen, L. & Sorvali, J., 2017. Diversity of high-latitude agricultural landscapes and crop rotations: Increased, decreased or back and forth? *Agricultural Systems*, 154, 25–33.

Popivanov, I., Petkova, T., Doycheva, V., Doychinova, T., Angelov, I. & Shalamanov, D., 2015. Vector control—Development and improvement of the modern chemical insecticides. *Journal of Pharmacy and Pharmacology*, 3, 1–8.

Prabha, S., Yadav, A., Kumar, A., Yadav, A., Yadav, H.K., Kumar, S., Yadav, R. & Kumar, R., 2016. Biopesticides—An alternative and eco-friendly source for the control of pests in agricultural crops. *Plant Arch*, 16, 902–906.

Pretty, J. & Bharucha, Z.P., 2015. Integrated pest management for sustainable intensification of agriculture in Asia and Africa. *Insects*, 6, 152–182.

Rajan, S., Parween, M. & Raju, N.J., 2023. Pesticides in the hydrogeo-environment: A review of contaminant prevalence, source and mobilisation in India. *Environmental Geochemistry and Health*, 45, 5481–5513.

Rajwade, J.M., Chikte, R.G. & Paknikar, K.M., 2020. Nanomaterials: New weapons in a crusade against phytopathogens. *Applied Microbiology and Biotechnology*, 104, 1437–1461.

Rasheed, A., Barqawi, A.A., Mahmood, A., Nawaz, M., Shah, A.N., Bay, D.H., Alahdal, M.A., Hassan, M.U. & Qari, S.H., 2022a. CRISPR/Cas9 is a powerful tool for precise genome editing of legume crops: A review. *Molecular Biology Reports*, 49, 5595–5609.

Rasheed, A., Raza, A., Jie, H., Mahmood, A., Ma, Y., Zhao, L., Xing, H., Li, L., Hassan, M.U. & Qari, S.H., 2022b. Molecular tools and their applications in developing salt-tolerant soybean (*Glycine max* L.) cultivars. *Bioengineering*, 9, 495.

Romeh, A.A., 2019. Integrated Pest Management for Sustainable Agriculture. *In* A.M. Negm & M. Abu-Hashim (eds.) *Sustainability of Agricultural Environment in Egypt: Part II: Soil-Water-Plant Nexus*. Cham: Springer International Publishing, 215–234.

Sachdev, S. & Singh, R.P., 2016. Current challenges, constraints and future strategies for development of successful market for biopesticides. *Climate Change and Environmental Sustainability*, 4, 129–136.

Saharan, V., Jain, D., Pareek, S., Pal, A., Kumaraswamy, R., Jakhar, S.K. & Singh, M., 2016. Viral, Fungal and Bacterial Disease Resistance in Transgenic Plants. *Advances in Plant Breeding Strategies: Agronomic, Abiotic and Biotic Stress Traits*, 6, 627–656.

Sarwar, M., 2015. Microbial insecticides-an ecofriendly effective line of attack for insect pests management. *International Journal of Engineering and Advanced Research Technology*, 1, 4–9.

Savary, S., Bregaglio, S., Willocquet, L., Gustafson, D., Mason D'croz, D., Sparks, A., Castilla, N., Djurle, A., Allinne, C., Sharma, M., Rossi, V., Amorim, L., Bergamin, A., Yuen, J., Esker, P., Mcroberts, N., Avelino, J., Duveiller, E., Koo, J. & Garrett, K., 2017. Crop health and its global impacts on the components of food security. *Food Security*, 9, 311–327.

Scavo, A., Fontanazza, S., Restuccia, A., Pesce, G.R., Abbate, C. & Mauromicale, G., 2022. The role of cover crops in improving soil fertility and plant nutritional status in temperate climates. A review. *Agronomy for Sustainable Development*, 42, 93.

Schröder, P., Sauvêtre, A., Gnädinger, F., Pesaresi, P., Chmeliková, L., Doğan, N., Gerl, G., Gökçe, A., Hamel, C., Millan, R., Persson, T., Ravnskov, S., Rutkowska, B., Schmid, T., Szulc, W., Teodosiu, C. & Terzi, V., 2019. Discussion paper: Sustainable increase of crop production through improved technical strategies, breeding and adapted management – A European perspective. *Science of the Total Environment*, 678, 146–161.

Shah, K.K., Modi, B., Pandey, H.P., Subedi, A., Aryal, G., Pandey, M. & Shrestha, J., 2021. Diversified crop rotation: An approach for sustainable agriculture production. *Advances in Agriculture*, 2021, 8924087.

Sharma, G., Shrestha, S., Kunwar, S. & Tseng, T.-M., 2021. Crop diversification for improved weed management: A review. *Agriculture*, 11, 461.

Sheikh, A.A., Wani, M.A., Bano, P., Un, S., Nabi, T.B., Bhat, M.A. & Dar, M.S., 2017. An overview on resistance of insect pests against Bt crops. *Journal of Entomology and Zoology Studies*, 5, 941–948.

Singh, O., Rathore, H.S. & Nollet, L.M., 2015. Biochemical pesticides. *Biopesticides Handbook* (Florida), 5, 171–183.

Sparks, T.C., Storer, N., Porter, A., Slater, R. & Nauen, R., 2021. Insecticide resistance management and industry: The origins and evolution of the insecticide resistance action committee (IRAC) and the mode of action classification scheme. *Pest Management Science*, 77, 2609–2619.

Tamanreet, K. & Mandeep, K., 2020. Ch. 1. Integrated Pest Management: A Paradigm for Modern Age. *In* K. Dimitrios, K. Anna & M. Kassio Ferreira (eds.) *Pests, Weeds and Diseases in Agricultural Crop and Animal Husbandry Production*. Rijeka: IntechOpen.

Tanveer, A., Arshad, M.S., Ayub, M., Javaid, M.M. & Yaseen, M., 2012. Effect of temperature, light, salinity, drought stress and seeding depth on germination of Cucumis melo var. agrestis. *Pakistan Journal of Weed Science Research*, 18.

Tanveer, A., Javaid, M.M., Irfan, M., Khaliq, A. & Yaseen, M., 2015. Yield losses in chickpea with varying densities of dragon spurge (*Euphorbia dracunculoides*). *Weed Science*, 63, 522–528.

Tanveer, M., Mahmood, A., Sarfraz, B., Zia, M.A., Javaid, M.M., Bibi, S., Naqve, M., Nadeem, M.A., Azeem, M. & Jabbar, A., 2023. Mechanism and Approaches to Enhancing Heat Stress Tolerance in Crop Plants. *In* M. Hasanuzzaman (ed.) *Climate-Resilient Agriculture, Vol 2: Agro-Biotechnological Advancement for Crop Production*. Cham: Springer International Publishing, 499–520.

Thomine, E., Mumford, J., Rusch, A. & Desneux, N., 2022. Using crop diversity to lower pesticide use: Socio-ecological approaches. *Science of the Total Environment*, 804, 150156.

Torres, J.B. & Bueno, A.D.F., 2018. Conservation biological control using selective insecticides – A valuable tool for IPM. *Biological Control*, 126, 53–64.

Trivedi, P., Mattupalli, C., Eversole, K. & Leach, J.E., 2021. Enabling sustainable agriculture through understanding and enhancement of microbiomes. *New Phytologist*, 230, 2129–2147.

Ullah, H., Mahmood, A., Ijaz, M., Tadesse, B. & Honermeier, B., 2013. Evaluation of anise (*Pimpinella anisum* L.) accessions with regard to morphological characteristics, fruit yield, oil contents and composition. *Journal of Medicinal Plants Research*, 7, 2177–2186.

van Bruggen, A.H., Gamliel, A. & Finckh, M.R., 2016. Plant disease management in organic farming systems. *Pest Management Science*, 72, 30–44.

van Esse, H.P., Reuber, T.L. & van Der Does, D., 2020. Genetic modification to improve disease resistance in crops. *New Phytologist*, 225, 70–86.

van Lenteren, J.C. & Nicot, P.C., 2020. Integrated Pest Management Methods and Considerations Concerning Implementation in Greenhouses. *In* M.L. Gullino, R. Albajes & P.C. Nicot (eds.) *Integrated Pest and Disease Management in Greenhouse Crops*. Cham: Springer International Publishing, 177–193.

Vats, S., 2015. Herbicides: History, Classification and Genetic Manipulation of Plants for Herbicide Resistance. *In* E. Lichtfouse (ed.) *Sustainable Agriculture Reviews: Volume 15*. Cham: Springer International Publishing, 153–192.

Verma, P., Hiremani, N.S., Pande, R. & Shamim, M., 2019. Plant Growth Promoting Rhizobacteria for the Control of Soil-Borne Diseases. *Biofertilizers and Biopesticides in Sustainable Agriculture*. New Jersey and Canada: Apple Academic Press, 113–140.

Waheed, H., Javaid, M.M., Shahid, A., Ali, H.H., Nargis, J. & Mehmood, A., 2019. Impact of foliar-applied Hoagland's nutrient solution on growth and yield of mash bean (*Vigna mungo* L.) under different growth stages. *Journal of Plant Nutrition*, 42, 1133–1141.

Wang, L., Zhao, Y., Al-Kaisi, M., Yang, J., Chen, Y. & Sui, P., 2020. Effects of seven diversified crop rotations on selected soil health indicators and wheat productivity. *Agronomy*, 10, 235.

Wasaya, A., Abbas, T., Yasir, T.A., Sarwar, N., Aziz, A., Javaid, M.M. & Akram, S., 2021. Mitigating drought stress in sunflower (*Helianthus annuus* L.) through exogenous application of β-aminobutyric acid. *Journal of Soil Science and Plant Nutrition*, 21, 936–948.

Wiedemann, R., Stamm, C. & Staudacher, P., 2022. Participatory knowledge integration to promote safe pesticide use in Uganda. *Environmental Science & Policy*, 128, 154–164.

Wuest, S.E., Peter, R. & Niklaus, P.A., 2021. Ecological and evolutionary approaches to improving crop variety mixtures. *Nature Ecology & Evolution*, 5, 1068–1077.

Yadav, I.C., Devi, N.L., Syed, J.H., Cheng, Z., Li, J., Zhang, G. & Jones, K.C., 2015. Current status of persistent organic pesticides residues in air, water, and soil, and their possible effect on neighboring countries: A comprehensive review of India. *Science of the Total Environment*, 511, 123–137.

Yu, T., Mahe, L., Li, Y., Wei, X., Deng, X. & Zhang, D., 2022. Benefits of crop rotation on climate resilience and its prospects in China. *Agronomy*, 12, 436.

9 Revitalizing Irrigation, Soil Moisture Management in Dryland Agriculture

*Filza Ishfaq**, *Muhammad Mansoor Javaid, Babar Ali, Rashad Mukhtar Balal, Ikram ul Haq, and Muhammad Awais*

9.1 INTRODUCTION

A specific type of agriculture known as s "dryland farming" is practiced in arid and semi-arid areas where annual rainfall amounts to just 20–35% of the potential water transpiration (Zhang *et al.*, 2022; Ameen *et al.*, 2024). This method relies entirely on rainfall, making efficient water conservation crucial for sustaining crops in challenging environments. The main objectives are to retain precipitation on the land, use drought-tolerant crops that align with rainfall patterns, and reduce soil evaporation to maximize water available for plant transpiration (Li *et al.*, 2020; Minhas *et al.*, 2020).

The modern era presents significant challenges for global agriculture, including the need to implement environmentally friendly practices, adapt to changing climates, and increase food production by 70% by 2050 to feed a growing population (O'Neill *et al.*, 2020). Additionally, dietary preferences are expected to shift from grains to meat and dairy products. A major concern is the slower growth in yield of major grain crops over the past 30 years (Fischer and Connor, 2018). For example, the annual increase in wheat production has fallen to less than 1%. To meet the future global food demand, the primary strategy is to enhance yield and growth by maximizing potential yield through the best varieties, agronomy, and overcoming limitations, thereby reducing the gap between actual farm yields and potential yield (Röös *et al.*, 2018).

Drylands, which make up approximately 41% of the planet's land area, must contribute their fair share to increase the crop production (Taseska *et al.*, 2023). These regions exhibit a gradient of primary productivity, ranging from hyper-arid, arid, and semi-arid to dry sub-humid areas, based on varying degrees of aridity or moisture deficit. The spatial extent of these areas is a gradient of primary productivity that ranges from hyper-arid, arid, and semi-arid to dry sub-humid, corresponding to the level of aridity or moisture deficit. Dryland farming is practiced mainly in the arid or dry sub-humid areas of the world where the annual rainfall is between 300 and 600 mm. These regions do not receive ample precipitation along with a high rate of evaporation, which creates moisture scarcity (Devkota *et al.*, 2022; Mukhtiar *et al.*, 2023). Innovative water management and soil conservation techniques are essential to maximizing the productivity of dryland farming. Research and development of crop varieties and farming practices adapted to these conditions are also important to improve yields and ensure sustainable agriculture in these areas.

In the semi-arid tropics, between 15 and 20 degrees South latitudes, dryland farming has been studied (Prăvălie *et al.*, 2019). These efforts have not been sufficient, however, to overcome the numerous technological, social, and economic barriers that have prevented the development of the commercial agriculture sector in these regions, with large areas being mainly for extensive livestock production (Dövényi-Nagy *et al.*, 2020; Aamer *et al.*, 2021). Similar agroclimates include that of southern China, the Mediterranean, southern Africa, central South America, and southern North

* Corresponding author: shahfilzashah9@gmail.com

DOI: 10.1201/9781003498698-9

America. Transforming breakthrough technologies, supportive policies, and facilitating community engagement will be needed to build sustainable agricultural practices that overcome the existing challenges. Innovations and research into climate-adaptive techniques are also paramount to continue developing productivity and resilience in these regions with technologies in water and soil management. So, in order to succeed for the long-term in global dryland farming, appropriate agricultural strategies need to be implemented that factor in local conditions and constraints.

In dryland areas, plant growth is often erratic, low-yielding, and unpredictable (Jaenisch *et al.*, 2022; Ahmed *et al.*, 2023). Plant growth indices in the most productive 16-week period range from 0.2 to 0.4 and in the optimal 13-week period range from 0.35 to 0.7. Potential grain yields where biotic or abiotic stresses are absent can in some of these regions be as much as two-fold higher than farm yields (Hochman and Horan, 2018; Khan *et al.*, 2023). Consequently, effective risk management and optimal use of local resources are prerequisites for profitable agricultural production to narrow the gap between the actual farm production and potential yield limited by net available water (da S. Andrea *et al.*, 2018). Boosting yields will require improving water use efficiency (WUE) and adopting drought-resistant crop varieties. Also, bringing in precision agriculture practices and working on robust farming may annihilate this gap between possible and actual yield. Investments in research and technologies in the dryland contexts are critical to improve agricultural productivity in the context of drylands.

Variables affecting soil water retention, including infiltration rates, storage capacity, and roots water absorption capabilities, play an important role in dryland farming crop and pastures yield (Abdallah *et al.*, 2021; Aslam *et al.*, 2023). However, crop yields often do not reach their water-limited potential due to additional environmental or management factors, such as evaporative loss of water held in fallows during northern cropping zones. Dryland agriculture, in particular, is susceptible to increasing temperatures, as yields of preferred crops may decrease over successive generations, while weeds and pests may thrive even more (Marengo *et al.*, 2020). Consequently, additional pesticide application will likely be required to achieve the desired level of pest control, though pest management becomes less effective at elevated temperatures (Deguine *et al.*, 2021; Rasheed *et al.*, 2022). Heat waves leading to extreme heat stress can also further reduce agricultural output. Indeed, in many regions, climate change is predicted to have a net negative effect on farming, exacerbating global food insecurity, a situation that policymakers must urgently address (Sun *et al.*, 2019). This requires investing in climate-resilient crop varieties, better water management practices, and improved pest management strategies.

9.2 CLIMATE CHANGE IMPACTS ON DRYLAND AGRICULTURE

Plant nutrient decomposition is restricted in dryland soils containing low concentrations of organic matter. Moreover, the soil is salt-affected and has poor structure, and limited moisture retention (Ayangbenro and Babalola, 2021). The interaction between agriculture and climate change is complex and involves a variety of interconnected elements such as discrepancies in average temperatures, precipitation patterns, and climate extremes like heat waves. In addition, shifts in pest and disease dynamics, changes in atmospheric carbon dioxide and ground-level ozone, both of which affect the food nutritional quality, the sea level variations also change farming systems (Rasheed *et al.*, 2021; Rocha *et al.*, 2022). Overcoming these challenges must involve a whole-systems approach, from adapting farming practices to a changing climate and managing soils sustainably, to breeding for resilient crop varieties. Thus, developing research in this area along with the house of innovation types of agricultural technology could help mitigate these impacts and improve the sustainability of dryland agriculture.

Dryland agriculture is already affected by climate change, although this effect is not uniform across the globe (O'Leary *et al.*, 2018). Crop yields are projected to decrease due to the negative impacts of climate change in low-latitude regions, while the impact in higher latitudes can be positive or negative. Certain vulnerable populations, particularly those in impoverished areas, are likely to face increased risks of food insecurity as a result of these changes. For example, 1–21% of the arable land area in South America, 1–18% in Africa, 1–17% in Europe, and 20–40% in India may disappear (Fischer and Connor, 2018; Boretti and Rosa, 2019). Since the late 1950s, new crop varieties, increased use of fertilizers, and the expansion of irrigated agriculture (Hemathilake and Gunathilake, 2022). However, ongoing climate change poses a significant threat to these gains, highlighting the need for adaptive strategies and sustainable practices to mitigate the impacts and ensure continued agricultural productivity. Addressing these challenges requires an integrated approach that considers both current and future climate conditions and emphasizes resilience and adaptation in agricultural systems (Chattha *et al.*, 2017; He *et al.*, 2019; Mead, 2022).

Elevated temperatures gradually diminish the yields of preferable crops while promoting the growth of weeds and pests (Kumari *et al.*, 2022). Higher pesticide rates will be required to maintain the same levels of control as pest management becomes less effective. Extreme heat stress from heat waves may reduce crop yields if they happen at certain stages of the plant's life cycle, such as pollination, pod, or fruit set (Nadeem *et al.*, 2018). Additionally, heat waves may cause plants to wilt owing to higher transpiration rates; if irrigation is not used to mitigate this effect, yield loss may follow. Excessive precipitation, which often leads to floods, may also be harmful to soil structure and crops (Figure 9.1) (Lv *et al.*, 2020). Since most plants' roots need air to breathe, they cannot endure extended periods of flooding. It is anticipated that climate change would have a negative overall influence on farming, endangering the world's food security (Muruganandam *et al.*, 2023).

Securing enough water for rainfed agricultural production is a major challenge for many developing nations situated in arid or semi-arid areas. In dry areas, issues with water shortages are only the consequence of insufficient rainfall (Gebru *et al.*, 2021). Though the yearly rainfall in semi-arid locations may be sufficient to sustain crops, the uneven distribution of the rainfall across time and place makes rainfed agriculture unfeasible. A reduction of one standard deviation from the mean annual rainfall often results in the total loss of the crop because of the large rainfall variance in

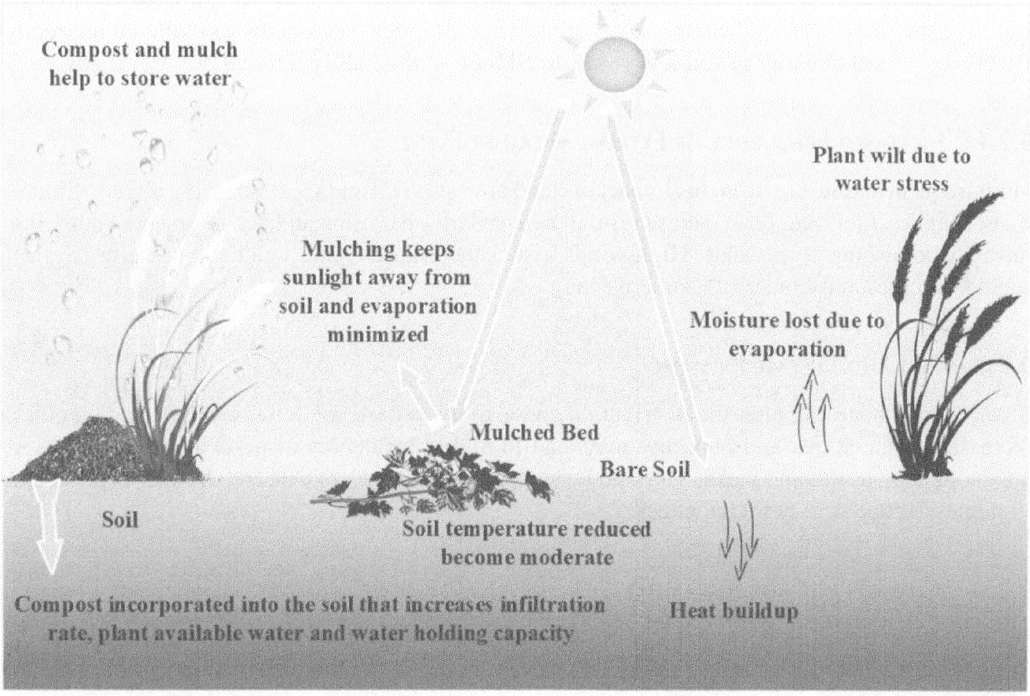

FIGURE 9.1 Mulches help to store water and reduce evaporation of water.

semi-arid locations (Ayanlade *et al.*, 2018). The issues of water shortage in these regions are made worse by water loss from evaporation and runoff. While regions receiving less than 300 mm of annual rainfall must turn to other means of securing enough water to support crop production, areas receiving low rainfall, between 300 and 600 mm annually, may be able to partially combat these problems through the use of supplemental irrigations (Ayanlade *et al.*, 2018).

Soil erosion is a major cause of soil degradation in drylands, driven by factors such as topography, rainfall, land use, and insufficient plant cover (Chalise *et al.*, 2019; Jianing *et al.*, 2022). Soil erosion, which is one of the most important environmental issues confronting human civilization, is the movement and transport of soil caused by many agents, mainly water, wind, and mass movement, leading to substantial soil loss (Owens, 2020). Annually, approximately 10 million hectares of farmland are rendered less productive due to soil erosion. Generally, soil is being lost 10–40 times faster than it is replenished, posing serious threats to future food security and environmental quality (Azadi *et al.*, 2018). Since over 98% of the calories that people consume come from the land, the productivity and health of soils are vital for sustaining and enhancing the global food supply. Through the loss of organic matter, reduced nutrient availability, and compromised hydrologic function, increased water erosion. These factors collectively diminish soil quality and agricultural productivity, highlighting the urgent need for effective soil conservation and management strategies to combat erosion and ensure long-term agricultural sustainability in dryland regions (Sanaullah *et al.*, 2020).

Consequences of climate change are the extremes of unpredictable rainfall, which may exacerbate issues with land degradation since excessive rain leads to runoff and water erosion, while insufficient rain results in a lack of plant cover and deterioration (Abrol and Gupta, 2019). Raindrops can raindrop force, surface and subsurface runoff, and river floods all erode soil. The most significant element influencing soil erosion brought on by rain is the strength of the rainfall. Soil particles are often transported away in proportion to the strength of the rainfall (Ni *et al.*, 2021).

Raindrops are essential for separating soil particles. They cause sheet erosion, the most common kind of soil deterioration from erosion, by removing a thin layer of soil from the land's surface. The soil particles become vulnerable to being washed away by runoff once they are loosed. The permeability of the soil is a crucial component that impacts soil erosion caused by rainfall and indirectly affects the overall quantity of soil loss (Martínez-Mena *et al.*, 2020) (Figure 9.2).

9.2.1 Increased Frequency of Extreme Weather Events

More frequent and severe droughts impact dryland crop survival and yield, stressing already limited water supplies. Flooding from intense rainfall can lead to soil erosion and loss of essential nutrients, further diminishing crop yields. High winds associated with extreme weather can cause physical damage to crops and exacerbate soil erosion.

9.2.2 Pest and Disease Pressure

Changes in climate can alter the distribution and activity of pests and diseases in dryland regions. Increased temperatures and humidity may lead to higher incidences of pests and plant diseases, impacting crop health and yields. Dryland farmers may need to adapt pest and disease management strategies to cope with new challenges.

9.2.3 Changes in Crop Suitability

Shifts in climate conditions may render traditional dryland crops less viable, necessitating the adoption of drought and heat-resistant varieties. Changes in crop suitability may require modifications in crop and land management practices to optimize productivity. Exploration of alternative crops and farming techniques becomes essential to adapt to new climatic realities.

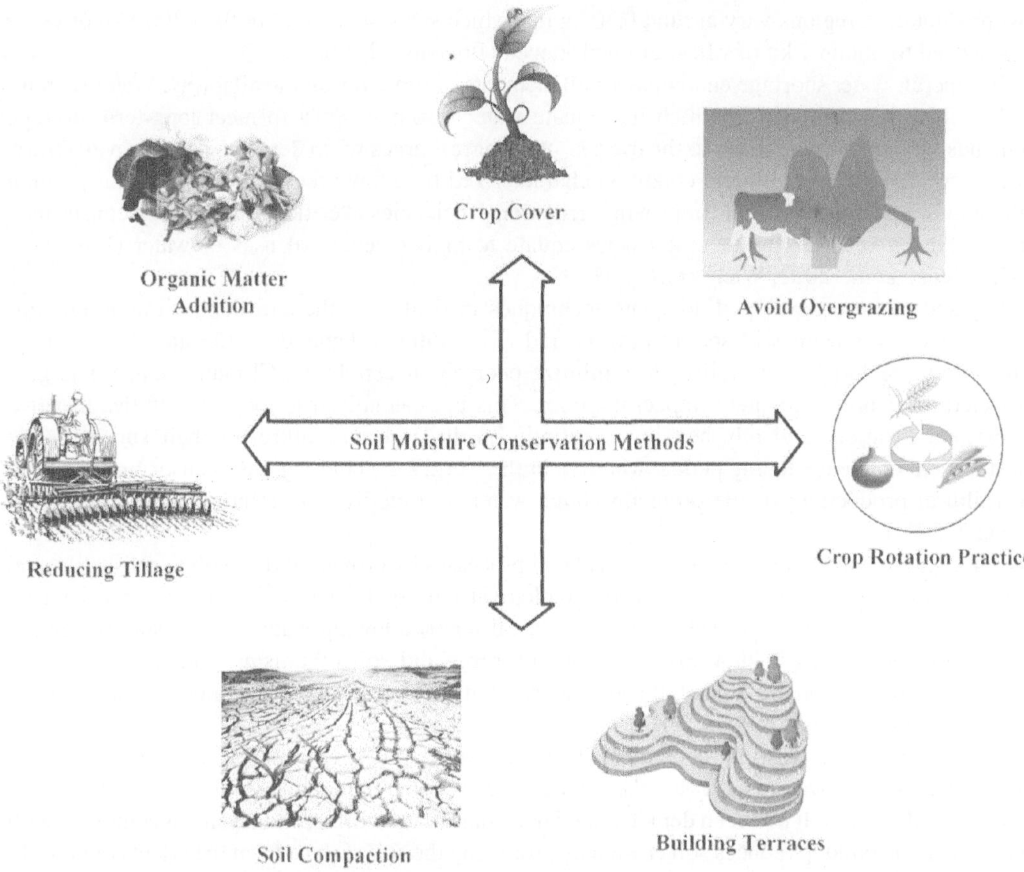

FIGURE 9.2 Soil moisture conservation methods.

9.2.4 Economic Impacts

Increased costs for water, irrigation, and inputs due to climate change may strain the financial stability of dryland farms. Reduced yields and higher risks of crop failure can impact profitability and economic sustainability. Investments in adaptation technologies and practices are necessary to mitigate economic challenges and ensure long-term viability.

9.2.5 Socioeconomic Effects

Changes in agricultural productivity may affect the livelihoods of communities dependent on dryland farming, leading to economic instability. Migration from increasingly unviable dryland areas may occur as farmers seek more sustainable opportunities. Impacts on food security and local economies due to reduced crop production and increased food cost can affect overall community well-being.

9.2.6 Water-Use Efficiency Problems in Dryland Agriculture

Since water is the key limitation for dryland farming, it is the main constraint for dryland production, and WUE improvement is essential for the success of dryland cropping systems (Han *et al.*, 2023). For instance, WUE estimates reported for wheat (*Triticum aestivum* L.) grown in low-precipitation regions vary around 0.30 kg m^3, which suggests that more than 3000 kg of water was needed to obtain 1 kg of wheat grain produced (Boersma, 2018).

In general, water shortage encompasses all elements of limited water availability. Water scarcity is described as a scenario in which inadequate water resources exist to meet long-term average demands. Water shortage refers to the overuse of water resources when demand exceeds availability. In contrast, aridity is a climatic condition characterized by a low precipitation evapotranspiration ratio and is a permanent event. Improving irrigation efficiencies is critical under water-constrained circumstances because greater efficiencies equate to higher beneficial uses of water (Liu *et al.*, 2022; Mcnee *et al.*, 2022; Wang *et al.*, 2024).

Drylanders have developed adequate techniques to deal with their difficult climatic circumstances. However, with widespread poverty and rising human demand on the unstable resource base, coping techniques are failing to minimize people's susceptibility. Climate change is largely expected to disproportionately impact the poor. This is especially true for communities that live in dryland locations and rely heavily or entirely on dryland agriculture for their survival. The most critical challenge facing people who live in the world's dryland regions is ensuring long-term agricultural productivity while protecting land, water, and biodiversity (Hatfield and Dold, 2019; Brendel, 2021).

Soil organic matter sustains many critical soil processes by providing the energy, substrates, and biological diversity necessary for supporting biological activity that has a direct effect on soil aggregation and water infiltration. Aggregation is crucial for (i) allowing water to infiltrate, (ii) giving soil organisms enough habitat space, (iii) giving roots and soil organisms enough oxygen, and (iv) halting soil erosion. An essential soil property that regulates runoff, leaching, and agricultural water availability is infiltration.

Poor aggregation, crusting, decreased plant-water availability, erosion, off-site effects of sedimentation, and poor water quality are the outcomes of exposing the soil to high-intensity rainfall and lacking residue cover. It has been demonstrated that conservation tillage management combined with surface residue buildup reduces soil erosion by protecting the soil surface from the effects of rainfall.

In dryland agriculture, transpiration rates are significantly higher due to the hot, windy, and dry climates characteristic of these regions. Field tests indicate that the atmospheric saturation deficit closely matches the observed gradient in dryland regions. As a result, transpiration rates are significantly higher compared to the cooler, calmer, and more humid conditions found in more favorable cropping zones. Reducing soil evaporation by maintaining soil cover is a crucial management strategy for enhancing WUE. This practice increases the amount of water available for transpiration. Evapotranspiration, the total water used by a crop, includes both the water transpired through plant leaves and the water evaporated from the soil surface between sowing and harvest (Zhang *et al.*, 2021a).

In dryland agriculture, evaporation and WUE are critical factors that influence crop productivity. Nearly all of the water consumed by the crop is lost via evaporation from the soil surface in the early stages of the growth season when plants are small and only partially shade the soil. Growing crops need more water for transpiration since their canopy covers a larger area of the soil, which means that more water is needed for transpiration overall. During the peak growth season, up to 90% or more of the water used may come from transpiration (Lüttschwager and Jochheim, 2020). The percentage of evapotranspiration that is used for transpiration over the whole growing season usually falls between 50 and 75%. Dryland crops lose a substantially larger percentage of their evapotranspiration via evaporation because they need less water, which leads to fewer plants overall and more bare ground available for evaporation unless the soil is covered (Farooq *et al.*, 2019).

Reduced soil evaporation may be achieved by leaving plant remains on the soil surface as mulch (Mbangiwa *et al.*, 2019). However, in dryland areas, plant residues are either removed for fuel or feed, or their production is insufficient. Plastic film has been employed recently, especially in China, to cover the soil surface in between plant rows; however for big fields, this is usually not viable due to several considerations including labor, cost, and water management. Despite these challenges, transpiration accounts for at least 90% of evapotranspiration, meaning that water consumption efficiency may be significantly boosted. The rapid growth and canopy development of fertilizer-applied plants may also have a favorable impact on water usage efficiency by shading the soil surface (Dou *et al.*, 2022). Additionally, altering plant population and row spacing can affect WUE, although the results may vary depending on specific conditions (Fischer *et al.*, 2019).

There is mounting evidence that the climate is changing. There are still a lot of unknowns about whether or not this is related to global warming. The hydrological cycle may become more intense as a result of increased climatic variability. Extreme event frequency and magnitude increases are already being seen in certain areas. Furthermore, it is possible that future extremes will be even more terrible than those that have already occurred. Certain areas and communities will be more susceptible to modifications in hydrological responses as a result of climate change and increased climatic variability. It will be necessary to use proactive management techniques to assist deal with and adjusting to these developments (Hunt *et al.*, 2018; Qaderi *et al.*, 2019; Zhang *et al.*, 2019).

Deficit irrigation is defined as water applications that fall short of the evapotranspiration level needs. To increase WUE, a number of agricultural techniques are being examined, including mulching, crop residue assimilation, low tillage, and the effects of modern irrigation techniques (Figure 9.3). However, the kind of soil and the surrounding environment determine each technique's potential and applicability. Improved agricultural and agronomic management techniques are the only way to produce more food with less water. Water resources in the Mediterranean, Europe, Central and Southern America, and Southern Africa are all significantly impacted by climate change. In water-stressed areas like southern and eastern Asia, climate change causes an increase in runoff. Thus, it is essential to use more efficient water management strategies in such situations. In agro-ecosystems, crop development and yield can frequently be impeded by inadequate water and restricted nutrient availability. This is because throughout their vegetative and reproductive phases, a large number of cereals and fodder crops are extremely vulnerable to water and nutrient deficits.

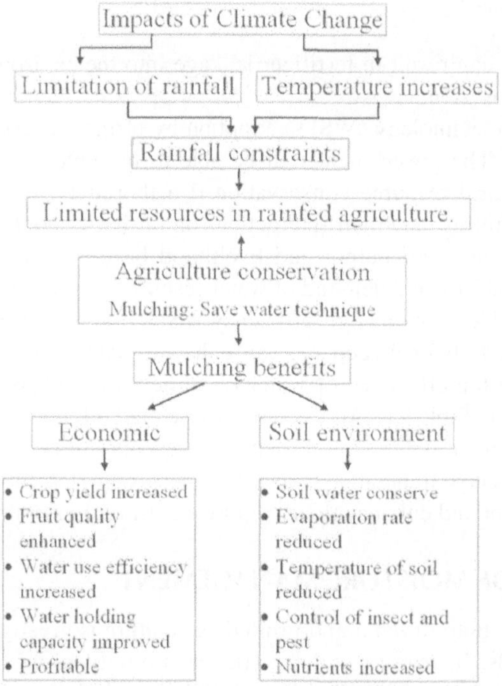

FIGURE 9.3 Revitalizing irrigation, soil moisture management in dryland agriculture.

TABLE 9.1
Soil and Water Conservation Problems in Various Agroclimatic Zones

Region	Rainfall (mm)	Areas	Main Issues	References
North Himalayan (excluding cold desert)	500–2000	Jammu and Kashmir, Himachal Pradesh	Erosion along hillslopes	Rawat *et al.* (2023)
Northeastern Himalayan	1500–2500	Northeastern hills	Shift from traditional to sustainable cultivation	Chapagain *et al.* (2019)
Indo-Gangetic alluvial soils	700–1000	Punjab, Haryana, Rajasthan, Uttar Pradesh, Bihar	Erosion, ravines, floods	Kumar *et al.* (2020)
Assam Valley and Gangetic delta	1500–2500	Assam, Tripura, West Bengal	Gully and streambank erosion	Bhattacharyya *et al.* (2016)
Desert area	150–500	Western Rajasthan, Haryana, Gujarat	Sand dune shifts, wind erosion	Kar (2022)
Mixed red, black, and yellow soils	600–700	Rajasthan, Uttar Pradesh, Madhya Pradesh	Ravine formation	Singh *et al.* (2018)
Black soils	00–7005	Rajasthan, Madhya Pradesh, Maharashtra	Erosion, groundwater decline	Periasamy and Shanmugam (2022)
Black soils (deep and medium deep)	800–1300	Maharashtra, Madhya Pradesh, Andhra Pradesh	Excessive soil loss	Srinivasarao *et al.* (2022)
Eastern red soils	1000–1500	West Bengal, Bihar, Orissa, Madhya Pradesh	Erosion, land mismanagement	Mandal *et al.* (2019)
Southern red soils	~750 (Kerala), >2500	Tamil Nadu, Kerala, Karnataka, Andhra Pradesh	Soil erosion from excessive rainfall	Jinger *et al.* (2023)
East-west coasts	~1000	Orissa to Saurashtra coast	Coastal salinity, soil loss	Ramasamy and Saravanavel (2020)

However, too much water can result in fertilizer leakage into the environment, raising production costs and polluting the ecosystem.

Water-saving irrigation technology (WSIT) adoption by farmers is crucial to attaining superior agricultural development. The government will be able to promote WSIT to support sustainable development and agricultural resource conservation if a thorough examination of the effects of risk aversion, technical training, and their interactions on farmers' adoption of WSIT is conducted. Furthermore, farmers' agricultural output and livelihood have been negatively impacted by the conflict between the availability and demand of water resources, which also poses a serious danger to their long-term sustainable growth. High-efficiency WSITs, such as drip irrigation systems, droppers, micro-sprinklers, and channel seepage, can enhance planting intensity, improve cultivated land quality, and increase the efficiency of water resource utilization when compared to conventional irrigation techniques (Table 9.1). Additionally, those methods can promote the double benefits of enhancing the environment and raising agricultural revenue. In order to resolve the aforementioned inconsistencies, it is now imperative to assist farmers in implementing high-efficiency water-saving irrigation equipment and enhance their cognitive abilities to minimize water loss.

9.3 IMPORTANCE OF MOISTURE MANAGEMENT

Production of crops under both dryland and rainfed agriculture is greatly influenced by soil moisture conservation methods. In regions with insufficient rainfall, using the leftover soil moisture from the previous crop can be successfully planted. If managed well, sufficient water adds greatly to the yield of rainfed crops (Li *et al.*, 2018).

Relying on the moisture-conserving technique employed, conservation techniques can also offer additional benefits such as weed management, protection of the soil surface from the detrimental effects of excessive rainfall and erosion, and improved nutrient distribution in the soil (Kumar *et al.*, 2022). These techniques can also help to reduce the amount of water used and increase the efficiency of water usage. By reusing organic matter and utilizing plant and crop residues that decompose in the soil, the need for waste management is lessened (Sharma *et al.*, 2019). Furthermore, these practices enhance soil fertility and structure, leading to better root development and overall plant health. Implementing comprehensive soil moisture conservation strategies is crucial for sustaining crop production in dryland and rainfed agricultural systems.

9.3.1 Barriers of Soil Moisture Conservation

In some regions, crop residues are periodically used to feed farm animals, thus raising the price of ecological preservation. Due to the challenges of producing new crops either through mulch or the utilization of diverse crop residue and the lack of mechanization of plant-producer systems, it is difficult to develop a conventional implementation of these soil moisture conservation mechanisms (Pereira *et al.*, 2021). Field residues are also burnt which limits the use of in-situ soil moisture conservation. Furthermore, not all crops are compatible with some forms of moisture conservation (Si *et al.*, 2017). For small-scale cultivators, the highly mechanized approaches are not affordable. One of the major challenges in the implementation of moisture preservation measures is the ignorance of farmers about conservation agricultural practices (Gangopadhyay *et al.*, 2023). Education and outreach programs are essential to raise awareness and knowledge about these practices. Low-cost, accessible conservation technologies can also address these challenges to promote sustainable farming systems.

9.3.2 Techniques of Soil Moisture Conservation

Among the many various techniques, only the better-known methods for conserving soil moisture are covered below.

9.3.3 Deep Tillage

Tillage, the most important practice to conserve soil moisture, adsorbs moisture and maximizes the infiltration of rainwater into the deeper layers, reduces surface runoff and nutrient loss, and enhances productivity (Bhattacharyya *et al.*, 2022). In dense soils, the solid pan is typically broken up with extensive tillage which also enhances the soil's physical properties further. Deep tillage is suggested for every two to three years at a depth of 25–30 cm. During the off-season and summer, dryland and rainfed areas become harder due to the soil being left fallow (Sigler *et al.*, 2020). Rainfall can facilitate heightened surface runoff and soil loss in them. It increases soil moisture conservation for crops during the monsoon by permitting deeper percolation of water during rainfall (Gupta *et al.*, 2020) (Figure 9.3). Moreover, the practice helps in deep root penetration and makes the plants stronger. Paired with the right tillage practices, soil structure, and water infiltration can be greatly improved (Table 9.2).

9.3.4 Conservation Tillage

Conservation tillage is an essential agricultural practice used in dryland agriculture which has been widely employed in order to conserve soil moisture, lower erosion and improve soil health. This technique uses very little disruption of the soil, and it keeps the surface of the soil covered with the remains of crops. This way reduces water, soil structure, and organic material content preserves wood. In conservation cultivation, for instance, only the seeding area is tilled (single furrow)

TABLE 9.2
Recommended Soil and Water Conservation Measures for Various Rainfall Zones Worldwide

Rainfall Range (mm)	Agricultural Practices	References
Less than 500	Contour bunds, small basins, tied ridges, inter row water harvesting system, scoops, khadin, ridging, mulching, contour cultivation with conservation furrows, sowing across slopes, off-season tillage, field bunds	Madhu (2022)
500–700	Zingg terrace, contour cultivation with conservation furrows, modified contour bunds, ridging, tide ridges, scoops, small basins, BBF (vertisols), inter row water harvesting system, field bunds, khadin, mulching, sowing across slopes	Bhattacharyya *et al.* (2016)
750–1000	graded bunds, lock and spill drains, nadi, vegetative bunds, conservation furrows, field bunds, small basins, BBF (vertisols), Tillage, sowing across slopes, Zingg terrace	Rao *et al.* (2022)
More than 1000	Chos, field bunds, BBF (vertisols), level terrace, vegetative bunds	Madhu (2022)

(Jat *et al.*, 2019). This only plows the land where the crop will be sown, thus resulting in less soil disturbance. Residual stubbles of the last harvest are left on the field. Tillage ought to be kept to a minimum, or, in extraordinary circumstances, dispensed by altogether, as this departure assists save resources and limits evaporate loss by not disturbing the settled harvested crop stubbles leaving them undisturbed, building up dirt good organic levels. Conservation agriculture may improve the soil absorption and retention capacity of water (Lateef *et al.*, 2019). Research shows that one such technique of conservation toward the soil moisture content in the land is applying conservation tillage in the soil; conservation tillage is one of the best tillage practices being practiced nowadays especially if it concerns to no-tillage operations (Hussain *et al.*, 2021). Additionally, the conventional farming practice thus reduces soil compaction, improving the root growth of the plants, ultimately leading to the better yield of crops. These methods allow farmers to reach sustainable levels of agricultural production, even in difficult dryland conditions.

9.3.5 Compartmental Bonding

One of the low-cost rainwater harvesting methods for rainfed vertisols with slope less than 1% is compartmental bunding. Pre-sizing fields divided the entire field into small-sized sections, to hold precipitation where it falls, while suppressing soil erosion (Racette *et al.*, 2020). These bunds limit the flow of water to concentrated parts, thus preventing runoff. This helps to improve rainwater infiltration into the soil profile, which is crucial for cultivating rabi crops using the stored moisture. Several previous studies have shown that compartmental bunding is advantageous for trapping in-situ moisture and ameliorating soil moisture dynamics in vertisols. In dryland and rainfed agriculture situations, compartmental bunding has been shown by research to be more effective for increasing yields (Tripathi, 2014). This technique will conserve the amount of water required and contribute to sustainable agriculture through soil and crop improvement.

9.3.6 Mulching

Mulching is an important agronomic method influencing the hydro-thermal cycles of soils through suppression of weeds, conservation of soil moisture, reduction of water evaporation, alter the physical environment of soil with an increase in soil fertility (Kader *et al.*, 2019). By providing a better growing environment, it is critical to improving soil health and crop productivity. This approach increases moisture retention and modifies soil thermal regime and the nutrient pool available for crop development and growth. Mulches are either inorganic (plastic sheeting) or organic (peat, wood chips, straw). Mulching enhances agricultural and veggie crop development, blooming, fruit output, and limited weed proliferation according to Ullah *et al.* (2021). It is successful at increasing water usage efficiency also because organic mulch can reduce soil evaporation (El-Beltagi *et al.*, 2022). Mulching can also enhance soil structure and increase microbial activity, thus promoting sustainable agriculture.

9.3.7 Basin Listing

Basin listing is a soil and water conservation method in dryland agriculture to improve infiltration and decrease soil erosion. In this method, we need a specific instrument that is called a basin lister, and it produces a series of basins on the slope of a hill (Elkadiri *et al.*, 2023). The basins facilitate seepage of rainwater into the soil for a long time, which is beneficial for more water availability to crops. The basin lister performed 11.0% better in crop yield compared to traditional summer plowing and was more effective than other methods like the wide bed former and chisel plow. This method is especially helpful to help preserve soil moisture in tomb months, leading to increased and more consistent food production (Sittig and Sur, 2023). Basin listing can promote such practices for sustainable agriculture by saving water and improving soil by preventing erosion.

9.3.8 Crop Rotation

The process of planting many crop varieties consecutively on the same piece of land over the course of a year is known as crop rotation (Stein and Steinmann, 2018). This sets out the way of improving soil quality, handling pests, and increasing nutrient accessibility. Crop rotation helps improve soil structure and fertility and reduces soil degradation. Crop rotation can also contribute to sustainable farming by diversifying production so that all available resources are used efficiently. When added to cereal crops, legumes as cover crops can retain soil moisture, enhance soil fertility, and reduce soil erosion (Bekele *et al.*, 2021). It optimizes soil nutrients, minimizes insect and weed infestation, and creates healthy soil. Soil moisture at different depths can also be managed by planting a sequence of deep- and shallow-rooted crops. Crop rotation will increase the moisture content of the soil, thus increasing the yield of rainfed agriculture, according to some research (Li *et al.*, 2022).

9.3.9 Strip Cropping

To reduce soil erosion and moisture loss, various crops are grown in alternating strips in this cropping technique (Jaskulska *et al.*, 2020). Close-growing segments in intertilled rows reduce runoff as well as filter out soil particles that are rinsed from the ground. Another kind of agronomic technique is strip cropping, which involves planting regular crops in the shape of comparatively thin strips down the slope of the field. Because of the way these strips are constructed, there should always be strips of close-growing, erosion-resistant crops between the strip crops (Verma *et al., 2020*). The probability of water penetration is also boosted by this runoff control, which raises the soil's moisture content. Strip cropping will boost yields compared to other methods of farming since it grows crops closer together and with fewer gaps between them. It will also raise soil moisture levels (Fahad *et al.*, 2022).

9.3.10 Hydrogels

Physically or chemically cross-linked polymers, known as hydrophilic gels or hydrogels, are able to absorb huge volumes of water without dissolving in soil and producing a vast volume of aqueous solution (Singh *et al.*, 2021). Because they have carboxylic groups and are hydrophilic by nature, hydrogels may adsorb water and bind cations. The ability of hydrogels to absorb water is influenced by a variety of variables, one of which is their ability to tolerate ionic solutions (Zhang *et al.*, 2021b). These gels reduce surface leaching, agricultural runoff, and soil erosion while increasing water retention capacity (Bashir *et al.*, 2020). Hydrogel seed coating produced the greatest amounts of grain and stover while also using the most water and using it most efficiently (Abou-Baker *et al.*, 2020). Incorporating hydrogels into agricultural practices can significantly improve soil moisture management and crop performance.

9.4 IMPLICATIONS FOR POLICY AND RESEARCH PRIORITIES FOR DRYLAND AGRICULTURE

Use of innovative technology and productivity development in rainfed regions are becoming more and more important for food security and agricultural productivity growth in Asia and sub-Saharan Africa (Rajaram and Van Ginkel, 2019). The underutilized potential of rainfed regions has to be investigated, and appropriate methods and policies need to be created to boost productivity development (Li *et al.*, 2018), if future technological advancements are to help the poor. It is necessary to reorient governmental policies and better focus development efforts toward dryland farmers (Rejekiningrum *et al.*, 2022).

9.4.1 Implications for Policy

Agricultural irrigation is the first step toward any policy action promoting dryland agriculture. It is thus critical to address the unfavorable policy results that dryland farmers experience (Jellason *et al.*, 2021). The subsequent compilation of policy suggestions was developed by a comprehensive examination of prior research, including the evaluation of micro-level VLS data and a national poverty assessment using the National Sample Surveys of India. Determining the key policy concerns that must be resolved in order to improve lives in the dryland areas (Li *et al.*, 2021).

9.4.2 Raise Public Investment in Technology and Infrastructure

Dryland agriculture is characterized by low productivity and low input utilization, hence it's critical to increase both public and private investment in advanced technology. Inadequate infrastructure in the drylands limits agricultural and non-farm earnings. Limitations in the supply of seeds and other inputs highlight how crucial it is for the public and private sectors to work together effectively in order to reach the rural poor. The marginal returns on technology and infrastructure investments are larger in arid regions than they are in irrigated ones. Food security will be directly impacted in particular by investments made in rural infrastructure (Balana *et al.*, 2020).

9.4.3 Rationalize Subsidies on Agricultural Inputs

The subsidy on fertilizer is a big problem. Though most fertilizers have been utilized for irrigated crops, they may be employed in both dryland and irrigated locations. Because there isn't any moisture available, dryland farmers see applying fertilizers as very dangerous and only use modest amounts of them. Nonetheless, banks and governments are offering farmers low-interest loans for irrigated crops (Martínez-Dalmau *et al.*, 2021). Two additional significantly subsidized inputs are irrigation water and energy, similar to fertilizers. Although everyone in society should profit from irrigation projects, this approach regrettably discriminates against dryland farmers (Vadez *et al.*, 2023).

9.4.4 Cover More Crops under Minimum Support Price Schemes

The government's rules regarding public distribution and purchase significantly discriminate against crops that get rain. For rainfed crops, minimum prices for support are also set, although procurement procedures seldom support them. For example, the highly supported PDS (Public Distribution System) and the Indian rice and wheat markets have changed market pricing ratios and reduced the price competitiveness of coarse grains. When PDS is replaced with a food stamp program, recipients still have the freedom to purchase the grains of their choosing. If these legislative measures are not actively embraced, rainfed crops and the farmers who cultivate them risk greater marginalization, which would compel them to look for other sources of income (Rai, 2018).

9.4.5 Cover More Households under Crop and Livestock Insurance

Farmers are concerned about their investments and the anticipated returns because of the growing expenses of cultivation as well as the dangers and uncertainties associated with dryland agriculture (Ahmad *et al.*, 2022). Thus, crop insurance in dryland regions is a critical policy effort that is required. Coverage for livestock and crop insurance should be made available to all dryland farmers at reduced costs, similar to the National Agricultural Insurance Scheme introduced by the Indian Ministry of Agriculture (Birthal *et al.*, 2022). Resolve the long-standing oilseed and pulse trade imbalances. Chronic issues include dependence on imports and the scarcity of oilseeds and pulses, particularly in India. Although imports of edible oil were reduced for a while by the Technology Oilseeds Mission (1986), imports have been steadily increasing during the 1990s. Since most of these commodities are cultivated in arid regions, putting more of a focus on oilseed and pulse production may help cut down on imports that are needless and deplete foreign currency (Birthal *et al.*, 2022).

9.4.6 Increase Inflow of Institutional Credit to Dryland Agriculture

Compared to irrigated regions, dryland farmers get far less institutional credit per hectare. Even while dryland agriculture may be a losing endeavor in any one year, it has been shown to be lucrative over a three- to five-year period. Therefore, even if dryland farmers fail after one or more years, it would be very beneficial to implement a new cyclical credit strategy that will satisfy all of their credit needs throughout this time (Ascui and Cojoianu, 2019).

9.4.7 Institute Measures to Cope with Globalization and Marginalization

In the agricultural sector, market reforms that promote production efficiency, the competitiveness of agricultural goods on both local and international markets, and the coordination and regulation of both import and export markets are becoming a significant policy concern. Dryland agriculture as a source of income, many small-scale farmers rely on it becoming more competitive via lower production costs (Xie *et al.*, 2021).

9.4.8 Facilitate Migration

Migration—seasonal, semi-permanent, or permanent—is a common coping mechanism used by the impoverished to break free from poverty (Das *et al.*, 2020). The SAT villages in India are home to informal marketplaces for migrant labor, which are crucial in maintaining a balance between the supply and demand of casual labor in the area. If certain dryland areas were to establish an institutionalized system for gathering and disseminating data on supply, demand, and pay rates, the effectiveness of unofficial migrant labor markets may be further enhanced. The percentage of pay that female employees get is 50% less than that of male employees. It is necessary to implement policies to address gender imbalance and inequality (O'Connor, 2019).

9.5 IMPLICATIONS FOR RESEARCH PRIORITIES

Prioritizing these previously disregarded areas when it comes to of technology, institutions, and policy is necessary, as evidence is mounting that public investments in less-favored regions, particularly in roads, markets, and research, have greater impacts on poverty and higher marginal productivity gains (Tesfaw *et al.*, 2022). A review of scientific and technological priorities in dryland regions is necessary in light of the literature's evidence of significant changes in village economies over the last several decades. Priority in regions of aridity. In order to maintain the development of the standard of living and livelihoods in the drylands, the framework focuses on the efficient and cost-effective use of water and land assets without inflicting irreversible harm to the environment. Governments, markets, institutions, and technology all have significant roles to play. Within this framework, strategic initiatives should prioritize maximizing biomass per drop of water by combining the right crops, livestock, grasses, shrubs, and trees; involving communities in these efforts and granting the most vulnerable sections access to water and land resources through a suitable framework of "rights"; and investing in the development and refinement of alternative technologies, as well as in physical infrastructure improvements and domestic market reforms (du Preez and van Huyssteen, 2020).

In dryland agriculture, one must go beyond advances in output and productivity. The relevance and application of research products to broader contexts are limited by the inability to consider important development restrictions and the failure to combine research and market difficulties (Eigenbrode *et al.*, 2018). Research has to be focused on producing better germplasm, using biotechnology, safeguarding crops from harmful pests and diseases, creating watersheds, improving the relationship between crops and animals, and coming up with novel ways to connect farmers with markets (both inputs as well as markets). It is necessary to simplify participatory research methods that are output-oriented and focused. Strategies like the commercialization of agriculture, the protection of intellectual property, free trade agreements, and the advancement of biotechnology must be pursued (Boersma, 2018). Table 9.1 provides an overview of the dryland agricultural research priorities (Barrett, 2021; Rehman *et al.*, 2023; Shehzadi *et al.*, 2023).

9.6 CONCLUSION

The physical, chemical, and biological properties of soils are all directly related to the conservation of soil moisture as moisture is needed to deliver an adequate supply of water and nutrients to crop plants. The methods depend on local conditions including soil type, geographical setting, climate, and farmer preferences. These roles are also crucial for agricultural sustainability in dryland and rainfed systems. Conservation measures benefit and limit different conservation strategies including mulching, tillage, and hydrogel systems. Mulching, for instance, reduces evaporation and improves the stock of minerals in the soil, yet, in places where plant residues are sparse or cannot be used, the mulching technique can be less effective. Likewise, soil structure can be improved with tillage or soil erosion if not managed correctly. Studies have indicated that the application of several soil moisture conservation techniques is more effective than the use of a single method. Farmers can combine these techniques such as compartmental bunding with mulching or application of hydrogel, to tackle multiple challenges and increase overall WUE. Overall, this integrated approach builds on the positive aspects of each technology and yields a more flexible strategy to the conditions faced in a range of dryland and rainfed cropping practices.

LIST OF ABBREVIATIONS

PDS public distribution system
WUE water use efficiency

REFERENCES

Aamer, M., Bilal Chattha, M., Mahmood, A., Naqve, M., Hassan, M.U., Shaaban, M., Rasul, F., Batool, M., Rasheed, A. & Tang, H., 2021. Rice residue-based biochar mitigates N_2O emission from acid red soil. *Agronomy*, 11, 2462.

Abdallah, A.M., Jat, H.S., Choudhary, M., Abdelaty, E.F., Sharma, P.C. & Jat, M.L., 2021. Conservation agriculture effects on soil water holding capacity and water-saving varied with management practices and agroecological conditions: A review. *Agronomy*, 11, 1681.

Abou-Baker, N.H., Ouis, M., Abd-Eladl, M. & Ibrahim, M.M., 2020. Transformation of lignocellulosic biomass to cellulose-based hydrogel and agriglass to improve beans yield. *Waste and Biomass Valorization*, 11, 3537–3551.

Abrol, I. & Gupta, R., 2019. Climate change-land degradation-food security nexus: Addressing India's challenge. *Journal of Agronomy Research*, 2, 17–35.

Ahmad, M.M., Yaseen, M. & Saqib, S.E., 2022. Climate change impacts of drought on the livelihood of dryland smallholders: Implications of adaptation challenges. *International Journal of Disaster Risk Reduction*, 80, 103210.

Ahmed, Z., Gui, D., Murtaza, G., Yunfei, L. & Ali, S., 2023. An overview of smart irrigation management for improving water productivity under climate change in drylands. *Agronomy*, 13, 2113.

Muruganandam, M., Rajamanickam, S., Sivarethinamohan, S., Gaddam, M.K.R., Velusamy, P., Gomathi, R., Ravindiran, G., Gurugubelli, T.R. and Muniasamy, S.K., 2023. Impact of climate change and anthropogenic activities on aquatic ecosystem–A review. *Environmental Research, 238*, 117233.

Ameen, M., Mahmood, A., Sahkoor, A., Zia, M.A. & Ullah, M.S., 2024. The role of endophytes to combat abiotic stress in plants. *Plant Stress*, 100435.

Ascui, F. & Cojoianu, T.F., 2019. Implementing natural capital credit risk assessment in agricultural lending. *Business Strategy and the Environment*, 28, 1234–1249.

Aslam, A., Mahmood, A., Ur-Rehman, H., Li, C., Liang, X., Shao, J., Negm, S., Moustafa, M., Aamer, M. & Hassan, M.U., 2023. Plant adaptation to flooding stress under changing climate conditions: Ongoing breakthroughs and future challenges. *Plants*, 12, 3824.

Ayangbenro, A.S. & Babalola, O.O., 2021. Reclamation of arid and semi-arid soils: The role of plant growth-promoting archaea and bacteria. *Current Plant Biology*, 25, 100173.

Ayanlade, A., Radeny, M., Morton, J.F. & Muchaba, T., 2018. Rainfall variability and drought characteristics in two agro-climatic zones: An assessment of climate change challenges in Africa. *Science of the Total Environment*, 630, 728–737.

Azadi, H., Keramati, P., Taheri, F., Rafiaani, P., Teklemariam, D., Gebrehiwot, K., Hosseininia, G., Van Passel, S., Lebailly, P. & Witlox, F., 2018. Agricultural land conversion: Reviewing drought impacts and coping strategies. *International Journal of Disaster Risk Reduction*, 31, 184–195.

Balana, B.B., Bizimana, J.-C., Richardson, J.W., Lefore, N., Adimassu, Z. & Herbst, B.K., 2020. Economic and food security effects of small-scale irrigation technologies in northern Ghana. *Water Resources and Economics*, 29, 100141.

Barrett, C.B., 2021. Overcoming global food security challenges through science and solidarity. *American Journal of Agricultural Economics*, 103, 422–447.

Bashir, S., Hina, M., Iqbal, J., Rajpar, A.H., Mujtaba, M.A., Alghamdi, N.A., Wageh, S., Ramesh, K. & Ramesh, S., 2020. Fundamental concepts of hydrogels: Synthesis, properties, and their applications. *Polymers*, 12, 2702.

Bekele, B., Ademe, D., Gemi, Y. & Habtemariam, T., 2021. Evaluation of intercropping legume covers with maize on soil moisture improvement in Misrak azerinet Berbere woreda, SNNPR, Ethiopia. *Water Conservation Science and Engineering*, 6, 145–151.

Bhattacharyya, R., Ghosh, B.N., Dogra, P., Mishra, P.K., Santra, P., Kumar, S., Fullen, M.A., Mandal, U.K., Anil, K.S. & Lalitha, M., 2016. Soil conservation issues in India. *Sustainability*, 8, 565.

Bhattacharyya, S.S., Leite, F.F.G.D., France, C.L., Adekoya, A.O., Ros, G.H., De Vries, W., Melchor-Martínez, E.M., Iqbal, H.M.N. & Parra-Saldívar, R., 2022. Soil carbon sequestration, greenhouse gas emissions, and water pollution under different tillage practices. *Science of the Total Environment*, 826, 154161.

Birthal, P.S., Hazrana, J., Negi, D.S. & Mishra, A.K., 2022. Assessing benefits of crop insurance vis-a-vis irrigation in Indian agriculture. *Food Policy*, 112, 102348.

Boersma, S.J., 2018. *Double-Crop Soybean Response to Maturity Group and Plant Population in the Ohio River Valley Region*. Purdue University.

Boretti, A. & Rosa, L., 2019. Reassessing the projections of the world water development report. *npj Clean Water*, 2, 15.

Brendel, O., 2021. The relationship between plant growth and water consumption: A history from the classical four elements to modern stable isotopes. *Annals of Forest Science*, 78, 47.

Chalise, D., Kumar, L. & Kristiansen, P., 2019. Land degradation by soil erosion in Nepal: A review. *Soil Systems*, 3, 12.

Chapagain, P.S., Ghimire, M. & Shrestha, S., 2019. Status of natural springs in the Melamchi region of the Nepal Himalayas in the context of climate change. *Environment, Development and Sustainability*, 21, 263–280.

Chattha, M.B., Khan, I., Mahmood, A., Chattha, M.U., Anjum, S.A., Ashraf, U., Zain, M. & Bilal, U., 2017. Seed priming with thiourea improves the performance of late sown wheat. *Journal of Agricultural Research*, 55(1), 29–39.

Das, P., Saha, J. & Chouhan, P., 2020. Effects of labor out-migration on socio-economic set-up at the place of origin: Evidence from rural India. *Children and Youth Services Review*, 119, 105512.

da S. Andrea, M.C., Boote, K.J., Sentelhas, P.C. & Romanelli, T.L., 2018. Variability and limitations of maize production in Brazil: Potential yield, water-limited yield and yield gaps. *Agricultural Systems*, 165, 264–273.

Deguine, J.-P., Aubertot, J.-N., Flor, R.J., Lescourret, F., Wyckhuys, K.A.G. & Ratnadass, A., 2021. Integrated pest management: Good intentions, hard realities. A review. *Agronomy for Sustainable Development*, 41, 38.

Devkota, M., Singh, Y., Yigezu, Y.A., Bashour, I., Mussadek, R. & Mrabet, R., 2022. Chapter Five – Conservation Agriculture in the Drylands of the Middle East and North Africa (MENA) Region: Past Trend, Current Opportunities, Challenges and Future Outlook. *In* D.L. Sparks (ed.) *Advances in Agronomy*. Academic Press, 253–305.

Dou, X., Wang, R., Zhou, X., Gao, F., Yu, Y., Li, C. & Zheng, C., 2022. Soil water, nutrient distribution and use efficiencies under different water and fertilizer coupling in an apple–maize alley cropping system in the Loess Plateau, China. *Soil and Tillage Research*, 218, 105308.

Dövényi-Nagy, T., Rácz, C., Molnár, K., Bakó, K., Szláma, Z., Jóźwiak, Á, Farkas, Z., Pócsi, I. & Dobos, A.C., 2020. Pre-harvest modelling and mitigation of aflatoxins in maize in a changing climatic environment—A review. *Toxins*, 12, 768.

du Preez, C.C. & van Huyssteen, C.W., 2020. Threats to soil and water resources in South Africa. *Environmental Research*, 183, 109015.

Eigenbrode, S.D., Binns, W.P. & Huggins, D.R., 2018. Confronting climate change challenges to dryland cereal production: A call for collaborative, transdisciplinary research, and producer engagement. *Frontiers in Ecology and Evolution*, 5.

El-Beltagi, H.S., Basit, A., Mohamed, H.I., Ali, I., Ullah, S., Kamel, E.A.R., Shalaby, T.A., Ramadan, K.M.A., Alkhateeb, A.A. & Ghazzawy, H.S., 2022. Mulching as a sustainable water and soil saving practice in agriculture: A review. *Agronomy*, 12, 1881.

Elkadiri, R., Momm, H.G., Bingner, R.L. & Moore, K., 2023. Spatial optimization of conservation practices for sediment load reduction in ungauged agricultural watersheds. *Soil Systems*, 7, 4.

Fahad, S., Chavan, S.B., Chichaghare, A.R., Uthappa, A.R., Kumar, M., Kakade, V., Pradhan, A., Jinger, D., Rawale, G., Yadav, D.K., Kumar, V., Farooq, T.H., Ali, B., Sawant, A.V., Saud, S., Chen, S. & Poczai, P., 2022. Agroforestry systems for soil health improvement and maintenance. *Sustainability*, 14, 14877.

Farooq, M., Hussain, M., Ul-Allah, S. & Siddique, K.H.M., 2019. Physiological and agronomic approaches for improving water-use efficiency in crop plants. *Agricultural Water Management*, 219, 95–108.

Fischer, R.A. & Connor, D.J., 2018. Issues for cropping and agricultural science in the next 20 years. *Field Crops Research*, 222, 121–142.

Fischer, R.A., Moreno Ramos, O.H., Ortiz Monasterio, I. & Sayre, K.D., 2019. Yield response to plant density, row spacing and raised beds in low latitude spring wheat with ample soil resources: An update. *Field Crops Research*, 232, 95–105.

Gangopadhyay, S., Chowdhuri, I., Das, N., Pal, S.C. & Mandal, S., 2023. The effects of no-tillage and conventional tillage on greenhouse gas emissions from paddy fields with various rice varieties. *Soil and Tillage Research*, 232, 105772.

Gebru, T.A., Brhane, G.K. & Gebremedhin, Y.G., 2021. Contributions of water harvesting technologies intervention in arid and semi-arid regions of Ethiopia, in ensuring households' food security, Tigray in focus1. *Journal of Arid Environments*, 185, 104373.

Gupta, R., Tyagi, N.K. & Abrol, I., 2020. Rainwater management and Indian agriculture: A call for a shift in focus from blue to green water. *Agricultural Research*, 9, 429–443.

Han, T., Lu, H., Lü, Y., Zhu, Y. & Fu, B., 2023. Crop switching could be a win-win solution for improving both the productivity and sustainability in a typical dryland farming region-Loess Plateau, China. *Journal of Cleaner Production*, 384, 135456.

Hatfield, J.L. & Dold, C., 2019. Water-use efficiency: Advances and challenges in a changing climate. *Frontiers in Plant Science*, 10, 429990.

He, G., Zhao, Y., Wang, L., Jiang, S. & Zhu, Y., 2019. China's food security challenge: Effects of food habit changes on requirements for arable land and water. *Journal of Cleaner Production*, 229, 739–750.

Hemathilake, D.M.K.S. & Gunathilake, D.M.C.C., 2022. Chapter 31 – Agricultural Productivity and Food Supply to Meet Increased Demands. *In* R. Bhat (ed.) *Future Foods*. Academic Press, 539–553.

Hochman, Z. & Horan, H., 2018. Causes of wheat yield gaps and opportunities to advance the water-limited yield frontier in Australia. *Field Crops Research*, 228, 20–30.

Hunt, J.R., Hayman, P.T., Richards, R.A. & Passioura, J.B., 2018. Opportunities to reduce heat damage in rain-fed wheat crops based on plant breeding and agronomic management. *Field Crops Research*, 224, 126–138.

Hussain, S., Hussain, S., Guo, R., Sarwar, M., Ren, X., Krstic, D., Aslam, Z., Zulifqar, U., Rauf, A., Hano, C. & El-Esawi, M.A., 2021. Carbon sequestration to avoid soil degradation: A review on the role of conservation tillage. *Plants*, 10, 2001.

Jaenisch, B.R., Munaro, L.B., Jagadish, S.V.K. & Lollato, R.P., 2022. Modulation of wheat yield components in response to management intensification to reduce yield gaps. *Frontiers in Plant Science*, 13.

Jaskulska, I., Romaneckas, K., Jaskulski, D. & Wojewódzki, P., 2020. A strip-till one-pass system as a component of conservation agriculture. *Agronomy*, 10, 2015.

Jat, R.K., Singh, R.G., Kumar, M., Jat, M.L., Parihar, C.M., Bijarniya, D., Sutaliya, J.M., Jat, M.K., Parihar, M.D., Kakraliya, S.K. & Gupta, R.K., 2019. Ten years of conservation agriculture in a rice–maize rotation of Eastern Gangetic Plains of India: Yield trends, water productivity and economic profitability. *Field Crops Research*, 232, 1–10.

Jellason, N.P., Conway, J.S., Baines, R.N. & Ogbaga, C.C., 2021. A review of farming challenges and resilience management in the Sudano-Sahelian drylands of Nigeria in an era of climate change. *Journal of Arid Environments*, 186, 104398.

Jianing, G., Yuhong, G., Yijun, G., Rasheed, A., Qian, Z., Zhiming, X., Mahmood, A., Shuheng, Z., Zhuo, Z. & Zhuo, Z., 2022. Improvement of heat stress tolerance in soybean (*Glycine max* L), by using conventional and molecular tools. *Frontiers in Plant Science*, 13, 993189.

Jinger, D., Kaushal, R., Kumar, R., Paramesh, V., Verma, A., Shukla, M., Chavan, S.B., Kakade, V., Dobhal, S. & Uthappa, A.R., 2023. Degraded land rehabilitation through agroforestry in India: Achievements, current understanding, and future prospectives. *Frontiers in Ecology and Evolution*, 11, 1088796.

Kader, M.A., Nakamura, K., Senge, M., Mojid, M.A. & Kawashima, S., 2019. Soil hydro-thermal regimes and water use efficiency of rain-fed soybean (*Glycine max*) as affected by organic mulches. *Agricultural Water Management*, 223, 105707.

Kar, A., 2022. Adaptation of the Agri-Based Society to Environmental Changes in Thar Desert. *In Climate Change, Disaster and Adaptations: Contextualising Human Responses to Ecological Change*. Springer, 151–171.

Khan, I., Ali, S.M., Chattha, M.U., Barbanti, L., Calone, R., Mahmood, A., Albishi, T.S., Hassan, M.U. & Qari, S.H., 2023. Neem and castor oil–coated urea mitigates salinity effects in wheat by improving physiological responses and plant homeostasis. *Journal of Soil Science and Plant Nutrition*, 23, 3915–3931.

Kumar, K.A., Thayalan, S., Reddy, R., Lalitha, M., Kalaiselvi, B., Parvathy, S., Sujatha, K., Hegde, R., Singh, S. & Mishra, B.B., 2020. Geology and geomorphology. *The Soils of India*, World Soils Book Series. 57–79. Springer, Cham. https://doi.org/10.1007/978-3-030-31082-0_4

Kumar, N., Chhokar, R.S., Meena, R.P., Kharub, A.S., Gill, S.C., Tripathi, S.C., Gupta, O.P., Mangrauthia, S.K., Sundaram, R.M., Sawant, C.P., Gupta, A., Naorem, A., Kumar, M. & Singh, G.P., 2022. Challenges and opportunities in productivity and sustainability of rice cultivation system: A critical review in Indian perspective. *Cereal Research Communications*, 50, 573–601.

Kumari, A., Lakshmi, G.A., Krishna, G.K., Patni, B., Prakash, S., Bhattacharyya, M., Singh, S.K. & Verma, K.K., 2022. Climate change and its impact on crops: A comprehensive investigation for sustainable agriculture. *Agronomy*, 12, 3008.

Lateef, A., Nazir, R., Jamil, N., Alam, S., Shah, R., Khan, M.N., Saleem, M. & Rehman, S.-U., 2019. Synthesis and characterization of environmental friendly corncob biochar based nano-composite – A potential slow release nano-fertilizer for sustainable agriculture. *Environmental Nanotechnology, Monitoring & Management*, 11, 100212.

Li, B., Zhang, X., Morita, S., Sekiya, N., Araki, H., Gu, H., Han, J., Lu, Y. & Liu, X., 2022. Are crop deep roots always beneficial for combating drought: A review of root structure and function, regulation and phenotyping. *Agricultural Water Management*, 271, 107781.

Li, C., Li, Y., Li, Y. & Fu, G., 2018. Cultivation techniques and nutrient management strategies to improve productivity of rain-fed maize in semi-arid regions. *Agricultural Water Management*, 210, 149–157.

Li, F.M., Li, X.G., Javaid, M.M., Ashraf, M. & Zhang, F., 2020. Ridge-furrow plastic film mulching farming for sustainable dryland agriculture on the Chinese loess plateau. *Agronomy Journal*, 112, 3284–3294.

Li, T., Jia, Y., Fava, F., Xu, Z., Zhu, J., Yang, Y., Tang, L., Wang, Y., Hao, Y. & Cui, X., 2021. A geographic identification of sustainable development obstacles and countermeasures in drylands: A case study in Inner Mongolia, China. *Ecological Indicators*, 132, 108257.

Liu, X., Wang, Y., Zhang, Y., Ren, X. & Chen, X., 2022. Can rainwater harvesting replace conventional irrigation for winter wheat production in dry semi-humid areas in China? *Agricultural Water Management*, 272, 107852.

Lüttschwager, D. & Jochheim, H., 2020. Drought primarily reduces canopy transpiration of exposed beech trees and decreases the share of water uptake from deeper soil layers. *Forests*, 11, 537.

Lv, H., Zhao, Y., Wang, Y., Wan, L., Wang, J., Butterbach-Bahl, K. & Lin, S., 2020. Conventional flooding irrigation and over fertilization drives soil pH decrease not only in the top- but also in subsoil layers in solar greenhouse vegetable production systems. *Geoderma*, 363, 114156.

Madhu, M., 2022. Soil and water conservation strategies for sustainable agriculture in changing climate scenario. *Community Based Climate Risk Management through Watershed Development*, 1.

Mandal, D., Mandal, C., Prasad, J. & Bhattacharyya, T., 2019. Acid soils in agro-ecological sub-regions of India: A revisit. *Indian Journal of Fertilisers*, 15, 1156–1166.

Marengo, J.A., Cunha, A.P.M.A., Nobre, C.A., Ribeiro Neto, G.G., Magalhaes, A.R., Torres, R.R., Sampaio, G., Alexandre, F., Alves, L.M., Cuartas, L.A., Deusdará, K.R.L. & Álvala, R.C.S., 2020. Assessing drought in the drylands of northeast Brazil under regional warming exceeding 4°C. *Natural Hazards*, 103, 2589–2611.

Martínez-Dalmau, J., Berbel, J. & Ordóñez-Fernández, R., 2021. Nitrogen fertilization. A review of the risks associated with the inefficiency of its use and policy responses. *Sustainability*, 13, 5625.

Martínez-Mena, M., Carrillo-López, E., Boix-Fayos, C., Almagro, M., García Franco, N., Díaz-Pereira, E., Montoya, I. & De Vente, J., 2020. Long-term effectiveness of sustainable land management practices to control runoff, soil erosion, and nutrient loss and the role of rainfall intensity in Mediterranean rainfed agroecosystems. *CATENA*, 187, 104352.

Mbangiwa, N.C., Savage, M.J. & Mabhaudhi, T., 2019. Modelling and measurement of water productivity and total evaporation in a dryland soybean crop. *Agricultural and Forest Meteorology*, 266–267, 65–72.

Mcnee, M.E., Rose, T.J., Minkey, D.M. & Flower, K.C., 2022. Effects of dryland summer cover crops and a weedy fallow on soil water, disease levels, wheat growth and grain yield in a Mediterranean-type environment. *Field Crops Research*, 280, 108472.

Mead, J., 2022. Power differentials in carbon emission negotiations – A world systems analysis to explore the sociological foundations of climate change research. *Social Science Chronicle*, 2, 1–15.

Minhas, P.S., Ramos, T.B., Ben-Gal, A. & Pereira, L.S., 2020. Coping with salinity in irrigated agriculture: Crop evapotranspiration and water management issues. *Agricultural Water Management*, 227, 105832.

Mukhtiar, A., Lihong, W., Mahmood, A., Ameen, M., Zia, M.A., Arshad, T., Naqve, M., Wahab, H.A., Rasheed, A. & Asghar, S., 2023. The role of endophytes and rhizobacteria to combat drought stress in wheat. *Notulae Botanicae Horti Agrobotanici Cluj-Napoca*, 51, 13453–13453.

Nadeem, M., Li, J., Wang, M., Shah, L., Lu, S., Wang, X. & Ma, C., 2018. Unraveling field crops sensitivity to heat stress: Mechanisms, approaches, and future prospects. *Agronomy*, 8, 128.

Ni, S.M., Zhang, D.Q., Cai, C.F., Wilson, G.V., Zhang, J.H. & Wang, J.G., 2021. Exploring rainfall kinetic energy induced erosion behavior and sediment sorting for a coarse-textured granite derived soil of south China. *Soil and Tillage Research*, 208, 104915.

O'Connor, P., 2019. Gender imbalance in senior positions in higher education: What is the problem? What can be done? *Policy Reviews in Higher Education*, 3, 28–50.

O'Leary, G.J., Aggarwal, P.K., Calderini, D.F., Connor, D.J., Craufurd, P., Eigenbrode, S.D., Han, X. & Hatfield, J.L., 2018. Challenges and responses to ongoing and projected climate change for dryland cereal production systems throughout the world. *Agronomy*, 8, 34.

O'Neill, B.C., Carter, T.R., Ebi, K., Harrison, P.A., Kemp-Benedict, E., Kok, K., Kriegler, E., Preston, B.L., Riahi, K., Sillmann, J., Van Ruijven, B.J., Van Vuuren, D., Carlisle, D., Conde, C., Fuglestvedt, J., Green, C., Hasegawa, T., Leininger, J., Monteith, S. & Pichs-Madruga, R., 2020. Achievements and needs for the climate change scenario framework. *Nature Climate Change*, 10, 1074–1084.

Owens, P.N., 2020. Soil erosion and sediment dynamics in the Anthropocene: A review of human impacts during a period of rapid global environmental change. *Journal of Soils and Sediments*, 20, 4115–4143.

Pereira, L.S., Paredes, P., Hunsaker, D.J., López-Urrea, R. & Mohammadi Shad, Z., 2021. Standard single and basal crop coefficients for field crops. Updates and advances to the FAO56 crop water requirements method. *Agricultural Water Management*, 243, 106466.

Periasamy, S. & Shanmugam, R.S., 2022. Agricultural Land Degradation in India. *In Impact of Agriculture on Soil Degradation I: Perspectives from Africa, Asia, America and Oceania*. Springer, 219–257.

Prăvălie, R., Bandoc, G., Patriche, C. & Sternberg, T., 2019. Recent changes in global drylands: Evidences from two major aridity databases. *CATENA*, 178, 209–231.

Qaderi, M.M., Martel, A.B. & Dixon, S.L., 2019. Environmental factors influence plant vascular system and water regulation. *Plants*, 8, 65.

Racette, K., Zurweller, B., Tillman, B. & Rowland, D., 2020. Transgenerational stress memory of water deficit in peanut production. *Field Crops Research*, 248, 107712.

Rai, P., 2018. The labor of social change: Seasonal labor migration and social change in rural western India. *Geoforum*, 92, 171–180.

Rajaram, S. & Van Ginkel, M., 2019. Achieving global food security for all: Focus on sub-Saharan Africa and Asia. *Le Déméter 2019*. IRIS éditions, Aachen, Germany, 297–309.

Ramasamy, S. & Saravanavel, J., 2020. Remote sensing revealed geomorphic anomalies and recent earth movements in Cauvery delta, Tamil Nadu, India. *Journal of the Indian Society of Remote Sensing*, 48, 1809–1827.

Rao, B.K., Annapurna, S., Rani, B.R., Rao, Z.S., Sunitha, K., Schindutt, M., Jamanal, S. & Ramesh, V., 2022. *Soil and Water Conservation Techniques in Rainfed Areas* [E-book]. Hyderabad: National Institute of Agricultural Extension Management (MANAGE) & Water and Land Management Training and Research Institute (WALAMTARI), Hyderabad. Hyderabad." (2022): 10–12.

Rasheed, A., Li, H., Nawaz, M., Mahmood, A., Hassan, M.U., Shah, A.N., Hussain, F., Azmat, S., Gillani, S.F.A. & Majeed, Y., 2022. Molecular tools, potential frontiers for enhancing salinity tolerance in rice: A critical review and future prospective. *Frontiers in Plant Science*, 13, 966749.

Rasheed, A., Seleiman, M.F., Nawaz, M., Mahmood, A., Anwar, M.R., Ayub, M.A., Aamer, M., El-Esawi, M.A., El-Harty, E.H. & Batool, M., 2021. Agronomic and genetic approaches for enhancing tolerance to heat stress in rice: A review. *Notulae Botanicae Horti Agrobotanici Cluj-Napoca*, 49, 12501–12501.

Rawat, Y.S., Negi, V.S., Moussa, I.M., Zaman, W. & Elansary, H.O., 2023. Diversity, distribution and vegetation assessment of woody plant species in the cold desert environment, North-Western Himalaya, India. *Sustainability*, 15, 10429.

Rehman, H.U., Mahmood, A., Ishfaq, F., Javaid, M.M., Haider, A., Asif, M., Nadeem, M.A. & Nargis, J., 2023. Sustainable Crop Management for Drylands. *In Climate-Resilient Agriculture, Vol 1: Crop Responses and Agroecological Perspectives*, Springer Nature Switzerland AG, 435–464.

Rejekiningrum, P., Apriyana, Y., Sutardi, Estiningtyas, W., Sosiawan, H., Susilawati, H.L., Hervani, A. & Alifia, A.D., 2022. Optimising water management in drylands to increase crop productivity and anticipate climate change in Indonesia. *Sustainability*, 14, 11672.

Rocha, J., Oliveira, S., Viana, C.M. & Ribeiro, A.I., 2022. Chapter 8 – Climate Change and Its Impacts on Health, Environment and Economy. *In* J.C. Prata, A.I. Ribeiro & T. Rocha-Santos (eds.) *One Health*. Academic Press, 253–279.

Röös, E., Mie, A., Wivstad, M., Salomon, E., Johansson, B., Gunnarsson, S., Wallenbeck, A., Hoffmann, R., Nilsson, U., Sundberg, C. & Watson, C.A., 2018. Risks and opportunities of increasing yields in organic farming. A review. *Agronomy for Sustainable Development*, 38, 14.

Sanaullah, M., Usman, M., Wakeel, A., Cheema, S.A., Ashraf, I. & Farooq, M., 2020. Terrestrial ecosystem functioning affected by agricultural management systems: A review. *Soil and Tillage Research*, 196, 104464.

Sharma, B., Vaish, B., Monika, Singh, U.K., Singh, P. & Singh, R.P., 2019. Recycling of organic wastes in agriculture: An environmental perspective. *International Journal of Environmental Research*, 13, 409–429.

Shehzadi, N., Ishfaq, F., Mahmood, A., Javaid, M.M., Naqve, M., Jabbar, A. & Aziz, A., 2023. Crop Improvement in the Desert. *In Climate-Resilient Agriculture, Vol 1: Crop Responses and Agroecological Perspectives*. Springer, 465–485.

Si, W.-X., Li, P.-F., Ashraf, M., Mo, F., Hui, X.-J., Batool, A., Mansoor, J., Afzal, M., Zhang, J. & Li, F.-M., 2017. Micro-field rain-harvesting farming system in semiarid Pakistan: A critical topic to address. *Pakistan Journal of Botany*, 49, 1217–1221.

Sigler, W.A., Ewing, S.A., Jones, C.A., Payn, R.A., Miller, P. & Maneta, M., 2020. Water and nitrate loss from dryland agricultural soils is controlled by management, soils, and weather. *Agriculture, Ecosystems & Environment*, 304, 107158.

Singh, K., Kumar, S., Gupta, K., Pradheep, K., Rajasekharan, P., Pandey, C., Ranjan, P., Gupta, S. & Gangopadhyay, K., 2018. Genetic resource management and utilisation of horticultural crops in India – A perspective. *International Journal of Innovative Horticulture*, 7, 71–103.

Singh, N., Agarwal, S., Jain, A. & Khan, S., 2021. 3-Dimensional cross linked hydrophilic polymeric network "hydrogels": An agriculture boom. *Agricultural Water Management*, 253, 106939.

Sittig, S. & Sur, R., 2023. Runoff mitigation on agricultural fields via micro-dams – Literature review and derivation of runoff curve number reductions. *Environmental Research*, 238, 117128.

Srinivasarao, C., Jasti, V.N.S.P., Kondru, V.R., Bathineni, V.S.K., Mudigiri, R., Venati, G.V., Priyadarshini, P., Abhilash, P.C. & Chaudhari, S.K., 2022. Land and water conservation technologies for building carbon positive villages in India. *Land Degradation & Development*, 33, 395–412.

Stein, S. & Steinmann, H.-H., 2018. Identifying crop rotation practice by the typification of crop sequence patterns for arable farming systems – A case study from Central Europe. *European Journal of Agronomy*, 92, 30–40.

Sun, Q., Miao, C., Hanel, M., Borthwick, A.G.L., Duan, Q., Ji, D. & Li, H., 2019. Global heat stress on health, wildfires, and agricultural crops under different levels of climate warming. *Environment International*, 128, 125–136.

Taseska, T., Yu, W., Wilsey, M.K., Cox, C.P., Meng, Z., Ngarnim, S.S. & Müller, A.M., 2023. Analysis of the scale of global human needs and opportunities for sustainable catalytic technologies. *Topics in Catalysis*, 66, 338–374.

Tesfaw, A., Alemu, D., Senbeta, F. & Teferi, E., 2022. Eucalyptus succession on croplands in the highlands of northwestern Ethiopia: Economic impact analysis using farm household model. *Resources*, 11, 71.

Tripathi, A., 2014. Resource Conservation Technologies: A Novel Paradigm to Magnify Resource Use Efficiency in Rainfed and Drylands of India. *Resource Conservation Technology in Pulses*. Scientific Publishers, 58–69.

Ullah, I., Lizalo, A., Emre Dogan, D. & Uray, Y. & Rashid, H., 2021. Relevance of mulching in strawberry cultivationed. *International Cappadocia Scientific Research Congress, 15–17 December 2021 / Cappadocia-Nevsehir,* 318–330.

Vadez, V., Pilloni, R., Grondin, A., Hajjarpoor, A., Belhouchette, H., Brouziyne, Y., Chehbouni, G., Kharrou, M.H., Zitouna-Chebbi, R. & Mekki, I., 2023. Water use efficiency across scales: From genes to landscapes. *Journal of Experimental Botany*, 74, 4770–4788.

Verma, M.N.K. & Chaudhary, V.R. & Sachan, D., 2020. Damini mobile app will give warning before lightning. *Editorial Members*, 68-69.

Wang, X., Wang, J., Zou, Y., Bie, Y., Mahmood, A., Zhang, L., Liao, L., Song, Z., Liu, G. & Zhang, C., 2024. Urea fertilization increased CO_2 and CH_4 emissions by enhancing c-cycling genes in semi-arid grasslands. *Journal of Environmental Management*, 356, 120718.

Xie, H., You, L., Dile, Y.T., Worqlul, A.W., Bizimana, J.-C., Srinivasan, R., Richardson, J.W., Gerik, T. & Clark, N., 2021. Mapping development potential of dry-season small-scale irrigation in sub-Saharan African countries under joint biophysical and economic constraints – An agent-based modeling approach with an application to Ethiopia. *Agricultural Systems*, 186, 102987.

Zhang, C., Dong, Z., Guo, Q., Hu, Z., Li, J., Wei, T., Ding, R., Cai, T., Ren, X., Han, Q., Zhang, P. & Jia, Z., 2022. Ridge–furrow rainwater harvesting combined with supplementary irrigation: Water-saving and yield-maintaining mode for winter wheat in a semiarid region based on 8-year in-situ experiment. *Agricultural Water Management*, 259, 107239.

Zhang, G., Ming, B., Shen, D., Xie, R., Hou, P., Xue, J., Wang, K. & Li, S., 2021a. Optimizing grain yield and water use efficiency based on the relationship between leaf area index and evapotranspiration. *Agriculture*, 11, 313.

Zhang, G., Shen, D., Ming, B., Xie, R., Jin, X., Liu, C., Hou, P., Xue, J., Chen, J., Zhang, W., Liu, W., Wang, K. & Li, S., 2019. Using irrigation intervals to optimize water-use efficiency and maize yield in Xinjiang, northwest China. *The Crop Journal*, 7, 322–334.

Zhang, W., Wang, P., Liu, S., Chen, J., Chen, R., He, X., Ma, G. & Lei, Z., 2021b. Factors affecting the properties of superabsorbent polymer hydrogels and methods to improve their performance: A review. *Journal of Materials Science*, 56, 16223–16242.

10 Implications and Potential of Marginal Land

Amna Ghafoor, Safura Bibi, Hina Nazir, Muhammad Talha Bin Yousaf, Muhammad Mansoor Javaid, and Muhammad Ather Nadeem*

10.1 INTRODUCTION

10.1.1 Marginal Lands

Marginal lands refer to areas with low agricultural productivity due to various factors such as poor soil fertility, water scarcity, steep slopes, and extreme climatic conditions (Chadha *et al.*, 2020). Due to poor soil and insufficient water supply conditions, marginal lands have little intrinsic value or potential for profit in agricultural or industrial activity (Khan *et al.*, 2024). Moreover, marginal land often degraded lands presents significant environmental challenges globally (Javaid, 2018). Therefore, increasing environmental protection and optimizing land management have been two major concerns in the distribution of crops to marginal regions.

According to current knowledge, the term "marginal lands" refers to regions that are geographically inaccessible, restricted by soil or climate, at high risk for environmental harm, or with weak ecosystem services and very little productivity, making them unprofitable (Haberzettl *et al.*, 2021). Marginal lands are a land resource in a transitory state and are extremely susceptible to natural processes and different management. Land degradation could be brought on by improper management of productive lands (Hussain *et al.*, 2019). Economic growth and food demand have substantially impacted marginal areas. Because of rising labor costs and agricultural intensification, marginal land has been disappearing across Europe (Kang *et al.*, 2013; Figure 10.1).

Widespread attention has been given to marginal lands because of their potential to increase food security and the production of pulses (Chattha *et al.*, 2017). The production of pulses in Pakistan is far less than what is needed and imports make up the difference. Pakistan's low production and yield of pulses is due to a lack of inventive crop development and a seed distribution system. Currently, the farmers utilize their own stored seed to grow 80% of the pulses (Ullah *et al.*, 2020).

10.2 MARGINAL LAND USE MANAGEMENT; CHALLENGES AND NEEDS

Marginal land has grown to encompass a wide range of issues. Marginal lands are often susceptible to soil erosion, desertification, and degradation of biodiversity (Arshad *et al.*, 2023). Due to their insufficient water-holding capacity and ensuring summer moisture deficits, many hillslope soils as well as sandy "excessively drained" soils are marginal and are frequently relegated to pasture/forage agriculture (Ullah *et al.*, 2020). Soil erosion brought on by poor previous management, natural fertility, loss of organic matter, and soil depth (such as growing annual crops that aren't appropriate to the soil) may also convert soils to marginal soils (Nazir and Lohano, 2022). To address these issues, a wide range of requirements are also required.

* Corresponding author: safura.bibi@uaf.edu.pk

DOI: 10.1201/9781003498698-10

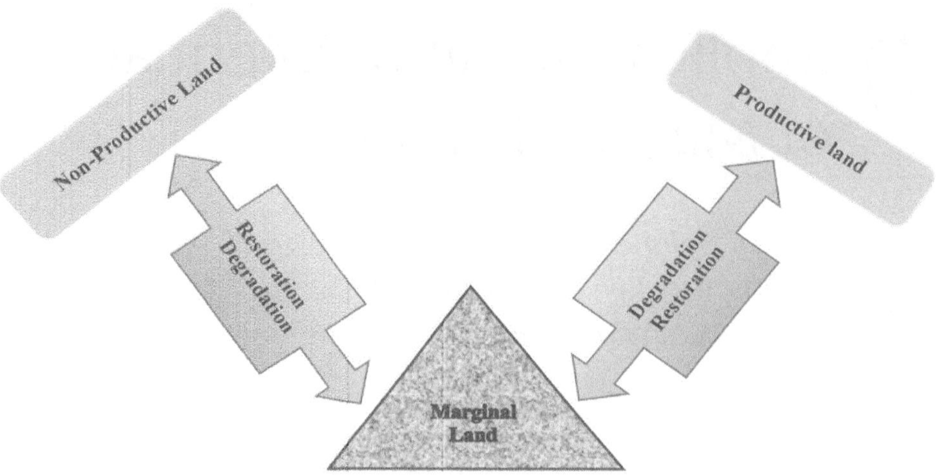

FIGURE 10.1 A transitional state of land uses to marginal lands.

10.3 CRITERIA FOR MARGINAL LAND MANAGEMENT AND ASSESSMENT

Managing marginal lands involves considering various criteria to determine their suitability for various uses. Soil quality, topography, climate conditions, water resources, biodiversity and ecological values, economic viability, legal and regulatory considerations are some of the factors that are responsible for determining the criteria of marginal lands (Li *et al.*, 2020). It is impossible to accurately assess and manage land resources without reliable criteria for defining marginal lands (Ahmad *et al.*, 2020). For instance, the amount of marginal land ranging from 100 million hectares to 1 million hectares varies on a global scale (The World Watch Institute), or 1.3 billion ha, or 36% of agricultural lands (Strijker, 2005). These figures were calculated using various standards for defining marginal land. Different management objectives across regions and countries may make it difficult to define unified marginal land criteria, but such criteria should be similar and adaptable for a variety of land use planning and policy-making activities (Van Dam *et al.*, 2010). Domestic marginal land should be defined based on current demands and problems, considering it is the new horizon defined by existing standards and evaluation techniques related to the quality of soil, water, air, and land. The criteria should also take into account how other land uses, such as cropland uses, wooded land, grasslands or rangelands, and other potentially useful land resources interact with each other. As a result, no single index or criterion can completely satisfy these demands (Kang *et al.*, 2013).

10.4 LIMITATIONS OF MARGINAL LAND FOR PULSES

Pulses are predominantly grown on marginal soils with limited inputs (Banerjee *et al.*, 2022). Pulses are subjected to a variety of abiotic and biotic challenges as a result of their growing needs, which results in significant production losses (Mahmood and Honermeier, 2019). Climate-change induced by global warming has likewise increased their susceptibility to new emerging insects, pests, abiotic and biotic stresses, and this could be more problematic in the years to come (Hassan *et al.*, 2018). Breeding and developing climate-resilient smart pulses is a challenging task as we are working in a shifting climate scenario. Pulses are, however, overcoming the climate change challenge while mitigating its effects, with low genetic diversity always a prominent challenge to the extent of adaptability (Oluyisola *et al.*, 2022). However, because of the genetic variance that already exists, there are still methods of exploiting innovative characteristics for the growth of climate-resilient cultivars. Accelerating the mining and utilization of adaptive traits that confer tolerance or resistance to climate-smart pulse through the use of state-of-the-art biotechnology approaches such as transgenic, genome editing, and epigenetics (Birthal *et al.*, 2014; Zsögön *et al.*, 2022).

10.5 ABIOTIC STRESSES

The most detrimental factor affecting the productivity and growth of crops is abiotic stress, which is largely inevitable (Nadeem *et al.*, 2022). Effective toleration requires confronting these stresses a complex phenomenon resulting from a variety of interactions (Bera *et al.*, 2021).

10.5.1 Drought

Drought stress affects the growth of pulses in many ways. For instance, the growth of chickpea growth, photosynthetic rate, and photosystem-II efficiency were all impacted by drought stress (Ullah *et al.*, 2019). Although pulses use very little water, water shortages during crucial stages of crop growth significantly lower yields. Even though among the pulses, chickpea is a reasonably drought-tolerant crop, they nevertheless need enough moisture during two crucial growth stages, namely planting and flowering/pod formation (Singh *et al.*, 2021). Similar to other plants, lentils need enough moisture to grow and create pods. Mung beans and mash beans, which are summer pulses, need at least four to five irrigations throughout the crop life cycle, although the flowering and pod-formation stages are the most important ones (Islam *et al.*, 2021). Although drought can influence the growth and yield production of pulses at any time, it is more severe when it happens during the reproductive stages and leads to production losses (Farooq *et al.*, 2017; Table 10.1).

Under extreme drought conditions a 42–70% decrease in chickpea seeds output has been reported (Leport *et al.*, 1998). Yield losses caused by drought stress occur in mash beans at the flowering and reproductive phases and range from 31 to 57 and 26%, respectively (Islam *et al.*, 2021). Chickpea and lentils frequently experience water stress in various growth stages (Singh *et al.*, 2022a).

10.5.2 High Temperature

Global climate models predict an increase of 4°C as well as an increase in heat stress by the end of this century. Although, pulses have the ability to tolerate high temperatures extremely high temperature can result in poor production of pulses. Winter pulses may experience a significant yield drop as a result of heat stress during grain fling and reproductive development (Hassan *et al.*, 2021).

TABLE 10.1
Effect of Drought Stress on Germination, Plant Growth, Water Relation and Nutrient

Crop	Growth Stage	Effect	Reduction Yield (%)	References
Chickpea	Flowering onwards	Reduced photosynthesis, decreased rate and duration of seed growth	52–70	De Oliveira *et al.* (2013)
Lentil	Podding onwards	Reduced leaf area, flower number, pod and seed number	70	Chelladurai and Erkinbaev (2020)
Mung bean	First flowering to 100 days of sowing	Reduced pollen viability and germination	87	Allahmoradi *et al.* (2011)

For the chickpea reproductive stage, the threshold temperature is 25°C. Similarly, the range of temperatures required for mungbean flowering and pod formation is 10–30°C (Farooq *et al.*, 2017). Beyond this point, temperature increases result in a decrease in yield. During the strength and durability, vegetative, and reproductive growth stages, the growth of pulses is inhibited by heat stress. Heat stress affects seedling germination and decreases seedling survival during the stand-establishing stages. Heat stress, however, lowers the weight and number at the reproductive stage and may result in grain abortion (Hassan *et al.*, 2022). Even soil bacteria that are in symbiosis with pulses survive in soil temperatures above (50°C). As seen in a variety of pulses, including chickpea (George *et al.*, 2022), alfalfa (Peng *et al.*, 2022), and mung bean (Hanumantharao *et al.*, 2016), high temperatures are detrimental to seed germination and seedling growth (Kamboj *et al.*, 2022). In mean square seedling germination, heat tolerance, and growth index, an elevated temperature of 50°C for 2 hours in mung bean and 35–50°C for four hours in lentil harms seedlings. The pulses are significantly impacted by high temperatures during the stand establishment and reproductive stages ((Javaid *et al.*, 2022).

10.5.3 Chilling Stress

Chilling stress is defined as a temperature below freezing (between 0 and 15°C). This stress also reduces the growth and yield of pulse crops. Tropical and subtropical plants are vulnerable to chilling temperatures (0–10°C) and unable to acclimate to the cold (Basu *et al.*, 2022) Chilling stress reduces the growth of chickpea by reducing the stand establishment during germination and seedling emergence stages. In pulses, the early stages of pod production and flowering are extremely susceptible to low temperatures (Khanpit *et al.*, 2022).

10.5.4 Salinity Stress

Salt stress is one of the main hurdles to the commercial production of pulses in Pakistan. Due to specific ion toxicity, osmotic stress, and nutritional imbalances, salt stress affects crop plant growth, germination, and yield formation (Verma *et al.*, 2022). Salt stress reduced seed output per plant due to lower seed per pod and 100-seed weight (Afzal *et al.*, 2022). The buildup of too much salt in pulses increases the coloring of anthocyanins in stems and leaves which hinders seedling establishment and seed germination. Salt stress affects biological nitrogen fixation and nitrogen uptake (Palma *et al.*, 2013). In chickpeas, 3.8 ds m^{-1} of salinity stress results in yield losses of up to 72% (Van Hoorn *et al.*, 2001). Similar to this, salt stress of 250 mM NaCl results in more than 80–100% yield losses in mung bean. In Pakistan, there are 10 Mha of salinized land, but there is no data on the saline land used for growing pulses (Hasanuzzaman *et al.*, 2022). Pulses are produced in different marginal areas with lower to moderate saline levels, which drastically reduces growth and production, percent (%) yield loss in lentils (Donato *et al.*, 2022).

10.5.5 Waterlogging Stress

Compared to other crops, pulses are more susceptible to waterlogging. Flooding may enhance the prevalence of soil-borne fungi illnesses in addition to anaerobic bacteria converting SO_4^{2-} to H_2S. Due to anaerobic metabolism and the demand for adenosine triphosphate (ATP) molecules exceeding the supply, growth is halted and death occurs due to waterlogging (Meirelles, 2022). High-yielding plants, intensive farming without the addition of micronutrients, (Ali *et al.*, 2014) deficiency of organic manures, leaching of nutrients in unavailable form such as potassium (K), nitrogen (N), calcium (Ca), sulfur (S), manganese (Mn), and boron (B) into unavailable forms in waterlogged soils, use of high-analysis fertilizers and negative interactions of all contribute to nutrient stress (Rana *et al.*, 2016a; Table 10.2).

TABLE 10.2
Yield Loss in Different Pulses Due to Biotic Stresses

Crop and Yield Loss (%)	Disease	Weeds	Yield Loss (%)	Insect-Pests or Yield (%)	References
Chickpea (20–35)	*Fusarium wilt, Ascochyta blight, Botrytis grey mould, Stunt virus*	*Chenopodium album, Melilotus indica, Avena ludoviciana, Medicago* spp., etc.	50–100	Pod-borer and cut worm (10–90)	Li *et al.* (2015)
Pigeon pea (30–90)	*Sterility mosaic virus, fusarium wilt*	*Celosia argentea, Portulaca ghalensis, Eclipta alba*	20–70	Pod-borer and leaf roller (70–80)	Raju *et al.* (2010)
Cow pea (50–90)	*Yellow mosaic virus*	*Cynodon dactylon, Cyperus rotundus*	10–100	White fly (20–55)	Yadav *et al.* (2017)
Mung bean (50–90)	*Cercosra leaf spot, powdery mildew*	*Physalis minima*	20–70	Jassid and pod-borer (20–55)	Shafiq *et al.* (1994)
Lentil (~50)	*Rust, wilt, sclerotia blight collor rot*	*Phalaris* spp., *Guizotia scabra, Avena* spp., *C. album*	20–70	Pod-borer	Talukdar (2013)
Field pea (15–67)	Powdery mildew, rust, downy mildew, wilt	*Avena* spp., *Cirsium arvense, Anagallis arvensis, C. album*, etc.	10–30	Stem and pod-borer, leaf minor	Raju *et al.* (2010)

10.6 BIOTIC STRESSES

Damage to plants from other living species, such as weeds, insects, pests, and nematodes results in biotic stress (Rana *et al.*, 2016a). The biggest and most significant categories among these are fungi and viruses, which have an impact on every part of the plant and at every stage of food legume growth (Pande *et al.*, 2009). Since intensive farming methods depend more and more on chemical pesticides to control insects and other pests, they are no longer viable (Rana *et al.*, 2016). As pesticides are used more frequently, the target pests develop resistance, which reduces biodiversity. More importantly, plants will be more susceptible to pests as a result of climate change. In these conditions, improved knowledge of plant physiology might lead to long-term management of biotic stresses (Cowie *et al.*, 2011).

10.6.1 Weeds

Yield losses of 20–90% are incurred due to weeds in different pulse crops (Cowie *et al.*, 2011). Higher weed infestation in the early vegetative growth stage is more sensitive in pulses. The development of the leaf area is slow for the first three to four weeks in cowpea, urdbean, and mungbean (Cowie *et al.*, 2011), and could take six weeks or longer in pigeonpea, chickpea, peas, and lentil. Nevertheless, greater row spacing and slow early development make weeds a significant problem in pigeon pea, potentially reducing productivity by as much as 80% (Cowie *et al.*, 2011). Pigeon pea being a perennial crop, in later stages several flushes of weed emerge to compete with the crop (Padmaja *et al.*, 2013). The weeds also act as hosts for diseases, nematodes, and insect pests that attack the pulses. In pigeon pea and peas, lambs quarters (*Chenopodium album* L.) serves as an alternative host for the gram caterpillar, while the greasy cut-worm (*Agrotis ipsilon* (Hufnagel) also lives in gram and peas (Cowie *et al.*, 2011).

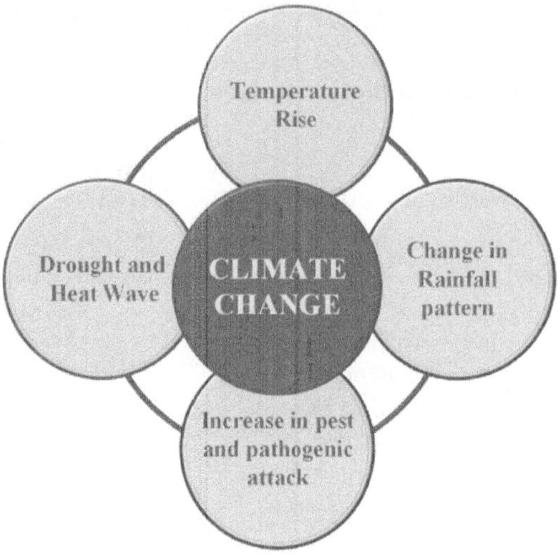

FIGURE 10.2 Biotic and abiotic stresses due to climate change.

10.6.2 Diseases and Insect-Pests

Depending on the crop variety, productivity losses brought on by a lack of plant protection measures range from 46 to 96%. The occurrence of pests, particularly the pod-borer in chickpeas, increases as a result of the of monsoons delayed arrival, which causes late pulse sowing in dryland areas. Pigeon pea plants infected with the sterility mosaic virus before 45 days of age lose 95–100% of their output, but older plants lose just 26–97% of their yield (Rana *et al.*, 2016b; Figure 10.2).

10.7 PULSES

Pulses emerge as a beacon of hope amidst the constraints of marginal lands. Their remarkable ability to grow in diversified agroclimatic zones and low input requirements make them ideal candidates for cultivation in challenging environments. In the context of food security pulses have a huge role as they are the primary supply of protein and minerals for vegetarians all across the world. Moreover, the use of marginal lands to grow pulses can be very helpful in the environmental and economic stability of any country.

Despite the fact that different varieties of pulses are grown in practically every climate, the adaptability of each cultivated pulse species is restricted to its region of origin (Kumar *et al.*, 2019). Chickpea, pigeon pea, green gram, lentil, and black gram are the five main pulse crops that are produced on vast scales, primarily in Asia and Africa, with a variety of soil types (Dagel, 1997). Chickpea (*Cicer arietinum* L.), mung bean (*Vigna radiata* (L.), lentil (*Lens culinaris* Medic.) u0014 and mash bean (*Vigna mungo* L.) are the dominant pulse crops grown on marginal land. 15% cereals and 30% pulses are protein content (Hall *et al.*, 2016). For example, protein content (Kuang *et al.*, 2022) in chickpeas ranges from 17 to 19% (Sreerama *et al.*, 2012); lentils from 23 to 31% (Ghumman *et al.*, 2016); mungbeans from 21 to 31% (Anwar *et al.*, 2007); and mash beans from 21 to 28% (Kole *et al.*, 2002). Pulses can help in restoring damaged soils by fixing nitrogen biologically, mobilizing nutrients such as phosphorus, increasing organic matter through root biomass and leaf fall, preventing soil erosion with cover, and stimulating soil aggregate formation through their more profound root systems (Zuazo and Pleguezuelo, 2009). Marginal lands provide for about one-third of the world's population, covering an estimated 36% of the total agricultural land of the world (1.3 billion ha) (Kuang *et al.*, 2022) and because food demand is rising in developing nations, it is necessary to cultivate marginal lands. Marginal lands have become a primary management objective in order to maintain crop production (Sahoo *et al.*, 2022). In various developing nations in Asia, Africa, and Latin America, pulses represent a significant crop group. Pulses lay 5.8% of worldwide arable marginal area and mainly cultivated as a variety (intercrop) with cereals and oil seeds on less productive and marginal fields (Fu *et al.*, 2022).

Pulses can adapt to a variety of latitudes, longitudes, and environmental conditions. Therefore, pulse crops contribute to enhancing ecological resilience, reducing poverty and hunger, and enhancing health and nutrition. Pulses yield in developed countries is about twice as high as in developing countries, pulses are also grown there making up just about an eighth of the total world average but slightly greater than a fifth of its production (Joshi *et al.*, 2017). The majority of the crop is used as animal forage, although in the past two to three decades, significant amounts have been grown for export to underdeveloped nations. Pulses are increasingly being used for domestic food use, especially in affluent nations like North America, as a result of rising knowledge of their nutritional worth and a trend toward vegetarianism. The year 2016 has been designated the International Year of Pulses in recognition of the significance of pulses in the human diet (Carrino *et al.*, 2020).

10.7.1 Pigeon Pea

More than 350 different local names have been given to the pigeon pea (*Cajanus cajan* L.) with arhar, tur, Congo pea, guandu, guandu, catjang pea, yellow dhal, and ambrevade being the most well-known (Ram and Choudhary, 2022). The term "pigeon pea" has its roots most likely in America, where pigeons are said to have preferred their seeds. The crop ranks as the fourth-largest edible legume crop worldwide (Munaweera *et al.*, 2022; Table 10.3).

10.7.1.1 Growth Regions

Pigeon peas are widely cultivated between 30°N and 35°S latitude in the tropics, subtropics, and warmer equatorial zones of Asia, East Africa, and Central America. The countries with the largest producing areas are India, Myanmar, Kenya, Malawi, Uganda, and Tanzania (Porkodi *et al.*, 2022). Three-fourths of the world's harvested land is occupied by India alone and it contributes about the same amount to output (Vsardana *et al.*, 2010).

10.7.1.2 Growth Conditions

Pigeon pea thrives in a temperature range of 25–35°C, if the soil is sufficiently moist to withstand temperatures as high as 45°C. Frost and cold temperatures can harm the crop (Aravani *et al.*, 2022). It is grown in India in regions with annual rainfall ranging from 600 to 1400 mm. Early and short-duration cultivars, however, do poorly in regions with heavy rainfall (Fowler *et al.*, 2021). Poor pod setting and seed filling caused by cloudy or rainy conditions during blooming and fruiting and pod borers may cause more damage. Pigeon peas are typically grown as dry land or rainfed crops since they are moderately drought tolerant. Despite having a deep root system, the crop gets half of the

TABLE 10.3
Production of Pigeon Pea and Productivity Rate with Area in Different Countries

Country	Area (million ha)	Production (million ha)	Productivity (kg/ha)	References
India	3.728 (76.7)	3.076	825	Singh *et al.* (2022b)
Myanmar	0.540 (11.1)	0.600	1111	Pawar *et al.* (2021)
Kenya	0.195 (4.0)	0.084	430	Balasubramanian and Sathiyamurthi (2021)
Malawi	0.168 (3.4)	0.150	893	Yohane *et al.* (2021)
Uganda	0.088 (1.8)	0.090	1023	Yohane *et al.* (2021)
Tanzania	0.067 (1.4)	0.048	718	Lemoine *et al.* (2022)

water the crop requires comes from the upper 50 cm of soil. The best times to provide water are during branching, blossom initiation, and pod filling (Kumar *et al.*, 2021). Pigeon peas can be grown in varied soils, include marginal soils and heavy clay loam (Bewley *et al.*, 2006; Vsardana *et al.*, 2010).

10.7.2 Chickpea

Chickpea (*C. arietinum* L.) also commonly known as Bengal gram and garbanzo bean is the world's third most important food grain legume after field pea and common bean (*Phaseolus vulgaris* L.) (Ahmad *et al.*, 2005). Chickpeas are grown for their seeds. Chickpeas are an important nutrition source and their demands are increasing. It has a high protein content of between 17 and 20%. People who live in the semi-arid tropics and cannot afford animal proteins or choose to follow a vegetarian diet rely heavily on chickpeas in their diets (Stoilova *et al.*, 2005). The low-cholesterol chickpea is a beneficial protein and carbohydrate source. Additionally, it is an excellent source of vitamins, dietary fiber, and minerals (Maurya *et al.*, 2018).

10.7.2.1 Growth Regions

The chickpea is a significant pulse crop that is farmed and eaten throughout the world, particularly in Africa and Asia. One of the most vital pulse crops cultivated and eaten in India is also known as Bengal gram. Chickpea output in India makes for about 45% of all pulse production (O-Maurya *et al.*, 2018). Like other pulses, chickpeas are primarily produced in India, which accounts for more than 75% of global production (Rao *et al.*, 2010). Pakistan's main pulse crop is the chickpea, which is grown on 73% of the country's territory dedicated to pulses. Due to a lack of irrigation water and low fertility in the Thal Desert, chickpeas are the only crop that can be profitably grown there (Nguli *et al.*, 2022).

10.7.2.2 Growth Conditions

It is most effectively suited for certain ranges in temperatures and low to moderate amounts of rain. Excess rains soon after seeding or during the flowering period hurt the crop. A severe cold can be seriously damaging (Jannat *et al.*, 2022). It is best suited in regions with annual rainfall between 600 and 900 mm. Because of its indefinite growth habit, the growth cycle lasts as long as there is moisture available (O-Maurya *et al.*, 2018; Table 10.4).

10.7.3 Mung Bean

The third-most significant pulse in Asia is the mung bean. Additionally, it goes by the names mung, green gram, moong, golden gram, and mungo. This crop's short growing season (90–120 days) ability to fix nitrogen, prevention of soil erosion, soil enrichment, low input needs, and extensive adaptability are just a few of its many advantageous qualities (Nguli *et al.*, 2022). For many Asian people mung bean is a high-quality protein source that accounts for up to 25% of the dry weight of the seed and is also a good source of micronutrients such as phosphorus, calcium, and vitamins (Pratap *et al.*, 2022).

TABLE 10.4
Chickpea Production, Average Yield, and Area of Production with Various Countries

Country	Area (1000 ha)	Production (1000 tons)	Share of World Production	Average Yield (t/ha)	References
India	7163	71.65	69.91	0.69	Khinea *et al.* (2022)
Pakistan	1043	10.43	8.02	0.54	Kumar and Malik (2022)
Turkey	533	5.33	8.43	1.1	Varma (2022)
Myanmer	203	2.03	2.56	0.88	Bhat *et al.* (2022)
Ethiopia	180	1.80	1.92	0.75	Khinea *et al.* (2022)

10.7.3.1 Growth Regions

In developing nations like Asia, Latin America, and Africa where it is cultivated primarily. Because of its high nutritional profile, brief growth season, and ability to restore soil fertility, mung bean is regarded as a cash crop (Ranawake *et al.*, 2011). Typically, it is grown in Pakistan during both of the growing seasons in marginal areas (Rabi and Kharif). Although inputs required to give developing plants the best conditions are insufficient in marginal areas, marginal lands support the growth of mung beans (Akram *et al.*, 2022).

10.7.3.2 Growth Conditions

As a warm-season vegetable, the mung bean with numerous types requires 90–120 days of frost-free weather from planting to its maturity (Aliku *et al.*, 2022). The ideal temperature range for growth of this crop is (27–30°C) thus it is often cultivated in summer (Schiavi *et al.*, 2022).

10.7.4 Lentils

Lentils are an essential rotational crop in agricultural systems because they increase soil fertility. The main impediment is increasing research funding and awareness on improving lentils in nations where the crop is an important component of the food system (Neda, 2020). The production of lentils has expanded globally at a pace of 8.6 kg/ha/year between 1961 and 2008. The main abiotic challenges where plants are now grown include a lack of rainfall, high spring temperatures, and freezing winters at high altitudes (Skendžić *et al.*, 2021). For those who cannot afford animal products, especially the poor, lentil grain is a vital source of protein. Additionally, lentil seed is rich in fiber, carbohydrates, Ascorbic acid, vitamins (inositol, niacin, and vitamin A), minerals, macro and micronutrients (Ca, Zn, K, Fe, and P) for a balanced diet. Additionally, it has a lot of lysine, a necessary amino acid that is scarce in grain protein. Lentils and cereals are complementary in terms of nutrition (Priego-Poyato *et al.*, 2021).

10.7.4.1 Growth Regions

A total of 94% of the world's output of lentils is produced by the top ten producers, indicating a considerable concentration of both export and production (Sidhu *et al.*, 2022). The top importing nations of lentils are Iran, Sri Lanka, Pakistan, India, Algeria, Turkey, Sri Lanka, Bangladesh, Egypt, United Arab Emirates, Saudi Arabia, Colombia, Peru, France and Morocco (FAO, 2010).

10.7.4.2 Growth Conditions

Depending on the variety, lentils typically require 90–120 days of frost-free, warm-season weather to go from planting to maturity (Chung, 2022). Consequently, the crop is usually grown in summer although the range of optimum temperatures for growth is 27°–30° (Chelladurai and Erkinbaev, 2020; Table 10.5).

TABLE 10.5
Lentils Production, Area, and Productivity Rate

Country	Area (1000 ha)	Production (1000t)	Productivity (kg/ha)	References
Canada	887.11	985.30	1391	Malik *et al.* (2022)
Mexico	708.20	7.36	953	Singh *et al.* (2022c)
USA	171.19	234.60	1370	Pramanik *et al.* (2022)
Australia	141.23	126.95	899	Sharma *et al.* (2022)
New Zealand	139.63	124.00	888	Malik *et al.* (2022)

10.8 MITIGATING STRATEGIES AGAINST STRESS IN MARGINAL LAND FOR PULSES

Field crop productivity is under pressure from the growing negative effects of climatic changes. Pulses are more susceptible to fluctuations brought on by climate change and its associated factors (Cowie *et al.*, 2011). To lessen the effects of climatic changes and variability in such a situation, adaptation and mitigation strategies must be developed and put into practice (Hulme, 2005). Due to the diversity of climate change effects, not a single method can effectively address the problem (Hayes *et al.*, 2018). To deal with these risks, a combination of technological and policy-related actions is needed. There are following strategies to overcome limitations in marginal land for pulses

10.8.1 Breeding Strategies

The creation of crop types that are climate-robust could be a viable defense against climatic risks. High osmotic adjustment varieties are being produced. These high osmotic adjustment types have been found by ICAR-IIPR (Indian Institute of Pulses Research-Innovative Industrial Properties, Inc.) (Cheng *et al.*, 2022; Table 10.6).

These varieties can be grown in regions with a lot of climate hazards (Hayes *et al.*, 2018). It is strongly advised to create pulse varieties resistant to many harmful conditions including salinity, alkalinity, acidity, drought, and biotic stress like insect and disease infestation. Therefore, the fundamental requirement to manage climatic risk has been the development of photo and thermos-sensitive genotypes (Hayes *et al.*, 2018).

10.8.2 Agronomic Strategies

Farming diversification is a successful strategy for lowering the risk involved in farming in unstable conditions. It is absolutely crucial in dryland environments where crop failure is common (Gull *et al.*, 2020). Crop diversification through mixed cropping and intercropping is a very promising strategy that will improve cropping intensities and create effective resource utilization. Agroforestry, alley cropping, ley farming, and other alternative land use systems can improve resource utilization while stabilizing pulse production (Palmo *et al.*, 2022).

10.8.2.1 Fallow and Conservation Tillage

The fallow method is meant to keep soil moisture under check. Other benefits include increased soil nutrient availability of certain soil-borne pests and their elimination. Standard agricultural practice in dry land (marginal land) farming involves using the fallow system to increase soil moisture storage, either with or without conservation tillage (M Tahat *et al.*, 2020).

Any tillage and planting method known as conservation covers 30% or more of the crop residue on the soil's surface following planting. Preserving ground cover and soil structure with mulch, like stubble, are the major goals of conservation tillage, which is primarily intended for decreased tillage operations (such as zero tillage, minimum tillage, and mulching). These procedures lessen runoff

TABLE 10.6
Varieties of Pulses Best Suited for Marginal Lands

Pulses	Varieties of Pulses	References
Pigeon pea	TGT501, Bennur, Local, RVK275, ICP1673, Bahar, GRG815	Cheng *et al.* (2022)
Black gram	PKG-U03, GP-15, IPU 06-12	Choudhary *et al.* (2022)
Green gram	Ganga-8, HUM1, ML5	Pradipa *et al.* (2022)

and allow for greater water penetration. Raised beds are a great alternative for planting in locations with a lot of rain (Kasprzyk *et al.*, 2022). Carbon sequestration in soil agriculture practices that raise soil organic carbon (SOC) must be used in order to lower greenhouse gas emissions and boost carbon sequestration in soils. The ever-changing climatic scenario has also been contributing to the decline in SOC content (Basso *et al.*, 2018). Soil physical, chemical, and biological environment is improved with the increase of SOC. The nitrogen-fixing ability of pulses makes them less fertilizer requirement leading to lower greenhouse gas emissions (Yuvaraj *et al.*, 2020).

10.8.2.2 Intercropping

Inter-cropping also known as multi-cropping, where two or more crops grow is described as growing more than one crop at the same time in the same field (Bamboriya *et al.*, 2022). Because of their deeper and larger root systems, pulse crops utilize more water for root development (Bell and De Oliveira, 2022).

Because of this root structure and their lag in initial growth, pulses are better adapted for rain-sheltered intercropping with cereals and oilseeds (Jadav *et al.*, 2022), and because these intercrops' reproductive growth does not correspond with that of the primary crop, the primary crop's yield is not adversely affected. (Ullah *et al.*, 2020). Leguminous crops including black bean, cowpea, mungbean, and pigeonpea can be seeded alongside cereal crops like mustard (*Brassica juncea* L.), wheat, maize (*Zea mays* L.), vegetables, and pearl millet (*Pennisetum glaucum*) (FAO, 2013). During the kharif season, pearl millet, pigeon pea, sesame (*Sesamum indicum* L.), maize, and cotton are interplanted with green gram (Swaminathan *et al.*, 2012). Green gram is produced as an intercrop with sugarcane that was sown in the springtime (Bil, 2022). In India, intercropping covers more than 70% of the land. In India's rainfed regions, winter pulses like chickpeas and lentils are regularly interplanted with oil seeds. Wheat and barley are also planted with chickpeas (*Hordeum vulgare* L.) (Kumawat *et al.*, 2022). Phosphorus and nitrogen use efficiency, wheat yields, chickpea yields, and net income were greater in wheat and chickpea intercropping systems than in monoculture systems of both crops (Akhtar *et al.*, 2010). Winter legumes like chickpeas and lentils can grow successfully alongside wheat and sugarcane in our agricultural system. As summer legumes, mungbean, and sesame can be planted together (Chelladurai and Erkinbaev, 2020; Table 10.7).

10.8.2.3 Seed Distribution System

The seed is a necessary component for producing something on a plot of land. By using high-quality seed, crop yields can be increased by 20–25% (Brooks, 2022). Farmer varietal selection to increase the creation of innovative low-cost seed systems and the cultivars they choose. Many countries have well-established seed farmer organizations that seek to supply good chickpea seeds. Similar methods can be used in Pakistan to increase seed output and distribution. Nearly 43,000 metric tonnes of seed are required for the country's pulse planting, but just 1400 metric tonnes are now available. Cooperatives that take breeder seeds and multiply them by working with contract farmers

TABLE 10.7
Grain Yield Production with Intercropping System

Inter Cropping System		Grain Yield of Main Crop (t/ha)		Increase in Yield (%)	References
Main Crop	**Inter Crop**	**Sole**	**Inter Cropping**		
Rice	Cow pea	4.02	5.08	26	Tanveer *et al.* (2022)
Rice	Sesbania	4.02	4.92	22	Tanveer *et al.* (2022)
Wheat	Chickpea (1:1)	1.51	1.72	14	Ullah and Hassan (2022)
Wheat	Chickpea (1:1)	1.51	1.67	11	Zeng *et al.* (2022)

to generate high-quality seeds for the widest possible distribution can manage seed distribution (Ullah *et al.*, 2022a).

10.8.2.4 Moisture and Water Conservation

The use of appropriate cultivation techniques is the prerequisite for preserving moisture for optimizing pulse yield since pulses are generally grown under moisture-stress conditions. Moisture conservation is the main restriction in dryland soils lacking irrigation resources (Ullah *et al.*, 2022b).

Greater focus should be placed on water conservation because poor germination and plant stand are caused by sub-optimal soil moisture at the time of sowing. As applied at a rate of 8 t/ha to red lateritic soils in western *Odisha*, *grass*, *gliricidia*, and *lantana mulch* dramatically increased soil moisture retention, increasing chickpea output by two to three times when in comparison to mulch while it's raining. Since pulses are sensitive to waterlogging, heavy summer plowing together with leveling is necessary for drylands to save moisture and to promote quick drainage. Similar to this, in instances of moisture stress, additional irrigation throughout the pre-rainy season would be beneficial for the stability of production or improvement. Therefore, it is preferable to ensure that water harvesting as well as in-situ rainfall conservation are maximized by fusing production and conservation methods (Nithisha *et al.*, 2022).

10.8.2.5 Irrigation Management

Only 12% of India's irrigated land is used for growing pulses (Rao *et al.*, 2022). In times of low soil moisture, such as during rabi and summer, irrigation is necessary for pulse crops. A better pulse production and more efficient use of nutrients can result from scientific irrigation scheduling and the use of water-saving irrigation techniques (Islam *et al.*, 2020).

In comparison to the traditional flood method, sprinkler irrigation in pea at an IW:CPE (irrigation water:cumulative pan evaporation) of 1.0 (2 irrigations) saved 4 cm of irrigation water (Singh *et al.*, 2013). When there is a dry spell, kharif pulses also require irrigation at crucial growth stages of blooming and pod development (Das *et al.*, 2022). In many areas of the country with scarce water supplies, micro-irrigation techniques like drip irrigation and spray irrigation are becoming more and more popular. Fertigation offers hope for widely spread crops like pigeon peas, allowing for the irrigation of 30–50% additional land (Singh *et al.*, 2013). When it comes to managing the limited water resources they must combine with other inputs for pulse crops. It appears that well-planned deficit-irrigation methods supplemented by micro-irrigation or slow water delivery devices are the most efficient way to conserve water. These schedules reduced the irrigation water need for mung bean by 43–50% and for common beans by 15–22% (Webber *et al.*, 2006).

10.8.2.6 Weed Management

More than 70% of land in India is used to grow pulses, which are more productive, especially during rains. Pulses play a significant role in intercrop systems with cereals, oilseeds, and commercial crops as they, apart from providing the base crop with additional nitrogen, have a significant smothering effect on weeds, and bind soil particles and thus, prevent soil degeneration, among other advantages (Barnsley, 2022). Pigeon pea yield is increased soil moisture is conserved and soil temperature is moderated when sugarcane garbage is mulched at 8 t/ha (Tayade *et al.*, 2016).

As a result, mulching previous crop leftovers helps pulse plants by lowering stress from weeds, contemporaneous insect pests, and disease. Nowadays, there are a number of reasonably priced herbicides available that may be widely applied to effectively manage weeds and increase pulse yield. According to studies, the pre-emergence treatment of pendimethalin (1.0 kg/ha) increased the grain production of pigeon peas, kidney beans, cowpeas, chickpeas, and lentils in Himachal Pradesh (Choudhary and Suri, 2014). Since weed stress has a significant impact on the very diversified agricultural sector of pulses, creative herbicide treatments are needed for better weed control. (Birthisel *et al.*, 2021).

10.8.2.7 Integrated Disease Management

A comprehensive integrated disease management (IDM) scheme is one that suppresses losses of grain legumes due to diseases to levels that are acceptable and do not compromise human health or the environment significantly over the long term (Brodt *et al.*, 2011). The IDM which integrates individual agronomic practices with balanced use of fungicides, pesticides, and bio-pesticides for effective management of diseases is synergistic and can be applied together in the fields of farmers to provide farmers with maximum economic benefits (Drechsel *et al.*, 2022).

By using cultural methods, the prevalence of virus-related diseases can be decreased, such as adaptation of early-maturing cultivars, healthy seeds, rouging, and altered sowing dates. Other successful measures include exterminating ratoon pigeon peas, uprooting and burning diseased plants, crop rotation to decrease inoculum levels and vector populations, chemical control methods such as seed treatment of 10% aldicarb or 25% carbofuran (3 g/kg seed), and spraying acaricide/insecticides (karathane, metasystox) to target the mite vector in the young crop phase (Ioriatti *et al.*, 2022).

10.9 CONCLUSION

Pulses are a gift from nature since they improve soil fertility and health through symbiotic nitrogen fixation, increase nutrition, food security, and more environmentally friendly. They also help to lessen the consequences of climate change. More than 80% of the world's pulses are produced using saved grain, which is of inferior quality since it was handled and stored improperly. High-quality seed manufacturing packages must be available in accordance with seed distribution (by non-governmental organizations (NGOs), public-private partnerships, and govt. institutes) and the sites/climate. Pulses are typically produced on marginal soils, where germination and seed establishment are key problems that can be solved by seed augmentation (priming, pelleting, or coating). The seed improvement increased grain quality while also improving germination, yield, and growth. Pulses being used as catch crops and intercropping are two more strategies to boost production. Between the main cereal crops, the short-duration pulses (such as mung bean and green gram) can be interplanted. Only a few locations use the drill method for sowing; most areas primarily use broadcasting. Using manual methods, insecticides, pesticides, and harvests are applied to the environment. Drills and mechanical harvesting sprayers mounted on tractors can all significantly increase productivity. Additionally, increasing the marketing system's support prices can help to raise the output of pulse. In conclusion, although it is challenging to grow pulses in marginal lands, we can achieve our goal of sustainable agriculture and food security by cultivating pulses on marginal lands. We can improve the yield of pulse cultivation in these areas by the implementation of proper management strategies to reduce the harmful effects of biotic and abiotic stress. This chapter has provided insights into the implications, challenges, and promising strategies for maximizing the potential of marginal lands for pulse cultivation.

LIST OF ABBREVIATIONS:

ATP	adenosine triphosphate
FAO	food and agriculture organization
ICAR-IIPR	Indian Institute of Pulses Research- Innovative Industrial Properties, Inc.
IDM	integrated disease management
IW:CPE	irrigation water: cumulative pan evaporation
NGO	non-governmental organization
SOC	soil organic carbon

REFERENCES

Afzal, M., Alghamdi, S.S., Migdadi, H.H., El-Harty, E., & Al-Faifi, S.A., 2022. Agronomical and physiological responses of faba bean genotypes to salt stress. *Agriculture*, 12, 235.

Ahmad, F., Gaur, P., & Croser, J., 2005. Chickpea (*Cicer arietinum* l.). *In* Genetic Resources, Chromosome Engineering, and Crop Improvement-Grain Legumes, 1, 187–217.

Ahmad, J., Iqbal, A., Mahmood, A., Iqbal, M.A., Khan, H.Z., Abbas, R.N., Akbar, N., & Maqsood, M., 2020. Effect of cutting management, seeding rates and sowing method on seed yield of alfalfa (*Medicago sativa* L.). *Pakistan Journal of Botany*, 52, 1449–1454.

Akram, H., Levia, D.F., Herrick, J.E., Lydiasari, H., & Schütze, N., 2022. Water requirements for oil palm grown on marginal lands: A simulation approach. *Agricultural Water Management*, 260, 107292.

Ali, B., Ali, A., Tahir, M., & Ali, S., 2014. Growth, seed yield and quality of mungbean as influenced by foliar application of iron sulfate. *Pakistan Journal of Life and Social Sciences*, 12, 20–25.

Aliku, O., Oshunsanya, S.O., & Aiyelari, E.A., 2022. Estimation of crop evapotranspiration of Okra using drainage Lysimeters under dry season conditions. *Scientific African*, 16, e01189.

Allahmoradi, P., Ghobadi, M., Taherabadi, S., & Taherabadi, S., 2011. Physiological aspects of mungbean (*Vigna radiata* L.) in response to drought stress. IPCBEE vol. 9, IACSIT Press. Singapore, *Int. Conf. Food Eng. Biotechnol*, 272–275.

Aravani, V.P., Sun, H., Yang, Z., Liu, G., Wang, W., Anagnostopoulos, G., Syriopoulos, G., Charisiou, N.D., Goula, M.A., & Kornaros, M., 2022. Agricultural and livestock sector's residues in Greece & China: Comparative qualitative and quantitative characterization for assessing their potential for biogas production. *Renewable and Sustainable Energy Reviews*, 154, 111821.

Arshad, T., Naqve, M., Mukhtiar, A., Javaid, M.M., Mahmod, A., Nadeem, M.A., & Khan, B.A., 2023. Conservation Tillage for Sustainable Agriculture. *Climate-Resilient Agriculture, Vol 1: Crop Responses and Agroecological Perspectives*. Cham: Springer International Publishing, 313–327.

Balasubramanian, A., & Sathiyamurthi, S., 2021. Effects of Nipping and Micronutrient Foliar Application on Dry Matter Production, Yield, Agronomic Efficiency and Nutrient Uptake of Short Duration Pigeonpea (*Cajanus Cajan* L.) Co (Rg)-7.

Bamboriya, S., Bana, R.S., Kuri, B.R., Kumar, V., Bamboriya, S.D., & Meena, R.P., 2022. Achieving higher production from low inputs using synergistic crop interactions under maize-based polyculture systems. *Environmental Sustainability*, *5*, 1–15.

Banerjee, J., Das, A., Parihar, A., Sharma, R., Pramanik, K., & Barpete, S., 2022. Genomic Designing towards Development of Abiotic Stress Tolerant Grass Pea for Food and Nutritional Security. *In Genomic Designing for Abiotic Stress Resistant Pulse Crops*. Springer, 345–381.

Barnsley, I., 2022. *Quantifying the Benefits of Natural Flood Management Methods in Groundwater-Dominated River Systems*. University of Southampton, UK.

Basso, B., Dumont, B., Maestrini, B., Shcherbak, I., Robertson, G.P., Porter, J.R., Smith, P., Paustian, K., Grace, P., & Asseng, S., 2018. Soil organic carbon and nitrogen feedbacks on crop yields under climate change. *Agricultural & Environmental Letters*, 3, 180026.

Basu, P.S., Chaturvedi, S.K., Gaur, P.M., Mondal, B., Meena, S.K., Das, K., Kumar, V., & Tewari, K., & Sharma, K., 2022. Physiological Mechanisms of Tolerance to Drought and Heat in Major Pulses for Improving Yield under Stress Environments. *Plant Response Mechanisms to Abiotic Stresses*.

Bell, R.W., & De Oliveira, T.S., 2022. Subsoil Constraints for Crop Production: Recent Advances, New Technologies, and Priorities for Further Research. *Subsoil Constraints for Crop Production*, 397–422.

Bewley, J.D., Black, M., & Halmer, P., 2006. *The Encyclopedia of Seeds: Science, Technology and Uses*. CABI, Wallingford, Oxfordshire, UK.

Bhat, S., Aditya, K., Kumari, B., Acharya, K.K., & Sendhil, R., 2022. Pulses Production, Trade and Policy Imperatives: A Global Perspective. *Advances in Legumes for Sustainable Intensification*, 639–656.

Bil, G., 2022. The Representation of Plants. *In* A Cultural History of Plantsin the Nineteenth Century: Volume 6, 6, 171.

Birthal, P.S., Khan, T., Negi, D.S., & Agarwal, S., 2014. Impact of climate change on yields of major food crops in India: Implications for food security. *Agricultural Economics Research Review*, 27, 145–155.

Birthisel, S.K., Clements, R.S., & Gallandt, E.R., 2021. How will climate change impact the 'many little hammers' of ecological weed management? *Weed Research*, 61, 327–341.

Brodt, S., Six, J., Feenstra, G., Ingels, C., & Campbell, D., 2011. Sustainable agriculture. *National Educational Knowledge*, 3.

Brooks, R.A., 2022. Testing the reliability of civic science data collection of plant traits in a perennial sunflower domestication experiment. Master's project, California Polytechnic State University, San Luis Obispo, CA. Available at: https://digitalcommons.calpoly.edu/nres_rpt/28

Carrino, L., Visconti, D., Fiorentino, N., & Fagnano, M., 2020. Biofuel production with castor bean: A win-win strategy for marginal land. *Agronomy*, 10, 1690.

Chadha, A., Florentine, S., Javaid, M., Welgama, A., & Turville, C., 2020. Influence of elements of climate change on the growth and fecundity of *Datura stramonium*. *Environmental Science and Pollution Research*, 27, 35859–35869.

Chattha, M., Hassan, M., Khan, I., Chattha, M., Munir, H., Nawaz, M., Mahmood, A., Usman, M., & Kharal, M., 2017. Alternate skip irrigation strategy ensure sustainable sugarcane yield. *JAPS: Journal of Animal & Plant Sciences*, 27.

Chelladurai, V., & Erkinbaev, C., 2020. Lentils. In *Pulses*. Springer, International Publishing, 129–143.

Cheng, M., Mccarl, B., & Fei, C., 2022. Climate change and livestock production: A literature review. *Atmosphere*, 13, 140.

Choudhary, A.K., & Suri, V., 2014. Scaling up of pulse production under frontline demonstration technology transfer program in Himachal Himalayas, India. *Communications in Soil Science and Plant Analysis*, 45, 1934–1948.

Choudhary, J.R., Kumar, S., & Jha, P.K. & Tripathi, A., Pulses as the climate smart crops.

Cowie, A., Penman, T., Gorissen, L., Winslow, M., Lehmann, J., Tyrrell, T., Twomlow, S., Wilkes, A., Lal, R., & Jones, J., 2011. Towards sustainable land management in the drylands: Scientific connections in monitoring and assessing dryland degradation, climate change and biodiversity. *Land Degradation & Development*, 22, 248–260.

Dagel, K.C., 1997. Defining drought in marginal areas: The role of perception. *The Professional Geographer*, 49, 192–202.

Das, A., Babu, S., Datta, M., Kumar, S., Singh, R., Avasthe, R., Rathore, S., Yadav, S., & Singh, V.K., 2022. Restoring soil carbon in marginal land of Indian Himalayas: Impact of crop intensification and conservation tillage. *Journal of Environmental Management*, 318, 115603.

De Oliveira, A.B., Alencar, N.L.M., & Gomes-Filho, E., 2013. Comparison between the water and salt stress effects on plant growth and development. *Responses of Organisms to Water Stress*, 4, 67–94.

Donato, C.J.R., Cendoya, E., Demonte, L.D., Repetti, M.R., Chulze, S.N., & Ramirez, M.L., 2022. Influence of abiotic factors (water activity and temperature) on growth and aflatoxin production by *Aspergillus flavus* in a chickpea-based medium. *International Journal of Food Microbiology*, 379, 109841.

Drechsel, P., Qadir, M., & Baumann, J., 2022. Water reuse to free up freshwater for higher-value use and increase climate resilience and water productivity. *Irrigation and Drainage*, *71*, 100–109.

Fowler, H.J., Lenderink, G., Prein, A.F., Westra, S., Allan, R.P., Ban, N., Barbero, R., Berg, P., Blenkinsop, S., & Do, H.X., 2021. Anthropogenic intensification of short-duration rainfall extremes. *Nature Reviews Earth & Environment*, 2, 107–122.

Fu, T., Xu, Y., Li, M., Xue, S., Duan, Z., & Xie, G.H., 2022. Bioenergy production potential of 16 energy crops on marginal land in China. *BioEnergy Research*, *15*(2), 998–1016.

George, F.E., Prema, A., Hema, M., & Prameela, P., 2022. Farm level damage and loss assessment of natural disasters: How realistic are the assessments? *Journal of Tropical Agriculture*, 60.

Gull, R., Bhat, T.A., Sheikh, T.A., Wani, O.A., Fayaz, S., Nazir, A., Saad, A., Jan, S., Nazir, I., & Nisa, R., 2020. Climate change impact on pulse in India—A review. *Journal of Pharmacognosy and Phytochemistry*, 9, 3159–3166.

Haberzettl, J., Hilgert, P., & Von Cossel, M., 2021. A critical review on lignocellulosic biomass yield modeling and the bioenergy potential from marginal land. *Agronomy*, 11, 2397.

Hanumantharao, B., Nair, R.M., & Nayyar, H., 2016. Salinity and high temperature tolerance in mungbean [*Vigna radiata* (L.) Wilczek] from a physiological perspective. *Frontiers in Plant Science*, 7, 957.

Hasanuzzaman, M., Parvin, K., Anee, T.I., Masud, A.a.C., & Nowroz, F., 2022. Salt Stress Responses and Tolerance in Soybean. *In Plant Stress Physiology-Perspectives in Agriculture*. IntechOpen, UK.

Hassan, M., Chattha, M., Chattha, M., Mahmood, A., & Sahi, S.T., 2018. Impact of harvesting times on chemical composition and methane productivity of *Sorghum* (*Sorghum bicolor* Moench L.). *Applied Ecology & Environmental Research*, 16.

Hassan, M.U., Chattha, M.U., Khan, I., Chattha, M.B., Barbanti, L., Aamer, M., Iqbal, M.M., Nawaz, M., Mahmood, A., & Ali, A., 2021. Heat stress in cultivated plants: Nature, impact, mechanisms, and mitigation strategies—A review. *Plant Biosystems-An International Journal Dealing with all Aspects of Plant Biology*, 155, 211–234.

Hassan, M.U., Ghareeb, R.Y., Nawaz, M., Mahmood, A., Shah, A.N., Abdel-Megeed, A., Abdelsalam, N.R., Hashem, M., Alamri, S., & Thabit, M.A., 2022. Melatonin: A vital pro-tectant for crops against heat stress: Mechanisms and prospects. *Agronomy*, 12, 1116.

Hayes, K., Blashki, G., Wiseman, J., Burke, S., & Reifels, L., 2018. Climate change and mental health: Risks, impacts and priority actions. *International Journal of Mental Health Systems*, 12, 1–12.

Hulme, P.E., 2005. Adapting to climate change: Is there scope for ecological management in the face of a global threat? *Journal of Applied Ecology*, 42, 784–794.

Hussain, M.I., Muscolo, A., Farooq, M., & Ahmad, W., 2019. Sustainable use and management of non-conventional water resources for rehabilitation of marginal lands in arid and semiarid environments. *Agricultural Water Management*, 221, 462–476.

Ioriatti, C., Mazzoni, V., Anfora, G., Cristofaro, M., Profeta, E., Roselli, G., & Sforza, R., 2022. Co-evolution and biological control: Multitrophic interactions between the tree of heaven, brown marmorated stink bug and their natural enemiesed. *ICE 2022: XXVI International Congress of Entomology: Entomology for our planet, Helsinki, Finland, July 17-22, 2022*FI, 361.

Islam, M.R., Kamal, M.M., Alam, M.A., Hossain, J., Soufan, W., Skalicky, M., Brestic, M., Habib-Ur-Rahman, M., El Sabagh, A., & Islam, M.S., 2021. Physiochemical changes of mung bean [*Vigna radiata* (L.) R. Wilczek] in responses to varying irrigation regimes. *Horticulturae*, 7, 565.

Islam, S.F.-U., Sander, B.O., Quilty, J.R., De Neergaard, A., Van Groenigen, J.W., & Jensen, L.S., 2020. Mitigation of greenhouse gas emissions and reduced irrigation water use in rice production through water-saving irrigation scheduling, reduced tillage and fertiliser application strategies. *Science of the Total Environment*, 739, 140215.

Jadav, M.L., Raidas, D.K., Kumawat, N., Girothia, O., Bhagat, D., & Choudhary, S., 2022. Pigeonpea (*Cajanus cajan* L.) Growth, yield and monetary influence by drip irrigation and mulch in vertisols of Madhya Pradesh. *Legume Research-An International Journal*, 45, 853–859.

Jannat, S., Shah, A.H., Ul Hassan, M., Sher, A., Fiaz, S., Elesawy, B.H., Ismail, K.A., El Askary, A., Gharib, A.F., & Qayyum, A., 2022. Genetic diversity of common bean (*Phaseolus vulgaris* L.) Ecotypes from Pakistan using simple sequence repeats. *Saudi Journal of Biological Sciences*, 29, 103300.

Javaid, M.M., 2018. Environmental factors affecting the germination and emergence of white horehound (*Marrubium vulgare* L.): A weed of arid-zone areas. *The Rangeland Journal*, 40, 47–54.

Javaid, M.M., Wang, X., Florentine, S.K., Ashraf, M., Mahmood, A., Li, F.-M., & Fiaz, S., 2022. Effects on photosynthetic response and biomass productivity of *Acacia longifolia* ssp. longifolia under elevated CO_2 and water-limited regimes. *Frontiers in Plant Science*, 13, 817730.

Kamboj, S., Kaur, M., & Sidhu, G.B.K., 2022. Seed quality as affected under different temperature regimes using different packings in pigeonpea (*Cajanus cajan* (L.) Millsp.) during storage. *The Pharma Innovation Journal*, 802–2808.

Kang, S., Post, W.M., Nichols, J.A., Wang, D., West, T.O., Bandaru, V., & Izaurralde, R.C., 2013. Marginal lands: Concept, assessment and management. *Journal of Agricultural Science*, 5, 129.

Kasprzyk, M., Szpakowski, W., Poznańska, E., Boogaard, F.C., Bobkowska, K., & Gajewska, M., 2022. Technical solutions and benefits of introducing rain gardens—Gdańsk case study. *Science of the Total Environment*, 835, 155487.

Khan, B.A., Nadeem, M.A., Najeeb Alawadi, H.F., Ayub, M.A., Mahmood, A., Abbas, T., Nijabat, A., Ameen, M., Abdullah, F., & Oraby, H., 2024. An overview of the role of nanoherbicides in tackling challenges of weed management in wheat: A novel approach. *Green Processing and Synthesis*, 13, 20240021.

Khanpit, V.V., Tajane, S.P., & Mandavgane, S.A., 2022. Extrusion for soluble dietary fiber concentrate: Critical overview on effect of process parameters on physicochemical, nutritional, and biological properties. *Food Reviews International*, 1–22.

Khinea, N.A., Kundub, K., Malikc, D., & Papangd, J.S., 2022. Production and trade of chickpea (*Cicer aneityum*) in India and Myanmar. *Indian Journal of Applied Business and Economic Research*, 3, 81–97.

Kuang, W., Liu, J., Tian, H., Shi, H., Dong, J., Song, C., Li, X., Du, G., Hou, Y., & Lu, D., 2022. Cropland redistribution to marginal lands undermines environmental sustainability. *National Science Review*, 9, nwab091.

Kumar, J., Choudhary, A.K., Gupta, D.S., & Kumar, S., 2019. Towards exploitation of adaptive traits for climate-resilient smart pulses. *International Journal of Molecular Sciences*, 20, 2971.

Kumar, V., & Malik, D., 2022. Trends and economic analysis of chickpea (*Cicer arietinum*) cultivation in India with Special Reference to Haryana. *Legume Research-An International Journal*, 1, 7.

Kumar, Y., Tiwari, K., Singh, T., & Raliya, R., 2021. Nanofertilizers and their role in sustainable agriculture. *Annals of Plant and Soil Research*, 23, 238–255.

Kumawat, A., Bamboriya, S.D., Meena, R.S., Yadav, D., Kumar, A., Kumar, S., Raj, A., & Pradhan, G., 2022. Legume-Based Inter-Cropping to Achieve the Crop, Soil, and Environmental Health Security. In *Advances in Legumes for Sustainable Intensification*. Elsevier, Amsterdam, 307–328.

Lemoine, M.D., Fink, T., Mencke, C., Schleberger, R., My, I., Rottner, L., Obergassel, J., Muenkler, P., Bergau, L., & Moser, F., 2022. CA-536-01 pulsed-field ablation based pulmonary vein isolation: Acute safety and efficacy in a multi-center real world scenario. *Heart Rhythm*, 19, S91.

Li, F.M., Li, X.G., Javaid, M.M., Ashraf, M., & Zhang, F., 2020. Ridge-furrow plastic film mulching farming for sustainable dryland agriculture on the Chinese loess plateau. *Agronomy Journal*, 112, 3284–3294.

Li, H., Rodda, M., Gnanasambandam, A., Aftab, M., Redden, R., Hobson, K., Rosewarne, G., Materne, M., Kaur, S., & Slater, A.T., 2015. Breeding for biotic stress resistance in chickpea: Progress and prospects. *Euphytica*, 204, 257–288.

Mahmood, A. & Honermeier, B., 2019. Effect of row spacing and cultivars on biomass yield and chemical composition of Sorghum bicolor L. *53. Jahrestagung*, 157.

Malik, D., Devi, M., & Reddy, A.A., 2022. Global status of lentil production with special reference to India. *Indian Journal of Agricultural Sciences*, 92, 474–479.

Meirelles, L.A., 2022. *The Nuanced Effects of Redox-Active Metabolites on Bacterial Physiology and Antibiotic Susceptibility*. California Institute of Technology, California.

M Tahat, M., M Alananbeh, K., A Othman, Y., & I Leskovar, D., 2020. Soil health and sustainable agriculture. *Sustainability*, 12, 4859.

Munaweera, T., Jayawardana, N., Rajaratnam, R., & Dissanayake, N., 2022. Modern plant biotechnology as a strategy in addressing climate change and attaining food security. *Agriculture & Food Security*, 11, 1–28.

Nadeem, M.A., Khan, B.A., Chadar, A.R., Maqbool, R., Raza, A., Javaid, M.M., Mahmood, A., Khan, M.I., Ahmad, M., & Irfan, M., 2022. Weed control and sustainable rice production through rice intensification system and conventional practices of weed competition periods and age of transplanted seedlings. *Semina: Ciências Agrárias*, 43, 2271–2292.

Nazir, A., & Lohano, H.D., 2022. Resilience through Crop Diversification in Pakistan. *In Climate Change and Community Resilience*. Singapore: Springer, 431–442.

Neda, E.K., 2020. Grain legumes production in Ethiopia: A review of adoption, opportunities, constraints and emphases for future interventions. *Turkish Journal of Agriculture-Food Science and Technology*, 8, 977–989.

Nguli, M., Gatari, M.J., Shepherd, K., Njenga, L., & Boman, J., 2022. Assessment of essential micronutrient levels in common beans (*Phaseolus vulgaris*) in Kenya by total reflection X-ray fluorescence. *X-Ray Spectrometry*, 51, 198–203.

Nithisha, A., Bokado, K., & Charitha, K.S., 2022. Mulches: Their impact on the crop production.

Oluyisola, O.E., Bhalla, S., Sgarbossa, F., & Strandhagen, J.O., 2022. Designing and developing smart production planning and control systems in the industry 4.0 era: A methodology and case study. *Journal of Intelligent Manufacturing*, 33, 311–332.

Palma, F., López-Gómez, M., Tejera, N., & Lluch, C., 2013. Salicylic acid improves the salinity tolerance of *Medicago sativa* in symbiosis with *Sinorhizobium meliloti* by preventing nitrogen fixation inhibition. *Plant Science*, 208, 75–82.

Palmo, T., Bahar, F., Singh, L., Mehdi, S., Bhat, M., Saad, A., Mir, M.S., Bashir, M., Rehman, U., & Bhat, T.A., 2022. Alternate land use systems for resource conservation.

Pawar, N., Bishnoi, N., & Malik, D., 2021. Resource use and economic potential of pigeonpea cultivation in Haryana. *International Journal of Agricultural Sciences*, 17, 239–244.

Peng, Y., Li, Z., Sun, T., Zhang, F., Wu, Q., Du, M., & Sheng, T., 2022. Modeling long-term water use and economic returns to optimize alfalfa-corn rotation in the corn belt of northeast China. *Field Crops Research*, 276, 108379.

Porkodi, G., Shanmugasundaram, R., Saravanapandian, P., Swaminathan, C., & Kumutha, K., 2022. Impact of different sources and levels of iron on yield and quality of groundnut.

Pradipa, C., Panneerselvam, S., Geethalakshmi, V., Bhuvaneeswari, K., & Maragatham, N., 2022. Potential impact of future climate change on spatial variability of blackgram yield over Tamil Nadu. *Journal of Agrometeorology*, 24, 157–164.

Pramanik, S., Bhattacharya, D., Kundu, A., Debangshi, U., & Munshi, C., 2022. Critical evaluation of production–import–export of lentils in India: an issue of food security.

Pratap, A., Dhaliwal, I., Singh, C.M., Mahalingam, A., Manivannan, N., Basavaraja, T., Rathore, M., Singh, Y., Tiwari, P., & Yadav, S., 2022. Biofortification of Mungbean. *In Biofortification of Staple Crops*. Springer, Cham, Switzerland, 295–333.

Priego-Poyato, S., Rodrigo-Garcia, M., Escudero-Feliu, J., Garcia-Costela, M., Lima-Cabello, E., Carazo-Gallego, A., Morales-Santana, S., Leon, J., & Jimenez-Lopez, J.C., 2021. Current Advances Research in Nutraceutical Compounds of Legumes, Pseudocereals and Cereals. *In Grain and Seed Proteins Functionality*. IntechOpen, UK.

Raju, N.L., Gnanesh, B.N., Lekha, P., Jayashree, B., Pande, S., Hiremath, P.J., Byregowda, M., Singh, N.K., & Varshney, R.K., 2010. The first set of EST resource for gene discovery and marker development in pigeonpea (*Cajanus cajan* L.). *BMC Plant Biology*, 10, 1–22.

Ram, S., & Choudhary, V., 2022. An economic analysis, production and marketing of pigeon pea in Surguja district of Chhattisgarh. *The Pharma Innovation Journal*, 11, 1123–1127.

Rana, A., Kalla, P., Verma, H., & Mohnot, J., 2016a. Recycling of dimensional stone waste in concrete: A review. *Journal of Cleaner Production*, 135, 312–331.

Rana, D., Dass, A., Rajanna, G., & Kaur, R., 2016b. Biotic and abiotic stress management in pulses. *Indian Journal of Agronomy*, 61, S238–S248.

Rao, P.P., Birthal, P., Bhagavatula, S., & Bantilan, M., 2010. Chickpea and pigeonpea economies in Asia: facts, trends and outlook.

Rao, P.V., Reddy, A.S., & Ramana, M., 2022. Effect of land configuration and weed management on performance of urdbean [*Vigna mungo* (L.) Hepper]. *Legume Research-An International Journal*, 1, 5.

Sahoo, K., Khatri, P., Kanwar, A., Singh, H.P., Mani, S., Bergman, R., Runge, T., & Kumar, D., 2022. Integrated environmental and economic assessments of producing energy crops with cover crops for simultaneous use as biofuel feedstocks and animal fodder. *Industrial Crops and Products*, 179, 114681.

Schiavi, D., Francesconi, S., Taddei, A.R., Fortunati, E., & Balestra, G.M., 2022. Exploring cellulose nanocrystals obtained from olive tree wastes as sustainable crop protection tool against bacterial diseases. *Scientific Reports*, 12, 1–14.

Shafiq, M., Hassan, A., Ahmad, N., & Rashid, A., 1994. Crop yields and nutrient uptake by rainfed wheat and mungbean as affected by tillage, fertilization, and weeding. *Journal of Plant Nutrition*, 17, 561–577.

Sharma, N., Sahu, J.K., Joshi, S., Khubber, S., Bansal, V., Bhardwaj, A., Bangar, S.P., & Bal, L.M., 2022. Modulation of lentil antinutritional properties using non-thermal mediated processing techniques—A review. *Journal of Food Composition and Analysis*, *109*(3), 104498.

Sidhu, J.S., Zafar, T., Benyathiar, P., & Nasir, M., 2022. Production, processing, and nutritional profile of chickpeas and lentils. In *Dry Beans and Pulses: Production, Processing, and Nutrition*, Hoboken, New Jersey, 383–407.

Singh, B., Padhy, A.K., Ambreen, H., Yadav, M., Bhardwaj, S., Singh, G., Pandey, V., Chakraborty, A., & Bhatia, S., 2022a. Understanding Abiotic Stress Responses in Lentil under Changing Climate Regimes. In *Developing Climate Resilient Grain and Forage Legumes*. Springer, Cham, Switzerland, 179–204.

Singh, J.M., Kaur, A., Chopra, S., Kumar, R., Sidhu, M., & Kataria, P., 2022b. Dynamics of production profile of pulses in India. *Legume Research-An International Journal*, 45, 565–572.

Singh, O., Singh, D.K., Singh, A., Singh, R.P., Pandey, S., & Bajpai, A.K., 2022c. Increasing productivity of lentil (*Lens culinaris*) using improved varieties under alluvial soil of Uttar Pradesh by cluster front line demonstrations. *Legume Research-An International Journal*, 1, 5.

Singh, V., Chauhan, Y., Dalal, R., & Schmidt, S., 2021. Chickpea. *In The Beans and the Peas*. Elsevier, Amsterdam, Netherlands, 173–215.

Skendžić, S., Zovko, M., Živković, I.P., Lešić, V., & Lemić, D., 2021. The impact of climate change on agricultural insect pests. *Insects*, 12, 440.

Stoilova, T., Pereira, G., De Sousa, M., & Carnide, V., 2005. Diversity in common bean landraces (*Phaseolus vulgaris* L.) from Bulgaria and Portugal. *Journal of Central European Agriculture*, 6, 443–448.

Strijker, D., 2005. Marginal lands in Europe—Causes of decline. *Basic and Applied Ecology*, 6, 99–106.

Talukdar, D., 2013. Selenium priming selectively ameliorates weed-induced phytotoxicity by modulating antioxidant defense components in lentil (*Lens culinaris* Medik.) and grass pea (*Lathyrus sativus* L.). *Annual Research & Review in Biology*, Hooghly, India, 195–212.

Tanveer, M.H., Abdullah, M., Ghafoor, I., Makhdoom, M., & Ahmad, J., 2022. JAU-17: A first high yielding and malting barley variety suitable for marginal lands of Punjab. *Journal of Agricultural Research*, 60, 11–16.

Tayade, A., Geetha, P., Dhanapal, R., & Hari, K., 2016. Effect of in situ trash management on sugarcane under wide row planting system. *Journal of Sugarcane Research*, 6, 35–41.

Ullah, A., Farooq, M., Qadeer, A., & Sanaullah, M., 2022a. Impact of zinc and plant growth-promoting bacteria on soil health as well as aboveground biomass of Desi and Kabuli chickpea under arid conditions. *Journal of the Science of Food and Agriculture*, 102, 2262–2269.

Ullah, A., & Hassan, M.U., 2022. Ahmed, 3 Muhammad 1, 4, 5 AL-Mayahi Farooq. *Soil Constraints on Crop Production*, 389, 1–11.

Ullah, A., Shah, T.M., & Farooq, M., 2020. Pulses production in Pakistan: Status, constraints and opportunities. *International Journal of Plant Production*, 14, 549–569.

Ullah, M., Khan, M., Das, G., Faruq, M., & Hossain, M., 2022b. International Journal of Agricultural Sciences and Technology.

Van Dam, J., Junginger, M., & Faaij, A.P., 2010. From the global efforts on certification of bioenergy towards an integrated approach based on sustainable land use planning. *Renewable and Sustainable Energy Reviews*, 14, 2445–2472.

Varma, P., 2022. An Overview of Pulses Economy. *Pulses for Food and Nutritional Security of India*, 9–25.

Verma, T., Bhardwaj, S., Singh, J., Kapoor, D., & Prasad, R., 2022. Triacontanol as a versatile plant growth regulator in overcoming negative effects of salt stress. *Journal of Agriculture and Food Research*, 10, 100351.

Webber, H., Madramootoo, C., Bourgault, M., Horst, M., Stulina, G., & Smith, D., 2006. Water use efficiency of common bean and green gram grown using alternate furrow and deficit irrigation. *Agricultural Water Management*, 86, 259–268.

Yadav, T., Nisha, K.C., Chopra, N., Yadav, M., Kumar, R., Rathore, D., Soni, P., Manarana, G., Tamta, A., & Kushwah, M., 2017. Weed management in cowpea—A review. *International Journal of Current Microbiology and Applied Sciences*, 6, 1373–1385.

Yohane, E.N., Shimelis, H., Laing, M., Shayanowako, A., Mathew, I., & Chintu, J.M., 2021. Pigeonpea production constraints and farmers' trait preferences in Malawi: Implications for variety design. *South African Journal of Plant and Soil*, 38, 326–337.

Yuvaraj, M., Pandiyan, M., & Gayathri, P., 2020. Role of legumes in improving soil fertility status. *Legume Crops-Prospects, Production and Uses.*

Zeng, Y., Ran, L., Fang, N., Wang, Z., Xu, Z., Lu, X., Yu, Q., Wang, L., Yu, S., & Shi, Z., 2022. How to balance green and grain in marginal mountainous areas? *Earth's Future*, 5, e2021EF002552.

Zsögön, A., Peres, L.E., Xiao, Y., Yan, J., & Fernie, A.R., 2022. Enhancing crop diversity for food security in the face of climate uncertainty. *The Plant Journal*, 109, 402–414.

Zuazo, V.C.H.D., & Pleguezuelo, C.R.O.R., 2009. Soil-erosion and runoff prevention by plant covers: A review. *Sustainable Agriculture*, 785–811.

11 Impact of Climate Change on Plant Physiology

Muhammad Saad Ullah, Umair Gull, Jameel Mohammed Al-Khayri, Umar Raza, and Ali Mardan*

11.1 INTRODUCTION

Extreme events are occurring more often, and meteorological conditions are changing. Climate change relates to the fluctuations in weather patterns around the globe, the increased probability of extreme weather occurrences, such as storms and droughts, and disruptions in natural processes and ecosystems. Climate change is the changing of the weather patterns, both regionally and universally over prolonged periods of time, and has shown to be disruptive in terms of soil and the processes that plants partake in. The Intergovernmental Panel on Climate Change (IPCC) has warned that even small degrees of change in temperature and precipitation can throw ecosystems out of kilter, leading to loss or degradation of habitat (Grigorieva *et al.*, 2023). Weather patterns will alter as a result of climate change, and severe weather events like storms and droughts will occur more often. Ecosystems and natural processes will also be disrupted. Even the increased CO_2 has the ability to increase agricultural productivity, more intense drought has the opposite impact. Climate change induces the occurrence of many abiotic stress factors, as shown in research with chickpeas and sorghum subjected to different stressors. The generation of secondary metabolites in plants is also impacted by elevated temperatures, which change metabolic pathways and produce a variety of metabolites (Ávila *et al.*, 2023).

Crop production is directly impacted by greenhouse gases (CO_2, N_2O, O_3, and CH_4) that cause climate change (Chadha *et al.*, 2020). It is estimated that atmospheric concentrations of N_2O have increased from approximately 270 parts per billion (ppb) in the pre-industrial era to 319 ppb in 2005 and that emissions of N_2O from the surface of the Earth have increased, as a result of human activity, by approximately 40%–50% compared to pre-industrial levels. Climate change is altering the patterns of rainfall. There will be times of heavy rain causing floods and droughts and off-season rain. It has been observed that in winter oilseed rape, increased rainfall during the fall and winter may favor disease development, in maize and soybean, more abundant precipitation during the spring damages juvenile plants. Another potential consequence of heavy rainfall is the heightened occurrence of flooding. These high floods increase the possibility of high soil moisture and accumulation of silt that causes farming areas to be destroyed or crop sowing to be delayed, endangering the food security of these nations (Ashraf *et al.*, 2023). Ecosystems are responsible for providing critical services such as the supply of habitat, the production of food, the purification of water, and the sequestration of carbon. Many species, however, may find it difficult to adjust to these shifting environments, which might cause changes in population size and distribution and even threaten their extinction. Drought often leads to water shortage, which in turn is often followed by salt stress. Soil salinity reduces agricultural yields in every salt-resistant crops like cotton, barley, and sugar beet and crops that are highly sensitive to salt compounds like sweet potato, wheat, or maize (Kopecká *et al.*, 2023). The ion toxicity and reduced soil water potential result in substantial suppression of the vegetative growth (Hassan *et al.*, 2022). Beyond timing and synchrony, climate change can influence phenology of important biological events, such as bloom, hibernation, and migration (Hanif *et al.*, 2017; John *et al.*, 2024).

* Corresponding author: jkhayri@kfu.edu.sa

DOI: 10.1201/9781003498698-11

Climate change impacts plant physiology. Plant physiology is crucial and necessary for plant adaptation, species persistency, and helping maintain ecosystem equilibrium. Climate change massively disturbs physiological processes, including photosynthesis, respiration, water balance, and mineral uptake. Now their ability to minimize water loss by manipulating osmotic potential and that of taking up water are the functions that are reflected in the adaptive responses to water limitation. Warming may also change the rate at which plants respire, potentially raising it enough to offset the benefits of enhanced photosynthesis (Javaid *et al.*, 2022; Vignesh *et al.*, 2024). Plant physiology governs mechanisms of growth dependent on temperature. However, the relationship between photosynthesis and respiration determines whether it behaves as a carbon sink or source. In addition, climate change alters the timing of flowering and leafing. It impacts various aspects of plant physiology that are vital for their adaptation, survival, and maintaining ecological balance (Singh *et al.*, 2023b). The plant's physiological response to water stress affects its ability to minimize water loss and improve water absorption through adjustments in its osmotic potential. Respiration rates of the plants might also be influenced by higher temperatures, whereby high temperatures lead to the increase in respiration (Tanveer *et al.*, 2012; Wahab *et al.*, 2023).

Temperature plays a pivotal role in coordinating growth processes regulated by plant physiology. We can confidently say that the Earth is getting warmer, as all model simulations fairly consistently expect a global movement to warmer temperature. A plant's ability to absorb or release carbon dioxide depends on how well photosynthesis and respiration work together. Moreover, climatic change modifies the timing of flowering and appearance of leaves in plants, possibly disturbing plant-pollinator interactions. The study conducted by Subedi *et al.* (2023) sheds light on the process by which plants adjust to changing temperatures. Projections indicate that by 2050, the average global temperature could rise by 1.5–2°C. This would lead to additional changes in the climate, including more frequent heatwaves, a decrease in freezing days, reduced rainfall but heavier precipitation, and an increased occurrence of droughts and other extreme weather events. These changes would have a detrimental effect on agricultural production. Projections suggest that 70% increase in total food production would be needed by the year 2050 to adequately nourish the global population. Nevertheless, the present patterns indicate that the rates of worldwide production of crucial crops are not rising enough to fulfill these requirements, resulting in an expanding disparity between supply and demand and giving rise to worries over future food security (Forslund *et al.*, 2023). Moreover, climate change is forecasted to result in reduced agricultural yields, which presents a significant risk to the world food supply. The regional variability of climate change's influence on agricultural output will undeniably provide obstacles to global food security. Hence, it is essential to preserve and enhance agricultural yield despite these climate limitations. Although stress leads to adverse effects on crops, there are some crops that have better adaptability under changing climatic conditions, which ensure hope for sustainable agriculture (Chadha *et al.*, 2020; Fagbemi *et al.*, 2023). Combined stress tests provide insights into nutrient status and consequences of changes within the context of nutritional security. This insight into tolerance and susceptibility is established in processes that have significance in developing climate-resilient crops for maintaining agricultural productivity, food security, and nutritional quality in the context of climate change. Climatic and meteorological conditions affect the physical and biological properties of agricultural, horticultural, and medicinal plants. It offers the advantages and disadvantages of these impacts (Greff *et al.*, 2023).

11.2 IMPACTS OF CLIMATE CHANGE ON PLANT DEVELOPMENT AND MORPHOLOGY

Elevated carbon dioxide has a positive effect on crop production by increasing photosynthesis and stomatal conductance, rather than just putting CO_2 into the air. Temperature plays an important role in determining the results of these responses, as differences in temperature can induce various responses in plants depending on the plant type, the extent of temperature increase, and the existing environmental conditions. Apart from this, the rise of various extreme weather has posed major challenges to the farming systems particularly which has become more frequent with climate change (Chadha *et al.*, 2019; Greff *et al.*, 2023).

In addition, strong winds and high storms can physically damage crops by bending and uprooting them from the ground, which has a negative effect on agricultural production. Moreover, precipitation changes can lead to water stress and potentially restrict crop transpiration, resulting in production losses. On the other hand, inundation and flooding due to extended rainfall lead to oxygen and nutrient deficiency in plants (Florentine *et al.*, 2016; Begna, 2020). Increasing atmospheric carbon dioxide concentration is seen as one of the key drivers of climate change. Its effect on crop physiology can be inconsistent. Higher levels of CO_2 stimulate photosynthesis, and subsequently, the increase in biomass as a result. On the other hand, they can also disrupt the balance of the carbon and nutrients inside the plants, thus impairing their nutritional quality and also their vulnerability against pests and diseases.

11.2.1 Root Developmental Responses

The response of roots to raised temperatures is crucial to determine crop's ability to tolerate climate change impacts. For instance, root lengths of maize and oilseeds increase with increasing temperatures, while cotton and sunflower have specific temperature ranges that provide optimal root growth. Both heat stress tolerance and associated root responses require effective communication from above- to below-ground. This differential effect on aquaporins and membrane fluidity translates into the heat-induced water uptake in crops. Elevated CO_2 enhances primary root length growth and lateral root formation in Arabidopsis and stimulates root meristem activity in other plants (Calleja-Cabrera *et al.*, 2020). A wider root system enables plants to absorb more water, allowing for greater transpiration and temperature regulation, although higher soil temperatures may reduce carbon allocation to roots, negatively impacting development.

Reproductive stages influence root development, where elevated soil temperature induces structural and functional changes in roots, helping plants adapt to climatic changes (Figure 11.1).

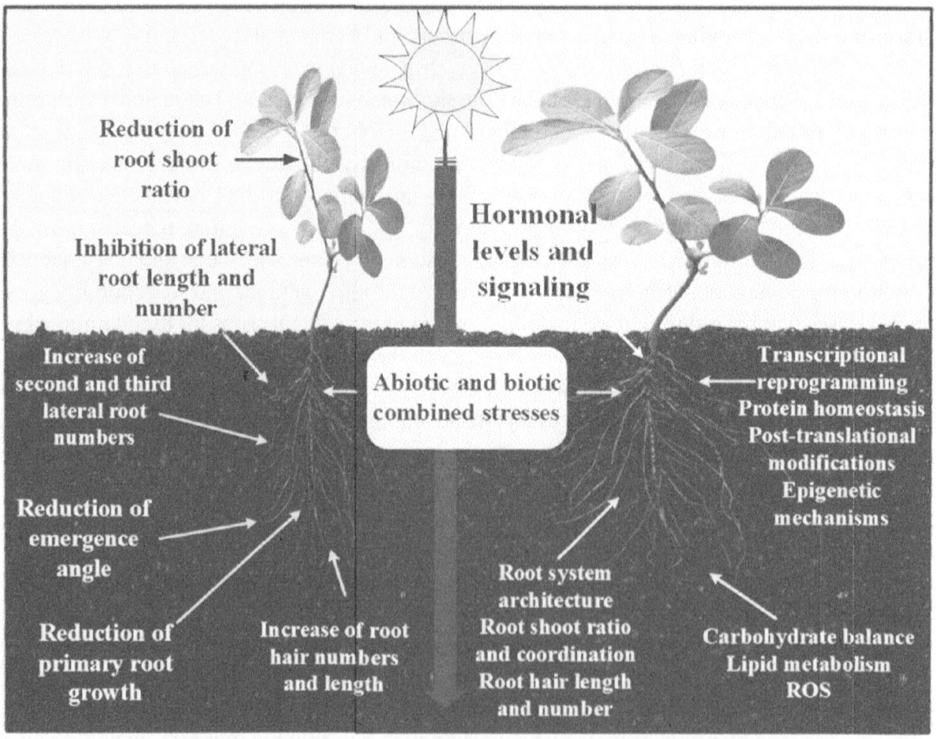

FIGURE 11.1 Root developmental responses to changing climate.

Root developmental responses to CO_2 enrichment significantly increase root biomass across various species, leading to a higher root-shoot ratio (Mehmood *et al.*, 2018; Vereecken *et al.*, 2022). For example, soybean has exhibited longer roots in superficial and intermediate layers under increased atmospheric CO_2. Genetic variation in density response to elevated CO_2 is observed in Populus species, highlighting gaps in understanding molecular mechanisms of root responses to CO_2 enrichment. Limited studies on gene expression in root tissues underscore the need for more research on genetic diversity and molecular pathways influencing root responses to climate change. Root developmental responses to drought stress often differ from shoot growth, with root growth frequently maintained or stimulated. For example, drought stress increases root growth rates in soybean during late vegetative and early reproductive stages and at greater soil depths. Enhanced root growth and deeper root distribution under drought suggest plasticity in root stress responses. Drought stress affects lateral root initiation and development in various species, with abscisic acid (ABA) suppressing lateral root initiation. Changes in root cellular anatomy, such as aerenchyma formation and cortical cell reduction, are linked to improved drought tolerance in maize. Genetic studies in rice highlight DRO1's role in regulating rooting depth under drought conditions, which affects grain yield and illustrates diverse drought adaptation strategies in plants (Shoaib *et al.*, 2022).

11.2.2 Leaf Developmental Responses to Elevated CO_2

Elevated CO_2 increases aboveground biomass and stimulates larger leaf growth due to enhanced cell proliferation and expansion in soybean and poplar. This response varies among cell types, as seen in growing leaves of *Populus euramericana*, where epidermal cells proliferated (Oksanen *et al.*, 2019). Elevated CO_2 reduces stomatal index and density, regulated by genetic factors like high CO_2 in Arabidopsis. Alterations in shoot morphology manifest as increased tillering in soybean, wheat, and rice. Elevated CO_2 also induces gene expression changes that enhance carbon uptake during photosynthesis and increase nighttime respiration rates. This effect has been found in soybean, Arabidopsis, and rice. Cell division and cell expansion define leaf development and are significantly affected by drought stress across species. Generally, drought reduces leaf expansion, particularly in Arabidopsis, where rosette area, leaf size, and epidermal cell dimensions decrease under mild drought (Nelissen *et al.*, 2018). Drought impact on final leaf size varies with leaf developmental stage; in *Ricinus communis*, rehydration rescues leaf expansion only if drought occurs after leaves reach a certain size, indicating the importance of early cell division in determining final leaf size. Conserved transcriptional responses to mild drought in Arabidopsis involve genes associated with ABA signaling, proline metabolism, and cell wall adjustments, with different gene sets activated depending on leaf age. Upregulated genes in young leaves include those related to cell wall synthesis and remodeling, likely facilitating growth maintenance under low turgor pressure (Chen *et al.*, 2021).

11.2.3 Reproductive Developmental Responses

Flowering time is crucial for a plant's reproductive success in dry environments. Two main strategies are recognized: drought escape, where plants flower early to complete reproduction before late-season droughts, and drought avoidance, where increased water-use efficiency helps plants avoid dehydration during early-season droughts. The impact of these strategies depends on the severity and timing of drought stress. Reproductive development, particularly male floral organs and gametophytes is highly sensitive to drought. In cereals like wheat and rice, water deficit during inflorescence development can hinder or entirely inhibit flowering (Aziz *et al.*, 2018; Sinha *et al.*, 2021). Similarly, drought stress in Arabidopsis causes early arrest of floral development and abnormal anther phenotypes. Arabidopsis mutants lacking MYB DOMAIN PROTEIN 21 (myb21) display reduced male filament elongation under drought stress, highlighting its importance in floral development during adverse conditions. Drought predominantly affects pollen viability in various species, leading to reduced seed set in cereals. Female reproductive structures and gametophytes are less impacted by drought compared to males, possibly due to prioritized female reproductive development (Wang *et al.*, 2023). Drought stress from flowering to seed maturation increases ovule abortion rates in Arabidopsis. Moderate drought has minimal effects on flowering, whereas severe drought reduces various reproductive parameters accompanied by distinct gene expression patterns. Tissue-specific sensitivity to drought is evident, with fertilized maize ovary tissue exhibiting stronger transcriptional responses compared to leaf tissue. Ovary-specific upregulated genes in maize are associated with ABA and sucrose metabolism, suggesting hormonal and sugar signaling involvement in kernel responses to drought (Singh *et al.*, 2023a). These findings highlight the heightened sensitivity of reproductive processes to drought, with male reproductive development being particularly susceptible compared to female development.

Plant responses to elevated temperature vary by tissue and developmental stage. Elevated temperature may advance reproductive events or damage reproductive structures. Reproductive events tend to occur earlier at higher temperatures. Accelerated flowering under elevated temperature may compromise resource accumulation needed for successful gamete production. Lohani *et al.*, (2019) revealed that fluctuating warm temperatures prompt earlier flowering in Arabidopsis compared to constant warmth, indicating a role of temperature range in regulating flowering time. Rising temperatures can disrupt crucial winter-chilling conditions for synchronous flowering and fruit set in fruit and nut trees, posing agricultural challenges. Temperature primarily affects crop production through male gametophyte sensitivity, reducing pollen viability, and impacting fertilization success. Heat stress also affects female floral organs, decreasing ovule number and increasing ovule abortion. Heat stress, whether constant or transient, impacts plant reproductive development, with variable effects on seed yield depending on the developmental stage. Both consistent exposure to elevated temperature and brief exposure to extreme temperature significantly influence plant reproductive development and yield (Santiago and Sharkey, 2019; Abdullah *et al.*, 2022). Molecular and cellular responses underlie the morphological and developmental changes induced by elevated temperature. While much of this work has been conducted in Arabidopsis, similar pathways likely operate in crop species, contributing to the detrimental impacts of elevated temperature on reproductive tissue development. Further research elucidating molecular mechanisms could aid in breeding climate-resistant crops.

Extreme temperatures or heat waves pose more immediate and severe threats to reproductive tissues compared to high ambient temperature. Both male and female reproductive organs are affected by high temperature, with a particular focus on the highly intricate male gametophyte life cycle, known for its responsiveness to environmental changes (Santiago and Sharkey, 2019; Krieg and Chambers, 2022).

11.3 FUNDAMENTAL PHYSIOLOGIC MECHANISMS IN CROPS

11.3.1 Photosynthesis

Extreme weather events such as storms, hail, and droughts can damage plant structures and disrupt photosynthesis. Strong winds and hail can physically damage leaves, disrupting photosynthetic processes. Drought stress exacerbates water shortages by closing stomata and reducing CO_2 absorption, further impeding photosynthesis. Severe weather can also induce oxidative stress and cellular damage, compromising overall plant health and resilience. High temperatures can denature photosynthetic enzymes and close stomata, limiting CO_2 absorption and hindering photosynthesis. The initial positive response to elevated CO_2 levels may be overshadowed by long-term detrimental effects. Increased atmospheric CO_2 concentrations initially enhance photosynthesis in C_3 plants by boosting rubisco enzyme activity (Tanveer *et al.*, 2013; Li *et al.*, 2020). However, prolonged exposure leads to photosynthetic acclimation, reducing photosynthesis rates. CO_2 enrichment boosts wheat and rice yields, while high temperatures decrease grain production by disrupting photosynthesis and increasing reactive oxygen species (ROS) production. Sustained temperatures of 45°C or higher suppress photosynthesis in rice, leading to plant death. Higher atmospheric CO_2 levels at elevated temperatures promote plant growth by enhancing photosynthetic pigment production. Elevated CO_2 levels increase protein levels associated with photosystems PSI (Photosystem I) and PSII (Photosystem II), while reducing stomatal conductance (Li *et al.*, 2021). C_4 plants exhibit greater resilience to high temperatures compared to C_3 plants. Changes in precipitation patterns directly impact water availability for plants, affecting photosynthesis. Water scarcity triggers stomatal closure, limiting CO_2 absorption, while excess rain may lead to waterlogging, hindering nutrient uptake and oxygen absorption by roots, thus affecting photosynthetic activity. Maize shows optimal photosynthetic temperature at 40°C, suggesting global warming may not severely impact yield. Rice under high temperatures exhibits increased respiratory enzyme activity but suffers from reduced water-use efficiency and transpiration rate, affecting photosynthesis negatively (Radha *et al.*, 2023). Heat and drought stress lead to a slight decrease in photosynthetic rates in *Hierochloe spicata* and *Valeriana jatamansi Jones.* Soil moisture deficiency reduces chlorophyll levels and photosynthesis rate, accompanied by changes in light saturation points and dark respiration rates. High temperatures disrupt photosynthesis by altering protein expression and enzyme activity, ultimately hindering plant development and productivity. Temperature variations also affect critical growth phases like blooming and fruit formation, closely tied to photosynthetic activities.

11.3.2 Rubisco Activase

The most prevalent protein in the world, rubisco, is also the essential enzyme for absorption of CO_2. It is extraordinarily heat-tolerant, continuing to operate at temperatures above 50°C. Raised temperatures, however, have the potential to quickly inactivate rubisco; rubisco activase may counteract this by causing an ATP-dependent activity known as carboxylation. Even though rubisco activase is necessary to sustain photosynthetic rates at high temperatures, it is very heat-sensitive; in some plant species, its activity is adversely impacted by temperatures exceeding 30°C (Amaral *et al.*, 2024). There are two variants of rubisco activase. The bigger form has two cysteine residues in the C-terminal extension, which allows for redox control via the thioredoxin system. It may arise from the alternate splicing of a single gene or from two separate genes. The heat sensitivity of photosynthesis is caused by thermal denaturation of rubisco activase rather than oxidation of the cysteine residues in their bigger form. Although studies are ongoing, the control of CO_2 fixation and rubisco activase is poorly understood in most key crops. Heat tolerance of rubisco activase varies among species, suggesting it as a target for developing heat-tolerant crop varieties. Stressful conditions further complicate rubisco's functioning due to the modulation of its inhibitor levels.

11.3.3 Respiration

Drought stress significantly impacts plant respiration by reducing water availability essential for metabolic activity. Stomatal closure to conserve water limits carbon dioxide intake, necessary for respiration. Under prolonged drought, plants resort to anaerobic respiration, which yields less ATP and produces harmful byproducts like ethanol. Anaerobic respiration is less efficient than aerobic respiration, reducing energy available for growth and maintenance. High temperatures increase energy expenditure on heat dissipation, such as transpiration, reducing resources for cellular functions, including respiration. Extended high-temperature exposure can harm cellular components and enzymes vital for respiration (Moore *et al.*, 2021).

Precipitation pattern changes influence plant respiration by altering soil moisture. Drought reduces respiration rates in water-stressed plants due to limited water. Stomatal closure decreases both transpiration and CO_2 assimilation, leading to reduced respiration rates and reliance on anaerobic respiration for energy (Chikov and Akhtyamova, 2019).

11.4 OTHER METABOLIC PROCESSES

11.4.1 Oxidative Stress and Protein Denaturation

Proteins and enzymes essential for biological processes are vulnerable to high temperatures, which can cause denaturation, impairing their structure and function. Cells naturally produce ROS during functions like mitochondrial energy production, and their levels are typically regulated by antioxidant defense mechanisms; however, external stressors like heat can disrupt this balance, leading to oxidative stress (Table 11.1). Elevated levels of carbon dioxide (CO_2) often enhance photosynthesis in C_3 plants, the mainstay of agricultural crops globally, through the CO_2 fertilization effect that stimulates photosynthesis, fostering plant growth and biomass production, particularly in fast-developing species (Waheed *et al.*, 2019; Bazinet *et al.*, 2022). However, elevated amounts of carbon dioxide (CO_2) might impact the process of absorbing and distributing nutrients, which could possibly lead to changes in the availability of nutrients in the soil. Increased carbon dioxide (CO_2) levels may also influence the way plants interact with insects, diseases, and herbivores, which may influence plant chemistry and decrease the nutritional value for certain herbivores. The disturbed plant-water relations arising due to the altered pattern of precipitation and drought stress disturb all the physiological system and health of the plants as a whole. In general, changes may include water availability, soil moisture, or plant ability concerning coping with the additional stress of water. Drought stress reduces plant cell water potential, prompting osmotic adjustments to maintain turgor pressure by absorbing solutes. This stress can impede cellular replication, leading to stunted growth and reduced leaf size, as well as disrupt reproductive processes, lowering flower and fruit yields. To cope with drought, plants undergo physiological and biochemical adjustments, known as drought acclimation, to enhance water-use efficiency and stress tolerance. However, recovery rates and extents vary depending on water availability, impacting plant resilience to drought stress. Elevated levels of carbon dioxide (CO_2) often enhance photosynthesis in C_3 plants, the mainstay of agricultural crops globally, through the "CO_2 fertilization effect." This phenomenon, most pronounced with sufficient nutrients and water, stimulates photosynthesis, fostering plant growth and biomass production, particularly in fast-developing species. Increased CO_2 levels may alter nutrient assimilation and distribution, impacting soil nutrient availability (Jin *et al.*, 2015).

TABLE 11.1
ROS in Plants, Their Site of Production, and Function

ROS	Subcellular Location	Role	References
Hydrogen peroxide (H_2O_2)	Chloroplasts, peroxisomes, mitochondria	Heat stress, UV-B radiation, pathogen infection	Xie *et al.* (2019a)
Hydroxyl radical (•OH)	Cytosol, chloroplasts, cell wall, membranes	Heavy metals, pathogen attack, wounding	Afzal *et al.* (2014), Das and Roychoudhury (2014)
Nitric oxide (NO)	Cytosol, chloroplasts, mitochondria, cell wall	Fungal infection, nitrogen stress, hypoxia	Dordas *et al.* (2003)
Peroxyl radical (ROO•)	Chloroplasts, mitochondria, peroxisomes	Cold stress, drought, oxidative burst	Bhattacharjee (2005)
Singlet oxygen (O_2)	Chloroplasts, peroxisomes, cell membranes	High light intensity, photooxidative stress	Tripathy and Pattanayak (2010)
Superoxide ($O_2^{\cdot-}$)	Chloroplasts, mitochondria, peroxisomes	Drought, salinity, high light, heavy metals	Gill and Tuteja (2010)

11.4.2 Grains Quality

The impact of global environmental modifications on crop quality, particularly food quality, is often overlooked. Studies have shown a general reduction in nitrogen and protein levels in response to increased carbon dioxide levels, but the nutritional composition and quality of food and forage crops remain largely unexplored. Existing evidence is sometimes conflicting, hindering our understanding of the effects of present and ongoing climate changes on food quality for both human and animal nutrition. Moreover, while small temperature differences seem to have a greater impact on quality than increased CO_2 levels, our understanding of the interactions between temperature and CO_2 on food quality is limited. This is due to a lack of knowledge about how CO_2 and temperature affect the connections between assimilating supply and demand.

The impact of temperature and elevated CO_2 levels on carbohydrate content in food crops is complex and influenced by experimental conditions. Soybean seeds showed changes in sucrose concentration with temperature elevation, while total soluble sugars and starch decreased with higher temperatures, minimally affected by elevated CO_2. In wheat, temperature changes had a greater effect on grain quality than CO_2 concentration, affecting starch content and granule composition. Under different temperature and CO_2 conditions, red kidney bean seed exhibited altered glucose, sucrose, and raffinose concentrations. High temperatures can expedite starch depositions in wheat grains, affecting the types of granules formed and consequently affecting malting processes (Wang *et al.*, 2021b). CO_2 levels both directly and indirectly influence the biosynthetic and degradation pathways of amylose in rice grains, and the processing suitability and the appearance quality have been reported to be degradative in some cases and there is no recorded effect in others. Genotype-specific factors under elevated CO_2 may drive these differences. The study by Alsajri *et al.* (2020) identified a positive correlation between temperature and soybean oil content, particularly within the range of 25–36°C. According to Thomas *et al.* (2003) maximum amount of oil in soybean seeds was accumulated at a daytime temperature of 32°C and a nighttime temperature of 22°C. Similar findings were observed in sunflower by Izquierdo *et al.* (2002), highlighting the significance of fatty acid composition for nutritional quality and food storage. Notably, increasing levels of carbon dioxide (CO_2) were found to have no significant impact on the composition and edible quality of maize grains and soybean seeds in various studies. However, quantitative variations in oil content in both non-starch and starch lipid fractions in wheat were more influenced by increased temperature (+4°C over ambient temperature) than elevated CO_2 levels. Elevated temperatures led to a decrease in stored lipids, particularly non-polar lipids, and qualitative alterations, such as an apparent increase in the conversion of oleate to linoleate (Xie, 2015; Mumtaz *et al.*, 2021). These modifications could significantly affect the milling characteristics of wheat and the baking attributes of flour. Taub *et al.* (2008) conducted a comprehensive meta-analysis involving 228 papers to examine the impact of increased atmospheric carbon dioxide concentration (ranging from 540 to 958 parts per million) on protein levels in key food crops. They found that wheat, barley, and rice experienced a decrease in grain protein content ranging from 10% to 15% of ambient CO_2 levels (315–400 ppm). Similarly, potatoes and soybeans showed reductions of 14% and 1.4%, respectively, in tuber protein concentration due to elevated CO_2 levels.

Attempts to mitigate the decrease in protein concentration by increasing nitrogen fertilizer application, even beyond current levels, have shown limited utilities (Nath Bhowmik and Das, 2018). However, excessive nitrogen supply may adversely affect rice grain quality, leading to increased chalkiness and decreased eating and cooking quality. In wheat, higher nitrogen fertilization levels may boost biomass and yield production at the expense of grain protein concentration. Moreover, elevated CO_2 levels may affect not only the quantity but also the composition of grain proteins in crops like rice and wheat (Ben Mariem *et al.*, 2021). The concentration and characteristics of gluten storage proteins, crucial for dough and bread quality in wheat, have been observed to decrease at high CO_2 levels in some studies, although this response appears to be cultivar dependent.

While there is limited research on the impact of high CO_2 levels on protein quality, it is anticipated that such conditions may lead to alterations in dough characteristics and rheological qualities crucial for bread-making in wheat analysis of wheat data also revealed a notable reduction in amino acid levels, ranging from 7.7% to 23.9%, due to CO_2 enrichment in over-the-counter experiments. However, certain essential amino acids may increase in proportion under elevated CO_2 conditions, mainly due to significant decreases in glutamate levels offsetting reductions in lysine and arginine.

11.5 ADAPTATION MECHANISMS IN PLANTS

Adaptation mechanisms in plants include many techniques and reactions that plants have developed over time to deal with changes in their surroundings, particularly when confronted with stressors, including adverse climatic conditions, nutritional insufficiencies, illnesses, and herbivore assaults. Physiological, biochemical, and genetic responses are the three main categories into which adaptation processes fall.

11.5.1 Drought Adaptation

Osmotic adjustment refers to an internal process by which living organisms regulate their osmotic pressures according to the changes the surrounding environment is undergoing. Such conditions occurring with drought stress and changing patterns of precipitation may have very marked effects on altering plant-water relations, hence giving rise to several disorders at physiological levels and deterioration in general plant health. These may modify plant-water-stress tolerance as well as water and soil moisture availability. Changes in precipitation patterns, such as increased frequency of droughts or changes in the timing of rainfall throughout the year, have a direct impact on the amount of water in the soil. Plants may be subjected to stressful conditions in order to obtain water for their growth and metabolic activities due to a decline in soil moisture as a result of scanty rainfall or prolonged drought spells (Govender *et al.*, 2022; Rasheed *et al.*, 2022b). In adapting to this, plants regulate water loss through transpiration by adjusting the size of the apertures called stomata on their leaves. Plants often shut their stomata to preserve water when water is in shortage. This may prevent the intake of carbon dioxide, causing a limit on photosynthetic rates. Water potential of plant cells is reduced by drought stress, developing a gradient that allows the intake of water from the soil to the plant. Consequently, some plants control their osmotic pressure by taking up solutes that decrease their water potential while maintaining turgor pressure (Anjum *et al.*, 2017).

Drought stress can elicit oxidative stress in the plant cell as a result of increased ROS production. ROS can inflict damage to cell membranes, proteins, and DNA and may even threaten survival and proper plant function. Thus, plants have developed the ability of drought acclimation for coping with drought stress, which is considered an internal adjustment of the physiological and biochemical processes of the plants toward enhancing water-use efficiency by the plant, as well as promoting tolerance against such stress. It follows, therefore, that plants can recuperate once water becomes accessible; in such a state of rejuvenation, recuperation might still be variable with respect to rate (Hussain *et al.*, 2019).

11.5.2 Heat and Cold Stress Adaptation

Heat shock proteins (HSPs), also known as stress proteins, are a specific group of proteins that help during high thermal temperature to prevent or dampen the heat-generated loosening of secondary, tertiary, and, therefore, structural integrity of subcellular macromolecular machinery such as misfolding/aggregation of cells. These proteins are induced by different types of stressors, including high temperatures, and their main role is to assist cells in coping and surviving under adverse conditions (Zain, 2017; Ghosh *et al.*, 2018).

HSPs are commonly referred to as "molecular chaperones." Chaperones are a class of proteins that are involved in the correct folding of other proteins and reduce unwanted interactions. The raised temperature of the cells may disturb the sensitive interlinks that are responsible for the three-dimensional structure of proteins due to the rise in thermal energy. This can lead to the denaturation of proteins, a process in which the protein loses its functional conformation and becomes biologically inert. Cells increase the synthesis of HSPs in response to heat stress (Park and Seo, 2015). HSPs play a very important role in maintaining a pool of proteins that readily become denatured in conditions of elevated temperature; thus, one of the crucial roles of HSPs in an organism is the prevention of heat shock-induced protein aggregation by binding with partly or fully unfolded or denatured proteins to facilitate subsequent refolding into correct functional forms. Essentially, the HSPs enable the plant cell to survive unfavorable stress in a normally physiological active state due to undenatured critical proteins.

However, besides their functions concerning prevention against denaturation, the HSP family proteins perform their functions rather broadly. They also play a crucial role in the protection of cellular components from stress damage, such as membranes and organelles. This helps in the maintenance of structural and functional integrity in cells and their components. In addition to that, HSPs facilitate the routing of misfolded or otherwise damaged proteins into pathways leading to their degradation. It is important to eliminate potentially toxic or nonfunctional proteins from the cell, and this process accomplishes that effectively. If we want to keep the cell healthy, we want to make sure that dysfunctional components do not accumulate.

HSPs are central to the cellular stress response and a fundamental adaptive mechanism. Cells induce a multitude of HSPs in response to certain durations and magnitudes of stress, allowing the cell to customize its response to the nuances of the stimuli. In some plants adapted to cold environments, antifreeze proteins are produced. These proteins prevent ice crystal growth within the cells of the plants, rescuing the plant from frost (Mogk *et al.*, 2019).

There are also some other mechanisms by which the effect of other stresses can be alleviated in plants. This helps maintain proper ion balance and reduces the incidence of cell damage. Besides, under salt stress, plants accumulate compatible solutes to maintain cellular water potential, similar to their adaptation under drought. These changes in plants improve their tolerance and resilience to stresses, hence allowing them to thrive under harsh conditions.

11.5.3 Molecular Responses to Climate Stress

Plants have several receptors at the cell membranes, which may sense the change in ambient conditions. Such receptors may perceive the stresses related to drought, high temperature, diseases, and nutritional deficiencies. The receptors, such as receptor-like kinases or pattern recognition receptors, may recognize pathogen-associated molecular patterns (PAMPs) and initiate an immune response. A series of events, triggered by the perception of stress, ultimately lead to the generation of secondary messengers, including Ca^{2+} and ROS (Couto and Zipfel, 2016). These secondary messengers act as signaling molecules and thus spread the stress signal within the cell. Secondary messengers induce the activation of a number of protein kinases, which include mitogen-activated protein kinases (MAPKs), calcium-dependent protein kinases (CDPKs), and receptor kinases, leading to the activation of a series of signaling pathways. These kinases in turn phosphorylate other downstream components, which activate transcription factors (TFs) and other regulatory proteins. The result of signaling pathways is the activation of TFs that bind selectively to DNA sequences in the promoters of stress-responsive genes. These TFs modulate the gene expression of those responsible for stress responses, including the genes encoding enzymes, chaperones, antioxidants, and other protective proteins (Manna *et al.*, 2021; Rasheed *et al.*, 2022a). The activation of TFs leads to a change in gene expression that allows synthesis of the stress-related proteins in plants. These proteins help the plant cope with stress through osmotic adjustment, heat shock response, detoxification of ROS, and other protectant mechanisms. Many stress signaling pathways may intersect and thus share common components. For example, a number of components of the ABA signaling pathway are also involved in other stress responses and thus enable the integration of the responses to various stressors. Plants use stress-responsive genes, TFs, and hormones to perceive and respond to environmental challenges. The interconnections of components inside the plants are highly advanced to coordinate adaptable reactions that ensure their long-lasting survival and strength. Such stress-responsive genes will, therefore, be expressed upon contact with environmental stresses; such a process eventually promotes the expression of appropriate proteins to confer protection indirectly or directly against stress-causing agents on the plant (Auge *et al.*, 2023). Indeed, the function of all stress-responsive genes falls broadly into categories, including detoxification, antioxidant defense, osmotic adjustment, protein stability, and signaling. Stress-responsive genes synthesize enzymes neutralizing injurious substances or scavenging ROS developed during the stress period. superoxide dismutase (SOD), CAT, and POD enzymes help with the detoxification process resulting in a reduction of oxidative stress (Mishra and Sharma, 2019). Besides that, they assist in osmotic adjustment resulting in enzymatic production that results in appropriate solutes development like proline and sugar. These chemicals help to maintain cellular water potential and turgor during stress.

Genes coding for HSPs and chaperones are induced under temperature-stress conditions. These proteins help in folding of proteins and prevent denaturation. They protect the cellular structures by preventing the aggregation of proteins. TFs are regulatory proteins that bind to particular DNA sequences in the promoter region of genes, hence activating or repressing the transcription rate of genes. TFs play an important role in the orchestration of stress responses through the activation and/or suppression of stress-responsive genes. Various stressors induce the activity of selective TFs, which differentially bind to specific DNA motifs associated with stress-responsive genes (Xie *et al.*, 2019b).

Dehydration-responsive element-binding proteins or DREBs modulate gene activities in response to drought and low-temperature conditions. Specific TFs promote the interactions between many stress pathways so that the plant can integrate and coordinate its responses to the different stressors. It ensures a harmonious and effective response to numerous environmental dangers.

Plant hormones or signaling molecules are key in growth, development, and response to environmental stimuli, including stress. They serve as crucial mediators in stress signaling pathways, synchronizing physiological and biochemical changes in the plant. ABA has a vital function as a hormone in stress responses, especially under conditions of water scarcity and high salt levels. It regulates the process of stomatal closure, reduces water loss, and initiates the activation of stress-responsive genes. Jasmonic acid (JA) and salicylic acid (SA) are two important plant hormones in the response of the immune system against infections and herbivores. JA modulates the body's response to injuries and its defense against plant-eating animals, while SA is linked with the body's overall acquired immunity to illnesses (Ullah *et al.*, 2013; Wang *et al.*, 2021a).

Ethylene plays a significant role in many stress responses, including those induced by flooding, physical damage, and pathogenic invasion. It has the capacity to regulate programmed cell death (PCD) and the development of roots in response to stress. These hormones have a role in stress responses related to nutrition availability and growth regulation. They may have an impact on root growth and nutrient absorption. Antioxidants and ROS are essential for plants to respond to environmental stress. ROS are chemical molecules composed of oxygen radicals and their derivatives, such as superoxide (O_2^-), hydrogen peroxide (H_2O_2), and hydroxyl radicals (OH^-). ROS are produced as a normal byproduct of metabolic processes (Pandhair and Sekhon, 2006). Nevertheless, their levels increase considerably in reaction to stressful circumstances. Antioxidants are substances that counteract the detrimental impact of ROS. It is crucial to maintain a precise balance between the production of ROS and the protection offered by antioxidant defense systems in order to successfully adjust to stress. Environmental conditions such as aridity, strong illumination, extreme temperatures, and invasion by pathogens may trigger the excessive production of ROS in plant cells. This disorder occurs due to disturbances in the balance between the production of ROS and the mechanisms that remove them. Elevated levels of ROS in the cell can lead to oxidative stress, a pathological reaction, damaging biological molecules, including proteins or lipids and even DNA. ROS also serve as signaling molecules in stress responses. ROS, in appropriate concentrations, can act as secondary messengers, activating a number of stress-responsive genes and pathways. ROS are involved in stress signaling, activating a variety of pathways such as the MAPKs and the CDPKs, which eventually modulate gene expression (Singh *et al.*, 2019). Plants have developed a complex antioxidant defense system to protect themselves from the damaging effects of ROS.

This system consists of enzymatic and non-enzymatic sources of antioxidants. For example, enzymes such as SOD, CAT, peroxidase (POD), and ascorbate peroxidase (APX) play a vital role in eliminating ROS. SOD catalyzes the conversion of SOD radicals to hydrogen peroxide, while CAT and POD are responsible for hydrogen peroxide detoxification. Vitamin C, GSH, tocopherols, and flavonoids are proteolytic antioxidants. These antioxidants work against ROS directly and assist in the regeneration of other antioxidants. This regulatory response is governed by plant hormones, including ABA, SA, and jasmonate (JA). These hormones can also induce the expression of genes encoding antioxidant enzymes and compounds, thereby enhancing the plants' ability to scavenge ROS (Zaib *et al.*, 2023). Elevated concentrations of ROS can trigger PCD, a regulated process whereby cells die to prevent additional harm to the plant. PCD is an important mechanism that protects against infection or harmful effects due to the release of pro-inflammatory cytokines while also preventing damage from cell death-associated stress through the clearance of infected or damaged cells.

11.5.4 Plant Hormones and Climate Adaptation

Plant hormones are highly central to the adaptation to climate change because they regulate a range of physiological, developmental, and stress responses that allow plants to thrive in ever-changing environments. Hormones play a key role in some climate adaptation, helping plants change their growth, development, and responses to different challenges. The relationship between plant hormones and stress responses is a complex and dynamic process based on the process of adaptation and development of plants in different environmental conditions. Plant hormones also work as signaling molecules, regulating a wide array of physiological, biochemical, and morphological adjustments to help plants adapt to environmental stimuli (Mahmood *et al.*, 2013; Sadiq *et al.*, 2020).

11.5.5 Abscisic Acid (ABA) and Stress Response

ABA plays a vital role as a hormone modulating stress response, particularly regarding drought and salt stress. Its important functions include the control of stomata closure and water balance. Plant ABA levels are elevated under water deficit, and stomata close, decreasing transpiration and water loss. This method not only conserves water but also avoids dehydration. ABA also induces the expression of stress-responsive genes involved in osmotic adjustment, antioxidant defense, and stress tolerance. Thus, ABA plays an important role in keeping seeds dormant and also in preventing the seeds from germination if environmental conditions are not suitable (Zhang *et al.*, 2006). Its key functions include regulating water uptake, inhibiting shoot growth, and triggering genes involved in stress response. ABA is vital in the regulation of seed dormancy especially in extreme environmental conditions. Seeds possess this adaptive characteristic that enables them to stay inactive during adverse times, such as drought, harsh temperatures, or other stressful situations, and then start growing when conditions become more conducive for plant development. ABA acts as a stimulus for plants to sense and react to environmental stress (Bhattacharyya, 2022). In response to stressors such as drought, salt, or extreme temperatures, seeds ramp up their production of ABA. The increased concentration of ABA therefore enhances dormancy and suppresses genes involved in germination.

ABA is involved in the acquisition of desiccation tolerance, an important trait for seeds to persist in drought-prone environments. It reinforces desiccation tolerance, triggers the rise of reserves, and increases the generation of antioxidants to lessen ROS. ABA also regulates resource distribution, as well as gene activation. The accumulation of components, such as nitrogen and phosphorus, with the capacity to support the initial growth of seedlings post-germination, is actively mediated by ABA (Chen *et al.*, 2020). ABA governs physiological and molecular responses to environmental stress. This is an important aspect in response to drought and salinity as it regulates gene expression, prevents excessive water loss, and promotes adaptation and stress resistance.

11.5.6 Auxins and Stress Response

Auxins mediate diverse stress responses, such as light, gravity, and nutrient deficiency signaling. They play a role in tropic responses such as phototropism and gravitropism as well as root growth. Auxins could potentially play a role in root growth in response to stressful conditions, enabling the plant to access deeper zones of the soil for water and nutrients (Leftley *et al.*, 2021). They also affect lateral root growth, which allows plants to increase nutrient uptake. Figure 11.2 showed the adaptive mechanisms of plants against various types of stresses.

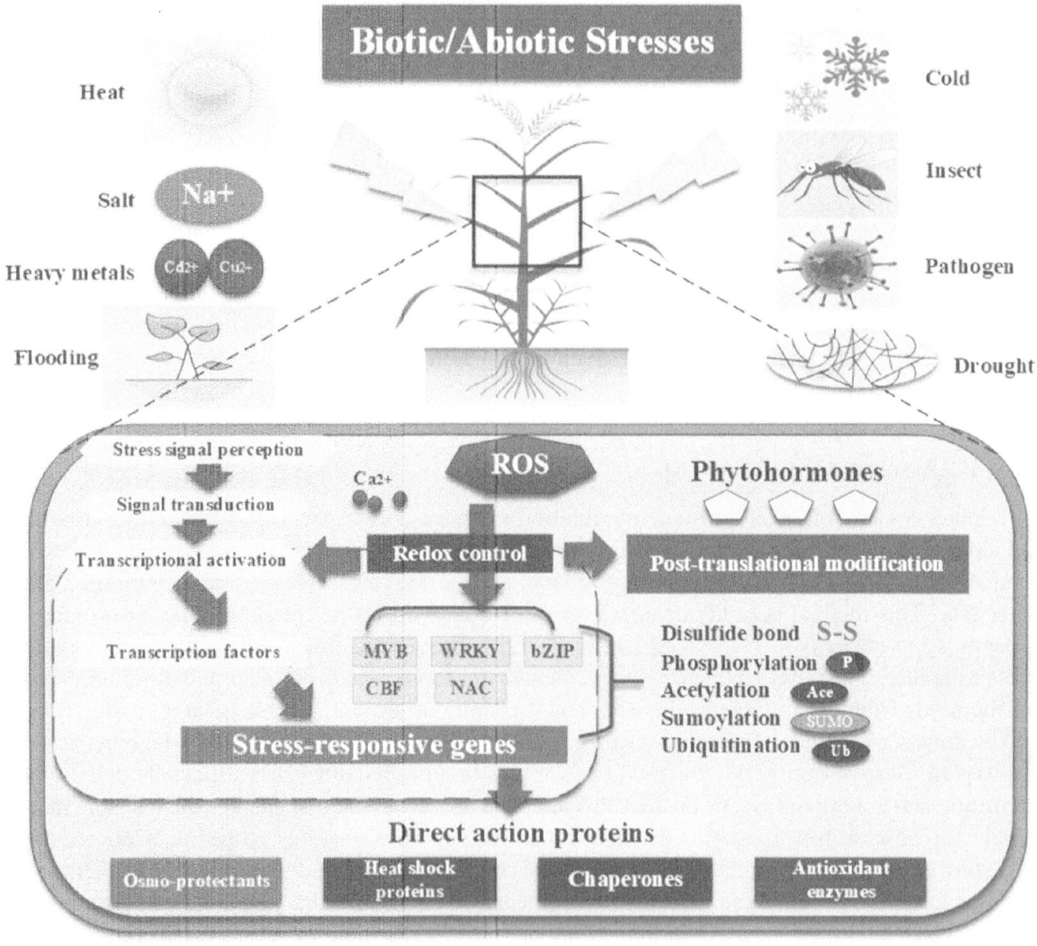

FIGURE 11.2 Plants adaptive mechanisms against various types of stresses.

11.6 ACCUMULATION AND DETOXIFICATION OF REACTIVE OXYGEN SPECIES

Plants under stress have challenges in using ATP and reduction equivalents for biosynthetic activities, resulting in the buildup of ROS. Although ROS has the ability to cause damage, they also function as signaling molecules. Some ROS, including superoxide anion radical, hydroxyl radical, and hydrogen peroxide, play a role in PCD. Production and detoxifying processes are critically important under stressful conditions like drought and high ambient temperatures. Antioxidant enzymes, including catalases, peroxidases, and SOD, play an important role in removing ROS. Moreover, hydrophilic as well as lipophilic antioxidant molecules serve as free radical scavengers and are implicated in the antioxidant response (Dumanović *et al.*, 2021). High levels of these chemicals promote rapid removal of ROS and also have a protective effect on subcellular components. When focusing on the detoxification of ROS, one must also consider the enzymes responsible for the production of antioxidant molecules, their expression prior to and during abiotic stress, their compartmentalization within the cell, and how their activity is controlled (Gawlik-Dziki, 2012).

11.7 CONCLUSIONS

Considering breeding, physiology, agronomists, and policy makers altogether find climate change a huge threat; different plant species bear in themselves different levels of their tolerance to drought and rise in temperature. In case of temperature, small elevation of it could be benefited at large for some crops like maize that is susceptible from low temperatures. The aim of gaining as broad an understanding as possible of key processes at the whole plant level is to select appropriate genetic types and to plan breedings for use in the future. Climate change has a number of impacts at the level of organism response, including organ development, assimilation, morphological adaptation, transport, senescence, and seed maturation. However, our current knowledge of these mechanisms is limited. Moreover, within a species, different genotypes may have quite different abilities to tolerate abiotic conditions such as drought and heat. When choosing genotypes, it is crucial to take into account their performance not only during periods of stress but also during recovery stages. Climate change and the increasing occurrence of severe events provide significant challenges for breeding, genotype selection, and agronomic techniques in monocultures. In heterogeneous ecosystems such as grasslands, interspecies competition emerges as an extra influential element to take into account. Varied plants may be affected differently by times of stress, resulting in changes in the species composition that may have a detrimental effect on the competition between cultivated plants and weeds.

ACKNOWLEDGMENT

The authors extend their appreciation for the support of the Deanship of Scientific Research, Vice Presidency for Graduate Studies and Scientific Research, King Faisal University, Saudi Arabia [Grant No. KFU252570].

LIST OF ABBREVIATIONS

ABA	abscisic acid
CDPKs	calcium-dependent protein kinases
H_2O_2	hydrogen peroxide
HSPs	heat shock proteins
JA	jasmonic acid
MAPKs	mitogen-activated protein kinases
O_2^-	superoxide
OH^-	hydroxyl radicals
PCD	programmed cell death
POD	peroxidase
ROS	reactive oxygen species
SA	salicylic acid
SOD	superoxide dismutase
TFs	transcription factors

REFERENCES

Abdullah, Mahmood, A., Bibi, S., Naqve, M., Javaid, M.M., Zia, M.A., Jabbar, A., Ud-Din, W., Attia, K.A. & Khan, N., 2022. Physiological, biochemical, and yield responses of linseed (*Linum usitatissimum* L.) in α-tocopherol-mediated alleviation of salinity stress. *Frontiers in Plant Science*, 13, 867172.

Afzal, F., Khurshid, R., Ashraf, M. & Gul Kazi, A., 2014. Reactive Oxygen Species and Antioxidants in Response to Pathogens and Wounding. *In* P. Ahmad (ed.) *Oxidative Damage to Plants*. Academic Press, San Diego, USA 397–424.

Alsajri, F.A., Wijewardana, C., Irby, J.T., Bellaloui, N., Krutz, L.J., Golden, B., Gao, W. & Reddy, K.R., 2020. Developing functional relationships between temperature and soybean yield and seed quality. *Agronomy Journal*, 112, 194–204.

Amaral, J., Lobo, A.K. & Carmo-Silva, E., 2024. Regulation of Rubisco activity in crops. *New Phytologist*, 241, 35–51.

Anjum, S.A., Ashraf, U., Zohaib, A., Tanveer, M., Naeem, M., Iftikhar Ali, I.A., Tahira Tabassum, T.T. & Usman Nazir, U.N., 2017. Growth and developmental responses of crop plants under drought stress: A review, 104, 267–276.

Ashraf, M., Niaz, Y., Hafeez, M., Ahmad, M.U.D. & Shah, A.A., 2023. Water Management Strategies for Agricultural Disasters. *In* M. Ahmed & S. Ahmad (eds.) *Disaster Risk Reduction in Agriculture*. Springer, Singapore, 67–89.

Auge, G., Hankofer, V., Groth, M., Antoniou-Kourounioti, R., Ratikainen, I. & Lampei, C., 2023. Plant environmental memory: Implications, mechanisms and opportunities for plant scientists and beyond. *AoB PLANTS*, 15, 1–15.

Ávila, R.G., Magalhães, P.C., Vitorino, L.C., Bessa, L.A., De Souza, K.R.D., Queiroz, R.B., Jakelaitis, A. & Teixeira, M.B., 2023. Chitosan induces sorghum tolerance to water deficits by positively regulating photosynthesis and the production of primary metabolites, osmoregulators, and antioxidants. *Journal of Soil Science and Plant Nutrition*, 23, 1156–1172.

Aziz, M., Ashraf, M. & Javaid, M.M., 2018. Enhancement in cotton growth and yield using novel growth promoting substances under water limited conditions. *Pakistan Journal of Botany*, 50, 1691–1701.

Bazinet, Q., Tang, L. & Bede, J.C., 2022. Impact of future elevated carbon dioxide on C_3 plant resistance to biotic stresses. *Molecular Plant-Microbe Interactions*, 35, 527–539.

Begna, T., 2020. Effects of drought stress on crop production and productivity. *International Journal of Research Studies in Agricultural Sciences*, 6, 34–43.

Ben Mariem, S., Soba, D., Zhou, B., Loladze, I., Morales, F. & Aranjuelo, I., 2021. Climate change, crop yields, and grain quality of C_3 cereals: A meta-analysis of [CO_2], temperature, and drought effects. *Plants*, 10, 1052.

Bhattacharjee, S., 2005. Reactive oxygen species and oxidative burst: Roles in stress, senescence and signal transduction in plants. *Current Science*, 1113–1121.

Bhattacharyya, S., 2022. Mechanism of Temperature Stress Acclimation and the Role of Transporters in Plants. *In* T. Aftab & A. Roychoudhury (eds.) *Plant Perspectives to Global Climate Changes*. Academic Press, San Diego, California, USA, 413–457.

Calleja-Cabrera, J., Boter, M., Oñate-Sánchez, L. & Pernas, M., 2020. Root growth adaptation to climate change in crops. *Frontiers in Plant Science*, 11, 523645.

Chadha, A., Florentine, S., Chauhan, B.S., Long, B., Jayasundera, M., Javaid, M.M. & Turville, C., 2019. Environmental factors affecting the germination and seedling emergence of two populations of an emerging agricultural weed: Wild lettuce (*Lactuca serriola*). *Crop and Pasture Science*, 70, 709–717.

Chadha, A., Florentine, S., Javaid, M., Welgama, A. & Turville, C., 2020. Influence of elements of climate change on the growth and fecundity of *Datura stramonium*. *Environmental Science and Pollution Research*, 27, 35859–35869.

Chen, K., Li, G.J., Bressan, R.A., Song, C.P., Zhu, J.K. & Zhao, Y., 2020. Abscisic acid dynamics, signaling, and functions in plants. *Journal of Integrative Plant Biology*, 62, 25–54.

Chen, Y., Dubois, M., Vermeersch, M., Inzé, D. & Vanhaeren, H., 2021. Distinct cellular strategies determine sensitivity to mild drought of Arabidopsis natural accessions. *Plant Physiology*, 186, 1171–1185.

Chikov, V. & Akhtyamova, G., 2019. Photorespiration and its role in the regulation of photosynthesis and plant productivity. *American Journal of Plant Sciences*, 10, 2179–2202.

Couto, D. & Zipfel, C., 2016. Regulation of pattern recognition receptor signalling in plants. *Nature Reviews Immunology*, 16, 537–552.

Das, K. & Roychoudhury, A., 2014. Reactive oxygen species (ROS) and response of antioxidants as ROS-scavengers during environmental stress in plants. *Frontiers in Environmental Science*, 2, 53.

Dordas, C., Rivoal, J. & Hill, R.D., 2003. Plant haemoglobins, nitric oxide and hypoxic stress. *Annals of Botany*, 91, 173–178.

Dumanović, J., Nepovimova, E., Natić, M., Kuča, K., & Jaćević, V., 2021. The significance of reactive oxygen species and antioxidant defense system in plants: A concise overview. *Frontiers in plant science*, 11, 552969.

Fagbemi, F., Oke, D.F. & Fajingbesi, A., 2023. Climate-resilient development: An approach to sustainable food production in sub-Saharan Africa. *Future Foods*, 7, 100216.

Florentine, S.K., Weller, S., Graz, P.F., Westbrooke, M., Florentine, A., Javaid, M., Fernando, N., Chauhan, B.S. & Dowling, K., 2016. Influence of selected environmental factors on seed germination and seedling survival of the arid zone invasive species tobacco bush (*Nicotiana glauca* R. Graham). *The Rangeland Journal*, 38, 417–425.

Forslund, A., Tibi, A., Schmitt, B., Marajo-Petitzon, E., Debaeke, P., Durand, J.-L., Faverdin, P. & Guyomard, H., 2023. Can healthy diets be achieved worldwide in 2050 without farmland expansion? *Global Food Security*, 39, 100711.

Gawlik-Dziki, U., 2012. Changes in the antioxidant activities of vegetables as a consequence of interactions between active compounds. *Journal of Functional Foods*, 4, 872–882.

Ghosh, S., Sarkar, P., Basak, P., Mahalanobish, S. & Sil, P.C., 2018. Role of Heat Shock Proteins in Oxidative Stress and Stress Tolerance. *In* A.A.A. Asea & P. Kaur (eds.) *Heat Shock Proteins and Stress*. Springer, Cham, Switzerland, 109–126.

Gill, S.S. & Tuteja, N., 2010. Reactive oxygen species and antioxidant machinery in abiotic stress tolerance in crop plants. *Plant Physiology and Biochemistry*, 48, 909–930.

Govender, T., Dube, T. & Shoko, C., 2022. Remote sensing of land use-land cover change and climate variability on hydrological processes in sub-Saharan Africa: Key scientific strides and challenges. *Geocarto International*, 37, 10925–10949.

Greff, B., Sáhó, A., Lakatos, E. & Varga, L., 2023. Biocontrol activity of aromatic and medicinal plants and their bioactive components against soil-borne pathogens. *Plants*, 12, 706.

Grigorieva, E., Livenets, A. & Stelmakh, E., 2023. Adaptation of agriculture to climate change: A scoping review. *Climate*, 11, 202.

Hanif, Z., Naeem, M., Ali, H.H., Tanveer, A., Javaid, M.M., Peerzada, A.M. & Chauhan, B.S., 2017. Effect of environmental factors on germination of *Salsola foetida*: Potential species for rehabilitation of degraded rangelands. *Rangeland Ecology & Management*, 70, 638–643.

Hassan, M.U., Nawaz, M., Mahmood, A., Shah, A.A., Shah, A.N., Muhammad, F., Batool, M., Rasheed, A., Jaremko, M. & Abdelsalam, N.R., 2022. The role of zinc to mitigate heavy metals toxicity in crops. *Frontiers in Environmental Science*, 10, 990223.

Hussain, S., Rao, M.J., Anjum, M.A., Ejaz, S., Zakir, I., Ali, M.A., Ahmad, N. & Ahmad, S., 2019. Oxidative Stress and Antioxidant Defense in Plants under Drought Conditions. *In* M. Hasanuzzaman, K.R. Hakeem, K. Nahar & H.F. Alharby (eds.) *Plant Abiotic Stress Tolerance: Agronomic, Molecular and Biotechnological Approaches*. Springer, Cham, Switzerland, 207–219.

Izquierdo, M., Häkkinen, K., Gonzalez-Badillo, J.J., Ibanez, J. & Gorostiaga, E.M., 2002. Effects of long-term training specificity on maximal strength and power of the upper and lower extremities in athletes from different sports. *European Journal of Applied Physiology*, 87, 264–271.

Javaid, M.M., Florentine, S., Mahmood, A., Wasaya, A., Javed, T., Sattar, A., Sarwar, N., Kalaji, H.M., Ahmad, H.B. & Worbel, J., 2022. Interactive effect of elevated CO_2 and drought on physiological traits of *Datura stramonium*. *Frontiers in Plant Science*, 13, 929378.

Jin, J., Tang, C. & Sale, P., 2015. The impact of elevated carbon dioxide on the phosphorus nutrition of plants: A review. *Annals of Botany*, 116, 987–999.

John, A., Riat, A.K., Ahmad Bhat, K., Ganie, S.A., Endarto, O., Nugroho, C., Handoko, H. & Wani, A.K., 2024. Adapting to climate extremes: Implications for insect populations and sustainable solutions. *Journal for Nature Conservation*, 79, 126602.

Kopecká, R., Kameniarová, M., Černý, M., Brzobohatý, B. & Novák, J., 2023. Abiotic stress in crop production. *International Journal of Molecular Sciences*, 24, 6603.

Krieg, C.P. & Chambers, S.M., 2022. The ecology and physiology of fern gametophytes: A methodological synthesis. *Applications in Plant Sciences*, 10, e11464.

Leftley, N., Banda, J., Pandey, B., Bennett, M. & Voß, U., 2021. Uncovering how auxin optimizes root systems architecture in response to environmental stresses. *Cold Spring Harbor Perspectives in Biology*, 13, a040014.

Li, S., Li, X., Wei, Z. & Liu, F., 2020. ABA-mediated modulation of elevated CO_2 on stomatal response to drought. *Current Opinion in Plant Biology*, 56, 174–180.

Li, T.-Y., Shi, Q., Sun, H., Yue, M., Zhang, S.-B. & Huang, W., 2021. Diurnal response of photosystem I to fluctuating light is affected by stomatal conductance. *Cells*, 10, 3128.

Lohani, N., Singh, M.B. & Bhalla, P.L., 2019. High temperature susceptibility of sexual reproduction in crop plants. *Journal of Experimental Botany*, 71, 555–568.

Mahmood, A., Ullah, H., Shahzad, A., Ali, H., Ahmad, S., Zia-Ul-Haq, M., Honermeier, B. & Hasanuzzaman, M., 2013. Dry matter yield and chemical composition of sorghum cultivars with varying planting density and sowing date. *Sains Malaysiana*, 42, 1529–1538.

Manna, M., Thakur, T., Chirom, O., Mandlik, R., Deshmukh, R. & Salvi, P., 2021. Transcription factors as key molecular target to strengthen the drought stress tolerance in plants. *Physiologia Plantarum*, 172, 847–868.

Mehmood, A., Naeem, M., Khalid, F., Saeed, Y., Abbas, T., Jabran, K., Sarwar, M.A., Tanveer, A. & Javaid, M.M., 2018. Identification of phytotoxins in different plant parts of *Brassica napus* and their influence on mung bean. *Environmental Science and Pollution Research*, 25, 18071–18080.

Mishra, P. & Sharma, P., 2019. Superoxide Dismutases (SODs) and Their Role in Regulating Abiotic Stress Induced Oxidative Stress in Plants. *In* M. Hasanuzzaman, M. Fujita, M. N. Oku, & J. A. Teixeira da Silva (eds.) *Reactive Oxygen, Nitrogen and Sulfur Species in Plants: Production, Metabolism, Signaling and Defense Mechanisms*, Hoboken, NJ: Wiley, 53–88.

Mogk, A., Ruger-Herreros, C. & Bukau, B., 2019. Cellular functions and mechanisms of action of small heat shock proteins. *Annual Review of Microbiology*, 73, 89–110.

Moore, C.E., Meacham-Hensold, K., Lemonnier, P., Slattery, R.A., Benjamin, C., Bernacchi, C.J., Lawson, T. & Cavanagh, A.P., 2021. The effect of increasing temperature on crop photosynthesis: From enzymes to ecosystems. *Journal of Experimental Botany*, 72, 2822–2844.

Mumtaz, S., Saleem, M.H., Hameed, M., Batool, F., Parveen, A., Amjad, S.F., Mahmood, A., Arfan, M., Ahmed, S. & Yasmin, H., 2021. Anatomical adaptations and ionic homeostasis in aquatic halophyte *Cyperus laevigatus* L. under high salinities. *Saudi Journal of Biological Sciences*, 28, 2655–2666.

Nath Bhowmik, S. & Das, A., 2018. Biofertilizers: A Sustainable Approach for Pulse Production. *In* R.S. Meena, A. Das, G.S. Yadav & R. Lal (eds.) *Legumes for Soil Health and Sustainable Management*. Singapore: Springer Singapore, 445–485.

Nelissen, H., Sun, X.H., Rymen, B., Jikumaru, Y., Kojima, M., Takebayashi, Y., Abbeloos, R., Demuynck, K., Storme, V. & Vuylsteke, M., 2018. The reduction in maize leaf growth under mild drought affects the transition between cell division and cell expansion and cannot be restored by elevated gibberellic acid levels. *Plant Biotechnology Journal*, 16, 615–627.

Oksanen, E., Lihavainen, J., Keinänen, M., Keski-Saari, S., Kontunen-Soppela, S., Sellin, A. & Sõber, A., 2019. Northern Forest Trees under Increasing Atmospheric Humidity. *In* F.M. Cánovas, U. Lüttge, R. Matyssek & H. Pretzsch (eds.) *Progress in Botany Vol. 80*. Springer, Cham, Switzerland, 317–336.

Pandhair, V. & Sekhon, B.S., 2006. Reactive oxygen species and antioxidants in plants: An overview. *Journal of Plant Biochemistry and Biotechnology*, 15, 71–78.

Park, C.-J. & Seo, Y.-S., 2015. Heat shock proteins: A review of the molecular chaperones for plant immunity. *The Plant Pathology Journal*, 31, 323.

Radha, B., Sunitha, N.C., Sah, R.P., Tp, M.A., Krishna, G., Umesh, D.K., Thomas, S., Anilkumar, C., Upadhyay, S. & Kumar, A., 2023. Physiological and molecular implications of multiple abiotic stresses on yield and quality of rice. *Frontiers in Plant Science*, 13, 996514.

Rasheed, A., Barqawi, A.A., Mahmood, A., Nawaz, M., Shah, A.N., Bay, D.H., Alahdal, M.A., Hassan, M.U. & Qari, S.H., 2022a. CRISPR/Cas9 is a powerful tool for precise genome editing of legume crops: A review. *Molecular Biology Reports*, 49, 5595–5609.

Rasheed, A., Mahmood, A., Maqbool, R., Albaqami, M., Sher, A., Sattar, A., Bakhsh, G., Nawaz, M., Hassan, M.U. & Al-Yahyai, R., 2022b. Key insights to develop drought-resilient soybean: A review. *Journal of King Saud University-Science*, 34, 102089.

Sadiq, Y., Zaid, A. & Khan, M.M.A., 2020. Adaptive Physiological Responses of Plants under Abiotic Stresses: Role of Phytohormones. *In* M. Hasanuzzaman (ed.) *Plant Ecophysiology and Adaptation under Climate Change: Mechanisms and Perspectives I: General Consequences and Plant Responses*. Springer, Singapore, 797–824.

Santiago, J.P. & Sharkey, T.D., 2019. Pollen development at high temperature and role of carbon and nitrogen metabolites. *Plant, Cell & Environment*, 42, 2759–2775.

Shoaib, M., Banerjee, B.P., Hayden, M. & Kant, S., 2022. Roots' drought adaptive traits in crop improvement. *Plants*, 11, 2256.

Singh, A., Kumar, A., Yadav, S. & Singh, I.K., 2019. Reactive oxygen species-mediated signaling during abiotic stress. *Plant Gene*, 18, 100173.

Singh, A., Pandey, H., Pandey, S., Lal, D., Chauhan, D., Aparna, Antre, S.H., B, S. & Kumar, A., 2023a. Drought stress in maize: Stress perception to molecular response and strategies for its improvement. *Functional & Integrative Genomics*, 23, 296.

Singh, B.K., Delgado-Baquerizo, M., Egidi, E., Guirado, E., Leach, J.E., Liu, H. & Trivedi, P., 2023b. Climate change impacts on plant pathogens, food security and paths forward. *Nature Reviews Microbiology*, 21, 640–656.

Sinha, R., Fritschi, F.B., Zandalinas, S.I. & Mittler, R., 2021. The impact of stress combination on reproductive processes in crops. *Plant Science*, 311, 111007.

Subedi, B., Poudel, A., & Aryal, S., 2023. The impact of climate change on insect pest biology and ecology: Implications for pest management strategies, crop production, and food security. *Journal of Agriculture and Food Research*, 14, 100733.

Tanveer, A., Arshad, M.S., Ayub, M., Javaid, M.M. & Yaseen, M., 2012. Effect of temperature, light, salinity, drought stress and seeding depth on germination of *Cucumis melo* var. agrestis. *Pakistan Journal of Weed Science Research*, 18.

Tanveer, A., Mumtaz, K., Javaid, M., Chaudhry, M., Balal, R. & Khaliq, A., 2013. Effect of ecological factors on germination of horse purslane (*Trianthema portulacastrum*). *Planta Daninha*, 31, 587–597.

Taub, G.E., Keith, T.Z., Floyd, R.G. & Mcgrew, K.S., 2008. Effects of general and broad cognitive abilities on mathematics achievement. *School Psychology Quarterly*, 23, 187.

Thomas, P.D., Campbell, M.J., Kejariwal, A., Mi, H., Karlak, B., Daverman, R., Diemer, K., Muruganujan, A. & Narechania, A., 2003. PANTHER: A library of protein families and subfamilies indexed by function. *Genome Research*, 13, 2129–2141.

Tripathy, B.C. & Pattanayak, G.K., 2010. Singlet Oxygen-Induced Oxidative Stress in Plants. *In* C.A. Rebeiz, C. Benning, H.J. Bohnert, H. Daniell, J.K. Hoober, H.K. Lichtenthaler, A.R. Portis & B.C. Tripathy (eds.) *The Chloroplast: Basics and Applications*. Springer, Dordrecht, Netherlands, 397–412.

Ullah, H., Mahmood, A., Ijaz, M., Tadesse, B. & Honermeier, B., 2013. Evaluation of anise (*Pimpinella anisum* L.) accessions with regard to morphological characteristics, fruit yield, oil contents and composition. *Journal of Medicinal Plants Research*, 7, 2177–2186.

Vereecken, H., Amelung, W., Bauke, S.L., Bogena, H., Brüggemann, N., Montzka, C., Vanderborght, J., Bechtold, M., Blöschl, G., Carminati, A., Javaux, M., Konings, A.G., Kusche, J., Neuweiler, I., Or, D., Steele-Dunne, S., Verhoef, A., Young, M. & Zhang, Y., 2022. Soil hydrology in the Earth system. *Nature Reviews Earth & Environment*, 3, 573–587.

Vignesh, A., Amal, T.C., Sivalingam, R., Selvakumar, S. & Vasanth, K., 2024. Unraveling the impact of nanopollution on plant metabolism and ecosystem dynamics. *Plant Physiology and Biochemistry*, 210, 108598.

Wahab, A., Muhammad, M., Munir, A., Abdi, G., Zaman, W., Ayaz, A., Khizar, C. & Reddy, S.P.P., 2023. Role of arbuscular mycorrhizal fungi in regulating growth, enhancing productivity, and potentially influencing ecosystems under abiotic and biotic stresses. *Plants*, 12, 3102.

Waheed, H., Javaid, M.M., Shahid, A., Ali, H.H., Nargis, J. & Mehmood, A., 2019. Impact of foliar-applied Hoagland's nutrient solution on growth and yield of mash bean (*Vigna mungo* L.) under different growth stages. *Journal of Plant Nutrition*, 42, 1133–1141.

Wang, Y., Mostafa, S., Zeng, W. & Jin, B., 2021a. Function and mechanism of jasmonic acid in plant responses to abiotic and biotic stresses. *International Journal of Molecular Sciences*, 22, 8568.

Wang, Y., Zhou, H., He, Y., Shen, X., Lin, S. & Huang, L., 2023. MYB transcription factors and their roles in the male reproductive development of flowering plants. *Plant Science*, 335, 111811.

Wang, Z., Ma, S., Sun, B., Wang, F., Huang, J., Wang, X. & Bao, Q., 2021b. Effects of thermal properties and behavior of wheat starch and gluten on their interaction: A review. *International Journal of Biological Macromolecules*, 177, 474–484.

Xie, J., 2015. *Evidence for Multiple Oxidation Pathways from Non-Volatile Products of Methyl Linoleate*: Rutgers, The State University of New Jersey, School of Graduate Studies.

Xie, X., He, Z., Chen, N., Tang, Z., Wang, Q. & Cai, Y., 2019a. The roles of environmental factors in regulation of oxidative stress in plant. *BioMed Research International*, 2019.

Xie, Z., Nolan, T.M., Jiang, H. & Yin, Y., 2019b. AP2/ERF transcription factor regulatory networks in hormone and abiotic stress responses in Arabidopsis. *Frontiers in Plant Science*, 10, 437723.

Zaib, M., Zeeshan, A., Akram, H., Amjad, W., Aslam, S. & Qasim, S., 2023. Impact of climate change on crop physiology and adaptation strategies: A review. *International Research Journal of Education and Technology*, 5, 15–38.

Zain, M., 2017. Foliar applied thiourea at different growth stages modulated late sown wheat. *Pakistan Journal of Science*, 69.

Zhang, J., Jia, W., Yang, J. & Ismail, A.M., 2006. Role of ABA in integrating plant responses to drought and salt stresses. *Field Crops Research*, 97, 111–119.

12 Modernizing Irrigation Management

*Humma Anwar Rana, Abrar Hussain**, *Rizwana Kanwal, Zahid Ali, and Basra Manzoor*

12.1 INTRODUCTION

Irrigation is the process of applying exact amounts of water to agricultural land to support the growth of lawns, crops, and landscape plants. Irrigation uses a variety of tube, pump, and spray technologies to artificially transport water to the land. In areas with irregular rainfall, dry spells, or an oncoming drought, irrigation is typically used (Rao *et al.*, 2022; Zafar *et al.*, 2023). Different crops require varied amounts, rates, timings, and quantities of irrigation, which also vary depending on the seasons and types of soil. As a case study, summer crops need more water than winter crops do (Sivarasan *et al.*, 2022). Agriculture suffers from insufficient and unpredictable rainfall. Low rainfall is the root cause of both famine and drought. Even with little rainfall, irrigation contributes to higher output. Irrigated land is more productive than non-irrigated land (Ashraf *et al.*, 2013). Irrigation contributes to stabilizing output and yield levels and increases farmers' income because it makes water readily available. Maximum irrigation is necessary although excessive irrigation can reduce crop yields (Ungureanu *et al.*, 2020; Rasheed *et al.*, 2022a). Too much water causes waterlogging, inhibits germination, raises the concentration of salt, and causes uprooting since roots cannot tolerate standing water. Therefore, for optimal cultivation, the appropriate procedure must be applied.

There are two approaches to irrigation: traditional methods and modern technologies. Irrigation is done manually where farmers take water from boreholes or channels and transport it to farmland, either with livestock or by themselves. Depending on the region, this approach may vary. The main advantage of this strategy is its low cost (Rasheed *et al.*, 2022d; Bhat *et al.*, 2023). However, the unequal distribution of water results in low efficiency. Additionally, there is a very significant probability of water loss. A few types of typical systems are the chain pump, lever system, and pulley system. The pump mechanism is the most frequently utilized and prevalent of them (Ali *et al.*, 2013; Ashraf, 2023).

The modern approach uses two systems: one is sprinkler setup, and the other is drip scheme. Sprinkler irrigation is an affordable method of assessing irrigation water that replicates natural rainfall. Water is typically pumped through a network of pipes for distribution. It then gets sprayed as mist into the air using sprinklers, breaking into small water droplets that fall to the soil (Ullah *et al.*, 2013; Zhang *et al.*, 2023). The majority of row, field, and tree crops can be irrigated with sprinklers, which allow water to be sprinkled either above or below the crop canopy. Enormous sprinklers are not advised, though, for watering sensitive plants like lettuce, since the enormous water droplets they produce could harm the plant (Tunio *et al.*, 2022). Sprinkler irrigation can be applied to both uniform and undulating farmable slopes. Every effort should be made to position the lateral pipes that deliver water to the nozzles along the contour of the land. By doing this, uniform irrigation can be achieved, as well as variations in pressure at the sprinklers can be reduced. Sprinklers can be used in most soil types but perform best on sandy soils with high infiltration rates (Eisenhauer *et al.*, 2021). To prevent surface runoff and capturing, set an average sprinkler application rate (in mm/hour) below the soil's fundamental infiltration rate. Sprinkler irrigation is not

* Corresponding author: abrar.hussain@cuisahiwal.edu.pk

 DOI: 10.1201/9781003498698-12

appropriate for crusting-prone soils. If sprinkler watering is the only alternative, rely on soft, fine sprays. Furthermore, it is better to avoid larger sprinklers that produce larger water droplets, which may exacerbate the problem (Tomas *et al.*, 2019).

Drip irrigation is the best way to deliver water and nutrients to crops. Thin tubes called "driplines," with small devices called "drippers" attached, are laid on top of the ground to carry water and fertilizers around a field. Each dripper releases drops laden with both fertilizer and water, providing a steady supply of nutrients and moisture directly to each plant in a field at its root zone (Manda *et al.*, 2021). Additionally, drip irrigation protects against disease by reducing the amount of water that comes into contact with plant leaves, stems, and fruit. Keeping the areas between plants dry increases accessibility, decreases weed development, and lowers effort. The system's efficiency results in significant water, money, and time savings. Furthermore, it performs well on uneven surfaces and reduces nutrient and water loss beneath the root zone, making irrigation more sustainable and effective (Tanveer *et al.*, 2013; Olamide *et al.*, 2022).

12.2 PRINCIPLES OF MODERN IRRIGATION MANAGEMENT

Looking for ways to reduce water consumption in irrigated agriculture, the greatest water user in the world has raised the issue due to factors such as climate change, limited water supplies, and continuous population development. Ecosystems and economies depend on water that is provided in the proper quality, quantity, timing, and place (Pérez-Blanco *et al.*, 2020; Hassan *et al.*, 2022). Water is needed for irrigation, which is a major component of global food production. The patterns of stream release, evaporation, and precipitation are adapted to natural ecosystems. Methods are exclusively considered to enhance the effective utilization of water by better management and modern irrigation technologies, hence reducing the amount of water used for crops while preserving or even improving total crop yield (different from the economic worth of water) (Bwambale *et al.*, 2022). The world's irrigated regions are facing a growing scarcity of water. In such cases, the latest irrigation strategy uses following two systems:

- Drip strategy (DS)
- Sprinkler system (SS)

12.2.1 Drip Irrigation System

Approximately two million hectares are currently irrigated using drip irrigation, which is the most efficient irrigation method currently in use (Figure 12.1). This represents less than 1% of all irrigated land worldwide (Chadha *et al.*, 2019; Ambomsa, 2020). A typical drip irrigation system's main requirements are the network of drip lines that irrigate each plant row, the hundreds of 25-cent plastic emitters that supply water at each drip point, and the filtering mechanism at the water receiving point. Nowadays, at least 4% of farmers utilize drip irrigation systems to irrigate their land (Shareef *et al.*, 2019). Drip irrigation systems also eliminate climate issues like salt and waterlogging that are connected to surface irrigation systems. Drip irrigation contributes to lower irrigation water consumption, more effective use of water, less wasteful tool usage, improved product quality, higher agricultural yields, and more efficient use of fertilizer (Kumar and Kumar, 2020; Rasheed *et al.*, 2022c). There are five classes into which the components of drip irrigation items can be divided.

12.2.2 The Primary Drip Irrigation Source

The most popular alternative is an irrigation system with drip emitters, which delivers water straight to the plant's deep roots. The ideal irrigation system for baskets, trees, bushes, etc. is one with drip emitters. Because of the low water release rate, the watering cycle typically lasts longer (Chirgwin and Sutton, 2019).

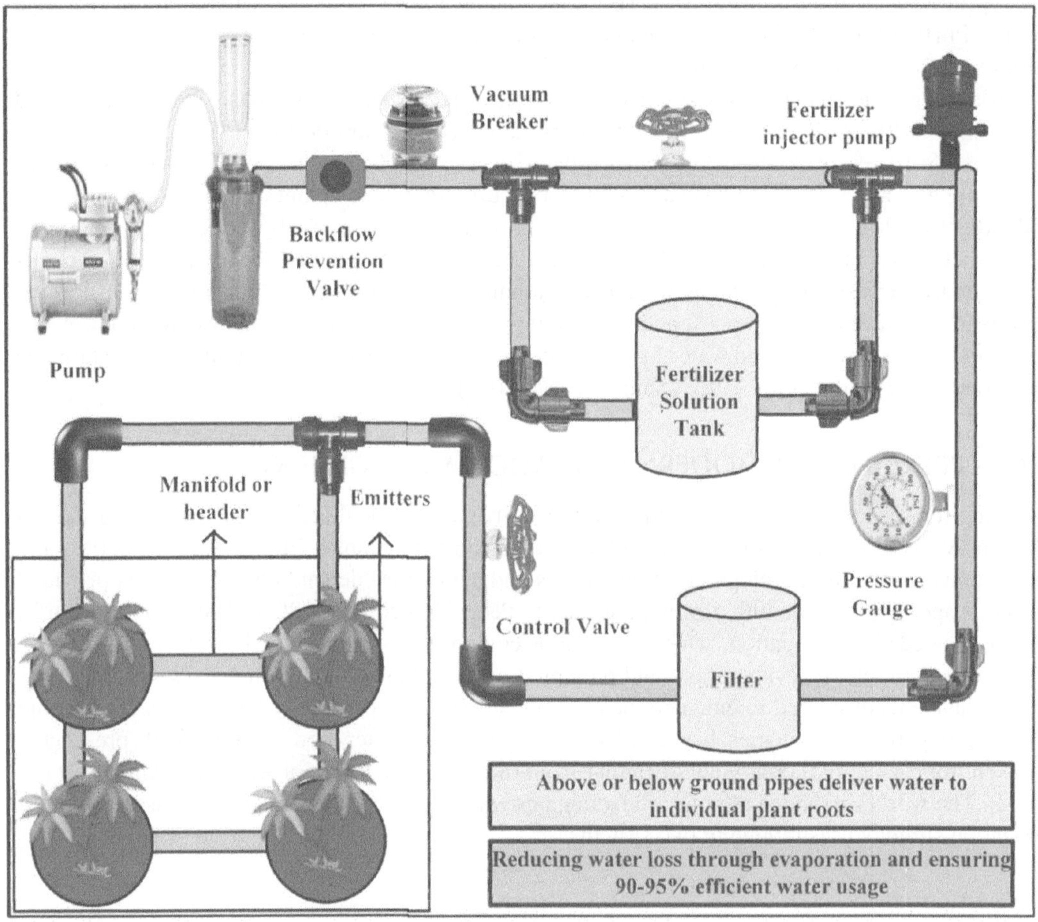

FIGURE 12.1 Drip irrigation system.

12.2.2.1 Bubbler

This drainage system is mostly used for shrubs and trees. Bubblers have slow watering periods and high discharge rates. Bubblier irrigation systems are especially useful for switching from traditional sprayers to low-volume systems because of their ability to operate at high pressures (Martínez de Azagra Paredes *et al.*, 2022).

12.2.2.2 Micro-Spray Irrigation System

A micro-spray irrigation system works well for covering ground and plants like ferns that require some humidity, as well as for low-volume drip irrigation applications (Kaur *et al.*, 2020).

12.2.2.3 Dripper Line Irrigation System

A drip irrigation system is the fast-growing and easiest to install method for watering landscapes, growing vegetables, shrubs, and so on. Other drip irrigation systems have rings for massive pots and garden boxes and loops hanging over big trees and containers (Shareef *et al.*, 2019).

12.2.2.4 The Mister Irrigation System

This kind of irrigation system works great for adding humidity, and the misters' irrigation system works best when it is upside down and sprays downward to water-hanging baskets. The quick evaporation of water in a mister can significantly lower the ambient temperature; that's why misters are being utilized more and more for outdoor cooling (Froiz-Míguez *et al.*, 2020).

12.2.2.5 Enter Drip Irrigation

A perfect irrigation system would water the plant as needed, at the right time, and irrigate the roots deeply rather than just the soil. Water is monitored at the source, which could be a tank or reservoir, and each time the valve is opened, the soil surrounding the plant acquires a tiny, steady, as well as equal amount of water. This kind of irrigation system is utilized via plastic "drip lines" that set both over the ground and below the soil root (Ayars *et al.*, 2023).

There are two primary kinds of drip irrigation systems.

12.2.2.5.1 Surface Irrigation System

For narrow emitter spacing (10"–16") and thin walls (9–12 mm) where the drip line is injected 1"–6" below the soil's surface, a surface drip irrigation system is helpful. Since the drip line for this kind of irrigation system is recycled and collected annually, it is frequently advised to refer to it as "temporary." Permanent or temporary sub-mains are both possible. Because these systems are labor-intensive, time-consuming, and require yearly reinstallation, they are typically utilized on high-value crops (Salahou *et al.*, 2023).

12.2.2.5.2 Sub-Surface Irrigation System

For a slightly thicker wall (12–16 mL), utilize the subsurface drip irrigation method (SDI) with emitter spacing of 22–28 inches. The drip line is injected 8–16 inches under the soil surface. These systems are critical for assuring longevity, since they are consistent in their design and installation. Although these are mostly used for row crops, subsurface irrigation systems are making their way into several high-value field crops. Subsurface drip irrigation systems have been Eco-Drip's focus since installation of the first system in 1980 (Waseem, 2020).

12.2.2.6 Sprinkler System

As the name suggests, it is a system of pipes or tubing that has nozzles or sprinkler heads attached to it so that water can be sprayed all over the ground.

These systems can be categorized as field, farm, or project systems depending on the region they cover (Wass and Pe, 2020).

12.2.2.7 Field Sprinkler Network

The field sprinkler network is intended to be used on several fields within one or more farm units. Pre-irrigation, fall or spring irrigation, seedling germination, or a specialized crop occurring in a crop rotation are the starting points for facilitating the conditions stated. For field SSs, power economy, distribution efficiency, and labor usage may be completely different from those of full farm systems (Chadha *et al.*, 2020; van Delden *et al.*, 2021).

12.2.2.8 Farm Sprinkler System

For a specific design region or farm unit where sprinkler irrigation will be the main technique of applying water, a whole farm SS has been planned specifically. Complete system planning considers the soils, crop sequences, and crops present in the designated design region. All sprinkling lines, main and sub-main lines, pumping plants as well as boosters, controllers for operation equipment, and other accessories needed for the water application necessary for effective irrigated farming are included in a farm SS (Patel and Prajapati, 2020).

12.2.2.9 Project Irrigation Systems

The U.S. Bureau of Reclamation, irrigation districts, and private citizens construct the water supply infrastructure needed to run individual field or farm SSs for irrigation projects that are now being created. The project systems service anything from 2 to 100 farms or more. A portion of these systems use closed pipe systems to transfer water drawn from a stream or well source to each farm turnout at a pressure sufficient to run farm SSs. Water is delivered from the source via other project systems into open canals, where a booster pump, a component of the agricultural SS, applies pressure to the water. Several channels from a running water source supply one or more agriculture systems (Aziz *et al.*, 2018; Gany *et al.*, 2019).

12.2.2.10 Components of the Sprinkler System

SSs consist of numerous parts. Nozzle heads, pressure regulators, riser pipes, couplers, tubing, and fittings are all part of the system, which starts at the nozzle's water spray and continues to the water source. The sprinkler lateral is the section of the irrigation system that is made up of these parts. A valve links the main pipeline, which is connected to the water supply source, to the sprinkler lateral (Bakhsh *et al.*, 2022) (Figure 12.2).

12.3 INTEGRATION OF TECHNOLOGY

Technology-enabled irrigation systems can maximize crop yields, save water, and operate more efficiently (Mpanga and Idowu, 2021). A brief description of the main technologies frequently utilized in present-day irrigation systems is provided below.

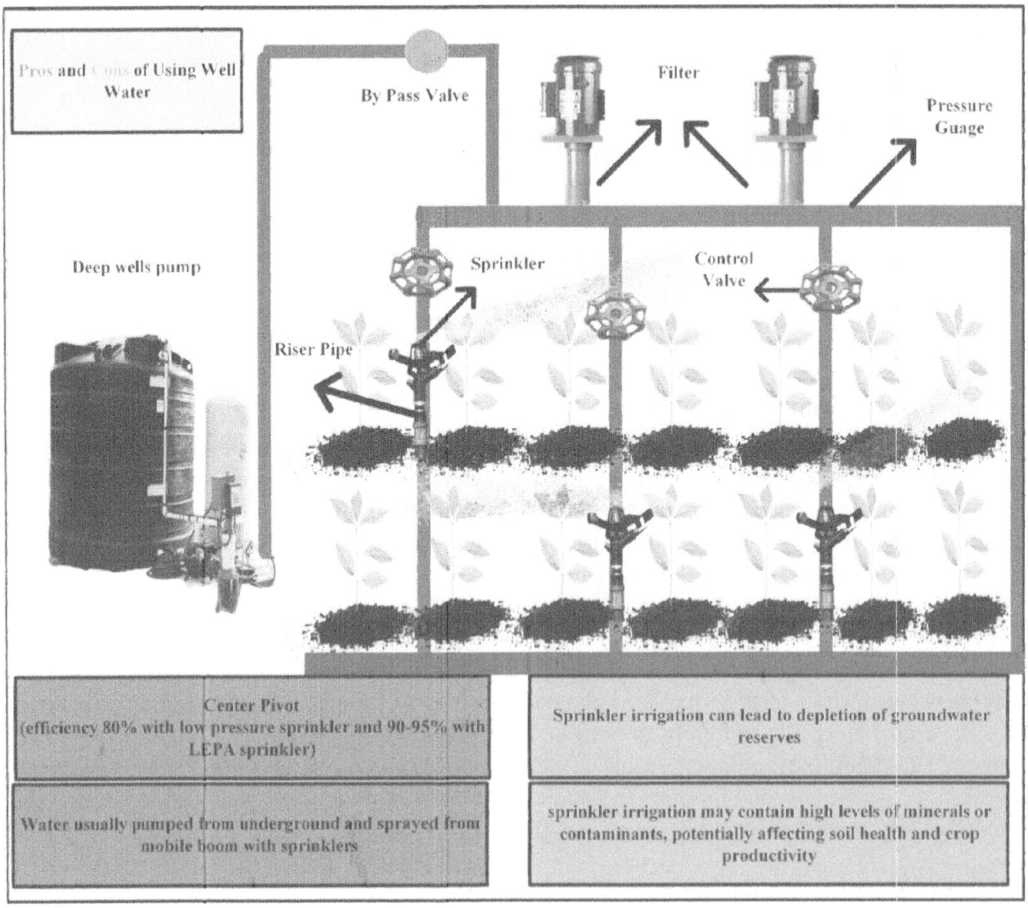

FIGURE 12.2 Sprinkler irrigation system.

12.3.1 Sensors for Measuring Soil Moisture

These sensors provide irrigation managers with real-time information regarding soil moisture levels. Farmers may save water and improve plant health by specifically measuring soil moisture levels and avoiding oversupply or underwatering their crops (Liao *et al.*, 2021).

12.3.2 Weather Monitoring and Forecasting

Through the installation of weather sensors and forecasting technologies into farmers' irrigation systems, farmers are able to adjust their watering schedules based on predicted environmental variables, including temperature, humidity, and rainfall (Sui and Vories, 2020).

12.3.3 Modern Irrigation Controller

Smart controllers automatically adjust the timing of irrigation and settings based on data collected by sensors, weather predictions, and soil conditions. However, these controllers help optimize water usage, save the cost of labor, and increase overall system efficiency (Bwambale *et al.*, 2022).

12.4 PLANNING AND DESIGNING MODERN IRRIGATION SYSTEMS

This requires a surface system for irrigation delivery that provides uniform and high-efficiency root zone reservoir filling to avoid crop stress and conserve resources such as energy, water, nutrients, and labor. It can also heat or cool the surrounding air to prevent frost damage to sensitive fruits and vegetable crops (Javaid *et al.*, 2012; Shakeel *et al.*, 2022). An irrigation system must always be capable of releasing salts that accumulate in the root zone. It can also be used to fertilize the land, apply pesticides, and prepare the soil for better cultivation. Design of surface irrigation system is a balancing act between water supply capacity and water supply availability and the appropriate amount and depth of irrigation (Lalehzari and Kerachian, 2021) (Table 12.1). There are two stages of this process.

12.4.1 Preliminary Design (Initial Design)

The system should be capable of operating with enough flexibility to give the crop water in different amounts and at different times, giving the farmer some flexibility in managing the moisture in the soil for maximum yields and saving labor, water, and energy. Water can be delivered continuously or sporadically, with comparatively constant flow rates and durations (Mehmood *et al.*, 2018; Obaideen *et al.*, 2022). Under such circumstances, irrigation flexibility in scheduling is confined to the level that each farmer or group of farmers can make collective decisions within their command zones. The water supply's limits in reaching an ideal watering schedule should be evaluated from the first stages of planning. To predict crop water demands, the next step in the design process is to collect and analyze data on local meteorology, soil, and cropping patterns. It is possible to determine how much groundwater the system should provide for the duration of the season using this approach (Pereira *et al.*, 2020). By evaluating the net crops demands with the water delivery system's capacity to supply water by a changeable schedule, a provisional timetable may be created. Compared to continuous or rotating water schedules, which are frequently challenging to fit crop requirements, on-demand systems ought to be more flexible. The farm should thoroughly consider the options before choosing a surface irrigation system (Davarpanah and Ahmadi, 2021; Jianing *et al.*, 2022). Row crops, modest farm flows and applications, and a high bi-directional slope are all favorable characteristics for furrow systems. Large field discharges, deeper application depths during most irrigations, and flatter terrain all favor border and basin systems. Where frequency and depth can be adjusted, quite a bit of leadership can be used (Stevens *et al.*, 2020; Naqve *et al.*, 2021).

TABLE 12.1
Techniques for Applying Water

Method Use	Protective Characteristics	Adjusted for	Efficiency (%)
Graded borders	Ensures even wetting and effective water usage. Uses powerful water streams safely, cutting down on the amount of time needed to cover an area.	Grain or hay can be placed on even slopes as much as 3%, established pasture can be placed on even slopes up to 6%. Ideal for mild soil types.	60–80
Basins and level border	Allows for effective water application management, beneficial for alkali control.	Crops grown closely together on level land with sandy soils.	60–80
Furrows	Offers no conservation characteristics unless water is applied very carefully, and furrows are placed on level ground.	Vegetable gardens, commercial crops, berries, orchards, and vineyards with all except coarse-textured soils are located on mild inclines.	60–80
Bends in corrugations	Ensures even wetting and stops erosive water buildup on terrain that is too steep or rolling for borders or basins and utilizes little streams.	Crops that grow closely together on sloped fields with slowly absorbing soil. Applying water to slopes more than 2% requires great care.	40–55
Wild flood	Gives uneven and sloping terrain little or no water control.	Without the use of boundaries, furrows, or other structures, water is permitted to flow across the ground.	25–40
Controlled flooding	In situations where soil can't be used for other purposes, it offers water control and rather consistent wetness.	Crops grown closely together on undulating terrain; pasture sod created with sprinklers or corrugations.	65–80

12.4.2 Thorough Design (In-Depth Layout)

The field's slope, the furrow, the border or basin's release and period, the location and size of headland structures and other facilities, and the availability of surface drainage facilities either for gathering tailwater for reuse or disposal are all determined during the detailed design process. Computations for land leveling are the first and last steps in the detailed design process. To ascertain the overall land slope by the direction of projected water flow, the actual topography is first assessed (Remund *et al.*, 2021; Rasheed *et al.*, 2022b). This does not have to be the in-depth analysis that's required to shift the earth. The detailed design process then proceeds to the decision on rates of flow and time frames that optimize application efficiency, tempered by a continuous review of the practical issues involved in farming the field later. This information is combined with an emphasis on the implementation of a distance based on an evaluation of crop water requirements. The next stage of thorough planning is to balance the water supply's flows and times against the field's allotted total flow and duration. If the entire supply is utilized to irrigate the entire field at once, it may be sufficient for small fields (Javaid *et al.*, 2022; Jain and Singh, 2023). Fields must be divided into sets and irrigated segment by segment, border by border, basin by basin, etc. The land and its supply of water must be compatible with these subdivisions, or "sets." Thus, the last stage of land leveling is started after the subdivisions are constructed. The field size and flow attributes must be determined before the top-layer irrigation system can be architecturally designed. Tubes or canals containing relevant control components must be the appropriate size for the field where the water is to be applied. If tailwater is allowed, there needs to be a way to remove these flows. The system's operation should also be considered by the designer (Brahmanand and Singh, 2022). To create the optimal strategy, the planner and the irrigator must work together. Understanding the needs for management, procedure, and maintenance must be a part of the conversation about the ideal system type (Rodorff *et al.*, 2019; Tang *et al.*, 2022).

12.4.3 Water Sources

Knowing the volume, quality, availability, flow rate, and source, whether above or below the ground, is essential to selecting the appropriate kind of system. Regardless of the planned system type, electricity is necessary for many groundwater sources. An underground water source could merely need an inline screen when employing micro irrigation, while a surface water supply would need a sophisticated filtration system (Condon *et al.*, 2021). Because of their high salinity (EC), certain sources may not be suitable for sprinkler irrigation. A micro-irrigation system works best with a steady supply of water, while a surface system may operate with longer intervals between water applications. On the other hand, surface systems need a sufficient flow for the most effective application, while spray or micro-irrigation techniques may function well with lesser quantities of application (Chaudhuri *et al.*, 2021).

12.4.4 Soil Properties

Design of an irrigation system should consider different soil properties. Permeability, transmissivity, and water-holding capacity (WHC) can be accurately predicted via the texture of the soil. This is an important consideration when considering a surface system due to the time between irrigations. Permeability is critical to the design of surface systems, and less so sprinklers. Another important water additional characteristic, especially in concerns of certain types of irrigation i.e. drip emitters, is the ability of the water to transmit through the soil, known as transmissivity. Water must be able to enter and move within the root zone (Baradaran and Tavazoei, 2022). The quantity and quality of water may limit crop selection. Depending on the type of crop cultivated, excessive salinity (EC) may reduce yield or even lead to crop failure (Zaman *et al.*, 2018).

12.4.5 Growing Season

Growing season can affect the choice of crops, and the information may be used to calculate the profitability of any system under consideration (Adhikari *et al.*, 2018).

12.4.6 Location

Various system structures can respond to future changes in temperature, humidity, precipitation, and wind. This also suggests that appropriate safety measures be used in close proximity between humans, cattle, and wildlife (Nayak *et al.*, 2020). Techniques for applying water are as follows.

12.5 BUILDING FARMER CAPABILITIES AND TRAINING

Farmers should have knowledge and skills about modern irrigation methods and irrigation technology to maximize water utilization, improve crop yields, and increase agricultural output. First of all, discuss the modern irrigation methods such as sprinklers, drip, and exactness irrigation. Explain the fundamental principles of these strategies and the advantages they possess over classical approaches (Karki *et al.*, 2020). Describe the ways in which these technologies can improve crop production and maximize irrigation efficiency. Facilitate hands-on training workshops so that farmers can receive practical information on establishing, operating, and maintaining modern irrigation systems (Neupane and Guo, 2019). Give examples of how to test irrigation equipment, modify sprinkler heads, and set up drip lines. Support farmers in realizing the financial advantages that come from utilizing current irrigation methods. Advise how to estimate the possible improvement in agricultural yields and earnings as well as the costs associated with putting these approaches into practice. Once the educational sessions are over, provide farmers with continuing guidance. Make instruments like telephone lines, internet discussion boards, and expert advice available. Observe how farmers who accepted current irrigation methods are doing and offer more support as required (Allani *et al.*, 2022).

12.6 BENEFITS TO THE ENVIRONMENT AND ECONOMY

Improved yields and higher-quality products result from effective irrigation management, ensuring that the cultivation receives sufficient water at the appropriate time. Food security and farmer income both may benefit from this. Compared to conventional techniques, these irrigation systems frequently use less energy. Farmers can save money on operating expenses and energy consumption by switching to drip irrigation instead of flood irrigation, which uses less energy and pressure (Raza *et al.*, 2022). The use of precision irrigation reduces overflow into water bodies, reduces water pollution, and protects aquatic ecosystems by enabling better application of pesticides as well as fertilizers. Overall, soil fertility and structure can be preserved by effective irrigation management, which can also stop soil erosion and salinization. Higher crop productivity is supported by healthy soils, which also minimize the need for expensive soil remediation methods (Shah and Wu, 2019). The initial financial investment needed to buy equipment and install new systems is one of the biggest barriers to using current irrigation techniques. The government may focus on building affordable technology, irrigation systems appropriate to rural farmers. There is also a need to work on building partnerships among banks, governments, and development organizations to provide affordable financing mechanisms for projects targeting modernization of irrigation (Ara *et al.*, 2021).

12.7 CONCLUSION

Modern irrigation technologies and agricultural practices help farmers to increase efficiency to meet the food needs around the world as compared to traditional old irrigation technologies. Modern irrigation technology reduces crop and soil erosion and water loss which helps farmers increase production. This chapter addresses practices for modernizing irrigation management methods, including combining high technology with sustainable methods aimed at maximizing the usability of resources, increasing the efficiency of water application, and minimizing environmental effects. Automation, data analytics, and smart sensors can refine irrigation systems to meet crop needs more closely, helping to minimize water waste and maximize productivity. In addition, the creation of systems such as drip irrigation, precision farming, and reuse of water contributes to reduce the use of resources and the establishment of resistance in the face of climate change. Under the circumstances, the objectives of improving food security, promoting sustainable agricultural practices, and ensuring the availability of water resources for future generations can be met only through improved irrigation management. Emerging technologies like remote sensing and satellite imaging enable farmers to continuously survey vast areas of land, producing data that informs targeted irrigation according to specific moisture levels and crop needs. Drones and aerial mapping also helps that, as in locating specific areas of need of irrigation, allowing to lower water use by only saturating the areas that require it. AI and machine learning algorithms analyze historical weather data, soil moisture, and crop type to forecast how much water is needed, helping keep irrigation in sync with crop growth cycles.

LIST OF ABBREVIATIONS

IoT Internet of Things
SS sprinkler system

REFERENCES

Adhikari, P., Araya, H., Aruna, G., Balamatti, A., Banerjee, S., Baskaran, P., Barah, B., Behera, D., Berhe, T. & Boruah, P., 2018. System of crop intensification for more productive, resource-conserving, climate-resilient, and sustainable agriculture: Experience with diverse crops in varying agroecologies. *International Journal of Agricultural Sustainability*, 16, 1–28.

Ali, H., Tanveer, A., Nadeem, M., Javaid, M., Kashif, M. & Chadhar, A., 2013. Allelopathic effects of *Rhynchosia capitata* on germination and seedling growth of mungbean. *Planta Daninha*, 31, 501–509.

Allani, M., Frija, A., Nemer, R., Ribbe, L. & Sahli, A., 2022. Farmers' perceptions on an irrigation advisory service: Evidence from Tunisia. *Water*, 14, 3638.

Ambomsa, A., 2020. Role of drip irrigation system as increasing water use efficiency over furrow irrigation system. *Academic Research Journal of Agricultural Science and Research*, 8, 252–262.

Ara, I., Turner, L., Harrison, M.T., Monjardino, M., Devoil, P. & Rodriguez, D., 2021. Application, adoption and opportunities for improving decision support systems in irrigated agriculture: A review. *Agricultural Water Management*, 257, 107161.

Ashraf, E., Shah, F., Luqman, M., Younis, M., Aziz, I. & Farooq, U., 2013. Use of untreated wastewater for vegetable farming: A threat to food safety. *International Journal of Applied Agricultural Sciences*, 5.

Ashraf, U., 2023. *Investigation of Design Factors of Drive Systems for Hydraulic Pumps*. Thesis. https://hdl.handle.net/2437/365405.

Ayars, J.E., Zaccaria, D. & Bali, K.M., 2023. *Microirrigation for Crop Production: Design, Operation, and Management*. Elsevier.

Aziz, M., Ashraf, M. & Javaid, M.M., 2018. Enhancement in cotton growth and yield using novel growth promoting substances under water limited conditions. *Pakistan Journal of Botany*, 50, 1691–1701.

Bakhsh, A., Choudhry, M.R., Khan, I.A. & Farooq, M., 2022. *Applied Irrigation Engineering: Endowment Fund Secretariat*. University of Agriculture.

Baradaran, A.A. & Tavazoei, M.S., 2022. Fuzzy system design for automatic irrigation of agricultural fields. *Expert Systems with Applications*, 210, 118602.

Bhat, S., Kulkarni, S., Deshmukh, R., Bhopal, S., Zwarteveen, M. & Sumbre, S., 2023. Conjunctive use of canal water and groundwater: An analysis based on farmers' practices in Ravangaon, Maharashtra. *Water Alternatives*, 16, 65–86.

Brahmanand, P. & Singh, A.K., 2022. Precision irrigation water management-current status, scope and challenges. *Indian Journal of Fertilisers*, 18, 372–380.

Bwambale, E., Abagale, F.K. & Anornu, G.K., 2022. Smart irrigation monitoring and control strategies for improving water use efficiency in precision agriculture: A review. *Agricultural Water Management*, 260, 107324.

Chadha, A., Florentine, S., Chauhan, B.S., Long, B., Jayasundera, M., Javaid, M.M. & Turville, C., 2019. Environmental factors affecting the germination and seedling emergence of two populations of an emerging agricultural weed: Wild lettuce (*Lactuca serriola*). *Crop and Pasture Science*, 70, 709–717.

Chadha, A., Florentine, S., Javaid, M., Welgama, A. & Turville, C., 2020. Influence of elements of climate change on the growth and fecundity of *Datura stramonium*. *Environmental Science and Pollution Research*, 27, 35859–35869.

Chaudhuri, S., Parakh, D., Roy, M. & Kaur, H., 2021. Groundwater-sourced irrigation and agro-power subsidies: Boon or bane for small/marginal farmers in India? *Groundwater for Sustainable Development*, 15, 100690.

Chirgwin, G.A. & Sutton, B., 2019. A low-cost, high-precision drip emitter suitable for low-pressure microirrigation systems. *Irrigation Science*, 37, 725–735.

Condon, L.E., Kollet, S., Bierkens, M.F., Fogg, G.E., Maxwell, R.M., Hill, M.C., Fransen, H.J.H., Verhoef, A., Van Loon, A.F. & Sulis, M., 2021. Global groundwater modeling and monitoring: Opportunities and challenges. *Water Resources Research*, 57, e2020WR029500.

Davarpanah, R. & Ahmadi, S.H., 2021. Modeling the effects of irrigation management scenarios on winter wheat yield and water use indicators in response to climate variations and water delivery systems. *Journal of Hydrology*, 598, 126269.

Eisenhauer, D.E., Martin, D.L., Heeren, D.M. & Hoffman, G.J., 2021. *Irrigation Systems Management*. American Society of Agricultural and Biological Engineers (ASABE).

Froiz-Míguez, I., Lopez-Iturri, P., Fraga-Lamas, P., Celaya-Echarri, M., Blanco-Novoa, Ó, Azpilicueta, L., Falcone, F. & Fernández-Caramés, T.M., 2020. Design, implementation, and empirical validation of an IoT smart irrigation system for fog computing applications based on Lora and Lorawan sensor nodes. *Sensors*, 20, 6865.

Gany, A.H.A., Sharma, P. & Singh, S., 2019. Global review of institutional reforms in the irrigation sector for sustainable agricultural water management, including water users' associations. *Irrigation and Drainage*, 68, 84–97.

Hassan, M.U., Nawaz, M., Mahmood, A., Shah, A.A., Shah, A.N., Muhammad, F., Batool, M., Rasheed, A., Jaremko, M. & Abdelsalam, N.R., 2022. The role of zinc to mitigate heavy metals toxicity in crops. *Frontiers in Environmental Science*, 10, 990223.

Jain, S.K. & Singh, V.P., 2023. *Water Resources Systems Planning and Management*. Elsevier.

Javaid, M., Tanveer, A., Ahmad, R., Yaseen, M. & Khaliq, A., 2012. Optimizing activity of herbicides at reduced rate on *Emex spinosa* Campd. with adjuvants. *Planta Daninha*, 30, 425–435.

Javaid, M.M., Florentine, S.K., Ashraf, M., Mahmood, A., Sattar, A., Wasaya, A. & Li, F.M., 2022. Photosynthetic activity and water use efficiency of *Salvia verbenaca* L. under elevated CO_2 and water-deficit conditions. *Journal of Agronomy and Crop Science*, 208, 536–551.

Jianing, G., Yuhong, G., Yijun, G., Rasheed, A., Qian, Z., Zhiming, X., Mahmood, A., Shuheng, Z., Zhuo, Z. & Zhuo, Z., 2022. Improvement of heat stress tolerance in soybean (*Glycine max* L), by using conventional and molecular tools. *Frontiers in Plant Science*, 13, 993189.

Karki, S., Burton, P. & Mackey, B., 2020. Climate change adaptation by subsistence and smallholder farmers: Insights from three agro-ecological regions of Nepal. *Cogent Social Sciences*, 6, 1720555.

Kaur, P., Kaur, K. & Singh, H., 2020. Role of micro-irrigation in vegetable crops. *International Journal of Agriculture Innovations and Research*, 9, 2319–1473.

Kumar, R. & Kumar, M., 2020. Effect of drip irrigated mulch on soil properties and water use efficiency – A review. *Journal of Soil and Water Conservation*, 19, 300–309.

Lalehzari, R. & Kerachian, R., 2021. An integrated framework for optimal irrigation planning under uncertainty: Application of soil, water, atmosphere and plant modeling. *Iranian Journal of Science and Technology, Transactions of Civil Engineering*, 45, 429–442.

Liao, R., Zhang, S., Zhang, X., Wang, M., Wu, H. & Zhangzhong, L., 2021. Development of smart irrigation systems based on real-time soil moisture data in a greenhouse: Proof of concept. *Agricultural Water Management*, 245, 106632.

Manda, R.R., Addanki, V.A. & Srivastava, S., 2021. Role of drip irrigation in plant health management, its importance and maintenance. *Plant Archives*, 21, 1294–1302.

Martínez de Azagra Paredes, A., Del Río San José, J., Reque Kilchenmann, J., Diez Hernández, J.M. & Sanz Ronda, F.J., 2022. Methods for watering seedlings in arid zones. *Forests*, 13, 351.

Mehmood, A., Naeem, M., Khalid, F., Saeed, Y., Abbas, T., Jabran, K., Sarwar, M.A., Tanveer, A. & Javaid, M.M., 2018. Identification of phytotoxins in different plant parts of *Brassica napus* and their influence on mung bean. *Environmental Science and Pollution Research*, 25, 18071–18080.

Mpanga, I.K. & Idowu, O.J., 2021. A decade of irrigation water use trends in southwestern USA: The role of irrigation technology, best management practices, and outreach education programs. *Agricultural Water Management*, 243, 106438.

Naqve, M., Wang, X., Shahbaz, M., Fiaz, S., Naqvi, W., Naseer, M., Mahmood, A. & Ali, H., 2021. Foliar spray of alpha-tocopherol modulates antioxidant potential of okra fruit under salt stress. *Plants*, 10, 1382.

Nayak, P., Kavitha, K. & Mallikarjuna Rao, C., 2020. IoT-enabled agricultural system applications, challenges and security issues. In *IoT and Analytics for Agriculture*. Springer, 139–163.

Neupane, J. & Guo, W., 2019. Agronomic basis and strategies for precision water management: A review. *Agronomy*, 9, 87.

Obaideen, K., Yousef, B.A., Almallahi, M.N., Tan, Y.C., Mahmoud, M., Jaber, H. & Ramadan, M., 2022. An overview of smart irrigation systems using IoT. *Energy Nexus*, 7, 100124.

Olamide, F.O., Olalekan, B.A., Tobi, S.U., Adeyemi, M.A., Julius, J.O. & Oluwaseyi, F.K., 2022. Fundamentals of Irrigation Methods and Their Impact on Crop Production. *In Irrigation and Drainage-Recent Advances*. IntechOpen.

Patel, N.H. & Prajapati, C.R., 2020. Agricultural sprinkler for irrigation system. *International Journal of Engineering and Technical Research*, 9, 162–166.

Pereira, L., Paredes, P. & Jovanovic, N., 2020. Soil water balance models for determining crop water and irrigation requirements and irrigation scheduling focusing on the FAO56 method and the dual K_c approach. *Agricultural Water Management*, 241, 106357.

Pérez-Blanco, C.D., Hrast-Essenfelder, A. & Perry, C., 2020. Irrigation technology and water conservation: A review of the theory and evidence. *Review of Environmental Economics and Policy*, 14, 216–239.

Rao, B.K., Annapurna, S., Rani, B.R., Rao, Z.S., Sunitha, K., Schindutt, M., Jamanal, S. & Ramesh, V., 2022. *Soil and Water Conservation Techniques in Rainfed Areas* [E-book]. Hyderabad: National Institute of Agricultural Extension Management (MANAGE) & Water and Land Management Training and Research Institute (WALAMTARI), Hyderabad.(manage.gov.in).

Rasheed, A., Barqawi, A.A., Mahmood, A., Nawaz, M., Shah, A.N., Bay, D.H., Alahdal, M.A., Hassan, M.U. & Qari, S.H., 2022a. CRISPR/Cas9 is a powerful tool for precise genome editing of legume crops: A review. *Molecular Biology Reports*, 49, 5595–5609.

Rasheed, A., Li, H., Nawaz, M., Mahmood, A., Hassan, M.U., Shah, A.N., Hussain, F., Azmat, S., Gillani, S.F.A. & Majeed, Y., 2022b. Molecular tools, potential frontiers for enhancing salinity tolerance in rice: A critical review and future prospective. *Frontiers in Plant Science*, 13, 966749.

Rasheed, A., Mahmood, A., Maqbool, R., Albaqami, M., Sher, A., Sattar, A., Bakhsh, G., Nawaz, M., Hassan, M.U. & Al-Yahyai, R., 2022c. Key insights to develop drought-resilient soybean: A review. *Journal of King Saud University-Science*, 34, 102089.

Rasheed, A., Raza, A., Jie, H., Mahmood, A., Ma, Y., Zhao, L., Xing, H., Li, L., Hassan, M.U. & Qari, S.H., 2022d. Molecular tools and their applications in developing salt-tolerant soybean (*Glycine max* L.) cultivars. *Bioengineering*, 9, 495.

Raza, F., Tamoor, M., Miran, S., Arif, W., Kiren, T., Amjad, W., Hussain, M.I. & Lee, G.-H., 2022. The socio-economic impact of using photovoltaic (PV) energy for high-efficiency irrigation systems: A case study. *Energies*, 15, 1198.

Remund, D., Liebisch, F., Liniger, H., Heinimann, A. & Prasuhn, V., 2021. The origin of sediment and particulate phosphorus inputs into water bodies in the Swiss Midlands–A twenty-year field study of soil erosion. *Catena*, 203, 105290.

Rodorff, V., Siegmund-Schultze, M., Guschal, M., Hölzl, S. & Köppel, J., 2019. Good governance: A framework for implementing sustainable land management, applied to an agricultural case in Northeast-Brazil. *Sustainability*, 11, 4303.

Salahou, M.K., Zhi, W., Chen, X., Zhang, Y., Lü, H. & Jiao, X., 2023. Improvement of the estimation of the infiltration function in surface irrigation systems. *PLoS One*, 18, e0291578.

Shah, F. & Wu, W., 2019. Soil and crop management strategies to ensure higher crop productivity within sustainable environments. *Sustainability*, 11, 1485.

Shakeel, Q., Shaheen, M.R., Ali, S., Ahmad, A., Raheel, M. & Bajwa, R.T., 2022. Postharvest Management of Fruits and Vegetables. *In Applications of Biosurfactant in Agriculture*. Elsevier, 1–16.

Shareef, T.M.E., Ma, Z. & Zhao, B., 2019. Essentials of drip irrigation system for saving water and nutrients to plant roots: As a guide for growers. *Journal of Water Resource and Protection*, 11, 1129.

Sivarasan, S., Vijayalakshmi, R., Kumar, V. & Surendran, U., 2022. Development of deficit irrigation program for improving the water use efficiency, growth, and yield of groundnut in the semiarid tropical region of India. *Journal of Soil and Water Conservation*, 77, 30–44.

Stevens, J., Sanewe, A., Steyn, J., Annandale, J. & Stirzaker, R., 2020. *Improving On-Farm Irrigation Water and Solute Management Using Simple Tools and Adaptive Learning*. Report No. TT821/20 Water Research Commission: Pretoria, South Africa (www.wrc.org.za).

Sui, R. & Vories, E.D., 2020. Comparison of sensor-based and weather-based irrigation scheduling. *Applied Engineering in Agriculture*, 36, 375–386.

Tang, H., Meng, G., Xiang, J., Mahmood, A., Xiang, G., Sanaullah, Liu, Y. & Huang, G., 2022. Toxic effects of antimony in plants: Reasons and remediation possibilities—A review and future prospects. *Frontiers in Plant Science*, 13, 1011945.

Tanveer, A., Mumtaz, K., Javaid, M., Chaudhry, M., Balal, R. & Khaliq, A., 2013. Effect of ecological factors on germination of horse purslane (*Trianthema portulacastrum*). *Planta Daninha*, 31, 587–597.

Tomas, S., Molle, B., Chevarin, C. & Serra-Wittling, C., 2019. Transport modeling in sprinkler irrigation. *Journal of Irrigation and Drainage Engineering*, 145, 04019014.

Tunio, M.H., Gao, J., Qureshi, W.A., Sheikh, S.A., Chen, J., Chandio, F.A., Lakhiar, I.A. & Solangi, K.A., 2022. Effects of droplet size and spray interval on root-to-shoot ratio, photosynthesis efficiency, and nutritional quality of aeroponically grown butterhead lettuce. *International Journal of Agricultural and Biological Engineering*, 15, 79–88.

Ullah, H., Mahmood, A., Ijaz, M., Tadesse, B. & Honermeier, B., 2013. Evaluation of anise (*Pimpinella anisum* L.) accessions with regard to morphological characteristics, fruit yield, oil contents and composition. *Journal of Medicinal Plants Research*, 7, 2177–2186.

Ungureanu, N., Vlăduț, V. & Voicu, G., 2020. Water scarcity and wastewater reuse in crop irrigation. *Sustainability*, 12, 9055.

van Delden, S., Sharathkumar, M., Butturini, M., Graamans, L., Heuvelink, E., Kacira, M., Kaiser, E., Klamer, R., Klerkx, L. & Kootstra, G., 2021. Current status and future challenges in implementing and upscaling vertical farming systems. *Nature Food*, 2, 944–956.

Waseem, M., 2020. *Improving Design Practices for Drip Irrigation Systems Installed in the Punjab Province, Pakistan.* University of Southern Queensland.

Wass, H.S., Jr & Pe, R.P.F., 2020. *Sprinkler Hydraulics: A Guide to Fire System Hydraulic Calculations.* Springer Nature.

Zafar, S., Afzal, H., Ijaz, A., Mahmood, A., Ayub, A., Nayab, A., Hussain, S., Maqsood, U.-H., Sabir, M.A. & Zulfiqar, U., 2023. Cotton and drought stress: An updated overview for improving stress tolerance. *South African Journal of Botany*, 161, 258–268.

Zaman, M., Shahid, S.A., Heng, L., Zaman, M., Shahid, S.A. & Heng, L., 2018. *Irrigation Water Quality.* Guideline for salinity assessment, mitigation and adaptation using nuclear and related techniques. Springer Nature, 113–131.

Zhang, R., Liu, Y., Zhu, D., Wu, P., Zheng, C., Zhang, X., Khudayberdi, N. & Liu, C., 2023. Simulating droplet distribution characteristics for sprinkler irrigation with a modified ballistic model. *Computers and Electronics in Agriculture*, 215, 108437.

13 Integrated Approaches for Sustainable Forage Production Systems

Umair Gull, Sher Afghan, Asif Iqbal, Muhammad Abdullah Saleem, Athar Mahmood, Hakki Akdeniz, and Ayesha Arshad*

13.1 INTRODUCTION

Fodder has a significant role to play in sustaining the livestock sector, and by extension humanity itself (Lukyanova *et al.*, 2021). It contributes to meet the increasing hunger for meat, dairy, and other agricultural products globally, regardless of being ignored in favor of major cereal crops (Makkar, 2018). It creates a relationship between livestock and sustainability in agriculture. Production of high-quality fodder is necessary at a larger scale. The fodder production should not be limited due to the environment, whether it's the rich pastures of temperate regions or well-managed forage crops in arid lands (Wright Morton and Miller, 2007). It completes the gaps between the foods that are served on our tables and the forage that keeps the livestock sector sustainable. However, the path toward sustainability and food security is not possible without challenges. Like other agricultural activities, fodder production faces several challenges, which are made worse in a time of climate change (Iñiguez, 2011). There are multiple ways in which the fundamental structure of livestock sector is at risk.

Modern fodder production systems get affected by multiple issues like weed infestations, disease and pest breakouts, land limitations, degraded soil, and a lack of water (Javaid *et al.*, 2012; Ullah *et al.*, 2013; Duressa *et al.*, 2014). Moreover, climate change is affecting every component of the agriculture sector. The fodder production is affected by changes in temperature, precipitation patterns, and extreme weather due to changes in the climate (Izaurralde *et al.*, 2011).

Traditional fodder production methods are affected due to climate change. It adds a degree of risk that farmers don't find easy to manage. In order to meet food demands, it is necessary for agriculture sector to adapt to climate changes (Jones *et al.*, 2017). Here is a need to further develop fodder production systems to address the impacts of climate change and increase the supply of animal feed (Cooper *et al.*, 2008). In order to combat the climate change, it is going to ensure the resilient and sustainable future.

13.2 IMPORTANCE OF FODDER PRODUCTION SYSTEMS

Fodder is a major component of the agricultural system that underpins livestock productivity and dairy development. It's a big source of protein, vitamins, and fiber. These nutrients play vital roles in the health, growth, and reproduction of livestock (Pond *et al.*, 2004; Waheed *et al.*, 2019). For example, the livestock sector is highly influenced by the quality as well as quantity of feed (Tyasi *et al.*, 2015; Mahmood *et al.*, 2021). Dairy cows should consume high-quality fodder along with all the vital nutrients in milk products like butter, cheese, and yogurt (Vandehaar and St-Pierre, 2006). Large-scale livestock production requires access to nutrient-rich feed. Fodder not only provides

* Corresponding author: umairgull@uaf.edu.pk

DOI: 10.1201/9781003498698-13

vital nutrients but also helps to convert feed into valuable products such as milk and meat (Ahmad *et al.*, 2010; Tyasi *et al.*, 2015). On the other hand, the population continues to increase and with it the demand for animal-based protein, and meeting this demand involves fodder; fodder plays an essential role in this scenario.

13.2.1 The Role of Fodder in Ensuring Food Security and Sustainability

Fodders create a sustainable cycle in our food systems by bridging the gap between crop and animal agriculture. There are numerous interwoven benefits of fodder systems that help ensure a resilient and diverse food supply through the conversion of plant material into high-quality animal products in an efficient manner (Sanginga *et al.*, 2003; Ashraf *et al.*, 2016). Fodder farming saves land, which is essential in a world where arable land is becoming more and more scarce. We can produce more food on the same size of land by utilizing the resources to their maximum potential (Berndes *et al.*, 2003). One of the sustainable elements of holistic farming systems is pasture, which is generated when animals graze over forages (Rodríguez-Ortega *et al.*, 2014). It is also important for carbon sequestration and enhancing microbial activity in the soil (Sarkar *et al.*, 2020).

Climate change is resulting complications in fodder production. Global warming (GS) and extreme climatic catastrophes (EC) pose huge threats to fodder availability (Turner *et al.*, 2011).

The less availability of water in drought and extreme heat waves causes to inhibit the growth and development of plants, which in turn leads to less fodder production (Parker *et al.*, 2020). As a result, the quality of the forage is decreased, making it less palatable to livestock. Long-term drought may result in fodder shortage (Speranza *et al.*, 2008; Ashraf *et al.*, 2013). In contrast, extreme rains and floods can result in low quality/low quantity of fodder crops (Maity *et al.*, 2016). Pest and disease outbreaks, resulting from climate changes, also affect fodder quality and availability (Sundström *et al.*, 2014). To ensure food security and viability of the livestock, it is important to address the unpredictable aspects of climate change in the fodder production systems.

13.3 TRADITIONAL FODDER PRODUCTION METHODS AND CHALLENGES

Traditional fodder production methods have supported livestock for centuries. Generally, Forages and pasturelands are grown and managed for animal feed. Here is an overview of some common traditional techniques.

13.3.1 Traditional Foraging Techniques

13.3.1.1 Grazing

Grazing livestock on natural pastures is one of the oldest and most basic methods of producing fodder. Animals are free to roam and eat grasses, legumes, and other natural forages.

13.3.1.2 Cut-and-Carry

Forage crops are grown, harvested, and transported to the animals for feeding in this method (Mengistu, 1997). Crops such as grasses, legumes, and cereal straws are involved.

13.3.1.3 Silage and Hay Making

Grass or legume crops are harvested at their peak maturity and either dried to make hay or ensiled to make silage. During seasons when fresh forage is scarce, these preserved forages provide feed (Ohshima *et al.*, 1997; Rasheed *et al.*, 2022c).

13.3.2 Limitations and Challenges Associated with These Methods

Traditional fodder production methods, while historically effective, have several limitations and challenges.

13.3.2.1 Seasonal Variability

Natural pastures and rain-fed crops are subject to seasonal fluctuations in quantity and quality. During the dry season, fodder may be scarce or of poor quality.

13.3.2.2 Crop Residues

After harvesting a primary crop, such as maize, the leftover plant material, or crop residues, can be used as livestock fodder (Devi *et al.*, 2017).

13.3.2.3 Land and Resource Constraints

Traditional methods tend to require large land areas and significant water resources. This may be unsustainable in areas with land degradation and scarcity of water (Desta, 2000).

13.3.2.4 Quality Control

It can be challenging to ensure the nutritional value of traditional fodder. Forages may not be sufficient enough to fulfill the dietary demands of modern livestock, which produce high-yields (Hazra, 2014).

13.3.2.5 Land Degradation

The long-term sustainability of traditional fodder production can be compromised by overgrazing and poor land management, which leads to land degradation and soil erosion (Assefa and Hans-Rudolf, 2016).

13.3.2.6 Weather Dependency

Traditional methods highly rely on the weather conditions (Forbes, 1998), making them vulnerable to the effects of climate change, such as extreme drought, high temperatures, and unpredictable rainfall.

13.4 CHANGING CLIMATE PATTERNS IMPACT TRADITIONAL FODDER PRODUCTION SYSTEMS

Traditional fodder production systems are at risk due to climate change. Climate change affects these systems in multiple ways. Forage crops can experience disturbance in their traditional growing seasons due to climate change. Crop growth and livestock feeding requirements cannot be matched due to changes in temperature and precipitation patterns. Increase in temperature, prolonged drought, and shift in precipitation patterns can reduce the yields of natural pastures and cultivated fodder crops, which cause in less feed availability (Ifejika Speranza, 2010). High level of carbon dioxide and temperature can reduce the nutritional value of forages. The health and productivity of livestock are affected by the consumption of low protein and nutrient feed (Lee *et al.*, 2017). The pest and disease outbreaks are more common in warm and humid environments, which harm both crops and livestock. The water scarcity is increasing due to climate change, which creates challenges in crop irrigation and maintenance of pastures (Mendoza-Grimón *et al.*, 2015). This has a direct impact on the quality and quantity of traditional fodder. Hence, traditional fodder production systems are facing more issues. Our fodder production systems need to make an urgent shift toward sustainable practices.

13.5 MODERN AGRICULTURAL TECHNOLOGIES FOR FODDER PRODUCTION

To maintain agricultural productivity, novel solutions will be necessary for modern era fodder production. New fodder cultivation techniques from modern agriculture technologies provide an insight and are essential for overcoming the challenges of traditional fodder cultivation. The following are key modern technologies.

13.5.1 Hydroponic Fodder Production

One of the present-day technologies that provide soil-less/soil-free production of nutrient-rich fodder is hydroponic fodder production. This method involves the growing of different fodders in trays under controlled light, temperature, and humidity (Sneath and Mcintosh, 2003). Hydroponic growth principles are the basis for this technique for the production of fodder in which they emphasize in efficient delivery of nutrients to the roots of plant. Rapid growth and effective nutrient supply make it a good feed for stock in this way. Recent findings claim that fodders produced using hydroponics may mature in a shorter period that allows its frequent harvest (Bhat, 2021). This is a critical practice as it ensures the provision of an adequate food supply for livestock (Fazaeli *et al.*, 2012). Continuous supply of fodder enhances the productivity and growth of livestock. Hydroponically grown fodder contains protein and vitamins (Abdula, 2022). According to a research done in Iran, protein levels in barley grown using hydroponic culture are significantly higher than that from conventional cultivation methods (Fazaeli *et al.*, 2012). Hydroponic feed produced higher milk output and overall dairy cow health.

Since this method uses far less water than regular field agriculture, it's an option that could particularly work in areas that find water scarce. Hydroponic fodder production has become popular throughout the dry areas of the Middle East, thanks to its water-saving capacities. Compared to traditional practices, a Qatar-based farm noted a 90% reduction in the water requirement for fodder production (Karanisa *et al.*, 2021). Hydroponics systems require less land area for cultivation, making it suitable from an urban or small-space perspective. In Meru, Kenya, an urban agriculture initiative employs hydroponic fodder systems to provide fresh feed for urban livestock as well as to maximize the use of limited areas in the city (Kibiti, 2017). Controlled environments protect crops from adverse climates, which makes hydroponic fodder production less susceptible to the consequences of climate change. For example, a farm in South Africa used hydroponic systems during a severe drought to provide consistent feed for their livestock, helping the animals stay healthy despite rapidly changing climate patterns (Shikwambana *et al.*, 2022). It is a high-quality feed that enhances animal health and productivity. Such systems can be installed in urban areas to provide fresh fodder, reducing transport and warehousing needs. It is a novel approach for regions affected by drought (Pomoni *et al.*, 2023). It provides support to the food security by maintaining a continuous supply of feed for livestock sector.

13.5.2 Vertical Farming

Vertical farming is a type of agriculture that uses straw bales, pots, or rows of ground. It is a modern agricultural process that can grow crops such as wheat, maize, and rice in vertical stacks, usually in controlled environments. Usually, it is used to relate with the vegetable and fruit production. However, it is now being used for fodder farming. It is an effective approach always available to fulfill food demand to livestock by enhancing fodder output (Mir *et al.*, 2022). In the vertical fodder farming, wheat, barley, and maize are commonly used as seeds to grow fodder plants (Yeşil and Tatar, 2020). A controlled environment is essential for the successful growth of crops as different factors (nutrient supply, light, humidity, temperature) must be monitored and controlled (Yeşil *et al.*, 2020). By cultivating the land efficiently, using the available land to its full potential, this method delivers year-round fodder production within a small footprint. It helps in parts of urban and space-heavy places. Controlled environments shield crops from impacts of severe droughts, drastic temperature swings, and unpredictable rainfall patterns. This way, fodder crops will always be available and less affected by the effects of climate change. The availability of fresh feed enhances the health and productivity of livestock (Seré *et al.*, 1996).

Vertical farming can save a significant amount of water compared to traditional methods by recirculating hydroponic systems (Al-Kodmany, 2018). However, Vertical farming has huge benefits in areas with limited water supply due to prolonged drought caused by climate change. Vertical farms are considered best-suited for urban areas or lands with less land and water availability. It contributes to land conservation and reduces pressure to expand the agricultural frontier, ultimately helping to preserve natural ecosystems. Use of pesticides and herbicide application in the controlled environments is lesser which facilitate sustainable fodder production and environment friendly (Ali *et al.*, 2013; Schütte *et al.*, 2017).

Vertical farming, integrated with automation and artificial intelligence (AI), offers a climate-resilient option for sustainable fodder production. This integration becomes a cornerstone of the long-term sustainability of the livestock industries.

13.5.3 Precision Agriculture

Precision agriculture, or precision farming, is the use of technology to help optimize crop production. There are many applications of this in fodder production. Precision agriculture allows farmers to determine the fertilization requirements of crops according to real-time data on soil nutrients (Hedley, 2015). They provide the precise nutrients required by the crop and as a result, maximum yield and minimal losses of nutrients occur. With the help of sensors and data analysis in a specific area, farmers can promptly fit their irrigation schedule per crop (Adeyemi *et al.*, 2017). Particularly as climate patterns change, effective management of water resources is becoming key. There is a vast amount of data that precision agriculture needs, collected from a variety of sources, including satellites, sensors, and ground-based equipment. This knowledge is utilized for optimization in real-time for planting, harvesting, as well as resource management to achieve sustainable fodder production (Mahmood *et al.*, 2015; Hakkim *et al.*, 2016).

The optimum path is the uses of resources through data by using technologies like sensors, satellites, and drones in precision agriculture. It is beneficial in fodder production due to its exact sowing, irrigation, and fertilization, resulting in higher yields and efficient resource utilization (Rasheed *et al.*, 2022b; Chandra, 2023). Sensors, drones, and IoT are used to increase fodder production in changing climates. Sensors can be used for real-time measurement of temperature, humidity, moisture, and other environmental conditions. This data helps the farmers at the time of managing their crop irrigation, crop residual management, fertilization, and pest control, resulting in sustainable fodder production (Cardell-Oliver *et al.*, 2004). Field imagery can be captured in high resolution by the help of drones with cameras and sensors. This helps to check crop health, detect pests and diseases, and assess the stresses like drought or erratic rainfall (Aziz *et al.*, 2018; Olson and Anderson, 2021). Drones offer a bird's eye view that addresses field problems as they arise. Environmental sensors and weather stations gather data and send them to a central platform. Farmers can use this data to inform their decisions and modify their farming practices as necessary when the environment changes (Mahmood *et al.*, 2013; Evans *et al.*, 2017). Similar precision agriculture practices with GPS-based equipment have also increased fodder crop yields in the US to a great extent. This assists in the adaptation to change climate patterns and enhances resource management.

13.5.4 Data Analytics in Fodder Farming

Data analytics include climate data that are of weather forecasts, historical climatological events, and projections of climate conditions. Through that analysis, the farmers will be able to best decide how to adapt to a changing climate. Data analytics can be used to detect climate-related issues, such as drought and pest outbreaks, at an early stage. Based on this data, farmers can take pre-emptive measures and formulate plans to subdue the loss that occurs due to the fluctuation in the climate. Analytics allows farmers to understand how to manage resources better for fodder production and account for climate conditions. For example, in drought conditions, analytics can suggest more effective irrigation practices to save water (Nazari *et al.*, 2018). By applying data analytics to historical data along with environmental factors, yield predictions can be made (Elavarasan *et al.*,

2018). Farmers can adjust their seed sowing and harvesting schedule to achieve the maximum fodder output under changing climate scenarios.

In precision agriculture, data analytics enables to form real-time decision resulting in more resilient and sustainable fodder. Automation and AI is being playing a crucial role in vertical farming for fodder production in time of decision-making. AI systems and sensors monitor and adjust environmental conditions such as temperature, humidity, and moisture levels (Mehmood *et al.*, 2019). Control of this is crucial to reducing the effects of climate change on plant development. AI can accurately deliver nutrients for crops, resulting in optimal use of resources with minimal waste (Mustafa *et al.*, 2022; Kumar *et al.*, 2023). This is essential for sustainable forage production, especially in the era of climate change. AI considers climatic conditions, live data, and crop output and provides the best solution to the farmers. Using robotics and AI, crop health and growth rate can be measured anytime (Jha *et al.*, 2019). It helps in identifying early symptoms of stress or disease and identifying solutions for quality of fodder.

13.5.5 Fodder Crop Breeding

Forage crops with improved yield, nutritional content, and pest and disease resistance are created using modern breeding techniques. These crops are engineered to thrive in a changing climate (Kole *et al.*, 2015).

13.5.6 Greenhouse Fodder Production

Greenhouses offer a controlled environment for growing fodder all year. They shield crops from severe weather and allow for customized climate conditions.

13.5.7 Fodder Conservation

Modern methods of preserving fodder as silage or hay preserve nutritional value, ensuring a year-round supply of high-quality feed. In Australia, advanced fodder conservation techniques such as silage and hay production have ensured a consistent supply of high-quality fodder during extreme weather events (Lean and Moate, 2021). This has been critical in assisting livestock farmers during droughts and bushfires.

13.6 TECHNICAL APPROACHES FOR HEALTHY FODDER PRODUCTION

13.6.1 Nutrients and Grazing Management

Nutrient management is pivotal for the persistence, productivity, and nutritive value of crops (Balal *et al.*, 2017; Saleem *et al.*, 2020). Fertilizer requirements differ between grazing pastures and hay production systems. Four principles of efficient fertilization program are as follows: using fertilizer source that is suitable and environmentally safe; applying the recommended rate based on soil tests; applying techniques at the right time; and applying at the right place (Hanif *et al.*, 2017). Consideration of environmental conditions is crucial to prevent volatilization losses (Zhen *et al.*, 2006). Timing is critical; for example, nitrogen application to forage grasses should occur after germination and at least 3 inches of growth to enhance root uptake (Babar *et al.*, 2009). Furthermore, applying nutrients at the right place is vital. Legumes, for instance, require phosphorus and potassium but no nitrogen, as nitrogen application can hinder biological nitrogen fixation (Franzluebbers *et al.*, 2014; Hassan *et al.*, 2022). Incorporating legumes into forage crops offers numerous benefits, enhancing productivity, nutritive value, animal performance, reducing nitrogen fertilizer costs, and extending the grazing season. However, unregulated grazing on diminishing areas has led to degradation of pastures. Demand for cropland, primarily for food production, reduces the available land for forage production and grazing (Wassie, 2020).

13.6.2 Cover Crop Grazing

Winter cover crops offer timely, nutritious forage for grazing, with good average daily gain for growing cattle. Grazing impacts on soil, including compaction, depend on factors like trampling intensity and soil characteristics (Kelly *et al.*, 2021). Evidence suggests that bulk density increases slightly with repeated cattle trampling on well-vegetated land. Negative impacts may occur, potentially due to increased water evaporation (Crowell *et al.*, 2022). However, enhanced fertility status and increased soil organic matter can contribute to greater subsequent crop growth. While some studies show little difference in short-term crop growth, various ecosystem services can be simultaneously increased with cover cropping. A balance is needed between reducing nitrate leaching and potential negative impacts on subsequent crop growth (Blanco-Canqui *et al.*, 2020; Meyer *et al.*, 2022).

13.6.3 Farm Integration

Historically, integrated crop-livestock systems have declined due to economic, social, and policy factors. The transition back to integrated systems faces challenges like expensive infrastructure investments and technical skills. However, organizing integrated systems beyond the farm level through collaborations among farms can be a viable alternative (Asai *et al.*, 2018). Various forms of collaboration, from material exchange to coordinated land use and shared resources, offer ecological, agronomic, and economic benefits. Integrating crops, livestock, and pastures allows fuel savings, cooperative markets to spread risks, and efficient resource use. It stimulates knowledge exchange, adaptive capacity, and shared labor and equipment (Ryschawy *et al.*, 2021). However, socio-technical constraints such as lack of knowledge, governance models, logistical costs, and suitable equipment pose challenges to achieving more integrated systems. Overcoming these constraints is crucial for promoting sustainable, circular agricultural practices (Stark *et al.*, 2021).

13.6.4 Backyard Forage Production

Backyard forage production is crucial as it develops within the home, providing convenient access to animals (Papanastasis *et al.*, 2008; Mengistu *et al.*, 2016). Additionally, tall-growing tropical grasses can be effectively incorporated into backyard forage development. Among the various approaches, tree legumes have been widely adapted to diverse climates with strong benefits of backyard forage production. Leveraging this strategy can serve as an enticement for forage development initiatives. This strategy holds significant potential for enhancing livestock productivity and overall agricultural sustainability. Backyard forage offers significant forage production and access for utilization (Mengistu *et al.*, 2017).

13.6.5 Intercropping and under-Sowing

Under-sowing and intercropping involves establishing both annual and perennial forage species. This practice is a short-medium term and a reliable pathway to farms with a constant supply of fresh forage (Reta-Sánchez *et al.*, 2016). The legume inclusion in the crop rotation system facilitates better soil fertility and soil structure in cropping soil. The on-farm trial alone introduces farmers to the benefits and methods of under-sowing and intercropping (Abera, 2012).

13.6.6 Over Sowing

This is one of the simplest forage improvement options, which can be of extremely low-cost depending on seed-rates used (Atsbha *et al.*, 2020). The method consists essentially of broadcasting or sowing improved forage species into common grazing lands, native pastures, and degraded areas without any cultivation or other input. To this end, this strategy works particularly well with pioneer legume species that establish rapidly and exhibit high rates of seed production (Garba *et al.*, 2022).

13.6.7 Aerial Sowing

Aerial sowing is a realistic and effective way to sow vast areas using improved forage seeds (Mengistu *et al.*, 2016). Choosing appropriate sites, especially ones with rough surfaces, is very important for the successful establishment of the stand. Aerial sowing is useful for widespread restoration in areas that are remote. Aerial sowing is mainly used to cover large areas by distributing a small quantity of seed in a short time and at a low cost (Campbell, 1992). This sowing method is good for high-density crops with rapid growth.

13.6.8 Contour Forage Production

Forage strips consist of the mixtures of herbaceous and legumes with grasses, planted on the contours (Nigus, 2017). It provides fuel wood, shelter, feed, soil stabilization, etc. Forage strips planted along the contour conserve soil by guiding plowing along the contour and lowering flow down the hill. This facilitates infiltration and reduces soil erosion, especially in the forage strip that is sown with a dense sward of grasses or herbaceous legumes. When woody leguminous species are combined with perennial, thick-rooted grasses, contour forage strips become especially effective (Cooper *et al.*, 1996). Farmers primarily value forage and feed products as the main advantages of forage strips, rather than their conservation features. Therefore, promoting forage strip strategies is essential to harnessing these benefits effectively. By highlighting the forage and feed advantages, farmers can be incentivized to adopt and implement forage strip strategies, ultimately improving their agricultural productivity and sustainability.

13.6.9 Alley Cropping

Incorporating long-lived browsing plants into the species mix for contour forage strips also supports alley farming initiatives. Alley farming requires careful placement and marking of contour strips, ensuring sufficient spacing for uninterrupted harvesting and plowing. Bare-rooted seedlings are recommended for alley farming establishment (Hill, 2017). Pioneer species like green leaf, Phalaris, and pigeon pea are ideal as understory plants, especially when combined with leucaena or tree lucerne. This method increases sustainability of agricultural systems by maximizing land use efficiency. It also promotes biodiversity and crop productivity.

13.6.10 Agroforestry

Agroforestry is a farming system in which crops and trees are integrated in the suitable manner. Most of the production strategies for forages can be adapted into agroforestry (Rao *et al.*, 2007). For example, contour forage banks and under-sowing of tree crops or forest plantations can be designed as agroforestry systems. Agroforestry maximizes the land utilization by enhancing productivity (Sobola *et al.*, 2015). The farmers with limited land resources can get more benefits by adapting to agroforestry. This can optimize the productivity of their lands. Moreover, many agroforestry strategies include leguminous; this is attractive to farmers having problem of declining soil productivity. Leguminous crops play an important role in agroforestry by increasing soil fertility through nitrogen fixation. By integrating trees and crops, agroforestry increases soil fertility, biodiversity, water retention and boosts the income of farmers (Moreno and Rolo, 2019).

13.6.11 Using Dual-Purpose Crops

The utilization of dual-purpose oilseeds and legumes presents challenges due to the lack of standardized protocols for maximizing both forage and seed production (Mcgrath *et al.*, 2021). Harvesting oilseed crops at peak maturity for seed production may result in reduced forage quality, while harvesting early for forage may compromise seed yield when the crop regrows (Neely *et al.*, 2015). Alternating between forage and seed harvests may mitigate this issue, allowing farmers to capitalize on market demand for either product each year. However, managing perennial oilseed crops for dual use also raises questions about potential trade-offs between traits optimized for forage versus seed production. For example, traits like plant height, which may enhance forage yield, could conflict with traits preferred for seed production, such as shorter, semi-dwarf stature to reduce lodging and allocate resources more efficiently to seed filling. Moreover, to increase diversity in agricultural systems, mixtures of species (polycultures) are essential. However, determining the optimal combination of species and cultivars for maximizing grain and forage yield, as well as nutritive value, remains a significant knowledge gap. Breeding crops for perennial mixtures necessitates evaluations conducted in the target environment, particularly within mixtures themselves (Tanveer *et al.*, 2013). This interdisciplinary approach combining breeding and agroecology poses a challenging but essential endeavor for developing optimal diverse perennial systems that can sustainably meet the needs of both forage and seed production (Snapp, 2017; Rasheed *et al.*, 2022a).

13.6.12 Harvesting and Storage

Balancing quality and quantity is crucial in forage crop management. Harvesting forage at the vegetative rather than the reproductive stage enhances crude protein, lowers fiber, and improves animal intake. Regularly testing hay for nutritive value allows for informed decisions in developing supplemental feeding programs tailored to specific livestock classes, impacting animal growth, breeding cycles, and reproductive rates.

13.6.13 Silage and Hay Making

Modern methods of preserving fodder as silage or hay preserve nutritional value, ensuring a year-round supply of high-quality feed. Advanced fodder conservation techniques such as silage and hay production have ensured a consistent supply of high-quality fodder during extreme weather events. This has been critical in assisting livestock farmers during droughts and extreme weather events. Grasses or legumes are harvested at their optimal maturity and dried into hay or ensiled into silage. The lean periods can also be tricky to navigate but can be overcome by preservation of excess fodder during a season. Farmers can store excess forage by using hay and silage production techniques when growth is above average. The preserved forage can then be used in times of scarcity, such as during droughts or winter months when access to fresh forage is limited (Coblentz *et al.*, 2021; Wasaya *et al.*, 2021). Fodder crops can be harvested and stored as hay or ensiled in the silage when in surplus. Gislon *et al.* (2020) reported increase in milk production in cows by feeding hay and silage.

13.6.14 Genetic Advancements in Fodder Crops

Genetic improvement for fodder crops has made considerable strides in recent years.

13.6.14.1 Drought Tolerance

Genetically modified (GM) fodders would be drought resistant, which are under research. Such fodders are able to resist water stress and produce a high biomass (Tanveer *et al.*, 2012; Jacobsen *et al.*, 2013).

13.6.14.2 Pest and Disease Resistance

GM fodders are made in order to control particular pests and diseases (Bawa and Anilakumar, 2013). This causes to decrease the requirement of chemical pesticides. It sustains a constant supply of fodder in this challenge.

13.6.14.3 Nutritional Enhancement

Improvement of nutritional value is the major focus of many scientific studies. Many studies are focused on protein enhancement and other nutritional contents, directly relating with forage digestibility (Sanginga *et al.*, 2003). GM fodders are more nutritious, which leads higher milk and meat production from the livestock. It also improves the health of dairy animals (Deb *et al.*, 2013).

13.6.14.4 Climate-Adaptive Traits

Environmental constraints (heat, drought, and salinity stress tolerance) are the greatest threats to forage production. Such adverse climatic conditions prompt the researchers for the use of genetic engineering techniques for improved cultivar development. Fodder production in the era of changing climate needs to be sustainable and a judicious choice of these traits will ensure it (Rao *et al.*, 2020).

13.7 SUSTAINABLE PRACTICES IN FODDER PRODUCTION SYSTEMS

Environmental challenges require sustainable practices for fodder production. These practices not only mitigate the effects of climate change but also make agricultural systems more resilient. Following are some examples of sustainable practices.

13.7.1 Organic Fodder Farming

Organic farming improves soil health, reduces chemical inputs, and promotes biodiversity. Organic fodder farming can improve the safety and quality of animal feed and reduce environmental impact (Hole *et al.*, 2005). Organic fodder farming relies on natural rather than synthetic chemicals to enhance soil fertility and forage quality. It reduces carbon emissions associated with making synthetic fertilizer and also increases biodiversity.

13.7.2 Conservation Tillage

Minimum or zero-tillage practices prevent soil disturbance ensuring soil structure preservation, decrease in erosion, and carbon sequestration (Bhatt, 2017). This helps the fight against climate change by sequestering carbon in the soil. Perennial fodder production further reduces the tillage operations which are beneficial for the soil health.

13.7.3 Crop Rotation

Rotating fodder crops with other crops has been demonstrated to disrupt disease cycles and increase soil fertility. Crop rotation also reduces climate risks by increasing agricultural commodities (Bowles *et al.*, 2020). Mixtures such as Italian ryegrass-red clover and perennial ryegrass-white clover are used in Europe for hay and grazed pastures (Peyraud *et al.*, 2014). The benefits of pasture-crop rotations include impacts on soil organic matter, soil fertility, and environmental quality (Moraine *et al.*, 2017). He increased the diversity of plant species in these rotations, helping wildlife populations, biological insect control, and watershed runoff quality (Blanco-Canqui *et al.*, 2020).

13.7.4 Agroforestry

The introduction of trees to fodder production systems provides multiple benefits. Trees provide shade, windbreak, and extra forage. They also assist in carbon monoxide sequestration and increase biodiversity (Jose, 2019).

13.7.5 Water Management

Sustainable water management is also significant in the sense that efficient irrigation as well as the harvesting of rainwater is helpful in conserving water while implementing strategies to adapt to climate variability and change.

13.8 ADVANTAGES OF ADOPTING MODERN AGRICULTURAL TECHNOLOGIES

Modernization of agricultural technologies has numerous benefits with respect to climate change. Controlled environments as with hydroponics and greenhouses protect crops from extreme weather, leading to increased yields and enhanced climate resilience (Kurade *et al.*, 2021). These systems generate premium green fodder throughout the year with minimum use of water. They have diminished the impacts of droughts and improved livestock nutrition. Drip irrigation and precision agriculture are resource-effective ways to lessen the environmental burden. Modern breeding and management practices boost crop yields, meaning that there is always enough fodder for all livestock, even when climate causes problems. Technologies facilitate the development of fodder that has better nutritional value, leading to healthier and more productive livestock (Balehegn *et al.*, 2020).

13.9 CHALLENGES AND FUTURE PROSPECTS

Multiple challenges under climate change context of introducing contemporary fodder production systems are the following:

- *High Initial Investment*: Smallholder farmers are reluctant to adopt modern technology, e.g., hydroponics, vertical farming, precision agriculture as they demand considerable outset at infrastructure and equipment (Swinton and Lowenberg-Deboer, 2001).
- *Technical Knowledge*: Operating and maintaining modern technologies require more technical knowledge to operate and maintain them. Efficient execution requires training, education, and extension services (Kitchen *et al.*, 2002).
- *Resource Constraints*: Water scarcity, cost of energy, and the availability of sufficient land can all limit the adoption of new technologies, particularly within climate-sensitive areas.
- *Market Access*: In remote areas, farmers may not have access to markets for modernized fodder products (Dixon *et al.*, 2001).
- *Regulatory Hurdles*: Each region has varying levels of strict rules and safety concerns associated with contemporary practices such as genetic engineering which helps to hide their adoption.
- *Digital Agriculture*: Integrated application of AI, big data, and blockchain technology will enable more precise monitoring on the basis of data-driven decision-making to optimize the use of resources and to adapt to climate (Basso and Antle, 2020).
- *Smart Farming Practices*: IoT devices, drones, and sensor technologies will become common and allow for real-time field monitoring and efficient resource management.
- *Diversification*: Integrated systems such as silvopasture will gain popularity due to their numerous benefits, including carbon sequestration and biodiversity support.
- *Adaptation to Climate Change*: There is also a need for research for the procedures and technologies that can adapt to changing weather patterns (Salinger *et al.*, 2000). This could include new crop varieties, improved water management strategies, or real-time monitoring tools.
- *Resource Efficiency*: Innovative practices can contribute to more resource-efficient practices, reducing the environmental impact of fodder production and conserving valuable resources like water and arable land.

- *Economic Viability*: Advanced technologies are out of reach for farmers, and research can provide them with cheap alternatives and make such advanced technologies economically viable and sustainable.
- *Environmental Conservation*: Innovations would reduce the environmental footprint of fodder production, while safeguarding ecosystems and biodiversity (Tittonell *et al.*, 2016).
- *Food Security*: Resilient fodder systems are fundamental for livestock and human food security. Fodder is a significant link in global food supply chains.

13.10 CONCLUSION

Integrated crop-livestock systems are gaining traction in the scientific community as crucial components for sustainable agricultural practices. Developing resilient systems within farms and regions requires both fieldwork and theoretical considerations. While agricultural systems are inherently complex, simplification is not the solution. Practical steps that have been envisioned to reconnect cropping and livestock operations include manure application, cover cropping, diverse cropping rotations incorporating perennial forages and watershed trading. Such measures help to cope with climate change, mitigate greenhouse gas emissions, improve soil health, protect the environment and biodiversity. Annual and perennial forages form the backbone of a circular, healthy, and sustainable farming and food system. Farmed herbivores, such as pasture-based livestock, provide animal welfare benefits as well as a positive image of livestock farming. When managed properly, grasslands can provide a wealth of ecosystem services, including biodiversity, hydrological function, biogeochemical cycling, reduced pesticide need, soil carbon sequestration, and preventing soil erosion.

REFERENCES

Abdula, A.H., 2022. Contribution of hydroponic feed for livestock production and productivity. *Science Frontiers*, 3, 1–7.

Abera, M., 2012. The effect of under sowing of forage legumes in maize on dry matter yield and nutritional value of the fodder in Baresa Watershed, Ethiopia. *International Journal of Science and Research*, 3, 1070–1075.

Adeyemi, O., Grove, I., Peets, S. & Norton, T., 2017. Advanced monitoring and management systems for improving sustainability in precision irrigation. *Sustainability*, 9, 353.

Ahmad, K., Ejaz, A., Khan, Z.I., Gondal, S., Fardous, A., Hussain, A., Sher, M., Valeem, E.E. & Ullah, S., 2010. Evaluation of dynamics of iron and manganese from pasture to buffaloes: A case study at rural livestock farms. *Pakistan Journal of Botany*, 42, 3415–3421.

Ali, H., Tanveer, A., Nadeem, M., Javaid, M., Kashif, M. & Chadhar, A., 2013. Allelopathic effects of *Rhynchosia capitata* on germination and seedling growth of mungbean. *Planta Daninha*, 31, 501–509.

Al-Kodmany, K., 2018. The vertical farm: A review of developments and implications for the vertical city. *Buildings*, 8, 24.

Asai, M., Moraine, M., Ryschawy, J., De Wit, J., Hoshide, A.K. & Martin, G., 2018. Critical factors for crop-livestock integration beyond the farm level: A cross-analysis of worldwide case studies. *Land Use Policy*, 73, 184–194.

Ashraf, E., Hayat, Z., Khan, M., Atif, F. & Haider, M., 2016. Impact of dairy farm management training workshop on the knowledge level of participants. *International Journal of Agriculture and Applied Sciences (Pakistan)*, 4.

Ashraf, E., Shah, F., Luqman, M., Younis, M., Aziz, I. & Farooq, U., 2013. Use of untreated wastewater for vegetable farming: A threat to food safety. *International Journal of Agricultural and Applied Sciences*, 5.

Assefa, E. & Hans-Rudolf, B., 2016. Farmers' perception of land degradation and traditional knowledge in Southern Ethiopia—Resilience and stability. *Land Degradation & Development*, 27, 1552–1561.

Atsbha, T., Wayu, S., Kidane, H., Degf, K. & Abreha, G., 2020. Improving the productivity of degraded pasture land through demonstration of legume forage over sowing: The case of Ayba pasture land, South Tigray, Ethiopia. *Cogent Environmental Science*, 6, 1778997.

Aziz, M., Ashraf, M. & Javaid, M.M., 2018. Enhancement in cotton growth and yield using novel growth promoting substances under water limited conditions. *Pakistan Journal of Botany*, 50, 1691–1701.

Babar, B.H., Tanveer, A., Tahir, M., Aziz, A., Ahmad, A.U.H., Nadeem, M.A. & Javaid, M.M., 2009. Allelopathic potential of wild onion (*Asphodelus tenuifolius*) on the germination and seedling growth of chickpea (*Cicer arietinum*). *Weed Biology and Management*, 9, 146–151.

Balal, R.M., Shahid, M.A., Javaid, M.M., Iqbal, Z., Liu, G.D., Zotarelli, L. & Khan, N., 2017. Chitosan alleviates phytotoxicity caused by boron through augmented polyamine metabolism and antioxidant activities and reduced boron concentration in *Cucumis sativus* L. *Acta Physiologiae Plantarum*, 39, 1–15.

Balehegn, M., Duncan, A., Tolera, A., Ayantunde, A.A., Issa, S., Karimou, M., Zampaligré, N., André, K., Gnanda, I. & Varijakshapanicker, P., 2020. Improving adoption of technologies and interventions for increasing supply of quality livestock feed in low-and middle-income countries. *Global Food Security*, 26, 100372.

Basso, B. & Antle, J., 2020. Digital agriculture to design sustainable agricultural systems. *Nature Sustainability*, 3, 254–256.

Bawa, A. & Anilakumar, K., 2013. Genetically modified foods: Safety, risks and public concerns—A review. *Journal of Food Science and Technology*, 50, 1035–1046.

Berndes, G., Hoogwijk, M. & Van Den Broek, R., 2003. The contribution of biomass in the future global energy supply: A review of 17 studies. *Biomass and Bioenergy*, 25, 1–28.

Bhat, A.G., 2021. *Development and Evaluation of Small-Scale Hydroponic Green Fodder Production System*. Tavanur: Department of Soil and Water Conservation Engineering, KCAET.

Bhatt, R., 2017. Zero tillage impacts on soil environment and properties. *Journal of Environmental and Agricultural Sciences*, 10, 01–19.

Blanco-Canqui, H., Drewnoski, M.E., Macdonald, J.C., Redfearn, D.D., Parsons, J., Lesoing, G.W. & Williams, T., 2020. Does cover crop grazing damage soils and reduce crop yields? *Agrosystems, Geosciences & Environment*, 3, e20102.

Bowles, T.M., Mooshammer, M., Socolar, Y., Calderón, F., Cavigelli, M.A., Culman, S.W., Deen, W., Drury, C.F., Y Garcia, A.G. & Gaudin, A.C., 2020. Long-term evidence shows that crop rotation diversification increases agricultural resilience to adverse growing conditions in North America. *One Earth*, 2, 284–293.

Campbell, M., 1992. Extending the frontiers of aerially sown pastures in temperate Australia: A review. *Australian Journal of Experimental Agriculture*, 32, 137–148.

Cardell-Oliver, R., Smettem, K., Kranz, M. & Mayer, K., 2004. Field testing a wireless sensor network for reactive environmental monitoring [soil moisture measurement]. In *Proceedings of the 2004 Intelligent Sensors, Sensor Networks and Information Processing Conference, 2004*. IEEE, 7–12.

Chandra, N., 2023. Significant advancements in agriculture science. *Research Trends in Agriculture Science*, II, 72.

Coblentz, W.K., Akins, M.S. & Cavadini, J.S., 2021. Nutritive value, silage fermentation characteristics, and aerobic stability of 3 round-baled, perennial-grass forages ensiled with or without a propionic-acid-based preservative. *Applied Animal Science*, 37, 239–255.

Cooper, P., Leakey, R., Rao, M. & Reynolds, L., 1996. Agroforestry and the mitigation of land degradation in the humid and sub-humid tropics of Africa. *Experimental Agriculture*, 32, 235–290.

Cooper, P.J., Dimes, J., Rao, K., Shapiro, B., Shiferaw, B. & Twomlow, S., 2008. Coping better with current climatic variability in the rain-fed farming systems of sub-Saharan Africa: An essential first step in adapting to future climate change? *Agriculture, Ecosystems & Environment*, 126, 24–35.

Crowell, H., Gamble, A.V., Feng, Y., Balkcom, K. & Yang, A., 2022. Impacts of winter grazing on soil health in southeastern cropping systems. *Agrosystems, Geosciences & Environment*, 5, e20240.

Deb, R., Sajjanar, B., Devi, K., Reddy, K., Prasad, R., Kumar, S. & Sharma, A., 2013. Feeding animals with GM crops: Boon or bane? *Indian Journal of Biotechnology*, 12, 311–322.

Desta, L., 2000. *Land Degradation and Strategies for Sustainable Development in the Ethiopian Highlands*. Amhara Region: ILRI (aka ILCA and ILRAD).

Devi, S., Gupta, C., Jat, S.L. & Parmar, M., 2017. Crop residue recycling for economic and environmental sustainability: The case of India. *Open Agriculture*, 2, 486–494.

Dixon, J.A., Gibbon, D.P. & Gulliver, A., 2001. *Farming Systems and Poverty: Improving Farmers' Livelihoods in a Changing World*. Food & Agriculture Org.

Duressa, D., Kenea, D., Keba, W., Desta, Z., Berki, G., Leta, G. & Tolera, A., 2014. Assessment of livestock production system and feed resources availability in three villages of Diga district Ethiopia. International Livestock Research Institute, Addis Ababa, Ethiopia.

Elavarasan, D., Vincent, D.R., Sharma, V., Zomaya, A.Y. & Srinivasan, K., 2018. Forecasting yield by integrating agrarian factors and machine learning models: A survey. *Computers and Electronics in Agriculture*, 155, 257–282.

Evans, K.J., Terhorst, A. & Kang, B.H., 2017. From data to decisions: Helping crop producers build their actionable knowledge. *Critical Reviews in Plant Sciences*, 36, 71–88.

Fazaeli, H., Golmohammadi, H., Tabatabayee, S. & Asghari-Tabrizi, M., 2012. Productivity and nutritive value of barley green fodder yield in hydroponic system. *World Applied Sciences Journal*, 16, 531–539.

Forbes, H., 1998. European agriculture viewed bottom-side upwards: Fodder-and forage-provision in a traditional Greek community. *Environmental Archaeology*, 1, 19–34.

Franzluebbers, A.J., Sawchik, J. & Taboada, M.A., 2014. Agronomic and environmental impacts of pasture–crop rotations in temperate North and South America. *Agriculture, Ecosystems & Environment*, 190, 18–26.

Garba, M., Gadaffi, S. & Idowu, W., 2022. Techniques of sown pasture establishment and management in Sudan Savanna zone Nigeria: Technical report. *Nigerian Journal of Animal Production*, 49, 266–274.

Gislon, G., Bava, L., Colombini, S., Zucali, M., Crovetto, G. & Sandrucci, A., 2020. Looking for high-production and sustainable diets for lactating cows: A survey in Italy. *Journal of Dairy Science*, 103, 4863–4873.

Hakkim, V.A., Joseph, E.A., Gokul, A.A. & Mufeedha, K., 2016. Precision farming: The future of Indian agriculture. *Journal of Applied Biology and Biotechnology*, 4, 068–072.

Hanif, Z., Naeem, M., Ali, H.H., Tanveer, A., Javaid, M.M., Peerzada, A.M. & Chauhan, B.S., 2017. Effect of environmental factors on germination of *Salsola foetida*: Potential species for rehabilitation of degraded rangelands. *Rangeland Ecology & Management*, 70, 638–643.

Hassan, M.U., Nawaz, M., Mahmood, A., Shah, A.A., Shah, A.N., Muhammad, F., Batool, M., Rasheed, A., Jaremko, M. & Abdelsalam, N.R., 2022. The role of zinc to mitigate heavy metals toxicity in crops. *Frontiers in Environmental Science*, 10, 990223.

Hazra, C., 2014. Feed and forage resources for sustainable livestock development. *Range Management and Agroforestry*, 35, 1–14.

Hedley, C., 2015. The role of precision agriculture for improved nutrient management on farms. *Journal of the Science of Food and Agriculture*, 95, 12–19.

Hill, J.E., 2017. Quality and yield responses of seven warm season forage grasses to partial shading in a simulated silvopasture in East Texas. https://scholarworks.sfasu.edu/etds/125

Hole, D.G., Perkins, A., Wilson, J., Alexander, I., Grice, P. & Evans, A.D., 2005. Does organic farming benefit biodiversity? *Biological Conservation*, 122, 113–130.

Ifejika Speranza, C., 2010. Drought coping and adaptation strategies: Understanding adaptations to climate change in agro-pastoral livestock production in Makueni district, Kenya. *The European Journal of Development Research*, 22, 623–642.

Iñiguez, L., 2011. The challenges of research and development of small ruminant production in dry areas. *Small Ruminant Research*, 98, 12–20.

Izaurralde, R.C., Thomson, A.M., Morgan, J., Fay, P., Polley, H. & Hatfield, J.L., 2011. Climate impacts on agriculture: Implications for forage and rangeland production. *Agronomy Journal*, 103, 371–381.

Jacobsen, S.-E., Sørensen, M., Pedersen, S.M. & Weiner, J., 2013. Feeding the world: Genetically modified crops versus agricultural biodiversity. *Agronomy for Sustainable Development*, 33, 651–662.

Javaid, M., Tanveer, A., Ahmad, R., Yaseen, M. & Khaliq, A., 2012. Optimizing activity of herbicides at reduced rate on *Emex spinosa* Campd. with adjuvants. *Planta Daninha*, 30, 425–435.

Jha, K., Doshi, A., Patel, P. & Shah, M., 2019. A comprehensive review on automation in agriculture using artificial intelligence. *Artificial Intelligence in Agriculture*, 2, 1–12.

Jones, J.W., Antle, J.M., Basso, B., Boote, K.J., Conant, R.T., Foster, I., Godfray, H.C.J., Herrero, M., Howitt, R.E. & Janssen, S., 2017. Toward a new generation of agricultural system data, models, and knowledge products: State of agricultural systems science. *Agricultural Systems*, 155, 269–288.

Jose, S., 2019. Environmental impacts and benefits of agroforestry. *In Oxford Research Encyclopedia of Environmental Science*. https://doi.org/10.1093/acrefore/9780199389414.013.195

Karanisa, T., Amato, A., Richer, R., Abdul Majid, S., Skelhorn, C. & Sayadi, S., 2021. Agricultural production in Qatar's hot arid climate. *Sustainability*, 13, 4059.

Kelly, C., Schipanski, M.E., Tucker, A., Trujillo, W., Holman, J.D., Obour, A.K., Johnson, S., Brummer, J.E., Haag, L. & Fonte, S.J., 2021. Dryland cover crop soil health benefits are maintained with grazing in the US High and Central Plains. *Agriculture, Ecosystems & Environment*, 313, 107358.

Kibiti, J.G., 2017. *Factors Influencing Adoption of Urban Hydroponic Farming. A Case of Meru Town, Meru County*. Kenya: University of Nairobi.

Kitchen, N., Snyder, C., Franzen, D. & Wiebold, W., 2002. Educational needs of precision agriculture. *Precision Agriculture*, 3, 341–351.

Kole, C., Muthamilarasan, M., Henry, R., Edwards, D., Sharma, R., Abberton, M., Batley, J., Bentley, A., Bryant, J. & Cai, H., 2015. Application of genomics-assisted breeding for generation of climate resilient crops: Progress and prospects. *Frontiers in Plant Science*, 6, 153364.

Kumar, J., Chawla, R., Katiyar, D., Chouriya, A., Nath, D., Sahoo, S. & Ali, A., 2023. Optimizing irrigation and nutrient management in agriculture through artificial intelligence implementation. *International Journal of Environment and Climate Change*, 13, 4016–4022.

Kurade, N., Chavan, P., Pawar, S., Gaikwad, B., Nirmale, A., Kumar, N. & Mp, B., 2021. Impacts and management of abiotic stresses on livestock in the drought prone areas of Maharashtra. *Technical Bulletin*, 30.

Lean, I. & Moate, P., 2021. Cattle, climate and complexity: Food security, quality and sustainability of the Australian cattle industries. *Australian Veterinary Journal*, 99, 293–308.

Lee, M.A., Davis, A.P., Chagunda, M.G. & Manning, P., 2017. Forage quality declines with rising temperatures, with implications for livestock production and methane emissions. *Biogeosciences*, 14, 1403–1417.

Lukyanova, M., Kovshov, V., Zalilova, Z., Lukyanov, V. & Araslanbaev, I., 2021. A systemic comparative economic approach efficiency of fodder production. *Journal of Innovation and Entrepreneurship*, 10, 1–17.

Mahmood, A., Hussain, A., Shahzad, A.N. & Honermeier, B., 2015. Biomass and biogas yielding potential of sorghum as affected by planting density, sowing time and cultivar. *Pakistan Journal of Botany*, 47, 2401–2408.

Mahmood, A., Ullah, H., Shahzad, A., Ali, H., Ahmad, S., Zia-Ul-Haq, M., Honermeier, B. & Hasanuzzaman, M., 2013. Dry matter yield and chemical composition of sorghum cultivars with varying planting density and sowing date. *Sains Malaysiana*, 42, 1529–1538.

Mahmood, A., Wang, X., Shahzad, A.N., Fiaz, S., Ali, H., Naqve, M., Javaid, M.M., Mumtaz, S., Naseer, M. & Dong, R., 2021. Perspectives on bioenergy feedstock development in Pakistan: Challenges and opportunities. *Sustainability*, 13, 8438.

Maity, A., Vijay, D., Mukherjee, A. & Lamichaney, A., 2016. Potential Impacts of Climate Change on Quality Seed Production: A Perspective of Hill Agriculture. *In* J.K. Bisht, V.S. Meena, P.K. Mishra & A. Pattanayak (eds.) *Conservation Agriculture: An Approach to Combat Climate Change in Indian Himalaya*. Singapore: Springer Singapore, 459–485.

Makkar, H., 2018. Feed demand landscape and implications of food-not feed strategy for food security and climate change. *Animal*, 12, 1744–1754.

Mcgrath, S., Behrendt, R., Friend, M. & Moore, A., 2021. Utilising dual-purpose crops effectively to increase profit and manage risk in meat production systems. *Animal Production Science*, 61, 1049–1061.

Mehmood, M.U., Chun, D., Han, H., Jeon, G. & Chen, K., 2019. A review of the applications of artificial intelligence and big data to buildings for energy-efficiency and a comfortable indoor living environment. *Energy and Buildings*, 202, 109383.

Mendoza-Grimón, V., Hernández-Moreno, J.M. & Palacios-Díaz, M.D.P., 2015. Improving water use in fodder production. *Water*, 7, 2612–2621.

Mengistu, A., 1998. Conservation-based forage development for Ethiopia. Self Help Development International and Institute for Sustainable Development. Berhanena Selam Printing Press, Addis Ababa, Ethiopia, 197.

Mengistu, A., Kebede, G., Assefa, G. & Feyissa, F., 2016. Improved forage crops production strategies in Ethiopia: A review. *Academic Research Journal of Agricultural Science and Research*, 4, 285–296.

Mengistu, A., Kebede, G., Feyissa, F. & Assefa, G., 2017. Review on major feed resources in Ethiopia: Conditions, challenges and opportunities. *Academic Research Journal of Agricultural Science and Research*, 5, 176–185.

Meyer, N., Bergez, J.-E., Justes, E. & Constantin, J., 2022. Influence of cover crop on water and nitrogen balances and cash crop yield in a temperate climate: A modelling approach using the STICS soil-crop model. *European Journal of Agronomy*, 132, 126416.

Mir, M.S., Naikoo, N.B., Kanth, R.H., Bahar, F., Bhat, M.A., Nazir, A., Mahdi, S.S., Amin, Z., Singh, L. & Raja, W., 2022. Vertical farming: The future of agriculture: A review. *The Pharma Innovation Journal*, 11, 1175–1195.

Moraine, M., Duru, M. & Therond, O., 2017. A social-ecological framework for analyzing and designing integrated crop–livestock systems from farm to territory levels. *Renewable Agriculture and Food Systems*, 32, 43–56.

Moreno, G. & Rolo, V., 2019. Agroforestry Practices: Silvopastoralism. *In Agroforestry for Sustainable Agriculture*. Burleigh Dodds Science Publishing, 119–164.

Mustafa, A., Athar, F., Khan, I., Chattha, M.U., Nawaz, M., Shah, A.N., Mahmood, A., Batool, M., Aslam, M.T. & Jaremko, M., 2022. Improving crop productivity and nitrogen use efficiency using sulfur and zinc-coated urea: A review. *Frontiers in Plant Science*, 13, 942384.

Nazari, B., Liaghat, A., Akbari, M.R. & Keshavarz, M., 2018. Irrigation water management in Iran: Implications for water use efficiency improvement. *Agricultural Water Management*, 208, 7–18.

Neely, C., Walsh, C., Davis, J., Hunt, C. & Brown, J., 2015. Investigation of early planted winter canola as a dual-purpose crop for silage and seed production. *Agronomy Journal*, 107, 1905–1914.

Nigus, A., 2017. Pasture management and improvement strategies in Ethiopia. *Journal of Biology, Agriculture and Healthcare*, 7, 69–78.

Ohshima, M., Kimura, E. & Yokota, H.-O., 1997. A method of making good quality silage from direct cut alfalfa by spraying previously fermented juice. *Animal Feed Science and Technology*, 66, 129–137.

Olson, D. & Anderson, J., 2021. Review on unmanned aerial vehicles, remote sensors, imagery processing, and their applications in agriculture. *Agronomy Journal*, 113, 971–992.

Papanastasis, V., Yiakoulaki, M., Decandia, M. & Dini-Papanastasi, O., 2008. Integrating woody species into livestock feeding in the Mediterranean areas of Europe. *Animal Feed Science and Technology*, 140, 1–17.

Parker, L.E., Mcelrone, A.J., Ostoja, S.M. & Forrestel, E.J., 2020. Extreme heat effects on perennial crops and strategies for sustaining future production. *Plant Science*, 295, 110397.

Peyraud, J.-L., Taboada, M. & Delaby, L., 2014. Integrated crop and livestock systems in Western Europe and South America: A review. *European Journal of Agronomy*, 57, 31–42.

Pomoni, D.I., Koukou, M.K., Vrachopoulos, M.G. & Vasiliadis, L., 2023. A review of hydroponics and conventional agriculture based on energy and water consumption, environmental impact, and land use. *Energies*, 16, 1690.

Pond, W.G., Church, D.B., Pond, K.R. & Schoknecht, P.A., 2004. *Basic Animal Nutrition and Feeding*. John Wiley & Sons.

Rao, C.S., Srinivas, T., Rao, R., Rao, N.S., Vinayagam, S.S. & Krishnan, P., 2020. *Climate Change and Indian Agriculture: Challenges and Adaptation Strategies*. ICAR-National Academy of Agricultural Research Management, Hyderabad, Telangana, India. p. 584.

Rao, K., Verchot, L.V. & Laarman, J., 2007. Adaptation to climate change through sustainable management and development of agroforestry systems. *Journal of SAT Agricultural Research*, 4, 1–30.

Rasheed, A., Barqawi, A.A., Mahmood, A., Nawaz, M., Shah, A.N., Bay, D.H., Alahdal, M.A., Hassan, M.U. & Qari, S.H., 2022a. CRISPR/Cas9 is a powerful tool for precise genome editing of legume crops: A review. *Molecular Biology Reports*, 49, 5595–5609.

Rasheed, A., Mahmood, A., Maqbool, R., Albaqami, M., Sher, A., Sattar, A., Bakhsh, G., Nawaz, M., Hassan, M.U. & Al-Yahyai, R., 2022b. Key insights to develop drought-resilient soybean: A review. *Journal of King Saud University-Science*, 34, 102089.

Rasheed, A., Raza, A., Jie, H., Mahmood, A., Ma, Y., Zhao, L., Xing, H., Li, L., Hassan, M.U. & Qari, S.H., 2022c. Molecular tools and their applications in developing salt-tolerant soybean (*Glycine max* L.) cultivars. *Bioengineering*, 9, 495.

Reta-Sánchez, D., Serrato-Corona, J., Quiroga-Garza, H., Gaytán-Mascorro, A. & Figueroa-Viramontes, U., 2016. Forage yield and chemical composition of canola (*Brassica napus* L.) as affected by sowing methods. *Grass and Forage Science*, 71, 281–290.

Rodríguez-Ortega, T., Oteros-Rozas, E., Ripoll-Bosch, R., Tichit, M., Martín-López, B. & Bernuès, A., 2014. Applying the ecosystem services framework to pasture-based livestock farming systems in Europe. *Animal*, 8, 1361–1372.

Ryschawy, J., Tiffany, S., Gaudin, A., Niles, M.T. & Garrett, R.D., 2021. Moving niche agroecological initiatives to the mainstream: A case-study of sheep-vineyard integration in California. *Land Use Policy*, 109, 105680.

Saleem, M.A., Tahir, M., Ahmad, T. & Tahir, M.N., 2020. Foliar application of boron improved the yield and quality of wheat (*Triticum aestivum* L.) in a calcareous field. *Soil & Environment*, 39, 59–66.

Salinger, M., Stigter, C. & Das, H., 2000. Agrometeorological adaptation strategies to increasing climate variability and climate change. *Agricultural and Forest Meteorology*, 103, 167–184.

Sanginga, N., Dashiell, K.E., Diels, J., Vanlauwe, B., Lyasse, O., Carsky, R., Tarawali, S., Asafo-Adjei, B., Menkir, A. & Schulz, S., 2003. Sustainable resource management coupled to resilient germplasm to provide new intensive cereal–grain–legume–livestock systems in the dry savanna. *Agriculture, Ecosystems & Environment*, 100, 305–314.

Sarkar, R., Corriher-Olson, V., Long, C. & Somenahally, A., 2020. Challenges and potentials for soil organic carbon sequestration in forage and grazing systems. *Rangeland Ecology & Management*, 73, 786–795.

Schütte, G., Eckerstorfer, M., Rastelli, V., Reichenbecher, W., Restrepo-Vassalli, S., Ruohonen-Lehto, M., Saucy, A.-G.W. & Mertens, M., 2017. Herbicide resistance and biodiversity: Agronomic and environmental aspects of genetically modified herbicide-resistant plants. *Environmental Sciences Europe*, 29, 1–12.

Seré, C., Steinfeld, H. & Groenewold, J., 1996. *World Livestock Production Systems.* Food and Agriculture Organization of the United Nations.

Shikwambana, S., Malaza, N. & Ncube, B., 2022. Enhancing the resilience and adaptive capacity of smallholder farmers to drought in the Limpopo Province, South Africa. *Conservation*, 2.

Snapp, S., 2017. Agroecology: Principles and Practice. *Agricultural Systems.* Elsevier, 33–72.

Sneath, R. & Mcintosh, F., 2003. *Review of Hydroponic Fodder Production for Beef Cattle.* Department of Primary Industries: Queensland Australia, 84, 54.

Sobola, O.O., Amadi, D. & Jamala, G., 2015. The role of agroforestry in environmental sustainability. *IOSR Journal of Agriculture and Veterinary Science*, 8, 20–25.

Speranza, C.I., Kiteme, B. & Wiesmann, U., 2008. Droughts and famines: The underlying factors and the causal links among agro-pastoral households in semi-arid Makueni district, Kenya. *Global Environmental Change*, 18, 220–233.

Stark, F., Ryschawy, J., Grillot, M., Mettauer, R., Martin, G., Cassagnes, A., Bertin, Y. & Moraine, M., 2021. Biotechnical and sociotechnical conditions required to engage an agroecological transition: A case-study of organic sheep crop integration in South-Western France. *Landscape 2021*: Diversity for Sustainable and Resilient Agriculture, Online Conference, 20–22 September 2021.

Sundström, J.F., Albihn, A., Boqvist, S., Ljungvall, K., Marstorp, H., Martiin, C., Nyberg, K., Vågsholm, I., Yuen, J. & Magnusson, U., 2014. Future threats to agricultural food production posed by environmental degradation, climate change, and animal and plant diseases–a risk analysis in three economic and climate settings. *Food Security*, 6, 201–215.

Swinton, S.M. & Lowenberg-Deboer, J., 2001. Global adoption of precision agriculture technologies: Who, when and why. *In Proceedings of the 3rd European conference on precision agriculture*, 557–562.

Tanveer, A., Arshad, M.S., Ayub, M., Javaid, M.M. & Yaseen, M., 2012. Effect of temperature, light, salinity, drought stress and seeding depth on germination of *Cucumis melo* var. agrestis. *Pakistan Journal of Weed Science Research*, 18.

Tanveer, A., Mumtaz, K., Javaid, M., Chaudhry, M., Balal, R. & Khaliq, A., 2013. Effect of ecological factors on germination of horse purslane (*Trianthema portulacastrum*). *Planta Daninha*, 31, 587–597.

Tittonell, P., Klerkx, L., Baudron, F., Félix, G.F., Ruggia, A., Van Apeldoorn, D., Dogliotti, S., Mapfumo, P. & Rossing, W.A., 2016. Ecological intensification: Local innovation to address global challenges. *Sustainable Agriculture Reviews*, 19, 1–34.

Turner, N.C., Molyneux, N., Yang, S., Xiong, Y.-C. & Siddique, K.H., 2011. Climate change in south-west Australia and north-west China: Challenges and opportunities for crop production. *Crop and Pasture Science*, 62, 445–456.

Tyasi, T.L., Gxasheka, M. & Tlabela, C.P., 2015. Assessing the effect of nutrition on milk composition of dairy cows: A review. *International Journal of Current Science*, 5, 56–63.

Ullah, H., Mahmood, A., Ijaz, M., Tadesse, B. & Honermeier, B., 2013. Evaluation of anise (*Pimpinella anisum* L.) accessions with regard to morphological characteristics, fruit yield, oil contents and composition. *Journal of Medicinal Plants Research*, 7, 2177–2186.

Vandehaar, M.J. & St-Pierre, N., 2006. Major advances in nutrition: Relevance to the sustainability of the dairy industry. *Journal of Dairy Science*, 89, 1280–1291.

Waheed, H., Javaid, M.M., Shahid, A., Ali, H.H., Nargis, J. & Mehmood, A., 2019. Impact of foliar-applied Hoagland's nutrient solution on growth and yield of mash bean (*Vigna mungo* L.) under different growth stages. *Journal of Plant Nutrition*, 42, 1133–1141.

Wasaya, A., Abbas, T., Yasir, T.A., Sarwar, N., Aziz, A., Javaid, M.M. & Akram, S., 2021. Mitigating drought stress in sunflower (*Helianthus annuus* L.) through exogenous application of β-aminobutyric acid. *Journal of Soil Science and Plant Nutrition*, 21, 936–948.

Wassie, S.B., 2020. Natural resource degradation tendencies in Ethiopia: A review. *Environmental Systems Research*, 9, 1–29.

Wright Morton, L. & Miller, L., 2007. Connecting sustainable agriculture to rural development: The case of pasture-based dairy grazing. *Community Development*, 38, 23–38.

Yeşil, V., Çakaloğullari, U. & Tatar, Ö., 2020. Effect of light intensity on dry matter accumulation of barley fodder in a vertical farming growth module. Agronomy LXIII, 2, 245–248.

Yeşil, V. & Tatar, Ö., 2020. An innovative approach to produce forage crops: Barley fodder in vertical farming system. *Scientific Papers. Series A. Agronomy*, 63.

Zhen, L., Zoebisch, M.A., Chen, G. & Feng, Z., 2006. Sustainability of farmers' soil fertility management practices: A case study in the North China Plain. *Journal of Environmental Management*, 79, 409–419.

14 Sustainable Fiber Production

Usman Tufail, Abrar Hussain, Rabia Zahid, and Ikram ul Haq*

14.1 INTRODUCTION

The adverse effects of utilizing materials derived from fossil fuels in large quantities, including heightened greenhouse gas emissions, expanding landfills, environmental contamination, and human health issues, have motivated materials scientists to explore innovative materials consistent with the principles of sustainability and circular economy. In an attempt to make goods better for the earth, more attention is being paid to natural resources as alternatives to materials made from oil. Sustainable fibers could be used in many end-uses because they are easy to find, long-lasting, eco-friendly, cause less tool wear, and irritate the skin less (Alemu Reta *et al.*, 2024). Since more than 2 millennia ago, reinforcement materials have frequently been natural fibers. Particularly common applications for cellulose-based natural fibers include the production of paper, polymeric composites, household textiles, waste-water treatment, and the spinning of regenerated cellulosic fibers (Frazier *et al.*, 2024). In addition to providing opportunities for the development of a variety of products, natural fibers have a substantial market share. For lightweight applications, including packaging, furniture, electronic appliances, sports equipment, and leisure products, cellulose-based natural fiber-reinforced polymer (FRP) composites are ideal. As a result of their composability, cellulose-based natural fibers are viable alternatives to synthetic fibers when used to reinforce polymers (Nyssanbek *et al.*, 2024). Conversely, cellulosic fibers possess certain limitations, including inadequate resistance to heat and microbes, fluctuations in quality that occur locally and seasonally, an irregular chemical composition, and a significant capacity for absorbing moisture. Due to the considerable variability in the characteristics of plant fibers, reliability should be prioritized in fiber cultivation and extraction, especially in composite materials and other engineering applications. The observed heterogeneity can be primarily attributed to variations in structure, chemical composition, and the specific plant part from which the fibers were isolated (Rasheed *et al.*, 2022a; Romanowska *et al.*, 2024).

In general, plant fibers are put into groups based on what part of the plant they come from. Because of this, fibers are mostly put into four groups: fruit, leaf, seed, and bast fibers. Plant fibers also include straw fibers (from corn, rice, and wheat), grass fibers (from bagasse and bamboo), and wood fibers (from softwood and hardwood). These fibers are very important in the textile and plastic industries (Dave and Macwan, 2024). Agricultural trash can also be used to make plastic products stronger or as an addition, which is good for the environment and getting rid of waste (Hanif *et al.*, 2017; Kumar *et al.*, 2024b). The very small structure of cellulosic fibers changes in big ways. The amount of these components differs depending on where the plant originated, how old it is, and which section of the plant is utilized to extract the fibers. Because of these variations, it is always important to consider the variety of the plant fibers. The way a substance's molecules are grouped usually reveals its basic structure. Polymeric fabric strands have a complicated structure made up of partly aligned, partially crystalline, and straight macromolecular chains. For a long time, researchers have been investigating how polymers are packed in plant fiber structures, and numerous potential solutions have been presented (Rasheed *et al.*, 2022b; Maraveas *et al.*, 2024).

The diversity of cellulosic fibers may be attributed to the chemical configuration of these fibers. All plants include cellulose as a significant constituent. Cellulose is a linear polymer composed of 1,4-β-glucan, which is composed of repeated units of D-anhydroglucose ($C_6H_{10}O_5$) with hydroxyl

* Corresponding author: abrar.hussain@cuisahiwal.edu.pk

 DOI: 10.1201/9781003498698-14

groups. The fiber has a pronounced hydrophilic character due to the presence of hydroxyl groups. Consequently, the presence of many hydrogen bonds among cellulose chains has the potential to enhance the crystallinity of the fiber, thus leading to an enhancement in its mechanical characteristics (Feng *et al.*, 2024). Hemicellulose, lignin, and pectin are some of the other parts of cellulosic fibers. It's possible to think of lignocellulosic fibers as hybrid materials made up of hemicellulose and cellulose microfibrils that are arranged in a random pattern along the fibers' length. A low-molecular-weight carbohydrate called hemicellulose is made up of hexoses, pentoses, and uronic acid groups. To put it another way, hemicellulose is a mixed bunch of different kinds of sugars that have a backbone made up of glucose, mannose, or xylose (Piazza *et al.*). The chains in hemicellulose are short and spread out. The amount of water in the fiber is controlled by hemicellulose, which is unstructured. Lignin is the second-most prevalent polymer in the world. It is made up of a lot of different branches and phenylpropane units that are arranged in a three-dimensional way. Lignin functions as a glue in cell walls, binding basic fibrils together and preventing microorganisms from breaking them down. Lignin also protects cellulose and hemicellulose from the outer world. Plant cell walls also have pectins, which are a type of protein. It makes up most of the middle lamella and can also be found in the primary and secondary cell walls. Cellulose, hemicellulose, and lignin are all linked together by this polymer (Delmer *et al.*, 2024). The wax affects the fiber's surface wetness and the way the fiber and matrix interact during production. When extracting fibers, it is very important to keep the structure of the fiber bundles so that the plant fibers' mechanical properties don't change (Bahrami *et al.*, 2024).

A major issue is that a growing quantity of utilized clothing is ending up in dumps as municipal solid waste (MSW). These garments are discarded by both consumers and fashion corporations. In 2018, the US Environmental Protection Agency (USEPA) estimated that textile trash accounted for 5.8% of all MSW that was discarded, or approximately 17 Mt (Sigman and Strow, 2024). The fashion and textile industries have been forced to adopt and use eco-friendly methods when making, using, and throwing away textiles because of the growing environmental issues caused by the buildup of non-biodegradable textile waste and the lack of landfill space. To keep the huge amount of cloth trash out of dumps, experts had to come up with new, long-lasting ways to add value and use it in useful ways (Rasheed *et al.*, 2022c; Azad *et al.*, 2024). We can lessen the damage that textile waste does to the earth by coming up with new ways to collect, recycle, and reuse it. This will help them get their money back and give people an easy way to get rid of their trash other than putting it in a dump. A circular textile economy, also known as a closed-loop system, is a more environmentally friendly way to deal with textile trash than the current open-loop method (Thacharodi *et al.*, 2024).

14.2 TYPES OF SUSTAINABLE FIBERS

14.2.1 Natural Fibers

Depending on their origin, natural fibers are classified as seed hair, bast fibers, or leaf fibers. Cotton (seed filaments), ramie, jute, and flax (bast fibers), sisal, and abaca (leaf fibers) are some examples. Jute, ramie, flax, and sisal are the fibers that are most frequently employed in the production of polymer composites. Additionally, wood flour, which is a form of natural fiber, has frequently been employed in the fabrication of natural fiber composites (Feng *et al.*, 2024). In Table 14.1, the characteristics of these fibers are detailed. The tensile strength of glass fibers is significantly greater than that of natural fibers, even though their modulus values are identical (Table 14.1). On the other hand, it is evident from examining the specific modulus of natural fibers (calculated as modulus/specific gravity) that the values displayed by natural fibers are either equal to or higher than those of glass fibers. For applications in which weight reduction is also a desired property, one of the primary benefits of utilizing natural fiber composites is their increased specific properties. Three primary varieties of fiber exist (Khan *et al.*, 2024).

TABLE 14.1
Mechanical Properties of Some Natural Fibers

Fiber	Length (mm)[a]	Tensile Strength (MPa)[a]	Elongation at Break (%)[a]	Moisture Content (%)[a]	Diameter (μm)[a]	Young's Modulus (GPa)[a]	Density (g/cm³)[a]	Reference
Flax	10–65 (37.5)	343–1035 (689)	1.2–3 (2.1)	7	5–38 (21.5)	50–70 (60)	1.38	Peças *et al.* (2018)
Banana	0.4–0.9 (0.65)	529–914 (721.5)	5–6 (5.5)	10–11 (10.5)	12–30 (21)	27–32 (29.5)	1.35	Peças *et al.* (2018)
Ramie	40–250 (145)	400–938 (669)	2–4 (3)	12–17 (14.5)	18–80 (49)	61.4–128 (94.7)	1.44	Thapliyal *et al.* (2023)
Sisal	0.8–8 (4.4)	507–855 (981)	1.9–3 (2.45)	11	7–47 (27)	9–22 (15.5)	1.2	Dias (1992)
Pineapple	3–8 (5.5)	170–1627 (898.5)	1–3 (2)	14	8–41 (24.5)	60–82 (71)	1.56	Ntenga *et al.* (2022)
Bamboo	1.5–4 (2.75)	270–862 (566)	1.6–8 (4.65)	11–17 (14)	25–88 (56.5)	18–89 (53)	0.6–1.1 (0.85)	Paul (2013)
Hemp	5–55 (30)	580–1110 (845)	1.6–4.5 (3.05)	8	10–51 (30.5)	30–60 (45)	1.47	Baran (2023)
Cotton	15–56 (35.5)	287–597 (442)	2–10 (6)	33–34 (33.5)	12–35 (23.5)	6–10 (8)	1.21	Baker *et al.* (2014)
Kenaf	1.4–11 (6.2)	295–930 (612.5)	2.7–6.9 (4.8)	6.2–12 (9.1)	12–36 (24)	22–60 (41)	1.2	Thapliyal *et al.* (2023)
Abaca	4.6–5.2 (4.9)	430–813 (621.5)	2.9	14	10–30 (20)	31.1–33.6 (32.35)	1.5	Baran (2023)
Jute	0.8–6 (3.4)	187–773 (480)	1.5–3.1 (2.3)	12	5–25 (15)	20–55 (37.5)	1.23	Weber-Kamin (2020)

[a] Denotes the unit of measurement for different parameters.

14.2.2 Hemp Fiber

As the demand for biodegradable, sustainable, and recyclable materials has increased in recent years, there has been a surge in the utilization of hemp fibers as reinforcement in composite materials. The stem of the plant contains hemp fibers, which impart strength and rigidity, a fundamental prerequisite for reinforcing composite materials, to it (Akter *et al.*, 2024). Comparable mechanical properties distinguish hemp fibers from glass fibers. However, the variability of their properties is their greatest drawback. Hemp fiber composites comprising thermoplastic, thermosetting, and biodegradable matrices have demonstrated favorable mechanical characteristics. Several hemp fiber surface treatments, which were implemented to enhance the interfacial bonding between the fiber and matrix, have yielded significant enhancements in the mechanical properties of the composites (Fahmi *et al.*, 2024).

14.2.3 Kapok Fiber

A light fiber with a porous tubular structure, the kapok fiber measures between 1 and 2 cm in length. Cotton fibers have a mean external diameter of 16.8 μm and a wall thickness of 3.9 μm, whereas the fibers consist of microtubes with a mean external diameter of 10 μm and a wall thickness of 0.1 μm (Qing *et al.*, 2024).

These traits that result in reduced vigor have also been observed in other *Ceiba* species. Kapok fiber accumulates a similar amount of lint as cotton due to its origins in endocarp cells. Cotton fiber is made from the seed coat's epidermal cells. As documented in different parts of the globe, the most common application of the fiber produced by any of the *Ceiba* species is as a padding material for mattresses and cushions (Daubert, 2020). On the contrary, kapok fiber aggregates possess advantageous characteristics, such as high porosity and superhydrophobicity, which render them well suited for life-saving supplies owing to their maneuverability and enhanced buoyancy. Furthermore, these aggregates exhibit biodegradability and resistance to acid and alkali and are abundant in nature, all of which are not found in artificial buoyancy materials (Zaharia *et al.*, 2024). Kapok fibers are short and lightweight, making them unsuitable for textile applications. To enhance their stability, kapok fibers must be incorporated with other cellulosic fibers, such as cotton or rayon, prior to their use in fabrics or filaments (Chowdhury *et al.*, 2024). In contrast to a comparable product constructed solely from cotton or synthetic fibers, kapok blended with other fibers or cotton in the production of fabrics or yarns may reduce the amount of water and resources used during manufacturing, thereby diminishing the product's carbon footprint and environmental impact (Mahmood *et al.*, 2021; Chen *et al.*, 2024). The apparel and textile sectors are among the most ecologically detrimental industries and encounter numerous obstacles in their pursuit of sustainability throughout the entire production lifecycle. Thus, the diversification of primary materials and environmentally benign manufacturing alternatives will contribute to the mitigation of present and future effects (Ullah *et al.*, 2013; Castellanos *et al.*, 2024).

Kapok fibers, which contain as much as 69% cellulose, have the potential to serve as a source of cellulose and nanocellulose. Additionally, the fibers of *Chaenomeles speciosa* and *Christensenia aesculifolia* have been reported to contain a significant amount of cellulose (Gómez-Maqueo and Gamboa-Debuen, 2022). This indicates that the fibers of various *Ceiba* species may serve as a valuable reservoir of cellulose and nanocellulose, which are polymers that are widely employed in the biotechnology sector. Due to its hydrophobic composition and high acetyl group content (approximately 13%), kapok fiber is an exceptional oil-absorbent (Zerga and Tahir, 2022). Oil that has been released into water may be recoverable using this fiber, according to one theory. Additionally, kapok fiber, being a naturally occurring substance characterized by its comparatively low cost and enhanced biodegradability, may present a more favorable alternative in comparison to conventional synthetic products (Cao *et al.*, 2023).

14.2.4 Organic Cotton

One of the most natural fabrics available, organic cotton tops our inventory of sustainable fabrics. Organic cotton, in contrast to conventional cotton cultivation (also known as the "dirtiest crop in the world"), is processed without the use of chemicals and pesticides, resulting in an overall energy and water consumption reduction of 62 and 88%, respectively (Harymawan *et al.*, 2022). Sustainable and ethical cotton is accompanied by a number of certifications that attest to the following: (A) cultivation and processing devoid of any chemicals or mechanical harvesting; and (B) final garment formulation devoid of any chemical substances. The most prevalent variety of organic cotton is GOTS-certified, or organic cotton that has received official approval from the World Health Organization (WHO) (Ervast, 2024). Additional relevant certifications, such as fair trade, guarantee equitable compensation and secure working conditions for farmers (although the absence of chemical exposure in the field is a significant factor in this regard as well). Our selection of sustainable fabrics is topped with organic cotton, which is among the most natural materials available. Organic cotton is processed without the use of chemicals and pesticides, resulting in a 62 and 88% reduction in overall energy and water use, respectively, when compared to traditional cotton cultivation (also known as the "dirtiest crop in the world") (Harymawan *et al.*, 2022). The prevailing variety of organic cotton is GOTS-certified, which signifies official endorsement from the WHO. Further pertinent certifications, such as fair trade, ensure that farmers are provided with secure working conditions and equitable compensation (although the absence of chemical exposure in the field is also a substantial factor in this regard) (Psychogiou and Tsoulfas, 2024).

14.2.5 Regenerated Cellulose Fiber

Regenerated cellulose is an artificial fiber manufactured through the process of organic solvent-catalyzed degradation of natural cellulose present in wood pulp, followed by fiber extrusion. Particular attention is paid to Lyocell, marketed under the brand name Tencel™, on account of the recyclable solvent utilized in its production. As the majority of its production is carried out by the Austrian manufacturer Lenzing, this initiative investigates whether a circular economy in the United States could be applied to its manufacturing process on a larger scale. In order to transition from waste disposal as in a linear economy to waste recycling and reuse, a primary objective of this project is to determine whether regenerated cellulose fibers, including lyocell, can be expanded in scale throughout the United States, recycled at a higher rate, and emerge as a viable long-term substitute for synthetic fibers. This group of materials is highly adaptable and is recognized for its biodegradability, textile-grade fineness, and capacity to mitigate environmental damage associated with the manufacturing of synthetic fibers. In general, the principal categories of these fibers consist of viscose, modal, and lyocell, all of which are rayon derivatives and are characterized by greater absorption than cotton.

14.2.6 Recycled Fibers

Recycled materials are those that have undergone reprocessing from reclaimed materials via manufacturing processes, thereby achieving incorporation into the ultimate product or a component thereof. The following fibers are frequently recycled: nylon, wool, polyester, and cotton. The recycling process in the textile industry appears to be falling behind, as only 12% of the world's textiles are recycled at present. Paper, plastic PET bottles, and glass are currently at 66, 29, and 27%, respectively, in comparison to it (Boutros *et al.*, 2021). This increases the focus on the modifications that must be implemented in order to reduce the amount of refuse sent to landfills and enable the recycling of materials that are recyclable.

14.3 CHEMICAL COMPOSITION OF NATURAL FIBERS

Particular types of natural fibers have distinct chemical compositions. The principal components of fibers are cellulose, hemicellulose, pectin, and lignin. The aggregate properties of the fiber are influenced by the properties of each constituent. Lignin is thermally stable but is accountable for UV degradation (Shafiq *et al.*, 2024), whereas hemicellulose is moisture absorption-resistant and causes biodegradation (Periyasamy *et al.*, 2023), as well as thermal and moisture absorption and thermal degradation of the fiber, respectively. The proportional distribution of each of these constituents differs among distinct types of fibers (Table 14.2). In general, fiber composition consists of 60–80% cellulose, 5–20% lignin, and a maximum of 20% moisture (Bi *et al.*, 2024).

TABLE 14.2
Chemical Composition of Plant Fibers

Fiber	Hemicellulose (%)	Pectin (%)	Wax (%)	Cellulose (%)	Lignin (%)	Refs.
Alfa	38.5	–	2	45.4	38.5	Derstine *et al.* (2018)
Bagasse	21	10	–	37	22	Kumar *et al.* (2021)
Banana	12.5	4	–	62.5	7.5	Bauri *et al.* (2016)
Barley	27–38	–	2–7	31–45	14–19	Hansson *et al.* (2024)
Eucalyptus	32.56	8.2	0.22	41.7	25.4	Hassan *et al.* (2023)
Cotton	4	6	0.6	89	0.75	Makamov *et al.* (2023)
Hemp	20	0.9	0.8	81	4	Ekren *et al.* (2023)
Flax	14.5	0.9	Flax	72.5	2.5	Avci *et al.* (2023)
Pineapple	17.5	4	–	80.5	8.3	Balo and Sua (2023)
Jute	16	0.2	0.5	67	9	Tareq *et al.* (2024)

14.4 SUSTAINABLE PRACTICES IN FIBER CULTIVATION

Biological diversity and crop rotation have historically been fundamental pillars of effective conventional agricultural production systems. An essential element of "integrated pest management" is the implementation of crop species rotations, which can effectively improve pest management (Tiwari, 2024). One way in which effective pest management can be improved, particularly with regard to insects and vegetation, is through a procedure known as "biological structuring" of systems. It is widely accepted in the farming community that crop rotations effectively mitigate the prevalence of nematodes, insects, plant pathogens, and vegetation. A stronger cultural control of pests can be anticipated in a rotation sequence where the disparities between crops are greater. By implementing rotational strategies, it has been possible to provide crop nutrients, incorporate diversity into agricultural systems, and control parasites in the field (Iqbal *et al.*, 2023). The operational mechanisms governing the interactions between plants and animals on an agricultural property may be referred to as the biological structuring of a system (Figure 14.1).

14.4.1 Organic Farming Techniques

14.4.1.1 Role of Crop Rotation

The monoculture agricultural technique may facilitate the proliferation of pathogenic microorganisms. Rotations that result in the growth of a plant from a different family disrupt the pathogen cycle, since the pathogen is incapable of infecting plants from various families. Consequently, the population of pathogens in the soil decreases significantly (Geoffrey, 2024). Consider a seven-year period of diverse crop rotation in which three years are devoted to alfalfa, two to chile pepper, and two to cotton, before returning to alfalfa. Cotton, alfalfa, and chile pepper are three plants that represent a range of goods (Jones and Congdon, 2024). The fact that these three plants—alfalfa, chile, and cotton—belong to distinct crop families enables them to assist in disrupting the disease cycle and increasing output. The Disease Control Rotation (DCR) contributes to pest management through the interference with the pest cycle, reduction of vegetation and diseases, enhancement of soil quality, and protection of the ecosystem (Lakhiar *et al.*, 2024) (Figure 14.1). The capacity of diversified crop rotation to disrupt the disease cycle within a soil profile is extraordinarily advantageous. Monoculture occurring on the same piece of land facilitates pathogen accumulation. The lack of rotation can facilitate the proliferation of these infections within the soil and impede their spread, thereby exacerbating the severity of plant disease outbreaks (Strydom *et al.*, 2024).

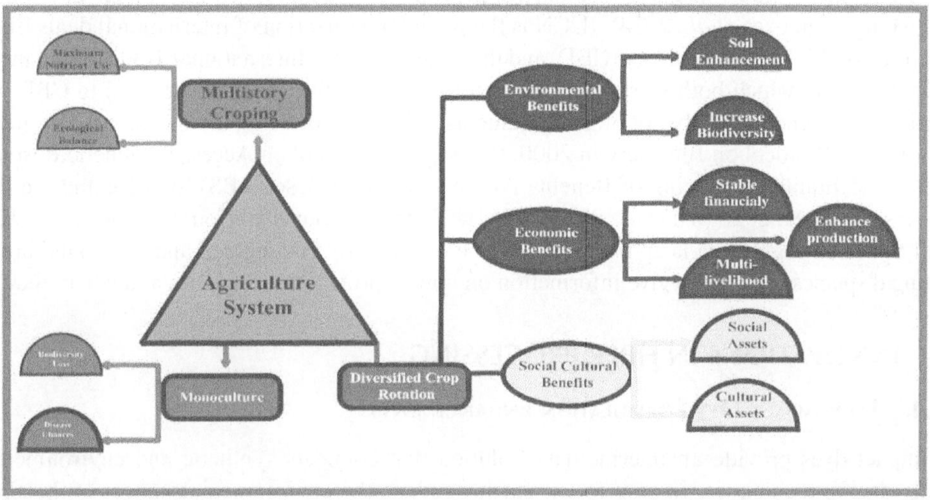

FIGURE 14.1 Conceptual framework of importance of DCR.

14.4.1.2 Soil Health Improvement

The fact that these three plants—alfalfa, chile, and cotton—belong to distinct crop families enables them to assist in disrupting the disease cycle and increasing output. The DCR contributes to pest management through the interference with the pest cycle, reduction of vegetation and diseases, enhancement of soil quality, and protection of the ecosystem. The capacity of diversified crop rotation to disrupt the disease cycle within a soil profile is extraordinarily advantageous. Monoculture occurring on the same piece of land facilitates pathogen accumulation. The lack of rotation can facilitate the proliferation of these infections within the soil and impede their spread, thereby exacerbating the severity of plant disease outbreaks (Harb *et al.*, 2024).

14.4.1.3 Organic Farming

Organic farming, on the contrary, uses closed-loop nutrient cycling, like biological nitrogen fixation and crop rotations, to keep the soil fertile by increasing the amount of organic matter in it. This means that synthetic fertilizers and pesticides are not used, and organic farming is better for the environment. Its current output per area of 0.037 Gha is still small compared to the world's 4.9 Gha of arable land, but it is steadily growing (Chadha *et al.*, 2020; Djedaiet *et al.*, 2024). Recent claims about organic farming include lower nitrogen losses to the environment and, more importantly, better soil carbon sequestration. These two benefits could cancel out 60–92% of current agricultural greenhouse gas emissions if all land were used for organic farming. In organic farming, food returns are often much lower than in standard farming because they depend on natural fertility (Ali *et al.*, 2013; Wu *et al.*, 2024). This is especially true when talking about standard methods that involve high inputs and high returns, where crops may be 16–72% bigger than in organic farming. Some good ways to deal with trash are to recycle, compost, and use organic waste as a resource. This lowers the damage that farming does to the earth (Šeremešić *et al.*, 2024).

14.4.1.4 Biodiversity Conservation

Most of the world's plant and animal life is assumed to live in warm climates. In contrast, studies revealed that tropical woodlands are decreasing at a pace of 0.8% each year. In the past 25 years, mankind has destroyed 10% of the world's tropical forests. (Martínez-Ruiz *et al.*, 2024). It is thought that tropical woods contribute more than 25% of the world's oxygen, and the seas make about 70% of all the oxygen that the earth needs (Bhuiyan *et al.*, 2024).

The International Union for Conservation of Nature and Natural Resources (IUCN) is the authority on the state of nature around the world and what needs to be done to protect it. Its major goal is to give governments, non-government organizations, companies, experts, and people from all over the world a place to meet and work together to protect wildlife and reach sustainable development goals (Monge-Ganuzas *et al.*, 2024). IUCN is the parent organization of international deals like the Convention on Biological Diversity (CBD) and the Convention for International Trade in Endangered Species (CITES), which both work to protect wildlife around the world. In addition to CBD, most countries have signed a number of other agreements aimed at protecting biodiversity. These include the Cartagena Protocol on Biosafety in 2000, the Nagoya Protocol on Access to Genetic Resources and Fair and Equitable Sharing of Benefits Arising from their Use (ABS) in 2010, and the Aichi Biodiversity Targets 2020 (Yu *et al.*, 2024). The IUCN Species Survival Commission and the IUCN Global Species Programme keep track of the state of critically endangered and rare plant, animal, and fungal species. They also give information on how to protect these species and where they live.

14.5 INNOVATIONS IN FIBER PROCESSING

14.5.1 Low-impact Dye Formulation Enhancements

Low-impact dyes provide an intermediary solution that connects synthetic and environmentally friendly dyeing techniques. In an effort to further reduce the environmental impact of their products, scientists are creating formulations that require fewer auxiliary compounds, including salt and alkali (Hernández-Pinto *et al.*, 2024). These chemicals, which are frequently utilized in traditional dyeing, produce harmful water effluents due to the substantial quantities required for the dyeing process. Sustainable coloring methods that are environmentally friendly are transforming the clothing industry. Recent improvements in a number of coloring methods show that the industry is committed to using environmentally friendly methods (Xu and Song, 2024). Scholars are currently involved in active efforts to improve the stability and color fastness of natural dyes, which have garnered attention due to their environmentally friendly characteristics. Co-pigmentation has been identified as a potentially effective method for enhancing color stability (Li *et al.*, 2024). Research has indicated that the addition of chitosan to natural dyes derived from *Caesalpinia* extract results in improved colorfastness, thereby enhancing their suitability for use in apparel applications (Xu *et al.*, 2024). This innovation possesses the capacity to broaden the application of natural dyes while preserving their long-lasting and vivid hues.

14.5.2 Advancements in Equipment Design and Bio-Based and Non-Toxic Inks

Methods of dyeing without water, like air-dye and supercritical carbon dioxide ($scCO_2$)-assisted dyeing (Khalil *et al.*, 2023), are getting better by designing and optimizing tools better. These new methods are much better than traditional ways. They make little trash, use less energy, work for shorter periods of time, and release fewer greenhouse gases. Continuous improvements in the design of tools are needed to make them more useful and economically viable as options for large-scale cloth production (Figure 14.2). In the printing business, using bio-based and non-toxic colors is becoming more popular. Bio-based mineral-oil-free (MOF) inkjet inks made for piezo printing

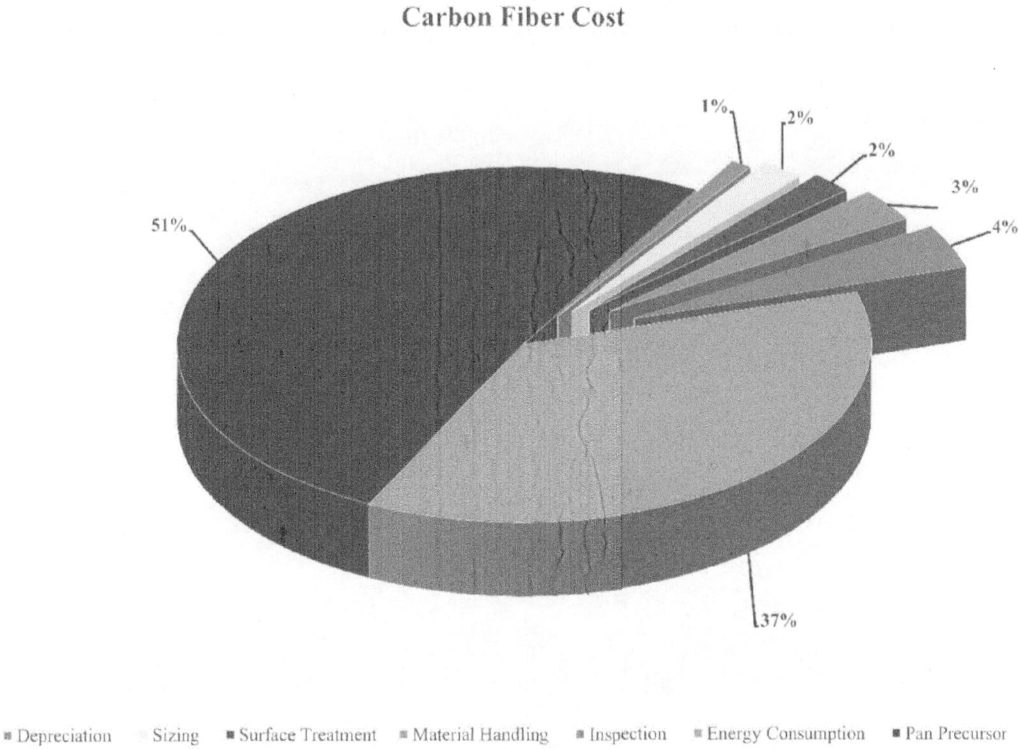

FIGURE 14.2 This figure shows how the usual costs of making carbon fiber are spread out, including the cost of the raw material (PAN). About 37% of the total cost of making PAN-based carbon fiber goes to energy costs, of which 21% are for the stability process.

technology are better for the earth than regular oil-based inks (Garcia, 2023). These bio-conscious paints meet the strict requirements set by environmental laws and rules around the world. This address worries about how printing processes affect the environment.

14.5.3 Energy-Efficient Manufacturing

The use of fibers, especially for fiber optic lines, plays an important role in energy conservation. The type of material utilized, such as glass or plastic, affects how much energy is needed. Cables made of glass normally utilize less energy than wires made of plastic. Microtrenching and tight buffering are two strategies to improve energy efficiency (Bado and Casas, 2021). Energy-efficient fiber production is important for cutting down on carbon emissions and energy use in many areas. When you use energy-efficient fiber building, you use less energy, save money, and get better results. Modern ways of making things with composites, like FRP composites, are stronger and lighter than steel. They help with clean energy efforts by using less energy and releasing fewer carbon emissions (Czapla *et al.*, 2021). Researchers are still looking into ways to make carbon fiber production more energy efficient by adding chemicals that come from living things to the base material (Figure 14.3). This would lower the amount of energy needed for the manufacturing process. These improvements in materials and production methods are necessary to reach fiber production goals for energy economy.

14.5.4 Benefits of Using Fiber in Energy-Efficient

Fiber-optic networks send data with less heat, which means that communication tools and data centers do not need to cool down as much. In turn, this saves even more energy and lowers carbon pollution. When compared to other technologies, fiber-optic lines require less power to send data.

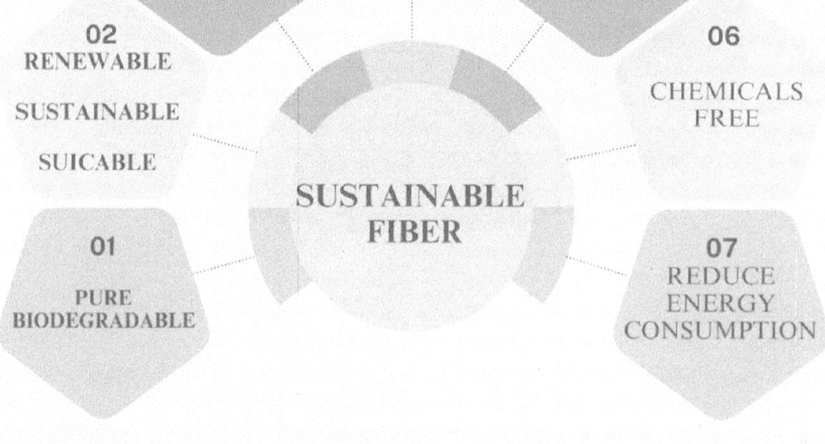

FIGURE 14.3 Advantages of sustainable production.

In fiber-optic communication, light signals replace electrical messaging, which utilizes significantly less energy. Adding fiber construction components, such as carbon fiber or fiberglass, to ordinary building materials strengthens them, improves their life, and uses less energy. Fiber-reinforced materials are extremely hard to break, so they last longer and require less maintenance. Fiber building that is energy-efficient uses less energy, which saves businesses money by lowering their power bills and making their operations more efficient. Fiber connections that use less energy have better total performance, with less signal loss, lower latency, and higher dependability, giving customers a smooth experience. Fiber construction materials are better at insulating than standard building materials. This means that homeowners will use less energy to heat and cool their homes, which will lower their utility bills (Molefe, 2022).

14.6 CASE STUDIES

Natural fibers that can be reused and are excellent for the environment have long been a valuable source of raw materials for producing eco-friendly products. Two of the most popular applications for natural fibers in Indonesia are eco-friendly composites, primarily lightweight composites, and fabrics. For textiles and car parts, ramie and kenaf show the most promise. To make eco-friendly composites, Malaysia wants to use natural fibers like kenaf, pineapple, and carbon waste from the oil palm business. Natural fibers have interesting properties, such as lower density, which means they are lighter, more expensive, biodegradable, easy to find, not too dangerous for your health when processed, having good specific strength and modulus, good thermal and acoustic insulation, good physical properties, and being easy to get. The physical, mechanical, chemical, biological, and anatomical qualities of natural fiber should be thought about in order to get the most out of it and make it last longer. Chemical factors, such as the presence of cellulose in the cell wall, have a significant impact on the tensile strength and mass of natural fiber. While there are certain advantages to using natural fibers, there are also some drawbacks, such as their short lifespan, uneven characteristics, poor adhesion, water absorption, and wettability. The hydrophilicity of natural fibers makes them poor adhesives for use in composites with slippery surfaces. These composites are not only difficult to deal with, but they also have poor dynamic properties. It follows that the appropriate technologies should be used to address the issues in order to expand the range of applications for these fibers.

Maiti *et al.* (2022) stated that using sustainable and natural FRP composites is growing quickly because they have many benefits, such as being good for the environment, biodegradable, cheap, and having good mechanical qualities. So, the natural FRP composites market should hit almost 3.7 million tons in 2022. It should also grow at a rate of more than 9% per year from 2022 to 2027. Trex Company, Inc., The AZEK Company, and Fiberon LLC are the main companies in the world that make natural FRP materials. Trex Company, Inc. has the biggest share of the market, with more than 10%. Asia-Pacific is the biggest market, with more than 30% of all production going there. The building and construction industry is likely to be the most significant market for natural FRP composites, followed by the automobile industry. Due to their eco-friendly uses in many fields, natural fibers are slowly replacing synthetic fibers as the main material used to reinforce composites. This is because natural fibers help clean up the environment in many ways.

Thyavihalli Girijappa *et al.* (2019) examined that chemical treatments on natural fibers mostly improve their qualities by changing their microstructure. These treatments also make the fibers more wettable and improve their surface shape, chemical groups, and tensile strength. The fiber's chemical treatment made the surface of the fiber stick better to the polymer matrix, which improved the thermomechanical properties of the composites. The tensile strength of ramie fibers gets better when they are treated with alkaline or salty chemicals, or when they are treated to get the best results. One of the most important ways to make natural fabrics less porous is to treat them chemically. This also makes them stick better to the matrix. Changes in the form and structure of the fibers can be observed after treatment. This is mainly due to the loss of non-cellulosic compounds from natural fibers. The qualities of the composites got a lot better after being treated with different chemicals. The heat stability of the composites strengthened with natural fibers also got better.

Natural fibers mitigate concrete cracking and enhance tensile strength. Fibers augment the internal distribution of stress and promote the stability of concrete; yet, they diminish the tensile strength (Castillo-Lara *et al.*, 2020; Zafar *et al.*, 2023). Fibers have the ability to improve the thermal insulation and noise reduction properties of concrete by decreasing its density. Overall, 85.7% of natural fiber concrete effectively contributes to lowering carbon dioxide emissions and facilitating the recycling of agricultural waste, resulting in significant environmental advantages (Chen *et al.*, 2023). Nevertheless, it is important to take into account the negative environmental impact caused by greenhouse gas emissions from the disposal of agricultural waste when considering natural fiber concrete. As a result, only 28.6% of natural fiber concrete materials provide adequate economic advantages (Ryłko-Polak *et al.*, 2022). The research indicates that 60% of natural fiber concrete is used in road construction and 30% is used in residential structures (Chen *et al.*, 2023).

14.7 ADVANTAGES OF SUSTAINABLE FIBERS

Supporting sustainable fiber production means using small-scale and local farming methods. This gives people in rural areas a way to make money and helps the local economy grow. Many fiber crops that are good for the environment, like flax and hemp, need less water than common crops like cotton (Barraco, 2021). Rainwater collection and drip watering are two methods that can be used to cut down on water use even more in sustainable fiber production (Wasaya *et al.*, 2021; Koudahe *et al.*, 2024). Hemp is an example of a renewable fiber crop that can take carbon dioxide out of the air as it grows, which further lowers greenhouse gas emissions. Unlike manufactured fibers made from fossil fuels, they come from plants, animals, or rocks that can be grown again or recycled. Organic or agroecological farming methods, which use less industrial fertilizers, pesticides, and herbicides, are often used to make sustainable fibers. This cuts down on chemical pollution of rivers and land (Hassan *et al.*, 2022; Kumar *et al.*, 2024a). There are usually fewer processes that use a lot of energy in sustainable fiber production, like making industrial fertilizers and pesticides. This means that the total carbon footprint is smaller. Using cover crops and rotating crops are examples of sustainable farming methods that help keep the soil's richness and structure (Tanveer *et al.*, 2012; Gentsch *et al.*, 2024). This stops erosion and makes farming more sustainable in the long run. Fair labor practices, like safe working conditions, fair pay, and involvement in the community, are often linked to sustainable fiber production. These practices are good for both workers and the areas where they are used (Naikwadi, 2024). Biodegradable fibers, like cotton, flax, and wool, can naturally break down at the end of their life cycle. This cuts down on trash and the damage they do to the earth (Ullah *et al.*, 2013; Ozek, 2024).

14.8 CHALLENGES IN SUSTAINABLE FIBER PRODUCTION

The textile manufacturing processes entail significant utilization of water, energy, and diverse chemicals, resulting in the production of waste either at the conclusion or as a secondary outcome (Stella *et al.*, 2024). The primary issue concerning water usage is the discharge of untreated effluents directly into adjacent water bodies. Water bodies are not the only cause of environmental damage (Chowdhary *et al.*, 2020). An essential aspect of the textile industry is the analysis and management of environmental pollution risks to prevent potential chemical disasters and their catastrophic consequences. Simultaneously, it is crucial to uphold the level of production (Tanveer *et al.*, 2015; Ostad-Ali-Askari, 2022). The fact that natural fibers can't grow very much is another problem with making sustainable fibers. Problems that might be hard to fix, like not having enough water, land loss, and city growth, limit the growth of natural products like cotton and wool. To solve this problem, sustainable fiber making needs to use cutting-edge tools and methods, such as hydroponics and vertical farms. These will make it easier to make natural fibers while also making them less harmful to the environment. Making things out of biodegradable fibers might cost more because it needs new tools and methods. To get around this problem, companies that make sustainable fibers need to use sustainable ways to get their raw materials, like recyclable materials, recycled polyester, organic cotton, and wool (Aziz *et al.*, 2018; Patti *et al.*, 2020). To solve this problem, sustainable fiber manufacturing needs to use methods that save money, like making production as efficient as possible and using recycled materials, that of the companies that supply feedstocks. The textile industry needs a lot of land and water to grow natural fibers like cotton and wool, which leads to soil loss and water shortages. The use of man-made fibers like nylon and polyester is another big problem for the earth. These fibers add to the problem of plastic waste. Because they need a lot of energy to make, some fibers, like carbon fiber, are very bad for the earth and release a lot of carbon dioxide when they are made. To solve this problem, sustainable fiber production must use clean energy sources like solar and wind power to cut down on carbon pollution and protect the earth as much as possible (Chadha *et al.*, 2019; Patti *et al.*, 2020).

14.9 FUTURE PERSPECTIVES

For ages, textiles have relied on natural fibers as their primary component. Clothes, household textiles, and, more recently, technological applications have all made use of natural fibers derived from cellulose and proteins. Hemp, one of the first plant fibers known, has recently gained popularity. The hemp industry, which comprises hemp fiber, seed, and oil, has grown dramatically since industrial hemp production resumed. This plant's fiber has the potential to be a green economy's alternative raw material, and growing hemp has the potential to launch a new supply chain that makes use of various hemp components in developing industries like the pharmaceutical and parapharmaceutical sectors. Rising consumer demand for eco-friendly products is expected to result in higher certified fiber output and lower ecological impacts in agriculture. Furthermore, GMOs (genetically modified organisms) and other new agricultural technologies hold considerable potential for addressing the world's future feed, fiber, and food demands. To be more specific, new biotech seeds have the potential to improve fertilizer use efficiency, crop output, and water use efficiency in cotton plants (Tedeschi *et al.*, 2022).

Composites made from natural fibers are becoming more popular in recent years. With the advent of biocomposites, or green composites, the composites sector is poised for rapid expansion and eventual incorporation into a circular economy (Khilji *et al.*, 2023). Reduced waste and environmental burden may result from improved recycling technologies and increased utilization of new components. It has been shown that using plant fiber composites in the car sector might help lower the potential for global warming. This is because the reinforcing substance, cellulose, has a light weighting impact. In order to improve characteristics and make them more recyclable, natural fiber reinforced composites with a polymer mix and/or biopolymer matrix will likely highlight the significance of fillers with high compatibility (Siakeng *et al.*, 2019). Natural fibers are raw materials that are renewable, recyclable, and reusable, as previously stated. Thus, transportation, filtration, geotextiles, and building are just a few of the many prospective uses for natural fibers (Rahman *et al.*, 2022).

LIST OF ABBREVIATIONS

CBD convention on biological diversity
DCR disease control rotation
FRP fiber-reinforced polymer
IUCN International Union for Conservation of Nature and Natural Resources
MSW municipal solid waste
WHO World Health Organization

REFERENCES

Akter, M., Uddin, M.H., & Anik, H.R., 2024. Plant fiber-reinforced polymer composites: A review on modification, fabrication, properties, and applications. *Polymer Bulletin*, 81, 1–85.

Alemu Reta, B., Babu, K.M., & Tesfaye, T., 2024. Studies on healthcare and hygiene textile materials treated with natural antimicrobial bioactive agents derived from plant extracts. *AATCC Journal of Research*, 24723444231215444.

Ali, H., Tanveer, A., Nadeem, M., Asghar, H., & Javaid, M., 2013. Germination ecology of *Rhynchosia capitata*: An emerging summer weed in Asia. *Planta Daninha*, 31, 249–257.

Avci, A., Eker, A.A., Bodur, M.S., & Candan, Z., 2023. Water absorption characterization of boron compounds-reinforced PLA/flax fiber sustainable composite. *International Journal of Biological Macromolecules*, 233, 123546.

Azad, A.K., Haq, U.N., Khairul Akter, M.M., & Uddin, M.A., 2024. Recycling Practices of Pre-Consumer Waste Generated from Textile Industry. *In Sustainable Manufacturing Practices in the Textiles and Fashion Sector.* Springer, 301–324.

Aziz, M., Ashraf, M., & Javaid, M.M., 2018. Enhancement in cotton growth and yield using novel growth promoting substances under water limited conditions. *Pakistan Journal of Botany*, 50, 1691–1701.

Bado, M.F., & Casas, J.R., 2021. A review of recent distributed optical fiber sensors applications for civil engineering structural health monitoring. *Sensors*, 21, 1818.

Bahrami, M., Butenegro, J.A., Abenojar, J., & Martinez, M.A., 2024. Adhesion characteristics of plasma-treated flax fabrics and elastoplastic properties of their biocomposites. *The Journal of Adhesion*, 100, 1–26.

Baker, A., Kenny, J., Kimball, J., & Belknap, T.H. Jr, 2014. *Time Charters*: Informa Law from Routledge.7th Edition.

Balo, F., & Sua, L.S., 2023. Application of Expert Decision Systems for Optimal Fiber Selection for Green Building Design Components. *In Pragmatic Engineering and Lifestyle: Responsible Engineering for a Sustainable Future.* Emerald Publishing Limited, 21–37.

Baran, I., 2023. *PULTRUSION: State-of-the-Art Process Models with Applications.* Elsevier.

Barraco, A., 2021. *Developing the New York Hemp Fiber Industry.*

Bauri, F., Avani, P., Sarkar, S., Bandyopadhyay, B., Misra, D., Debnath, S., & Dey, K., 2016. *Ex-Situ Conservation and Performance of Genetic Resources of Banana in New Alluvial Zone of West Bengal.*

Bhuiyan, M.M.U., Rahman, M., Naher, S., Shahed, Z.H., Ali, M.M., & Islam, A.R.M.T., 2024. Oxygen declination in the coastal ocean over the twenty-first century: Driving forces, trends, and impacts. *Case Studies in Chemical and Environmental Engineering*, 9, 100621.

Bi, H., Wei, Y., Wang, Z., & Chen, G., 2024. Fundamental investigation of micro-nano cellulose and lignin interaction for transparent paper: Experiment and electrostatic potential calculation. *International Journal of Biological Macromolecules*, 260, 129180.

Boutros, M., Saba, S., & Manneh, R., 2021. Life cycle assessment of two packaging materials for carbonated beverages (polyethylene terephthalate vs. glass): Case study for the Lebanese context and importance of the end-of-life scenarios. *Journal of Cleaner Production*, 314, 128289.

Cao, L., Wang, H., Wang, F., Xue, K., Cui, L., Shen, H., & Xu, G., 2023. Hollow reversible kapok fibrous membranes with amphiphilicity, natural antibacterial properties, and biodegradability. *Industrial Crops and Products*, 204, 117401.

Castellanos, H.G., Aryanfar, Y., Keçebaş, A., Assad, M.E.H., Islam, S., Naveed, Q.N., & Lasisi, A., 2024. A new paradigm for mining energy from industrial sludge: A low-cost fuel. *Journal of Water Process Engineering*, 59, 104987.

Castillo-Lara, J.F., Flores-Johnson, E.A., Valadez-Gonzalez, A., Herrera-Franco, P.J., Carrillo, J.G., Gonzalez-Chi, P., & Li, Q., 2020. Mechanical properties of natural fiber reinforced foamed concrete. *Materials*, 13, 3060.

Chadha, A., Florentine, S., Chauhan, B.S., Long, B., Jayasundera, M., Javaid, M.M., & Turville, C., 2019. Environmental factors affecting the germination and seedling emergence of two populations of an emerging agricultural weed: Wild lettuce (*Lactuca serriola*). *Crop and Pasture Science*, 70, 709–717.

Chadha, A., Florentine, S., Javaid, M., Welgama, A., & Turville, C., 2020. Influence of elements of climate change on the growth and fecundity of *Datura stramonium*. *Environmental Science and Pollution Research*, 27, 35859–35869.

Chen, L., Chen, Z., Xie, Z., Wei, L., Hua, J., Huang, L., & Yap, P.-S., 2023. Recent developments on natural fiber concrete: A review of properties, sustainability, applications, barriers, and opportunities. *Developments in the Built Environment*, 16, 100255.

Chen, Y., Bi, Y., Wu, X., & Ding, X., 2024. Impact of additional carbon storage of natural plant fiber on product carbon footprint: A case study of cotton/kapok blended T-shirt vs pure cotton T-shirt. *Journal of Cleaner Production*, 434, 140237.

Chowdhary, P., Bharagava, R.N., Mishra, S., & Khan, N., 2020. Role of industries in water scarcity and its adverse effects on environment and human health. *In* Environmental Concerns and Sustainable Development: Volume 1: Air, Water and Energy Resources. Singapore, Springer, 235–256.

Chowdhury, U.S., Rahman, F., Ehsan, M.F., Pabel, M.Y., & Islam, M.M., 2024. Catalytic Methods for Sustainable Textile Dyeing. In *Sustainable Manufacturing Practices in the Textiles and Fashion Sector.* Springer, 143–172.

Czapla, A., Ganesapillai, M., & Drewnowski, J., 2021. Composite as a material of the future in the era of green deal implementation strategies. *Processes*, 9, 2238.

Daubert, S., 2020. *Between the Rocks and the Stars: Narratives in Natural History.* Vanderbilt University Press.

Dave, P.N., & Macwan, P.M., 2024. Biocomposites with Natural Fibers. *In Advances in Biocomposites and their Applications.* Elsevier, 81–120.

Delmer, D., Dixon, R.A., Keegstra, K., & Mohnen, D., 2024. The plant cell wall—Dynamic, strong, and adaptable—Is a natural shapeshifter. *The Plant Cell*, 36, koad325.

Derstine, B.A., Holcombe, S.A., Ross, B.E., Wang, N.C., Su, G.L., & Wang, S.C., 2018. Skeletal muscle cutoff values for sarcopenia diagnosis using T10 to L5 measurements in a healthy US population. *Scientific Reports*, 8, 11369.

Dias, J., 1992. *Efeitos De Vantagens Preferenciais: o Impacto Da Convenção De Lomé Nas Exportações Dos Acp Para A Cee: 1975-1985.* Universidade Tecnica de Lisboa (Portugal).

Djedaiet, A., Ayad, H., & Ben-Salha, O., 2024. Oil prices and the load capacity factor in African oil-producing OPEC members: Modeling the symmetric and asymmetric effects. *Resources Policy*, 89, 104598.

Ekren, S., Ozturk, G., Gokçol, A., Ertekin, M., Geren, H., & Ilker, E., 2023. Effect of nitrogen application rate and harvesting stage on yield and agronomic parameters of industrial hemp (*Cannabis sativa* L.). *Crop Science*, 63, 3051–3064.

Ervast, M., 2024. *Tracing the value of traceability: Identifying the value of traceability within textile and apparel value chains.* MS Thesis at University of Turku.

Fahmi, F.F., Mahdi, T.H., Ferdousi, U.S., Uddin, M.B., Sarker, M.E., & Syduzzaman, M., 2024. Effect of Interfacial Bonding Characteristics on Fatigue Behavior of Hemp Fibre Reinforced Polymer Composites. *Interfacial Bonding Characteristics in Natural Fiber Reinforced Polymer Composites: Fiber-Matrix Interface in Biocomposites.* Springer, 155–201.

Feng, Y., Hao, H., Lu, H., Chow, C.L., & Lau, D., 2024. Exploring the development and applications of sustainable natural fiber composites: A review from a nanoscale perspective. *Composites Part B: Engineering*, 111369.

Frazier, R.M., Vivas, K.A., Azuaje, I., Vera, R., Pifano, A., Forfora, N., Jameel, H., Ford, E., Pawlak, J.J., & Venditti, R., 2024. Beyond cotton and polyester: An evaluation of emerging feedstocks and conversion methods for the future of fashion industry. *Journal of Bioresources and Bioproducts*, 9.

Garcia, M.a.R., 2023. *Influence of Inks, Coatings, and Varnishes on Paper Re-pulpability and the Environmental Performance of the Material.* Kauno technologijos universitetas.

Gentsch, N., Riechers, F.L., Boy, J., Schweneker, D., Feuerstein, U., Heuermann, D., & Guggenberger, G., 2024. Cover crops improve soil structure and change organic carbon distribution in macroaggregate fractions. *SOIL*, 10, 139–150.

Geoffrey, R.D.V., 2024. Pests and Pathogens. *In Vegetable Brassicas and Related Crucifers*, vol. 40. CABI, 295.

Gómez-Maqueo, X., & Gamboa-Debuen, A., 2022. The biology of the genus Ceiba, a potential source for sustainable production of natural fiber. *Plants*, 11, 521.

Hanif, Z., Naeem, M., Ali, H.H., Tanveer, A., Javaid, M.M., Peerzada, A.M., & Chauhan, B.S., 2017. Effect of environmental factors on germination of *Salsola foetida*: Potential species for rehabilitation of degraded rangelands. *Rangeland Ecology & Management*, 70, 638–643.

Hansson, M., Youssef, H.M., Zakhrabekova, S., Stuart, D., Svensson, J.T., Dockter, C., Stein, N., Waugh, R., Lundqvist, U., & Franckowiak, J., 2024. A guide to barley mutants. *Hereditas*, 161, 1–55.

Harb, H.E., El-Tabakh, M.A., Khattab, A.M., Mohamed, Y.A., Saleh, A.M., & El-Abeid, S.E., 2024. Recent Advances of Using Innovative Strategies in Management of Millet Plant Pathogens. In Genetic Improvement of Small Millets. Spinger, 297–328.

Harymawan, I., Irsyad, M.A., & Nurchoirotun, C., 2022. *Discovering the Potential of Ramie a Support to the Strategic Ramie Industry in Indonesia.* Airlangga University Press.

Hassan, M.N., Mydin, M.O., & Nawi, M.N.M., 2023. Secondary natural fibre-reinforced cementitious composites: A comprehensive review. *Journal of Advanced Research in Applied Sciences and Engineering Technology*, 29, 105–125.

Hassan, M.U., Nawaz, M., Mahmood, A., Shah, A.A., Shah, A.N., Muhammad, F., Batool, M., Rasheed, A., Jaremko, M., & Abdelsalam, N.R., 2022. The role of zinc to mitigate heavy metals toxicity in crops. *Frontiers in Environmental Science*, 10, 990223.

Hernández-Pinto, F.J., Miranda-Medina, J.D., Natera-Maldonado, A., Vara-Aldama, Ó, Ortueta-Cabranes, M.P., Del Mercado-Pardiño, J.a.V., El-Aidie, S.A., Siddiqui, S.A., & Castro-Muñoz, R., 2024. Arabinoxylans: A review on protocols for their recovery, functionalities and roles in food formulations. *International Journal of Biological Macromolecules*, 259, 129309.

Iqbal, J., Kamran, M., Shafique, M.M., Jawad, M., Arshad, Z.N., & Raza, F., & Nawaz, M.S., 2023. *Harnessing Genetic Diversity for Sustainable Maize Production: A Contemporary Perspective.*

Jones, R.A., & Congdon, B.S., 2024. Australian cool-season pulse seed-borne virus research: 1. Alfalfa and cucumber mosaic viruses and less important viruses. *Viruses*, 16, 144.

Khalil, E., Sarkar, J., Rahman, M.M., Shamsuzzaman, M., & Das, D., 2023. Advanced Technology in Textile Dyeing. *In Advanced Technology in Textiles: Fibre to Apparel.* Springer, 97–138.

Khan, F., Hossain, N., Mim, J.J., Rahman, S.M., Iqbal, M.J., Billah, M., & Chowdhury, M.A., 2024. Advances of composite materials in automobile applications—A review. *Journal of Engineering Research*, 24.

Khilji, I.A., Chilakamarry, C.R., Surendran, A.N., Kate, K., & Satyavolu, J., 2023. Natural fiber composite filaments for additive manufacturing: A comprehensive review. *Sustainability*, 15, 16171.

Koudahe, K., Aguilar, J., Djaman, K., & Sheshukov, A.Y., 2024. Evapotranspiration, fiber yield and quality, and water productivity of cotton (*Gossypium hirsutum* L.) under different irrigation technologies in a semiarid climate. *Irrigation Science*, 1–20.

Kumar, A., Kumar, V., & Singh, B., 2021. Cellulosic and hemicellulosic fractions of sugarcane bagasse: Potential, challenges and future perspective. *International Journal of Biological Macromolecules*, 169, 564–582.

Kumar, R., Goyal, M.K., Surampalli, R.Y., & Zhang, T.C., 2024a. River pollution in India: Exploring regulatory and remedial paths. *Clean Technologies and Environmental Policy*, 9, 1–23.

Kumar, R.S., Sasikumar, R., & Dhilipkumar, T., 2024b. Exploiting agro-waste for cleaner production: A review focusing on biofuel generation, bio-composite production, and environmental considerations. *Journal of Cleaner Production*, 435, 140536.

Lakhiar, I.A., Yan, H., Zhang, J., Wang, G., Deng, S., Bao, R., Zhang, C., Syed, T.N., Wang, B., & Zhou, R., 2024. Plastic pollution in agriculture as a threat to food security, the ecosystem, and the environment: An overview. *Agronomy*, 14, 548.

Li, R., Feng, H., Wang, S., Zhuang, D., & Zhu, J., 2024. A colorimetry-enhanced tri-functional film with high stability by polyphenol-anthocyanin co-pigmentation/conjugate: New prospect for active intelligent food packaging. *Food Chemistry*, 138927.

Mahmood, A., Wang, X., Shahzad, A.N., Fiaz, S., Ali, H., Naqve, M., Javaid, M.M., Mumtaz, S., Naseer, M., & Dong, R., 2021. Perspectives on bioenergy feedstock development in Pakistan: Challenges and opportunities. *Sustainability*, 13, 8438.

Maiti, S., Islam, M.R., Uddin, M.A., Afroj, S., Eichhorn, S.J., & Karim, N., 2022. Sustainable fiber-reinforced composites: A review. *Advanced Sustainable Systems*, 6, 2200258.

Makamov, A., Shavkiev, J., Kholmuradova, M., Boyqobilov, U., Normamatov, I., Norbekov, J., Khusenov, N., Kushakov, S., Yuldasheva, Z., & Khoshimov, S., 2023. *Cotton Genotypes Appraisal for Morpho-Physiological and Yield Contributing Traits under Optimal and Deficit Irrigated Conditions.*

Maraveas, C., Kyrtopoulos, I.V., Arvanitis, K.G., & Bartzanas, T., 2024. The aging of polymers under electromagnetic radiation. *Polymers*, 16, 689.

Martínez-Ruiz, M., Arroyo-Rodríguez, V., Arasa-Gisbert, R., Hernández-Ruedas, M.A., & San-José, M., 2024. Maintenance of different life stages of old-growth forest trees in deforested tropical landscapes. *Ecology*, e4273.

Molefe, M., 2022. *Architectural Considerations and Resource Allocation in Energy Efficient Networking.*

Monge-Ganuzas, M., Guillén-Mondéjar, F., Díaz-Martínez, E., Herrero, N., & Brilha, J., 2024. Geoconservation at the International Union for Conservation of Nature (IUCN). *Philosophical Transactions of the Royal Society A*, 382, 20230053.

Naikwadi, S., 2024. Role of Governments and Consumers. *In Climate Action Through Eco-Friendly Textiles.* Springer, 163–178.

Ntenga, R., Saidjo, S., Wakata, A., Djoda, P., Tango, M., & Mfoumou, E., 2022. *Extraction, Applications and Characterization of Plant Fibers.* IntechOpen.

Nyssanbek, M., Kuzina, N., Kondrashchenko, V., & Azimov, A., 2024. Effects of plasma treatment on biodegradation of natural and synthetic fibers. *npj Materials Degradation*, 8, 23.

Ostad-Ali-Askari, K., 2022. Management of risks substances and sustainable development. *Applied Water Science*, 12, 65.

Ozek, H.Z., 2024. Sustainability, Biodegradability and Life Cycle Analysis of Wool. *In The Wool Handbook*. Elsevier, 401–440.

Patti, A., Cicala, G., & Acierno, D., 2020. Eco-sustainability of the textile production: Waste recovery and current recycling in the composites world. *Polymers*, 13, 134.

Paul, M., 2013. *Use of Woodchip for Agricultural Livestock Bedding*. Bangor University.

Peças, P., Carvalho, H., Salman, H., & Leite, M., 2018. Natural fibre composites and their applications: A review. *Journal of Composites Science*, 2, 66.

Periyasamy, R., Hemanth Kumar, M., Rangappa, S.M., & Siengchin, S., 2023. A comprehensive review on natural fillers reinforced polymer composites using fused deposition modeling. *Polymer Composites*, 44, 3715–3747.

Piazza, V., Da Silva, B., Frassoldati, A., Lietti, L., Chiara, G., Rajendran, K., Chen, D., Faravelli, T., & Beretta, A., 2024. Unravelling the Complexity of Hemicellulose Pyrolysis: A New Methodology for the Detailed Product Speciation for Xylan and Glucomannan in TGA and Fixed Bed Reactor. *Chemical Engineering Journal*, 497, 154579.

Psychogiou, E.-E., & Tsoulfas, G.T., 2024. Critical factors affecting trust in the wine supply chain in Greece: A grey dematel approach. *Logistics*, 8, 24.

Qing, A., Zhao, Y., & Zhang, Z., 2024. Microwave, Millimeter Wave and Terahertz Applications of Liquid Crystal. *In Microwaves, Millimeter Wave and Terahertz Liquid Crystals: Preparation, Characterization and Applications*. Springer, 95–187.

Rahman, S.S., Siddiqua, S., & Cherian, C., 2022. Sustainable applications of textile waste fiber in the construction and geotechnical industries: A retrospect. *Cleaner Engineering and Technology*, 6, 100420.

Rasheed, A., Barqawi, A.A., Mahmood, A., Nawaz, M., Shah, A.N., Bay, D.H., Alahdal, M.A., Hassan, M.U., & Qari, S.H., 2022a. CRISPR/Cas9 is a powerful tool for precise genome editing of legume crops: A review. *Molecular Biology Reports*, 49, 5595–5609.

Rasheed, A., Mahmood, A., Maqbool, R., Albaqami, M., Sher, A., Sattar, A., Bakhsh, G., Nawaz, M., Hassan, M.U., & Al-Yahyai, R., 2022b. Key insights to develop drought-resilient soybean: A review. *Journal of King Saud University-Science*, 34, 102089.

Rasheed, A., Raza, A., Jie, H., Mahmood, A., Ma, Y., Zhao, L., Xing, H., Li, L., Hassan, M.U., & Qari, S.H., 2022c. Molecular tools and their applications in developing salt-tolerant soybean (*Glycine max* L.) cultivars. *Bioengineering*, 9, 495.

Romanowska, B., Różańska, W., & Zimniewska, M., 2024. The influence of the chemical composition of flax and hemp fibers on the value of surface free energy. *Materials*, 17, 1104.

Ryłko-Polak, I., Komala, W., & Białowiec, A., 2022. The reuse of biomass and industrial waste in biocomposite construction materials for decreasing natural resource use and mitigating the environmental impact of the construction industry: A review. *Materials*, 15, 4078.

Šeremešić, S., Dolijanović, Ž, Tomaš Simin, M., Milašinović Šeremešić, M., Vojnov, B., Brankov, T., & Rajković, M., 2024. Articulating organic agriculture and sustainable development goals: Serbia case study. *Sustainability*, 16, 1842.

Shafiq, A., Bhatti, I.A., Amjed, N., Zeshan, M., Zaheer, A., Kamal, A., Naz, S., & Rasheed, T., 2024. Lignin derived polyurethanes: Current advances and future prospects in synthesis and applications. *European Polymer Journal*, 112899.

Siakeng, R., Jawaid, M., Ariffin, H., Sapuan, S., Asim, M., & Saba, N., 2019. Natural fiber reinforced polylactic acid composites: A review. *Polymer Composites*, 40, 446–463.

Sigman, H., & Strow, R., 2024. *China's Waste Import Ban and US Solid Waste Management: Effects of the Loss of a Waste Haven.*

Stella, F., Garofalo, S.F., Cavallini, N., Fino, D., & Deorsola, F.A., 2024. Closing the loop: Analysis of biotechnological processes for sustainable valorisation of textile waste from the fast fashion industry. *Sustainable Chemistry and Pharmacy*, 38, 101481.

Strydom, R., Wilson, C., Tegg, R., Balendres, M., & Van Der Waals, J., 2024. Advancements in *Spongospora subterranea*: Current knowledge, management strategies, and research gaps. *Potato Research*, 67, 1–41.

Tanveer, A., Arshad, M.S., Ayub, M., Javaid, M.M., & Yaseen, M., 2012. Effect of temperature, light, salinity, drought stress and seeding depth on germination of *Cucumis melo* var. agrestis. *Pakistan Journal of Weed Science Research*, 18.

Tanveer, A., Javaid, M.M., Irfan, M., Khaliq, A., & Yaseen, M., 2015. Yield losses in chickpea with varying densities of dragon spurge (*Euphorbia dracunculoides*). *Weed Science*, 63, 522–528.

Tareq, M.Z., Sadat, M.A., Islam, S.N., Fakir, M.S.A., & Hossain, M.A., 2024. Day-neutral jute plant: A hallmark for developing new jute genotype. *Plant Science Today*. DOI: 10.14719/pst.2356.

Tedeschi, A., Cerrato, D., & Menenti, M., 2022. Is the potential for multi-functional use of industrial hemp greater than maize under saline conditions? *Sustainability*, 14, 15646.

Thacharodi, A., Hassan, S., Meenatchi, R., Bhat, M.A., Hussain, N., Arockiaraj, J., Ngo, H.H., Sharma, A., Nguyen, H., & Pugazhendhi, A., 2024. Mitigating microplastic pollution: A critical review on the effects, remediation, and utilization strategies of microplastics. *Journal of Environmental Management*, 351, 119988.

Thapliyal, D., Verma, S., Sen, P., Kumar, R., Thakur, A., Tiwari, A.K., Singh, D., Verros, G.D., & Arya, R.K., 2023. Natural fibers composites: Origin, importance, consumption pattern, and challenges. *Journal of Composites Science*, 7, 506.

Thyavihalli Girijappa, Y.G., Mavinkere Rangappa, S., Parameswaranpillai, J., & Siengchin, S., 2019. Natural fibers as sustainable and renewable resource for development of eco-friendly composites: A comprehensive review. *Frontiers in Materials*, 6, 226.

Tiwari, A.K., 2024. IPM essentials: Combining biology, ecology, and agriculture for sustainable pest control. *Journal of Advances in Biology & Biotechnology*, 27, 39–47.

Ullah, H., Mahmood, A., Ijaz, M., Tadesse, B., & Honermeier, B., 2013. Evaluation of anise (*Pimpinella anisum* L.) accessions with regard to morphological characteristics, fruit yield, oil contents and composition. *Journal of Medicinal Plants Research*, 7, 2177–2186.

Wasaya, A., Abbas, T., Yasir, T.A., Sarwar, N., Aziz, A., Javaid, M.M., & Akram, S., 2021. Mitigating drought stress in sunflower (*Helianthus annuus* L.) through exogenous application of β-aminobutyric acid. *Journal of Soil Science and Plant Nutrition*, 21, 936–948.

Weber-Kamin, A., 2020. *Reinforced Concrete Coupling Beams with High-Strength Steel Bars*. University of Kansas.

Wu, Z., Feng, X., Zhang, Y., & Fan, S., 2024. Repositioning fertilizer manufacturing subsidies for improving food security and reducing greenhouse gas emissions in China. *Journal of Integrative Agriculture*, 23, 430–443.

Xu, F., Zhang, Z., Zhao, Z., Ji, X., Liu, J., & Song, X., 2024. Dyeing performance and color evaluation of cotton fabrics dyed with *Caesalpinia sappan* L. and *Galla chinensis* Mill. extract, and the evaluation of binary sequential dyeing method. *Fibers and Polymers*, 25, 1–24.

Xu, H., & Song, H.-Y., 2024. Key factors influencing Chinese consumers' demand for naturally dyed garments: Data analysis through KJ method and KANO model. *Sustainability*, 16, 1189.

Yu, D., Wang, L., & Hu, F., 2024. Agro-Biodiversity Conservation and Implications in China. In *Organic Agriculture and Biodiversity in China*. Elsevier, 25–45.

Zafar, S., Afzal, H., Ijaz, A., Mahmood, A., Ayub, A., Nayab, A., Hussain, S., Maqsood, U.-H., Sabir, M.A., & Zulfiqar, U., 2023. Cotton and drought stress: An updated overview for improving stress tolerance. *South African Journal of Botany*, 161, 258–268.

Zaharia, C., Musteret, C.-P., & Afrasinei, M.-A., 2024. The use of coagulation–flocculation for industrial colored wastewater treatment—(I) The application of hybrid materials. *Applied Sciences*, 14, 2184.

Zerga, A.Y., & Tahir, M., 2022. Biobased kapok fiber nano-structure for energy and environment application: A critical review. *Molecules*, 27, 8107.

15 Cutting-Edge Strategies for Environmental Stress

An Overview

Abdul Jabbar, Saima Asghar, Athar Mahmood, Muhammad Mazhar Iqbal, Rimsha Kanwal, and Komal Rehman*

15.1 INTRODUCTION

Climate change, alien species, ecosystem loss, high temperatures, contamination of heavy metals, higher levels of salinity, drought, and all kinds of wastes (in the air, water, and land) are just some of the things that can put stress on the environment (Mehmood *et al.*, 2017; Aziz *et al.*, 2018). Heavy metals are hazardous metals having atomic density ≥5 g cm^{-3}, which are found harmful to the growth and yield of crops. Alleged environmental stresses are factors that stop ecosystems from growing, reproducing, and making things (Mahmood *et al.*, 2022). The study of environmental stressors has its origins in studies of environmental variables, with the groundbreaking work of Liebig (Ullah *et al.*, 2013; Daiber *et al.*, 2022). Environmental stress may originate from natural processes or human activities like the production and discharge of new chemical compounds and extensive land-use changes, and it can also be anthropogenic (Smith *et al.*, 2016). The components of natural and anthropogenic stressors are often similar to one another (Hammond *et al.*, 2020). An individual may experience stress if they are subjected to social or environmental demands that have an impact on their mental or physical health and ability to carry out daily tasks (Kalia and Knauft, 2020). When stress levels rise beyond what most organisms can tolerate, it begins to play a role in lowering biotic potential. Ecosystem development may after that be limited or impeded. Stressors have some effect on every living thing, as well as their groups, escapes, and people (Koza *et al.*, 2022).

Land plants endure a wide range of unfavorable or even harmful environmental factors daily (Stevenson *et al.*, 2020). Biodiverse ecosystems are better able to withstand extreme environmental conditions, such as rising temperatures or salt concentrations, than less diverse ones (Shehzadi *et al.*, 2023). The agricultural sector continues to face the tremendous environmental pressures produced by the intensifying global warming and climatic changes. These stresses have a substantial impact on crop yields (Mahmood *et al.*, 2013; Tanveer *et al.*, 2023). Major crop yields have dropped by as much as 50% due to environmental stress caused by climate change, which includes extreme heat or cold, drought, salt, or metal toxicity (Tanveer *et al.*, 2013; Sharma *et al.*, 2023). The toxic effects of abiotic stressors on plant growth and productivity including salinity, drought, extreme temperature, cold, metal toxicity, ozone, UV radiation, and nutritional deficiencies are becoming more significant when taking into account the direct and indirect effects of environmental stress (Muhammad *et al.*, 2021; Rasheed *et al.*, 2022a). Both natural systems and humans are subject to environmental and ecological stress as a result of human activities such as deforestation, overfishing, and other forms of pollution. Ecosystems are essential for the survival of humans and plants owing to the provision of essential nutrients and the remove of harmful substances (Jabbar *et al.*, 2023). Ecosystems are under pressure due to current consumption and growth trends. Some of the symptoms that ecosystems are

* Corresponding author: rimshakanwal700@gmail.com

DOI: 10.1201/9781003498698-15

TABLE 15.1
The Proteome and Transcriptome Reactions of Plants to Nanoparticles (NPs) in Response to Environmental Stressors

Nanoparticles	Stress	Plant	Major Regulated Genes	References
AgNPs	Salinity	Tomato	RBOH1; APX2; MAPK2; ERF5; MAPK3; DDF2. – AREB; MAPK2; P5CS1; CRK1.	Almutairi (2019)
Al_2O_3NPs	Heavy metals	Mung bean	Proteins related to protein synthesis/degradation; glycolysis; lipid metabolism. NmrA-Like, PABP2	Hossain et al. (2015)
AgNPs	Flooding	Soya bean	Root proteins for stress signaling and cell metabolism Glyoxalase II 3 and fermentation-related proteins ADH1; PDC	Wang and Komatsu (2020)
TiO_2NPs	Drought	Maize	LRubisco; S Rubisco; Chlorophyll a/b-Binding Protein; PEPC	Al-Khayri et al. (2023)
SiNPs	Cadmium	Wheat	LCT1; NRAMP5 sHMA3; LSI1	Shi et al. (2018)
TiO_2NPs	Cold stress	Chickpea	Genes related to cellular defense, cell signaling, transcriptional regulation, and chromatin modification	Silva et al. (2022)

under stress include environmental deterioration, loss of biodiversity, deforestation, and the collapse of social and economic institutions (Watson *et al.*, 2018).

Global agriculture relies on a small number of crop species that have been cultivated to maximize yield within a relatively restricted range of environmental variables (Folberth *et al.*, 2020). It's generally recognized that advances in plant science could solve some of the problems with farming and food production (Movilla-Pateiro *et al.*, 2021). Sustainable agricultural production and resilience in the face of climate variation can currently be improved through a variety of measures (Douxchamps *et al.*, 2017; Mahmood *et al.*, 2021). The presence of endophytes and microbial symbionts like actinobacteria, fungi, or yeast can greatly enhance plant growth and development, especially when faced with challenging environmental Stress (Hanaka *et al.*, 2021). All environmental stress mentioned above can be mitigated by genetic engineering, biotechnology, habitat restoration, gene editing by stress tolerance, bioremediation, and phytoremediation, climate-smart agriculture, and nature-based solutions (NBS) (Naz *et al.*, 2022; Table 15.1).

15.2 MANAGEMENT STRATEGIES FOR ENVIRONMENTAL STRESS

15.2.1 Bioremediation and Phytoremediation

One most important methods for cleaning up contaminated areas is bioremediation, which makes use of microorganisms that either already exist in the environment or are exogenously introduced to feed on and decompose pollutants (Saeed *et al.*, 2021). Microorganisms and/or plants are used in biological remediation, which includes phytoremediation and bioremediation, to eliminate, break down, and/or immobilize pollutants in the environment (Raklami *et al.*, 2022). Biological remediation is widely used to restore contaminated places because of its low cost, easy operation, and environmental friendliness. Utilizing organic materials like microorganisms and plants are recognized as bioremediation (Ye *et al.*, 2017). This method's feasibility is contingent upon the type, extent, and context of the pollution. It is possible to use algae, fungi, bacteria, etc. in bioremediation, and being able to adapt to strange surroundings enhances their efficiency (Samuel *et al.*, 2022). For instance, psychrophiles flourish in cold climes, acidophiles live in acidic conditions, and halophiles live in saltwater regions (Zgonik *et al.*, 2021). Phytoremediation is the use of plants to remove, contain, or transform contaminants. This can be accomplished directly (e.g., by plants hyperaccumulating heavy metals) or indirectly (e.g., by plants stimulating microorganisms in the rhizosphere) (Hussain *et al.*, 2022; Rasheed *et al.*, 2022b). Phytoextraction, phytoaccumulation, phytotransformation, phytodegradation, stabilization, phytovolatilization, rhizofiltration, and rhizodegradation are some industry-wide approaches, to mitigate the environmental stress from plants (Jeevanantham *et al.*, 2019).

15.2.2 Ex-Situ Bioremediation and In-Situ Bioremediation

"Ex-situ" refers to transporting a mat for pollution removal to a distant treatment site (Quinnan *et al.*, 2022). Excavating contaminated soil and transporting it elsewhere is a major undertaking, which is why this categorization is not very common (Freschet *et al.*, 2021). The first step in ex-situ remediation is to create an environment away from the site that is optimal for soil conditions, including moisture, nutrients, and oxygen (Wang *et al.*, 2021). Land farming, artificial biopiles, and compost pile irrigation are all forms of ex-situ bioremediation as described in Figure 15.1.

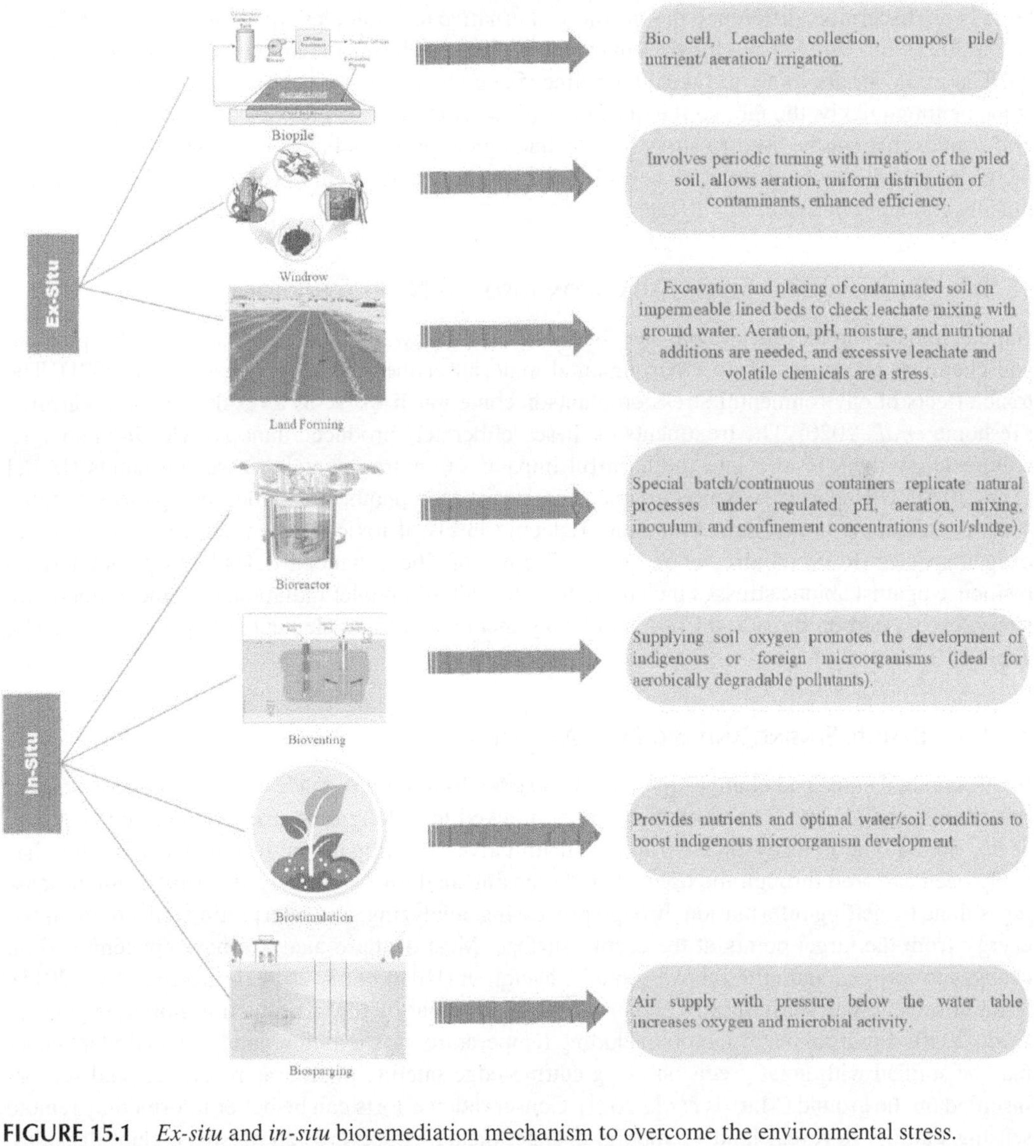

FIGURE 15.1 *Ex-situ* and *in-situ* bioremediation mechanism to overcome the environmental stress.

In situ bioremediation allows us to treat soil and groundwater simultaneously without excavating any further (Sales Da Silva *et al.*, 2020). Through the in-situ remediation process, a system can be transformed from partially closed to completely open. These include techniques such as bio-slurping, percolation, bioventing, air sparging, and pump-and-treat (Li *et al.*, 2023).

15.2.3 Innovations in Technology for Mitigation of Environmental Stress

Environmental scientists and policy makers, who care about the environment, are hopeful that discoveries and green tools will make climate change history (Loh *et al.*, 2020; Mackenthun, 2021). As technology improves, the positive link between new ideas and reducing the negative effects of different stresses on the environment is getting more and more attention (Li *et al.*, 2020; Zafar *et al.*, 2023). Climate change, resource loss, and pollution are all very serious problems that need to be looked at in detail to see how new innovative tools either make things worse or make them better (Erdelen and Richardson, 2021). More often than not, environmental technologies aim to save energy as a factor rather than improve quality. So, as one can see with "green" and "grey" power, there is no discernible difference in how these alternative technologies work for customers (Hassan *et al.*, 2023). So, big changes for the environment that will lower the amount of greenhouse gases (GHGs) in the air are likely to take a long time (Eskander and Fankhauser, 2020). Climate change is happening quickly, though, so the question isn't whether as a species, we can come up with several studies that can be used to show how technological progress has helped solve the problem of increased global warming and with new ideas in the future, but whether we can put them into action quickly enough (Nielsen *et al.*, 2021).

15.2.4 Environmental Stress Alleviation through Nanotechnology

Nanotechnology holds the crucial key because it is a new method that can alter the physical and chemical properties of an environmental material at the nanoscale level (Ariga, 2021). The toxic effects of environmental stress on plants become much worse as a result of global warming (Teshome *et al.*, 2020). The treatments of these deliberately produced nanoparticles in nano agriculture allow them to alleviate the harmful impacts of environmental stresses on plants (Mittal *et al.*, 2020). Nanoparticles exhibit unique characteristics in plants due to their charge-to-size ratio, such as enhancing antioxidant status and reducing levels of toxicant chemicals, such as reactive oxygen species (ROS) (Andrés *et al.*, 2018). The use of silicon nanoparticles helps plants develop resilience against abiotic stresses including drought, salt, ultraviolet radiation, extreme temperature stress, excess carbon dioxide (eCO_2), and heavy metals and toxic elements (Tripathi *et al.*, 2023; Table 15.2).

15.2.5 Remote Sensing and Big Data Analytics

Environmental signs and changes in ecosystem stress from human-made causes like deforestation, river dams, and air and water pollution can be tracked for a long time using remote sensing (Roy *et al.*, 2023). The potency of environmental initiatives to mitigate environmental pressures has lately been assessed through the use of remote sensing analysis (Werner *et al.*, 2019). Remote sensing is done by getting information through processing, analyzing, and interpreting reflected/emitted energy from the target points at the earth's surface. Most of these analyses have concentrated on changes to forested and other above-ground vegetation (Ullah *et al.*, 2014; Temesgen *et al.*, 2015). Remote sensing has substantially enhanced the assessment of forest protection initiatives (Yanai *et al.*, 2020). Environmental factors including temperature, precipitation, land use, and plant cover may be studied with great precision using cutting-edge satellite photography, drones, and sensors installed on the ground (Martos *et al.*, 2021). Conservation efforts can be better informed by remote sensing's ability to reveal land's temporal change, including soil erosion hotspots, plant densities,

TABLE 15.2
Effect of Environmental Stress on Plants with Their Mechanism to Mitigate the Proper Growth and Development

Environmental Stress	Effects on Plant	References
Heavy metals	Senescence, stunted growth and photosynthesis, chlorosis, poor biomass accumulation, changed water balance, and nutrient absorption are all factors that lead to plant mortality.	Oves *et al.* (2016)
Salinity	Causes an imbalance in ions and a lack of available water. Enhanced reactive oxygen species (ROS) generation, decreased stomata opening, decreased photosynthesis, and limited absorption and translocation of mineral nutrients and water were all symptoms of osmotic and oxidative stress.	Hasanuzzaman *et al.* (2021), Khondoker *et al.* (2023)
Drought	Does damage by increasing osmotic and oxidative stress. Oxidative stress and ion loss have been exacerbated. Reduced mineral nutrient uptake and transport. Enzyme activity decline, denaturation of proteins	Wahab *et al.* (2022)
Cold	Uneven nutrition, oxidative damage, and osmotic stress result. Result: less photosynthetic activity, ROS buildup, and enzyme activity suppression.	Kumar *et al.* (2018)
Heat	Produces oxidative stress, water shortage, and osmotic stress, all of which increase the generation of reactive oxygen species (ROS), which in turn accelerate protein misfolding and denaturation. Less CO_2 fixation, discolored leaves, accelerated foliar aging, and disrupted ion transport.	Sachdev *et al.* (2021)
Radiations	Reduced photosynthetic activity, chlorophyll breakdown, oxidative stress, and increased ROS generation and oxidative damage.	Hong *et al.* (2014)
Pesticides	Low levels of chlorophyll and protein, which causes plants to have less effective photosynthetic capacity.	Parween *et al.* (2016)
Water logging	Reduced photosynthetic activity, chlorophyll breakdown, oxidative stress, and increased ROS generation and oxidative damage.	Akhtar and Nazir (2013)

and other indicators (Dawson *et al.*, 2016). The data may be used by land managers to pinpoint the places most in danger and devise strategies to mitigate them (Cosgrove and Loucks, 2015).

15.3 CLIMATE ENGINEERING AND GEOENGINEERING

The deliberate and extensive use of technology to alter Earth's natural processes is known as geoengineering (Oomen and Meiske, 2021). Climate engineering is another name for this field that reflects its focus on technology as a tool to fight environmental stress (Moraci *et al.*, 2020). There are three main subfields under climate/geoengineering technologies: (1) so-called solar radiation management (i.e., directing sunlight into space) (Moayedi and Mosavi, 2021), (2) removal and sequestration of GHGs (García-Freites *et al.*, 2021), and (3) manipulation of weather patterns (Zhang *et al.*, 2020b). These innovations encompass a broad array of methods, such as releasing sulfuric acid into the stratosphere or "whitening" clouds to reflect sunlight; discharging iron particles into the oceans to support plankton that absorbs carbon dioxide; releasing silver iodide into clouds to create rain; or genetically modifying crops to have more photo reflective foliage (Su *et al.*, 2020; Abbatt and Ravishankara, 2023). Attempts to actively intervene in the climate by removing carbon dioxide from the atmosphere after it has been released constitute the majority of geoengineering technologies that are classified as carbon dioxide removal (CDR) (Faran and Olsson, 2018). Geoengineering might entail capturing atmospheric carbon dioxide and storing it in a different location, most likely underground, or it could involve reflecting a specific quantity of sunlight into space (Zhang *et al.*, 2015). The generation of GHGs through human activities, such as the combustion of fossil fuels, is the principal driver of global warming (Kweku *et al.*, 2018). Neither the generation of GHG emissions nor their effects on ecosystems nor human health and safety could be mitigated with solar geoengineering (Sovacool *et al.*, 2022).

15.4 TECHNOLOGIES TO CONSERVE AND PURIFY WATER

Environmental stress such as drought induces excessive floral abscission in yellow lupine, causing to mitigate the production and economic ramifications in agriculture (Florkiewicz *et al.*, 2020). Because excessive salt concentrations in soil change plant development through alternation in metabolism, ion toxicity, producing secondary oxidative stress, and osmotic stress, excess salinity poses a serious environmental risk to crops (Parihar *et al.*, 2015; Balal *et al.*, 2017). A major contributor to water loss is poor irrigation, which is particularly problematic in regions prone to drought, unpredictable precipitation, or high rates of evaporation (Pereira *et al.*, 2002). The watering schedule may be adjusted based on soil moisture, weather, and plant demands with the help of smart irrigation systems that employ sensors, controllers, and software (Bwambale *et al.*, 2022). While increasing crop yields and quality, this method can reduce water usage by as much as half when compared to traditional irrigation methods (Koech and Langat, 2018). Water treatment and recycling have been greatly improved by innovations in technology such as membrane filtration, biological treatment, UV disinfection, and reverse osmosis, among others (Joo and Tansel, 2015).

15.5 MODERN INNOVATIONS IN GENETICS AND BIOTECHNOLOGY TO OVERCOME ENVIRONMENTAL STRESS

Climate change is causing more frequent and severe droughts, water shortages, floods, degraded soils, and high temperatures, all of which pose serious problems for agriculture in the current era (Shahzad *et al.*, 2021). Most of the crops aren't exactly hardy in extreme environmental stress. Rising freshwater shortage at a time when the global population is projected to reach about 10,000,000,000 in 2050, it is increasing at an exponential rate (Zhang *et al.*, 2020a). For that decade, scientists predict that food production should quadruple from where it is now. As a result, new agricultural methods will need to be developed to deal with these challenges (Abegunde and Obi, 2022). Biotechnology can help to feed the world by increasing crop yields with less input (Durham and Mizik, 2021). Another way is to reduce the amount of agricultural chemicals needed by crops, resulting in less run-off into the environment (Khanna *et al.*, 2022). Lastly, biotech crops allow farmers to reduce tilling of farmland and apply pesticides less frequently (Anderson *et al.*, 2019).

15.6 STRESS-TOLERANCE GENE MODIFICATION

Transgenic plants improve resistance to abiotic stressors by manipulating genes (Anwar and Kim, 2020). Plants can be engineered to be more resistant to abiotic stress and to have better quality, texture, and shelf life by using genome modification methods like CRISPR/Cas9 (Sami *et al.*, 2021). Defensin, chitinase, osmotin, and glucanase genes, as well as genes associated with disease, are being introduced to several horticultural crops globally (Parmar *et al.*, 2017). Horticultural crop yields are negatively impacted by environmental stresses like heat, drought, and salt (Florentine *et al.*, 2016; Toscano *et al.*, 2019). To increase tolerance to these stresses, plants have been engineered to produce stress-protecting chemicals like mannitol, glycine betaine, and heat shock proteins (HSPs) (Gupta *et al.*, 2022). Other transcription factors that play a role in this process include DREB1, MAPK, WRKY, and many more (Erpen *et al.*, 2018). Some different transgenic strategies that aim to improve environmental stress in plants are shown in Figure 15.2.

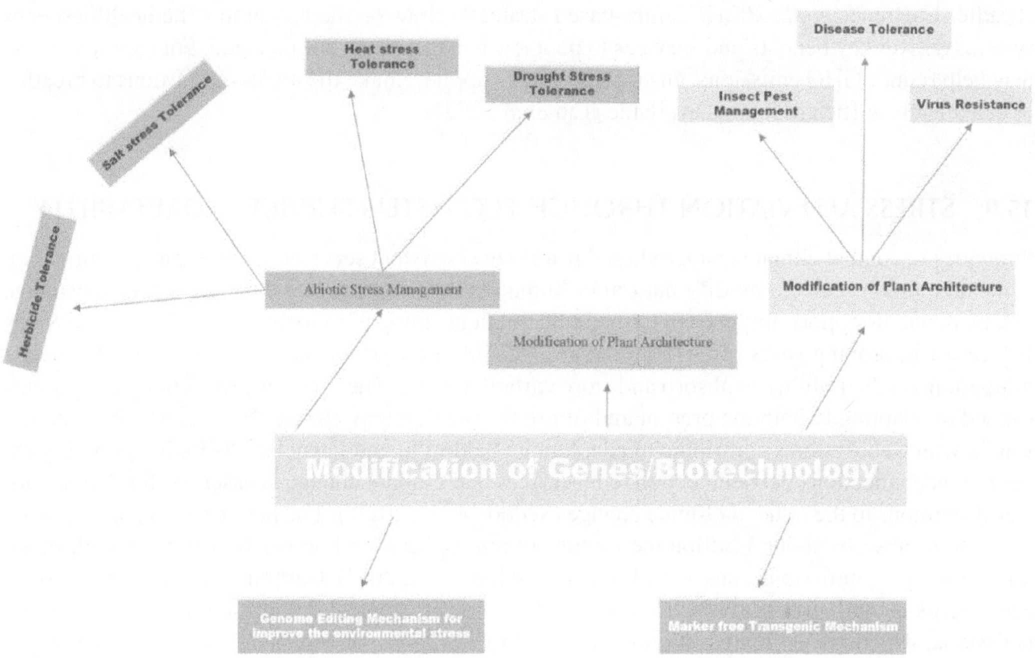

FIGURE 15.2 An investigation of the several transgenic approaches to crop improvement under environmental stress.

15.7 IMPROVEMENT OF CROP RESILIENCE UNDER ENVIRONMENTAL STRESS

In the presence of environmental stress, plants exhibit various behaviors, including short-term evasion or tolerance and long-term changes in paleontology, morphology, and physiochemistry (Rastogi *et al.*, 2014). Under stress, specific major lenience methods including ion transporters, osmoprotectants, numerous proteins during abnormal embryo formation, and characteristics entangled in response chutes and transcriptional switches are crucial (Raza, 2021). Excessive expression of some stress-related genes in plants has the potential to increase their stress resistance by decreasing their cytosolic uptake of harmful ions like Na^+ (Almeida *et al.*, 2017). Overexpressing the vacuolar H^+ pyrophosphatase (V-PPase) genes has helped scientists improve the external stress resistance and production of various crops, such as alfalfa, rice, wheat, sugarcane, cotton, peanuts, and tobacco (Schilling *et al.*, 2017; Afzal *et al.*, 2020). Plants that overexpress V-PPase have stronger root systems, that boost their capability to obtain nutrients and water since auxin polar transport is more active in these plants (Kriegel *et al.*, 2015).

15.8 SOLUTION STRATEGIES ADAPTED FROM NATURE

The term "nature-based solutions" describes methods that rely on the natural world to lessen the impact of climate variation, save species diversity, and improve people's standard of life (Seddon *et al.*, 2021b). These approaches include tree planting, agroforestry, regenerative agriculture, and community forest management (Aryal *et al.*, 2019). NBS can be implemented in urban settings to mitigate and minimize the impacts of climate variation (Seddon *et al.*, 2021b). New studies revealed that NBS and the land sector as a whole could reduce climate change by 30% by 2050, which is the cut required to meet the goal of limiting global warming in what is now the Paris Agreement

(Bradfer-Lawrence *et al.*, 2021). Nature-based strategies draw on the notion that the healthiest ecosystems deliver key benefits and services to people when they are well-managed. For example, these may help reduce GHG emissions, ensure safe water supplies, make the air less hazardous to breathe, or help ensure sufficient food is available (Liu *et al.*, 2021).

15.9 STRESS ALLEVIATION THROUGH ECOSYSTEM SERVICES ADAPTABILITY

Strategies to combat climate change should prioritize ecosystem services: adaptation and mitigation while adaptation seeks to modify natural or human systems to mitigate damage or take advantage of advantageous opportunities from climatic fluctuations, mitigation focuses on reducing emissions sources or improving sinks of GHGs (Gao *et al.*, 2018). Ecosystems have a role in climate change mitigation via their ability to absorb and store carbon dioxide. The services provided by ecosystems can aid in adapting to both the present and future threats posed by climate change, which is another way in which ecosystems contribute to adaptation (Baldocchi and Penuelas, 2019). Recognizing the numerous connections between ecosystem services and climate change is essential for ecosystem-based solutions to the issue of climate change (Seddon *et al.*, 2021a). The net expansion of terrestrial ecosystems absorbs about 3 billion metric tons of atmospheric carbon per year (Pg/yr), making up 30% of the CO_2 emissions caused by humans (Jakhar *et al.*, 2017). Carbon sequestration by forest ecosystems is vital, especially in tropical forests; however, tropical deforestation results in carbon emissions, which are estimated to range from 0.8 to 2.8 Pg/yr., or 6–17% of the world's anthropogenic CO_2 emissions. Managing ecosystems can help reduce the impact of climate change (Mcnicol *et al.*, 2018).

15.10 MANAGE BALANCED NUTRITION TO REDUCE ENVIRONMENTAL STRESS

The natural balance of the world environment has been upset by changing weather conditions in this modern age. Worldwide, agricultural output is severely constrained by environmental stress (Abbass *et al.*, 2022). Indirectly and directly, crop growth and environmental conditions are being affected by a multitude of abiotic stresses brought about by climate change (Ullah *et al.*, 2021). In many agricultural ecosystems, crop development and output can be impacted by abiotic stresses such as drought, temperature changes, soil salinity, soil alkalinity, and heavy metal stressors (Seleiman *et al.*, 2021). These factors can limit global food security. Heat or chilling/freezing stress can be caused by extremely high or low temperatures (10–15°C) compared to an ideal environment (Waraich *et al.*, 2012). Stress effects on plant processes including photosynthesis, seed germination, development, growth, and reproduction are far-reaching and detrimental to plant productivity (Hanif *et al.*, 2017; Dhankher and Foyer, 2018). Plant hormones are affected by heat and temperature stress (Babar *et al.*, 2009; Abdelrahman *et al.*, 2020). When plants are exposed to high temperatures, they produce more ROS while reducing enzyme activity, chlorophyll content, and gas exchange parameters (Hasanuzzaman *et al.*, 2013).

Reportedly, several plant nutrients reduce heat stress. Boron, for instance, is said to be a crucial micronutrient that reduces the harmful effects of ROS caused by heat stress and greatly enhances antioxidant system activity (Sachdev *et al.*, 2021). Chlorophyll, membrane fluidity, and tissue water content are all impacted by cold stress in plants. Similarly, selenium (Se) shields plants from ROS damage because of its structural function in glutathione peroxidase (GPX) formation (Al-Khayri *et al.*, 2023). Zinc is a crucial nutrient because it protects plants from the harmful impacts of heat stress and because it significantly preserves membrane permeability (Umair Hassan *et al.*, 2020). Figure 15.3 describes the various abiotic stressors and how they affect plants.

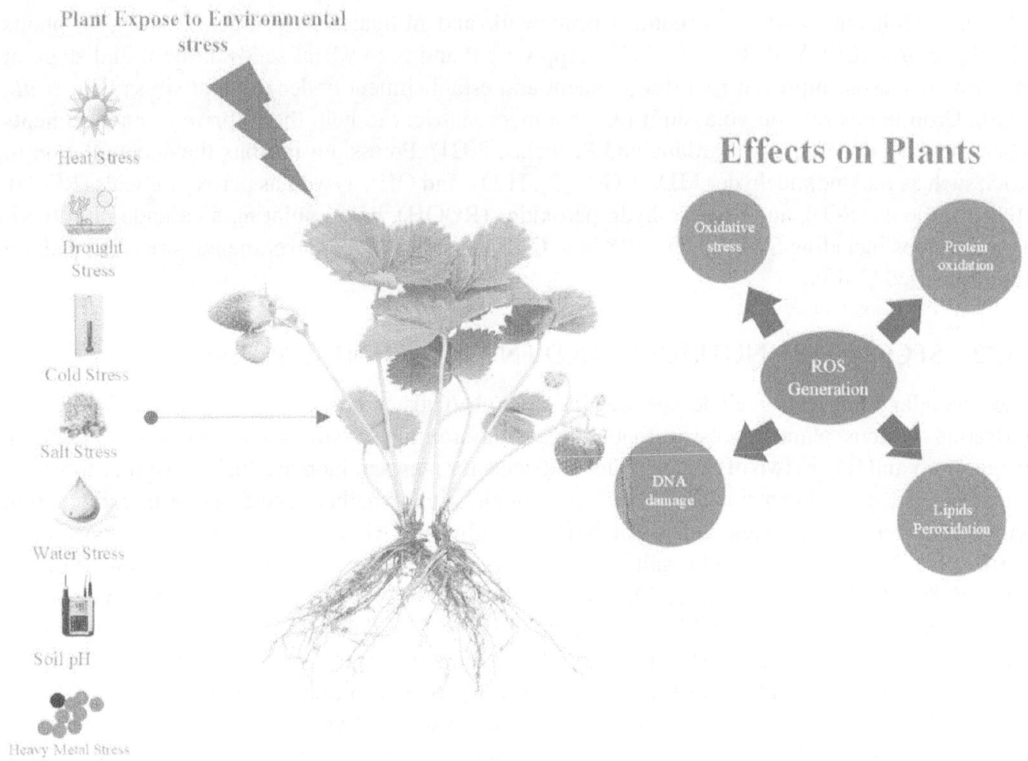

FIGURE 15.3 Effect of abiotic stress on crop plants.

15.11 PRIMARY NUTRIENTS

Nutrient management methods can ensure a successful harvest even in the most trying of environments (Rehman *et al.*, 2022). Nutrients in plants promote several defense mechanisms that plants use to overcome abiotic stress (Sachdev *et al.*, 2021). These include genes that deal with stress, antioxidant enzyme biosynthesis, Osmo protectants, HSPs, detoxification of ROS, enhancement of photosynthetic activity, stabilization of membranes, repair of DNA, protection of proteins' functional or structural aspects, decreased uptake of heavy metals, and countless more (Sachdev *et al.*, 2021). When properly implemented, nutrient management can increase agricultural output by decreasing environmental stress. Nitrogen (N) is the most important plant nutrient for structural health, growth, physiology, and stress alleviation (Sapkota *et al.*, 2021). It also helps plants deal with a wide range of biotic and abiotic pressures by working as a signaling molecule or free radical (Wani *et al.*, 2021). Increasing proline levels and the activities of enzymes that neutralize ROS, including superoxide dismutase (SOD), peroxidase (POD), catalase (CAT), and ascorbate peroxidase (APX), reduce the quantity of ROS in the body. Nitrogen can also interact with phytohormones like auxin and abscisic acid (ABA) and stop programmed cell death (PCD) (Azhar Mehmood *et al.*, 2018; El-Beltagi *et al.*, 2020).

Phosphoric fertilizer is strongly linked to plant development during times of stress, and it raises levels of proteins and molecules that help plants manage drought stress (Khanna *et al.*, 2022). In addition to facilitating optimal leaf water potential maintenance, stress tolerance, and cell membrane integrity, phosphorus (P) aids in this process by speeding up stomatal conductance and the rate of net photosynthesis (Khondoker *et al.*, 2023). For both the main and secondary plant metabolisms, biomolecules including ATP, NADPH, nucleic acids, phospholipids, and sugar-phosphates rely on P molecules as their structural framework and mitigate environmental stress in plants (Chadha *et al.*, 2019; Wahab *et al.*, 2023). Applying P and N to wheat seeds at the initial stage of the growth season improved root development and establishment under drought stress (Lin *et al.*, 2020). Crop plants rely on potassium (K) as a macronutrient to help them thrive in environments when they are under stress (Sardans and Peñuelas, 2021). Potassium inhibits the accumulation of ROS, such as malondialdehyde (MDA), O^{2-}, $^1O_2$, H_2O_2, and OH•, as well as peroxy radicals (ROO•), alkoxy radicals (RO•), and organic hydroperoxides (ROOH), by stimulating a cascade of antioxidant enzymes including SOD, POD, APX, and CAT in response to environmental stress (El-Beltagi and Mohamed, 2013).

15.12 SECONDARY NUTRIENTS AND ENVIRONMENTAL STRESS

The secondary nutrients include specifically calcium (Ca), magnesium (Mg), and sulfur (S), for activating various plant processes that mitigate environmental stresses (Kumari *et al.*, 2022). Magnesium and Ca are two of the most vital nutrients for plants to increase their resistance to stress. It makes a substantial contribution to reducing abiotic stress whether used alone or in combination with external plant hormones such as salicylic acid, ethylene, ABA, jasmonic acid, and polyamines (Arif *et al.*, 2021). Heat, drought, salt, and metal stress are known to lower the bioavailability of Ca^{2+}, Mg^{2+}, and K^+. Plants require Mg^{2+} for enzymes like ATPase, kinases, and polymerases to function properly. Additionally, Mg^{2+} helps to maintain cell wall cohesion by altering the hydration level of the plasma membrane (Farhat *et al.*, 2016). This, in turn, boosts the tolerance of cells to drought conditions by enhancing the viscosity of the protoplasm during drought stress. After the roots absorb sulfate (SO_4^{2-}), sulfur is converted into cysteine (Cys) Cysteine serves as a precursor for the important S compounds methionine (Met), S-adenosylmethionine, glutathione (GSH), homo-GSH (h-GSH), phytochelatins (PCs), sulfolipids, iron-sulfur clusters, allylic, and glucosinolates that play critical roles in environmental stress. These compounds contribute to plant development and mitigate environmental stress (González-Morales *et al.*, 2017).

15.13 THE FUNCTION OF MICROBES IN RESOLVING ENVIRONMENTAL STRESS

Microorganisms can grow in a broad variety of environments and have remarkable metabolic capabilities, which contribute to their widespread distribution in the biosphere (Lynch and Neufeld, 2015). Biodegradation of contaminants can also make use of microbes' dietary flexibility. Microbes can help plants grow better in abiotic stress situations by increasing their production of amino acids, organic acids, mineral phosphate solubilization, ACC-deaminase, chitinase, and glucanase, and low-molecular-weight osmolytes like proline and glycine betaine (Shaikh *et al.*, 2021). Due to these factors, environmental biodegradation is less effective, thus less preferable. Bioremediation is beneficial only in settings in which microbial growth and function are possible (Medfu Tarekegn *et al.*, 2020).

15.14 MANAGEMENT OF ENVIRONMENTAL STRESSORS THROUGH BIOAUGMENTATION AND BIOSTIMULATION

Bioaugmentation involves the addition of microorganisms capable of degrading resistant molecules in a contaminated environment. Essentially, it is the introduction of a culture into a contaminated site to assist in biodegradation. The process of biostimulation, which is a mechanism of removing elements that make the degradation of pollutants in the presence of oxygen more difficult, is commonly known (Dehnavi and Ebrahimipour, 2022). Biostimulation is a nature-friendly method using cheap chemicals with reliable technology and low energy (Udume *et al.*, 2023). This mechanism is dependent on the chemical characteristics of the pollutants, the concentration of the pollutants, the physiochemical properties of the surrounding environment, and the availability of the required microorganisms (Gorovtsov *et al.*, 2020). The preference for the biostimulation method is a result of the higher competitiveness of native bacteria as opposed to foreign ones and the capacity to maintain the natural diversity of the microbial community (Trivedi *et al.*, 2021). The efficacy of *Bacillus* sp., *Rhodococcus* sp., and *Citrobacter* sp. in the process of bioremediation of heavy metals using the biostimulation approach (Gayathiri *et al.*, 2022).

15.15 BIOLEACHING AND BIOSORPTION

Bioleaching, also known as biomining, is a mining and biohydrometallurgy process that involves the use of microorganisms (such as bacteria or archaea) to extract valuable metals from low-grade ore (Nkuna *et al.*, 2022). It takes advantage of the natural interactions between these microbes and minerals. Whereas, biotech is a conventional method, called biosorption, whereby biological materials serve as biosorbents that can remove heavy metals from effluent (Ali Redha, 2020). Bioleaching is an extremely beneficial process to eliminate iron or sulfur contamination. Microorganisms create an acidic environment and dissolve heavy metals during this process, which immobilizes them in a liquid solution (Rahman and Singh, 2020). This is the process of removing heavy metals from pollution through the displacement of protons and ions, forming complexes and chelates, and physically interacting with electrostatic forces (Chandran *et al.*, 2023). Certain bacteria, including *Streptomyces* sp., have shown the potential for biosorption leading to bioremediation. By using biochar and complexation led to the purification of wastewater where K11, *Bacillus anthracis,* and *Rhodococcus erythropolis*, removed copper (II) oxide and hexavalent chromium (Cr). Species involved in complexation (*Rhodobacter blasticus* and *Bacillus licheniformis*) (Ambika *et al.*, 2022; Jia *et al.*, 2022).

15.16 ROLE OF ACTINOBACTERIA IN OVERWHELMING ENVIRONMENTAL STRESSES

Actinobacteria is a large phylum of gram-positive bacteria on plains of soil. They also are known to produce various secondary metabolites with different bioactivities, and they belong to high-G+C bacteria (Salam *et al.*, 2020). The relationships between the microbial population with agricultural plants are the most important reason for the adaptation and survival of both against environmental stress. Actinobacteria have evolved intricate strategies to cope with various biotic and abiotic stresses in their habitat, such as competition for food (Abdul Rahman *et al.*, 2021). Micromolecules undergo evolution among environmental stresses through some of the established genetic mechanisms leading to affect the chromosomal alterations in their genome (Sánchez-Romero and Casadesús, 2020). Changes in gene expression form the majority of bacterial responses to stress and environmental changes, among the many diverse mechanisms bacteria employ to cope with environmental stress (Pal *et al.*, 2022). Actinobacteria live in the rhizospheres of different plants and also produce compounds that promote plants, including under stressful conditions like salinity, drought, heavy metal-polluted soils, and low nutrient availability (Barra Caracciolo and Terenzi, 2021). Several prospectuses make use of actinobacteria to help plants deal with salt stress or become more resistant to it by the formation of an antioxidant defense system, the preservation of osmotic balance, and the production of 1-aminocyclopropane-1-carboxylic acid (ACC) deaminase (Kumar *et al.*, 2023).

15.17 CONCLUSION AND FUTURE PROSPECTUS

Agriculture serves as the fundamental support for both developed and developing countries, with more than 60% of the population in developing countries relying on agriculture for their means of living. The growing global population necessitates a heightened focus on augmenting food production. The agriculture industry faces diverse types of abiotic stresses. Furthermore, the significant climatic fluctuations caused by global warming contribute to a rise in the adverse effects of abiotic stressors. It was widely conceded that abiotic stress conditions such as salt, drought, temperatures, heavy metal poisonous, and pollution (both organic and inorganic) have significant adverse impacts on crop plants. These impacts include reduced plant growth and productivity, physiological changes, alterations in osmotic balance, and ion cytotoxicity. Climate change exacerbates the already harmful impact of environmental stress on plant life and mitigates agricultural production (Chadha *et al.*, 2020). The combined impact of plants and the bacteria they connect with is crucial to the successful implementation of phytoremediation technologies.

Microorganisms that live in harsh conditions possess useful qualities that can help farmland by keeping crops valuable and making them more productive. An effective method for reducing organic pollutant biodegradation and environmental stress is also available. The ecosystem approach is a strategy for the integrated management of land, water, and living resources that promotes conservation and sustainable use in an equitable way. The ecosystem approach will support a balance of three goals of the convention, as specified in this chapter. However, new technology can also promote environmental restoration in ways such as green energy and less reliance on fossil fuels. If nations desire to streamline their industrial processes, they could reap the advantages of cutting-edge technologies. Environmental engineers play a critical role in seeking, developing renewable energy sources, improving energy efficiency as well as assessing possible climate threats to help overcome these impacts.

We must continue to invest in research and development, raise awareness in various sectors, and empower communities and institutions to take action. Through inculcating the latest techniques in research and technology, not only do we strengthen our capacity to respond to existing environmental pressures, but we also lay the groundwork for a sustainable and successful future for future generations. We must collectively strive to answer that call, as we continue on the road to resiliency for our environment.

REFERENCES

Abbass, K., Qasim, M.Z., Song, H., Murshed, M., Mahmood, H., & Younis, I., 2022. A review of the global climate change impacts, adaptation, and sustainable mitigation measures. *Environmental Science and Pollution Research*, 29, 42539–42559.

Abbatt, J.P., & Ravishankara, A.R., 2023. Opinion: Atmospheric multiphase chemistry—Past, present, and future. *Atmospheric Chemistry and Physics*, 23, 9765–9785.

Abdelrahman, M., Burritt, D.J., Gupta, A., Tsujimoto, H., & Tran, L.-S.P., 2020. Heat stress effects on source–sink relationships and metabolome dynamics in wheat. *Journal of Experimental Botany*, 71, 543–554.

Abdul Rahman, N.S.N., Abdul Hamid, N.W., & Nadarajah, K., 2021. Effects of abiotic stress on soil microbiome. *International Journal of Molecular Sciences*, 22, 9036.

Abegunde, V.O., & Obi, A., 2022. The role and perspective of climate smart agriculture in Africa: A scientific review. *Sustainability*, 14, 2317.

Afzal, M.Z., Jia, Q., Ibrahim, A.K., Niyitanga, S., & Zhang, L., 2020. Mechanisms and signaling pathways of salt tolerance in crops: Understanding from the transgenic plants. *Tropical Plant Biology*, 13, 297–320.

Akhtar, I., & Nazir, N., 2013. Effect of waterlogging and drought stress in plants. *International Journal of Water Resources and Environmental Science*, 2, 34–40.

Ali Redha, A., 2020. Removal of heavy metals from aqueous media by biosorption. *Arab Journal of Basic and Applied Sciences*, 27, 183–193.

Al-Khayri, J.M., Rashmi, R., Surya Ulhas, R., Sudheer, W.N., Banadka, A., Nagella, P., Aldaej, M.I., Rezk, A.a.-S., Shehata, W.F., & Almaghasla, M.I., 2023. The role of nanoparticles in response of plants to abiotic stress at physiological, biochemical, and molecular levels. *Plants*, 12, 292.

Almeida, D.M., Oliveira, M.M., & Saibo, N.J., 2017. Regulation of Na^+ and K^+ homeostasis in plants: Towards improved salt stress tolerance in crop plants. *Genetics and Molecular Biology*, 40, 326–345.

Almutairi, Z.M.E.A., 2019. Plant molecular defense mechanisms promoted by nanoparticles against environmental stresses. *International Journal of Biological*, 21, 259–270.

Ambika, S., Kumar, M., Pisharody, L., Malhotra, M., Kumar, G., Sreedharan, V., Singh, L., Nidheesh, P., & Bhatnagar, A., 2022. Modified biochar as a green adsorbent for removal of hexavalent chromium from various environmental matrices: Mechanisms, methods, and prospects. *Journal of Chemical Engineering*, 439, 135716.

Anderson, J.A., Ellsworth, P.C., Faria, J.C., Head, G.P., Owen, M.D., Pilcher, C.D., Shelton, A.M., & Meissle, M., 2019. Genetically engineered crops: Importance of diversified integrated pest management for agricultural sustainability. *Frontiers in Bioengineering and Biotechnology*, 7, 24.

Andrés, N.C., Sieben, J.M., Baldini, M.N., Rodríguez, C.H., Famiglietti, A.N., & Messina, P.V., 2018. Electroactive Mg^{2+}-hydroxyapatite nanostructured networks against drug-resistant bone infection strains. *ACS Applied Materials & Interfaces*, 10, 19534–19544.

Anwar, A., & Kim, J.-K., 2020. Transgenic breeding approaches for improving abiotic stress tolerance: Recent progress and future perspectives. *International Journal of Molecular Sciences*, 21, 2695.

Arif, Y., Singh, P., Bajguz, A., Alam, P., & Hayat, S., 2021. Silicon mediated abiotic stress tolerance in plants using physio-biochemical, omic approach and cross-talk with phytohormones. *Journal of Physiology and Biochemistry*, 166, 278–289.

Ariga, K., 2021. Nanoarchitectonics: What's coming next after nanotechnology? *Nanoscale Horizons*, 6, 364–378.

Aryal, K., Thapa, P.S., & Lamichhane, D., 2019. Revisiting agroforestry for building climate resilient communities: A case of package-based integrated agroforestry practices in Nepal. *Emerging Science Journal*, 3, 303–311.

Azhar Mehmood, A.M., Muhammad Naeem, M.N., Farhan Khalid, F.K., Yousaf Saeed, Y.S., Tasawer Abbas, T.A., Jabran, K., Sarwar, M., Asif Tanveer, A.T., & Javaid, M., 2018. Identification of phytotoxins in different plant parts of *Brassica napus* and their influence on mung bean.

Aziz, M., Ashraf, M., & Javaid, M.M., 2018. Enhancement in cotton growth and yield using novel growth promoting substances under water limited conditions. *Pakistan Journal of Botany*, 50, 1691–1701.

Babar, B.H., Tanveer, A., Tahir, M., Aziz, A., Ahmad, A.U.H., Nadeem, M.A., & Javaid, M.M., 2009. Allelopathic potential of wild onion (*Asphodelus tenuifolius*) on the germination and seedling growth of chickpea (*Cicer arietinum*). *Weed Biology and Management*, 9, 146–151.

Balal, R.M., Shahid, M.A., Javaid, M.M., Iqbal, Z., Liu, G.D., Zotarelli, L., & Khan, N., 2017. Chitosan alleviates phytotoxicity caused by boron through augmented polyamine metabolism and antioxidant activities and reduced boron concentration in *Cucumis sativus* L. *Acta Physiologiae Plantarum*, 39, 1–15.

Baldocchi, D., & Penuelas, J., 2019. The physics and ecology of mining carbon dioxide from the atmosphere by ecosystems. *Global Change Biology*, 25, 1191–1197.

Barra Caracciolo, A., & Terenzi, V., 2021. Rhizosphere microbial communities and heavy metals. *Microorganisms*, 9, 1462.

Bradfer-Lawrence, T., Finch, T., Bradbury, R.B., Buchanan, G.M., Midgley, A., & Field, R.H., 2021. The potential contribution of terrestrial nature-based solutions to a national 'net zero' climate target. *Journal of Applied Ecology*, 58, 2349–2360.

Bwambale, E., Abagale, F.K., & Anornu, G.K., 2022. Smart irrigation monitoring and control strategies for improving water use efficiency in precision agriculture: A review. *Agricultural Water Management*, 260, 107324.

Chadha, A., Florentine, S., Chauhan, B.S., Long, B., Jayasundera, M., Javaid, M.M., & Turville, C., 2019. Environmental factors affecting the germination and seedling emergence of two populations of an emerging agricultural weed: Wild lettuce (*Lactuca serriola*). *Crop and Pasture Science*, 70, 709–717.

Chadha, A., Florentine, S., Javaid, M., Welgama, A., & Turville, C., 2020. Influence of elements of climate change on the growth and fecundity of *Datura stramonium*. *Environmental Science and Pollution Research*, 27, 35859–35869.

Chandran, D.G., Muruganandam, L., & Biswas, R., 2023. A review on adsorption of heavy metals from wastewater using carbon nanotube and graphene-based nanomaterials. *Environmental Science and Pollution Research*, 30, 110010–110046.

Cosgrove, W.J., & Loucks, D.P., 2015. Water management: Current and future challenges and research directions. *Water Resources Research*, 51, 4823–4839.

Daiber, A., Frenis, K., Kuntic, M., Li, H., Wolf, E., Kilgallen, A.B., Lecour, S., Van Laake, L.W., Schulz, R., & Hahad, O., 2022. Redox regulatory changes of circadian rhythm by the environmental risk factors traffic noise and air pollution. *Antioxidants & Redox Signaling*, 37, 679–703.

Dawson, T.P., Cutler, M., & Brown, C., 2016. The role of remote sensing in the development of SMART indicators for ecosystem services assessment. *Biodivers*, 17, 136–148.

Dehnavi, S., & Ebrahimipour, G., 2022. Comparative remediation rate of biostimulation, bioaugmentation, and phytoremediation in hydrocarbon contaminants. *International Journal of Environmental Science and Technology*, 19, 11561–11586.

Dhankher, O.P., & Foyer, C.H., 2018. *Climate Resilient Crops for Improving Global Food Security and Safety*. Wiley Online Library, 877–884.

Douxchamps, S., Debevec, L., Giordano, M., & Barron, J., 2017. Monitoring and evaluation of climate resilience for agricultural development—A review of currently available tools. *World Development Perspectives*, 5, 10–23.

Durham, T.C., & Mizik, T., 2021. Comparative economics of conventional, organic, and alternative agricultural production systems. *Economies*, 9, 64.

El-Beltagi, H.S., & Mohamed, H.I., 2013. Reactive oxygen species, lipid peroxidation and antioxidative defense mechanism. *Notulae Botanicae Horti Agrobotanici*, 41, 44–57.

El-Beltagi, H.S., Mohamed, H.I., & Sofy, M.R., 2020. Role of ascorbic acid, glutathione and proline applied as singly or in sequence combination in improving chickpea plant through physiological change and antioxidant defense under different levels of irrigation intervals. *Molecules*, 25, 1702.

Erdelen, W.R., & Richardson, J.G., 2021. A world after COVID-19: Business as usual, or building bolder and better? *Global Policy*, 12, 157–166.

Erpen, L., Devi, H.S., Grosser, J.W., & Dutt, M., 2018. Potential use of the DREB/ERF, MYB, NAC and WRKY transcription factors to improve abiotic and biotic stress in transgenic plants. *Plant Cell, Tissue and Organ Culture (PCTOC)*, 132, 1–25.

Eskander, S.M., & Fankhauser, S., 2020. Reduction in greenhouse gas emissions from national climate legislation. *Nature Climate Change*, 10, 750–756.

Faran, T.S., & Olsson, L., 2018. Geoengineering: Neither economical, nor ethical—A risk–reward nexus analysis of carbon dioxide removal. *International Environmental Agreement*, 18, 63–77.

Farhat, N., Zorrig, W., Salhi, T., Abdelly, C., Smaoui, A., & Rabhi, M., 2016. Do medium-and long-term magnesium deficiencies affect potassium and calcium nutrition in *Sulla carnosa*. *Agrochimica*, 60, 10–12871.

Florentine, S.K., Weller, S., Graz, P.F., Westbrooke, M., Florentine, A., Javaid, M., Fernando, N., Chauhan, B.S., & Dowling, K., 2016. Influence of selected environmental factors on seed germination and seedling survival of the arid zone invasive species tobacco bush (*Nicotiana glauca* R. Graham). *The Rangeland Journal*, 38, 417–425.

Florkiewicz, A.B., Kućko, A., Kapusta, M., Burchardt, S., Przywieczerski, T., Czeszewska-Rosiak, G., & Wilmowicz, E., 2020. Drought disrupts auxin localization in abscission zone and modifies cell wall structure leading to flower separation in yellow lupine. *International Journal of Molecular Sciences*, 21, 6848.

Folberth, C., Khabarov, N., Balkovič, J., Skalský, R., Visconti, P., Ciais, P., Janssens, I.A., Peñuelas, J., & Obersteiner, M., 2020. The global cropland-sparing potential of high-yield farming. *Nature Sustainability*, 3, 281–289.

Freschet, G.T., Pagès, L., Iversen, C.M., Comas, L.H., Rewald, B., Roumet, C., Klimešová, J., Zadworny, M., Poorter, H., & Postma, J.A., 2021. A starting guide to root ecology: Strengthening ecological concepts and standardising root classification, sampling, processing and trait measurements. *New Phytologist*, 232, 973–1122.

Gao, J., Hou, H., Zhai, Y., Woodward, A., Vardoulakis, S., Kovats, S., Wilkinson, P., Li, L., Song, X., & Xu, L., 2018. Greenhouse gas emissions reduction in different economic sectors: Mitigation measures, health co-benefits, knowledge gaps, and policy implications. *Environmental Pollution*, 240, 683–698.

García-Freites, S., Gough, C., & Röder, M., 2021. The greenhouse gas removal potential of bioenergy with carbon capture and storage (BECCS) to support the UK's net-zero emission target. *Biomass Bioenergy*, 151, 106164.

Gayathiri, E., Prakash, P., Karmegam, N., Varjani, S., Awasthi, M.K., & Ravindran, B., 2022. Biosurfactants: Potential and eco-friendly material for sustainable agriculture and environmental safety—A review. *Agronomy*, 12, 662.

González-Morales, S., Pérez-Labrada, F., García-Enciso, E.L., Leija-Martínez, P., Medrano-Macías, J., Dávila-Rangel, I.E., Juárez-Maldonado, A., Rivas-Martínez, E.N., & Benavides-Mendoza, A., 2017. Selenium and sulfur to produce Allium functional crops. *Molecules*, 22, 558.

Gorovtsov, A.V., Minkina, T.M., Mandzhieva, S.S., Perelomov, L.V., Soja, G., Zamulina, I.V., Rajput, V.D., Sushkova, S.N., Mohan, D., & Yao, J., 2020. The mechanisms of biochar interactions with microorganisms in soil. *Environmental Geochemistry and Health*, 42, 2495–2518.

Gupta, A., Mishra, R., Rai, S., Bano, A., Pathak, N., Fujita, M., Kumar, M., & Hasanuzzaman, M., 2022. Mechanistic insights of plant growth promoting bacteria mediated drought and salt stress tolerance in plants for sustainable agriculture. *International Journal of Molecular Sciences*, 23, 3741.

Hammond, T.T., Ortiz-Jimenez, C.A., & Smith, J.E., 2020. Anthropogenic change alters ecological relationships via interactive changes in stress physiology and behavior within and among organisms. *Integrative and Comparative Biology*, 60, 57–69.

Hanaka, A., Ozimek, E., Reszczyńska, E., Jaroszuk-Ściseł, J., & Stolarz, M., 2021. Plant tolerance to drought stress in the presence of supporting bacteria and fungi: An efficient strategy in horticulture. *Hortic*, 7, 390.

Hanif, Z., Naeem, M., Ali, H.H., Tanveer, A., Javaid, M.M., Peerzada, A.M., & Chauhan, B.S., 2017. Effect of environmental factors on germination of *Salsola foetida*: Potential species for rehabilitation of degraded rangelands. *Rangeland Ecology & Management*, 70, 638–643.

Hasanuzzaman, M., Nahar, K., Alam, M.M., Roychowdhury, R., & Fujita, M., 2013. Physiological, biochemical, and molecular mechanisms of heat stress tolerance in plants. *International Journal of Molecular Sciences*, 14, 9643–9684.

Hasanuzzaman, M., Raihan, M.R.H., Masud, A.C., Rahman, K., Nowroz, F., Rahman, M., Nahar, K., & Fujita, M., 2021. Regulation of reactive oxygen species and antioxidant defense in plants under salinity. *International Journal of Molecular Sciences*, 22, 9326.

Hassan, A., Yang, J., Usman, A., Bilal, A., & Ullah, S., 2023. Green growth as a determinant of ecological footprint: Do ICT diffusion, environmental innovation, and natural resources matter? *PLoS One*, 18, e0287715.

Hong, M.J., Kim, J.-B., Yoon, Y.H., Kim, S.H., Ahn, J.-W., Jeong, I.Y., Kang, S.-Y., Seo, Y.W., & Kim, D.S., 2014. The effects of chronic gamma irradiation on oxidative stress response and the expression of anthocyanin biosynthesis-related genes in wheat (*Triticum aestivum*). *International Journal of Radiation Biology*, 90, 1218–1228.

Hossain, Z., Mustafa, G., & Komatsu, S., 2015. Plant responses to nanoparticle stress. *International Journal of Molecular Sciences*, 16, 26644–26653.

Hussain, A., Rehman, F., Rafeeq, H., Waqas, M., Asghar, A., Afsheen, N., Rahdar, A., Bilal, M., & Iqbal, H.M., 2022. In-situ, ex-situ, and nano-remediation strategies to treat polluted soil, water, and air—A review. *Chemosphere*, 289, 133252.

Jabbar, A., Ibrahim, H., Mahmood, A., Javaid, M.M., Wasaya, A., Nadeem, M.A., Ismael, M., Awais, M., & Khan, S.R., 2023. Modern Agronomic Measurement for Climate-Resilient Agriculture. *In Climate-Resilient Agriculture, Vol 2: Agro-Biotechnological Advancement for Crop Production*. Springer, 81–105.

Jakhar, R.R., Yadav, S., Jakhar, R.K., Devra, P., Ram, H., & Kumar, R., 2017. Potential and importance of carbon sequestrations in agricultural soils. *International Journal of Current Microbiology and Applied Sciences*, 6, 1776–1788.

Jeevanantham, S., Saravanan, A., Hemavathy, R., Kumar, P.S., Yaashikaa, P., & Yuvaraj, D., 2019. Removal of toxic pollutants from water environment by phytoremediation: A survey on application and future prospects. *Environmental Technology and Innovation*, 13, 264–276.

Jia, Q.-Y., Chen, C.-M., Li, B.-Z., & Wang, L., 2022. Construction and characterization of an engineered recombinant *Rhodopseudomonas palustris* to remove Cd^{2+}, Zn^{2+} and Cu^{2+}. *Journal of Environmental Chemical Engineering*, 10, 107698.

Joo, S.H., & Tansel, B., 2015. Novel technologies for reverse osmosis concentrate treatment: A review. *Journal of Environmental Management*, 150, 322–335.

Kalia, V., & Knauft, K., 2020. Emotion regulation strategies modulate the effect of adverse childhood experiences on perceived chronic stress with implications for cognitive flexibility. *PloS One*, 15, e0235412.

Khanna, M., Atallah, S.S., Kar, S., Sharma, B., Wu, L., Yu, C., Chowdhary, G., Soman, C., & Guan, K., 2022. Digital transformation for a sustainable agriculture in the United States: Opportunities and challenges. *Agricultural Economics*, 53, 924–937.

Khondoker, M., Mandal, S., Gurav, R., & Hwang, S., 2023. Freshwater shortage, salinity increase, and global food production: A need for sustainable irrigation water desalination—A scoping review. *Earth*, 4, 223–240.

Koech, R., & Langat, P., 2018. Improving irrigation water use efficiency: A review of advances, challenges and opportunities in the Australian context. *Water*, 10, 1771.

Koza, N.A., Adedayo, A.A., Babalola, O.O., & Kappo, A.P., 2022. Microorganisms in plant growth and development: Roles in abiotic stress tolerance and secondary metabolites secretion. *Microorganisms*, 10, 1528.

Kriegel, A., Andrés, Z., Medzihradszky, A., Krüger, F., Scholl, S., Delang, S., Patir-Nebioglu, M.G., Gute, G., Yang, H., & Murphy, A.S., 2015. Job sharing in the endomembrane system: Vacuolar acidification requires the combined activity of V-ATPase and V-PPase. *The Plant Cell*, 27, 3383–3396.

Kumar, A., Behera, I., Langthasa, M., & Prakashnaroju, S., 2023. Effect of plant growth-promoting rhizobacteria on alleviating salinity stress in plants: A review. *Journal of Plant Nutrition*, 46, 2525–2550.

Kumar, R., Singh, P., & Singh, S., 2018. A review report: Low temperature stress for crop production. *International Journal of Pure & Applied Bioscience*, 6, 575–598.

Kumari, V.V., Banerjee, P., Verma, V.C., Sukumaran, S., Chandran, M.a.S., Gopinath, K.A., Venkatesh, G., Yadav, S.K., Singh, V.K., & Awasthi, N.K., 2022. Plant nutrition: An effective way to alleviate abiotic stress in agricultural crops. *International Journal of Molecular Sciences*, 23, 8519.

Kweku, D.W., Bismark, O., Maxwell, A., Desmond, K.A., Danso, K.B., Oti-Mensah, E.A., Quachie, A.T., & Adormaa, B.B., 2018. Greenhouse effect: Greenhouse gases and their impact on global warming. *Journal of Scientific Research*, 17, 1–9.

Li, H., Zheng, J., Zheng, W., Zu, H., Chen, H., Yang, J., Qu, W., Leng, L., Feng, Y., & Yang, Z., 2023. In situ acid etching boosts mercury accommodation capacities of transition metal sulfides. *Nature Communications*, 14, 1395.

Li, Z., Liao, G., & Albitar, K., 2020. Does corporate environmental responsibility engagement affect firm value? The mediating role of corporate innovation. *BSE*, 29, 1045–1055.

Lin, Y., Watts, D.B., Kloepper, J.W., Feng, Y., & Torbert, H.A., 2020. Influence of plant growth-promoting rhizobacteria on corn growth under drought stress. *Communications in Soil Science and Plant Analysis*, 51, 250–264.

Liu, H.-Y., Jay, M., & Chen, X., 2021. The role of nature-based solutions for improving environmental quality, health and well-being. *Sustainability*, 13, 10950.

Loh, S., Foth, M., Caldwell, G.A., Garcia-Hansen, V., & Thomson, M., 2020. A more-than-human perspective on understanding the performance of the built environment. *Architectural Science Review*, 63, 372–383.

Lynch, M.D., & Neufeld, J.D., 2015. Ecology and exploration of the rare biosphere. *Nature Reviews Microbiology*, 13, 217–229.

Mackenthun, G., 2021. Sustainable stories: Managing climate change with literature. *Sustainability*, 13, 4049.

Mahmood, A., Ullah, H., Shahzad, A., Ali, H., Ahmad, S., Zia-Ul-Haq, M., Honermeier, B., & Hasanuzzaman, M., 2013. Dry matter yield and chemical composition of sorghum cultivars with varying planting density and sowing date. *Sains Malaysiana*, 42, 1529–1538.

Mahmood, A., Wang, X., Shahzad, A.N., Fiaz, S., Ali, H., Naqve, M., Javaid, M.M., Mumtaz, S., Naseer, M., & Dong, R., 2021. Perspectives on bioenergy feedstock development in Pakistan: Challenges and opportunities. *Sustainability*, 13, 8438.

Mahmood, I., Arabnejad, H., Suleimenova, D., Sassoon, I., Marshan, A., Serrano-Rico, A., Louvieris, P., Anagnostou, A., Je Taylor, S., & Bell, D., 2022. FACS: A geospatial agent-based simulator for analysing COVID-19 spread and public health measures on local regions. *Journal of Simulation*, 16, 355–373.

Martos, V., Ahmad, A., Cartujo, P., & Ordoñez, J., 2021. Ensuring agricultural sustainability through remote sensing in the era of agriculture 5.0. *Applied Sciences*, 11, 5911.

Mcnicol, I.M., Ryan, C.M., & Mitchard, E.T., 2018. Carbon losses from deforestation and widespread degradation offset by extensive growth in African woodlands. *Nature Communications*, 9, 3045.

Medfu Tarekegn, M., Zewdu Salilih, F., & Ishetu, A.I., 2020. Microbes used as a tool for bioremediation of heavy metal from the environment. *Cogent Food & Agriculture*, 6, 1783174.

Mehmood, M.A., Ibrahim, M., Rashid, U., Nawaz, M., Ali, S., Hussain, A., & Gull, M., 2017. Biomass production for bioenergy using marginal lands. *Sustainable Production and Consumption*, 9, 3–21.

Mittal, D., Kaur, G., Singh, P., Yadav, K., & Ali, S.A., 2020. Nanoparticle-based sustainable agriculture and food science: Recent advances and future outlook. *Frontiers in Nanotechnology*, 2, 579954.

Moayedi, H., & Mosavi, A., 2021. An innovative metaheuristic strategy for solar energy management through a neural networks framework. *Energies*, 14, 1196.

Moraci, F., Errigo, M.F., Fazia, C., Campisi, T., & Castelli, F., 2020. Cities under pressure: Strategies and tools to face climate change and pandemic. *Sustainability*, 12, 7743.

Movilla-Pateiro, L., Mahou-Lago, X., Doval, M., & Simal-Gandara, J., 2021. Toward a sustainable metric and indicators for the goal of sustainability in agricultural and food production. *Critical Reviews in Food Science and Nutrition*, 61, 1108–1129.

Muhammad, I., Shalmani, A., Ali, M., Yang, Q.-H., Ahmad, H., & Li, F.B., 2021. Mechanisms regulating the dynamics of photosynthesis under abiotic stresses. *Frontiers in Plant Science*, 11, 615942.

Naz, M., Benavides-Mendoza, A., Tariq, M., Zhou, J., Wang, J., Qi, S., Dai, Z., & Du, D., 2022. CRISPR/Cas9 technology as an innovative approach to enhancing the phytoremediation: Concepts and implications. *Journal of Environmental Management*, 323, 116296.

Nielsen, K.S., Clayton, S., Stern, P.C., Dietz, T., Capstick, S., & Whitmarsh, L., 2021. How psychology can help limit climate change. *American Psychologist*, 76, 130.

Nkuna, R., Ijoma, G.N., Matambo, T.S., & Chimwani, N., 2022. Accessing metals from low-grade ores and the environmental impact considerations: A review of the perspectives of conventional versus bioleaching strategies. *Minerals*, 12, 506.

Oomen, J., & Meiske, M., 2021. Proactive and reactive geoengineering: Engineering the climate and the lithosphere. *Wiley Interdisciplinary Reviews: Climate Change*, 12, e732.

Oves, M., Saghir Khan, M., Huda Qari, A., Nadeen Felemban, M., & Almeelbi, T., 2016. Heavy metals: Biological importance and detoxification strategies. *Journal of Bioremediation and Biodegradation*, 7, 1–15.

Pal, A., Bhattacharjee, S., Saha, J., Sarkar, M., & Mandal, P., 2022. Bacterial survival strategies and responses under heavy metal stress: A comprehensive overview. *Critical Reviews in Microbiology*, 48, 327–355.

Parihar, P., Singh, S., Singh, R., Singh, V.P., & Prasad, S.M., 2015. Effect of salinity stress on plants and its tolerance strategies: A review. *Environmental Science and Pollution Research*, 22, 4056–4075.

Parmar, N., Singh, K.H., Sharma, D., Singh, L., Kumar, P., Nanjundan, J., Khan, Y.J., Chauhan, D.K., & Thakur, A.K., 2017. Genetic engineering strategies for biotic and abiotic stress tolerance and quality enhancement in horticultural crops: A comprehensive review. *3 Biotech*, 7, 1–35.

Parween, T., Jan, S., Mahmooduzzafar, S., Fatma, T., & Siddiqui, Z.H., 2016. Selective effect of pesticides on plant—A review. *Critical Reviews in Food Science and Nutrition*, 56, 160–179.

Pereira, L.S., Oweis, T., & Zairi, A., 2002. Irrigation management under water scarcity. *Agricultural Water Management*, 57, 175–206.

Quinnan, J., Morrell, C., Nagle, N., & Maynard, K.G., 2022. Ex situ soil washing to remove PFAS adsorbed to soils from source zones. *Remediation Journal*, 32, 151–166.

Rahman, Z., & Singh, V.P., 2020. Bioremediation of toxic heavy metals (THMs) contaminated sites: Concepts, applications and challenges. *Environmental Science and Pollution Research*, 27, 27563–27581.

Raklami, A., Meddich, A., Oufdou, K., & Baslam, M., 2022. Plants—Microorganisms-based bioremediation for heavy metal cleanup: Recent developments, phytoremediation techniques, regulation mechanisms, and molecular responses. *International Journal of Molecular Sciences*, 23, 5031.

Ramamurthy, P.C., Singh, S., Kapoor, D., Parihar, P., Samuel, J., Prasad, R., Kumar, A., & Singh, J., 2021. Microbial biotechnological approaches: Renewable bioprocessing for the future energy systems. *Microbial Cell Factories*, 20, 1–11.

Rasheed, A., Barqawi, A.A., Mahmood, A., Nawaz, M., Shah, A.N., Bay, D.H., Alahdal, M.A., Hassan, M.U., & Qari, S.H., 2022a. CRISPR/Cas9 is a powerful tool for precise genome editing of legume crops: A review. *Molecular Biology Reports*, 49, 5595–5609.

Rasheed, A., Mahmood, A., Maqbool, R., Albaqami, M., Sher, A., Sattar, A., Bakhsh, G., Nawaz, M., Hassan, M.U., & Al-Yahyai, R., 2022b. Key insights to develop drought-resilient soybean: A review. *Journal of King Saud University-Science*, 34, 102089.

Rastogi, R.P., Sinha, R.P., Moh, S.H., Lee, T.K., Kottuparambil, S., Kim, Y.-J., Rhee, J.-S., Choi, E.-M., Brown, M.T., & Häder, D.-P., 2014. Ultraviolet radiation and cyanobacteria. *Journal of Photochemistry and Photobiology*, 141, 154–169.

Raza, A., 2021. Eco-physiological and biochemical responses of rapeseed (Brassica napus l.) To abiotic stresses: Consequences and mitigation strategies. *Journal of Plant Growth Regulation*, 40, 1368–1388.

Rehman, A., Saba, T., Kashif, M., Fati, S.M., Bahaj, S.A., & Chaudhry, H., 2022. A revisit of internet of things technologies for monitoring and control strategies in smart agriculture. *Agronomy*, 12, 127.

Roy, P., Pal, S.C., Chakrabortty, R., Saha, A., & Chowdhuri, I., 2023. A systematic review on climate change and geo-environmental factors induced land degradation: Processes, policy-practice gap and its management strategies. *Geological Journal*, 58, 3487–3514.

Sachdev, S., Ansari, S.A., Ansari, M.I., Fujita, M., & Hasanuzzaman, M., 2021. Abiotic stress and reactive oxygen species: Generation, signaling, and defense mechanisms. *Antioxidants*, 10, 277.

Saeed, H.I., Alyas, J., Saeed, A., Majeed, T., & Anjum, I., 2021. Role of microorganisms in bioremediation. *PJBB*, 2, 140–176.

Salam, N.A., Jiao, J.-Y., Zhang, X.-T., & Li, W.-J., 2020. Update on the classification of higher ranks in the phylum Actinobacteria. *International Journal of Systematic and Evolutionary Microbiology*, 70, 1331–1355.

Sales Da Silva, I.G., Gomes De Almeida, F.C., Padilha Da Rocha E Silva, N.M., Casazza, A.A., Converti, A., & Asfora Sarubbo, L., 2020. Soil bioremediation: Overview of technologies and trends. *Energies*, 13, 4664.

Sami, A., Xue, Z., Tazein, S., Arshad, A., He Zhu, Z., Ping Chen, Y., Hong, Y., Tian Zhu, X., & Jin Zhou, K., 2021. CRISPR–Cas9-based genetic engineering for crop improvement under drought stress. *Bioengineered*, 12, 5814–5829.

Samuel, M.S., Ravikumar, M., John J, A., Selvarajan, E., Patel, H., Chander, P.S., Soundarya, J., Vuppala, S., Balaji, R., & Chandrasekar, N., 2022. A review on green synthesis of nanoparticles and their diverse biomedical and environmental applications. *Catalysts*, 12, 459.

Sánchez-Romero, M.A., & Casadesús, J., 2020. The bacterial epigenome. *Nature Reviews Microbiology*, 18, 7–20.

Sapkota, T.B., Jat, M.L., Rana, D.S., Khatri-Chhetri, A., Jat, H.S., Bijarniya, D., Sutaliya, J.M., Kumar, M., Singh, L.K., & Jat, R.K., 2021. Crop nutrient management using nutrient expert improves yield, increases farmers' income and reduces greenhouse gas emissions. *Scientific Reports*, 11, 1564.

Sardans, J., & Peñuelas, J., 2021. Potassium control of plant functions: Ecological and agricultural implications. *Plants*, 10, 419.

Schilling, R.K., Tester, M., Marschner, P., Plett, D.C., & Roy, S.J., 2017. AVP1: One protein, many roles. *Trends in Plant Science*, 22, 154–162.

Seddon, N., Smith, A., Smith, P., Key, I., Chausson, A., Girardin, C., House, J., Srivastava, S., & Turner, B., 2021a. Getting the message right on nature-based solutions to climate change. *Global Change Biology*, 27, 1518–1546.

Seddon, N., Smith, A., Smith, P., Key, I., Chausson, A., Girardin, C., House, J., Srivastava, S., & Turner, B., 2021b. Getting the message right on nature-based solutions to climate change. *Global Change Biology*, 27, 1518–1546.

Seleiman, M.F., Al-Suhaibani, N., Ali, N., Akmal, M., Alotaibi, M., Refay, Y., Dindaroglu, T., Abdul-Wajid, H.H., & Battaglia, M.L., 2021. Drought stress impacts on plants and different approaches to alleviate its adverse effects. *Plants*, 10, 259.

Shahzad, A., Ullah, S., Dar, A.A., Sardar, M.F., Mehmood, T., Tufail, M.A., Shakoor, A., & Haris, M., 2021. Nexus on climate change: Agriculture and possible solution to cope future climate change stresses. *Environmental Science and Pollution Research*, 28, 14211–14232.

Shaikh, S., Yaqoob, M., & Aggarwal, P., 2021. An overview of biodegradable packaging in food industry. *Current Opinion in Food Science*, 4, 503–520.

Sharma, J., Kumar, S., Kumar, V., Singh, P., Khyalia, P., Verma, S., Saini, S., & Sharma, A., 2023. Foliar application of glycine betaine to ameliorate lead toxicity in barley plants by modulating antioxidant enzyme activity and biochemical parameters. *Environmental Research Communications*, 5, 075002.

Shehzadi, N., Ishfaq, F., Mahmood, A., Javaid, M.M., Naqve, M., Jabbar, A., & Aziz, A., 2023. Crop Improvement in the Desert. *In Climate-Resilient Agriculture, Vol 1: Crop Responses and Agroecological Perspectives*. Springer, 465–485.

Shi, Z., Yang, S., Han, D., Zhou, Z., Li, X., Liu, Y., & Zhang, B., 2018. Silicon alleviates cadmium toxicity in wheat seedlings (*Triticum aestivum* L.) by reducing cadmium ion uptake and enhancing antioxidative capacity. *Environmental Science and Pollution Research*, 25, 7638–7646.

Silva, S., Dias, M.C., & Silva, A.M., 2022. Titanium and zinc based nanomaterials in agriculture: A promising approach to deal with (a) biotic stresses? *Toxics*, 10, 172.

Smith, P., House, J.I., Bustamante, M., Sobocká, J., Harper, R., Pan, G., West, P.C., Clark, J.M., Adhya, T., & Rumpel, C., 2016. Global change pressures on soils from land use and management. *Global Change Biology*, 22, 1008–1028.

Sovacool, B.K., Baum, C.M., & Low, S., 2022. Climate protection or privilege? A whole systems justice milieu of twenty negative emissions and solar geoengineering technologies. *Political Geography*, 97, 102702.

Stevenson, P.C., Bidartondo, M.I., Blackhall-Miles, R., Cavagnaro, T.R., Cooper, A., Geslin, B., Koch, H., Lee, M.A., Moat, J., & O'hanlon, R., 2020. The state of the world's urban ecosystems: What can we learn from trees, fungi, and bees? *Plants, People, Planet*, 2, 482–498.

Su, H., Cheng, Y., & PöSchl, U., 2020. New multiphase chemical processes influencing atmospheric aerosols, air quality, and climate in the anthropocene. *Accounts of Chemical Research*, 53, 2034–2043.

Tanveer, A., Mumtaz, K., Javaid, M., Chaudhry, M., Balal, R., & Khaliq, A., 2013. Effect of ecological factors on germination of horse purslane (*Trianthema portulacastrum*). *Planta Daninha*, 31, 587–597.

Tanveer, M., Mahmood, A., Sarfraz, B., Zia, M.A., Javaid, M.M., Bibi, S., Naqve, M., Nadeem, M.A., Azeem, M., & Jabbar, A., 2023. Mechanism and Approaches to Enhancing Heat Stress Tolerance in Crop Plants. *In Climate-Resilient Agriculture, Vol 2: Agro-Biotechnological Advancement for Crop Production*. Springer, 499–520.

Temesgen, H., Affleck, D., Poudel, K., Gray, A., & Sessions, J., 2015. A review of the challenges and opportunities in estimating above ground forest biomass using tree-level models. *Scandinavian Journal of Forest Research*, 30, 326–335.

Teshome, D.T., Zharare, G.E., & Naidoo, S., 2020. The threat of the combined effect of biotic and abiotic stress factors in forestry under a changing climate. *Frontiers in Plant Science*, 11, 601009.

Toscano, S., Trivellini, A., Cocetta, G., Bulgari, R., Francini, A., Romano, D., & Ferrante, A., 2019. Effect of preharvest abiotic stresses on the accumulation of bioactive compounds in horticultural produce. *Frontiers in Plant Science*, 10, 468818.

Tripathi, S., Mahra, S., Tiwari, K., Rana, S., Tripathi, D.K., Sharma, S., & Sahi, S., 2023. Recent advances and perspectives of nanomaterials in agricultural management and associated environmental risk: A review. *Nanomater*, 13, 1604.

Trivedi, P., Mattupalli, C., Eversole, K., & Leach, J.E., 2021. Enabling sustainable agriculture through understanding and enhancement of microbiomes. *New Phytologist*, 230, 2129–2147.

Udume, O.A., Abu, G.O., Stanley, H.O., Vincent-Akpu, I.F., Momoh, Y., & Eze, M.O., 2023. Biostimulation of petroleum-contaminated soil using organic and inorganic amendments. *Plants*, 12, 431.

Ullah, A., Bano, A., & Khan, N., 2021. Climate change and salinity effects on crops and chemical communication between plants and plant growth-promoting microorganisms under stress. *Frontiers in Sustainable Food Systems*, 5, 618092.

Ullah, H., Mahmood, A., & Honermeier, B., 2014. Essential oil and composition of anise (*Pimpinella anisum* L.) with varying seed rates and row spacing. *Pakistan Journal of Botany*, 46, 1859–1864.

Ullah, H., Mahmood, A., Ijaz, M., Tadesse, B., & Honermeier, B., 2013. Evaluation of anise (*Pimpinella anisum* L.) accessions with regard to morphological characteristics, fruit yield, oil contents and composition. *Journal of Medicinal Plants Research*, 7, 2177–2186.

Umair Hassan, M., Aamer, M., Umer Chattha, M., Haiying, T., Shahzad, B., Barbanti, L., Nawaz, M., Rasheed, A., Afzal, A., & Liu, Y., 2020. The critical role of zinc in plants facing the drought stress. *Agriculture*, 10, 396.

Wahab, A., Abdi, G., Saleem, M.H., Ali, B., Ullah, S., Shah, W., Mumtaz, S., Yasin, G., Muresan, C.C., & Marc, R.A., 2022. Plants' physio-biochemical and phyto-hormonal responses to alleviate the adverse effects of drought stress: A comprehensive review. *Plants*, 11, 1620.

Wahab, A., Muhammad, M., Munir, A., Abdi, G., Zaman, W., Ayaz, A., Khizar, C., & Reddy, S.P.P., 2023. Role of arbuscular mycorrhizal fungi in regulating growth, enhancing productivity, and potentially influencing ecosystems under abiotic and biotic stresses. *Plants*, 12, 3102.

Waheed, H., Javaid, M.M., Shahid, A., Ali, H.H., Nargis, J., & Mehmood, A., 2019. Impact of foliar-applied Hoagland's nutrient solution on growth and yield of mash bean (*Vigna mungo* L.) under different growth stages. *Journal of Plant Nutrition*, 42, 1133–1141.

Wang, L., Cheng, Y., Naidu, R., & Bowman, M., 2021. The key factors for the fate and transport of petroleum hydrocarbons in soil with related in/ex situ measurement methods: An overview. *Frontiers in Environmental Science*, 9, 756404.

Wang, X., & Komatsu, S., 2020. Proteomic techniques for the development of flood-tolerant soybean. *International Journal of Molecular Sciences*, 21, 7497.

Wani, K.I., Naeem, M., Castroverde, C.D.M., Kalaji, H.M., Albaqami, M., & Aftab, T., 2021. Molecular mechanisms of nitric oxide (NO) signaling and reactive oxygen species (ROS) homeostasis during abiotic stresses in plants. *International Journal of Molecular Sciences*, 22, 9656.

Waraich, E., Ahmad, R., Halim, A., & Aziz, T., 2012. Alleviation of temperature stress by nutrient management in crop plants: A review. *Journal of Soil Science and Plant Nutrition*, 12, 221–244.

Wasaya, A., Abbas, T., Yasir, T.A., Sarwar, N., Aziz, A., Javaid, M.M., & Akram, S., 2021. Mitigating drought stress in sunflower (*Helianthus annuus* L.) through exogenous application of β-aminobutyric acid. *Journal of Soil Science and Plant Nutrition*, 21, 936–948.

Watson, J.E., Evans, T., Venter, O., Williams, B., Tulloch, A., Stewart, C., Thompson, I., Ray, J.C., Murray, K., & Salazar, A., 2018. The exceptional value of intact forest ecosystems. *Nature Ecology and Evolution*, 2, 599–610.

Werner, T., Bebbington, A., & Gregory, G., 2019. Assessing impacts of mining: Recent contributions from GIS and remote sensing. *Extractive Industries and Society*, 6, 993–1012.

Yanai, R.D., Wayson, C., Lee, D., Espejo, A., Campbell, J.L., Green, M.B., Zukswert, J.M., Yoffe, S., Aukema, J., & Lister, A., 2020. Improving uncertainty in forest carbon accounting for REDD+ mitigation efforts. *Environmental Research Letters*, 15, 124002.

Ye, S., Zeng, G., Wu, H., Zhang, C., Dai, J., Liang, J., Yu, J., Ren, X., Yi, H., & Cheng, M., 2017. Biological technologies for the remediation of co-contaminated soil. *Critical Reviews in Biotechnology*, 37, 1062–1076.

Zafar, S., Afzal, H., Ijaz, A., Mahmood, A., Ayub, A., Nayab, A., Hussain, S., Maqsood, U.-H., Sabir, M.A., & Zulfiqar, U., 2023. Cotton and drought stress: An updated overview for improving stress tolerance. *South African Journal of Botany*, 161, 258–268.

Zgonik, V., Mulec, J., Eleršek, T., Ogrinc, N., Jamnik, P., & Ulrih, N.P., 2021. Extremophilic microorganisms in central Europe. *Microorganisms*, 9, 2326.

Zhang, D., Sial, M.S., Ahmad, N., Filipe, A.J., Thu, P.A., Zia-Ud-Din, M., & Caleiro, A.B., 2020a. Water scarcity and sustainability in an emerging economy: A management perspective for future. *Sustainability*, 13, 144.

Zhang, J., Zhang, Y., Xing, J., Yu, H., Zhang, R., Chen, Y., Zhang, D., Yin, P., Tian, X., & Wang, Q., 2020b. Introducing selective agrochemical manipulation of gibberellin metabolism into a cereal crop. *Nature Plants*, 6, 67–72.

Zhang, Z., Moore, J.C., Huisingh, D., & Zhao, Y., 2015. Review of geoengineering approaches to mitigating climate change. *Journal of Cleaner Production*, 103, 898–907.

16 Recent Advances in Seed Enhancement

Asma Zafar, Umair Gull, Ayesha Arshad, Iqra Anwar, and Muhammad Mansoor Javaid*

16.1 INTRODUCTION

High-quality seed has a substantial capacity to augment on-farm production and improve food security. The quality of seeds is crucial for achieving profitability in both production and marketing (Moulick *et al.*, 2024). High-quality seeds possess genetic and physical integrity, displaying robustness and freedom from insect pests and infections. Seeds of superior quality, which have been strengthened, account for around 30% of the overall yield (Mahmood *et al.*, 2021; Basavaraja *et al.*, 2023). Plant uniformity is a measure of seed quality that is attained by the presence of very vigorous seeds. Seed quality is determined by several aspects during the process of seed development, including processing, gathering, desiccation, purification, categorization, packing, and preservation (Sundareswaran *et al.*, 2023). Agricultural practitioners are always seeking superior seeds to guarantee consistent field establishment and enhanced productivity (Ullah *et al.*, 2013; Kedisso *et al.*, 2023).

The worldwide output and, therefore, food security are influenced by the availability, quality, and cost of seeds (Maity *et al.*, 2023). The informal seed systems, which are established and maintained by farmers without legal paperwork, account for 75–90% of their food crop farming. Informal seed networks continue to be the primary source of seed for smallholder farmers in the developing globe (Westengen *et al.*, 2023). The inadequate storage conditions, vulnerability of sprouting seeds, and early seedlings to dehydration stress result in the loss of ability to tolerate drying out when the seeds are hydrated. Additionally, anticipated climate change, characterized by irregular rainfall patterns and unexpected temperature extremes, could affect the quality of seeds (Rasheed *et al.*, 2022b; Sine *et al.*, 2024). Different seed technologies could improve the genetic quality of low-vigor seeds, helping them to thrive under subsistence farming conditions and thus increase the supply of high-quality seeds to the local seed industry (Bello and Mbhele, 2024).

Efficient and consistent germination of seedlings is critical for their successful establishment, developing a deep root system before the top layers of soil become dry, compact, or overheated (Carnevale Zampaolo *et al.*, 2023). In commercial agriculture, optimizing seed germination and the early establishment of seedlings is relevant because they are considered the most vulnerable stages of the crop plant life cycle (Naqve *et al.*, 2021; Uebersax *et al.*, 2023). Such methods include postharvest processing necessary for preparing seeds for planting and treatments commonly known as 'seed enhancements' or 'seed treatments' in agriculture and within the research community (Bhabani *et al.*, 2024). Germination begins with the uptake of water which culminates in the emergence of the radicle. In reaction to changing environmental conditions and limitations in seedbeds (Jarrar *et al.*, 2023). Several approaches are currently being employed to improve seed placement and to optimize seedling establishment and growth. Multiple reports are accessible such as Pagano *et al.* (2023b) stated that seed treatments were developed to interrupt dormancy, thus improving germinability, processes that add to stress resistance. Methods to improve crop germination in the field, considering the seed's temperature and soil moisture response have been proposed by

* Corresponding author: iqraanwar84@gmail.com

DOI: 10.1201/9781003498698-16

several scientists. Ijaz *et al.* (2023) focused on the practical aspects of seed treatment methods which include conditioning, protection, and physiological benefits. The strategies may be grouped into physiological domains such as seed priming, coating, and pelleting and biological factors such as seed enhancement (Shelar *et al.*, 2023).

This chapter aims to explore novel trends in seed enhancement technology, particularly precision coating, genetic priming, or adding a minimal amount of other materials such as nanopolymers. The goal is to understand these treatments and their impact on improving seed performance, such as increased germination rates, better stress resistance, and improved nutrient uptake. Moreover, the chapter underscores the implications of these advances for broader global agricultural challenges, including food security and increased resilience to climate change. In general, it presents a nuanced and multifaceted perspective on how seed improvement innovations can help enable a more resilient, efficient, and sustainable future for world agriculture.

16.2 TRADITIONAL SEED ENHANCEMENT TECHNIQUES

16.2.1 Seed Priming

Seed priming is a simple, cost-effective technology that improves seed emergence uniformity, seedling vigor, and increases crop yields. The regulation of water in this way acts to initiate metabolic processes in the first phase of germination associated with the emergence of the radicle (Zaghdoud *et al.*, 2023). The seeds would be primed for a certain time by immersing them in liquids containing either plant hormones or beneficial microbes (Pagano *et al.*, 2023a). Because the prepared seeds were moistened, it is important to bring them back to a suitable moisture content for packing, transport, and sowing (Kavale *et al.*, 2023). Seed priming is one of the important techniques to enhance the physiological traits of seeds. In this technique, the controlled uptake and the initiation of pregerminated metabolism in seeds while inhibiting the growth of the radicle (Paul *et al.*, 2024).

Seed priming refers to this pre-hydration of seeds, as it has the potential for improving germination and seedling establishment through dormancy alleviation, desiccation tolerance, and development (Rasheed *et al.*, 2022a; Chandel *et al.*, 2024). The improved germination of primed seeds is mainly due to a decrease in the time taken to imbibe water, enzyme activation, accumulation of germination-provoking factors, restoration of metabolism during water uptake, and osmotic equilibrium adjustments (Javaid *et al.*, 2022; Narejo *et al.*, 2023). Priming activates the cellular defense systems of plants and thus increases their ability to withstand future exposure to environmental stimuli. Seed priming is a well-known technique that can improve the germination process of weaker, damaged, or fully mature seeds (Anwar *et al.*, 2023).

Seeds show favorable effects of priming owing to the metabolic alterations in seeds, when seeds absorb water through the imbibition process (Tang *et al.*, 2022; Khan *et al.*, 2023). Hence enhanced agricultural production and crop yield. The activities initiated during the rehydration of primed seeds at the time of sowing drive multiple biological processes (Boamah *et al.*, 2023) and include mobilization of nucleic acids and new proteins, generation of ATP molecules, accumulation of phospholipids and sterols, activation of antioxidant system, and repair of DNA. Under extreme conditions, primed seeds depend on repair processes, e.g., DNA repair, to conserve seed viability. Moreover, DNA damage repair, especially the optimal repair is a prerequisite for cell cycle progression and DNA duplication in embryonic cells (Lv *et al.*, 2024). There are a number of environmental challenges that seeds must face during development, after harvest, and during germination in the field. Various environmental stressors can lead to the damage of a variety of biological macromolecules like lipids, proteins, and nucleic acids (Saud *et al.*, 2023).

During the first phases of seed germination DNA repair mechanisms such as Nucleotide Excision Repair (NER), Base Excision Repair (BER), and even Base and NER are triggered to preserve genome integrity (Szurman-Zubrzycka *et al.*, 2023). In primed seeds, the activation of pregerminated metabolism, along with water uptake and uncontrolled synthesis of reactive oxygen species (ROS), is primarily triggered by DNA repair pathways (Li *et al.*, 2022). Seed DNA repair mechanisms respond differently to priming chemicals, and improvement of the priming process for various seeds necessitates an evaluation of these chemicals. Due to the priming process, the seeds' storage potential diminishes, and the primed seeds require low temperatures to preserve their quality (Paul *et al.*, 2022). Priming was reported to enhance the activity of antioxidant enzymes such as ascorbate peroxidase (APX, EC 1.11.1.11), superoxide dismutase (SOD, EC 1.15.1.1), catalase (CAT, EC. 1.11.1.6), and glutathione reductase (GR, EC 1.8.1.7) (Yadav *et al.*, 2021). The timing of priming treatment needs to be controlled to improve the seed quality and to avoid the loss of viability due to overdosing the time of soaking. The prolonged time of seed exposure to priming treatment causes increased oxidative damage to DNA and hence will have a long-lasting effect on seed viability (Adetunji *et al.*, 2021). Hydropriming, osmopriming, hormonal-priming, halopriming, matrix priming, and pregerminated seeds are approaches to seed priming.

16.2.1.1 Hydropriming

A soaking process of seeds in water prior to application for priming. This method means; introducing smaller quantities of water, but repeatedly, to the seeds. A typical apparatus for this purpose is a drum, and the presence of wet air may be taken advantage of to promote additional evaporative flow of water (Hanif *et al.*, 2017; Gabhane *et al.*, 2020). On-farm steeping process, referred to as hydropriming, is an extremely effective and cost-efficient hydropriming technology. In this method, cereal and legume seeds are soaked in a warm water bath for a limited period of time. This seed priming technique helps to enhance the ability of the crop to withstand the abiotic stresses that it faces in the cultivated environment (Oyebamiji *et al.*, 2024).

16.2.1.2 Osmopriming

Osmopriming also called osmoconditioning involves soaking seeds in solutions with adequate aeration and low water potential followed by washing and drying. A promising technique (halopriming procedure) is the submersion of seeds in osmotic solutions, like potassium nitrate (KNO_3), potassium chloride (KCl), and sodium chloride (NaCl). Thus, it resulted in higher uniformity and stronger resilience of the crops leading to improved overall productivity (Sang *et al.*, 2023).

16.2.1.3 Hormonal-Priming

The use of gibberellic acid (GA_3) and PEG priming produced increased photosynthetic potential, enhanced antioxidant properties, advanced seedling emergence, and promoted the growth of white clover affected in soils contaminated with heavy metals (Aziz *et al.*, 2018; Mitra *et al.*, 2023). Utilizing GA_3 on spring wheat seedlings led to increased grain yield and enhanced resistance to salt. The previous result was attained by the regulation of hormone equilibrium and the alteration of ion absorption and accumulation between the shoots and roots (Parul *et al.*, 2023).

16.2.1.4 Matrix Priming

The process of matrix conditioning involves the incubation of seeds in matrices that are moist, solid, and insoluble, such as vermiculite, cross-linked hydrogels, and diatomaceous earth. However, this approach includes a gradual process of absorption. Pregerminated seeds may be used as a means of enhancing seeds, however, this is only feasible for a limited number of crop species (Noronha *et al.*, 2023). Unlike typical priming, seeds are permitted to exhibit radicle protrusion. This approach entails the immersion of seeds until the physical process of germination begins, after which the wet seeds are planted (Chadha *et al.*, 2019; Pompelli *et al.*, 2023). In contrast, priming comprises briefly soaking the seeds, followed by drying and storing them. Pregerminated seeds enable rapid and consistent germination and development of seedlings (Paul *et al.*, 2023). The seeds that have been primed are then dried to get the optimal moisture content required for safe storage. Primed seeds often exhibit superior performance under challenging settings compared to non-primed seeds (Ali *et al.*, 2013a; Mamedi *et al.*, 2024).

Priming improves the overall vitality of seeds and helps to overcome many constraints that seeds may have, including photo-dormancy, postharvest dormancy, and thermo-dormancy. On-farm seed priming is a process that allows farmers to prime their own seeds by soaking them before planting (Mehmood *et al.*, 2018; Ashraf *et al.*, 2023). This requires knowing the maximum period or safe limits for soaking the seeds. Oversoaking might injure the seeds and the budding plant. Overwatering may also damage the seeds and the emerging plant. To stop germination when soaking seeds in water, it should be done according to the provided safe limits for every variety accordingly During this process, farmers often soak the seeds in water for 8–10 hours and then dry on the surface and will be planted immediately. However, for optimal results, many seeds like rice and maize need to be soaked for a longer duration, about 16–18 hours. On-farm seed priming is an age-old agricultural practice familiar to farmers for centuries. This seed priming technique is highly effective for many grains and legumes producing better yields (Ali *et al.*, 2023).

16.2.2 Seed Coating

The need to address the challenge of increasing agricultural production at low environmental costs cannot be emphasized enough when we consider the provision of sufficient and nutritious food to the global growing human population (Aziz *et al.*, 2018; Paudel *et al.*, 2023). Sowing seeds and their subsequent germination and emergence are critical to the growth and productivity of all plants that participate in sexual reproduction. At its most basic level, a seed is a tiny, living plant whose biological function is to protect and nurture an embryo. In efforts to improve the seed itself, it is important to heed the soil. Seed's biological, chemical, and physiological characteristics greatly affect plant performance and resistance to adverse environmental conditions, which influences their final yields. According to a recent study, seed disease is considered a contributing factor in the reduction of germination percentage and seedling establishment (Zulfiqar *et al.*, 2023). Pre-seed treatment might be used to treat these illnesses. One way to do this is by applying pesticides right onto the seed. Seed coating is a process in which different chemicals such as fertilizers, nutrients, moisture adsorbing or repellant agents, plant growth regulators, rhizobium inoculum, chemicals, and pesticides adhere together on seeds. This process aims at maximizing seed capacities and breaking seed dormancy (Kavusi *et al.*, 2023).

The process, first employed in 1930 for treating cereal seeds, consists of coating insecticide as a thin and permeable layer into the surface of the seeds. The benefits become even greater in the seed industry, where seed coating allows seeds to realize their natural vigor. And indeed, the method is employed for propagating seeds of various garden plants, as well as common n-haus crops including maize, sunflower, canola, alfalfa, and perennial grass types (Javaid *et al.*, 2012; Campos *et al.*, 2023). This application acts as a shield against damage from seeded infections. As the seed takes up moisture and warms up to a certain temperature, the outermost layer melts or cracks, allowing the radical to escape the seed. This strategy aims to increase the anchor effectiveness of materials and seeds together so as to minimize pesticide waste from evaporation and leakage. Moreover, it has to cope with the adverse effects of some pesticides on seeds (Ali *et al.*, 2013b; Karimi-Maleh *et al.*, 2024). These features further facilitate the accuracy and efficacy of pesticides enabling reduced usage, environmental pollution, and costs (Tanveer *et al.*, 2013; Vasileiou *et al.*, 2024).

16.3 EMERGING TECHNOLOGIES IN SEED ENHANCEMENT

16.3.1 Nanotechnology

Nanotechnology, which is the manipulation of material within a nanoscale range, holds enormous promise for modern agriculture. Nanopriming means you should apply a nanoscale coating to your surface to optimize its performance or character. Diverse artificial nanoparticles (NPs), have witnessed a surge in popularity over the last decade. This is a novel method that can change current agriculture and increase crop yield (Periakaruppan *et al.*, 2023). Nanotechnology in agriculture is shown with the structures of nano-pesticides, nano-fertilizers, nano-formulations, and nanosensors, and plant growth regulators are demonstrated by recent studies. To enhance the effectiveness of various agricultural inputs (herbicides, plant growth regulators and fertilizers, etc.) and to promote environmentally sustainable agriculture, NPs are added to different agricultural inputs (Yadav *et al.*, 2023).

NPs such as gold (AuNPs), silver AgNPs, multi-walled carbon nanotubes (MWCNTs), and others are a new technology for seed preparation that have a positive impact on germination, vigor, and growth of seedlings (Bhardwaj *et al.*, 2023). Impact of NPs on the performance of seeds and seedlings is presented in Table 16.1.

Silver NPs (AgNPs) improve rice seed amylase enzyme activity and thus enhance seed germination and growth. Seed priming has a significant impact on the imbibition of early germination seedlings. Nanopriming demonstrated improved seed performance due to higher β-amylase enzyme activity, soluble sugar content, aquaporin gene expression, stress tolerance, water uptake, antioxidant activation, and hydroxyl radical generation (Imtiaz *et al.*, 2023). Primed seeds uptake water faster than non-primed seeds (Mondal *et al.*, 2024). Nanopriming alters water and starch metabolism of rice seedlings. Plant nanopriming results in higher concentrations of soluble sugars in seedlings as a result of starch hydrolysis by β-amylase. Applying silver NPs (AgNPs) to rice seedlings enhances their dehydrogenase activity, which is a measure of seed viability. In rice, nanopriming with AgNPs increases aquaporin gene expression. Priming upregulated aquaporin genes PIP 1;1 and PIP 1;2, increasing water absorption. Nanopriming regulates SOD and CAT activity (Mondal *et al.*, 2024).

Chemically manufactured NPs are costly, environmentally damaging, and uneconomical for agriculture. In-vitro bioreduction of metal ions by plant extracts produces green NPs. This affordable

TABLE 16.1
Summary of the Impact of Nanoparticles (NPs) on the Performance of Seeds and Seedlings

Nanoparticles (NPs) Applied	Species	Effects Described	Reference
300, 600, 900, 1200 mg/L	*Triticum aestivum* L.	Seed priming with silicon nanoparticles (SiNPs)-enhanced wheat growth, productivity, and chlorophyll levels and alleviated oxidative stress through increased activity of antioxidant enzyme.	Raza *et al.* (2023)
20, 40, 80, and 160 mg/L	*Citrullus lanatus* (Thunb.) Matsum and Nakai	The use of Iron oxide nanoparticles (Fe-NPs) during seed priming resulted in alterations to the antioxidant capacity and jasmonate-linked defenses of watermelon seedlings.	Chouhan and Mandal (2021)
20 mg/L	*Oryza sativa* L.	Zero-valent iron (nZVI) nanoparticles boosted fragrant rice cultivar Gobindabhog seedling development.	Guha *et al.* (2018)
0.02%	*Vigna radiata* L.	Using (TiO_2) nanoparticles to prime green gram seeds boosted germination and growth.	Rhaman *et al.* (2022)
30 mg/mL	*Petroselinum crispum*	Nano-anatase (TiO_2) increased germination parameters, seedling growth, vigor, and chlorophyll	
25 ppm	*Cajanus cajan* L.	The application of Fe nanoparticles (NPs) to seeds produced enhanced germination metrics, including germination percentage, germination rate, seedling growth, and vigor index.	Gupta *et al.* (2022)
100 mg/L	*Alnus subcordata*	The application of multi-walled carbon nanotubes (MWCNTs) to seeds resulted in improved seed germination and increased tolerance to drought stress.	Yousefi *et al.* (2017)
30 µg/mL	*Boswellia ovalifoliolata*	Silver nanoparticles (SNPs) increased the process of germination and promoted the development of seedlings.	Guleria *et al.* (2023)

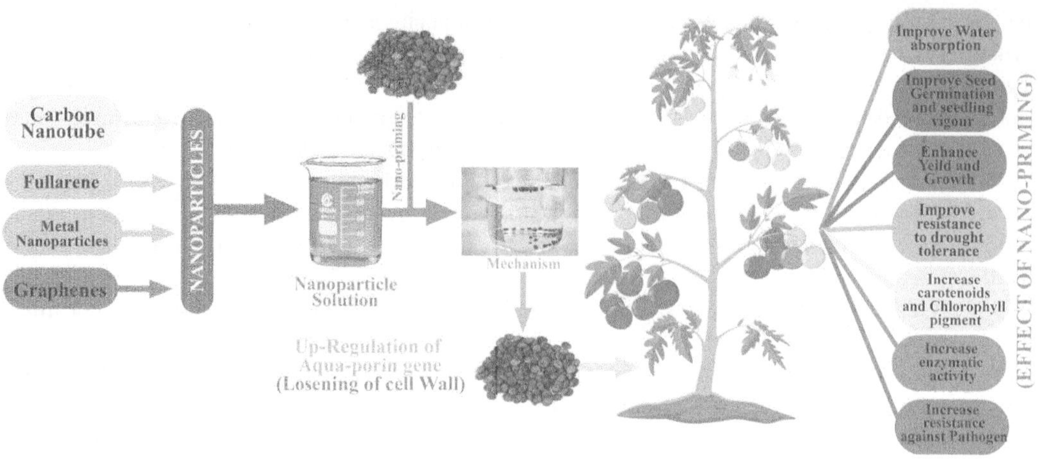

FIGURE 16.1 Effect of nanoparticles on seeds and seedlings.

NP manufacturing method uses plant extracts and metal acids and salts including gold, copper, iron, silver, and platinum. Green gold and silver NPs are good in priming seeds. Nanopriming works in rice, maize, onion, green gram, and other crops (Bhandari *et al.*, 2023). Seed and seedling response to NPs is presented in Figure 16.1.

16.3.2 Biotechnology

In natural conditions, obtaining high-quality seeds by spontaneous variation is quite rare. Conventional breeding has achieved significant advancements in enhancing crops. Although this approach exhibits a slow and the existing genetic resources for enhancing seed vigor are limited. In recent decades, the advancement of global omics methods, such as transcriptomics, genomics, and proteomics, has led to the identification of multiple prospective candidates (genes/proteins) associated with seed vigor (Roychowdhury *et al.*, 2023). This has been accomplished with exceptional efficiency. The possibilities have the potential to be used in breeding programs and/or biotechnological approaches aimed at improving seed vigor and increasing agricultural yields. Late embryogenesis abundant (LEA) proteins and heat shock proteins (HSPs), which are protective molecules, are often linked to the ability to withstand dehydration and live for a long time. These molecules are stored throughout the maturity phase of seed development. These stress-induced proteins may potentially contribute to seed vitality (Lah *et al.*, 2023).

Arabidopsis seedlings that have been genetically modified to produce an excess of (heat shock factor (HSF)) show increased levels of HSPs and higher resistance to aging. A knockout mutation in ATEM6, a member of the Arabidopsis group 1 LEA family, led to an early manifestation of traits, indicating that the ATEM6 protein plays a role in regulating water retention or loss during seed maturity (Wang *et al.*, 2023). Dehydrins are LEA proteins that build up during the development of seeds and in reaction to non-living environmental pressures in non-reproductive tissues. RNAi was used to decrease the expression of seed-specific dehydrin genes (LEA14, XERO1, and RAB18) in Arabidopsis. As a consequence, there was a 50% decrease in the duration and capacity of the seeds to grow. Phospholipase D plays a role in the first phases of seed decay by breaking down phospholipids and producing phosphatidic acid (PA). PA affects membranes and the storage lipids of the seeds. The deletion of the Arabidopsis PLDα1 gene, which encodes a member of the lipid-hydrolyzing phospholipase D family, resulted in seeds with decreased levels of lipid peroxides and increased resistance to aging (Bansal *et al.*, 2023).

Currently, the problem of achieving food security with limited arable land has now become a prominent global issue primarily due to changing climatic patterns and the growing population worldwide. It is reasonable to generate high-vigor seeds for agricultural production by engineering the PIMT in seed tissues. This technology also potentially could have a vast effect on food security worldwide when applied to major cereal crops such as wheat, rice, and maize. While there has been considerable progress, several questions remain (Kibria *et al.*, 2023). It is still unclear if the enhanced seed vigor observed from the over-expression of PIMT and other proteins correlated with crop nutritional quality. There still has to be established a comparison between the health and nutritional features of genetically manipulated seed with improving vigor with its unaltered version (Animasaun *et al.*, 2023).

Although there are many possible candidates (genes/proteins) linked to seed vigor, their effectiveness in enhancing seed vigor has to be confirmed by large-scale reverse genetics experiments before they can be used for agriculturally important crop species. The rapid development of novel genome-editing methods allows for the accurate adjustment of desired features with unparalleled precision and effectiveness (Animasaun *et al.*, 2023). Clustered regularly interspaced short palindromic repeats (CRISPR), one of the current genome-editing techniques, is characterized by its simplicity, speed, and affordability. It has a wide range of applications in modifying plant genomes to create customized crops. However, it is crucial to bear in mind that the proper use of GM food or feed necessitates a thorough evaluation of potential health hazards and environmental impacts (Nawaz *et al.*, 2023).

Currently, there is a lack of information on the use of CRISPR technology to modify seed vigor in plants. Genome-editing methods provide a viable means of controlling the build-up of proteins linked to seed vigor in a seed-specific way. This approach has the potential to significantly shorten the time required to develop accessible crop varieties. Therefore, the development of transgenic seeds and their subsequent use in agriculture is essential for effectively nourishing a rapidly expanding population in a changing environment (Albahri *et al.*, 2023).

The foundation of agricultural productivity is the quality of seeds. Every seed manufacturer relentlessly strives to get high-quality seeds. GM technology is a highly efficient, cost-effective, and environmentally friendly method for enhancing the strength and vitality of seeds, altering their color or form, and increasing the nutritional content and other desirable characteristics of crops. The use of GM technologies will significantly transform the landscape of agriculture (Chaudhry *et al.*, 2023).

16.3.3 Microbial Seed Coating

Seed coating is considered a very effective technique for advancing sustainable agriculture. It is the process of improving the physiological traits of seeds, allowing for planting, enhancing growth indices, and reducing abiotic and biotic stress. Seed coating includes a few techniques, which ensure consistent and strong adhesion (Jarrar *et al.*, 2023). Several plant species with diverse dimensions, shapes, textures, and germination features have been studied for seed coating. Seed inoculations with plant-beneficial microorganisms (PBM, e.g., rhizobia, bacteria, and fungi) can increase seed germination and growth and stress resilience (biotic, e.g., biotic (pathogens and pests) as well as abiotic (salt, drought, and heavy metals, etc.) stressors) of plants, while limiting agrochemical use (Wahab *et al.*, 2023).

Seed coating is a method employed to improve the growth and development of plants by coating the seeds. It protects plants from diseases, disorders, pests, salinity, drought, and temperature. This method provides a safe habitat for future generations of plants. One such viable technology now under viewpoint in sustainable agriculture is the seed coating procedures. Microbial seed coating is an emerging biotechnological approach to enhance agricultural quality and production under environmental stress (Shelar *et al.*, 2023).

TABLE 16.2
Effects of Different Methods of PBM Inoculation on Plant Growth

Plant	Inoculation	PBM	Effect Inoculation	References
Triticum aestivum L.	Soil	*Plymuthica Bacillus subtilis*	Soil inoculation reduced the toxicity of chromium.	Naz *et al.* (2023)
Glycine max	Seed	*Bacillus megaterium*, *Trichoderma longibrachiatum*, and *Trichoderma simmonsii*	The process of seed inoculation resulted in enhanced germination and seedling indices in the presence of cold stress.	Jha and Mohamed (2023)
Zea mays	Root	*Pseudomonas putida*	Leaf necrosis was decreased by root injection.	Giordano *et al.* (2023)
Arabidopsis thaliana	Root	*Pseudomonas fluorescens*	Leaf protection against the oomycete *Peronospora* was achieved by root injection.	Zhu *et al.* (2024)
Oryza sativa	Soil	*Pseudomonas aeruginosa*, *Bacillus amyloliquefaciens*, and *Trichoderma harzianum*	Inoculation of soil may effectively control diseases in aerobic rice.	Liu *et al.* (2023)
Triticum sp.	Seed	*Streptomyces*, *Aspergillus*, *Bacillus*	Seed the infusion enhanced crop productivity and decreased *Rhizoctonia* presence.	Geetha *et al.* (2023)
O. sativa	Seed	*Rhizobia*	The process of seed injection demonstrated efficacy in enhancing plant development.	Balusamy *et al.* (2023)
Ryegrass	Seed, soil, root	*Burkholderia phytofirmans*	The biomass of plants output enhanced by soil inoculation.	Naseer *et al.* (2024)

Seed coating has the potential to create a profitable industry through the emergence of the use of seed coating and high-performing PBM strains in agricultural output. In addition, even the microbial seed coating application in agriculture appears to be effective, but it may not be consistent because it is affected by species of plant, growth conditions, and experimental size. The PBM formulations produced by indigenous strains in particular geographical environments require further investigation. Seed coating also provides a way for the future with formulations needing to be tailored to local conditions and agricultural practices including herbicides, fertilizers, and irrigation management (Mcgregor *et al.*, 2024). But there are need for more exploration and determination through further studies regarding several aspects such as survival and viability of microorganisms, selection of ingredients and optimized formulation, and the manufacturing cost, which are the factors limiting the wide field performance and large-scale adoption of seed coating (Hari *et al.*, 2024).

PBM has been implicated in the reduction of both biotic and abiotic stress caused by climate change. These optimized formulas improve viability with PBM, as well as the general development and productivity of plants. PBM seed coating is one of the highly energy efficient processes for sustainable agriculture. Thus, the application of PBM in seed coating is a promising approach and will offer great potential for application in resource- and environment-friendly agriculture (Table 16.2). However, in order to achieve broad use and integration into agricultural management methods, it requires further growth and investment (Imtiaz *et al.*, 2023).

16.4 SUSTAINABLE SEED ENHANCEMENT PRACTICES

16.4.1 Eco-Friendly Coatings and Treatments

Seed coating is the procedure that involves the application of an outer layer of material surrounding the first seed coat (Figure 16.2). The primary objective of seed coating is to modify the physical

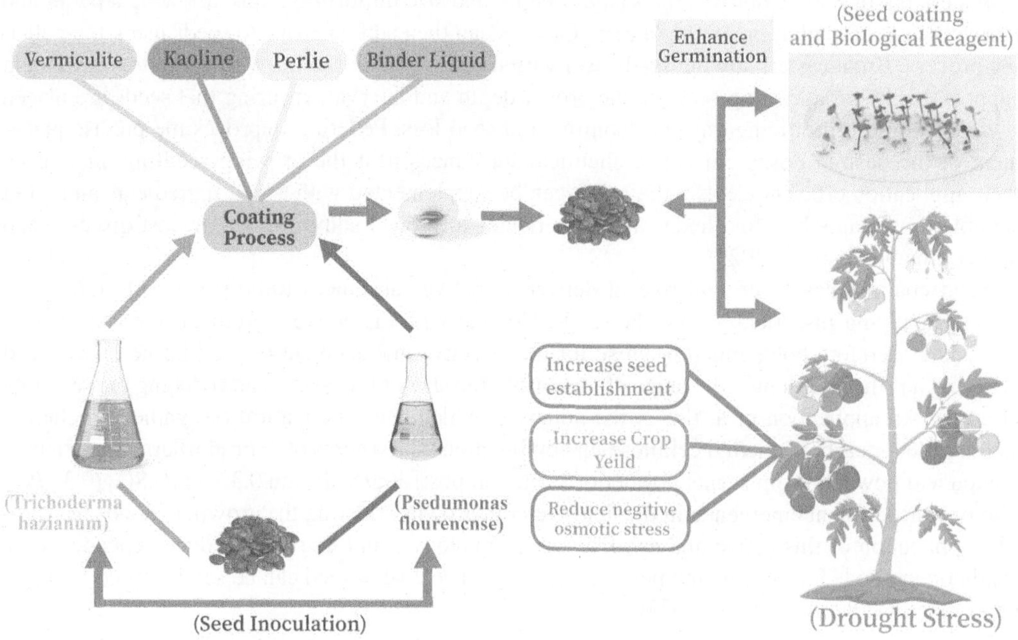

FIGURE 16.2 Ecofriendly seed coating.

attributes of seeds with the aim of improving their manageability via the standardization of their weight and size. Furthermore, the application of the coating facilitates the transportation of diverse bioactive components, including nutrients, protective agents, symbionts, soil adjuvants, phytoactive compounds, growth regulators, pigments, and trace elements. Seed coating may be classified into three distinct categories: film coating, encrusting, and pelleting. These categories are determined based on the physical attributes of the coatings, including weight, size, and sorting qualities (Briassoulis, 2023).

Film coating is the process of applying a thin coating of supplementary substance (typically less than 10% of the seed's weight) around the seed. On the other hand, encrusting involves applying a thick coating around the seed, which significantly increases its weight and volume without changing its original shape. By using film coating, a thin coating of premium polymer is applied to the seed, causing little changes to the seeds' size and weight. Both encrusting and pelleting include the addition of a substantial exterior coating to the seed. However, it should be noted that encrusting and pelleting are separate processes. Encrusting enhances seed size without changing its shape similar to the original variety, while pelleting does change the shape of coated seeds from the original (Pedrini *et al.*, 2023).

Generally, the weights are used to distinguish encrusted and film-coated seeds, whereas pelleted seeds are separated according to their diameter. In seed coating, synthetic polymers are utilized as binders to enhance the adhesion and retention of active compounds onto the seed as well as regulate the absorption of water to induce early germination. The temperature-responsive polymers regulate the amount of water that is absorbed by the seed at specific temperature thresholds ensuring that seed germination will only take place when the environmental conditions are optimal for continued growth (El Idrissi *et al.*, 2024) Seed pelleting, on the other hand, is the practice of adding a coating on top of seed forming a sizable amount of seed coating over a seed which can make an irregularly shaped seed become spherical. The synthetic coating is thick, so it's difficult to tell the original shape of the seed. Mechanical sowing faces challenges due to the seeds of some crops (e.g., lettuce and sugar beet) not maintaining a uniform shape, and the seeds of plants like Begonia, Calceolaria, and many other horticultural crops being small (Jarrar *et al.*, 2023).

Pelletizing of seeds improves the shape, weight, and size uniformity; thus, allowing a better and more efficient crop planting with planters. Osseins can then add nutrients to seeds using the pelleting process. Broadcasters sowing seeds over varied width configurations with variable rows should support this by broadcasting seeds at the proper depth and interval, ensuring that seeds are placed accurately, and minimizing any shortcoming and seed loss. Pelleting improves the precise placement of the desired dosage of active chemical substances near the growing seedling, increasing their application efficiency. Also, the pellet can be supplemented with active ingredients and other suitable protectants like fungicides and insecticides to protect some species against diseases and pests (Alhammad *et al.*, 2023).

Seed coating plays a crucial role in delivering active substances to crop pest infestations by directly applying insecticides onto the seeds. Coating with the active ingredient or encapsulation procedure therefore holds much promise for the effective management of seed-borne disease and pelleting has the additional advantage of increasing the size of tiny seeds and reducing the seed rate of crops. An application of a film of the adhesive, which could be natural or synthetic including starch gruel or carboxy methyl cellulose, in combination with 0 acres of natural fillers (if you added arappu leaf powder, neem kernel powder or vermicompost) and/or thiram 0.3% or $ZnSO_4$ (0.3 g/kg) also enhanced (plant emergence in the field) veteran growth affecting the growth of soybean crops. The application of this active and eco-friendly technology could markedly reduce dependence on synthetic methods for disease and pest management. The coated seed can be safely stored for up to three months (Okonkwo *et al.*, 2023).

16.4.2 Utilization of Waste Materials in Seed Enhancement

Potential ways to improve either crop performance or seed quality could come from using a variety of waste items. Utilizing waste materials for seed enhancement provides an innovative and environmentally friendly approach to advancing agricultural practices. Biochar is a beneficial seed-coating material produced by the pyrolysis of biomass that can improve soil fertility and moisture retention. For example, organic waste in the form of compost and agricultural residues can be converted into biofertilizers or applied as seed coating so as to facilitate germination and provide nutrients (Hamidzadeh *et al.*, 2023).

Seed pelleting can be done using recycled paper pulp to give a controlled environment for growing immature seedlings. Waste-based biostimulants, such as plant or seaweed extracts, can help seedlings to become more resistant to stress and also grow. Growth performance is further improved with extracts obtained from organic waste sources used to treat seedlings, providing for faster and more consistent germination. It is necessary to study and test extensively to ensure that waste products positively impact soil health and do not introduce any harmful substances (Jat Baloch *et al.*, 2023). Coffee grounds, aquaponic water, and biopolymers sourced from waste can be used to enrich seed mixes. Any of these measures can only guarantee that agricultural practices are environment-friendly if local legislation is followed. For the latter, they offer benefits like improved soil structure and more nutrients available for the plant. Anaerobic digestate, which is produced after anaerobic digestion (the decomposition of organic materials due to lack of oxygen), is an excellent nutrient source that may be used as seed treatment (Chozhavendhan *et al.*, 2023).

16.5 FUTURE PERSPECTIVES AND CHALLENGES OF SEED ENHANCEMENT

Seed invigoration, or seed enhancements, is a term for treatments applied to seed after harvesting to enhance the germination and develop seedlings, which is an essential stage that occurs during sowing. For the last 24 years, indiscriminate methods have been used for the improvement of seeds. This information is capable of guidance on seed priming methods, especially if stress resistance continues, and eventually preconditioned seeds that can survive against adverse environmental conditions can be produced (Khan *et al.*, 2023). This is especially important in areas where salinity, drought, or other stress may occur. The germination and seedling establishment are usually poor in small seeds; the coating facilitates this process, thus avoiding losses. In this eco-friendly method, the seed is treated to improve its physical characteristics for reliable germination and protection from soil predators. These biological agents improve plant physiology and seed performance, providing a safer alternative than chemical coatings. Yield success relies on the ultimate seeds, and seed coating is a good approach to reinforce abiotic challenges such as drought (Jarrar *et al.*, 2023).

Coated seeds show lower storability and require immediate utilization after coating and early change over to field planting. The key aspects contributing to the viability of agricultural seed coating applications are superabsorbent polymers and the optimal population of microbes (Paravar *et al.*, 2023). Tools had been developed from biotechnological approaches with the potential to develop seeds with better traits, post-biological engineering, and latest genetics. This includes features linked to improved nutritional value, drought resilience, and disease resistance. A lot of research is done on making the seeds that fight against the evils of climate change. Future improvements in seeds may emphasize traits that can help withstand extreme heat, droughts, and other climate-related stressors (Grigorieva *et al.*, 2023).

16.6 CONCLUSION

Various methods have been used to increase seed quality, with the aim of improving germination, reducing seedling emergence time, and enhancing stand establishment and production. Many research investigations have provided evidence of the advantageous impacts of quality improvement approaches on various field crops. The optimal resolution for germination-related issues, particularly in cases when crops are cultivated under harsh circumstances, significantly enhances the market value of seeds. Numerous approaches for enhancing seed quality have been developed and are now being used in various crops. It has the potential to improve the rates and percentages of germination and seedling emergence, hence facilitating the development of a suitable stand under various environmental circumstances.

LIST OF ABBREVIATIONS

AgNPs silver nanoparticles
AuNPs gold nanoparticles
BER base excision repair
CRISPR clustered regularly interspaced short palindromic repeats
GA_3 gibberellic acid
GR glutathione reductase
HSF heat shock factor
HSPs heat shock proteins
LEA late embryogenesis abundant
MWCNTs multi-walled carbon nanotubes
NER nucleotide excision repair
PA phosphatidic acid
PBM plant-beneficial microorganisms

REFERENCES

Adetunji, A.E., Adetunji, T.L., Varghese, B., Sershen, & Pammenter, N.W., 2021. Oxidative stress, ageing and methods of seed invigoration: An overview and perspectives. *Agronomy*, 11, 2369.

Albahri, G., Alyamani, A.A., Badran, A., Hijazi, A., Nasser, M., Maresca, M., & Baydoun, E., 2023. Enhancing essential grains yield for sustainable food security and bio-safe agriculture through latest innovative approaches. *Agronomy*, 13, 1709.

Alhammad, B.A., Abdel-Aziz, H.M.M., Seleiman, M.F., & Tourky, S.M.N., 2023. How can biological and chemical silver nanoparticles positively impact physio-chemical and chloroplast ultrastructural characteristics of *Vicia faba* seedlings? *Plants*, 12, 2509.

Ali, H., Tanveer, A., Nadeem, M., Asghar, H., & Javaid, M., 2013a. Germination ecology of *Rhynchosia capitata*: An emerging summer weed in Asia. *Planta Daninha*, 31, 249–257.

Ali, H., Tanveer, A., Nadeem, M., Javaid, M., Kashif, M., & Chadhar, A., 2013b. Allelopathic effects of *Rhynchosia capitata* on germination and seedling growth of mungbean. *Planta Daninha*, 31, 501–509.

Ali, K., Mubasher, H.M., Sher, A., Sattar, A., & Manaf, A., 2023. Seed Priming for Abiotic Stress Tolerance. *In* M. Hasanuzzaman (ed.) *Climate-Resilient Agriculture, Vol 2: Agro-Biotechnological Advancement for Crop Production.* Cham: Springer International Publishing, 641–665.

Animasaun, D.A., Adedibu, P.A., Shkryl, Y., Emmanuel, F.O., Tekutyeva, L., & Balabanova, L., 2023. Modern plant biotechnology: An antidote against global food insecurity. *Agronomy*, 13, 2038.

Anwar, T., Munwwar, F., Qureshi, H., Siddiqi, E.H., Hanif, A., Anwaar, S., Gul, S., Waheed, A., Alwahibi, M.S., & Kamal, A., 2023. Synergistic effect of biochar-based compounds from vegetable wastes and gibberellic acid on wheat growth under salinity stress. *Scientific Reports*, 13, 19024.

Ashraf, U., Kiran, M., Shahid, M.N., Anjum, S.A., & Khan, I., 2023. Biofortification Through Seed Priming in Food Crops: Potential Benefits and Future Scope. *In* M. Hasanuzzaman, M.S. Tahir, M. Tanveer & A.N. Shah (eds.) *Mineral Biofortification in Crop Plants for Ensuring Food Security.* Singapore: Springer Nature Singapore, 261–296.

Aziz, M., Ashraf, M., & Javaid, M.M., 2018. Enhancement in cotton growth and yield using novel growth promoting substances under water limited conditions. *Pakistan Journal of Botany*, 50, 1691–1701.

Balusamy, S.R., Joshi, A.S., Perumalsamy, H., Mijakovic, I., & Singh, P., 2023. Advancing sustainable agriculture: A critical review of smart and eco-friendly nanomaterial applications. *Journal of Nanobiotechnology*, 21, 372.

Bansal, S., Sundararajan, S., Shekhawat, P.K., Singh, S., Soni, P., Tripathy, M.K., & Ram, H., 2023. Rice lipases: A conundrum in rice bran stabilization: A review on their impact and biotechnological interventions. *Physiology and Molecular Biology of Plants*, 29, 985–1003.

Basavaraja, T., Husain, A., & Tiwari, P., 2023. Chapter 14: Quality Seed Production Techniques in Pulses. *Kaptan Baboo Syed Tazeen Zaidi Dr. Manoj Kumar Mishra Prabhas Kumar Shukla*, 151.

Bello, A.O., & Mbhele, T.P., 2024. A fuzzy-AHP multi-criteria decision-making approach for a sustainable supply chain of rice farming stakeholders in Edu-Patigi LGA, Kwara State, Nigeria. *Sustainability*, 16, 1751.

Bhabani, M.G., Shams, R., & Dash, K.K., 2024. Microgreens and novel non-thermal seed germination techniques for sustainable food systems: A review. *Food Science and Biotechnology*, 33, 1541–1557.

Bhandari, G., Dhasmana, A., Chaudhary, P., Gupta, S., Gangola, S., Gupta, A., Rustagi, S., Shende, S.S., Rajput, V.D., Minkina, T., Malik, S., & Slama, P., 2023. A perspective review on green nanotechnology in agro-ecosystems: Opportunities for sustainable agricultural practices & environmental remediation. *Agriculture*, 13, 668.

Bhardwaj, I., Kumar, V., Bhardwaj, N., Verma, R., Bhardwaj, Y., & Kumari, T., 2023. A glimpse into the performance and synthesis of microbial nanoparticles and its new advances in soil enrichment and plant nutrition: A review. *Nanotechnology for Environmental Engineering*, 8, 943–964.

Boamah, P.O., Onumah, J., Aduguba, W.O., & Santo, K.G., 2023. Application of depolymerized chitosan in crop production: A review. *International Journal of Biological Macromolecules*, 235, 123858.

Briassoulis, D., 2023. Agricultural plastics as a potential threat to food security, health, and environment through soil pollution by microplastics: Problem definition. *Science of the Total Environment*, 892, 164533.

Campos, E.V.R., Ratko, J., Bidyarani, N., Takeshita, V., & Fraceto, L.F., 2023. Nature-based herbicides and micro-/nanotechnology fostering sustainable agriculture. *ACS Sustainable Chemistry & Engineering*, 11, 9900–9917.

Carnevale Zampaolo, F., Kassam, A., Friedrich, T., Parr, A., & Uphoff, N., 2023. Compatibility between conservation agriculture and the system of rice intensification. *Agronomy*, 13, 2758.

Chadha, A., Florentine, S., Chauhan, B.S., Long, B., Jayasundera, M., Javaid, M.M., & Turville, C., 2019. Environmental factors affecting the germination and seedling emergence of two populations of an emerging agricultural weed: Wild lettuce (*Lactuca serriola*). *Crop and Pasture Science*, 70, 709–717.

Chandel, N.S., Tripathi, V., Singh, H.B., & Vaishnav, A., 2024. Breaking seed dormancy for sustainable food production: Revisiting seed priming techniques and prospects. *Biocatalysis and Agricultural Biotechnology*, 55, 102976.

Chaudhry, A., Hassan, A.U., Khan, S.H., Abbasi, A., Hina, A., Khan, M.T., & Abdelsalam, N.R., 2023. The changing landscape of agriculture: Role of precision breeding in developing smart crops. *Functional & Integrative Genomics*, 23, 167.

Chouhan, D., & Mandal, P., 2021. Applications of chitosan and chitosan based metallic nanoparticles in agrosciences—A review. *International Journal of Biological Macromolecules*, 166, 1554–1569.

Chozhavendhan, S., Karthigadevi, G., Bharathiraja, B., Praveen Kumar, R., Abo, L.D., Venkatesa Prabhu, S., Balachandar, R., & Jayakumar, M., 2023. Current and prognostic overview on the strategic exploitation of anaerobic digestion and digestate: A review. *Environmental Research*, 216, 114526.

El Idrissi, A., Channab, B.-E., Essamlali, Y., & Zahouily, M., 2024. Superabsorbent hydrogels based on natural polysaccharides: Classification, synthesis, physicochemical properties, and agronomic efficacy under abiotic stress conditions: A review. *International Journal of Biological Macromolecules*, 258, 128909.

Gabhane, J.W., Bhange, V.P., Patil, P.D., Bankar, S.T., & Kumar, S., 2020. Recent trends in biochar production methods and its application as a soil health conditioner: A review. *SN Applied Sciences*, 2, 1307.

Geetha, N., Sunilkumar, C.R., Bhavya, G., Nandini, B., Abhijith, P., Satapute, P., Shetty, H.S., Govarthanan, M., & Jogaiah, S., 2023. Warhorses in soil bioremediation: Seed biopriming with PGPF secretome to phytostimulate crop health under heavy metal stress. *Environmental Research*, 216, 114498.

Giordano, D.F., Pastor, N.A., Rouws, L.F.M., De Freitas, K.M., Erazo, J.G., Del Canto, A., Da Silva Coelho, I., Oddino, C.M., & Torres, A.M., 2023. *Trichoderma harzianum* ITEM 3636 colonizes peanut roots as an endophyte and protects the plants against late leaf spot. *Symbiosis*, 89, 337–352.

Grigorieva, E., Livenets, A., & Stelmakh, E., 2023. Adaptation of agriculture to climate change: A scoping review. *Climate*, 11, 202.

Guha, T., Ravikumar, K.V.G., Mukherjee, A., Mukherjee, A., & Kundu, R., 2018. Nanopriming with zero valent iron (nZVI) enhances germination and growth in aromatic rice cultivar (*Oryza sativa* cv. Gobindabhog L.). *Plant Physiology and Biochemistry*, 127, 403–413.

Guleria, G., Thakur, S., Shandilya, M., Sharma, S., Thakur, S., & Kalia, S., 2023. Nanotechnology for sustainable agro-food systems: The need and role of nanoparticles in protecting plants and improving crop productivity. *Plant Physiology and Biochemistry*, 194, 533–549.

Gupta, N., Jain, S.K., Tomar, B.S., Anand, A., Singh, J., Sagar, V., Kumar, R., Singh, V., Chaubey, T., Abd-Elsalam, K.A., & Singh, A.K., 2022. Impact of foliar application of ZnO and Fe_3O_4 nanoparticles on seed yield and physio-biochemical parameters of cucumber (*Cucumis sativus* L.) seed under open field and protected environment vis a vis during seed germination. *Plants*, 11, 3211.

Hamidzadeh, Z., Ghorbannezhad, P., Ketabchi, M.R., & Yeganeh, B., 2023. Biomass-derived biochar and its application in agriculture. *Fuel*, 341, 127701.

Hanif, Z., Naeem, M., Ali, H.H., Tanveer, A., Javaid, M.M., Peerzada, A.M., & Chauhan, B.S., 2017. Effect of environmental factors on germination of *Salsola foetida*: Potential species for rehabilitation of degraded rangelands. *Rangeland Ecology & Management*, 70, 638–643.

Hari, A., Doddapaneni, T.R.K.C., & Kikas, T., 2024. Common operational issues and possible solutions for sustainable biosurfactant production from lignocellulosic feedstock. *Environmental Research*, 251, 118665.

Ijaz, M., Khan, F., Ahmed, T., Noman, M., Zulfiqar, F., Rizwan, M., Chen, J., H.M. Siddique, K., & Li, B., 2023. Nanobiotechnology to advance stress resilience in plants: Current opportunities and challenges. *Materials Today Bio*, 22, 100759.

Imtiaz, H., Shiraz, M., Mir, A.R., Siddiqui, H., & Hayat, S., 2023. Nano-priming techniques for plant physio-biochemistry and stress tolerance. *Journal of Plant Growth Regulation*, 42, 6870–6890.

Jarrar, H., El-Keblawy, A., Ghenai, C., Abhilash, P.C., Bundela, A.K., Abideen, Z., & Sheteiwy, M.S., 2023. Seed enhancement technologies for sustainable dryland restoration: Coating and scarification. *Science of the Total Environment*, 904, 166150.

Jat Baloch, M.Y., Zhang, W., Sultana, T., Akram, M., Shoumik, B.a.A., Khan, M.Z., & Farooq, M.A., 2023. Utilization of sewage sludge to manage saline–alkali soil and increase crop production: Is it safe or not? *Environmental Technology & Innovation*, 32, 103266.

Javaid, M., Tanveer, A., Ahmad, R., Yaseen, M., & Khaliq, A., 2012. Optimizing activity of herbicides at reduced rate on *Emex spinosa* Campd. with adjuvants. *Planta Daninha*, 30, 425–435.

Javaid, M.M., Florentine, S.K., Ashraf, M., Mahmood, A., Sattar, A., Wasaya, A., & Li, F.M., 2022. Photosynthetic activity and water use efficiency of *Salvia verbenaca* L. under elevated CO_2 and water-deficit conditions. *Journal of Agronomy and Crop Science*, 208, 536–551.

Jha, Y., & Mohamed, H.I., 2023. Inoculation with *Lysinibacillus fusiformis* strain YJ4 and *Lysinibacillus sphaericus* strain YJ5 alleviates the effects of cold stress in maize plants. *Gesunde Pflanzen*, 75, 77–95.

Karimi-Maleh, H., Ghalkhani, M., Saberi Dehkordi, Z., Mohsenpour Tehran, M., Singh, J., Wen, Y., Baghayeri, M., Rouhi, J., Fu, L., & Rajendran, S., 2024. Mof-enabled pesticides as developing approach for sustainable agriculture and reducing environmental hazards. *Journal of Industrial and Engineering Chemistry*, 129, 105–123.

Kavale, M.G., Veeragurunathan, V., Jasiwar, S., Rajai, J.V., Baraiya, M., Grace, P.G., & Mantri, V.A., 2023. Long-distance translocation of planting material of *Kappaphycus alvarezii* (Doty) Doty ex P. C. Silva (Rhodophyta) for commercial farming: Assessing metabolic response, viability, and vigor. *Aquaculture International*, 32, 1629–1645.

Kavusi, E., Shahi Khalaf Ansar, B., Dehghanian, Z., Asgari Lajayer, B., Nobaharan, K., Ma, Y., & Glick, B.R., 2023. Delivery of beneficial microbes via seed coating for medicinal and aromatic plant production: A critical review. *Journal of Plant Growth Regulation*, 42, 575–597.

Kedisso, E.G., Guenthner, J., Maredia, K., Elagib, T., Oloo, B., & Assefa, S., 2023. Sustainable access of quality seeds of genetically engineered crops in Eastern Africa-Case study of Bt Cotton. *GM Crops & Food*, 14, 1–23.

Khan, M.N., Fu, C., Li, J., Tao, Y., Li, Y., Hu, J., Chen, L., Khan, Z., Wu, H., & Li, Z., 2023. Seed nanopriming: How do nanomaterials improve seed tolerance to salinity and drought? *Chemosphere*, 310, 136911.

Kibria, M.G., Aspy, N.N., Ullah, E., Dewan, M.F., Hasan, M.A., Hossain, M.A., Haseeb, M., & Hossain, M.E., 2023. Quantifying the effect of agricultural greenhouse gas emissions, food production index, and land use on cereal production in South Asia. *Journal of Cleaner Production*, 432, 139764.

Lah, N.H.C., El Enshasy, H.A., Mediani, A., Azizan, K.A., Aizat, W.M., Tan, J.K., Afzan, A., Noor, N.M., & Rohani, E.R., 2023. An insight into the behaviour of recalcitrant seeds by understanding their molecular changes upon desiccation and low temperature. *Agronomy*, 13, 2099.

Li, R., Wu, L., Shao, Y., Hu, Q., & Zhang, H., 2022. Melatonin alleviates copper stress to promote rice seed germination and seedling growth via crosstalk among various defensive response pathways. *Plant Physiology and Biochemistry*, 179, 65–77.

Liu, D., Samtani, J., Johnson, C., Zhang, X., Butler, D.M., & Derr, J., 2023. Brewer's spent grain with yeast amendment shows potential for anaerobic soil disinfestation of weeds and *Pythium irregulare*. *Agronomy*, 13, 2081.

Lv, T., Li, J., Zhou, L., Zhou, T., Pritchard, H.W., Ren, C., Chen, J., Yan, J., & Pei, J., 2024. Aging-induced reduction in safflower seed germination via impaired energy metabolism and genetic integrity is partially restored by sucrose and DA-6 treatment. *Plants*, 13, 659.

Mahmood, A., Wang, X., Shahzad, A.N., Fiaz, S., Ali, H., Naqve, M., Javaid, M.M., Mumtaz, S., Naseer, M., & Dong, R., 2021. Perspectives on bioenergy feedstock development in Pakistan: Challenges and opportunities. *Sustainability*, 13, 8438.

Maity, A., Paul, D., Lamichaney, A., Sarkar, A., Babbar, N., Mandal, N., Dutta, S., Maity, P.P., & Chakrabarty Chakrabarty, S.K., 2023. Climate change impacts on seed production and quality: Current knowledge, implications, and mitigation strategies. *Seed Science and Technology*, 51, 7–38.

Mamedi, A., Sharifzadeh, F., Maali-Amiri, R., & Divargar, F., 2024. Physiological and biological responses of Ca^{2+}-primed quinoa seed longevity stored at different hermetic storage conditions. *Journal of Plant Growth Regulation*, 43, 1967–1984.

Mcgregor, A., Bojovic, M., Ghammachi, N., & Mihrshahi, S., 2024. Producing knowledge about the sustainability and nutritional values of plant and animal-based beef: Funding, metrics, geographies and gaps. *Journal of Cleaner Production*, 441, 140900.

Mehmood, A., Naeem, M., Khalid, F., Saeed, Y., Abbas, T., Jabran, K., Sarwar, M.A., Tanveer, A., & Javaid, M.M., 2018. Identification of phytotoxins in different plant parts of *Brassica napus* and their influence on mung bean. *Environmental Science and Pollution Research*, 25, 18071–18080.

Mitra, D., Pellegrini, M., Olatunbosun, A.N., Mondal, R., Del Gallo, M., Chattaraj, S., Chakroborty, D., Priyadarshini, A., Khoshru, B., Guerra Sierra, B.E., De Los Santos-Villalobos, S., Senapati, A., Djebaili, R., Das Mohapatra, P.K., & Panneerselvam, P., 2023. Chapter 5 – Seed Priming With Microbial Inoculants for Enhanced Crop Yield. *In* V.K. Sharma, A. Kumar, M.R.Z. Passarini, S. Parmar & V.K. Singh (eds.) *Microbial Inoculants*. Academic Press, 99–123.

Mondal, S., Yadav, K.N., Vikram, N., & Panda, D., 2024. Chapter 20 – Physiological and Molecular Basis of Seed Priming with Nanomaterials. *In* K.A. Abd-Elsalam (ed.) *Nanofertilizer Synthesis*. Elsevier, 345–358.

Moulick, D., Mukherjee, A., Das, A., Roy, A., Majumdar, A., Dhar, A., Pattanaik, B.K., Chowardhara, B., Ghosh, D., & Upadhyay, M.K., 2024. Selenium—An environmentally friendly micronutrient in agroecosystem in the modern era: An overview of 50-year findings. *Ecotoxicology and Environmental Safety*, 270, 115832.

Naqve, M., Wang, X., Shahbaz, M., Fiaz, S., Naqvi, W., Naseer, M., Mahmood, A., & Ali, H., 2021. Foliar spray of alpha-tocopherol modulates antioxidant potential of okra fruit under salt stress. *Plants*, 10, 1382.

Narejo, G.A., Mirbahar, A.A., Yasin, S., Sirohi, M.H., & Saeed, R., 2023. Effect of hydro and KNO_3 priming on seed germination of cotton (*Gossypium hirsutum* L.) under gnotobiotic conditions. *Journal of Plant Growth Regulation*, 42, 1592–1603.

Naseer, M., Yang, Y.-M., Zhu, Y., Zhao, L., Cao, J., Wang, S., Wang, W.-Y., & Xiong, Y.-C., 2024. Nano-iron and AM fungi inoculation in dryland wheat field: A sustainable alternative to plastic film mulching. *Field Crops Research*, 306, 109208.

Nawaz, M.A., Ali, M.A., Golokhvast, K.S., Tsatsakis, A.M., & Chung, G., 2023. Chapter 1 – GMOs: History, Economic Status, Risks, and Socio-Economic Regulatory Frameworks. *In* M.A. Nawaz, G. Chung, K.S. Golokhvast & A.M. Tsatsakis (eds.) *GMOs and Political Stance*. Academic Press, 1–13.

Naz, H., Sayyed, R.Z., Khan, R.U., Naz, A., Wani, O.A., Maqsood, A., Maqsood, S., Fahad, A., Ashraf, S., & Show, P.L., 2023. Mesorhizobium improves chickpea growth under chromium stress and alleviates chromium contamination of soil. *Journal of Environmental Management*, 338, 117779.

Noronha, C., Liu, S., Bahar, M.H., & Mosher-Gallant, N., 2023. Buckwheat (*Fagopyrum esculentum*): Its impact on wireworm development and survival. *Arthropod-Plant Interactions*, 17, 429–440.

Okonkwo, C.E., Hussain, S.Z., Onyeaka, H., Adeyanju, A.A., Nwonuma, C.O., Bashir, A.A., Farooq, A., Zhou, C., & Shittu, T.D., 2023. Lignin polyphenol: From biomass to innovative food applications, and influence on gut microflora. *Industrial Crops and Products*, 206, 117696.

Oyebamiji, Y.O., Adigun, B.A., Shamsudin, N.A., Ikmal, A.M., Salisu, M.A., Malike, F.A., & Lateef, A.A., 2024. Recent advancements in mitigating abiotic stresses in crops. *Horticulturae*, 10, 156.

Pagano, A., Macovei, A., Xia, X., Padula, G., Hołubowicz, R., & Balestrazzi, A., 2023a. Seed priming applied to onion-like crops: State of the art and open questions. *Agronomy*, 13, 288.

Pagano, A., Pagano, P., Dueñas, C., Griffo, A., Gaonkar, S.S., Messina, F., Balestrazzi, A., & Macovei, A., 2023b. Seed Quality Assessment and Improvement between Advancing Agriculture and Changing Environments. *In* Global Climate Change and Plant Stress Management. 317–334.

Paravar, A., Piri, R., Balouchi, H., & Ma, Y., 2023. Microbial seed coating: An attractive tool for sustainable agriculture. *Biotechnology Reports*, 37, e00781.

Parul, Kumar, A., Chourasia, S., Kumar, R., Devi, G., Kumar, A., Mann, A., Sheoran, P., & Sanwal, S.K., 2023. Physiological, Biochemical, and Molecular Responses to Salt Stress and Seed Priming Approach to Enhance Salt Tolerance in Bread Wheat. *In* A. Kumar, P. Dhansu & A. Mann (eds.) *Salinity and Drought Tolerance in Plants: Physiological Perspectives*. Singapore: Springer Nature Singapore, 61–86.

Paudel, D., Neupane, R.C., Sigdel, S., Poudel, P., & Khanal, A.R., 2023. COVID-19 pandemic, climate change, and conflicts on agriculture: A trio of challenges to global food security. *Sustainability*, 15, 8280.

Paul, A., Mondal, S., Chakraborty, K., & Biswas, A.K., 2024. Moving forward to understand the alteration of physiological mechanism by seed priming with different halo-agents under salt stress. *Plant Molecular Biology*, 114, 24.

Paul, A., Mondal, S., Pal, A., Biswas, S., Chakraborty, K., Mazumder, A., Biswas, A.K., & Kundu, R., 2023. Seed priming with NaCl helps to improve tissue tolerance, potassium retention ability of plants, and protects the photosynthetic ability in two different legumes, chickpea and lentil, under salt stress. *Planta*, 257, 111.

Paul, S., Dey, S., & Kundu, R., 2022. Seed priming: An emerging tool towards sustainable agriculture. *Plant Growth Regulation*, 97, 215–234.

Pedrini, S., Webber, Z., D'agui, H., Dixon, K., Just, M., Arya, T., & Turner, S., 2023. Customise the seeds, not the seeder: Pelleting of small-seeded species for ecological restoration. *Ecological Engineering*, 196, 107105.

Periakaruppan, R., Romanovski, V., Thirumalaisamy, S.K., Palanimuthu, V., Sampath, M.P., Anilkumar, A., Sivaraj, D.K., Ahamed, N.a.N., Murugesan, S., Chandrasekar, D., & Selvaraj, K.S.V., 2023. Innovations in modern nanotechnology for the sustainable production of agriculture. *ChemEngineering*, 7, 61.

Pompelli, M.F., Jarma-Orozco, A., & Rodriguez-Páez, L.A., 2023. Imbibition and germination of seeds with economic and ecological interest: Physical and biochemical factors involved. *Sustainability*, 15, 5394.

Rasheed, A., Li, H., Nawaz, M., Mahmood, A., Hassan, M.U., Shah, A.N., Hussain, F., Azmat, S., Gillani, S.F.A., & Majeed, Y., 2022a. Molecular tools, potential frontiers for enhancing salinity tolerance in rice: A critical review and future prospective. *Frontiers in Plant Science*, 13, 966749.

Rasheed, A., Raza, A., Jie, H., Mahmood, A., Ma, Y., Zhao, L., Xing, H., Li, L., Hassan, M.U., & Qari, S.H., 2022b. Molecular tools and their applications in developing salt-tolerant soybean (*Glycine max* L.) cultivars. *Bioengineering*, 9, 495.

Raza, M.a.S., Zulfiqar, B., Iqbal, R., Muzamil, M.N., Aslam, M.U., Muhammad, F., Amin, J., Aslam, H.M.U., Ibrahim, M.A., Uzair, M., & Habib-Ur-Rahman, M., 2023. Morpho-physiological and biochemical response of wheat to various treatments of silicon nano-particles under drought stress conditions. *Scientific Reports*, 13, 2700.

Rhaman, M.S., Tania, S.S., Imran, S., Rauf, F., Kibria, M.G., Ye, W., Hasanuzzaman, M., & Murata, Y., 2022. Seed priming with nanoparticles: An emerging technique for improving plant growth, development, and abiotic stress tolerance. *Journal of Soil Science and Plant Nutrition*, 22, 4047–4062.

Roychowdhury, R., Das, S.P., Gupta, A., Parihar, P., Chandrasekhar, K., Sarker, U., Kumar, A., Ramrao, D.P., & Sudhakar, C., 2023. Multi-omics pipeline and omics-integration approach to decipher plant's abiotic stress tolerance responses. *Genes*, 14, 1281.

Sang, L., Lv, X., & Wu, Y., 2023. $NaNO_3$-KNO_3-KCl/K_2CO_3 with the elevated working temperature for CSP application: Phase diagram calculation and machine learning. *Solar Energy*, 252, 322–329.

Saud, S., Jiang, Z., Chen, S., & Fahad, S., 2023. Interaction of melatonin on post-harvest physiology and quality of horticultural crops. *Scientia Horticulturae*, 321, 112286.

Shelar, A., Nile, S.H., Singh, A.V., Rothenstein, D., Bill, J., Xiao, J., Chaskar, M., Kai, G., & Patil, R., 2023. Recent advances in nano-enabled seed treatment strategies for sustainable agriculture: Challenges, risk assessment, and future perspectives. *Nano-Micro Letters*, 15, 54.

Sine, B., Ndiaye, A., Kanfany, G., Grondin, A., Ndiaye, A., Faye, A., Sy, O., Diack, O., Kane, N.A., & Laplaze, L., 2024. Physiological and Molecular Bases of Drought and Heat Tolerance in Pearl Millet. *In* V.A. Tonapi, N. Thirunavukkarasu, S.K. Gupta, P.I. Gangashetty & O.P. Yadav (eds.) *Pearl Millet in the 21st Century: Food-Nutrition-Climate Resilience-Improved Livelihoods.* Singapore: Springer Nature Singapore, 247–278.

Sundareswaran, S., Ray Choudhury, P., Vanitha, C., & Yadava, D.K., 2023. Seed quality: Variety development to planting—An overview. *Seed Science and Technology: Biology, Production, Quality*, 1–16.

Szurman-Zubrzycka, M., Jędrzejek, P., & Szarejko, I., 2023. How do plants cope with DNA damage? A concise review on the DDR pathway in plants. *International Journal of Molecular Sciences*, 24, 2404.

Tang, H., Meng, G., Xiang, J., Mahmood, A., Xiang, G., Sanaullah, Liu, Y., & Huang, G., 2022. Toxic effects of antimony in plants: Reasons and remediation possibilities—A review and future prospects. *Frontiers in Plant Science*, 13, 1011945.

Tanveer, A., Mumtaz, K., Javaid, M., Chaudhry, M., Balal, R., & Khaliq, A., 2013. Effect of ecological factors on germination of horse purslane (*Trianthema portulacastrum*). *Planta Daninha*, 31, 587–597.

Uebersax, M.A., Cichy, K.A., Gomez, F.E., Porch, T.G., Heitholt, J., Osorno, J.M., Kamfwa, K., Snapp, S.S., & Bales, S., 2023. Dry beans (*Phaseolus vulgaris* L.) as a vital component of sustainable agriculture and food security—A review. *Legume Science*, 5, e155.

Ullah, H., Mahmood, A., Ijaz, M., Tadesse, B., & Honermeier, B., 2013. Evaluation of anise (*Pimpinella anisum* L.) accessions with regard to morphological characteristics, fruit yield, oil contents and composition. *Journal of Medicinal Plants Research*, 7, 2177–2186.

Vasileiou, M., Kyrgiakos, L.S., Kleisiari, C., Kleftodimos, G., Vlontzos, G., Belhouchette, H., & Pardalos, P.M., 2024. Transforming weed management in sustainable agriculture with artificial intelligence: A systematic literature review towards weed identification and deep learning. *Crop Protection*, 176, 106522.

Wahab, A., Muhammad, M., Munir, A., Abdi, G., Zaman, W., Ayaz, A., Khizar, C., & Reddy, S.P.P., 2023. Role of arbuscular mycorrhizal fungi in regulating growth, enhancing productivity, and potentially influencing ecosystems under abiotic and biotic stresses. *Plants*, 12, 3102.

Wang, X., Zhu, Y., Tang, L., Wang, Y., Sun, R., & Deng, X., 2023. Arabidopsis HSFA9 acts as a regulator of heat response gene expression and the acquisition of thermotolerance and seed longevity. *Plant and Cell Physiology*, 65, 372–389.

Westengen, O.T., Dalle, S.P., & Mulesa, T.H., 2023. Navigating toward resilient and inclusive seed systems. *Proceedings of the National Academy of Sciences*, 120, e2218777120.

Yadav, N., Garg, V.K., Chhillar, A.K., & Rana, J.S., 2023. Recent advances in nanotechnology for the improvement of conventional agricultural systems: A review. *Plant Nano Biology*, 4, 100032.

Yadav, R., Juneja, S., & Kumar, S., 2021. Cross priming with drought improves heat-tolerance in chickpea (*Cicer arietinum* L.) by stimulating small heat shock proteins and antioxidative defense. *Environmental Sustainability*, 4, 171–182.

Yousefi, S., Kartoolinejad, D., & Naghdi, R., 2017. Effects of priming with multi-walled carbon nanotubes on seed physiological characteristics of Hopbush (*Dodonaea viscosa* L.) under drought stress. *International Journal of Environmental Studies*, 74, 528–539.

Zaghdoud, C., Ollio, I., Solano, C.J., Ochoa, J., Suardiaz, J., Fernández, J.A., & Martínez Ballesta, M.D.C., 2023. Red LED light improves pepper (*Capsicum annuum* L.) seed radicle emergence and growth through the modulation of aquaporins, hormone homeostasis, and metabolite remobilization. *International Journal of Molecular Sciences*, 24, 4779.

Zhu, F., Cao, M.-Y., Zhang, Q.-P., Mohan, R., Schar, J., Mitchell, M., Chen, H., Liu, F., Wang, D., & Fu, Z.Q., 2024. Join the green team: Inducers of plant immunity in the plant disease sustainable control toolbox. *Journal of Advanced Research*, 57, 15–42.

Zulfiqar, U., Haider, F.U., Maqsood, M.F., Mohy-Ud-Din, W., Shabaan, M., Ahmad, M., Kaleem, M., Ishfaq, M., Aslam, Z., & Shahzad, B., 2023. Recent advances in microbial-assisted remediation of cadmium-contaminated soil. *Plants*, 12, 3147.

17 Contemporary Nutrient Management

*Asif Mukhtiar, Wasim Javed, and Ayman El Sabagh**

17.1 INTRODUCTION

The growth of population has social and economic pressure, which have encouraged the pursuit of technologies by scientists on the frontier, in the quest for the holy grail of agricultural output by land area and agricultural output by time (Smaje, 2020). In contrast, achieving a balance between nutrient needs for crops and nutrients held by soil is vital for high yields and soil health and fertility, and efficient, sustainable production and preventing environmental devastation (Briat *et al.*, 2020). For this reason, nutrient management is an integral part of modern agriculture, consistent monitoring and application of the nutrients that are vital for plant production to achieve market demand while limiting the risk of polluting the environment.

Integrated nutrient management (INM) is a management strategy that deals with the quality of production and environmental conservation for future generations.

It promotes sustainable agriculture by maintaining soil fertility while minimizing reliance on chemical fertilizers, thereby increasing crop yields and profitability. It is based on the application and preservation of nutrients, innovative approaches that boost nutrient accessibility to plants, and knowledge propagation among farmers and researchers (Azhar Mehmood *et al.*, 2018; Selim, 2020). Previously, nutrient management was driven by the desire to maximize production. However, sustainable nutrient management now includes optimizing production, averting on-site soil degradation, and minimizing the outside influence of applied nutrients (Devi *et al.*, 2022).

Primarily, INM refers to integrating old and current techniques of managing nutrients into an environmentally friendly and economically ideal farming system that takes maximum advantage of all possible sources of organic, biological, and inorganic nutrient sources in a careful, productive, and effective manner (Ullah *et al.*, 2013; Shah and Wu, 2019). One key aspect of INM is the use of precision agriculture and other inputs in a targeted and efficient manner (Waheed *et al.*, 2019). It maximizes all aspects of nutrient cycling, including N, P, K, and input and output of various micro and macronutrients to integrate nutrient demand by the crops and discharge in the natural environment (Chadha *et al.*, 2019; Zhang *et al.*, 2020). INM practices reduce losses from volatilization, leaching, runoff, emission levels, and immobility while accomplishing substantial nutrient-use efficiency (Rehman *et al.*, 2019; Hassan *et al.*, 2022).

External inputs to agriculture, including chemical fertilizers, organic amendments, inoculating bacteria, and chemical pesticides, are implemented with the ultimate objective of optimizing productivity and financial benefits, whereas their negative impacts on soil ecosystems and organisms are often neglected (Soytong *et al.*, 2021). Mineral fertilizers have very few direct negative effects, but their use can improve the biological activity of soil by increasing system efficiency, crop residue return, and organic matter in the soil (Bhunia *et al.*, 2021). Another substantial repercussion, particularly of N fertilization, is acidification of the soil, which has significant adverse consequences for the organisms in the soil (Gull *et al.*, 2023). A fertilizer experiment by growing rice found that a balanced application of K, N, and P increased the biomass of microbes by boosting the diversity of the microbial community (Mukhtiar *et al.*, 2023b).

* Corresponding author: ayman.elsabagh@siirt.edu.tr

DOI: 10.1201/9781003498698-17

Organic amendments, including manure, organic matter, biosolids, compost, and humic substances, serve as both direct and indirect sources of nutrients and carbon (C) for soil microbes through increased plant growth and plant residue return (Goswami *et al.*, 2022). Bio-fertilizers are microbial inoculants containing beneficial microorganisms to provide nutrients (N, P, and S) such as N-fixing or P-solubilizing or plant growth-promoting microbial strains to increase the availability of nutrients in assimilable forms (Goldan *et al.*, 2023). Hence, the use of combinations of various nutrient sources is essential for efficient nutrient management in achieving sustainable and efficient agricultural production. Organic and unified production systems could provide alternatives to conventional systems. However, integrated agricultural practices have not gained the same level of acceptance among farmers compared to traditional production methods (Chadha *et al.*, 2020; Aftab and Hakeem, 2022).

Nutrient management has evolved over time, with contemporary nutrient management incorporating a range of advanced strategies and technologies to more effectively manage nutrient inputs and outputs. This chapter describes various aspects of contemporary nutrient management in relation to physiological soil fertility, health, and sustainable productivity, moreover the production of high-quality crops and various practices for managing plant nutrients in the soil.

17.2 NUTRIENT MANAGEMENT STRATEGIES

There are currently several nutritional management techniques in use. These are the conventional methods, including a cover crop (green manure) method and a livestock manure-based strategy (Carr *et al.*, 2020). A conventional nutrient management strategy is the one in which chemical fertilizers become the predominant source of N (and other essential nutrients such as P and K), and these products are typically applied according to supplier-recommended rates (Robertson and Groffman, 2024).

17.2.1 Cover Crop (Green Manure)

One standard cover crop approach involves using nutrient-rich crops, already present in the soil N pools, likely seeded in the fall and tilled into the ground in the spring, prior to sowing a primary crop. Cover crops have well-characterized decomposition rates (Scavo *et al.*, 2022). Cover crops may assist in adding nutritious biomass to bare soil and protect against soil erosion during winter months. Cover crops usually need considerably fewer commercial fertilizers when compared to conventionally grown crops (Farzadfar, 2022). Cover crops may include leguminous and non-leguminous crops or both. Commonly used cover crops include hairy vetch containing 4.62% N, crimson clover enriched with 3.67% N, winter pea having 4.56% N, and rye (Gareau, 2012).

17.2.2 Animal Manure

Animal dung is an invaluable asset with multiple benefits and difficulties. Manure is a nutrient-rich fertilizer that improves soil health and lessens waste disposal problems by containing vital nutrients, including K, P, and N. By encouraging carbon sequestration and lowering the manufacture of synthetic fertilizers, its use additionally may slow down the rate of global warming. Manure management does present certain challenges though, such as the potential of water contaminants, odor and air quality challenges, nutrient imbalance, and disease and contaminant hazards. A manure-based method uses raw animal manure such as that of dairy cattle, hogs, pigs, and chickens to provide plant nutrients in soil (Rayne and Aula, 2020). The typical nitrogen concentrations of fresh manures vary: hog manure contains 0.5%–0.7% (Kokulan *et al.*, 2021), chicken manure contains 0.5%–0.9% (Hsu and Lai, 2022), manure of dairy cows possesses 0.3%–0.5% (Pei *et al.*, 2023), and beef cattle manure contains 0.5%–0.6% nitrogen (Goldan *et al.*, 2023).

17.3 NUTRIENT MANAGEMENT IN SOIL USING AGRICULTURAL FERTILIZERS

For the best growth of crops, productivity, and environmental sustainability, soil nutrient management is important. Crop nutrients management is shown in Figure 17.1. Fixing soil nutrients, sustaining fertility, and securing crop nutrition are all made feasible by agricultural fertilizers. N, P, K, micronutrients, and organic fertilizers are among the several types of fertilizers. Fertilizer application is guided by the pH and nutrient shortages found by soil testing. Fresh farm waste is frequently treated prior to its application in a treatment facility, like an aerobic or anaerobic lagoon (Ogburn *et al.*, 2023). An organic nutrient management approach is one that uses organic agricultural fertilizers and does not use synthetic fertilizers (Zhao *et al.*, 2023). As a result, both manure-based systems and cover crop systems (without the use of commercial fertilizer) can be considered organic in nature.

17.3.1 Organic Agricultural Fertilizers

Organic fertilizers are derived from naturally occurring substances that are biodegradable, serving as soil enrichers and nutrient boosters. Essentially, any naturally sourced material that easily decomposes and contributes to soil fertility qualifies as organic fertilizer (Kayesh *et al.*, 2023). These substances undergo decomposition by various microorganisms, breaking down into soluble particles that roots can absorb for nourishment (Babar *et al.*, 2009; Zuo *et al.*, 2021). Manure, plant residues, seaweed, slurry, sewage, and worms represent naturally sourced options for green manure, while compost, biochar, bone meal, blood meal, and seaweed extracts are examples of manufactured organic fertilizers (Løes, 2023). Additionally, crops themselves can be cultivated to enrich the soil with nutrients (Ho *et al.*, 2022). Presently, every farmer seeks effective solutions to agricultural

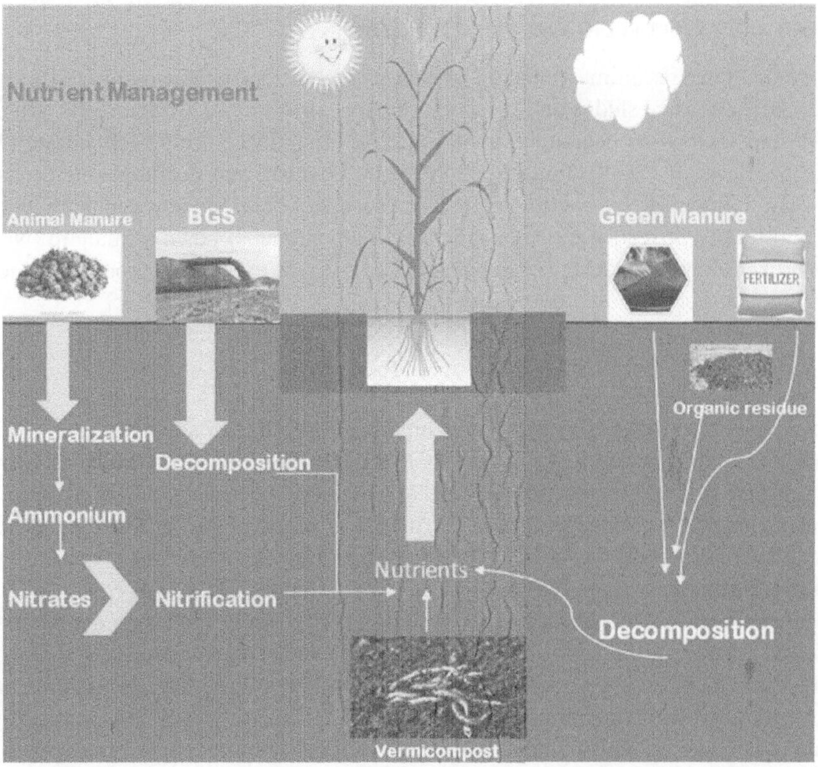

FIGURE 17.1 Showing the mechanism of different applied organic fertilizers to soil to maintain soil nutrients.

challenges without sacrificing yields. It is widely recognized that organic fertilization practices are environmentally friendly and yield gradual but consistent results. The use of fertilizers stands as one of the paramount innovations of the agricultural revolution (Lazcano *et al.*, 2021).

Organic fertilizers contribute to the refinement of soil structure by nourishing it, boosting productivity, and safeguarding against erosion (Liu *et al.*, 2021a).

In contrast, artificial and inorganic fertilizers pose drawbacks as they necessitate frequent reapplication to maintain soil fertility, essentially becoming a dependency for sustained productivity (Faloye *et al.*, 2023). They can be likened to a form of soil medication, sustaining productivity only as long as they are consistently applied. Excessive use of artificial fertilizers can lead to eutrophication and nutrient pollution (Liu *et al.*, 2021b).

17.3.2 Inorganic Agricultural Fertilizers

Inorganic agricultural fertilizers consist of chemical substances and are commonly known as synthetic or artificial fertilizers (Pahalvi *et al.*, 2021). Unlike organic, artificial fertilizers (urea, granular triple superphosphate) are typically non-biodegradable and are produced in factories using advanced technologies. The manufacturing processes result in a durable and highly efficient product with a coarse texture (Vanmare, 2021).

17.4 ECO-FRIENDLY NUTRIENT MANAGEMENT IN SOIL

There are various ways to improve or replenish nutrients in the soil using different organic fertilizers and wastes and employing microorganisms, including N-fixing and plant growth-promoting rhizobacteria (PGPR) (Mukhtiar *et al.*, 2023a).

17.4.1 Naturally Occurring Organic Fertilizers

This category encompasses animal manure (Bhunia *et al.*, 2021), peat (Misnikov, 2020), slurry waste (Chew *et al.*, 2019), sewage sludge and biosolids (Delibacak *et al.*, 2020), seaweeds (Raghunandan *et al.*, 2019), and various biodegradable materials. Additionally, mining rock sulfate (Mahmood *et al.*, 2013; Slama *et al.*, 2019), limestone, phosphate, potash, and decomposing crops (Yahyah *et al.*, 2023) also fall within this classification. The process of breaking down manure, with the help of bacteria and fungi, converts it into a beneficial soil amendment known as manure (Naqve *et al.*, 2023). This broken-down substance is used in agrarian practice to increase ground productivity and substantially increase output.

17.4.2 Bio-Compost

Bio-compost is a sugar industry waste-based organic fertilizer (Krishnaveni *et al.*, 2020). This process produces both nitrogen and phosphate-solubilizing bacteria and beneficial fungi such as decomposing fungi, through the decomposition carried out by bacteria and fungi that are harmless to both humans and plants (Li *et al.*, 2020). Farmers can increase the soil fertility and ultimately improve yields with bio-composting (De Corato, 2020). The use of sawdust (Fatima *et al.*, 2021), organic matter (Hoffland *et al.*, 2020), and manure from animals (Dadrasnia *et al.*, 2021) for nutrient recovery has the ability to significantly reduce the cost of importing inorganic fertilizers while also significantly reducing the environmental impact of current waste disposal methods (Hidalgo *et al.*, 2021). Dairy and livestock (cattle, sheep, goats, and pigs) generate a large amount of feces, which can be recycled through composting (Parihar *et al.*, 2019). During this process, a nutrient-rich substrate was obtained with an appropriate total nitrogen, carbon, carbon/nitrogen (C/N) ratio, nitrate nitrogen, ammonium nitrogen, electrical conductivity (EC), pH, and microbial respiration, which are best for agricultural cultivation (Farghali *et al.*, 2023).

According to de Souza Mendes *et al.* (2023), the bio-compost alters the physicochemical characteristics of the soil while serving as a nutrient source for plants. The nutrient elements (nitrogen, magnesium, phosphorus, potassium, zinc, manganese, iron, and copper) are more than in bio-compost where the raw materials, sawdust and goat manure, are there. Hence, the final product contained macro and micronutrients that will be beneficial for restoring soil nutrients. It is due to the highest value of nutrients in bio-compost which enable it to be used as a source of nutrient growth and productivity of plants.

In a similar way, a study conducted by Abdelfattah *et al.* (2021) found that bio-compost from legumes and 25% of NPK rates not only ensures the physicochemical criteria but also maintains mineral outlays and plant's physiological each of *Proteus vulgaris* on sand-based medium. Essentially, by adding leguminous bio-compost at 1.2 kg/m^2 along with 75% of the soil NPK dose and applying up to 25% of the NPK dosages in *Phaseolus vulgaris* is causing a considerable improvement in soil fertility, vegetation physiological as well as biochemical parameters facilitating in the improvement of plants growth parameters, overall productivity and yield.

17.4.3 Vermicompost

This is the organic fertilizer that comprises mainly K, N, organic carbon, P, S, hormones and enzymes. When it is applied dependably, it enhances the nutrient status of the soil by returning lost nutrients and ultimately resulting in lasting fertility. The term "vermicompost" comes from Latin and is comprised two words such as "virmi," representing worms, and "compost," being a fertilizer. Therefore, vermicompost is essentially worm compost, produced as worms consume and excrete organic waste (Al-Maamori *et al.*, 2023).

According to investigations, the waste produced by worms is less polluting and has higher nitrogen contents than organic matter (Waqar *et al.*, 2019; Singh and Sinha, 2022). Additionally, it contains high water-soluble nutrients (Kumar *et al.*, 2020). The presence of earthworms in soil amends its physical features by increasing water movement and porosity of soil (Le Bayon *et al.*, 2021). Organic vermicompost contains a lot of enzymes that uplift microbial activity and ultimately boost microbial biomass in the soil. Worms select a variety of household waste that are not damaging to them, including cardboard and various leaves, which serve to be both worm bedding and a source of carbon (Hřebečková *et al.*, 2019). Worms can also eat tree leaves and peels of fruits and vegetables. Various foodstuffs, such as rice, sugarcane, oats, corn, and kitchen waste, except citrus fruits, are all appropriate for composting (Bellitürk and Sundari, 2024).

According to Boruah *et al.* (2019), research was carried out to develop vermicompost. Worms, sludge, regular compost, and paper waste were retained in the containers. The various attributes of the soil, including carbon-nitrogen ratio, pH, EC, total K, total N, Ca, available P, and levels of the materials, were calculated before earthworms changed these into vermicompost. The experiment depicted a drop in pH (6.01), EC (1.89 dS/m), and C/N (10). In addition, the amounts of N, K, P, and calcium have been calculated to be 1.8%, 284.4 mg/kg, 242.7 mg/kg, and 0.034%, respectively.

Vermicompost provides many benefits in agriculture and horticulture. Table 17.1 shows the positive impacts of organic fertilizers on soil structure and crop productivity. It encourages the growth of beneficial soil microorganisms and enzymes, which are vital to nutrient cycling, combating soil-borne pathogens, and boosting soil quality overall (Tanveer *et al.*, 2012; Markam, 2021). Vermicompost nutrient composition is determined by the organic materials used during the composting process. A wide range of input biological substances results in a broader range of nutrients and trace elements in the vermicompost (Walia and Kaur, 2024).

17.4.4 Rhizobacteria as Bio-fertilizer

PGPRs are bacteria that cause nitrogen-fixing nodules to form on vegetable roots like peas and beans, and they play a crucial part in agriculture (Zope *et al.*, 2019). These PGPRs are widely

TABLE 17.1
Presenting the Positive Impacts of Organic Fertilizers on Soil Structure and Crop Productivity

Nutrient Source	Effect on Soil Quality	Crop Productivity	References
Poultry and cattle manure	Increase pH of soil Reduced the bulk density of soil	Significantly improved the yield of maize	Geng *et al.* (2019), Du *et al.* (2020)
Vermicompost	Soil physical structure	Increases Maize Productivity from 94.1 to 261%	Are *et al.* (2017)
Pig manure	Enhances soil organic carbon concentration Maintains soil pH	Improve grain and yield of straw in wheat and maize	Wang *et al.* (2019)
Rapeseed meal	Improves soil organic matter up to 54%	Enhances wheat yield	Cen *et al.* (2020)
Compost	Improves pH, SOC of soil Limits bulk density of soil Increases cation exchange rate	Straw and productivity of maize	Ejigu *et al.* (2021)
Biochar	Improves nutrients availability for plants	Improves soil organic carbon	Agegnehu *et al.* (2016)

distributed in the rhizosphere of plants; these bacteria stimulate the growth and development of plants using several mechanisms, such as their contribution to the increased availability of mineral nutrients, phytohormone regulation, and biocontrol of plant pathogens (Etesami and Adl, 2020). The achievement of PGPR inoculants depends on the establishing, persistence, and intricate interaction in rhizosphere (Thies and Grossman, 2023). The rhizo-microbiome is then established by the recruitment of diverse microbial settlement patterns, enriched by root exudates and plant cell debris, which are essential to farming (Jain *et al.*, 2020).

The rhizo-microbiome constructs and dominates extracellular molecules like hormones, secondary metabolites, antibiotics, and signaling molecules (Jain *et al.*, 2020). In addition, the rhizo-microbiome's microbial structure assists in enhancing soil texture. In studies by Huang *et al.* (2014) and Rosier *et al.* (2018), PGPR had the potential to enhance the growth of plants either through treatments or inoculation. PGPR modifies host plant characteristics that increase rates of nutrient acquisition and root function (Ramírez-Cariño *et al.*, 2024). Goyal *et al.* (2023) evaluated PGPR isolates regarding traits of plant growth promotion and other related facets such as ethylene emission, osmolyte accumulation, nutrient uptake, soil nutrient availability, and soil enzyme activity under drought stress. The results verified the positive effects of plant nutrient uptake.

17.4.5 Azotobacter

Azotobacter is very useful for increasing the rate of availability of atmospheric N in soil which is essential for plant growth (Aasfar *et al.*, 2021). It also safeguards the roots of plants from other soil pathogens. The first report of *Azotobacter* use as a bio-fertilizer was released by Gerlach and Voel in 1902 (Nongthombam *et al.*, 2021). Functions in phosphate solubilization are detrimental to plant growth and growth regulation such as auxins, cytokinins, gibberellins, amino acids, and vitamins (Farman *et al.*, 2019). *Azotobacter* spp. in soil are known to help in various aspects of plant growth like seed germination, including CGR (Crop Growth Rate) (Chennappa *et al.*, 2017; Sumbul *et al.*, 2020; Salimi and Hamedi, 2021). In addition, their abundance is positively correlated with organic matter, pH, soil moisture, temperature, and microbiological properties of the soils (Kurrey *et al.*, 2018). Soil profile depth also affects the abundance of these bacteria (Aasfar *et al.*, 2021). Moreover, microorganisms, compost, crop rotation, organic fertilizers, and intercropping work synergistically to enhance soil health and improve crop nutrient uptake, leading to more sustainable and productive agricultural systems (Figure 17.2).

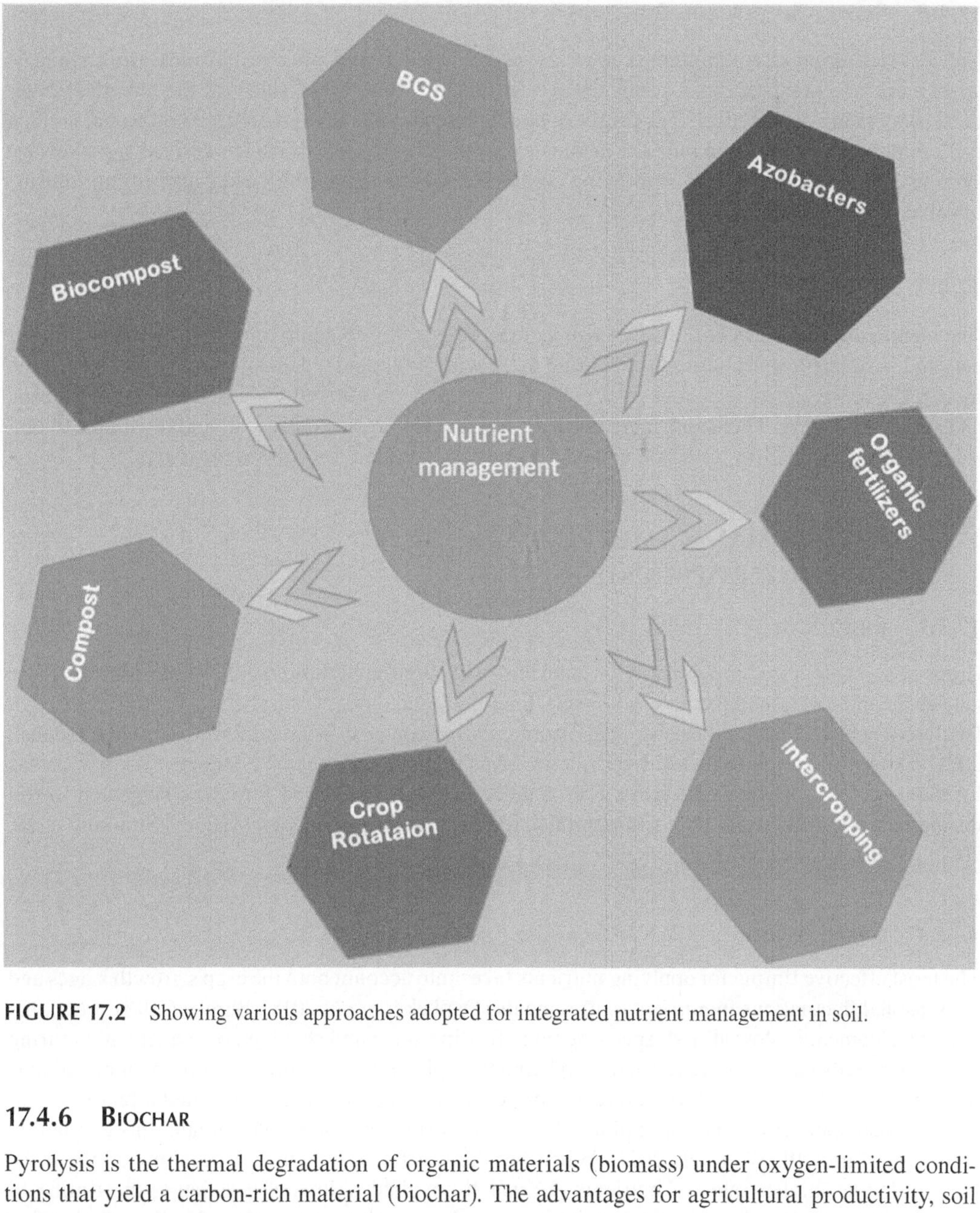

FIGURE 17.2 Showing various approaches adopted for integrated nutrient management in soil.

17.4.6 Biochar

Pyrolysis is the thermal degradation of organic materials (biomass) under oxygen-limited conditions that yield a carbon-rich material (biochar). The advantages for agricultural productivity, soil health, and environmental sustainability are many, and this is driving strong adoption of this practice in relation to modern fertilizer management.

17.4.7 Bioslurry

Anaerobic digestion is the biological process that occurs in biogas plants in which organic materials such as manure, crop residues, and food waste are digested without the presence of oxygen with the help of microbes to generate energy; the process yields bioslurry, a nutrient-dense sludge that is the end product of anaerobic digestion. The produced bioslurry is a great natural fertilizer due to its richness in nutrients, its ability to improve soil health, and its importance for sustainable agriculture and is increasingly applied in current nutrient management.

17.4.8 Sewage Sludge and Biosolids

Nutrient-rich organic compounds known as sewage sludge and biological solids are created at wastewater treatment facilities when sewage is treated. These by-products of the chemical treatment process are often recycled as fertilizer or soil amendment due to their high nutritional content and organic matter. Modern nutrient management involves the use of sludge derived from sewage systems and biosolids, thereby improving soil fertility, enhancing crop yields, and promoting sustainable farming techniques.

17.4.9 Seaweeds

The significance of seaweeds, also known as macroalgae, for contemporary management of nutrients and sustainable agriculture is becoming more widely recognized. Seaweeds are rich in organic matter, bioactive substances, and vital nutrients that may be utilized as biostimulants, fertilizers, and soil conditioners. They contribute to better crop development, soil health, and resistance to outside stresses by providing a natural and eco-friendly substitute for synthetic fertilizers.

17.5 NUTRIENT MANAGEMENT USING FOUR "R" OF FERTILIZER APPLICATION

17.5.1 Right Rate

Crops necessitate a specific quantity of nutrients to ensure the highest yields. While some of these nutrients can be obtained from the soil itself, the remaining amount must be supplemented through fertilizers, which can be either chemical or organic like compost from animal wastes (Me Trenkel, 2021). Determining the right fertilizer rate involves meticulous research conducted across various locations, crops, varieties, and years. For instance, a portion of N and P may be subjected to soil during bed preparation with the rest administered through drip irrigation systems (Simonne *et al.*, 2017; Abdullah *et al.*, 2022).

17.5.2 Right Timing

The most effective timing for applying nutrients takes into account both the crop's growth stages and the seasonal fluctuations in nutrient requirements (Dapkekar *et al.*, 2018). Bareke (2018) stated that crop development is slow at first, speeding up at fruiting stage and slowing down again at maturing stage after seeds have germinated and crop being transplanted. The timing of nutrient application is inherently connected to the correct rate and place. A case in point: in drip-irrigated tomato cultivation, nutrient application rates are tailored to crop maturity with lower rates of nutrients overlaid in the stages of growth (Wu *et al.*, 2021). Whenever the vegetative growth is maximum and the fruits are developing, higher rates of nutrients are usually applied. Although predicting the timing of rainfall can be challenging, every effort in this regard should be made to time fertilizer application accordingly, to minimize nutrient leaching from excessive future rainfall (Mahmud *et al.*, 2021).

17.5.3 Right Placement

For optimal nutrient efficiency, it is essential to place nutrients where the plant can easily access them. During fertilization, nutrients are always placed in the rhizoplane due to the fact that, in general, most of the nutrients flow upward through the root system (Marschner and Rengel, 2023). There are two common systems for nutrient placement: banding and broadcasting. Banding means applying nutrients tightly in bands in the soil, especially close to the plant in the plant-growing region (Hopkins *et al.*, 2020). In contrast, broadcasting means uniformly applying fertilizer to soil surface (Sutradhar *et al.*, 2017). The decision is predominantly based on the crop along with the root system which is developed and spread.

17.5.4 Right Source

The selection of fertilizer source and its application is an important consideration and is influenced by several factors such as convenience of application, costs per unit weight of nutrients, and nutrient-use efficiency (Sharma and Bali, 2017; Mustafa *et al.*, 2022). For example, controlled-release N supply can obtain more profitable applications, as it partitions available nutrients in packets throughout the growing season rather than a large single application from a soluble source at side-dressing (Kim Hocheol *et al.*, 2018). Manure can also be a beneficial fertilizer source if the farmer desires to utilize the organic matter with plant nutrients (Hassan *et al.*, 2020; Verma *et al.*, 2020). With this type of soil, this organic matter may improve the soil water holding capacity and nutrients substance.

17.6 AGRONOMIC PRACTICES TO MANAGE NUTRIENTS

17.6.1 Crop Rotation and Diversification

Crop rotation is one approach to resilient farm management that aims to increase soil organic matter while decreasing soil erosion (Franzluebbers and Gastal, 2019). Diversified crop rotation (DCR) is a well-known quality for enhancing global farming systems. The practices have great potential for enhancing system productivity and soil health. Increased water uptake and storage capacity as well as an increase of beneficial soil organisms are significant benefits due to improved soil characteristics that can increase crop resistance to drought and adverse growing environments over multiple crop cycles (Shah et al., 2021). DCR involves an ensemble of crops, reduces farmer increasing production risks and uncertainties along with the soil and environmental sustainability (Hanif *et al.*, 2017; Wang *et al.*, 2020). In addition, diverse crop rotations can provide farmers with alternative income (Wang *et al.*, 2021).

DCR are different from monocultures and double-farmed rotations because they contain three or more crop rotations (Shah et al., 2021). Well-structured crop rotation strategies can minimize trade-offs between crop productivity and environmental performance, promote long-term soil fertility, and break weed and pathogen cycles due to intrinsic nutrient cycling processes (Drinkwater *et al.*, 2017).

17.6.2 Intercropping

Intercropping systems improve crop quality and yield, as well as environmental quality, by maximizing the use of land, light, water, and soil nutrients (Glaze-Corcoran *et al.*, 2020). Although nutrient efficiency in cereal/legume intercropping systems has been studied, notably, when it comes to N, especially legume N_2 fixation, little focus has been paid to soil P or micronutrients like iron (Fe) and zinc (Zn). Nevertheless, few recent studies have been conducted on this topic; however, one Brazilian study found more soil P accumulated in cereals intercropped with legumes and more Fe and Zn accumulated in legume crops intercropped with cereals (Hossain *et al.*, 2022; Rasheed *et al.*, 2022).

17.6.3 Zero Tillage

The physical, chemical, and biological characteristics of soil are extensively affected by zero tillage (ZT) in contrast with conventional tillage (CT) (Arshad *et al.*, 2023). Agricultural conservation of zero till can improve soil quality through soil structure preservation and biological activity, nutrient cyclical mechanisms, and water absorption while enhancing the water use efficiency (Somasundaram *et al.*, 2020). Sowing wheat directly on the standing rice stubble through ZT leads to minimum soil disturbance, without an impact on seed germination (Mishra *et al.*, 2022). Under ZT conditions, the decomposition of organic substances is a primary source for nutrients that are readily available to the vegetative roots (Rahman *et al.*, 2020). Several studies show that experimental designs with ZT had much higher exchangeable K in the 0–5 cm depth than CT systems (van der Merwe, 2022).

Chaves *et al.* (2021) reported that changing tillage and residue integration increases the aeration of soil and increases the rate of residue decomposition, which impacts soil organic N. Nunes *et al.* (2020) noticed that soil-available P and K were unaffected during the first phase of tillage experiments for each crop sequence. ZT plots contained more organic carbon than the conventional tilled plots. Consequently, ZT enhances nutrient accessibility in the upper layers of the soil, which may be advantageous for both desired crops and weeds (Mondal *et al.*, 2020).

17.7 LIMITATIONS TO MANAGE SOIL FERTILITY

Several strategies are used to improve soil fertility, which includes the traditional methods of inorganic fertilizer application, as well as the use of organic fertilizing materials, such as plant residues (rice straw and husk), green manure, and animal manure (Bhunia *et al.*, 2021). Crop management systems include cover cropping, legumes, mulching, fallowing, agroforestry, and have been extensively studied (Liu *et al.*, 2021c). In addition, the adoption of high yielding and genetically advanced varieties of crops is one potential approach for enhancing yield productivity.

Plant waste and biomass in soils have a profound effect on soil nutrients, biological activity, crop potential, and physical structure (Mahmood *et al.*, 2015; Sharma *et al.*, 2017). Farmers may find organic materials such as green manure, rice straw, organic biomass, and husk less appealing than burning straw due to the limited effects of organic matter on crop productivity. Transporting and incorporating biomass into the field is also labor-intensive and expensive (Geetha Thanuja *et al.*, 2021).

Crop residues, on the other hand, have a high economic value and are commonly used for livestock feed; so they are rarely found in the field (Sarkar *et al.*, 2020; Mahmood *et al.*, 2021). Nonetheless, incorporating rice straw, which is widely available in rice fields, may effectively replenish and preserve maximum nutrients in soil, mainly N, K, P, S, and silicon over time.

Singh *et al.* (2018) have shown that incorporating rice husk into rice cultivation can better provide the required N, P, and K in comparison to chemical fertilizer application. Despite the potential benefits, crop management practices such as tree fallow systems are not widely adopted by farmers, who prefer alternative land use options, especially in areas with high population pressure, such as humid and tropical regions (Rahman *et al.*, 2017). Furthermore, the effectiveness of improved fallow systems in semi-arid tropical regions of Africa has yet to be proven (Partey *et al.*, 2017). Furthermore, fallow systems have limited potential for improving soil fertility in shallow and poorly drained soils.

Leguminous plants can be used as fallow crops before the cropping season or intercropped with other crops to store N for future crops (Diacono *et al.*, 2021). However, it is critical to recognize that plant residues' effects on soils and crops are influenced by factors such as their quality and decomposition which results in regulating nutrient release. Mupangwa *et al.* (2018) found that plant remains of low quality used to be mulch had the least impact on maize grain yield and protein concentration when compared with residues of high quality in soil. Animal wastes from pigs, cows, poultry, sheep, and goats are high in nutrients, making them valuable nutrients for maintaining the fertility of soil and agricultural productivity (Itelima *et al.*, 2018). However, collecting bulky dung from grazing livestock or gathering excreta (both dung and urine) is difficult, especially in rural areas where such manures are scarce and not readily available for sale (Adegbeye *et al.*, 2020). Poultry manure, on the other hand, may be more readily available in urban areas with high poultry production (Loyon, 2018). Poultry manure is currently in high demand, but the available supply may not meet farmers' needs. Due to this scarcity, farmers are frequently forced to pay in advance for manure delivery to their fields.

17.8 ADVANTAGES OF CONTEMPORARY NUTRIENT MANAGEMENT

1. Optimizing nutrient input-output efficiency
2. Reducing nutrient losses due to leaching, runoff, volatilization, emissions, and immobilization
3. Improving soil health
4. Promoting soil biodiversity
5. Enhancing crop growth and yield
6. Improve the resistance of plants to both biotic and abiotic stresses
7. Maximizing farm profitability by reducing total costs of production
8. Supporting sustainable agricultural practices
9. Minimizing environmental impacts

17.9 MODERN NUTRIENT MANAGEMENT STRATEGIES

17.9.1 Soil Testing and Plant Analysis

- Soil sampling and soil nutrient analysis,
- Accessing plant nutrient deficiency symptoms

17.9.2 Farm Waste Valorization and Commercialization

Nutrient-rich farm waste, such as crop residues, animal manure, and kitchen waste, would be recycled to produce value-added products to be used as organic fertilizers and amendments with standardized and reliable compositions to improve soil fertility.

Blending of organic and inorganic fertilizers to match crop needs can improve efficiency and reduce the risk of nutrient imbalances.

17.9.3 Sensor and Drone Technologies for Real-Time Monitoring

1. Various nutrient sensors, crop sensors, and weather sensors can be used to assess crop health and nutrient status in real time.
2. Drones equipped with multispectral or hyperspectral cameras can capture detailed images of fields, allowing farmers to assess soil health, detect nutrient deficiencies, and monitor crop health throughout the growing season.
3. Drones can be used for rapid crop scouting, allowing farmers to identify areas of nutrient stress, disease, or pest infestation.

17.9.4 Precision Agriculture Technologies

This includes the use of modern technologies to target fertilizer application to specific areas of a field.

It includes the following:

1. Soil mapping using GPS and GIS technologies.
2. Nutrient monitoring using various nutrient sensors, crop sensors, and weather sensors.
3. Various sensors can be used to assess crop health and nutrient status in real time.

4. *Variable Rate Application* (*VRA*): Through precision agriculture, farmers can apply fertilizers at variable rates across a field, based on soil nutrient levels and crop requirements. This ensures that nutrients are applied where they are most needed, optimizing their use and reducing environmental impact.
5. *Data Management and Analysis*: Precise agriculture generates huge volume of data, including soil test results, yield maps, sensor data, and others. Farmers can use advanced data management and analysis tools to assist their nutrient management decisions.
6. Use of crop models.
7. *Nutrient Budgeting*: To measure detailed nutrient inputs and outputs at farms that help estimating more reliable amounts and types of fertilizer to apply thus decreasing overuse and possible waste.

17.9.5 Remote Sensing Technologies

1. Remote sensing can be used to identify nutrient deficiencies.
2. Spectral remote sensing to estimate leaf chlorophyll content, vegetation health, and vigor.

17.9.6 Fluorescence Imaging for Accessing Photosynthesis

1. Thermal imaging to access canopy temperature
2. Drones, satellites, and other remote sensing technologies can provide valuable information about nutrient deficiencies and crop health
   - Various indices, such as the N reflectance index (NRI) or the chlorophyll index (CI) or normalized difference vegetation index (NDVI), are specifically designed and used to detect N or chlorophyll or other nutrient deficiencies.

By analyzing remote sensing data, we can identify areas within fields that may be experiencing nutrient deficiencies. This information can be used to implement targeted nutrient application strategies, such as variable rate fertilization and band placement, to address these deficiencies and improve crop health and productivity.

17.10 USE OF AI FOR NUTRIENT MANAGEMENT

AI has the potential to revolutionize nutrient management in agriculture by providing farmers with valuable insights and tools to optimize nutrient use, improve crop health, and protect the environment that are as follows:

1. AI algorithms can analyze soil and crop data, as well as weather and other environmental factors, to recommend accurate fertilizer amounts for specific areas of a field, preventing over-fertilization and nutrient waste.
2. AI can analyze real-time data from satellites, drones, and ground-based sensors to monitor crop health and nutrient status. This information can be used to detect nutrient deficiencies at an early stage and to initiate countermeasures.
3. AI-based decision support systems can integrate information from multiple sources, such as soil tests, satellite imagery, weather forecasts, and crop models, to provide farmers with real-time nutrient management recommendations.
4. AI algorithms can optimize irrigation scheduling based on soil moisture contents and crop nutrient needs, facilitating to minimize water and nutrient losses through leaching.
5. AI can be employed to develop crop models that simulate plant growth and nutrient uptake under different conditions. These models can help farmers make informed decisions about nutrient management strategies.

6. AI can analyze large datasets to identify patterns and trends related to nutrient management, helping to enhance overall nutrient-use efficiency.
7. AI-powered automation and robots can be used for precise fertilizer application based on real-time data, reducing labor and improving accuracy.

17.11 CONCLUSION

The sustainability challenges of modern agricultural practices have been well documented, and alternative nutrient management practices provide many potential solutions. In addition to these, several other approaches like precision agriculture, organic farming, bio-fertilizers, and crop rotation also help to improve nutrient uptake efficiency, mitigate environmental damage, and maintain soil health and productivity for the future. While these alternative solutions to fast-track sustainable agriculture are promising, their widespread adoption is stymied by technological barriers, knowledge gaps, and socioeconomic factors. Some of the key challenges that need to be resolved are insufficient information regarding the long-term effects of applying nutrients, lack of access to advanced technologies, limited incentives for farmers to adopt these practices as well as the requirement for education and support systems to promote behavior change. Also important is the recognition of these connections to the wider environmental and socioeconomic context in which agricultural systems reside. Effective nutrient management also has to consider water quality, biodiversity conservation, climate change resilience, and social equity. Thus, holistic as well as integrated approaches that can encompass these multifaceted realities are key to creating truly sustainable agri-food systems.

LIST OF ABBREVIATIONS

CI chlorophyll index
CT conventional tillage
DCR diversified crop rotation
INM integrated nutrient management
K potassium
N nitrogen
NDVI normalized difference vegetation index
P phosphorus
PGPR plant growth-promoting rhizobacteria
S sulfur
VRA variable rate application
ZT zero tillage

REFERENCES

Aasfar, A., Bargaz, A., Yaakoubi, K., Hilali, A., Bennis, I., Zeroual, Y. & Meftah Kadmiri, I., 2021. Nitrogen fixing *Azotobacter* species as potential soil biological enhancers for crop nutrition and yield stability. *Frontiers in Microbiology*, 12, 628379.

Abdelfattah, M.A., Rady, M.M., Belal, H.E., Belal, E.E., Al-Qthanin, R., Al-Yasi, H.M. & Ali, E.F., 2021. Revitalizing fertility of nutrient-deficient virgin sandy soil using leguminous biocompost boosts *Phaseolus vulgaris* performance. *Plants*, 10, 1637.

Abdullah, Mahmood, A., Bibi, S., Naqve, M., Javaid, M.M., Zia, M.A., Jabbar, A., Ud-Din, W., Attia, K.A. & Khan, N., 2022. Physiological, biochemical, and yield responses of linseed (*Linum usitatissimum* L.) in α-tocopherol-mediated alleviation of salinity stress. *Frontiers in Plant Science*, 13, 867172.

Adegbeye, M.J., Salem, A.Z.M., Reddy, P.R.K., Elghandour, M.M.M. & Oyebamiji, K.J., 2020. Waste Recycling for the Eco-Friendly Input Use Efficiency in Agriculture and Livestock Feeding. In M. M. M. Elghandour & A. Z. M. Salem (Eds.), *Resources Use Efficiency in Agriculture*, Springer, 1–45.

Aftab, T. & Hakeem, K.R., 2022. *Sustainable Plant Nutrition: Molecular Interventions and Advancements for Crop Improvement*. Academic Press.

Agegnehu, G., Nelson, P.N. & Bird, M., 2016. Crop yield, plant nutrient uptake and soil physicochemical properties under organic soil amendments and nitrogen fertilization on Nitisols. *Soil Tillage Research*, 160, 1–13.

Al-Maamori, H.A., Salman, A.D., Al-Budeiri, M., Al-Shami, Y.A.O. & Al-Shaabani, E.M., 2023. Effect of vermicompost production on some soil properties and nutrients in plants. *In IOP Conference Series: Earth and Environmental Science*. IOP Publishing, 012006.

Are, K.S., Adelana, A.O., Fademi, I.O. & Aina, O.A., 2017. Improving physical properties of degraded soil: Potential of poultry manure and biochar. *Agriculture Natural Resources*, 51, 454–462.

Arshad, T., Naqve, M., Mukhtiar, A., Javaid, M.M., Mahmood, A., Nadeem, M.A. & Khan, B.A., 2023. Conservation Tillage for Sustainable Agriculture. *In Climate-Resilient Agriculture, Vol 1: Crop Responses and Agroecological Perspectives.* Springer International Publishing, 313–327.

Azhar Mehmood, A.M., Muhammad Naeem, M.N., Farhan Khalid, F.K., Yousaf Saeed, Y.S., Tasawer Abbas, T.A., Jabran, K., Sarwar, M., Asif Tanveer, A.T. & Javaid, M., 2018. Identification of phytotoxins in different plant parts of *Brassica napus* and their influence on mung bean. *Environmental Science and Pollution Research*, 25, 18071–18080.

Babar, B.H., Tanveer, A., Tahir, M., Aziz, A., Ahmad, A.U.H., Nadeem, M.A. & Javaid, M.M., 2009. Allelopathic potential of wild onion (*Asphodelus tenuifolius*) on the germination and seedling growth of chickpea (*Cicer arietinum*). *Weed Biology and Management*, 9, 146–151.

Bareke, T., 2018. Biology of seed development and germination physiology. *Advances in Plants & Agriculture Research*, 8, 336–46.

Bellitürk, K. & Sundari, R.S., 2024. The Power of Earthworm: Vermicompost Drives to Sustainable Agriculture. In *Earthworm Technology in Organic Waste Management.* Elsevier, 307–321.

Bhunia, S., Bhowmik, A., Mallick, R. & Mukherjee, J., 2021. Agronomic efficiency of animal-derived organic fertilizers and their effects on biology and fertility of soil: A review. *Agronomy*, 11, 823.

Boruah, T., Barman, A., Kalita, P., Lahkar, J. & Deka, H., 2019. Vermicomposting of citronella bagasse and paper mill sludge mixture employing *Eisenia fetida. Bioresource Technology Reports*, 294, 122147.

Briat, J.-F., Gojon, A., Plassard, C., Rouached, H. & Lemaire, G., 2020. Reappraisal of the central role of soil nutrient availability in nutrient management in light of recent advances in plant nutrition at crop and molecular levels. *European Journal of Agronomy*, 116, 126069.

Carr, P.M., Cavigelli, M.A., Darby, H., Delate, K., Eberly, J.O., Fryer, H.K., Gramig, G.G., Heckman, J.R., Mallory, E.B. & Reeve, J.R.J.A.J., 2020. Green and animal manure use in organic field crop systems. *Agronomy Journal*, 112, 648–674.

Cen, Y., Guo, L., Liu, M., Gu, X., Li, C. & Jiang, G., 2020. Using organic fertilizers to increase crop yield, economic growth, and soil quality in a temperate farmland. *PeerJ*, 8, e9668.

Chadha, A., Florentine, S., Chauhan, B.S., Long, B., Jayasundera, M., Javaid, M.M. & Turville, C., 2019. Environmental factors affecting the germination and seedling emergence of two populations of an emerging agricultural weed: Wild lettuce (*Lactuca serriola*). *Crop and Pasture Science*, 70, 709–717.

Chadha, A., Florentine, S., Javaid, M., Welgama, A. & Turville, C., 2020. Influence of elements of climate change on the growth and fecundity of *Datura stramonium. Environmental Science and Pollution Research*, 27, 35859–35869.

Chaves, B., Redin, M., Giacomini, S.J., Schmatz, R., Léonard, J., Ferchaud, F. & Recous, S., 2021. The combination of residue quality, residue placement and soil mineral N content drives C and N dynamics by modifying N availability to microbial decomposers. *Soil Biology and Biochemistry*, 163, 108434.

Chennappa, G., Naik, M., Amaresh, Y., Nagaraja, H. & Sreenivasa, M., 2017. *Azotobacter*: A potential biofertilizer and bioinoculants for sustainable agriculture. *Microorganisms for Green Revolution*, 1, 87–106.

Chew, K.W., Chia, S.R., Yen, H.-W., Nomanbhay, S., Ho, Y.-C. & Show, P.L., 2019. Transformation of biomass waste into sustainable organic fertilizers. *Sustainability*, 11, 2266.

Dadrasnia, A., De Bona Muñoz, I., Yáñez, E.H., Lamkaddam, I.U., Mora, M., Ponsá, S., Ahmed, M., Argelaguet, L.L., Williams, P.M. & Oatley-Radcliffe, D.L., 2021. Sustainable nutrient recovery from animal manure: A review of current best practice technology and the potential for freeze concentration. *Journal of Cleaner Production*, 315, 128106.

Dapkekar, A., Deshpande, P., Oak, M.D., Paknikar, K.M. & Rajwade, J.M., 2018. Zinc use efficiency is enhanced in wheat through nanofertilization. *Scientific Reports*, 8, 6832.

De Corato, U., 2020. Agricultural waste recycling in horticultural intensive farming systems by on-farm composting and compost-based tea application improves soil quality and plant health: A review under the perspective of a circular economy. *Science of the Total Environment*, 738, 139840.

Delibacak, S., Voronina, L. & Morachevskaya, E., 2020. Use of sewage sludge in agricultural soils: Useful or harmful. *Eurasian Journal of Soil Science*, 9, 126–139.

de Souza Mendes, A.K., Vilhena, M.D.P.S.P., Silva, M.V.O., Berrêdo, J.F., Da Costa, M.L. & De Sousa Trindade, M.J., 2023. Solid bio-compost as a nutrient source for family farming. *Journal of Agriculture and Food Research*, 12, 100575.

Devi, L., Sheikh, S., Kumari, N. & Kumar, R., 2022. Role of biofertilizers in organic agriculture. *A Monthly Peer Reviewed Magazine for Agriculture and Allied Sciences*, 42.

Diacono, M., Persiani, A., Castellini, M., Giglio, L. & Montemurro, F., 2021. Intercropping and rotation with leguminous plants in organic vegetables: Crop performance, soil properties and sustainability assessment. *Biological Agriculture and Horticulture*, 37, 141–167.

Drinkwater, L.E., Schipanski, M., Snapp, S. & Jackson, L.E., 2017. Ecologically Based Nutrient Management. In *Agricultural Systems*. Elsevier, 203–257.

Du, Y., Cui, B., Wang, Z., Sun, J. & Niu, W., 2020. Effects of manure fertilizer on crop yield and soil properties in China: A meta-analysis. *Catena*, 193, 104617.

Ejigu, W., Selassie, Y.G., Elias, E. & Damte, M., 2021. Integrated fertilizer application improves soil properties and maize (*Zea mays* L.) yield on Nitisols in Northwestern Ethiopia. *Heliyon*, 7.

Etesami, H. & Adl, S.M., 2020. Plant Growth-Promoting Rhizobacteria (PGPR) and Their Action Mechanisms in Availability of Nutrients to Plants. In A. Rakshit, H. B. Singh, & A. Sen (Eds.), *Phyto-Microbiome in Stress Regulation*, Springer, 147–203.

Faloye, O.T., Ajayi, A.E., Babalola, T., Omotehinse, A.O., Adeyeri, O.E., Adabembe, B.A., Ogunrinde, A.T., Okunola, A. & Fashina, A., 2023. Modelling crop evapotranspiration and water use efficiency of maize using artificial neural network and linear regression models in biochar and inorganic fertilizer-amended soil under varying water applications. *Water*, 15, 2294.

Farghali, M., Mohamed, I.M., Osman, A.I. & Rooney, D.W., 2023. Seaweed for climate mitigation, wastewater treatment, bioenergy, bioplastic, biochar, food, pharmaceuticals, and cosmetics: A review. *Environmental Chemistry Letters*, 21, 97–152.

Farman, S., Mushtaq, A. & Azeem, M.W., 2019. Plant growth regulators (PGRs) and their applications: A review. *International Journal of Chemical and Biochemical Science*, 15, 94–103.

Farzadfar, S., 2022. *Role of Cover Crops in Regulating Nitrogen Dynamics, Productivity and Phytochemical Quality in Vegetable Production Systems*. Doctoral dissertation, University of Saskatchewan, Saskatoon, Canada.

Fatima, I., Ahmad, M., Vithanage, M. & Iqbal, S., 2021. Abstraction of nitrates and phosphates from water by sawdust-and rice husk-derived biochars: Their potential as N-and P-loaded fertilizer for plant productivity in nutrient deficient soil. *Journal of Analytical and Applied Pyrolysis*, 155, 105073.

Franzluebbers, A.J. & Gastal, F., 2019. Building Agricultural Resilience with Conservation Pasture-Crop Rotations. In *Agroecosystem Diversity*. Elsevier, 109–121.

Gareau, S.E., 2012. Economic Analysis of Conventional and Alternative Nutrient Management Approaches. In A. Srivastava (Ed.), *Advances in Citrus Nutrition*, Springer, Dordrecht, 415–424.

Geetha Thanuja, K., Marimuthu, S., Ramesh, D. & Karthikeyan, S., 2021. Paddy Straw-Based Circular Economy for Sustainable Waste Management. In *Handbook of Solid Waste Management: Sustainability through Circular Economy*. Springer, 1–28.

Geng, Y., Cao, G., Wang, L. & Wang, S., 2019. Effects of equal chemical fertilizer substitutions with organic manure on yield, dry matter, and nitrogen uptake of spring maize and soil nitrogen distribution. *PLoS One*, 14, e0219512.

Glaze-Corcoran, S., Hashemi, M., Sadeghpour, A., Jahanzad, E., Afshar, R.K., Liu, X. & Herbert, S.J., 2020. Understanding intercropping to improve agricultural resiliency and environmental sustainability. *Advances in Agronomy*, 162, 199–256.

Goldan, E., Nedeff, V., Barsan, N., Culea, M., Panainte-Lehadus, M., Mosnegutu, E., Tomozei, C., Chitimus, D. & Irimia, O., 2023. Assessment of manure compost used as soil amendment—A review. *Processes*, 11, 1167.

Goswami, L., Kushwaha, A., Singh, A., Saha, P., Choi, Y., Maharana, M., Patil, S.V. & Kim, B.S., 2022. Nano-biochar as a sustainable catalyst for anaerobic digestion: A synergetic closed-loop approach. *Catalysts*, 12, 186.

Goyal, V., Rani, D., Ritika, Mehrotra, S., Deng, C. & Wang, Y., 2023. Unlocking the potential of nano-enabled precision agriculture for efficient and sustainable farming. *Plants*, 12, 3744.

Gull, T., Mahmood, A., Shaheen, C., Javaid, M.M., Zia, M.A., Naqve, M., Bibi, S., Nadeem, M.A., Ameen, M. & Nargis, J., 2023. Climate Change and Nutrient Use Efficiency of Plants. In *Climate-Resilient Agriculture, Vol 1: Crop Responses and Agroecological Perspectives*. Springer, 291–312.

Hanif, Z., Naeem, M., Ali, H.H., Tanveer, A., Javaid, M.M., Peerzada, A.M. & Chauhan, B.S., 2017. Effect of environmental factors on germination of *Salsola foetida*: Potential species for rehabilitation of degraded rangelands. *Rangeland Ecology & Management*, 70, 638–643.

Hassan, M.U., Chattha, M.U., Barbanti, L., Mahmood, A., Chattha, M.B., Khan, I., Mirza, S., Aziz, S.A., Nawaz, M. & Aamer, M., 2020. Cultivar and seeding time role in sorghum to optimize biomass and methane yield under warm dry climate. *Industrial Crops and Products*, 145, 111983.

Hassan, M.U., Nawaz, M., Mahmood, A., Shah, A.A., Shah, A.N., Muhammad, F., Batool, M., Rasheed, A., Jaremko, M. & Abdelsalam, N.R., 2022. The role of zinc to mitigate heavy metals toxicity in crops. *Frontiers in Environmental Science*, 10, 990223.

Hidalgo, D., Corona, F. & Martín-Marroquín, J., 2021. Nutrient recycling: From waste to crop. *Biomass Conversion and Biorefinery*, 11, 207–217.

Ho, T.T.K., Le, T.H., Tran, C.-S., Nguyen, P.-T., Thai, V.-N. & Bui, X.-T., 2022. Compost to improve sustainable soil cultivation and crop productivity. *Case Studies in Chemical and Environmental Engineering*, 6, 100211.

Hoffland, E., Kuyper, T.W., Comans, R.N., Creamer, R.E. & Soil, 2020. Eco-functionality of organic matter in soils. *Plant*, 455, 1–22.

Hopkins, B.G., Stark, J.C. & Kelling, K.A., 2020. Nutrient Management. In J. Stark, M. Thornton, & P. Nolte (Eds), *Potato Production Systems*, Springer, Cham, 155–202.

Hossain, A., Maitra, S., Ahmed, S., Mitra, B., Ahmad, Z., Garai, S., Mondal, M., Adeel, M., Shankar, T. & Meena, R.S., 2022. Legumes for Nutrient Management in the Cropping System. In *Advances in Legumes for Sustainable Intensification*. Elsevier, 93–112.

Hřebečková, T., Wiesnerová, L. & Hanč, A., 2019. Changes of enzymatic activity during a large-scale vermicomposting process with continuous feeding. *Journal of Cleaner Production*, 239, 118127.

Hsu, C.-M. & Lai, H.-Y., 2022. Comprehensive assessment of the influence of applying two kinds of chicken-manure-processed organic fertilizers on soil properties, mineralization of nitrogen, and yields of three crops. *Agronomy*, 12, 2355.

Huang, X.-F., Chaparro, J.M., Reardon, K.F., Zhang, R., Shen, Q. & Vivanco, J.M., 2014. Rhizosphere interactions: Root exudates, microbes, and microbial communities. *Botany*, 92, 267–275.

Itelima, J., Bang, W., Onyimba, I., Sila, M. & Egbere, O., 2018. A review: biofertilizer; a key player in enhancing soil fertility and crop productivity. *Journal of Microbiology and Biotechnology Reports*, 2, 22–28.

Jain, S., Jain, J. & Singh, J., 2020. The Rhizosphere Microbiome: Microbial Communities and Plant Health. In A. Varma, S. Tripathi, & R. Prasad (Eds.), *Plant Microbiome Paradigm*, Springer, Cham, 175–190.

Kayesh, E., Gomasta, J., Bilkish, N., Koly, K.A. & Mallick, S.R., 2023. A Holistic Approach of Organic Farming in Improving the Productivity and Quality of Horticultural Crops. *In Organic Fertilizers-New Advances and Applications*. IntechOpen.

Kim Hocheol, K.H., Bae Kangsoon, B.K., Yun Seokkyu, Y.S. & Kim Taechoon, K.T., 2018. Shoot growth and fruit characteristics of 'Yumi' peach in response to side dressing with fused phosphate fertilizer. *Horticultural Science and Technology*, 36, 630–639.

Kokulan, V., Wadu, M.C.M., Akinremi, O.O. & Buckley, K., 2021. Nitrogen and phosphorus distribution in plant, soil, and leachate as affected by liquid hog manure and chemical fertilizers. *Canadian Journal of Soil Science*, 101, 402–414.

Krishnaveni, A., Chinnasamy, S., Elumalai, J. & Muthaiyan, P., 2020. Sugar industry wastes as wealth of organic carbon for soil. *Environmental Factors Affecting Human Health*, 137.

Kumar, R., Sharma, P., Gupta, R., Kumar, S., Sharma, M.M.M., Singh, S. & Pradhan, G., 2020. Earthworms for Eco-Friendly Resource Efficient Agriculture. In S. Kumar, R. S. Meena, & M. K. Jhariya (Eds.), *Resources Use Efficiency in Agriculture*, Springer, Singapore, 47–84.

Kurrey, D.K., Sharma, R., Lahre, M.K. & Kurrey, R.L., 2018. Effect of *Azotobacter* on physio-chemical characteristics of soil in onion field. *The Pharma Innovation Journal*, 7, 108–13.

Lazcano, C., Zhu-Barker, X. & Decock, C., 2021. Effects of organic fertilizers on the soil microorganisms responsible for N_2O emissions: A review. *Microorganisms*, 9, 983.

Le Bayon, R.C., Bullinger, G., Schomburg, A., Turberg, P., Brunner, P., Schlaepfer, R. & Guenat, C., 2021. Earthworms, Plants, and Soils. In A. Hunt, M. Egli, & B. Faybishenko (Eds.), *Hydrogeology, Chemical Weathering, and Soil Formation*, Wiley, 81–103.

Li, Y., Li, Q., Guan, G. & Chen, S., 2020. Phosphate solubilizing bacteria stimulate wheat rhizosphere and endosphere biological nitrogen fixation by improving phosphorus content. *PeerJ*, 8, e9062.

Liu, J., Shu, A., Song, W., Shi, W., Li, M., Zhang, W., Li, Z., Liu, G., Yuan, F. & Zhang, S., 2021a. Long-term organic fertilizer substitution increases rice yield by improving soil properties and regulating soil bacteria. *Geoderma*, 404, 115287.

Liu, L., Zheng, X., Wei, X., Kai, Z. & Xu, Y., 2021b. Excessive application of chemical fertilizer and organophosphorus pesticides induced total phosphorus loss from planting causing surface water eutrophication. *Scientific Reports*, 11, 23015.

Liu, R., Thomas, B.W., Shi, X., Zhang, X., Wang, Z. & Zhang, Y., 2021c. Effects of ground cover management on improving water and soil conservation in tree crop systems: A meta-analysis. *Catena*, 199, 105085.

Løes, A.-K., 2023. Fertilisers from aquaculture and other marine waste. *ESPP Scope Newsletter*, 149, 49–51.

Loyon, L., 2018. Overview of animal manure management for beef, pig, and poultry farms in France. *Frontiers in Sustainable Food Systems*, 2, 36.

Mahmood, A., Hussain, A., Shahzad, A.N. & Honermeier, B., 2015. Biomass and biogas yielding potential of sorghum as affected by planting density, sowing time and cultivar. *Pakistan Journal of Botany*, 47, 2401–2408.

Mahmood, A., Ullah, H., Shahzad, A., Ali, H., Ahmad, S., Zia-Ul-Haq, M., Honermeier, B. & Hasanuzzaman, M., 2013. Dry matter yield and chemical composition of sorghum cultivars with varying planting density and sowing date. *Sains Malaysiana*, 42, 1529–1538.

Mahmood, A., Wang, X., Shahzad, A.N., Fiaz, S., Ali, H., Naqve, M., Javaid, M.M., Mumtaz, S., Naseer, M. & Dong, R., 2021. Perspectives on bioenergy feedstock development in Pakistan: Challenges and opportunities. *Sustainability*, 13, 8438.

Mahmud, K., Panday, D., Mergoum, A. & Missaoui, A., 2021. Nitrogen losses and potential mitigation strategies for a sustainable agroecosystem. *Sustainability*, 13, 2400.

Markam, S., 2021. Vermicompost, its importance and benefit in agriculture. *The Pharma Innovation Journal*, 10, 3163–3167.

Marschner, P. & Rengel, Z., 2023. Nutrient Availability in Soils. In *Marschner's Mineral Nutrition of Plants*. Elsevier, 499–522.

Me Trenkel, T., 2021. *Slow- and Controlled-Release and Stabilized Fertilizers: An Option for Enhancing Nutrient Use Efficiency in Agriculture*. International Fertilizer Industry Association (IFA).

Mishra, J., Kumar, R., Mondal, S., Poonia, S., Rao, K., Dubey, R., Raman, R.K., Dwivedi, S., Kumar, R. & Saurabh, K., 2022. Tillage and crop establishment effects on weeds and productivity of a rice-wheat-mungbean rotation. *Field Crops Research*, 284, 108577.

Misnikov, O., 2020. The use of peat deposits of the Russian federation for organic fertilizers production. *In E3S Web of Conferences*. EDP Sciences, 01015.

Mondal, S., Chakraborty, D., Bandyopadhyay, K., Aggarwal, P. & Rana, D.S., 2020. A global analysis of the impact of zero-tillage on soil physical condition, organic carbon content, and plant root response. *Land Degradation and Development*, 31, 557–567.

Mukhtiar, A., Lihong, W., Mahmood, A., Ameen, M., Zia, M.A., Arshad, T., Naqve, M., Wahab, H.A., Rasheed, A. & Asghar, S., 2023a. The role of endophytes and rhizobacteria to combat drought stress in wheat. *Notulae Botanicae Horti Agrobotanici Cluj-Napoca*, 51, 13453–13453.

Mukhtiar, A., Mahmood, A., Zia, M.A., Ameen, M., Dong, R., Shoujun, Y., Javaid, M.M., Khan, B.A. & Nadeem, M.A., 2023b. Role of biogas slurry to reclaim soil properties providing an eco-friendly approach for crop productivity. *Bioresource Technology Reports*, 101716.

Mupangwa, W., Nyagumbo, I., Mutsamba, E., Mhlanga, B. & Mataruse, J., 2018. Maize responses to reduced tillage, different plant residue mulch and nitrogen fertiliser on granitic sandy soils of Zimbabwe. *South African Journal of Plant Soil*, 35, 367–376.

Mustafa, A., Athar, F., Khan, I., Chattha, M.U., Nawaz, M., Shah, A.N., Mahmood, A., Batool, M., Aslam, M.T. & Jaremko, M., 2022. Improving crop productivity and nitrogen use efficiency using sulfur and zinc-coated urea: A review. *Frontiers in Plant Science*, 13, 942384.

Naqve, M., Mukhtiar, A., Arshad, T., Zia, M.A., Mahmood, A., Javaid, M.M. & Aziz, A., 2023. Biostimulants in Sustainable Agriculture. In *Climate-Resilient Agriculture, Vol 1: Crop Responses and Agroecological Perspectives*. Springer, 535–548.

Nongthombam, J., Kumar, A., Sharma, S. & Ahmed, S., 2021. *Azotobacter*: A complete review. *Bulletin of Environment, Pharmacology and Life Sciences*, 10, 72–79.

Nunes, R.D.S., De Sousa, D.M.G., Goedert, W.J., De Oliveira, L.E.Z., Pavinato, P.S. & Pinheiro, T.D., 2020. Distribution of soil phosphorus fractions as a function of long-term soil tillage and phosphate fertilization management. *Frontiers in Earth Science*, 8, 350.

Ogburn, N.J., Duan, L., Subashchandrabose, S.R., Sorgeloos, P., O'connor, W., Megharaj, M. & Naidu, R., 2023. Agricultural wastes for brine shrimp *Artemia* production: A review. *Reviews in Aquaculture*, 15, 1159–1178.

Pahalvi, H.N., Rafiya, L., Rashid, S., Nisar, B. & Kamili, A.N., 2021. Chemical fertilizers and their impact on soil health. *Microbiota and Biofertilizers, Ecofriendly Tools for Reclamation of Degraded Soil Environs*, 2, 1–20.

Parihar, S., Saini, K., Lakhani, G., Jain, A., Roy, B., Ghosh, S., Aharwal, B. & Studies, Z., 2019. Livestock waste management: A review. *Journal of Entomology*, 7, 384–393.

Partey, S.T., Zougmoré, R.B., Ouédraogo, M. & Thevathasan, N.V., 2017. Why promote improved fallows as a climate-smart agroforestry technology in sub-Saharan Africa? *Sustainability*, 9, 1887.

Pei, Y., Chen, X., Niu, Z., Su, X., Wang, Y. & Wang, X., 2023. Effects of nitrogen fertilizer substitution by cow manure on yield, net GHG emissions, carbon and nitrogen footprints in sweet maize farmland in the Pearl River Delta in China. *Journal of Cleaner Production*, 399, 136676.

Raghunandan, B., Vyas, R., Patel, H. & Jhala, Y., 2019. Perspectives of Seaweed as Organic Fertilizer in Agriculture. In Panpatte, D. & Jhala, Y. (eds), *Soil Fertility Management for Sustainable Development*, Springer, Singapore, 267–289.

Rahman, M.M., Alam, M.S., Kamal, M.Z.U. & Rahman, G.M., 2020. Organic Sources and Tillage Practices for Soil Management. In S. Kumar, R. S. Meena, & M. K. Jhariya (Eds.), *Resources Use Efficiency in Agriculture*, Springer, Singapore, 283–328.

Rahman, S.A., Sunderland, T., Roshetko, J.M. & Healey, J.R., 2017. Facilitating smallholder tree farming in fragmented tropical landscapes: Challenges and potentials for sustainable land management. *Journal of Environmental Management*, 198, 110–121.

Ramírez-Cariño, H.F., Morales, I., Guadarrama-Mendoza, P.C., Tlilayatzi-Bello, A.A. & Valadez-Blanco, R., 2024. Beneficial Traits of Crop-Associated Rhizobacteria Toward Plant Growth Promotion, Biofertilization, and Nutrient Enhancement of Harvested Agro-Products. In *Bacterial Secondary Metabolites*. Elsevier, 299–334.

Rasheed, A., Barqawi, A.A., Mahmood, A., Nawaz, M., Shah, A.N., Bay, D.H., Alahdal, M.A., Hassan, M.U. & Qari, S.H., 2022. CRISPR/Cas9 is a powerful tool for precise genome editing of legume crops: A review. *Molecular Biology Reports*, 49, 5595–5609.

Rayne, N. & Aula, L., 2020. Livestock manure and the impacts on soil health: A review. *Soil Systems*, 4, 64.

Rehman, A., Ullah, A., Nadeem, F. & Farooq, M., 2019. Sustainable Nutrient Management. In M. Farooq & M. Pisante (Eds.), *Innovations in Sustainable Agriculture*, Springer, Cham, 167–211.

Robertson, G.P. & Groffman, P., 2024. Nitrogen Transformations. In *Soil Microbiology, Ecology and Biochemistry*. Elsevier, 407–438.

Rosier, A., Medeiros, F.H., Bais, H.P. & Soil, 2018. Defining plant growth promoting rhizobacteria molecular and biochemical networks in beneficial plant-microbe interactions. *Plant*, 428, 35–55.

Salimi, F. & Hamedi, J., 2021. Biofertilizers: Microbes for Agricultural Productivity. In A. N. Yadav (Ed.), *Soil Microbiomes for Sustainable Agriculture: Functional Annotation*, Springer, Cham, 407–469.

Sarkar, S., Skalicky, M., Hossain, A., Brestic, M., Saha, S., Garai, S., Ray, K. & Brahmachari, K., 2020. Management of crop residues for improving input use efficiency and agricultural sustainability. *Sustainability*, 12, 9808.

Scavo, A., Fontanazza, S., Restuccia, A., Pesce, G.R., Abbate, C. & Mauromicale, G., 2022. The role of cover crops in improving soil fertility and plant nutritional status in temperate climates. A review. *Agronomy for Sustainable Development*, 42, 93.

Selim, M.M., 2020. Introduction to the integrated nutrient management strategies and their contribution to yield and soil properties. *International Journal of Agronomy*, 2020.

Shah, K.K., Modi, B., Pandey, H.P., Subedi, A., Aryal, G., Pandey, M., & Shrestha, J. 2021. Diversified crop rotation: An approach for sustainable agriculture production. *Advances in Agriculture*, 2021(1), 8924087.

Shah, F. & Wu, W., 2019. Soil and crop management strategies to ensure higher crop productivity within sustainable environments. *Sustainability*, 11, 1485.

Sharma, B., Sarkar, A., Singh, P. & Singh, R.P., 2017. Agricultural utilization of biosolids: A review on potential effects on soil and plant grown. *Waste Management*, 64, 117–132.

Sharma, L.K. & Bali, S.K., 2017. A review of methods to improve nitrogen use efficiency in agriculture. *Sustainability*, 10, 51.

Simonne, E.H., Gazula, A., Ozores-Hampton, M., Devalerio, J. & Hochmuth, R.C., 2017. Localized Application of Fertilizers in Vegetable Crop Production. In F. Tei, S. Nicola, & P. Benincasa (Eds.), *Advances in Research on Fertilization Management of Vegetable Crops*, Springer, Cham, 149–181.

Singh, C., Tiwari, S., Gupta, V.K. & Singh, J.S., 2018. The effect of rice husk biochar on soil nutrient status, microbial biomass and paddy productivity of nutrient poor agriculture soils. *Catena*, 171, 485–493.

Singh, S. & Sinha, R.K., 2022. Vermicomposting of Organic Wastes by Earthworms: Making Wealth from Waste by Converting 'Garbage into Gold' for Farmers. In *Advanced Organic Waste Management*. Elsevier, 93–120.

Slama, M.I., Omar, A.E., Bassiony, S.S. & Ghoneem, 2019. Effect of compost and natural rocks as partial substitutes for NPK mineral fertilizers on yield and fruit quality of 'Flame' seedless grapevine grown in two different locations of Egypt. *Spanish Journal of Agricultural Research*, 17, e0903.

Smaje, C., 2020. *A Small Farm Future: Making the Case for a Society Built Around Local Economies, Self-Provisioning, Agricultural Diversity and a Shared Earth.* Chelsea Green Publishing.

Somasundaram, J., Sinha, N., Dalal, R.C., Lal, R., Mohanty, M., Naorem, A., Hati, K., Chaudhary, R., Biswas, A. & Patra, A., 2020. No-till farming and conservation agriculture in South Asia – Issues, challenges, prospects and benefits. *Critical Reviews in Plant Sciences*, 39, 236–279.

Soytong, K., Song, J. & Tongon, R., 2021. Agricultural inputs for organic agriculture. In *International Seminar on Promoting Local Resources for Sustainable Agriculture and Development (ISPLRSAD 2020)*. Atlantis Press, 523–532.

Sumbul, A., Ansari, R.A., Rizvi, R. & Mahmood, I., 2020. *Azotobacter*: A potential bio-fertilizer for soil and plant health management. *Saudi Journal of Biological Sciences*, 27, 3634–3640.

Sutradhar, A.K., Kaiser, D.E. & Behnken, L.M., 2017. Soybean response to broadcast application of boron, chlorine, manganese, and zinc. *Agronomy Journal*, 109, 1048–1059.

Tanveer, A., Arshad, M.S., Ayub, M., Javaid, M.M. & Yaseen, M., 2012. Effect of temperature, light, salinity, drought stress and seeding depth on germination of *Cucumis melo* var. agrestis. *Pakistan Journal of Weed Science Research*, 18.

Thies, J.E. & Grossman, J.M., 2023. The Soil Habitat and Soil Ecology. In N.T. Uphoff & J.E. Thies (Eds.), *Biological Approaches to Regenerative Soil Systems*, CRC Press, 69–84.

Ullah, H., Mahmood, A., Ijaz, M., Tadesse, B. & Honermeier, B., 2013. Evaluation of anise (*Pimpinella anisum* L.) accessions with regard to morphological characteristics, fruit yield, oil contents and composition. *Journal of Medicinal Plants Research*, 7, 2177–2186.

van der Merwe, A., 2022. *The Effects of Long-Term Tillage and Crop Rotation Practices on Nutrient Stratification in the Western Cape*. Stellenbosch University.

Vanmare, D.J.A.S.R.R., 2021. A review on the significance of organic fertilizers. *Agricultural Science Research & Review*, 1, 140–145.

Verma, B., Pramanik, P. & Bhaduri, D., 2020. Organic Fertilizers for Sustainable Soil and Environmental Management. In R. Meena (Ed.), *Nutrient Dynamics for Sustainable Crop Production*, Springer, Singapore, 289–313.

Waheed, H., Javaid, M.M., Shahid, A., Ali, H.H., Nargis, J. & Mehmood, A., 2019. Impact of foliar-applied Hoagland's nutrient solution on growth and yield of mash bean (*Vigna mungo* L.) under different growth stages. *Journal of Plant Nutrition*, 42, 1133–1141.

Walia, S.S. & Kaur, T., 2024. Beneficial Role of Vermicompost: Nutrient Content in Vermicompost and Success Stories. In *Earthworms and Vermicomposting: Species, Procedures and Crop Application*. Springer, 135–146.

Wang, H., Xu, J., Liu, X., Zhang, D., Li, L., Li, W. & Sheng, L., 2019. Effects of long-term application of organic fertilizer on improving organic matter content and retarding acidity in red soil from China. *Soil Tillage Research*, 195, 104382.

Wang, L., Zhao, Y., Al-Kaisi, M., Yang, J., Chen, Y. & Sui, P., 2020. Effects of seven diversified crop rotations on selected soil health indicators and wheat productivity. *Agronomy*, 10, 235.

Wang, T., Jin, H., Fan, Y., Obembe, O. & Li, D., 2021. Farmers' adoption and perceived benefits of diversified crop rotations in the margins of US Corn Belt. *Journal of Environmental Management*, 293, 112903.

Waqar, A., Shah, G.M., Bakhat, H.F., Shahid, M., Aslam, M., Ashraf, M.R., Hafeez, R., Murtaza, B. & Rashid, M.I., 2019. The earthworm species *Pheretima hawayana* influences organic wastes decomposition, nitrogen mineralization and maize N recovery. *European Journal of Soil Biology*, 90, 1–8.

Wu, Y., Yan, S., Fan, J., Zhang, F., Xiang, Y., Zheng, J. & Guo, J., 2021. Responses of growth, fruit yield, quality and water productivity of greenhouse tomato to deficit drip irrigation. *Scientia Horticulturae*, 275, 109710.

Yahyah, S.M., Mahmud, A.A., Abdullahi, M. & Haruna, A., 2023. Recent advances in the chemistry of nitrogen, phosphorus and potassium as fertilizers in soil: A review. *Pedosphere*, 33, 385–406.

Zhang, X., Davidson, E., Zou, T., Lassaletta, L., Quan, Z., Li, T. & Zhang, W., 2020. Quantifying nutrient budgets for sustainable nutrient management. *Global Biogeochemical Cycles*, 34, e2018GB006060.

Zhao, Z., Mao, Y., Gao, S., Lu, C., Pan, C. & Li, X., 2023. Organic carbon accumulation and aggregate formation in soils under organic and inorganic fertilizer management practices in a rice–wheat cropping system. *Scientific Reports*, 13, 3665.

Zope, V., El Enshasy, H.A. & Sayyed, R., 2019. Plant Growth-Promoting Rhizobacteria: An overview in agricultural perspectives. In R. Sayyed (Ed.), *Plant Growth Promoting Rhizobacteria for Sustainable Stress Management. Microorganisms for Sustainability*, Springer, Singapore, 345–361.

Zuo, W., Jincheng, W., Shiqiang, S., Pinhua, R., Runkai, W. & Shihui, L., 2021. Microencapsulated soil conditioner with a water-soluble core: Improving soil nutrition of crop root. *Journal of Microencapsulation*, 38, 22–35.

18 Cultivating Resilience

Crop Adapting to Elevated CO_2 and Temperature

*Fouzia Shabbir, Singarayer Florentine, Muhammad Mansoor Javaid**, *Zeki Erden, Çağdaş Can Toprak, and Hakki Akdeniz*

18.1 INTRODUCTION

Climate change refers to a long-term change in temperature, precipitation, wind, and other atmospheric conditions worldwide and will have the most severe impacts in certain regions (Rawat *et al.*, 2024). These actions have mainly been the result of humans emitting greenhouse gases like carbon dioxide (CO_2), methane (CH_4), and nitrous oxide (N_2O) into the atmosphere. These activities include fossil fuel combustion, deforestation converted for industrial and agricultural purposes, growing emissions from transportation, and the impact of more growing internal establishments that cover arable land surfaces. These greenhouse gases act as a buffer layer in the Earth's atmosphere causing a gradual rise in temperature over the years which triggers the occurrences collectively referred to as "global warming" (Irfan *et al.*, 2024).

The effects of climate change on ecosystems have already been shown, with dire and complex repercussions (Ashraf *et al.*, 2013; Javaid *et al.*, 2022b; Jacquemart *et al.*, 2024), and according to the Intergovernmental Panel on Climate Change, many plant and animal species may be more susceptible to extinction as a result of environmental changes. Critical habitat loss or degradation can result from instability in a range of ecosystems which will be induced by relatively small fluctuations in temperature and precipitation (Siddha and Sahu, 2022). As an illustration, the magnitude of coastal ecosystems, such as saline marshes and mangroves, may diminish as a result of rising sea levels, and in addition, there will be adverse effects of increasing temperatures on the structure of forests and grasslands (Gull *et al.*, 2023; Temmerman *et al.*, 2023). The timing of significant biological occurrences, including migration, flowering, and hibernation, may also be altered due to climate change (Ponti and Sannolo, 2023). It has already been noted that an inordinate quantity of carbon dioxide is being absorbed by the oceans from the atmosphere, resulting in an increase in acidity. In this respect, impeding the ability of organisms to develop their shells and skeletons, this may be detrimental to marine ecosystems, especially coral reefs and mollusk populations, ultimately resulting in the extinction of marine life. Of key importance in this regard are ecosystems that are natural carbon sinks, such as peatlands and forests, that absorb more carbon dioxide from the atmosphere than they release, and therefore must be protected at all costs (Miller *et al.*, 2022; Aslam *et al.*, 2023).

Adaptation strategies have, indeed, already been proposed to address some of the impacts of climate change. These actions have included both mitigating and adaptive measures to bring industrial and agricultural practices in line with present or expected climatic changes and adapter effects (Gomez-Zavaglia *et al.*, 2020). In the agricultural area, farm-level adaptations have included

* Corresponding author: mmansoorjavaid@gmail.com

DOI: 10.1201/9781003498698-18

changing planting and harvesting dates, adjusting crop rotation, improving irrigation water management, adapting fertilizer regimes, and adopting different tilling practices (Grigorieva *et al.*, 2023).

This chapter aims to examine the responses of the environment and various plants to enhanced levels of CO_2 and (B) rising temperatures (Badiani *et al.*, 2023). It also has been developed to examine the potential consequences of a wide range of species and assess how physiological processes of plants have changed and the functioning of ecosystems altered, so as to observe implications for how practices may be adapted in order to mitigate the impact of climate change on land-based ecosystems.

18.2 UNDERSTANDING THE IMPACT OF ELEVATED CO_2 AND TEMPERATURE ON CROPS

18.2.1 The Effects of Temperature Increase on Crop Growth and Development

Elevated temperatures significantly affect many pathophysiological and metabolic mechanisms in plants, having consequences for respiration and photosynthesis (Javaid *et al.*, 2022c; Zandalinas *et al.*, 2022). These effects will differ in magnitude as well as type, depending on the species of the plants, the extent of increase in temperature, and the concentration of atmospheric CO_2, among other environmental factors (Jianing *et al.*, 2022; Nunes, 2023). The following sections provide a referenced summary of the ways in which plants can be affected by environmental changes. Response of different crops to elevated CO_2 and temperature is presented in Table 18.1.

18.2.2 Photosynthesis

It has been found that the impacts of increased temperatures on photosynthesis can be seen to be both advantageous and detrimental to plant species. For example, mild temperature rises can often enhance photosynthetic rates by boosting enzyme activity and facilitating CO_2 transport in plant tissues (Chadha *et al.*, 2019; Lobo *et al.*, 2022). This is commonly referred to as the "temperature response" of photosynthesis and is frequently noticed in colder climates when a little rise in temperature can significantly improve photosynthetic efficiency. However, in other situations, increasing temperatures might lead to the negation of any favorable photosynthesis impacts due to the harmful consequences of heat stress (Zahra *et al.*, 2023). Elevated temperatures can often cause those enzymes responsible for photosynthesis to undergo denaturation, which clearly will hinder their activity. Heat stress can cause stomata closure, reducing CO_2 absorption and hindering photosynthesis, and these adverse consequences may surpass any beneficial outcomes of the original temperature increase (Chaudhry and Sidhu, 2022).

18.2.3 Respiration

It is observed that vascular respiration rates may be impacted by increasing temperature, and, in this respect, elevated temperatures typically result in greater respiration rates (Sadok *et al.*, 2021). It has been determined that an increase in temperatures accelerates the metabolic process responsible for respiration, resulting in elevated energy use. As a consequence, if respiration rates grow more than photosynthetic rates, the plant may experience a net loss of energy. This issue is especially troublesome in scenarios when the carbon supply is restricted by conditions such as drought or nutrient deficits (Weller *et al.*, 2020; Lynch, 2022).

18.2.4 Protein Denaturation

Proteins and enzymes are essential for biological functions, but elevated temperatures can lead to their denaturation, which seriously disrupts their structure and activity (Tian *et al.*, 2022). There are weak forces, such as hydrogen bonds, which maintain the three-dimensional configuration of proteins, but high temperatures can disturb these interactions, causing denaturation. Denatured

TABLE 18.1
Understanding the Impact of Elevated CO_2 and Temperature Levels on Crops

Crop	Impact of Elevated CO_2	Impact of Elevated Temperature	Combined Impact	References
Wheat	An elevation in the rate of photosynthesis results in improved growth and the accumulation of biomass.	Decreased grain quality caused by hastened maturity and reduced grain-filling time.	Overall productivity is impacted differently according to the extent of CO_2 growth and temperature rise.	Rehman *et al.* (2020) Ullah *et al.* (2022)
Maize	Improved water utilization efficiency, potentially alleviating the impacts of drought stress.	Pollination inefficiency results in decreased kernel setting and grain yield.	Reduced crop productivity caused by the adverse effects of increased carbon dioxide levels and higher temperatures.	Li *et al.* (2023)
Rice	Increased biomass output, particularly in C_3 cultivars, leading to higher tiller and panicle development.	Higher vulnerability to pests and diseases like blast and sheath blight in warmer climates.	The grain yield is affected variably by cultivar-specific reactions and interactions with other environmental conditions.	Hosahatti *et al.* (2021)
Soybean	Enhanced growth and development, along with increased branching and pod formation, were noted in response to increasing CO_2 levels.	Reduced nitrogen fixation caused by elevated temperatures results in decreased nutrient availability and potential crop output.	Decreased crop productivity consistency due to the potential negative effects of increased carbon dioxide being counteracted by heat stress and nitrogen deficiencies.	Santachiara *et al.* (2019)
Cotton	Increased photosynthesis and carbon assimilation rates led to enhanced fiber output and quality.	The fiber quality is compromised, as evidenced by a reduction in both fiber length and strength when subjected to high temperatures.	Variable effects on fiber production as a whole, contingent upon genotype-specific reactions and management strategies.	Zha *et al.* (2023)
Barley	Enhanced drought tolerance and water use efficiency are the results of stomatal closure and decreased transpiration rates in response to elevated CO_2.	A decrease in the duration of grain loading and kernel weight resulted in a reduction in both grain yield and quality.	Yield and quality parameters exhibit variable responses, wherein environmental interactions and genotype exert a substantial influence.	Balla *et al.* (2019)
Potato	Improved tuber growth and productivity, influenced by higher photosynthetic activity and enhanced distribution of carbon to root storage organs.	Reduced tuber quality is indicated by elevated levels of reducing sugars and heightened vulnerability to physiological issues such as heat sprouting.	Diverse impacts on total tuber yield, affected by unique cultivar reactions and agricultural techniques.	Stark *et al.* (2020)
Tomato	Increased fruit production and yield are due to improved carbon absorption and higher flower output in response to greater levels of carbon dioxide.	Reduced fruit quality, including decreased firmness, color intensity, and shelf life, in high-temperature environments.	Variable influence on total fruit production and quality, affected by genotype-specific reactions and agricultural methods	Jiang *et al.* (2020)

proteins are physiologically inert and, moreover, have the potential to interfere with cellular functions such as signaling cascades, gene expression, and nutrient transport (Peters *et al.*, 2023).

18.2.5 Crop Yield and Quality

Projected rising air temperatures in the next decades are expected to lead to a decline in grain yields for major crops due to increased temperature stress (Becker *et al.*, 2023). It is known that, once temperatures reach a specific threshold, plant growth, pollination, and reproductive systems are negatively impacted (Sinha *et al.*, 2021; Rasheed *et al.*, 2022). This restriction means that when air temperatures go beyond the ideal level, crop output losses accelerate rather than just decrease proportionally to the temperature rise. Experience has shown that yield growth for maize, soybean, and cotton, increases steadily as temperatures rise to 29°C–32°C, but then dramatically decreases with further temperature increases beyond this range. It has been estimated that such temperature increases may lead to production reductions ranging from 2.5% to 10% in various agronomic species during the course of the 21st century (Kabir *et al.*, 2023).

18.3 EFFECT OF ELEVATED CO_2 ON PLANT PHYSIOLOGY

18.3.1 Photosynthesis and Carbon Assimilation

Increased levels of CO_2 often boost photosynthesis in C_3 plants, which constitute the majority of the world's crops (Welgama *et al.*, 2019; Walsh and Lundgren, 2024), with the higher amounts of CO_2 contributing to enhanced carbon fixation which is required for photosynthesis (Chadha *et al.*, 2020; Correa *et al.*, 2023). This is referred to as the "CO_2 fertilization effect", and the enhanced levels of photosynthesis result in significantly increased plant growth and biomass output, particularly when nutrients and water are plentiful.

18.3.2 Nutrient Uptake and Allocation

Increased CO_2 levels have been shown to affect how plants absorb and distribute nutrients (Elbasiouny *et al.*, 2022). It appears that enhanced carbon fixation might necessitate the need for more nutrients by the plant to facilitate its development. This situation will clearly be impacted by the availability of minerals in the soil (Saleem *et al.*, 2023), and therefore under conditions of restricted nutrition availability, increased CO_2 levels may not be able to significantly enhance plant growth.

18.3.3 Water Use Efficiency

Higher CO_2 levels frequently result in increased water use efficiency in plants (Florentine *et al.*, 2016; Zhao *et al.*, 2020). This efficiency arises from the greater CO_2 levels to permit a portion of the stomata to close, thereby decreasing water loss via transpiration. In other studies, it has been observed that herbivores, pathogens, and insects may exhibit altered interactions with plants as CO_2 levels rise (Javaid *et al.*, 2022a; Welti and Kaspari, 2024), and it appears that increased CO_2 levels may significantly modify the chemistry of plants, rendering them less palatable to particular herbivores (Wari *et al.*, 2022; Khan *et al.*, 2023).

18.4 CROP ADAPTATION STRATEGIES

18.4.1 Genetic Diversity and Breeding for Resilience

In order to deliberately develop genetic responses to climate change, it is necessary to effectively exploit and evaluate existing genetic variability possibilities (Clark *et al.*, 2023). In this area, the traditional use of molecular markers for genetic studies is largely ineffective for predicting adaptive variability due to the lack of connections between quantitative genetic variation and molecular genetic diversity (Rasheed *et al.*, 2021; Chhina *et al.*, 2024). Additionally, heterozygosity at these markers is not systematically correlated with inbreeding levels. Nevertheless, the determination of complete genomic sequences may provide valuable insights for addressing several pressing concerns in this regard, such as (i) inbreeding depression, (ii) the extent and structure of adaptive variation, (iii) the degree of local adaptation, and (iv) the nature of genotype-environment interactions. Phenotypic variation in response to environmental stimuli may result from inheritable genetic evolution or plasticity (which is not inheritable). In addition, the diversity of epigenetic and other transgenerational effects, in conjunction with their function in adaptation and maladaptation, must be comprehended (Crino *et al.*, 2024). It has been recognized that breeding strategies have increased output but, at the same time, reduced genetic variety. In this regard, in addition to the wild relatives that exist for crop species, examples of agrobiodiversity are also found in farmers' fields. These sites of bio-reserves represent a gene pool that is continuously developing naturally, and this phenomenon may demonstrate species' normal adaptations to climate change. At the more systematic level, gene banks worldwide, such as the Consultative Group on International Agricultural Research (CGIAR), store extensive collections of genetically varied plants, which include enhanced crop varieties, traditional landraces, and wild crop species (Chattha *et al.*, 2017; Ghamkhar, 2023). When working in this area, the nature of the gene pool should be thoroughly documented before being made accessible to researchers who are adjusting agricultural systems to upcoming climate change.

It has been suggested that germplasm agents can be used as the primary material for developing genetic diversity. Some methods for conserving and making use of biodiversity are noted in the following descriptions of practical actions:

1. Attending to regional simulations which demonstrate how biodiversity reacts to climate change (Harrison *et al.*, 2024).
2. Methodical selection of conservation sites, using climate change considerations as a fundamental part of the selection process (Northrup *et al.*, 2022).
3. Implementation of climate change management as an explicit management parameter in the management of biodiversity across regional landscapes, including center-protected areas and their environs (Ridlon *et al.*, 2023).
4. Developing systems to facilitate both local integration of management and those spanning international boundaries, together with recognition of boundaries between protected park areas and non-park conservation areas.
5. Allocation of resources from nations with the most abundant resources and the most significant contribution to climate change, to regions where the impacts of climate change and biodiversity loss are most severe (Abbass *et al.*, 2022).

18.5 SUGGESTED STRATEGIES FOR AGRONOMIC PRACTICE TO DEVELOP CLIMATE RESILIENCE

To effectively manage climate change and extreme events, agriculture must undergo adaptation to become climate-smart and climate-resilient (Tunio *et al.*, 2024). To achieve this situation, a number of consequent actions are necessary: (i) the adoption of sustainable production methods, which encompass integrated livestock and cropping systems, in conjunction with managed agroforestry; (ii) the utilization of bio-based and organic fertilizers, crop rotations, crop associations, no-till practices, conservation agriculture, carbon sequestration, and microbial processes to augment soil organic matter; and (iii) the mitigation of greenhouse gas emissions and reduction of nutrient losses from agriculture, together with enhancement of nutrient use efficiency through precision agriculture.

Rotating crops has been shown to be a long-term and successful method to enhance the resistance of crops to stress (Dormatey *et al.*, 2020; Wang *et al.*, 2024). It has been shown that implementing a three-year rotation of wheat and legumes can substantially decrease the negative impacts of drought stress, and this action leads to an increase in crop output. Furthermore, crop rotation improves soil quality by promoting the breakdown of various crop residues over time, leading to favorable effects on soil water movement (Yu *et al.*, 2022). In general, crop rotation which is carried out for species with varying water needs is a viable approach to reduce negative impacts in regions prone to drought. In addition, cultivating crops with low water requirements during periods of abundant water availability can help save soil moisture for the following season's crop.

It is also well established that both diversification of crops and intercropping, influence the physical properties and water retention of soil, and this can provide crop plants with protection against thermal stress (Kumawat *et al.*, 2022). Involving a range of plant species can significantly contribute to ecological advantages via niche overlap, complementarity in resource utilization, and/or modifications in soil structure and chemistry. Notwithstanding this practice though, negative consequences may result from plant interspecific interaction, this being contingent upon (i) the species implicated in the overlap, (ii) the accessibility of resources, and (iii) uncontrollable environmental circumstances. Additionally, crop diversification induces phenotypic plasticity, which allows plants to exploit a greater quantity of accessible resources, such as nutrients, by increasing their root biomass.

The implementation of diversified cultivation systems also enhances the biomass and activity of soil microbes, in conjunction with the microbial community residing in the vicinity of the plant roots (Cai *et al.*, 2024). This action results in positive circumstances, including enhanced soil structure, organic matter accumulation, and water retention. In this context, the provision of shade by a variety of plantation types fosters subterranean microbial activity through (i) the mitigation of temperature extremes, (ii) obstruction of ultraviolet radiation penetration, and (iii) reduction of soil coalescence caused by diminished soil dehydration. In diverse plantations, elevated root densities, species-specific microbial communities, and root-specific interactions with the soil microbiome all contribute to increased and enhanced soil microbial activity (Lu *et al.*, 2024).

18.6 TECHNOLOGICAL INNOVATIONS FOR BETTER MANAGING THE EFFECTS OF CLIMATE CHANGE

A pivotal function that the use of technology can perform, is to assist producers in adjusting their management practices to the effects of climate change (Balasundram *et al.*, 2023). By using a range of technological aids, concerned farmers can maintain the productivity of their operations despite the impacts of extreme weather and shifting conditions.

18.6.1 Precision Agriculture

In order to conserve water and reduce chemical usage, precision agricultural technologies, including sensors, drones, and GPS mapping, can assist farmers in optimizing their irrigation, fertilization, and insect control procedures. An additional benefit is that these technologies can assist producers in real-time monitoring and therefore their response, to shifting weather patterns and soil conditions (Dhanaraju *et al.*, 2022).

18.6.2 Climate Modeling

The use of software that simulates climate change can assist farmers in forecasting forthcoming climate conditions and weather patterns, enabling them to prepare for and adjust to shifting circumstances (Anwar *et al.*, 2013).

18.6.3 Crop Monitoring

Utilizing crop monitoring technology such as satellite images and remote sensing, enables farmers to constantly monitor the health and growth of their crops, facilitating the early detection of issues and timely remedial measures (Getahun *et al.*, 2024).

18.6.4 Water Management

In the face of increasing water scarcity and drought, intelligent irrigation systems and water monitoring technologies can assist farmers in optimizing their water usage, which can both conserve their resources and ensure that plants obtain sufficient hydration (Lakhiar *et al.*, 2024).

18.6.5 Soil Health Management

Farmers can measure the health and productivity of their soil with the aid of soil health monitoring technologies, such as sensors and remote sensing. The compilation of this information will significantly enable farm managers to make informed decisions regarding fertilization and other practices that can enhance the resilience and health of the soil (Mathenge *et al.*, 2022).

18.6.6 Greenhouse Management

Technological hydroponics and greenhouse applications which include climate control systems, enable farm managers to cultivate crops in monitored environments (Farooq *et al.*, 2022a). This assists managers in overcoming the difficulties posed by variable climate-related issues and extreme weather events.

18.6.7 Livestock Management

By employing a range of monitoring systems, sensors, and other devices that can gauge changes in animals' diet, behavior, and physical health, the information obtained can assist producers in optimizing the health and productivity of their livestock (Džermeikaitė *et al.*, 2023).

18.6.8 Food Storage and Preservation

By utilizing technology-based food storage and preservation technologies, which include such necessary actions as vacuum encapsulation and refrigeration, farmers can reduce food waste and extend the expiration life of their products (Akinsemolu and Onyeaka, 2024).

18.6.9 Energy Efficiency

By installing energy-efficient lighting and solar panels on their farm properties, producers can reduce their carbon footprint and save money on energy costs (Minoofar *et al.*, 2023).

18.6.10 Online Marketplaces

Online marketplaces, such as farm-to-table websites, can facilitate direct sales between farmers and local clients. This action can markedly decrease their dependence on conventional supply chains, which are susceptible to climate-related and other industrial interruptions (Tchonkouang *et al.*, 2024).

18.7 CASE STUDIES AND BEST PRACTICES

18.7.1 Case Study 1: Wheat Cultivation in Drought-Prone Regions

Drought is an abiotic stressor that causes substantial reductions in cereal yields and has emerged in recent years as a significant impediment to agricultural progress (Fadiji *et al.*, 2022). Although drought can impact wheat at any time throughout its life cycle, its impact is most pronounced during the anthesis and grain-filling stages (Zahra *et al.*, 2021). This impact is termed "terminal drought", and results in a substantial reduction in yield. Research has shown that a prolonged decrease in relative water content, due to drought conditions, results in the closure of stomata and a subsequent reduction in stomatal conductance (Figure 8.1). Additionally, the influence of drought conditions affects photosynthesis, in that, there are altered levels of chloroplast, mitochondria, and chlorophyll within the interior structure (Hu *et al.*, 2023). Owing to oxidative stress exerted on chloroplasts with stomatal closure, net photosynthesis is diminished, resulting in inadequate grain development and grain set. Additional problems are that (i) a drought, which occurs during the reproductive stage, has the potential to induce sterility in developing spikelets and (ii) the occurrence of drought during the pollen development stage results in pollen abortion and a subsequent decrease in the number of grains (Zahra *et al.*, 2021). In conjunction with these problems, drought that occurs during the anthesis phase primarily results in particle size reduction.

To ameliorate the conflict between wheat production and water scarcity, numerous strategies have been implemented on a trial basis. These actions include (i) water-saving irrigation field management, (ii) the use of drought-resistant wheat varieties, and (iii) the installation of water-saving irrigation equipment (Juraev and Karimov, 2020). In this respect, the cultivation of wheat varieties that are resistant to drought serves as the cornerstone for enhancing both crop yield and water productivity in drought-prone areas. Indeed, global initiatives to develop drought-resistant cultivars have been ongoing for an extended period, although their development is critically affected by a variety of environmental factors. In the course of this work, it has been found that drought resistance is a multifaceted attribute of a grain species that is influenced by several genes. These genes arise from the interplay of various fundamental elements or adaptive characteristics of the grain, each of which can be affected by intricate genetic and environmental variations (Mao *et al.*, 2023). Hence, it is essential in the creation and assessment of drought-resistant crop varieties, to examine the drought-tolerant characteristics in order to guarantee sustainable food production under the anticipated drought effects caused by climate change (Figure 18.1).

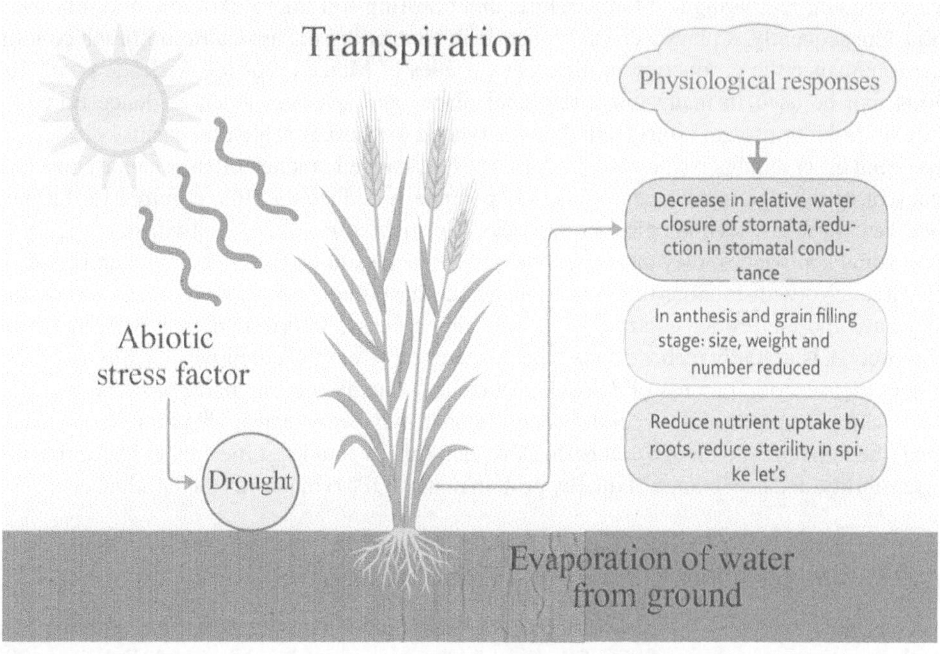

FIGURE 18.1 Abiotic stress factor from (drought) and the corresponding physiological response of winter wheat.

18.7.2 Case Study 2: Maize Production in Regions Vulnerable to Heat Stress

Cropping maize (*Zea mays* L.) accounts for 19.5% of the total global caloric intake, having now evolved into a significant industrial product (Hossain *et al.*, 2023). It is therefore of extreme concern that high and low-temperature extremes occur throughout the maize growth period, and pose a significant threat to maintaining the viability of maize yields. Maize plants are extremely susceptible to heat stress (>30°C), and their grain yield plummets to unacceptable levels when plants are exposed to heat stress for an extended period of time.

Currently, the demand for maize is consistently rising, due to its usage in food, feed, fodder, and as an industrial raw material. Records show that the expansion of the maize industry began when border areas, neighboring India, began cultivating maize hybrids in the 1980s by initially acquiring Indian seeds. However, open-pollinated varieties (OPVs) are the outcome of spontaneous pollination between plants, and, as such, their genetic variability may be restricted (Mamo *et al.*, 2023). Consequently, the grains may lack the required characteristics that make it receptive to additional inputs like fertilizers, water, and pesticides. Thus, the yield of OPVs cannot be raised beyond a specific limit despite using high initial inputs.

Hybrid maize research officially began in 1987 to enhance production, and National Maize Research Programs (NMRPs) have now developed and officially recorded seven single cross-hybrids in partnership with the International Maize and Wheat Improvement Center (CIMMYT). Various hybrids, such as Rampur Hybrid-2 (2012), Khumal Hybrid-2 (2014), Rampur Hybrid-4 (2016), Rampur Hybrid-6 (2016), Rampur Hybrid-8 (2018), and Rampur Hybrid-10 (RH-10) (2018), may be commonly found on the market, with Rampur Hybrid-8 and Rampur Hybrid-10 being heat-tolerant single cross-hybrids.

Drought, low soil moisture, excessive evaporation, weed competition, and pests during crucial crop growth phases, all lead to substantial output losses in the maize industry and require effective soil and water conservation methods (Begum and Momen, 2023). In this regard, mulching is a highly effective method that enhances crop microclimate by conserving soil moisture, reducing evaporation losses, promoting water uptake and transpiration in plants, preventing soil degradation and weed growth, regulating soil temperature, and boosting soil microbial growth to improve soil nutrition. Consequently, it plays a crucial function in controlling the agricultural growth conditions in order to enhance the production of different varieties of Maize crops (de Andrade *et al.*, 2023). Mulching can be used in many unconventional places and seasons which evidence limited soil moisture, in order to protect crops from damage related to unpredictable rain events.

It has been observed that participatory approaches to changed practice often generate considerable efficiency of knowledge transfer, and thus may potentially reduce the time required for farmers to embrace new varieties, technologies, or practices (Cornish *et al.*, 2023). Notwithstanding the successes of some approaches, they are expensive and throw a stain on the economic viability of crops in some areas. Notwithstanding this issue, farmers do have the opportunity to adopt a new variety that they have chosen through a participatory selection process, according to their preferred criteria. In this respect, it is urgently recommended that researchers strengthen their partnership with farmers throughout the selection process of a novel or enhanced-quality protein maize varieties, in order to facilitate their widespread and successful adoption once they become available for purchase. In such a context, the implementation of a comprehensive value chain could significantly enhance the prominence, feasibility, and acceptance of quality protein maize (QPM) materials (Senarathna *et al.*, 2024).

18.7.3 Case Study 3: Rice Cultivation in Flood Prone Areas

Rice is consumed by over half of the world's seven billion inhabitants as their staple sustenance, and Asia is the continent where over 90% of the world's rice is produced and ingested (Das *et al.*, 2023). Rainfed rice comprises roughly half of the 37% of the world's total rice area in South Asia, and recurrent flash floods can submerge plants entirely for 10–15 days. This is one of the most significant obstacles to rice production in the rainfed lowlands of South and Southeast Asia. In addition,

30% of the rice-growing area in India (12–14 M ha) is susceptible to catastrophic inundation, resulting in an average yield of merely 0.5–0.8 t ha^{-1}. Records show that Coastal Odisha, for example, is a region notorious for frequent flash floods that decimate paddy crops during the rainy season. As a result, farmers are compelled to either abandon paddy cultivation entirely or revert to low-yielding traditional landraces. These floods diminish the seasonal earnings of farmers, consequently impeding their propensity to engage in technological innovation, such as the cultivation of novel varieties.

It is predicted that the prolonged negative impacts of climate change will increase the likelihood of farmers being exposed to more severe flood occurrences (Bedeke, 2023). Anticipated flooding in many regions is expected to come from rising sea levels, erratic rainfall patterns, and projected escalation in the occurrence and severity of floods from extreme weather phenomena. Fortunately, the integration of stress tolerance into high-yielding varieties has been shown to be a practical method for producing varieties that can thrive in challenging conditions. This suggests that addressing the issue of climate change can be partially achieved by developing plant types that are adaptive to changing conditions and capable of providing robust harvests (Janni *et al.*, 2024).

In this respect, the International Rice Research Institute (IRRI) utilized marker-assisted breeding (MAB) to introduce the Sub1 (submergence tolerance 1) quantitative trait locus (QTL) from an Indian landrace into various well-known rice varieties, such as the widely cultivated Swarna mega variety in Odisha and other Indian states. The updated Swarna model, known as SS1, is capable of enduring full submersion for more than two weeks and can recover once the floodwater recedes. Following a recent thorough assessment and successful performance in farmers' fields, SS1 was approved for commercial production in India, and Odisha, a state with a long history of floods, is producing seeds of this kind for farmers in coastal regions prone to flooding (Sharma *et al.*, 2020).

Adaptation is a continuous process in agriculture and is a crucial strategy for mitigating the severity of environmental and socioeconomic impacts on agricultural value chains and farmer livelihoods (Marsden *et al.*, 2023). Molecular mechanism of quiescent and escape strategies is shown in Figure 18.2. Indeed, an increasing number of agricultural practitioners now implement various

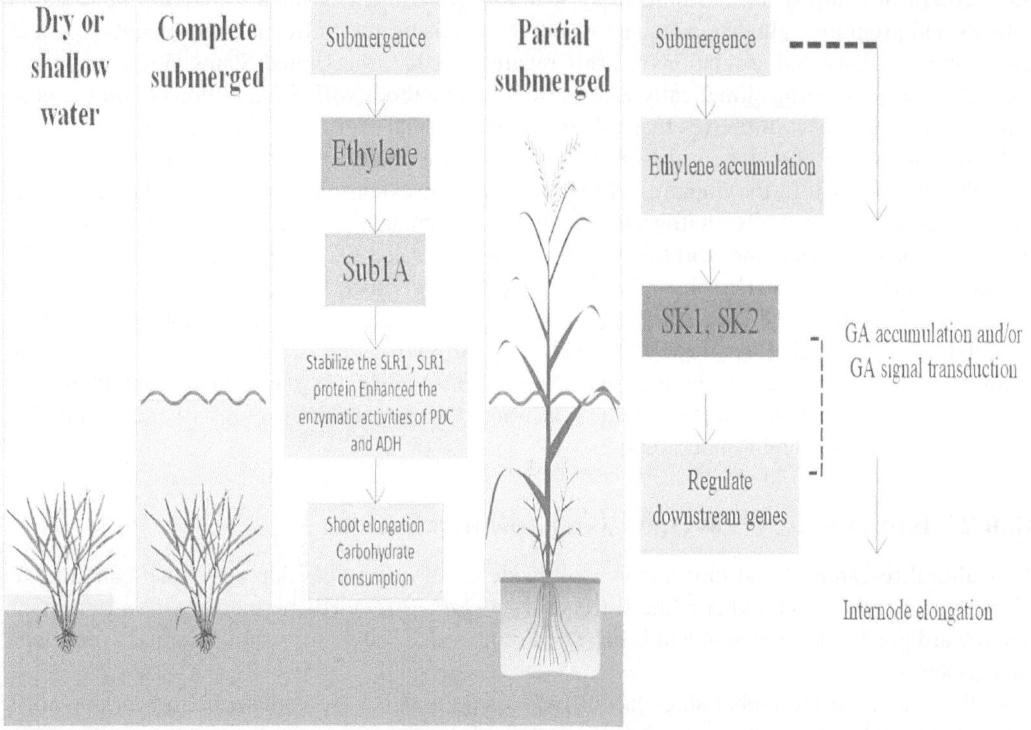

FIGURE 18.2 Molecular mechanism of quiescent and escape strategy.

agronomic modifications to their operations. These modifications include (i) the utilization of stress-tolerant varieties, (ii) the adjustment of sowing dates and irrigation schedules, (iii) the promotion of conservation agriculture practices such as multiple cropping and strategic nutrient management, and (iv) integrated pest management. In this respect, certain agronomic practices, such as "climate-smart agriculture", are presently packaged and marketed as adaptation packages. However, it is noted that, for farmers to implement these practices, site-specificity and the incorporation of local practices remain crucial.

18.8 POLICIES AND INSTITUTIONAL SUPPORT FOR RESILIENCE BUILDING

18.8.1 Government Policies and Incentives for Climate-Smart Agriculture

In the pursuit of mitigating and adapting to climate change, agricultural modifications have emerged as a critical domain, and the key objective of climate-smart agriculture (CSA) is to increase the resilience of agricultural systems to the impacts of global warming (Jat *et al.*, 2024). In an effort to increase farm productivity and efficiency with minimal environmental impact, precision agriculture platforms have contributed to the widespread adoption of climate-smart farming, but it has been seen that there is often considerable difficulty for farmers, particularly those in developing nations, to transition to climate-smart agriculture. Consequently, more widespread adoption of these CSA practices requires extensive support and promotion from the local government and non-government organizations.

It is recognized that numerous variables influence the adoption of climate-wise agriculture by agribusinesses. These include (i) the level of understanding and knowledge by the business, (ii) the accessibility of inputs, funding, and markets, (iii) the availability of resources for risk management, and (iv) the necessity for ensuring a suitable economic return.

In this development area, Governments may potentially persuade the public to adopt climate-wise agriculture practices by implementing incentive programs and eliminating obvious barriers (Lee, 2024). In addition, for sustainable agricultural growth, it is critical to develop new, sound policies and programs, allocate adequate resources to ensure their effective implementation, and amend existing laws and regulations. In this regard, in 2022, the United States Government has taken the lead in adopting climatically wise agricultural methods with a $22 billion commitment to Agriculture Department initiatives that adhere to climate-smart agricultural principles.

Participation of civil society is also crucial to the success of incentives to embrace climate-smart agricultural practices. In the domains of policy advocacy, extension services, and the evaluation and expansion of climatically intelligent agriculture systems, non-governmental organizations possess the capacity to effect meaningful change (Trautman *et al.*, 2023). As an illustration, Swedish non-governmental organizations have led a fruitful effort to reduce carbon emissions from Kenyan plantations. This inaugural carbon initiative in Africa's agricultural sector, brought together more than 60,000 small landholders, fostering the exchange of knowledge and enhancing resource management. This type of initiative implies that Governments and NGOs may aid in the facilitation of the development of climate-smart agriculture, leading to the establishment of robust and sustainable farming practices (Ma and Rahut, 2024).

18.8.2 International Cooperation and Knowledge Sharing

Agricultural researchers and universities worldwide are trying to develop crops that can survive climatic changes due to rapid climatic shifts (Sarma *et al.*, 2024). His partnership is an important step toward greater development and is expected to provide food security and sustainable development globally.

Collaboration on the global stage has already made possible the combination of cutting-edge research, specialized knowledge, and the expertise of developed geographies and scientific fields

(Rossoni *et al.*, 2024). Producers have had access to better crop varieties due to the diffusion of information, best practices and technologies, and novel agricultural strategies and climate-specific adaptive interventions. Writing about the actors involved, the systemic structures of knowledge exchange that are formed, and how this process then enables producers to be informed decision makers and adapt practices based on their ability to mitigate climate change.

The impacts of climate change know no borders and the measures to confront it are expensive and should be joint investments of our economies (Leichenko and O'brien, 2024). Through international cooperation, economies of scale enable nations to invest in sustainable agricultural practices, research and development, and infrastructural improvements. Investing in relevant initiatives like water-efficient irrigation systems, precision farming, and agroforestry leads to a higher local return on investment for countries and a more scalable global impact.

Experience has indicated that international cooperation allows nations to establish unified rules and criteria for sustainable agricultural and environmentally friendly activities (Melnyk, 2023). Aligning rules across borders can simplify market entry for agricultural products, encourage equitable trade agreements, and enhance the economies of all involved countries. Additionally, the incorporation of cross-border cooperation bolsters resilience to climate-induced shifts in markets whereby food supply chains become more stable and secure long-term due to established trading relations.

The strategy of cooperative endeavors in agriculture, can improve climate resilience through several methods (Zulfikri *et al.*, 2024). Regional cooperation could set up seed banks for climate-adaptive crops that are specifically suited to a local environment a step that could give farmers better options for crops that could survive harsh weather events. The other element, the coordination of research and technology transfers, can lead to drought-resistant cultivars, flood control techniques, and sustainable soil management practices, keeping food production strong despite climate-related challenges.

Moreover, the global partnership helps develop uniform laws and regulations with respect to sustainable agriculture and climate change mitigation (Wijerathna-Yapa and Pathirana, 2022). At the same time, international agreements and conventions can at least provide space where countries can work together, define common goals, and require each other to adhere to those commitments. In general, the world agricultural community has to deliver change on a larger scale, and the single voice of its executive can allow more ambitious climate and agriculture policies.

18.9 CHALLENGES AND FUTURE DIRECTIONS

A particular challenge in this area is that developing agriculture that is climate-smart is a complex and sophisticated topic. This complexity, therefore, requires interventions that are facilitated by practitioners with significant technical and functional skills (Keshavarz and Sharafi, 2023). Furthermore, ongoing in-service capacity development is imperative for individuals, organizations, and stakeholders who are desirous of operating within the wider enabling environment in order to provide assistance to producers.

At an operational level, farmers may not immediately observe the effects of some restoration initiatives or crop rotations. It is therefore essential to sustain farmers' enthusiasm for the program and ensure that emerging benefits are made visible throughout various time periods. In addition, efforts to tackle sustainable livelihoods, introduce climate change adaptation, and catalyze mitigation of environmentally poor practices, will involve operating at several geographic levels. These levels range from local plots and farms to entire farming systems, landscapes, or territories, and this wide brief for concern contributes to the intricacy of climate-smart actions.

Table 18.2 shows the previous efforts have indicated that community-level climate change interventions frequently necessitate the formation of multidisciplinary teams comprising stakeholders with varying educational backgrounds, working methodologies, and justifications (Ansari *et al.*, 2023). Overcoming this administrative obstacle may pose a substantial challenge since it concerns

TABLE 18.2
Description of Policies and Institutional Support for Resilience Building

Policy/Institution	Description	Role in Resilience Building
Climate change adaptation strategies and policies	National and international policies targeting the effects of climate change on agriculture. These may consist of adaptive strategies, financial mechanisms, and research projects aimed at improving resilience.	Support the incorporation of climate change factors into agricultural policy and offer tools for implementing adaptation measures.
Research and Development Institutions	Organizations dedicated to agricultural research and development, both public and private. These institutes research crop reactions to increased levels of carbon dioxide and temperature to create more robust agricultural methods.	Develop new technologies through research to identify adaptation strategies, and disseminate this information to farmers.
Agricultural extension services	Government agencies or groups that offer agricultural education and outreach to farmers. These programs distribute information on climate-smart agriculture practices and assist farmers in adopting resilient measures.	Provide farmers with information, training, and assistance to help them adopt climate-smart techniques and integrate them into their farming operations.
Insurance schemes	Governments or private insurers may provide crop insurance programs to mitigate the risks associated with the adverse effects of climate change, including crop failures and extreme weather events. Financial security provided by these programs may inspire producers to implement resilient agricultural practices.	Advocate for financial safeguards against losses attributable to climate change, encourage the implementation of resilient practices, and foster long-term investment and planning in resilience.
Sustainable agricultural policies	Advocate for sustainable agricultural techniques, such as conservation agriculture, agroforestry, or organic farming. The policies seek to boost resilience by enhancing soil health, water conservation, and biodiversity.	Promote the implementation of practices that improve soil health, preserve water, and decrease greenhouse gas emissions to promote resistance to the effects of climate change.

human capital, institutions, and policies, but it is nevertheless essential for ensuring sustainable livelihoods and enduring contributions to community resilience to climate change.

It is suggested that future research needs to address the issues related to the magnitude of this problem area and account for the modifications to climate-smart agricultural narratives. In this way, the substantial divide that exists between science and policy can be narrowed through the implementation of cross-disciplinary research objectives. In addition, the international scientific research agenda necessitates a realignment of investment priorities toward the "on-farm" and "off-farm" circumstances of smallholder farmers (Moojen *et al.*, 2024). This requires a reevaluation of the socioeconomic, political, and institutional aspects of the CSA discourse. One way to achieve this is by fostering interdisciplinary research that improves communication between the social, managerial, and economic research dimensions (Lindgreen *et al.*, 2024). Such a contribution would require an examination of the current state of knowledge linkages between the CSA and non-governmental organizations, farmer-led groups, and other non-traditional elitist development and research organizations that collaborate to generate knowledge.

18.10 CONCLUSION

While it is widely accepted that climate changes are causing damage globally by negatively impacting agriculture and its yields (Farooq *et al.*, 2022b), wide reliance on Industrialization, together with toxic emissions from many sources, are leading to further global warming and disruption of the Earth's ecology. In this respect, the effects of climate change are significantly impacting general plant growth and economic crop productivity. In this context, abiotic stressors are the primary form of stress emerging for climate change that plants experience. Thus, the primary present need is to investigate the genetic basis of plant responses to various abiotic environments to help reduce stress levels and maintain growth and productivity.

It is in this regard that certain molecular and physiological obstacles act as bottlenecks to plant development and must urgently be addressed to enable and enhance plant adaptability to abiotic settings. It is understood that varied precipitation patterns and temperature fluctuations are critical indicators of environmental stresses and that it is exceedingly challenging to rectify the agricultural imbalance caused by these changes. Indeed, the approach to addressing this issue and the particular strategies that ought to be implemented, currently remain uncertain (Yang *et al.*, 2024). As a consequence, it is imperative that scientists optimize the development and proliferation of plants under abiotic stresses, and carefully examine the outcomes of parallel investigations. It is likely that in order to safeguard agriculture in the future, novel cultural techniques, diverse cultivation schemes, and numerous conventional and non-conventional approaches will be needed to increase crop resistance to biotic and abiotic stresses. Most probably, innovative breeding strategies will be developed to contribute to the availability of more climate-resilient crops that show more tolerance to heat and drought. The utilization of genotyping strategies, genome-wide association studies (GWAS) and genomic selection with high-throughput phenotyping, will be crucial for identifying the various genes that contribute to crop enhancement in the face of climate change (Afzal *et al.*, 2023). Whilst the application of genetic engineering techniques in the development of transgenic plants with improved resistance to various biotic and abiotic stressors has been substantial (Kumar *et al.*, 2024), moving forward, it will be necessary to develop environmentally sustainable crops via CRISPR/Cas9-mediated genome editing in order to combat increasing climate change (Ambika *et al.*, 2024).

LIST OF ABBREVIATIONS

CGIAR Consultative Group on International Agricultural Research
GWAS genome-wide association studies
NMRP National Maize Research Program

REFERENCES

Abbass, K., Qasim, M.Z., Song, H., Murshed, M., Mahmood, H. & Younis, I., 2022. A review of the global climate change impacts, adaptation, and sustainable mitigation measures. *Environmental Science and Pollution Research*, 29, 42539–42559.

Afzal, S., Mubeen, M., Hussain, S., Ali, M., Javeed, H.M.R., Al-Ashkar, I., Soufan, W., Pandey, S., Islam, M.S. & El Sabagh, A., 2023. Modern Breeding Approaches for Climate Change. *In Climate Change Impacts on Agriculture: Concepts, Issues and Policies for Developing Countries.* Springer, 299–313.

Akinsemolu, A.A. & Onyeaka, H.N., 2024. Microorganisms Associated with Food Spoilage and Foodborne Diseases. *In Food Safety and Quality in the Global South.* Springer, 489–531.

Ambika, Bhati, S. & Kumar, R., 2024. Plant Breeding Using the CRISPR-Cas9 System for Food Security and Facing Climate Change. *In Plant Genome Editing Technologies: Speed Breeding, Crop Improvement and Sustainable Agriculture.* Springer, 149–181.

Ansari, D., Schönenberg, R., Abud, M., Becerra, L., Brahim, W., Castiblanco, J., De La Vega-Leinert, A.C., Dudley, N., Dunlop, M. & Figueroa, C., 2023. Communicating climate change and biodiversity loss with local populations: Exploring communicative utopias in eight transdisciplinary case studies. *UCL Open Environment*, 5.

Anwar, M.R., Liu, D.L., Macadam, I., & Kelly, G., 2013. Adapting agriculture to climate change: A review. *Theoretical and Applied Climatology*, 113, 225–245.

Ashraf, E., Shah, F., Luqman, M., Younis, M., Aziz, I. & Farooq, U., 2013. Use of untreated wastewater for vegetable farming: A threat to food safety. *International Journal of Agricultural and Applied Sciences*, 5.

Aslam, A., Mahmood, A., Ur-Rehman, H., Li, C., Liang, X., Shao, J., Negm, S., Moustafa, M., Aamer, M. & Hassan, M.U., 2023. Plant adaptation to flooding stress under changing climate conditions: Ongoing breakthroughs and future challenges. *Plants*, 12, 3824.

Badiani, M., Raschi, A., Paolacci, A.R. & Miglietta, F., 2023. Plants Responses to Elevated CO_2; A Perspective from Natural CO_2 Springs. *In Environmental Pollution and Plant Responses.* Routledge, 45–81.

Balasundram, S.K., Shamshiri, R.R., Sridhara, S. & Rizan, N., 2023. The role of digital agriculture in mitigating climate change and ensuring food security: An overview. *Sustainability*, 15, 5325.

Balla, K., Karsai, I., Bónis, P., Kiss, T., Berki, Z., Horvath, A., Mayer, M., Bencze, S. & Veisz, O., 2019. Heat stress responses in a large set of winter wheat cultivars (*Triticum aestivum* L.) depend on the timing and duration of stress. *PLoS One*, 14, e0222639.

Becker, R., Schüth, C., Merz, R., Khaliq, T., Usman, M., Aus Der Beek, T., Kumar, R. & Schulz, S., 2023. Increased heat stress reduces future yields of three major crops in Pakistan's Punjab region despite intensification of irrigation. *Agricultural Water Management*, 281, 108243.

Bedeke, S.B., 2023. Climate change vulnerability and adaptation of crop producers in sub-Saharan Africa: A review on concepts, approaches and methods. *Environment, Development and Sustainability*, 25, 1017–1051.

Begum, M.M. & Momen, M.N., 2023. Drought. *In The Palgrave Encyclopedia of Global Security Studies.* Springer, 332–337.

Cai, S., Xu, S., Zhang, D., Geisen, S. & Zhu, H., 2024. Soil microbial biomass and bacterial diversity enhanced through fallow cover cropping in rice–fish coculture. *Agronomy*, 14, 456.

Chadha, A., Florentine, S., Chauhan, B.S., Long, B., Jayasundera, M., Javaid, M.M. & Turville, C., 2019. Environmental factors affecting the germination and seedling emergence of two populations of an emerging agricultural weed: Wild lettuce (*Lactuca serriola*). *Crop and Pasture Science*, 70, 709–717.

Chadha, A., Florentine, S., Javaid, M., Welgama, A. & Turville, C., 2020. Influence of elements of climate change on the growth and fecundity of *Datura stramonium. Environmental Science and Pollution Research*, 27, 35859–35869.

Chattha, M.B., Khan, I., Mahmood, A., Chattha, M.U., Anjum, S.A., Ashraf, U., Zain, M. & Bilal, U., 2017. Seed priming with thiourea improves the performance of late sown wheat. *Journal of Agricultural Research*, 55, 29–39.

Chaudhry, S. & Sidhu, G.P.S., 2022. Climate change regulated abiotic stress mechanisms in plants: A comprehensive review. *Plant Cell Reports*, 41, 1–31.

Chhina, A.K., Abhari, N., Mooers, A. & Lewthwaite, J., 2024. Linking the spatial and genomic structure of adaptive potential for conservation management: A review. *Genome*, 67, 403–423.

Clark, R.D., Catalano, K.A., Fitz, K.S., Garcia, E., Jaynes, K.E., Reid, B.N., Sawkins, A., Snead, A.A., Whalen, J.C. & Pinsky, M.L., 2023. The practice and promise of temporal genomics for measuring evolutionary responses to global change. *Molecular Ecology Resources*, 25, 1–17.

Cornish, F., Breton, N., Moreno-Tabarez, U., Delgado, J., Rua, M., De-Graft Aikins, A. & Hodgetts, D., 2023. Participatory action research. *Nature Reviews Methods Primers*, 3, 34.

Correa, S.S., Schultz, J., Lauersen, K.J. & Rosado, A.S., 2023. Natural carbon fixation and advances in synthetic engineering for redesigning and creating new fixation pathways. *Journal of Advanced Research*, 47, 75–92.

Crino, O.L., Bonduriansky, R., Martin, L.B. & Noble, D.W., 2024. A conceptual framework for understanding stress-induced physiological and transgenerational effects on population responses to climate change. *Evolution Letters*, 8, 161–171.

Das, K., Roy, P. & Tiwari, R.K.S., 2023. Biofortification of Rice, an Impactful Strategy for Nutritional Security: Current Perspectives and Future Prospect. *In* A. B. Roy & S. Banerjee (Eds.), *Plant-Based Diet.* Springer, 22–25.

de Andrade, L.A., Santos, C.H.B., Frezarin, E.T., Sales, L.R. & Rigobelo, E.C., 2023. Plant growth-promoting rhizobacteria for sustainable agricultural production. *Microorganisms*, 11, 1088.

Dhanaraju, M., Chenniappan, P., Ramalingam, K., Pazhanivelan, S. & Kaliaperumal, R., 2022. Smart farming: Internet of things (IoT)-based sustainable agriculture. *Agriculture*, 12, 1745.

Dormatey, R., Sun, C., Ali, K., Coulter, J.A., Bi, Z. & Bai, J., 2020. Gene pyramiding for sustainable crop improvement against biotic and abiotic stresses. *Agronomy*, 10, 1255.

Džermeikaitė, K., Bačėninaitė, D. & Antanaitis, R., 2023. Innovations in cattle farming: Application of innovative technologies and sensors in the diagnosis of diseases. *Animals*, 13, 780.

Elbasiouny, H., El-Ramady, H., Elbehiry, F., Rajput, V.D., Minkina, T. & Mandzhieva, S., 2022. Plant nutrition under climate change and soil carbon sequestration. *Sustainability*, 14, 914.

Fadiji, A.E., Santoyo, G., Yadav, A.N. & Babalola, O.O., 2022. Efforts towards overcoming drought stress in crops: Revisiting the mechanisms employed by plant growth-promoting bacteria. *Frontiers in Microbiology*, 13, 962427.

Farooq, M.S., Riaz, S., Helou, M.A., Khan, F.S., Abid, A. & Alvi, A., 2022a. Internet of things in greenhouse agriculture: A survey on enabling technologies, applications, and protocols. *IEEE Access*, 10, 53374–53397.

Farooq, M.S., Uzair, M., Raza, A., Habib, M., Xu, Y., Yousuf, M., Yang, S.H. & Ramzan Khan, M., 2022b. Uncovering the research gaps to alleviate the negative impacts of climate change on food security: A review. *Frontiers in Plant Science*, 13, 927535.

Florentine, S.K., Weller, S., Graz, P.F., Westbrooke, M., Florentine, A., Javaid, M., Fernando, N., Chauhan, B.S. & Dowling, K., 2016. Influence of selected environmental factors on seed germination and seedling survival of the arid zone invasive species tobacco bush (*Nicotiana glauca* R. Graham). *The Rangeland Journal*, 38, 417–425.

Getahun, S., Kefale, H. & Gelaye, Y., 2024. Application of precision agriculture technologies for sustainable crop production and environmental sustainability: A systematic review. *The Scientific World Journal*, 2024, 2126734.

Ghamkhar, K., 2023. Plant Genetic Resources and Their Values. *In Plant Genetic Resources for the 21st Century*. Apple Academic Press, 3–7.

Gomez-Zavaglia, A., Mejuto, J.C. & Simal-Gandara, J., 2020. Mitigation of emerging implications of climate change on food production systems. *Food Research International*, 134, 109256.

Grigorieva, E., Livenets, A. & Stelmakh, E., 2023. Adaptation of agriculture to climate change: A scoping review. *Climate*, 11, 202.

Gull, T., Mahmood, A., Shaheen, C., Javaid, M.M., Zia, M.A., Naqve, M., Bibi, S., Nadeem, M.A., Ameen, M. & Nargis, J., 2023. Climate Change and Nutrient Use Efficiency of Plants. *In Climate-Resilient Agriculture, Vol 1: Crop Responses and Agroecological Perspectives*. Springer International Publishing, 291–312.

Harrison, S., Franklin, J., Hernandez, R.R., Ikegami, M., Safford, H.D. & Thorne, J.H., 2024. Climate change and California's terrestrial biodiversity. *Proceedings of the National Academy of Sciences*, 121, e2310074121.

Hosahatti, R., Jeevan, B., Mishra, K., Subbanna, A. & Kant, L., 2021. Blast Disease: Historical Importance, Distribution, and Host Infectivity across Cereal Crops. *In Blast Disease of Cereal Crops: Evolution and Adaptation in Context of Climate Change*. Springer, 1–13.

Hossain, F., Zunjare, R.U., Muthusamy, V., Kumar, A., Madhavan, J., Ikkurti, G., Katral, A., Talukder, Z.A., Chhabra, R. & Chand, G., 2023. Genetic Improvement of Specialty Corn for Nutritional Quality Traits. *In Maize Improvement: Current Advances in Yield, Quality, and Stress Tolerance under Changing Climatic Scenarios*. Springer, 235–257.

Hu, F., Zhang, Y. & Guo, J., 2023. Effects of drought stress on photosynthetic physiological characteristics, leaf microstructure, and related gene expression of yellow horn. *Plant Signaling & Behavior*, 18, 2215025.

Irfan, M., Musarat, M.A., Alaloul, W.S. & Ghufran, M., 2024. Radiative Forcing on Climate Change: Assessing the Effect of Greenhouse Gases on Energy Balance of Earth. *In Advances and Technology Development in Greenhouse Gases: Emission, Capture and Conversion*. Elsevier, 137–167.

Jacquemart, M., Weber, S., Chiarle, M., Chmiel, M., Cicoira, A., Corona, C., Eckert, N., Gaume, J., Giacona, F. & Hirschberg, J., 2024. Detecting the impact of climate change on alpine mass movements in observational records from the European Alps. *Earth-Science Reviews*, 258, 104886.

Janni, M., Maestri, E., Gullì, M., Marmiroli, M. & Marmiroli, N., 2024. Plant responses to climate change, how global warming may impact on food security: A critical review. *Frontiers in Plant Science*, 14, 1297569.

Jat, H.S., Sharma, P.C., Choudhary, M., Poonia, T., Dagar, J.C. & Parihar, C.M., 2024. Climate-Smart Agriculture (CSA) Practices for Sustainable Intensification in Major Agri-Food Systems of South Asia. *In Climate-Smart and Resilient Food Systems and Security*. Springer, 37–71.

Javaid, M.M., Florentine, S., Mahmood, A., Wasaya, A., Javed, T., Sattar, A., Sarwar, N., Kalaji, H.M., Ahmad, H.B. & Worbel, J., 2022a. Interactive effect of elevated CO_2 and drought on physiological traits of *Datura stramonium*. *Frontiers in Plant Science*, 13, 929378.

Javaid, M.M., Florentine, S.K., Ashraf, M., Mahmood, A., Sattar, A., Wasaya, A. & Li, F.M., 2022b. Photosynthetic activity and water use efficiency of *Salvia verbenaca* L. under elevated CO_2 and water-deficit conditions. *Journal of Agronomy and Crop Science*, 208, 536–551.

Javaid, M.M., Wang, X., Florentine, S.K., Ashraf, M., Mahmood, A., Li, F.-M. & Fiaz, S., 2022c. Effects on photosynthetic response and biomass productivity of *Acacia longifolia* ssp. longifolia under elevated CO_2 and water-limited regimes. *Frontiers in Plant Science*, 13, 817730.

Jiang, W., Li, N., Zhang, D., Meinhardt, L., Cao, B., Li, Y. & Song, L., 2020. Elevated temperature and drought stress significantly affect fruit quality and activity of anthocyanin-related enzymes in jujube (*Ziziphus jujuba* Mill. cv.'Lingwuchangzao'). *PLoS One*, 15, e0241491.

Jianing, G., Yuhong, G., Yijun, G., Rasheed, A., Qian, Z., Zhiming, X., Mahmood, A., Shuheng, Z., Zhuo, Z. & Zhuo, Z., 2022. Improvement of heat stress tolerance in soybean (*Glycine max* L), by using conventional and molecular tools. *Frontiers in Plant Science*, 13, 993189.

Juraev, F. & Karimov, G., 2020. Water saving technique and technology of subsurface irrigation. *In IOP Conference Series: Materials Science and Engineering.* IOP Publishing, 012095.

Kabir, M., Habiba, U.E., Khan, W., Shah, A., Rahim, S., Patricio, R., Ali, L. & Shafiq, M., 2023. Climate change due to increasing concentration of carbon dioxide and its impacts on environment in 21st century; A mini review. *Journal of King Saud University-Science*, 35, 102693.

Keshavarz, M. & Sharafi, H., 2023. Scaling up climate-smart regenerative agriculture for the restoration of degraded agroecosystems in developing countries. *Sustainable Production and Consumption*, 38, 159–173.

Khan, A., Riaz, A., Mahmood, A., Bibi, S., Florentine, S., Javaid, M.M. & Nargis, J., 2023. Phytoremediation of Atmospheric Pollutants in the Era of Climate Change. *In Climate-Resilient Agriculture, Vol 1: Crop Responses and Agroecological Perspectives.* Springer International Publishing, 257–272.

Kumar, T., Wang, J.-G., Xu, C.-H., Lu, X., Mao, J., Lin, X.-Q., Kong, C.-Y., Li, C.-J., Li, X.-J. & Tian, C.-Y., 2024. Genetic engineering for enhancing sugarcane tolerance to biotic and abiotic stresses. *Plants*, 13, 1739.

Kumawat, A., Bamboriya, S.D., Meena, R.S., Yadav, D., Kumar, A., Kumar, S., Raj, A. & Pradhan, G., 2022. Legume-Based Inter-Cropping to Achieve the Crop, Soil, and Environmental Health Security. *In Advances in Legumes for Sustainable Intensification.* Elsevier, 307–328.

Lakhiar, I.A., Yan, H., Zhang, C., Wang, G., He, B., Hao, B., Han, Y., Wang, B., Bao, R. & Syed, T.N., 2024. A review of precision irrigation water-saving technology under changing climate for enhancing water use efficiency, crop yield, and environmental footprints. *Agriculture*, 14, 1141.

Lee, W.Y., 2024. *The Impact of Climate Change on Personal Finances: Managing Food Expenditure.* Available at SSRN 4857042. https://ssrn.com/abstract=4857042

Leichenko, R. & O'brien, K., 2024. *Climate and Society: Transforming the Future.* John Wiley & Sons.

Li, H., Liu, K., Li, Z., Zhang, M., Zhang, Y., Li, S., Wang, X., Zhou, J., Zhao, Y. & Liu, T., 2023. Mixing trait-based corn (*Zea mays* L.) cultivars increases yield through pollination synchronization and increased cross-fertilization. *The Crop Journal*, 11, 291–300.

Lindgreen, A., Di Benedetto, C.A., Brodie, R.J., Van Der Borgh, M. & Schäfers, T., 2024. Engaging in Cross-Disciplinary Research. *In How to Achieve Societal Impact through Engaged and Collaborative Scholarship.* Edward Elgar Publishing, 137–148.

Lobo, A.K.M., Catarino, I.C., Silva, E.A., Centeno, D.C. & Domingues, D.S., 2022. Physiological and molecular responses of woody plants exposed to future atmospheric CO_2 levels under abiotic stresses. *Plants*, 11, 1880.

Lu, J., Liu, Y., Zou, X., Zhang, X., Yu, X., Wang, Y. & Si, T., 2024. Rotational strip peanut/cotton intercropping improves agricultural production through modulating plant growth, root exudates, and soil microbial communities. *Agriculture, Ecosystems & Environment*, 359, 108767.

Lynch, J.P., 2022. Edaphic stress interactions: Important yet poorly understood drivers of plant production in future climates. *Field Crops Research*, 283, 108547.

Ma, W. & Rahut, D.B., 2024. Climate-smart agriculture: Adoption, impacts, and implications for sustainable development. *Mitigation and Adaptation Strategies for Global Change*, 29, 44.

Mamo, T., Asheesh, S. & Mahama, A.A., 2023. Participatory plant breeding and participatory variety selection. *In* W. P.Suza & K. R.Lamkey (Eds.), *Crop Improvement.* Iowa State University Digital Press, 161–171.

Mao, H., Jiang, C., Tang, C., Nie, X., Du, L., Liu, Y., Cheng, P., Wu, Y., Liu, H. & Kang, Z., 2023. Wheat adaptation to environmental stresses under climate change: Molecular basis and genetic improvement. *Molecular Plant*, 16, 1564–1589.

Marsden, A.R., Zander, K.K. & Lassa, J.A., 2023. Smallholder farming during Covid-19: A systematic review concerning impacts, adaptations, barriers, policy, and planning for future pandemics. *Land*, 12, 404.

Mathenge, M., Sonneveld, B.G. & Broerse, J.E., 2022. Application of GIS in agriculture in promoting evidence-informed decision making for improving agriculture sustainability: A systematic review. *Sustainability*, 14, 9974.

Melnyk, T.Y., 2023. Tools for the formation of a "favorable environment" for organic production as a prerequisite for the sustainable development of agriculture in Ukraine. *In IOP Conference Series: Earth and Environmental Science*. IOP Publishing, 012124.

Miller, M.A., Tonoto, P. & Taylor, D., 2022. Sustainable development of carbon sinks? Lessons from three types of peatland partnerships in Indonesia. *Sustainable Development*, 30, 241–255.

Minoofar, A., Gholami, A., Eslami, S., Hajizadeh, A., Gholami, A., Zandi, M., Ameri, M. & Kazem, H.A., 2023. Renewable energy system opportunities: A sustainable solution toward cleaner production and reducing carbon footprint of large-scale dairy farms. *Energy Conversion and Management*, 293, 117554.

Moojen, F.G., Ryschawy, J., Wulfhorst, J., Archer, D.W., De Faccio Carvalho, P.C. & Hendrickson, J.R., 2024. Case study analysis of innovative producers toward sustainable integrated crop-livestock systems: Trajectory, achievements, and thought process. *Agronomy for Sustainable Development*, 44, 26.

Northrup, J.M., Vander Wal, E., Bonar, M., Fieberg, J., Laforge, M.P., Leclerc, M., Prokopenko, C.M. & Gerber, B.D., 2022. Conceptual and methodological advances in habitat-selection modeling: Guidelines for ecology and evolution. *Ecological Applications*, 32, e02470.

Nunes, L.J., 2023. The rising threat of atmospheric CO_2: A review on the causes, impacts, and mitigation strategies. *Environments*, 10, 66.

Peters, J., Oliva, R., Caliò, A., Oger, P. & Winter, R., 2023. Effects of crowding and cosolutes on biomolecular function at extreme environmental conditions. *Chemical Reviews*, 123, 13441–13488.

Ponti, R. & Sannolo, M., 2023. The importance of including phenology when modelling species ecological niche. *Ecography*, 2023, e06143.

Rasheed, A., Li, H., Nawaz, M., Mahmood, A., Hassan, M.U., Shah, A.N., Hussain, F., Azmat, S., Gillani, S.F.A. & Majeed, Y., 2022. Molecular tools, potential frontiers for enhancing salinity tolerance in rice: A critical review and future prospective. *Frontiers in Plant Science*, 13, 966749.

Rasheed, A., Seleiman, M.F., Nawaz, M., Mahmood, A., Anwar, M.R., Ayub, M.A., Aamer, M., El-Esawi, M.A., El-Harty, E.H. & Batool, M., 2021. Agronomic and genetic approaches for enhancing tolerance to heat stress in rice: A review. *Notulae Botanicae Horti Agrobotanici Cluj-Napoca*, 49, 12501–12501.

Rawat, A., Kumar, D. & Khati, B.S., 2024. A review on climate change impacts, models, and its consequences on different sectors: A systematic approach. *Journal of Water and Climate Change*, 15, 104–126.

Rehman, M., Fahad, S., Saleem, M.H., Hafeez, M., Rahman, M.U., Liu, F. & Deng, G., 2020. Red light optimized physiological traits and enhanced the growth of ramie (*Boehmeria nivea* L.). *Photosynthetica*, 58.

Ridlon, A.D., Grosholz, E.D., Hancock, B., Miller, M.W., Bickel, A., Froehlich, H.E., Lirman, D., Pollock, F.J., Putnam, H.M. & Tlusty, M.F., 2023. Culturing for conservation: The need for timely investments in reef aquaculture. *Frontiers in Marine Science*, 10, 1069494.

Rossoni, A.L., De Vasconcellos, E.P.G. & De Castilho Rossoni, R.L., 2024. Barriers and facilitators of university-industry collaboration for research, development and innovation: A systematic review. *Management Review Quarterly*, 74, 1841–1877.

Sadok, W., Lopez, J.R. & Smith, K.P., 2021. Transpiration increases under high-temperature stress: Potential mechanisms, trade-offs and prospects for crop resilience in a warming world. *Plant, Cell & Environment*, 44, 2102–2116.

Saleem, S., Mushtaq, N.U., Rasool, A., Shah, W.H., Tahir, I. & Rehman, R.U., 2023. Plant Nutrition and Soil Fertility: Physiological and Molecular Avenues for Crop Improvement. *In Sustainable Plant Nutrition*. Elsevier, 23–49.

Santachiara, G., Salvagiotti, F. & Rotundo, J.L., 2019. Nutritional and environmental effects on biological nitrogen fixation in soybean: A meta-analysis. *Field Crops Research*, 240, 106–115.

Sarma, H.H., Borah, S.K., Dutta, N., Sultana, N., Nath, H. & Das, B.C., 2024. Innovative approaches for climate-resilient farming: Strategies against environmental shifts and climate change. *International Journal of Environment and Climate Change*, 14, 217–241.

Senarathna, S., Mel, R. & Malalgoda, M., 2024. Utilization of cereal-based protein ingredients in food applications. *Journal of Cereal Science*, 103867.

Sharma, S., Rana, D.S., Jat, M.L., Biswal, S., Khatri-Chhetri, A. & Pathak, H., 2020. *A Compendium of Technologies, Practices, Services and Policies for Scaling Climate Smart Agriculture in Odisha (India)*. International Rice Research Institute, 1–85.

Siddha, S. & Sahu, P., 2022. Impact of Climate Change on the River Ecosystem. *In Ecological Significance of River Ecosystems*. Elsevier, 79–104.

Sinha, R., Fritschi, F.B., Zandalinas, S.I. & Mittler, R., 2021. The impact of stress combination on reproductive processes in crops. *Plant Science*, 311, 111007.

Stark, J.C., Love, S.L. & Knowles, N.R., 2020. Tuber Quality. *In* J. C.Stark, M.Thornton, & P.Nolte (Eds.), *Potato Production Systems*. Springer Nature, 479–497.

Tchonkouang, R.D., Onyeaka, H. & Nkoutchou, H., 2024. Assessing the vulnerability of food supply chains to climate change-induced disruptions. *Science of the Total Environment*, 171047.

Temmerman, S., Horstman, E.M., Krauss, K.W., Mullarney, J.C., Pelckmans, I. & Schoutens, K., 2023. Marshes and mangroves as nature-based coastal storm buffers. *Annual Review of Marine Science*, 15, 95–118.

Tian, J., Walayat, N., Ding, Y. & Liu, J., 2022. The role of trifunctional cryoprotectants in the frozen storage of aquatic foods: Recent developments and future recommendations. *Comprehensive Reviews in Food Science and Food Safety*, 21, 321–339.

Trautman, S., Chevallier, R., Cramer, L. & Mutamba, M., 2023. Scaling climate-smart agriculture for accelerated agri-food systems transformation in Africa. African Union Development Agency, NEPAD & Accelerating Impacts of CGIAR Climate Research for Africa (AICCRA).

Tunio, R.A., Li, D. & Khan, N., 2024. Maximizing farm resilience: The effect of climate smart agricultural adoption practices on food performance amid adverse weather events. *Frontiers in Sustainable Food Systems*, 8, 1423702.

Ullah, A., Nadeem, F., Nawaz, A., Siddique, K.H. & Farooq, M., 2022. Heat stress effects on the reproductive physiology and yield of wheat. *Journal of Agronomy and Crop Science*, 208, 1–17.

Walsh, C.A. & Lundgren, M.R., 2024. Nutritional Quality of Photosynthetically Diverse Crops under Future Climates. *Plants, People, Planet*, 6, 1272–1283.

Wang, X., Wang, J., Zou, Y., Bie, Y., Mahmood, A., Zhang, L., Liao, L., Song, Z., Liu, G. & Zhang, C., 2024. Urea fertilization increased CO_2 and CH_4 emissions by enhancing C-cycling genes in semi-arid grasslands. *Journal of Environmental Management*, 356, 120718.

Wari, D., Aboshi, T., Shinya, T. & Galis, I., 2022. Integrated view of plant metabolic defense with particular focus on chewing herbivores. *Journal of Integrative Plant Biology*, 64, 449–475.

Welgama, A., Florentine, S., Marchante, H., Javaid, M.M. & Turville, C., 2019. The germination success of *Acacia longifolia* subsp. longifolia (Fabaceae): A comparison between its native and exotic ranges. *Australian Journal of Botany*, 67, 414–424.

Weller, S.L., Javaid, M.M. & Florentine, S.K., 2020. Evaluation of the growth response of arid zone invasive species Salvia verbenaca cultivars to atmospheric carbon dioxide and soil moisture. *The Rangeland Journal*, 42, 45–53.

Welti, E.A. & Kaspari, M., 2024. Elevated CO_2, nutrition dilution, and shifts in Earth's insect abundance. *Current Opinion in Insect Science*, 101255.

Wijerathna-Yapa, A. & Pathirana, R., 2022. Sustainable agro-food systems for addressing climate change and food security. *Agriculture*, 12, 1554.

Yang, Y., Tilman, D., Jin, Z., Smith, P., Barrett, C.B., Zhu, Y.-G., Burney, J., D'odorico, P., Fantke, P. & Fargione, J., 2024. Climate change exacerbates the environmental impacts of agriculture. *Science*, 385, eadn3747.

Yu, T., Mahe, L., Li, Y., Wei, X., Deng, X. & Zhang, D., 2022. Benefits of crop rotation on climate resilience and its prospects in China. *Agronomy*, 12, 436.

Zahra, N., Hafeez, M.B., Ghaffar, A., Kausar, A., Al Zeidi, M., Siddique, K.H. & Farooq, M., 2023. Plant photosynthesis under heat stress: Effects and management. *Environmental and Experimental Botany*, 206, 105178.

Zahra, N., Wahid, A., Hafeez, M.B., Ullah, A., Siddique, K.H. & Farooq, M., 2021. Grain development in wheat under combined heat and drought stress: Plant responses and management. *Environmental and Experimental Botany*, 188, 104517.

Zandalinas, S.I., Balfagón, D., Gómez-Cadenas, A. & Mittler, R., 2022. Plant responses to climate change: Metabolic changes under combined abiotic stresses. *Journal of Experimental Botany*, 73, 3339–3354.

Zha, L., Jin, W., Fang, G., Guo, J., Liu, Z., Hu, W., Zhou, Z. & Meng, Y., 2023. Long-term residue retention affects seedcotton yield by regulating distribution of photosynthetic carbon, boll capacity of the root system and soil quality. *Field Crops Research*, 303, 109120.

Zhao, W., Liu, L., Shen, Q., Yang, J., Han, X., Tian, F. & Wu, J., 2020. Effects of water stress on photosynthesis, yield, and water use efficiency in winter wheat. *Water*, 12, 2127.

Zulfikri, A., Ningsih, E.M.N., Harsono, I. & Susanto, H., 2024. Agricultural adaptation strategies to weather fluctuations for improved agribusiness sustainability in West Java. *West Science Nature and Technology*, 2, 17–23.

19 Farming Systems for Sustainable Resource Utilization

*Rizwan Maqbool**, *Nazma Khan, Muhammad Ather Nadeem, Athar Mahmood, and Muhammad Junaid Nisar*

19.1 INTRODUCTION

Sustainable agriculture is one of the crucial components of modern farming systems. It focuses on responsibly utilizing resources so that current needs can be met without compromising the ability to meet future needs. Sustainable agriculture involves the proper organization of individual components and technologies within the farming system to optimize resource use efficiency, increase productivity, and promote environmental stewardship. Sustainable resource utilization refers to the efficient use and conservation of natural resources, including soil, water, and biodiversity, to ensure the long-term sustainability of agricultural production systems and environmental integrity.

Sustainable resource utilization is a concept in agriculture that includes practices that help to maintain or improve the productivity of agricultural systems while minimizing environmental harm, e.g., crop rotation, multiple cropping, conservation tillage, and integrated pest management (IPM) (Benlhabib *et al.*, 2014). Sustainable resource utilization includes crop residue, organic matter, and nutrient recycling that improves the health and fertility of soil, which is essential to the long-term sustainability of agriculture (Jadhav *et al.*, 2024).

Ultimately, farming systems help to foster sustainability as a network of enterprises that can be integrated to maximize resource utilization and improve system efficiency (overall production) (Tahat *et al.*, 2020) Multiple enterprises or enterprises in different compartments (for example, crops, livestock, or aquaculture) are part of an integrated farming system (IFS) that can maximize resource utilization and facilitate resource efficiency. By growing multiple enterprises or crops simultaneously, farmers can maximize inputs for multiple crops at the same time (e.g., cover crops that can be managed later, such as grazing or harvesting forage). Diversification of a farm's enterprises can also improve the system's natural resilience to environmental stressors (e.g., drought, floods) and increase farm income, both of which benefit the long-term sustainability of a farm operation overall (Rasheed *et al.*, 2022b).

Thus, the conclusion can be made that the sustainable utilization of resources in agriculture is crucial. Better utilization of resources may be obtained through system integration, technological innovation, and implementation of resource-saving technologies and measures, while it also contributes to enhancing production capacity, environmental sustainability, and the resilience of the agricultural systems.

19.2 OVERVIEW OF SUSTAINABLE FARMING SYSTEMS

19.2.1 Integrated Farming Systems

IFS is an ecologically, economically, and socially sustainable agriculture intervention involving crop, livestock, and aquaculture components synergistically integrated into the farming system (Singh *et al.*, 2020). The combination of different farming enterprises may lead to enhanced system

* Corresponding author: rizwan.maqbool@uaf.edu.pk

DOI: 10.1201/9781003498698-19

productivity and resource use efficiency and provide more employment potential. IFS involves the optimum and efficient utilization of resources to maximize land productivity and sustainability in agriculture. Several integrated components may provide more options to spread the risks. IFS increases farm productivity by combining various farming enterprises and utilizing the resource-rich land (Singh *et al.*, 2020).

19.2.2 Conservation Agriculture

Conservation agriculture is a production system for ensuring food security that evolved as a result of the negative consequences of intensive agriculture and a growing concern related to environmental degradation and climate change caused by traditional systems (Babar *et al.*, 2009). Shahbaz *et al.* (2022) wrote that the negative consequences of intensive agriculture include 'significantly reduced water retention and quality of soil; decreased soil organic matter; and increased soil erosion agriculture.' Most importantly, conservation agriculture replaces the unsustainable use of chemicals and fertilizers with crop management systems that use less land and labor. Employing conservation agriculture practices, such as IPM, agroforestry, and agroecology, can help farmers develop more sustainable and resilient food systems, reduce the reliance on chemical pesticides, enhance soil health, increase biodiversity, and mitigate environmental impact. For instance, Waheed *et al.* (2019) explained that conservation farming is 'a natural ecosystem management system that aims at sustainable crop productivity, economic profitability and food sufficiency through resource and environmental conservation.' It delivers sustainable food systems by implementing integrated water management plans, agroforestry, and promoting living soils. Rana (2023) stated that this farming system is based on the principle of the no-tillage method and 'improves productivity and reduces cost of crop agricultural production, where farmers can generate more crops with less labor and less cost.' Meanwhile, conservation agriculture 'minimizes farming costs and efforts' to protect and enhance 'soil productivity' (Javaid *et al.*, 2012). Unlike intensive agriculture, conservation agriculture proposes a range of conservation practices for an implementation of ecological improvements in farming system (EIFS). That is, for the best results, conservation agriculture essentially proposes organic farming. Chadha *et al.* (2019) stated that conservation agriculture proposes a range of conservation practices to be implemented in all farming systems so that ecological improvements can be achieved, and these practices can be adapted to different needs and contexts.

19.2.3 Organic Farming

Organic farming is a sustainable farming system that emphasizes the conservation and recycling of resources to develop more sustainable production systems (Mahmood *et al.*, 2013; Smith *et al.*, 2015). The prolific use of synthetic inputs and chemicals often degrades soils, reduces biodiversity, and leads to unsustainable use of resources. Organic farming promotes the natural health of soils, biodiversity, and sustainable use of resources by prohibiting synthetic inputs and chemicals. Organic farming plays a vital role in making agriculture more sustainable by reducing carbon footprints, conserving water, and ensuring natural pest controls.

19.2.4 Agroforestry

Agroforestry is the intentional incorporation of trees along with crops and/or pasture (livestock). Studies have demonstrated that agroforestry can represent an important route to sustainable agriculture. Agroforestry practices are argued to avoid soil and environmental degradation that has taken place in the past or is currently taking place, increase agricultural production, support carbon sequestration, and enhance soil and ecosystem functions and the services they provide for human welfare (Ali *et al.*, 2013). An important component of the related strategy of sustainable intensification of agricultural production, which benefits restoration of degraded landscapes for increased ecosystem functions and services, is integration of food production in the landscape for the conservation of biodiversity. Agroforestry provides many benefits, including carbon sequestration, soil quality, and biodiversity, but there are impediments to adoption. Such challenges include more thorough studies on effectiveness and stakeholders (Castle *et al.*, 2022; Park *et al.*, 2022; Tranchina *et al.*, 2024).

19.2.5 Permaculture for Sustainable Resource Utilization

Permaculture is an intensive, sustainable farming system involving complex multi-strata polycultures, perennial plants, crop-animal integration, habitat diversity, water management, and on-site energy production (Morel *et al.*, 2019). The system designed in such a way that it could replicate the integrated systems used by nature, harnessing the energy of the sun and minimizing inputs from the oceans and the winds so that more is produced than is put in, waste is minimized, and ecosystems are never compromised (i.e., resources are better utilized) to do a better job of increasing energy (and further the goal of improving soil quality and resilience at the same time) should be a significant improvement in mass farming (Meena *et al.*, 2024). Agriculture is facing environmental, economic, and social problems as a result of intensification, industrialization, and specialization, and permaculture has been seen as a way to offer another pathway, one more in harmony with ecological systems than the current 'industrial organic farming' model (Tanveer *et al.*, 2015). Studies show that permaculture's yield increases are higher than sustainable agriculture and organic farming and that farm yields and food production are two to ten times better than traditional forms of farming (Mahmood *et al.*, 2015). By better utilizing the sun's energy in natural cycles, agroforest systems, and developments of sustainable multispecies perennials, preserving and increasing biodiversity helps with the mitigation of the climate crisis (Brozovic, 2020; Rasheed *et al.*, 2022c). The design system not only offers a sustainable form of agriculture but also creates sustainable human environments, with the discipline's main goal being to make the most efficient use of the available resources and raw energy inputs of a particular place or region to create a sustainable, healthy, and fair living system for human and non-human entities (Szilágyi et al., 2021). While there is no strict principle to be followed when applying the system, the principles that guide it encourage the individual states of homeostasis for humans and their surrounding ecosystems and natural systems. Using both indigenous and modern agricultural knowledge, permaculture offers a way to design sustainable agricultural systems relating to farming, urban design, construction, and all other forms of living systems, offering an alternative to organic farming that ensures increased biodiversity and decreased food-water-energy nexus issues (Suh, 2014; Gashute and Hale, 2023). Overall, IFS, conservation agriculture, organic farming, agroforestry, and permaculture are considered integral parts of agroecological approach. It is extremely important to integrate these approaches into agricultural systems so that enhanced productivity can be achieved, as well as conserved natural resources and incorporated environmental sustainability into the agriculture.

19.3 WATER RESOURCE MANAGEMENT

19.3.1 Water Use Efficiency in Agriculture

Water use efficiency, which is a key sustainability aspect, is commonly used to refer to the efficient utilization of water resources to achieve maximum productivity of the crop while reducing the amount of water wasted.

Recent studies indicated that employing improved water use efficiency technologies, such as drip irrigation, mulching, and suitable timing of irrigation, can promote crop productivity (Zafar *et al.*, 2023). If water use efficiency technologies, such as precision irrigation systems and water-saving technologies, are adopted, the water consumption would be greatly reduced, even though the crop yield could stay the same or even be increased. Therefore, farmers could significantly promote water use sustainability through the adoption of water use efficiency technologies.

19.3.2 Irrigation Systems for Sustainable Water Use

Identification and implementation of the suitable fit of appropriate irrigation systems could commemorate sustainable use of water in the agricultural sector. Different efforts to control water supply in crops, like drip irrigation, sprinkler irrigation, and furrow irrigation. These strategies help to provide the appropriate distribution of water in different conditions and reduce water losses. Several studies focus on the correct supply of irrigation water through suitable technological systems specific to crops and their quality (Aziz *et al.*, 2018).

Additionally, reaching the equilibrium of the irrigation system with crops' needs and soil features creates an adept irrigation process, which leads to optimum utilization of water use efficiency and subsequently improves agricultural productivity in different conditions (Hanif *et al.*, 2017).

An integration of different technological systems in agriculture, whether of drip or sprinkler irrigation, helps to optimize it with sustainability measures and achieve savings in water use, reduction of environmental pollution, and improved sustainability of the agricultural farm.

19.3.3 Rainwater Harvesting Techniques

The second strategy that is considered important for sustainable water management in agriculture is rainwater harvesting. Capturing and storing rainwater is an approach that helps to balance irregular rainfall seasons, improve irrigation water supply, minimize dependency on groundwater, and increase water availability for irrigation, especially in dry periods. In many parts of the world, farmers depend entirely on rainfall for crop production. If rainfalls are erratic and not sufficient, farmers can face colossal yield loss. For instance, by implementing rainwater harvesting techniques, crop production can be significantly improved. In many dry regions of the world, rainfall can be highly erratic, resulting in a big challenge in meeting water demands in dry periods.

Over the past few decades, research shows that if rainwater harvesting systems are implemented depending on the local conditions, they could be a potential solution to handle water scarcity and improve agricultural resilience to environmental variability and change.

19.3.4 Drainage Management

Drainage management was seen as an important context for effective water utilization in agriculture by farmers and extension workers, while drainage systems provided an important tool for managing water flows in a sustainable manner. This enabled crop production and improved productivity by preventing waterlogging and soil erosion, enhancing soil structure, reducing nutrient loss, and leading to healthier crops. Establishing appropriate drainage systems was crucial to ensuring effective water flows, especially for carrying away excess water from the fields. Other studies also noted how drainage management could be combined with other water-saving practices and integrated into the overall water-saving approach of farmers. In summary, it is important to emphasize systems. Being efficient with water use, particularly using the appropriate irrigation methods, rainwater harvesting cropping systems, and drainage management, will ensure higher and maximum use of resources, better quality produce, and better farming overall.

19.4 NUTRIENT RESOURCE MANAGEMENT

19.4.1 Integrated Nutrient Management

Through this integrated nutrient management (INM) approach, farmers combine organic and inorganic features and bio-fertilizers to supply a balanced diet of nutrients to the crop and improve its nutrient use efficiency. INM also leads to improvement of soil health and reduces the use of chemical fertilizers, thereby inducing long-term sustainability of agriculture. Overuse of chemical fertilizers can lead to the contamination of soil and drinking water, as well as vast greenhouse gas emissions and wide-ranging environmental problems. Integrated nutrition management can help minimize these environmental problems and improve the health of the planet. According to Jain *et al.* (2024), an integrated nutrition management approach in agriculture leads to a balanced and optimum diet of nutrients being supplied to the crop. This improves the nutrient use efficiency of the crops and thus improves the soil health and productivity with reduced use of chemical fertilizers. Integrated nutrition management has been found to be successful in reducing greenhouse gas emissions because it relies on recycling of residues and greater use of organic fertilizers than traditional agriculture. It also results in better use of nitrogen by plants and thus reduction in nitrous oxide emissions.

19.4.2 Precision Agriculture for Nutrient Application

Precision agriculture techniques can help reduce nutrient loss by allowing for targeted application of nutrients based on data gathered in real time and variability across the field. For instance, through global positioning system (GPS)-guidance of machinery and utilization of soil sensors, farmers can apply the right amount of nutrients as needed by crop, thereby improving efficiency and reducing the environmental footprint (Bala *et al.*, 2024).

Precision agriculture improves nutrient management through the principle of '4Rs,' i.e., right source, right rate, right time, and right place of nutrient application.

19.4.3 Biological Nitrogen Fixation

Biological nitrogen fixation, in which plants and microorganisms convert atmospheric nitrogen to a form usable by plants, can be utilized by promoting nitrogen-fixing crops, such as legumes, or through the inoculation of soils with nitrogen-fixing bacteria. This enables farmers to use less synthetic nitrogen fertilizer in nutrient management (Gawande *et al.*, 2023), thus improving the sustainability of agriculture and reducing greenhouse gas emissions related to nitrogen fertilizer manufacturing (Shang *et al.*, 2019).

19.4.4 Composting and Organic Fertilizers

Composting and organic fertilization utilizes nutrients in organic waste for agricultural production by recycling crop residues, manure, kitchen scraps, and other wastes. The practice of composting can help farmers to fertilize their land by returning nutrients to the soil, thus improving soil fertility and structure. Organic fertilizer provides slow-release nutrition to plants, and by increasing soil biodiversity and microbial activity, it promotes the health of the soil and the sustainability of agricultural systems in the long run (Olivo *et al.*, 2024).

To sum up, nutrient resource management is an important factor of a sustainable crop production system. Consequently, nutrient management integration, precision agriculture for nutrient application, biological nitrogen fixation, composting, and the use of organic fertilizers are the significant ways for efficient resource utilization and sustainable environment for agriculture.

19.5 BIODIVERSITY AND ECOLOGICAL RESOURCE MANAGEMENT

19.5.1 Crop Diversification Strategies

Crop diversification involves including different crops in a given area, which plays an important role in the efficient use of resources in agricultural systems and enhances the sustainability of different farming systems. Indeed, crop diversification can help to reduce the presence of weeds, improve soil health, and increase the resilience of the farming system as a whole. Crop diversification, including crop rotation, the use of legumes, and reduced tillage, is an important strategy for sustainable agriculture, bringing a remarkable contribution in the reduction of synthetic inputs and increasing crop yields, according to a recent paper. Studies support the idea that diversified production systems can double food output in developed and developing countries while simultaneously reducing environmental impact, and therefore they have been recognized as a key component of ecological intensification in agriculture (Hayat *et al.*, 2010; Ullah *et al.*, 2013). The use of biodiversity-based production systems that promote positive feedbacks between species can boost productivity and avoid the excessive use of mineral fertilizers and pesticides.

19.5.2 Integrated Pest Management

IPM is one of the key cornerstones of any sustainable crop-production system. Many studies have reported additive and complementary synergies in integrated crop-livestock systems that eventually helped to improve the overall production activities on farm by promoting the integration of crops such as legumes with livestock and other crops (Asante *et al.*, 2020). The ecosystem network analysis in SICLS has also elucidated the role of IPM in sustaining crop-livestock agricultural diversity, reducing the service gaps in ecological functions that can improve ecosystem services, and restoring the validity of environmental sustainability (Paramesh *et al.*, 2020). In IPM, farmers and farm managers are encouraged to utilize a range of strategies in pest management to control the pests without depending on any chemical pesticides.

19.5.3 Pollinator Conservation

Habitat management for beneficial insects (including pollinators) can act as a non-chemical strategy to enhance crop pollination in all pollinated crops and thus may improve overall crop yield. Numerous studies establish that, where wild pollinators are concerned, where you have pollinator-friendly habitat in the farming matrix, you can be supportive of pollinators and consequently increase your crop yield. Farmers can enhance pollination services, improve crop yields, and make overall farming more sustainable by managing farm landscapes to attract pollinators.

19.5.4 Habitat Management for Beneficial Organisms

One of the major strategies for enhancing ecological servicing in farming systems is habitat management for 'third-order organisms.' Integrated crop-livestock systems enhancing ecosystem services sustain agricultural diversification and may improve soil health and restore ecological services, reported Western Kenya (Dauda *et al.*, 2022). The structural transition in crop composition across these regions has also taught us the role of social relationships in promoting sustainable farming practices and crop diversification (Ochieng *et al.*, 2020; Mahmood *et al.*, 2021). Providing the available habitat for 'third-order organisms' who provide ecosystem services improves the resilience of the farming system.

To summarize, the sustainable use of resources in farming systems is largely achieved through certain common crop diversification strategies, on-farm IPM, pollinator conservation, and even habitat management for beneficial organisms. When farming systems engage in these strategies, resources will be used in greater proportion to benefit people, productivity will increase, and sustainable use of resources is more likely.

19.6 ENERGY RESOURCE MANAGEMENT

19.6.1 Energy Efficiency in Farming Operations

Agricultural efficiency is an essential topic as it can increase agricultural output, resource conservation, and farm viability. Agricultural utilization of energy is important because it has widespread implications on agricultural yields, resource efficiency, and overall farming sustainability. Studies revealed that adopting renewable energy for farm operations could improve agricultural productivity, reduce environmental footprint, and strengthen the agriculture sector (Ahmad *et al.*, 2010; Fatima *et al.*, 2023). To promote agricultural viability, energy utilization within farms must be efficient by eliminating energy waste, optimizing whole-farm efficiency, and encouraging resource utilization. These can be achieved through recent agricultural technology, like precision agriculture and emerging smart farming technologies (Cofas and Bălăceanu, 2023). For instance, a growth mindset reduces the energy needed for crops and contributes to indoor farming. Moreover, in recent times, energy generation was seen as an important developmental goal, leading to both governments and farmers adopting various energy sources, including solar and wind power generation, to reduce depletion.

19.6.2 Renewable Energy Sources in Agriculture

Integrating renewable energy sources into agricultural systems plays a vital role in achieving energy efficiency and reducing dependence on non-renewable energy sources. Mahmood *et al.* (2015) reported that using agricultural holdings for biomass energy production would provide a feasible renewable energy source for the society and conservation of the environment. Furthermore, energy needs of the farmers could be met using bioenergy crops to harvest solar energy by plants and biogas from animal and other organic waste (Ghaly and Al-Hattab, 2012). In this regard, the future of farming links to a circular economy that involves 100 percent utilization of waste and biomass for energy generation, therefore reducing greenhouse gas emissions and other environmental effects of the agricultural sector. In terms of smart farming, the use of blockchain would provide authenticated wireless links between drones and the sensors in smart farming to enhance energy efficiency and automation in agriculture.

19.6.3 Bioenergy Crops and Their Role in Sustainable Farming

Bioenergy crops can play an important role in carbon emissions reductions, as renewable sources of bioenergy can replace traditional fossil fuels (Saidivaliyeva and Khalikova, 2024). The cultivation of energy crops, such as switchgrass and miscanthus, has led to more sustainable energy production with enhanced soil health and biodiversity (Saidivaliyeva and Khalikova, 2024). With the help of precision farming enabled by artificial intelligence and big data analytics, bioenergy crop cultivation can be more resource-efficient and environmentally friendly (Gupta and Gupta, 2024). Utilization of energy balance and energy indices, especially in crop production systems, can help the farmers make decisions and increase energy efficiency and decrease the environmental impact of growing crops for long-term sustainability in agriculture (Azarpour *et al.*, 2012; Ashraf *et al.*, 2016).

To conclude, the proper management of energy resources from a homeostatic viewpoint is an important factor in making our production systems more sustainable, leading to an increase in the quality of crops. In particular, increasing energy use efficiency, the use of renewable energy sources, and the production of bioenergy crops can be the winning strategies in the relationship between the farmer and the environment.

19.7 TECHNOLOGICAL INNOVATIONS FOR SUSTAINABLE RESOURCE USE

19.7.1 Precision Agriculture Technologies

Technologies such as precision agriculture affected contemporary practice by allowing resources to be used efficiently to maximize productivity and limit the environmental footprint. Precision agriculture tools encompass GPS guidance systems paired with variable rate technology and automated machinery, enabling fertilizer, pesticide, and water applications at variable rates contingent upon specific field conditions. The use of these technologies allows farmers to better manage and allocate resources, thereby leading to the achievement of higher crop yields at a lower input cost and with a smaller indentation on the environment (Usenko et al., 2024). The use of precision agriculture technologies in farming systems brings more efficiency to resource use in agricultural activities, contributing to making crop production more sustainable. The farmer can use resources responsibly and effectively. Sustainable crop production involves the use of limited resources efficiently and productively (Rivington *et al.*, 2007; Ashraf *et al.*, 2013).

19.7.2 Remote Sensing and GIS Applications

Remote sensing and geographic information system (GIS) applications are essential for sustainable utilization of resources for crop production. The reason for this is that data from remote sensing techniques on crop health, soil, and environmental conditions help farmers to take decisions on utilizing resources in a manner to ensure the sustainability of crops (Vogeler *et al.*, 2016). For instance, with the help of remote sensing techniques, such as satellite images and drones, farmers can monitor crop growth to detect abnormalities indicating pest infestation, as well as plan for interventions to avoid yield loss, improve crop growth, and mobilize inputs in places where they are necessitated (Mehmood *et al.*, 2018; Loures *et al.*, 2020). Further, GIS applications equip farmers with spatial analysis tools to help identify the best location for growing crops, devise the best irrigation plans to enhance water use efficiency, and minimize environmental risks. All these have helped make farming systems that can capitalize on precision agriculture for sustainable use of resources for crop production (Rivington *et al.*, 2007; Martin-Clouaire and Rellier, 2010).

19.7.3 Smart Farming and Internet of Things (IoT)

Smart farming with Internet of Things (IoT) is a paradigm shift toward a data-driven, automated approach to farm management. IoT devices like sensors, actuators, and drones are used to collect data on crop production, soil health, weather, and other factors relating to crop production (Dhinakaran *et al.*, 2023). The collected data are used for monitoring and managing farm operations remotely. Smart farming increases accuracy in resource use in crop production, reduces wastage, saves cost, and increases productivity by precisely managing farm operations (Almomani *et al.*, 2021). It uses IoT devices with data analytics and machine learning (ML) algorithms to model data and produce recommendations for farmers. High productivity and quality of crops can be obtained when a smart farming system is in operation, thus enhancing the sustainability of crop production (Chadha *et al.*, 2019; Reza *et al.*, 2022). Smart farming practices help to enhance the efficiency of resource use in crop production and, in turn, lead to environmental sustainability by reducing inputs and maximizing outputs (Hsiao and Hsu, 2014).

19.7.4 Decision Support Systems for Resource Management

Decision support systems (DSSs) are also important for helping farmers make informed decisions related to inputs and output allocations in crop production. These DSSs utilize data from sensors, weather forecasting services, and historical records to recommend input allocations and farming practices for farmers (Cabrera, 2012). The use of DSS in crop farming systems can help farmers to become more efficient, reduce production risks, and improve farm profitability (Flores *et al.*, 2010). DSS for resource management can help farmers optimize their irrigation scheduling, fertilizer applications, and pest-control activities, which can lead to more sustainable crop production practices (Rasheed *et al.*, 2022a). Farmers using DSS to make more informed decisions within their cropping systems can better adapt to changing weather patterns, market dynamics, and regulatory requirements that would otherwise affect the sustainability and profitability of their operations.

19.8 SOCIO-ECONOMIC ASPECTS OF SUSTAINABLE FARMING SYSTEMS

19.8.1 Economic Viability of Sustainable Farming Practices

Since viability is a crucial determinant of sustainability, economic considerations are vital for the long-term success and viability of any sustainable farming system. Some empirical studies indicate that when the government provides green-production subsidies and levies for non-green practices are enforced, the likelihood of the growers using sustainable practices significantly increases, highlighting the positive correlation between economic incentives and sustainable farming methods (Lu *et al.*, 2024). Market incentives or price premiums earmarked to reward producers for their use of sustainable production methods can further enhance the economic viability of sustainable farming systems. For example, some types of produce in the UK market, such as raspberries and potatoes, are eco-labeled, and empirical studies have examined whether producers will be compensated by the consumers in the form of a premium price when they produce food according to sustainable production methods (Roheim *et al.*, 2011).

19.8.2 Policy Support for Sustainable Agriculture

Even if the initial technological building blocks were available in a low-carbon form, their economic feasibility depends greatly on policy support. This can provide the regulatory framework and incentives for farmers to adopt ecologically sound practices. In China, 'identity-preserved production and marketing' systems that promote the direct and secure connection between producers and consumers have proliferated in recent decades due to changing tastes, technological innovations, and intensified regulation and control over food quality and safety. This shows that policy can play a key role in driving Chinese agriculture toward sustainability. For example, since China launched its carbon trading scheme in 2013, researchers have found that some Chinese companies have shifted their portfolio of projects to cheaper and more energy-efficient ones with the potential to reduce corporate emissions and induce corporate redesign of production processes.

19.8.3 Farmer Education and Extension Programs

Educating the farmers about sustainable farming techniques and building their capacity to put their ability into practice through extension programs is crucial in raising awareness (Wilkins, 2008). Extension programmers play an important role in agricultural practice, as researchers are able to transfer knowledge and tools to farmers by bridging the gap between agricultural research and agricultural practice (Mir *et al.*, 2022). Education and extension provide farmers with the needed knowledge and skills to implement sustainable farming systems.

19.8.4 Market Incentives for Sustainable Production

Market incentives can also be an important driver; they can encourage farmers to adopt sustainable production practices. To illustrate, tens of sustainable agriculture and forestry certification schemes have been devised to ameliorate the sustainability of environmental and social aspects of production processes, and producers self-report their adoption of sustainable practices (Newton *et al.*, 2015). Furthermore, institutionalizing market incentives that can accelerate smallholders' certification, as seen in the oil palm sector, has been shown to significantly improve the uptake of sustainable practices, supporting once more the role of market incentives to promote sustainability (Wibowo *et al.*, 2023). Production practices beneficial to the environment and society can be made more profitable to farmers through market incentives.

19.9 CASE STUDIES AND SUCCESS STORIES

19.9.1 Regional Examples of Successful Sustainable Farming Systems

An enquiry-based approach to the analysis of sustainable farming systems in different regions indicates there are distinct pathways through which agricultural sustainability has been advanced on a global scale. Long-term rotation trials with cereals in the Mediterranean region have contributed to the development of new technologies and crop-management systems for increased sustainability of crop production (Ryan *et al.*, 2008). Sustainability in these trials has also guaranteed enriched crop productivity, long-lasting effectiveness, and productivity from biological, environmental, and economic perspectives. Similarly, in the Himalayan region, mainstreaming cultivation of Barahnaja has come up as an important strategy to improve the food and nutritional security of the region (Gururani *et al.*, 2021). Here, farmers are combining traditional farming with modern scientific knowledge so as to optimize crop production in an ecologically sustainable manner.

Additionally, conservation tillage and nutrient practices are also found to enhance crop-water productivity and the economic profitability of wheat production. Through these practices, farmers can mitigate the loss of soil fertility, a major hazard to crop productivity and profitability. Crop-water productivity and the economic profitability of wheat production can also be improved under conservation agriculture, especially when considering diverse cropping systems. For instance, recent studies on conservation agriculture in Ethiopia have identified best maize-legume-based cropping systems for improving crop productivity sustainably under conservation agriculture practices. Through their investigation, Mekasha *et al.* (2021) pinpointed a plethora of practices with regard to double intercropping and double cropping under conservation agriculture to leverage the residual soil moisture to enhance productivity and sustainability.

Moreover, the selection of most profitable but sustainable farming crops in the nutrient-rich Wadi areas of Saudi Arabia using multiple-criteria decision analysis (MCDM) highlights the importance of sustainable crop production activities in managing the disparity between demand and supply (Tirth *et al.*, 2021). These studies categorically highlight that selection of crops for good profitability combined with sustainable criteria can improve the productivity of farms while maintaining ecological equilibrium. Moreover, the implementation of best management practices for maintaining agricultural production in response to climate change in the Choctawhatchee Watershed in Alabama has advocated for increasing production and decreasing fertilizer use through crop rotation and legume incorporation (Chadha *et al.*, 2020; Afroz *et al.*, 2021).

19.9.2 Lessons Learned and Best Practices

Meanwhile, to lay the groundwork for others to identify the lessons learned over time in different models of sustainable farming systems and best practices in order to replicate and/or improve upon them to achieve the objectives of sustainable, adequate, and accessible food production and environmental sustainability, examples shared below are worth highlighting.

For instance, the application of sustainable production practices among the farmers for yellow pepper production was evaluated in various research studies in Nigeria, revealing some key practices for sustainable crop production, as farmer field schools let farmers discover and adopt sustainable practices, including mass selection, poultry manure, and minimum use of agro-chemicals, as suggested for its high productivity with respect to consumers' preferences and low cost (Tanveer *et al.*, 2015; Nwonu *et al.*, 2022).

Overall, these regional cases of successful sustainable farming systems highlight the different approaches to enhancing agricultural sustainability around the world. Experiences and best practices in sustainable farming underline the importance of knowledge sharing and adoption of new, sustainable practices by farmers and many others involved in the agricultural sector to achieve sustainable agriculture in the long term.

19.10 FUTURE PERSPECTIVES AND CHALLENGES

19.10.1 Emerging Trends in Sustainable Farming Systems

Trends emerging from sustainable farming systems are shaping the future of agriculture toward cleaner and more efficient farming practices. Digital multi-precision technologies, such as artificial intelligence and sensors, could revolutionize farming approaches by taking decisions based on data measurements and therefore optimizing resources (de Oliveira *et al.*, 2024). However, vertical farming techniques, where plants are grown in vertically stacked layers, are increasingly employed as a sustainable approach for growing crops and solving problems related to land shortage and the environment (Mustafa *et al.*, 2022; Parameswari *et al.*, 2024). On this basis, we can argue that farming systems are shifting toward a sustainable and technology-based farming approach. Furthermore, sustainable agriculture also promotes the use of organic farming systems, which are known to be an emerging trend in sustainable farming systems. Organic farming focuses on reducing the use of chemical input, improving soil health, and increasing biodiversity, which eventually increases the sustainability of crop production (Hazmi *et al.*, 2023). Organic farming practices can help farmers to reduce the negative impact on the environment, increase soil health, increase land productivity, and improve the sustainability of agricultural systems in the long term.

19.10.2 Addressing Climate Change through Sustainable Resource Use

Since agricultural systems are especially sensitive to climate change, adaptation of agricultural production systems to the impacts of a changing climate needs an increased resilience of a system to climate change through sustainable resource use. Crop types and cropping practices, such as crop rotation, biological pest control, or soil conservation management, are the core of the multi-nutrient management systems in integrated agricultural systems, and they determine the efficiency as well as the resilience of the system. These practices can help improve soil health and agricultural productivity, lower greenhouse gas emissions, increase crop production, and enhance adaptations to climate change.

In addition, climate change can be reduced based on resource efficiency and environmental footprints for innovative implementation of sustainable special integrated livestock farming systems (Witjaksono and Indrasti, 2018). When the crop and livestock components of an IFS are synergistically managed, the operations involved can interdependently be optimized for best resource use, waste minimization, and overall production system sustainability. Transitioning farming toward organic management systems can improve the sustainability of food production in terms of climate change mitigation, especially through biodiversity enhancement (Khoiri *et al.*, 2021).

19.10.3 Scaling Up Sustainable Practices

This means that the main issue to expand the sustainable use in farm systems is the scale-up of those good practices. In order to do so, it is necessary to first increase the awareness of farmers on the use of sustainable agriculture practices. This increase in the average farmer's knowledge should induce an increase in the number of sustainable farming systems adopted. (In this case, the way that a measure can causally affect a specific outcome is called instrumentality.) Gebska *et al.* (2020) suggested that an increase in farmers' knowledge of sustainable agriculture practices and sustainable systems of production, as well as in their expectation to have a low use of chemicals, might trigger positive consequences on farming systems.

This also opens the door to different ways of measuring the economic performance of farming practices and assessing the sustainable performance of farms through tools like the Balanced Scorecard, balancing the environmental and social perspectives with the economic (see examples in Chen *et al.*, 2020; Chopin *et al.*, 2021). These aspects are also needed to be able to scale up sustainable farming systems. The key question is the benefits of producing foods and fibers sustainably and their contribution to economic viability greater than the costs to our wellbeing and natural capital that stem from the detrimental effects of agriculture? These are crucial common grounds for understanding how we may achieve more sustainable and equitable food systems.

Conclusively, sustainable farming system trends with better adoption of nutrients, herbicides, pesticides, and water management; climate change mitigation through sustainable use of resources; and scaling up of sustainable agriculture are the key factors in the direction of ensuring sustainable resource use in agriculture. These will help the agricultural industry by being more resilient, efficient, and environmentally friendly through adoptions of various advanced technologies, organic farming, and sustainable managing techniques.

19.11 CONCLUSION

19.11.1 Summary of Key Points

To explore different farm systems for sustainable utilization of resources, it is crucial to consider that IFS can be useful in developing an efficient use of resources and improving productivity, as the farming system can be integrated depending on the combination of crop, livestock, and aquaculture components. An IFS can be the best option in the utilization of resources in a farming system as compared to others, as a report reveals the ratio of land utilization, livestock feed and protein, fertilizer demand, and water demand, which will be efficient as compared to others. Moreover, there are many systems, but to make farming options efficient as discussed above, one should use animal manure and crop residue, should use technology like artificial intelligence and robotics, should be efficient in using resources in farming, should produce bio-fertilizer and bio-pesticide, should raise tourism and agro-tourism, demand generation, skill development, etc., which will ultimately help in achieving more sustainability in the production system. The incorporation of renewable technologies in farming practices can play a vital role in supporting the regeneration of resources and establishing a sustainable agriculture system in the long run, and also mankind can enjoy the benefits of nature. Supplying energy should be up to 100 percent through renewable energy sources, water should be conserved while maintaining crop productivity, glasshouses should be constructed and operated with maximum energy efficiency, heat should be recovered and used for further purposes, and wastes should be recycled rather than disposed off, as it helps in reducing carbon dioxide and odorous gases, along with maintaining the sustainability of an ongoing farming system. Organic agriculture can promote sustainable production systems as it is natural and preservative of resources and recycles all inputs without polluting nature. Moreover, it is commendable that eco-efficient approaches along with vertical farming techniques like hydroponics and aeroponics can help in promoting sustainability in farming that can harness agronomic and ecological benefits.

19.11.2 Call to Action for Implementing Sustainable Farming Systems

To realize sustainable farming systems, it is necessary for the collective effort of farmers, policymakers, researchers, and other stakeholders to implement the evolved IFS.

- Farmers have to switch to IFSs, which would promote the efficiency in utilization of resources and enhance production and productivity (Goswami *et al.*, 2023). Policymakers have to provide support, incentives, regulations, and legislation for promoting sustainable farming practices and for the adoption of new technology (Wilkins, 2008).
- Researchers should develop new methods and technologies to optimize the utilization and sustainability of crop production (Singh *et al.*, 2024).
- Stakeholders have to work together to raise awareness, provide training, and share knowledge to facilitate the evolution of the sustainable farming system.

To conclude, the path toward the development of sustainable farming systems through efficient use of resources is a collective effort and requires a comprehensive approach that combines synergistically the various elements, innovative useful technologies, and resource-based management approaches that, in unity, can help in developing environmentally sustainable and high-yielding farming systems that have the potential to feed the world in the future.

LIST OF ABBREVIATIONS

DSS decision support system
EIFS ecological improvements in farming system
GIS geographic information system
GPS global positioning system
IFS integrated farming system
INM integrated nutrient management
IPM integrated pest management
MCDM multiple-criteria decision analysis
ML machine learning

REFERENCES

Afroz, M.D., Li, R., Muhammed, K., Anandhi, A., & Chen, G., 2021. Best management practices for sustaining agricultural production at Choctawhatchee Watershed in Alabama, USA, in response to climate change. *Air, Soil and Water Research*, 14, 1178622121991789.

Ahmad, K., Ejaz, A., Khan, Z.I., Gondal, S., Fardous, A., Hussain, A., Sher, M., Valeem, E.E., & Ullah, S., 2010. Evaluation of dynamics of iron and manganese from pasture to buffaloes: A case study at rural livestock farms. *Pakistan Journal of Botany*, 42, 3415–3421.

Ali, H., Tanveer, A., Nadeem, M., Asghar, H., & Javaid, M., 2013. Germination ecology of *Rhynchosia capitata*: An emerging summer weed in Asia. *Planta Daninha*, 31, 249–257.

Almomani, A., Al-Nawasrah, A., Alomoush, W., Al-Abweh, M., Alrosan, A., & Gupta, B.B., 2021. Information management and IoT technology for safety and security of smart home and farm systems. *Journal of Global Information Management (JGIM)*, 29, 1–23.

Asante, B.O., Villano, R.A., & Battese, G.E., 2020. Evaluating complementary synergies in integrated crop-livestock systems in Ghana. *International Journal of Social Economics*, 47, 72–85.

Ashraf, E., Hayat, Z., Khan, M., Atif, F., & Haider, M., 2016. Impact of dairy farm management training workshop on the knowledge level of participants. *International Journal of Agriculture and Applied Sciences (Pakistan)*, 4.

Ashraf, E., Shah, F., Luqman, M., Younis, M., Aziz, I., & Farooq, U., 2013. Use of untreated wastewater for vegetable farming: A threat to food safety. *International Journal of Applied Agricultural Sciences*, 5.

Azarpour, E., Moraditochaee, M., & Bozorgi, H.R., 2012. Evaluating energy balance and energy indices of wheat production in rain-fed farming in northern Iran. *African Journal of Agricultural Research*, 7, 1950–1955.

Aziz, M., Ashraf, M., & Javaid, M.M., 2018. Enhancement in cotton growth and yield using novel growth promoting substances under water limited conditions. *Pakistan Journal of Botany*, 50, 1691–1701.

Babar, B.H., Tanveer, A., Tahir, M., Aziz, A., Ahmad, A.U.H., Nadeem, M.A., & Javaid, M.M., 2009. Allelopathic potential of wild onion (*Asphodelus tenuifolius*) on the germination and seedling growth of chickpea (*Cicer arietinum*). *Weed Biology and Management*, 9, 146–151.

Bala, J., Choudhary, K., Bochalya, R.S., Himanshu, & Sharda, A. (2024). Effect of Different Nitrogen Levels and Nano Urea on Yield and Economics of Wheat under Mid Hills of Himachal Pradesh, India. *International Journal of Plant & Soil Science*, 36(12), 68–74. DOI: 10.9734/ijpss/2024/v36i125184.

Benlhabib, O., Yazar, A., Qadir, M., Lourenço, E., & Jacobsen, S., 2014. How can we improve Mediterranean cropping systems? *Journal of Agronomy and Crop Science*, 200, 325–332.

Brozovic, D., 2020. Business model based on strong sustainability: Insights from an empirical study. *Business Strategy and the Environment*, 29, 763–778.

Cabrera, V.E., 2012. DairyMGT: A suite of decision support systems in dairy farm management. *Decision Support Systems*, 282.

Castle, S.E., Miller, D.C., Merten, N., Ordonez, P.J., & Baylis, K., 2022. Evidence for the impacts of agroforestry on ecosystem services and human well-being in high-income countries: A systematic map. *Environmental Evidence*, 11, 10.

Chadha, A., Florentine, S., Chauhan, B.S., Long, B., Jayasundera, M., Javaid, M.M., & Turville, C., 2019. Environmental factors affecting the germination and seedling emergence of two populations of an emerging agricultural weed: Wild lettuce (*Lactuca serriola*). *Crop and Pasture Science*, 70, 709–717.

Chadha, A., Florentine, S., Javaid, M., Welgama, A., & Turville, C., 2020. Influence of elements of climate change on the growth and fecundity of *Datura stramonium*. *Environmental Science and Pollution Research*, 27, 35859–35869.

Chen, N., Yang, X., & Shadbolt, N., 2020. The balanced scorecard as a tool evaluating the sustainable performance of Chinese emerging family farms—Evidence from Jilin province in China. *Sustainability*, 12, 6793.

Chopin, P., Mubaya, C.P., Descheemaeker, K., Öborn, I., & Bergkvist, G., 2021. Avenues for improving farming sustainability assessment with upgraded tools, sustainability framing and indicators. A review. *Agronomy for Sustainable Development*, 41, 1–20.

Cofas, E., & Bălăceanu, C.T., 2023. Evaluation of the biomass energy production potential in agricultural holdings in relation to their size. Case study for cop farms in Romania. *Romanian Agricultural Research*, 40, 655–666.

Dauda, I., Anthony, D., Ruben, J., & Kara, H., 2022. Profitability of watermelon production diversification strategy: Evidence from smallholder farmers in Taraba State, Nigeria. *Asian Journal of Agricultural Extension, Economics & Sociology*, 40, 861–868.

de Oliveira, F.M., Ferraz, G.E.S., André, A.L.G., Santana, L.S., Norton, T., & Ferraz, P.F.P., 2024. Digital and precision technologies in dairy cattle farming: A bibliometric analysis. *Animals*, 14, 1832.

Dhinakaran, D., Gopalakrishnan, S., Manigandan, M., & Anish, T., 2023. IoT-Based Environmental Control System for Fish Farms with Sensor Integration and Machine Learning Decision Support. *arXiv preprint arXiv:2311.04258.*

Fatima, S.A., Nasim, S.F., & Ahmed, S., 2023. Enhancing agricultural operations: Big data analytics using distributed and parallel computing. *International Journal of Emerging Engineering and Technology*, 2, 1–7.

Flores, C.I., Holzapfel, E.A., & Lagos, O., 2010. A dynamic decision support system for farm water management in surface irrigation: Model development and application. *Chilean Journal of Agricultural Research*, 70, 278–286.

Gashute, T.S., & Hale, T.K., 2023. The role of permaculture in the integration of indigenous and modern agricultural knowledge: Evidence from Konso, Ethiopia. *Sustainable Development*, 31, 1781–1789.

Gawande, V., Saikanth, D., Sumithra, B., Aravind, S.A., Swamy, G.N., Chowdhury, M., & Singh, B.V., 2023. Potential of precision farming technologies for eco-friendly agriculture. *International Journal of Plant & Soil Science*, 35, 101–112.

Gebska, M., Grontkowska, A., Swiderek, W., & Golebiewska, B., 2020. Farmer awareness and implementation of sustainable agriculture practices in different types of farms in Poland. *Sustainability*, 12, 8022.

Ghaly, A.E., & Al-Hattab, M., 2012. An innovative farm scale biogas/composting facility for a sustainable medium size dairy farm.

Goswami, R., Brodt, S., Patra, S., Dasgupta, P., Mukherjee, B., & Nandi, S., 2023. Resource interaction in smallholder farms is linked to farm sustainability: Evidence from Indian Sundarbans. *Frontiers in Sustainable Food Systems*, 7, 1081127.

Gupta, N., & Gupta, P., 2024. Robotics and Artificial Intelligence (AI) in Agriculture with Major Emphasis on Food Crops. *Digital Agriculture: A Solution for Sustainable Food and Nutritional Security*. Springer, 577–605.

Gururani, K., Sood, S., Kumar, A., Joshi, D.C., Pandey, D., & Sharma, A., 2021. Mainstreaming Barahnaja cultivation for food and nutritional security in the Himalayan region. *Biodiversity and Conservation*, 30, 551–574.

Hanif, Z., Naeem, M., Ali, H.H., Tanveer, A., Javaid, M.M., Peerzada, A.M., & Chauhan, B.S., 2017. Effect of environmental factors on germination of Salsola foetida: Potential species for rehabilitation of degraded rangelands. *Rangeland Ecology & Management*, 70, 638–643.

Hayat, Z., Pasha, T., Khattak, F., Jabbar, M., Nasir, Z., & Samiullah, S., 2010. Consumer's perception and willingness to buy nutrient enriched designer eggs in Pakistan. *Education*, 5, 0.067.

Hazmi, M., Suryaningrum, D.A., Umarie, I., & Swasono, M.a.H., 2023. Emerging trends and future directions in organic agriculture and environmentally-Friendly farming practices: A bibliometric analysis. *West Science Interdisciplinary Studies*, 1, 438–447.

Hsiao, T.Y., & Hsu, Y.Y., 2014. Modeling different scenarios for forecasting human resources requirements in Taiwan's recreational farms. *International Journal of Business Administration*, 5, 1.

Jadhav, N., Rajnivas, B., Subapriya, V., Sivaramakrishnan, S., & Premalatha, S., 2024. Enhancing crop growth efficiency through IoT-enabled smart farming system. *EAI Endorsed Transactions on Internet of Things*, 10.

Jain, S., Manasa, S., Ranganna, G., Mohapatra, S., & Verma, S. & Mishra, S., 2024. Cultivating sustainability: A comprehensive review of organic farming practices for nutrient-rich fruit production in India. *Plant Archives*, 24(1), 516–524.

Javaid, M., Tanveer, A., Ahmad, R., Yaseen, M., & Khaliq, A., 2012. Optimizing activity of herbicides at reduced rate on *Emex spinosa* Campd. with adjuvants. *Planta Daninha*, 30, 425–435.

Khoiri, A.N., Cheevadhanarak, S., Jirakkakul, J., Dulsawat, S., Prommeenate, P., Tachaleat, A., Kusonmano, K., Wattanachaisaereekul, S., & Sutheeworapong, S., 2021. Comparative metagenomics reveals microbial signatures of sugarcane phyllosphere in organic management. *Frontiers in Microbiology*, 12, 623799.

Loures, L., Chamizo, A., Ferreira, P., Loures, A., Castanho, R., & Panagopoulos, T., 2020. Assessing the effectiveness of precision agriculture management systems in Mediterranean small farms. *Sustainability*, 12, 3765.

Lu, Y., Sun, Z., Xu, J., & Yao, G., 2024. Unveiling agricultural shifts through stakeholder interactions and strategic dynamics in green rice production. *Agriculture*, 14, 724.

Mahmood, A., Hussain, A., Shahzad, A.N., & Honermeier, B., 2015. Biomass and biogas yielding potential of sorghum as affected by planting density, sowing time and cultivar. *Pakistan Journal of Botany*, 47, 2401–2408.

Mahmood, A., Ullah, H., Shahzad, A., Ali, H., Ahmad, S., Zia-Ul-Haq, M., Honermeier, B., & Hasanuzzaman, M., 2013. Dry matter yield and chemical composition of sorghum cultivars with varying planting density and sowing date. *Sains Malaysiana*, 42, 1529–1538.

Mahmood, A., Wang, X., Shahzad, A.N., Fiaz, S., Ali, H., Naqve, M., Javaid, M.M., Mumtaz, S., Naseer, M., & Dong, R., 2021. Perspectives on bioenergy feedstock development in Pakistan: Challenges and opportunities. *Sustainability*, 13, 8438.

Martin-Clouaire, R., & Rellier, J.-P., 2010. A Generic Framework for Simulating Agricultural Production Systems. *In Modelling Nutrient Digestion and Utilisation in Farm Animals*. Wageningen Academic, 11–21.

Meena, L., Kochewad, S., Kumar, D., Malik, S., & Meena, S., 2024. Development of sustainable integrated farming systems for small and marginal farmers and ecosystem services—A comprehensive review. *Agricultural Science Digest*, 44, 391–397.

Mehmood, A., Naeem, M., Khalid, F., Saeed, Y., Abbas, T., Jabran, K., Sarwar, M.A., Tanveer, A., & Javaid, M.M., 2018. Identification of phytotoxins in different plant parts of *Brassica napus* and their influence on mung bean. *Environmental Science and Pollution Research*, 25, 18071–18080.

Mekasha, G.M., Mupangwa, W.T., Seyoum, S.A., Lemma, A.Z., Senbeta, H.B., Onu, M.H., & Tello, M.A., 2021. Identification of best maize-legume based cropping systems under conservation agriculture practices for Central Rift Valley of Ethiopia. *Journal of Plant Breeding and Crop Science*, 13, 35–45.

Mir, M.S., Naikoo, N.B., Amin, Z., Bhat, T.A., Nazir, A., Kanth, R.H., Singh, P., Raja, W., Singh, L., & Fayaz, S., 2022. Integrated farming system: A tool for doubling farmer's income. *Journal of Experimental Agriculture International*, 44, 47–56.

Morel, K., Léger, F., & Ferguson, R.S., 2019. Permaculture. *Encyclopedia of Ecology, 2nd edition*, Oxford, Elsevier, 4, 559–567.

Mustafa, A., Athar, F., Khan, I., Chattha, M.U., Nawaz, M., Shah, A.N., Mahmood, A., Batool, M., Aslam, M.T., & Jaremko, M., 2022. Improving crop productivity and nitrogen use efficiency using sulfur and zinc-coated urea: A review. *Frontiers in Plant Science*, 13, 942384.

Newton, P., Alves-Pinto, H.N., & Pinto, L.F.G., 2015. Certification, forest conservation, and cattle: Theories and evidence of change in Brazil. *Conservation Letters*, 8, 206–213.

Nwonu, P.C., Nwobodo, C.E., Onwubuya, E.A., & Obazi, S.A., 2022. Farmers' use of sustainable production practices for yellow pepper crop in the Nsukka agricultural zone, Enugu State, Nigeria. *Acta Universitatis Sapientiae, Agriculture and Environment*, 14, 13–28.

Ochieng, J., Kirimi, L., Ochieng, D.O., Njagi, T., Mathenge, M., Gitau, R., & Ayieko, M., 2020. Managing climate risk through crop diversification in rural Kenya. *Climatic Change*, 162, 1107–1125.

Olivo, A.J., Godber, O.F., Perry, J., Wattiaux, M.A., & Ketterings, Q.M., 2024. Engaging secondary school students in sustainability assessments through farm nutrient mass balances. *Natural Sciences Education*, 53, e20147.

Paramesh, V., Sreekanth, G.B., Chakurkar, E.B., Chethan Kumar, H., Gokuldas, P., Manohara, K.K., Ramdas Mahajan, G., Rajkumar, R.S., Ravisankar, N., & Panwar, A.S., 2020. Ecosystem network analysis in a smallholder integrated crop–livestock system for coastal lowland situation in tropical humid conditions of India. *Sustainability*, 12, 5017.

Parameswari, P., Ragini, M., Singh, V., Nm, R., Tiwari, A.K., Belagalla, N., Pandey, S.R., & Kolekar, S.N., 2024. Vertical farming: Revolutionizing sustainable agriculture in the 21st century. *Journal of Scientific Research and Reports*, 30, 917–930.

Park, M.S., Baral, H., & Shin, S., 2022. *Systematic Approach to Agroforestry Policies and Practices in Asia.* MDPI, 635.

Rana, M.M., 2023. Conservation agriculture for sustainable crop productivity and economic return for the smallholders of Bangladesh: A systematic review. *Turkish Journal of Agriculture-Food Science and Technology*, 11, 2009–2015.

Rasheed, A., Barqawi, A.A., Mahmood, A., Nawaz, M., Shah, A.N., Bay, D.H., Alahdal, M.A., Hassan, M.U., & Qari, S.H., 2022a. CRISPR/Cas9 is a powerful tool for precise genome editing of legume crops: A review. *Molecular Biology Reports*, 49, 5595–5609.

Rasheed, A., Mahmood, A., Maqbool, R., Albaqami, M., Sher, A., Sattar, A., Bakhsh, G., Nawaz, M., Hassan, M.U., & Al-Yahyai, R., 2022b. Key insights to develop drought-resilient soybean: A review. *Journal of King Saud University-Science*, 34, 102089.

Rasheed, A., Raza, A., Jie, H., Mahmood, A., Ma, Y., Zhao, L., Xing, H., Li, L., Hassan, M.U., & Qari, S.H., 2022c. Molecular tools and their applications in developing salt-tolerant soybean (*Glycine max* L.) cultivars. *Bioengineering*, 9, 495.

Reza, A.W., Saymatul Jannat, K., Islam, M., & Das Barman, S., 2022. Smart pre-seeding decision support system for agriculture. *Journal of Information Systems and Telecommunication (JIST)*, 2, 102.

Rivington, M., Matthews, K.B., Bellocchi, G., Buchan, K., Stöckle, C.O., & Donatelli, M., 2007. An integrated assessment approach to conduct analyses of climate change impacts on whole-farm systems. *Environmental Modelling & Software*, 22, 202–210.

Roheim, C.A., Asche, F., & Santos, J.I., 2011. The elusive price premium for ecolabelled products: Evidence from seafood in the UK market. *Journal of Agricultural Economics*, 62, 655–668.

Ryan, J., Singh, M., & Pala, M., 2008. Long-term cereal-based rotation trials in the Mediterranean region: Implications for cropping sustainability. *Advances in Agronomy*, 97, 273–319.

Saidivaliyeva, B., & Khalikova, L., 2024. Renewable energy and agriculture: Developing a common language for technological integration. *BIO Web of Conferences, EDP Sciences*, 06006.

Shahbaz, P., Ul Haq, S., & Boz, I., 2022. Adoption of Conservation Agriculture as a Driver of Sustainable Farming: Opportunities, Constraints, and Policy Issues. *In Smart Farming-Integrating Conservation Agriculture, Information Technology, and Advanced Techniques for Sustainable Crop Production.* IntechOpen.

Shang, Y., Hasan, M.K., Ahammed, G.J., Li, M., Yin, H., & Zhou, J., 2019. Applications of nanotechnology in plant growth and crop protection: A review. *Molecules*, 24, 2558.

Singh, S., Singh, P., Kumar, A., Baheliya, A.K., & Patel, K.K., 2024. Promoting environmental sustainability through vertical farming: A review. *Journal of Advances in Biology & Biotechnology*, 27, 210–219.

Singh, V., Rathore, S., Singh, R., Upadhyay, P.K., & Shekhawat, K., 2020. Integrated farming system approach for enhanced farm productivity, climate resilience and doubling farmers' income. *The Indian Journal of Agricultural Sciences*, 90, 1378–1388.

Smith, L.G., Williams, A.G., & Pearce, B.D., 2015. The energy efficiency of organic agriculture: A review. *Renewable Agriculture and Food Systems*, 30, 280–301.

Suh, J., 2014. Towards sustainable agricultural stewardship: Evolution and future directions of the permaculture concept. *Environmental Values*, 23, 75–98.

Szilágyi, A., Plachi, E., Nagy, P., & Simon, B., & Centeri, C., 2021. Assessing earthworm populations in some Hungarian horticultural farms: Comparison of conventional, organic and permaculture farming. *Biology and Life Sciences Forum*, MDPI, 11.

Tahat, M. M., Alananbeh, K. M., Othman, Y. A., & Leskovar, D. I., 2020. Soil health and sustainable agriculture. *Sustainability*, 12, 4859.

Tanveer, A., Javaid, M.M., Irfan, M., Khaliq, A., & Yaseen, M., 2015. Yield losses in chickpea with varying densities of dragon spurge (*Euphorbia dracunculoides*). *Weed Science*, 63, 522–528.

Tirth, V., Singh, R.K., Islam, S., & Algahtani, A., 2021. Optimal selection of agricultural crops for sustainable farming in nutrient-rich Wadi areas of Saudi Arabia using MCDM. *Journal of Engineering Research*, 9.

Tranchina, M., Reubens, B., Frey, M., Mele, M., & Mantino, A., 2024. What challenges impede the adoption of agroforestry practices? A global perspective through a systematic literature review. *Agroforestry Systems*, 98, 1–21.

Ullah, H., Mahmood, A., Ijaz, M., Tadesse, B., & Honermeier, B., 2013. Evaluation of anise (*Pimpinella anisum* L.) accessions with regard to morphological characteristics, fruit yield, oil contents and composition. *Journal of Medicinal Plants Research*, 7, 2177–2186.

Usenko, L.N., Guzey, V.A., Usenko, N.M., & Usenko, A.M., 2024. Analysis of key technologies of digital transformation in agriculture. *BIO Web of Conferences, EDP Sciences*, 03001.

Vogeler, I., Cichota, R., & Beautrais, J., 2016. Linking land use capability classes and APSIM to estimate pasture growth for regional land use planning. *Soil Research*, 54, 94–110.

Waheed, H., Javaid, M.M., Shahid, A., Ali, H.H., Nargis, J., & Mehmood, A., 2019. Impact of foliar-applied Hoagland's nutrient solution on growth and yield of mash bean (*Vigna mungo* L.) under different growth stages. *Journal of Plant Nutrition*, 42, 1133–1141.

Wibowo, L.R., Erdi, E., Hutabarat, S., Nurfatriani, F.N., Utomo, M., Nawireja, I.K., Pramudya, E.P., Kurniasari, D.R., Cahyono, E., & Kurniadi, R., 2023. Accelerating certification of oil palm smallholders through institutionalization of various incentives. *Forest and Society*, 7, 263–294.

Wilkins, R.J., 2008. Eco-efficient approaches to land management: A case for increased integration of crop and animal production systems. *Philosophical Transactions of the Royal Society B: Biological Sciences*, 363, 517–525.

Witjaksono, J., & Indrasti, R., 2018. Developing integrated crop-livestock farming system as a strategy for green growth in low income countries: A brief review. *Animal and Veterinary Sciences*, 6, 12–16.

Zafar, S., Afzal, H., Ijaz, A., Mahmood, A., Ayub, A., Nayab, A., Hussain, S., Maqsood, U.-H., Sabir, M.A., & Zulfiqar, U., 2023. Cotton and drought stress: An updated overview for improving stress tolerance. *South African Journal of Botany*, 161, 258–268.

20 Eco-Friendly Pest Management

*Ayesha Sadeeqa Muhammad, Muhammad Ahmad Zeshan, Sami Ullah, and Muhammad Anjum Zia**

20.1 INTRODUCTION

20.1.1 What Is Eco-Friendly Pest Management?

Plant pests collectively cause 45% of crop losses of worth 290 billion annually, according to careful estimates. The expenditures incurred on chemical pesticides add up more monetary burden on the farmers that not only squeeze the cost-benefit ratio of the crop production but also deteriorate the environment (Tanveer *et al.*, 2013a; Tudi *et al.*, 2021). Although, chemical pesticides have significant positive effect on crop yield increment with efficient pest control, it has also deteriorated the soil, air, and water apart from targeting the non-target organisms. Pesticide residues cause food safety concerns among domestic consumers and trade barriers for export. In food production, one way to address this issue is to replace pesticides with biopesticides (pest control chemicals derived from living organisms); however, commercialization of these products is very slow. Biopesticides are obtained from beneficial microbes, plant extracts, and bio-chemicals produced by the microbes. Biopesticide usage is progressively increasing by 10% annually worldwide (Yadav, 2022).

The byproducts of biopesticides are eco-friendly, easily biodegradable, and cost-effective. Biopesticides are being used in agricultural settings since the domestication of plants has been started due to their non-toxic feature (Dar *et al.*, 2021). Biopesticides do not pollute soil, air, and water, as well as these are target specific i.e. don't harm non-target organisms (Baker *et al.*, 2020).

20.1.2 Multifaceted Impacts of Sustainable Pest Management

The extensive damages caused by pests force the growers to use synthetic chemical pesticides for quick relief that have multiple harmful effects i.e. environmental deterioration, development of resistance in pests, killing of beneficial microbes and insects (Sarkar *et al.*, 2021). An environmentally sustainable substitute is the use of microbial biopesticides, phyto-pesticides obtained from beneficial microbes and plant extracts, which are less dangerous both for the environment and public health. The most often utilized biopesticides are bio-insecticides (*Bacillus thuringiensis*), bio-fungicides (*Trichoderma* spp.), and bio-herbicides (*Phytophthora* spp.) (Fenibo *et al.*, 2022).

The benefits of biopesticides are as follows:

1. Eco-friendly
2. Kills only target organism
3. Effective in very low amounts and degrade quickly in the environment
4. These are integral part of integrated pest management (IPM) programs (Kiss, 2020).

Additionally, biopesticides offer the following advantages (Table 20.1).

* Corresponding author: anjum.zia@uaf.edu.pk

DOI: 10.1201/9781003498698-20

TABLE 20.1
Benefits of Biopesticides

Factors	Advantages of Biopesticides
Economical	More expensive but fewer applications
Durability and after-effect	Sustainable effects with lower environmental residues
Effect of knockdown	Gradual but long-lasting control
Managing and bulkiness	May require more careful handling; dependent on formulation/carrier
Return of pests	Less recurrence due to targeted action and reduced resistance development
Resilience	Biopesticides degrade faster, so slightly less persistent but safer
Impact on beneficial plant life	Less harmful to non-target and beneficial organisms
Specificity of target	Highly specific, minimizing damage to advantageous pests and natural enemies
Waiting period	Often shorter; usually host-specific and safer for immediate use
Type of control	Primarily preventive, supporting integrated pest management (IPM)
Durability of shelf	Usually, shorter shelf life compared to synthetic pesticides

20.2 UNDERSTANDING THE INSIGHTS AND IMPLICATIONS OF PEST BEHAVIOR

20.2.1 Behavioral Dynamics of Economically Important Pests

The interaction of pests with host plants affects the choice of IPM strategy that depends on four categories of pests: (1) subeconomic, (2) sporadic, (3) perennial, and (4) severe. A subeconomic pest's general equilibrium position is always lower than the economic harm threshold, even at the height of its population peak. This type of insect may result in direct losses; however, if the values of the host (crop) are low and the densities of pests are consistently low, it is not reasonable to start control measures whose expenses outweigh the value of host harm (Dent and Binks, 2020). If population peaks of pests occasionally reach above the economic harm threshold, their average equilibrium position is almost always below it. One extremely prevalent kind of pest is the occasional one. It may occur on or close to a host almost annually, yet it only infrequently results in financial loss. A perennial pest is a general equilibrium position that is below the point of economic injury, but its peak numbers occur so frequently that economic damage typically happens once a year. When a severe pest appears in or on a host, the end outcome is always economic damage since the pest's general equilibrium position is always above the level of economic injury. Perennial and harmful pests are the ones that do the most harm and present the biggest problems to an IPM program, as one might think (Singh *et al.*, 2023).

Identification of pests is a vital step in accurate diagnosis and treatment plans accordingly in a wider range of environments (Florentine *et al.*, 2016). So, the etiology, biology, systematics, ecology, and spatial distribution must serve as a strong basis for pest identification. Inaccurate identifications may render the treatments ineffective as overwatering may cause plant damage that is mistaken for a fungal infection since many fungal infections develop in moist environments. Many viral infections have close similarities with nutrient deficiencies; soil-borne pathogens cause wilting symptoms that may be misunderstood with drought or hard pan of soil (Hassan *et al.*, 2019). The misleading identification of pests may enhance spray expenses without any benefit (Debo, 2023).

20.2.2 The Life Cycles of Pests and Hosts

A fundamental aspect of IPM is to understand the crop growth and development. We can't ignore the pest entirely. The relationships between the environment, the crop, and the pest are crucial. It

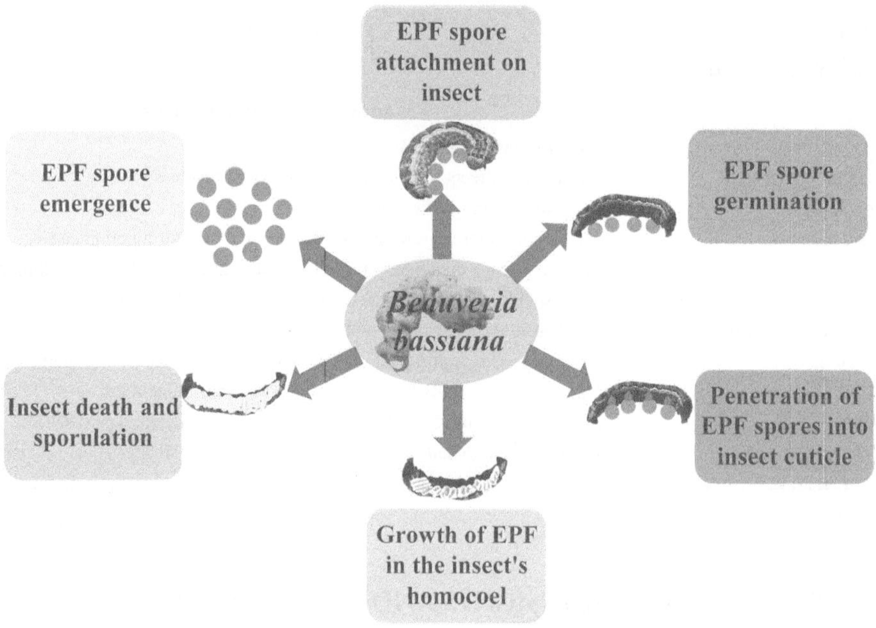

FIGURE 20.1 The life cycle of pests and hosts.

would be necessary to investigate pest life cycle, its mobility, distribution, host range, and fundamental biology to formulate an effective IPM program (Figure 20.1). When you notice a bug, it might be very late to take action rather than giving it a quick pesticide spray. Usually, preventive measures can be applied to a different phase of the life cycle. For instance, pre-emergent pesticide and mulching can stop weeds from growing from last year's seed; the life cycle of pest can also be interrupted by knowing its survival phase (Ali *et al.*, 2013b; Witcher and Poudel, 2020).

The devising of IPM strategies depends on estimation of pest density, infestation, and ecology (Tanveer *et al.*, 2013b; Tanveer *et al.*, 2015). The host and pest interaction has a crucial role in making treatment plans for the management (Gely *et al.*, 2020). A crop's ability to withstand stress from insects that take plant sap, such as aphids and leafhoppers, may be reduced while it is under stress, and this stress may cause the economic threshold to drop a little. When weed populations compete with crops for scarce water during a drought, they may result in financial losses even though they wouldn't usually (Tanveer *et al.*, 2013a). It is well known that the weather has an impact on the growth and survival of pests. Several weeds are more prevalent in the spring or fall because of the specific weather patterns that may have an impact on weed seed development (Mahmood *et al.*, 2013; Travlos *et al.*, 2020).

A certain pest level is tolerable under some circumstances such as soybeans can withstand defoliation rather well, no immediate action may be required if there are only a few caterpillars within the field and the overall population isn't growing much. On the other hand, there comes a time when cost control measures must be implemented i.e. the farmer can set costs to stop when the damage caused by the pest outweighs the expense of controlling it. Pest tolerance varies depending on whether the pest poses a health risk (low tolerance) or just causes cosmetic harm (high tolerance in non-commercial situations) (Gladman, 2022).

The term "integrated" means using a variety of strategies simultaneously or integrating them together with the goal of keeping one pest species below the point of economic harm. In practical terms, the integration happens gradually over time, even if in theory, a single strategy is the outcome of simultaneously integrating multiple approaches; while some of the strategies work well together, and some don't. Legal control, habitat modification, and biological control are undoubtedly complementary strategies. Host resistance is a method that can be used independently or in conjunction with the previously stated strategies. In general, chemical control and host resistance work well. A management strategy, thus, combines one or more workable techniques into a unified package. In IPM, site manager can make changes to the plan before the pests reach their action threshold and need action once more (Tanveer *et al.*, 2015; Hendrichs *et al.*, 2021).

20.3 ECO-FRIENDLY TECHNIQUES FOR PEST CONTROL

20.3.1 Biological Control: Pest Management through the Use of Natural Enemies

In order to reduce a pest organism's population density or influence and lessen its potential harm, biological control involves using of living species, such as natural enemies. Agriculture will face three challenges in the future: to maintain healthy food production for the growing global population, to reduce the negative impacts of agrochemicals on the environment and human health, and to ensure reasonable profit or return for the producer; therefore, one goal for the future is to change pest management techniques by enhancing biological pest control potential, including reducing the usage of pesticides. In ecologically safe pest management, biological control is a strategy that "uses a particular type of organism (the 'natural enemies') for control of another (the pest species)." Natural enemies encompass a variety of creatures, such as predators, parasitoids, pathogens, entomopathogenic nematodes, hyperparasites of plant diseases, competitive microbes, herbivores that consume weeds and their seeds, competitors for resources, and organisms that produce toxins, sometimes known as allelopathy or antibiosis (Balal *et al.*, 2017; Hassan *et al.*, 2022b). Releases of organisms that are augmented can be inoculative, creating an overall population that is anticipated to behave across several generations, with the organisms anticipated to produce immediate decreases (Maurya *et al.*, 2022).

In a broader sense, biological techniques also encompass the use of pheromones to track pest populations and prevent breeding, the release of sterile insects, and the production of pesticides called biopesticides that are derived from micro-organisms or their byproducts. Bio-stimulants can lessen the effects of pest activity because of an indirect mode of action or complex of constituents that increases the tolerance of plants to abiotic stresses, rather than just because of the existence of a recognized plant-protective compound that directly affects pests (Hassan *et al.*, 2022a). There are numerous strategies to use natural enemies in agriculture, including inundative, classical, and conservation biological control. The conservation biological control method that uses macroscopic natural enemies works by regulating the surroundings to encourage naturally occurring natural enemies (Mahmood *et al.*, 2021; Thomine *et al.*, 2022). Its primary goal is to "enhance the activities of currently present natural enemies in order to provide pest suppression," particularly by diversifying the plant community and giving them ongoing access to a variety of sources of food (nectar, pollen, prey, and alternate hosts) and shelter in spite of harvest, crop senescence, and in some cases, the use of pesticides.

20.3.2 Cultural Control: Altering the Surroundings to Keep Pests Away

Cultural practices are essential to the development of an IPM strategy because they modify the agricultural system's ecology. This method is known as "preventive pest management." Cultural practices, such as crop rotation, tillage, nutrient control, and planting date adjustments, intentionally change habitats to reduce the number of pest species. Crop rotation is a useful tool for reducing insect populations. A well-known method for changing the synchronization of pests and hosts to prevent infestation of the crop is to modify the planting date. Hessian flies in wheat, wheat streak mosaic virus, and boll weevils in cotton can be avoided by changing planting dates (Chauhan *et al.*, 2022).

20.3.3 Mechanical Control: Using Physical Obstacles or Traps to Keep Pests Away

This method pertains to the application of diverse physical or mechanical methods with the purpose of eliminating pests, removing them, or destroying them. In certain situations, trapping might be comparable to behavioral management. Physical or mechanical control techniques include pest exclusion using handpicking or vacuuming, row covers or netting to remove pests, rodent pest traps, mechanical tools for weed control, modifying environmental conditions like heat or humidity in greenhouses, visual or physical bird deterrents, and steam sterilization or solarization like sonic devices or reflective material. This strategy makes use of a range of mechanical and physical methods to exclude pests and to trap, remove, or destroy them. These treatments eliminate pests by using gadgets, equipment, obstacles, or extremely high or low temperatures (Rehman *et al.*, 2022).

Physical and mechanical controls consist of:

1. Agricultural techniques such as manual weeding, slash and burn, and tillage.
2. Defoliation of some crops and pruning of fruit and forest tree sections afflicted with pests.
3. Tilling the soil with a machine to eradicate weeds or insects that hibernate.
4. Soil is mechanically tilled to eradicate plants and insects that hibernate.
5. Putting up traps for pests such as mollusks, rats, and insects.
6. Using screens, netting, plant collars, vacuuming, or handpicking can help exclude pests.
7. Freezers to keep pests out from products that are stored.
8. For weed control, use infrared light, flame, or hot water.
9. Noise-producing equipment or other insect deterrents.
10. Altering environmental factors like solarization, steam sterilization, or the humidity or heat in greenhouses.
11. Bird perches are placed in fields to enable birds to perch and consume insect pests in their larval, pupal, and egg phases.
12. Installation of physical or visual bird deterrents in the field, such as scarers, sound devices, or reflecting materials, depending on the situation (Pruteanu *et al.*, 2023).

20.3.4 Utilizing Natural Materials to Manage Pests is Known as "Plant-Incorporated Protectants"

A potential strategy to mitigate crop damage caused by phytophagous arthropod pests involves genetically modifying plants to produce genes that encode poisons that kill insects. The utilization of genetically modified (GM) crops has experienced a significant surge in acceptance, and these plants carry one or more genes derived from separate species (Kumar *et al.*, 2020). Plant-derived pesticides and the genetic material required for the plant to produce the pesticide are known as plant-incorporated protectants. For instance, researchers can insert a gene for a particular *B. thuringiensis* (Bt) pesticidal protein into the genetic makeup of the plant. Afterward, the pest is controlled by the plant as it feeds upon it because the plant produces the pesticidal protein. *B. thuringiensis* mode of action is presented in Figure 20.2.

Plant-incorporated protectants and herbicide-tolerant crops are two main classifications for GM crops (Beckie *et al.*, 2021; Rasheed *et al.*, 2022). In each of these instances, the traits are attained by introducing particular genes into the organism's genetic makeup. First, 11 Bt plants were registered using the Bt plant-incorporated protectant for usage; however, five of these registrations are no longer effective. Bt-incorporated varieties of corn and cotton are marketed, and a Bt soy variety is registered. The production of transgenic plants (Bt plants), that indicate insecticidal δ-endotoxins that arise from soil bacteria and expressing these toxins, provides protection against insect-induced crop loss. Bt endotoxins are guaranteed to be inert in vertebrates, including humans, due to their strong reliance on the alkaline intestinal environment of the insect. Targeting the main pests of tobacco, cotton, tomato, rice, maize, and corn, these commercially manufactured proteins are

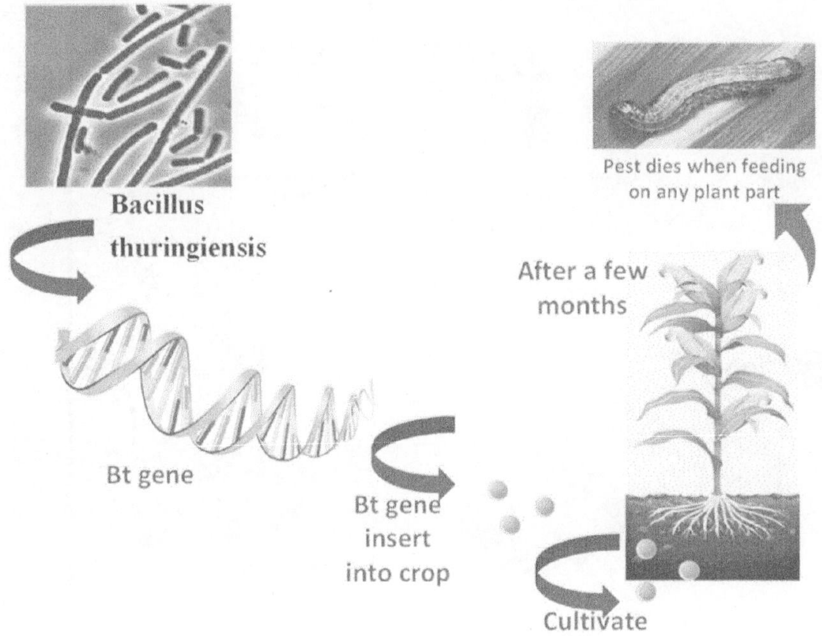

FIGURE 20.2 *Bacillus thuringiensis* mode of action.

notably used for higher coverage by reaching those parts of a plant that foliar sprays cannot (Chung *et al.*, 2022).

20.4 THE DEFINITION AND TENETS OF INTEGRATED PEST MANAGEMENT (IPM)

IPM is a method of controlling pests that uses physical, mechanical, cultural, biological, and educational strategies to keep pest populations below a point where they cause intolerable harm or annoyance. Regular monitoring is used to determine whether and when treatments are necessary. As a last option, least-toxic chemical controls are employed (Mumtaz *et al.*, 2023).

Treatments are only used in IPM programs when monitoring indicates that the pest will lead to unacceptable economic, medical, or esthetic harm. They are not applied in accordance with a predetermined schedule. The best and least disruptive treatments are those that don't interfere with natural pest control. The IPM idea serves as the foundation for the pest-specific management techniques that are discussed later in this book (Klassen and Vreysen, 2021).

IPM, which was first created for the agricultural industry, offers a method for locating and minimizing the causes of pest issues. It also helps assess if the expense of a certain pest control measure is justified by the outcome. When yield numbers and market returns are available to compare results against, estimating economic harm to a crop can be done quite simply (Benami *et al.*, 2021). However, quantifying the fear or retaliation brought on by a specific organism or the media-driven obsession with having a "perfect lawn" is far more challenging. You have to decide if it is worth it to expose your family and pets to the possibly harmful chemicals in "conventional" pesticides to stop the damage, or if you are ready to invest the necessary time and energy in less-toxic alternatives (Zaller and Zaller, 2020).

These are obviously not simple issues to answer, as it is more difficult to measure the costs and benefits than it is in traditional agriculture. Even in the agricultural industry, evaluation is getting harder as costs that were previously thought to be external to production—such as pesticide residues

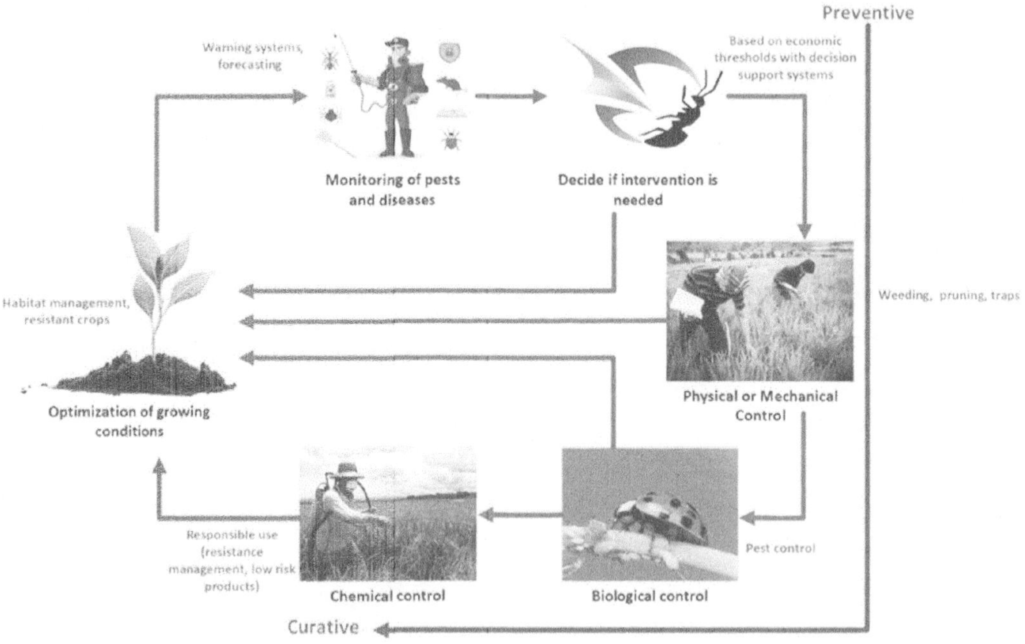

FIGURE 20.3 An introduction to integrated pest management.

on food and in groundwater, regulatory requirements, educational initiatives, the purchase and use of protective equipment, and work-related exposure to toxic materials—begin to enter the equation (Ostad-Ali-Askari, 2022).

The types of pest issues, the settings in which they arise, and the individual and collective values of the people dealing with the issues are all quite different and dynamic. The best we can hope to accomplish in this book is to provide you with some evaluation standards and information sources to enable you to determine, on your own, when and how pest control measures can be applied at their best. In the end, it is up to you to decide what is acceptable in terms of bug populations or being exposed to potentially dangerous substances (Alvarez-Herms *et al.*, 2024).

Since its establishment in the late 1950s, IPM has been the predominant crop protection paradigm (Figure 20.3). As per the Food and Agriculture Organization (FAO) of the United Nations, "The term IPM refers to the methodical evaluation of all the readily available pest control strategies, followed by the incorporation of suitable controls that prevent the growth of pest populations, limit the use of pesticides and other approaches to levels which are financially feasible, and lessen or eliminate risks to the environment and public health.

IPM supports natural pest management methods and the development of crops that are productive with the least amount of disturbance to agro-ecosystems". IPM, formerly referred to as integrated management, has improved crop protection globally, thanks to its broad guiding principles (Dyck *et al.*, 2021). These fundamental ideas and standards of IPM, as stated explicitly in the 1960s, are as follows:

1. There will be species that are potentially hazardous at manageable abundances.
2. The management unit is the ecosystem.
3. Maximizes the utilization of natural control agents.
4. Any control approach has the potential to have unanticipated and undesired impacts.
5. An interdisciplinary strategy is necessary.

Furthermore, the following are the primary guidelines:

1. Evaluate the state of the pests and set thresholds.
2. Create plans to bring equilibrium positions down.
3. In an emergency, look for corrective actions that will cause the least amount of environmental disturbance.
4. Create methods of observation.

20.5 THE FUNDAMENTALS OF INTEGRATED PEST CONTROL

Six fundamental tactics are recommended by IPM's main proponents to enhance pest management strategies.

20.5.1 Avoidance

Creating an atmosphere that deters pest activity and infestation is the most affordable and effective strategy to prevent many pest-related issues. These techniques include preventing dangerous organisms from becoming issues by organizing and overseeing a range of choices, including the following:

1. Intercropping and crop rotation.
2. Using suitable cultivation methods, such as direct sowing, conservation tillage, undersowing, pruning, controlling planting times and densities, and seedbed hygiene.
3. Using certified/standard seeds, planting material, and resistant cultivars.
4. Appropriate water management and a supply of nutrients that is well-balanced.
5. Using field hygiene and sanitation methods to stop the spread of pathogenic organisms.
6. Preserving and promoting advantageous organisms.

20.5.2 Identification of Pests

One of the most important methods for reducing the number of pests is pest identification. Furthermore, as a pest's species or strain may behave differently at multiple sites, a technique developed to manage a pest at one location cannot be used to another if the identity of the pest is unknown. Therefore, a strong basis in the etiology, systematics, taxonomy, spatial distribution, and taxonomy of pests must be established (Galvão, 2021).

20.5.3 Putting in Place a Routine Monitoring and Inspection Mechanism

Anthropomorphic classifications that are both advantageous and detrimental are called pests. For example, termites, which are seen to be organisms that are beneficial in forests because they turn dead trees into organic matter, are also regarded as pests because they feed on very valuable wood. In order to ascertain the types and extent of pest infestation at a specific location, pest inspection involves both routine site visits and traps (Vreysen, 2021). It also entails routinely monitoring the presence of species that have been correctly identified and classified as either beneficial organisms or pests, as well as the harm that pests do, agricultural characteristics, and environmental influences. A crucial component of IPM systems is the monitoring process, which aids in fast detection, rapid exchange of data about pest activity, infestation severity assessment, cause identification, and population projections for the future. Methods for monitoring the environment must be created to evaluate the instantaneous and dynamic elements of the pest's incidence, activity, or density (Kaur and Kaur, 2020). Crop advisors can respond to changing environmental circumstances and establish appropriate pest population levels, efficient reduction strategies, and breaches of the action threshold by having a thorough understanding of these environmental relationships (Cusworth and Lorimer, 2024).

20.5.4 Establish the Economic Harm and the Pest Activity Threshold Level

Bringing a pest within acceptable bounds is the main goal of IPM, instead of its complete eradication. According to the FAO, a pest is any strain, biotype, or species of animal, plant, or pathogenic agent that harms materials, plant products, plants, or habitats. This definition also covers animals that are nuisance to the public's health as well as vectors of parasites or pathogens that cause sickness in humans and animals. A small number of pests are permissible in IPM, but action must be taken when pest levels cross a threshold (Wolff, 2023). The term economic injury level (EIL) refers to the smallest number of pests that may cause economic harm, while the term "damage boundary" refers to the lowest degree of damage to agricultural crops at which the damage caused can be measured. The term economic threshold level (ETL) refers to the pest densities at which preventive measures ought to be implemented to keep an ever-growing pest population from attaining a point of economic harm (Sankhla *et al.*, 2022). The quantity of the population of pests that necessitates corrective action for reasons related to human health, economy, or esthetics is known as action threshold level (ATL) and it varies based on the layout of the site and its intended usage. Because ATL and ETL are site- and pest-specific, they could be appropriate at one location but inappropriate at another. The next phase is a decision-making process that uses timely and reliable information to make judgments about managing pests and preventing them. The economic threshold must be set beyond the EIL in order for controlling practice to be profitable, or at the very least break even.

20.5.5 Formulating Methods for Management

Mechanical, biological, cultural, chemical, or a mix of these controls are used in IPM treatment method execution (Abbas *et al.*, 2022). While employing a single tactic could work for a while, combining various techniques can offer precautions against disruptions to the environment (such as the eradication of natural enemies or insect resistance), which sometimes result from depending too heavily on a single tactic. Synthetic pesticides should only be used as a last resort when all other options exhaust, and the system for monitoring demonstrates that the population of pests is still over action thresholds. While using chemicals that are synthetic, the least hazardous materials should be selected to reduce exposure to any organisms that are not targeted. The ultimate objective is pest management with little environmental impact (Deguine *et al.*, 2021a).

20.5.6 Assessing and Maintaining Documentation

The final phase of IPM is evaluation, which is frequently seen as one of the most crucial. To find out if the pest management measures are working, a frequent evaluation program is crucial. It involves evaluating an IPM project and the outcomes it has produced (Bueno *et al.*, 2021). Furthermore, knowing how successful the IPM program is enables you to adjust the IPM strategy before pests approach the point of action threshold and need to be controlled once more. In order to identify patterns and trends in insect outbreaks, a system for keeping records is necessary. Every investigation or treatment should record information on the identification of the pest, its size and distribution, its population, suggestions for the future preventive measures, and a thorough account of its treatment.

20.6 HOW TO PUT AN IPM PROGRAM INTO ACTION

In order to reduce the number of pests on the resource or host that is protected, push-pull pest control techniques employ a variety of behavior-modifying cues to alter the abundance and distribution of beneficial insect populations or pests (Mishra *et al.*, 2022). The use of stimuli that conceal the host's appearance or that are repellent or deterrent causes pests to be deterred or repelled from the resource (push). Using highly visible and alluring stimuli such as trap crops, pests are simultaneously drawn to the area, where they are highly concentrated, making it easier to eradicate them. East African subsistence farmers devised foremost successful push-pull technique (Luttermoser, 2023). The two main foods of East Africa, sorghum (*Sorghum bicolor*) and maize (*Zea mays*), are targeted by lepidopteran stem borers, such as *Chilo partellus, Sesamia calamistis, Eldana saccharina*, and *Busseola fusca*, which result in yield losses of 10–50%. Farmers employ plants that are suitable for them along with natural enemies for combining the use of trap crops and intercrops. When non-hosts, like silverleaf desmodium (*Desmodium uncinatum*) molasses grass (*Melinis minutiflora*) and green-leaf desmodium (*Desmodium intortum*), are interplanted with sorghum or maize, they drive away stem borers (the push). Trap crops, primarily Sudan grass (*Sorghum vulgare* sudanense) and Napier grass (*Pennisetum purpureum*), are planted around the boundaries of the fields to draw and concentrate pests (the pull) (Njeru, 2020). These grasses serve two purposes because cattle graze them as well. In addition to increasing parasitic activity by a parasitoid wasp, molasses grass cultivated as an intercrop lowers populations of stem borer by producing volatiles that repel them. Similar repelling volatiles are also produced by desmodium, but it also generates sesquiterpenes that inhibit a parasitic African witchweed (*Striga hermonthica*), which is a significant farmland yield limitation in East Africa. The desmodium chemicals promote witchweed seed germination and the eventual mortality of plant seedlings (Babar *et al.*, 2009; Javaid *et al.*, 2022). East Africa's enhanced livestock and grain yields as a result of the push-pull method have had a major influence on food security.

20.7 OBSTACLES AND PROSPECTS FOR THE FUTURE

20.7.1 Barriers to Putting Eco-Friendly Pest Management Techniques into Practice

Principal Obstacles to Advancement: IPM is not widely used in forestry, public health, urban areas, agriculture, or any other industry, despite its considerable advancements. Agriculture has profited the most out of all the industries. Programs for operational control of pests are rarely fully integrated. Ecologically sound management of pests has not been applied to a wide range of resources and pests due to political, financial, cultural, and technical obstacles among others. Understanding limitations even though there has been a significant increase in research on IPM recently, much more work has to be done to gather the data and refine the monitoring methods necessary for IPM deployment (Moustafa, 2023). It is especially crucial that experts from every relevant field collaborate on this study. Integrating the research findings requires collaboration between different disciplines. IPM studies ought to employ computer models and system analysis more frequently. Before ecologically appropriate pest management can be effectively deployed and before workable IPM schemes are made accessible to a wide range of resources and pests a number of obstacles must be overcome.

If there is an IPM system, it is frequently very hard to market it to agriculturalists and other people who are used to the less complex chemical management approach. It must first be demonstrated to these people that IPM is a more cost-effective solution for controlling pests than chemical control (Fitri *et al.*, 2021). After that, they need to be instructed on how to gather and use the data required for IPM deployment. Livestock owners, farmers, and homeowners can implement several IPM techniques, even if it might feel strange at first. The usage of chemical controls may therefore persist until people receive training on an IPM alternative, even if the latter may be more efficient and less expensive.

Sources of User Information: Those that typically supply agriculturalists as well as other users with the information also support the usage of pesticides. Pesticide salespeople provide farmers in California, the wealthiest agricultural state in the country, with the majority of their pest management advice (Van den Bosch, 2023). Only 1% of the knowledge used by cotton farmers in the area to manage insect pests comes from farm advisers employed by the Cooperative Extension Service

(CES), the official organization in charge of informing the public about pest control. It is to be expected that information from the chemical industry is used far more frequently by homeowners as well as farmers than information from the CES or independent pest management advisors. The number of representatives from the chemical industry is much higher than that of independent advisors and CES pest control specialists. In contrast to the 119 extension employees who worked full- or part-time on agricultural pest control education programs in Iowa in 1973, approximately 4000 or more people were involved in the retail selling of agricultural pesticides.

Lack of Enough Skilled Workers: There are thought to be 200,000 certified commercial applicators of pesticides in the country, including owners of pest control companies and aerial applicators (Dhananjayan *et al.*, 2020). In contrast, there were only roughly 1120 CES professionals working on agricultural and animal health issues in 1977, including IPM. A little more than 500 private consultants are employed by farmers' cooperatives and farm service companies as well as on their own.

Institutional Limitations: In addition to encouraging the continued use of pesticides that are chemical-based, government policies and initiatives also obstruct IPM advancement in other ways. While eating small plant-feeding insects does not appear to pose a health risk, the Food and Drug Administration (FDA) has generally decreased the amount of insects and insect components allowed in food since the 1930s (Hervet and Morrison III, 2021). Furthermore, the esthetic quality of fruits and vegetables (such as being free of insect blemishes) has become more important to food processors, wholesale merchants, and retailers. The end consequence is higher losses because more of the crop is now deemed unfit for commercial use, and fruit and vegetable producers must apply 10–20% more pesticides to comply with FDA requirements and standards for cosmetic appearance. The rules may be more onerous than required, but consumers have a right to be shielded against contaminated items that pose a health risk. Commercial production of certain alternative control measures required in IPM systems is discouraged by a 1972 modification to the Federal Fungicide, Insecticide and Rodenticide Act. The amended law demands that before being registered by the EPA, commercially produced insects and weed disease-causing agents (bacteria, fungi, viruses), insect hormones, and insect sex attractants must undergo the same fundamental testing processes that apply to conventional pesticides (Javaid *et al.*, 2022). These options are typically highly selective and occasionally successful against a single or small number of insect species, in contrast to wide-spectrum pesticides made from chemicals. Due to the restricted possibility of financial gain, industry is hesitant to allocate funds for the necessary investigation, creation, and marketing (Stefani *et al.*, 2020).

20.7.2 New Developments and Technology in Environmentally Friendly Pest Management

Innovative, long-lasting control strategies must be developed in order for IPM to be effective. This could be accomplished by utilizing information from various evolutionary study fields. A well-researched method for managing pest insects in IPM techniques is mating disruption, which involves decreasing the frequency of mating by dispersing sex pheromones that influence the location of the pests' partners (Rizvi *et al.*, 2021). Utilizing the wealth of genetic, ecological, and behavioral knowledge on sexual selection and mating may open up a number of new avenues for investigation into lowering the fitness of pests; one such avenue might be the creation of all-male insect populations by using sex transformation genes. Additionally, it has been demonstrated lately that when applying pesticides, it is important to take into account the difference in mating patterns among the targeted populations of insects, since sexual selection may accelerate the development of resistance (Jehan *et al.*, 2020). Examining sexual selection in oomycetes and harmful fungus may also be important. For instance, it has been proposed that effectors—small proteins released by pathogens that are crucial to infection processes—may have additional purposes, like modifying the host microbiome and regulating relationships with other bacteria. Effectors may be sexually chosen, just as they may impact relationships within the same species. Understanding the relationships between microbes, plants, and insects has led to the development of another intriguing strategy (Ayaz *et al.*, 2023). Plant-associated microbial communities can enhance nutrient intake, growth, and host tolerance against herbivores and diseases (Waheed *et al.*, 2019). High plant microbiome diversity may therefore be a crucial component of a plant's defense mechanism. From an evolutionary standpoint, it makes sense that the host would be chosen to exercise control over the microbiota, while the bacteria would be picked for their strong capacity for competition and any beneficial impacts on them as byproducts. It may be useful to understand how plants can optimize their fitness through the development and maintenance of their microbiome in order to reduce disease, for example, by the use of beneficial bacteria in treatments or plant breeding based on the ability to regulate the microbiota. Furthermore, a deeper comprehension of the relationships within plant innate resistance and the microbiome may aid in enhancing biological pest control agents' efficacy, which is frequently quite variable in field settings (Francis *et al.*, 2020).

Although using live biological control agents to reduce pest populations and the harm they create is not a unique idea, evolutionary research may be able to support and enhance its effectiveness and long-term viability. Given that these agents are living things, they might change in reaction to the conditions of their breeding program or the facility in which they are raised. As a result, they can acquire characteristics that could harm their introduction into the wild or their relationships with the pest they are trying to eradicate (O'Mahony, 2020). Before being released into the field, maladaptations or inbreeding of the biocontrol organisms could be reduced or eliminated using an evolutionary-based framework. Furthermore, pests have the ability to adapt and develop ability to resist biocontrol organisms. Thus, selection for features that are effective against the inherent variety of defense qualities found in the targeted pests should be possible when breeding biocontrol agents. Therefore, it will be essential to do research on the interactions among biocontrol organisms and their prey in order to determine which genes and features to focus on during breeding efforts. Plant breeding and biocontrol organism breeding can be combined to create plant varieties with complementary effects on biocontrol (such as volatiles which attract predators) and to produce biocontrol organisms which are well-suited to the crop they are supposed to protect (Torracchi C *et al.*, 2020).

20.7.3 Suggestions for Expanding the Use of Environmentally Friendly Pest Control Techniques

Suggested Course of Action: IPM cannot be successfully implemented by an individual private or public organization or governmental program (Deguine *et al.*, 2021b). Certain roles are suitable for the U.S. Department of Agriculture, the National Science Foundation, the Environmental Protection Agency, the Executive Office, the Departments of Health, Welfare, and Education, Defense, State and Interior, as well as other federal, state, and privately held organizations. Numerous opportunities exist for collaboration with foreign agencies and scientists on specific challenges as well as policy formulation related to control of pests and regulation of pesticide. This chapter suggests a number of federal government programs that will eventually result in national pest control legislation, which will encourage the advancement and application of IPM both domestically and globally. The recommendations aim to enhance public understanding of the principles of ecology underlying IPM, promote monitoring and research, improve the coherence and efficacy of federal policies, and increase the number of certified advisors for IPM who are available to all sector users (Sotirov *et al.*, 2020).

20.7.4 Coordination and Execution of Federal-State Relations

Issuance of a presidential directive or executive order mandating the installation of IPM technology on all federally owned, controlled, or leased lands, buildings, and infrastructure, the creation of an Executive Interagency Group on IPM has been initiated to oversee worldwide and national pest management policies and initiatives to launch a nationwide monitoring effort to look for resistance to herbicides in species of weeds and assessment of the viability of a "early warning system" to spot issues with using chemical pesticides to control major pests (Ali *et al.*, 2013a). Production of information systems, including computers, for use in IPM initiatives. Establishment of protocols to expedite the licensing of insect pheromones, pest disease agents, and other non-traditional pest management techniques.

20.7.5 Promoting Use and Eliminating Barriers

For setting up standards for approving, sponsoring, reviewing, and assessing Extension IPM demonstrations, an assessment of the rules has been established by the food advertising and processing sectors, government, and government tolerance criteria for pest components in food and the esthetic appearance of vegetables and fruits (Travičić *et al.*, 2023). Assessment is meant for government pest eradication initiatives, both current and proposed, and development of standards for program approval, review, funding, and assessment. Certification standards for independent pest control consultants and helping certified persons start IPM consulting businesses have been created. Bank policies for loans related to agriculture in order to ascertain whether terms included in the loans stimulate the use of excessive pesticide methods are examined. The viability of pest-specific hazards protection strategies for farmers taking part in IPM initiatives is evaluated. International search for natural enemies and fundamental research on harmful imported pests in their natural environments has been increased. Development and management of natural ecosystems and plant germplasm reserves for scientific research and conservation in other countries where potentially dangerous pests exist or where U.S. crops and their hazardous pests originated are now coevolving (Oberemok *et al.*, 2023).

20.8 EDUCATION

Establishment of teaching resources for the general public on the ecological fundamentals of integrated pest control funding initiatives for public education, research, and demonstration to promote IPM in cities.

20.9 BASIC RESEARCH AND EVALUATION

The research has enhanced national usage of pesticides survey frequency and methodology. Funding has been increased for multidisciplinary research on pesticide resistance, improved pesticide application techniques, insect monitoring and detection, and the effects of pesticides on the environment. Assessment of pest management for rangelands, cattle, stored goods, and rights of way has been initiated. A project has been launched to create tools, methods, and oversight protocols that reduce the risks of using chemical pesticides to the environment and public health when are employed through programs like IPM. The start of an interdisciplinary study designed to create integrated mosquito control that is sustainable for the environment (Jacquet *et al.*, 2022).

20.10 CONCLUSION

The earlier models may have been inadequate in effectively promoting and implementing IPM, because they did not provide enough room to incorporate the business, social, human, and communication parts of the whole equation. Numerous instances covered in this study demonstrated how these elements affect the creation, dissemination, and effective use of IPM methods globally. In the modern era, when sophisticated farming techniques and tools for communication are essential to the production and consumption of food, IPM is redefined for the agricultural sector, that is a consumer-centered enterprise, and for the global trade sector, which is influenced by multiple factors. The two outermost layers of the new model are the force that drives behind farming activities and have a substantial impact on IPM across the whole crop production process, even though they are not limited to pest control alone. A vital role in the production of food is played by economists, educators, agricultural researchers, sociologists, economists, managers, producers, pest management specialists, makers of agricultural inputs, retailers, and consumers (Sjah and Zainuri, 2020). The latest IPM model offers a framework for concentrating on the different aspects of the paradigm and promoting cooperation across disciplines by rearranging the components and integrating a variety of elements which affect them. With the goal of ensuring affordability for consumers, profitability for growers, and food security for the world's expanding population, this novel approach is anticipated to direct IPM tactics at a global scale (Han *et al.*, 2024).

20.11 FUTURE PERSPECTIVES AND CHALLENGES

Crop protection by chemical pesticides leads to ineffective control over time and deteriorated environmental components (Sánchez-Bayo, 2021). The requirement was to protect crops without disturbing eco-diversity over extended periods in a sustainable manner. The production of pesticides from the natural sources would serve the purpose without leaving harmful residues thereafter. The pesticides obtained from any of the natural sources are target specific, efficient, and easily degradable, leaving no harmful matter for the end user as well (Mishra *et al.*, 2020). The natural pesticides have abundant quantity of growth hormones that boost plant defense system, enhance growth rate apart from targeting the pests. The farming community has an ideal opportunity to cope with the issues related to climate change by get ridding from synthetic chemicals and promoting organic farming (Chadha *et al.*, 2020; Rehman *et al.*, 2023).

Generally, it is considered that organic amendments could not provide quick and efficient pest management. Nematodes are deleterious pest of vegetables which deter the absorption of water and nutrients apart from reducing the root anchorage in soil and predispose the plant to other biotic and abiotic stresses (Oladosu *et al.*, 2022). Nematodes are effectively controlled by using fumigants consisted of synthetic chemicals that have adverse effects on soil. Seaweed and marigold extract are substitutes for these since they slow down nematode development and eventually wipe out the population (Abd-Elgawad, 2021).

Sustainable and effective crop protection task can be accomplished by using IPM that relies on combined and rational use of all available pest management strategies. IPM relies upon less damage to the environment and people while using it in agricultural or urban settings. IPM can ensure safe food production by modifying cultural practices, encouraging disease and pest resistant genotypes (Angon *et al.*, 2023).

The recent shifts in government and public views will undoubtedly encourage the development and application of IPM going forward (Deguine *et al.*, 2021b). Major challenges for IPM in future are gene technology constraints i.e. crops may be GM to be more compatible with the IPM practices. As the pests can adapt to a wider range of environments due to their elastic nature, it would be challenging for the plant breeders to produce resistant and immune cultivars under climate change scenario.

New and emerging trends in IPM: Information technology and genetic studies (Bioinformatics); it would be effective in understanding the mechanisms of pathogen or pest action and the route of plant defense system which would provide accurate basis for diagnosis and management. The analysis of pest and host genomes can be used to explore all the critical pathways of pest and host interaction. Precision farming tools such as remote sensing, global positioning system (GPS), and image analysis could be very much decisive in efficient IPM, transgenic plants. Nano-pesticides are one of the emerging trends that can create the difference in future of IPM; the key future of nano-pesticides is to allow only specific quantity of active ingredients in response to target.

LIST OF ABBREVIATION

ATL action threshold level
Bt *B. thuringiensis*
CES Cooperative Extension Service
EIL economic injury level
ETL economic threshold level
FAO Food and Agriculture Organization
FDA Food and Drug Administration
GM genetically modified
GPS global positioning system
IPM integrated pest management

REFERENCES

Abbas, M., Saleem, M., Hussain, D., Ramzan, M., Jawad Saleem, M., Abbas, S., Hussain, N., Irshad, M., Hussain, K., & Ghouse, G., 2022. Review on integrated disease and pest management of field crops. *International Journal of Tropical Insect Science*, 42, 3235–3243.

Abd-Elgawad, M.M.M., 2021. Optimizing safe approaches to manage plant-parasitic nematodes. *Plants*, 10, 1911.

Ali, H., Abid, S.A., Ahmad, S., Sarwar, N., Arooj, M., Mahmood, A., & Shahzad, A.N., 2013a. Integrated weed management in cotton cultivated in the alternate-furrow planting system. *Journal of Food, Agriculture and Environment*, 11, 1664–1669.

Ali, H., Tanveer, A., Nadeem, M., Asghar, H., & Javaid, M., 2013b. Germination ecology of *Rhynchosia capitata*: An emerging summer weed in Asia. *Planta Daninha*, 31, 249–257.

Alvarez-Herms, J., González-Benito, A., Corbi, F., & Odriozola, A., 2024. What if gastrointestinal complications in endurance athletes were gut injuries in response to a high consumption of ultra-processed foods? Please take care of your bugs if you want to improve endurance performance: A narrative review. *European Journal of Applied Physiology*, 124, 383–402.

Angon, P.B., Mondal, S., Jahan, I., Datto, M., Antu, U.B., Ayshi, F.J., & Islam, M.S., 2023. Integrated pest management (IPM) in agriculture and its role in maintaining ecological balance and biodiversity. *Advances in Agriculture*, 2023, 5546373.

Ayaz, M., Li, C.-H., Ali, Q., Zhao, W., Chi, Y.-K., Shafiq, M., Ali, F., Yu, X.-Y., Yu, Q., Zhao, J.-T., Yu, J.-W., Qi, R.-D., & Huang, W.-K., 2023. Bacterial and fungal biocontrol agents for plant disease protection: Journey from lab to field, current status, challenges, and global perspectives. *Molecules*, 28, 6735.

Babar, B.H., Tanveer, A., Tahir, M., Aziz, A., Ahmad, A.U.H., Nadeem, M.A., & Javaid, M.M., 2009. Allelopathic potential of wild onion (*Asphodelus tenuifolius*) on the germination and seedling growth of chickpea (*Cicer arietinum*). *Weed Biology and Management*, 9, 146–151.

Baker, B.P., Green, T.A., & Loker, A.J., 2020. Biological control and integrated pest management in organic and conventional systems. *Biological Control*, 140, 104095.

Balal, R.M., Shahid, M.A., Javaid, M.M., Iqbal, Z., Liu, G.D., Zotarelli, L., & Khan, N., 2017. Chitosan alleviates phytotoxicity caused by boron through augmented polyamine metabolism and antioxidant activities and reduced boron concentration in *Cucumis sativus* L. *Acta Physiologiae Plantarum*, 39, 1–15.

Beckie, H.J., Busi, R., Lopez-Ruiz, F.J., & Umina, P.A., 2021. Herbicide resistance management strategies: How do they compare with those for insecticides, fungicides and antibiotics? *Pest Management Science*, 77, 3049–3056.

Benami, E., Jin, Z., Carter, M.R., Ghosh, A., Hijmans, R.J., Hobbs, A., Kenduiywo, B., & Lobell, D.B., 2021. Uniting remote sensing, crop modelling and economics for agricultural risk management. *Nature Reviews Earth & Environment*, 2, 140–159.

Bueno, A.D.F., Panizzi, A.R., Hunt, T.E., Dourado, P., Pitta, R., & Gonçalves, J., 2021. Challenges for adoption of integrated pest management (IPM): The soybean example. *Neotropical Entomology*, 50, 5–20.

Chadha, A., Florentine, S., Javaid, M., Welgama, A., & Turville, C., 2020. Influence of elements of climate change on the growth and fecundity of *Datura stramonium*. *Environmental Science and Pollution Research*, 27, 35859–35869.

Chauhan, C., Sharma, V., Thakur, A., Verma, S.K., Tewari, S.K., Panwar, R.K., Kumar, R., & Gautam, A., 2022. Understanding Environmental Associated Biotic Stress Response in Plants. *In Molecular Response and Genetic Engineering for Stress in Plants, Volume 2: Biotic Stress*. Bristol: IOP Publishing, 1-1–1-28.

Chung, Y.H., Church, D., Koellhoffer, E.C., Osota, E., Shukla, S., Rybicki, E.P., Pokorski, J.K., & Steinmetz, N.F., 2022. Integrating plant molecular farming and materials research for next-generation vaccines. *Nature Reviews Materials*, 7, 372–388.

Cusworth, G., & Lorimer, J., 2024. On disease configurations, black-grass blowback, and probiotic pest management. *Annals of the American Association of Geographers*, 114, 462–480.

Dar, S.A., Wani, S.H., Mir, S.H., Showkat, A., Dolkar, T., & Dawa, T., 2021. Biopesticides: Mode of action, efficacy and scope in pest management. *Journal of Advanced Research in Biochemistry and Pharmacology*, 4, 1–8.

Debo, M.C., 2023. *Evaluation of Herbicide Spray Dates against Noxious Weeds of Sugarcane (Saccharum officinarum L.) under Irrigation Condition at Wonji Shoa Sugar Estate* (Master's thesis). Haramaya University, Haramaya. Ethiopia.

Deguine, J.-P., Aubertot, J.-N., Flor, R.J., Lescourret, F., Wyckhuys, K.A., & Ratnadass, A., 2021a. Integrated pest management: Good intentions, hard realities. A review. *Agronomy for Sustainable Development*, 41, 38.

Deguine, J.-P., Aubertot, J.-N., Flor, R.J., Lescourret, F., Wyckhuys, K.a.G., & Ratnadass, A., 2021b. Integrated pest management: Good intentions, hard realities. A review. *Agronomy for Sustainable Development*, 41, 38.

Dent, D., & Binks, R.H., 2020. *Insect Pest Management*. Cabi.

Dhananjayan, V., Jayakumar, S., & Ravichandran, B., 2020. Conventional methods of pesticide application in agricultural field and fate of the pesticides in the environment and human health. *Controlled release of pesticides for sustainable agriculture*, 1–39.

Dyck, V.A., Hendrichs, J., & Robinson, A.S., 2021. *Sterile Insect Technique: Principles and Practice in Area-Wide Integrated Pest Management*. Taylor & Francis.

Fenibo, E.O., Ijoma, G.N., & Matambo, T., 2022. Biopesticides in sustainable agriculture: Current status and future prospects. *New and future development in biopesticide research: Biotechnological exploration*, 1–53.

Fitri, I.R., Hanum, F., Kusnanto, A., & Bakhtiar, T., 2021. Optimal pest control strategies with cost-effectiveness analysis. *The Scientific World Journal*, 2021, 1–17.

Florentine, S.K., Weller, S., Graz, P.F., Westbrooke, M., Florentine, A., Javaid, M., Fernando, N., Chauhan, B.S., & Dowling, K., 2016. Influence of selected environmental factors on seed germination and seedling survival of the arid zone invasive species tobacco bush (*Nicotiana glauca* R. Graham). *The Rangeland Journal*, 38, 417–425.

Francis, F., Jacquemyn, H., Delvigne, F., & Lievens, B., 2020. From diverse origins to specific targets: Role of microorganisms in indirect pest biological control. *Insects*, 11, 533.

Galvão, C., 2021. Taxonomy. *In Triatominae-The Biology of Chagas Disease Vectors*. Springer, 15–38.

Gely, C., Laurance, S.G., & Stork, N.E., 2020. How do herbivorous insects respond to drought stress in trees? *Biological Reviews*, 95, 434–448.

Gladman, A.K., 2022. *Understanding Interactions between Biopesticides and Partial Crop Resistance for Management of Aphid Pests of Brassica Crops*. University of Warwick.

Han, P., Rodriguez-Saona, C., Zalucki, M.P., Liu, S.-S., & Desneux, N., 2024. A theoretical framework to improve the adoption of green integrated pest management tactics. *Communications Biology*, 7, 337.

Hassan, M.U., Chattha, M.U., Barbanti, L., Chattha, M.B., Mahmood, A., Khan, I., & Nawaz, M., 2019. Combined cultivar and harvest time to enhance biomass and methane yield in sorghum under warm dry conditions in Pakistan. *Industrial Crops and Products*, 132, 84–91.

Hassan, M.U., Ghareeb, R.Y., Nawaz, M., Mahmood, A., Shah, A.N., Abdel-Megeed, A., Abdelsalam, N.R., Hashem, M., Alamri, S., & Thabit, M.A., 2022a. Melatonin: A vital protectant for crops against heat stress: Mechanisms and prospects. *Agronomy*, 12, 1116.

Hassan, M.U., Nawaz, M., Mahmood, A., Shah, A.A., Shah, A.N., Muhammad, F., Batool, M., Rasheed, A., Jaremko, M., & Abdelsalam, N.R., 2022b. The role of zinc to mitigate heavy metals toxicity in crops. *Frontiers in Environmental Science*, 10, 990223.

Hendrichs, J., Vreysen, M., Enkerlin, W., & Cayol, J., 2021. Strategic Options in Using sterile Insects for Area-Wide Integrated Pest Management. *In Sterile Insect Technique*. CRC Press, 841–884.

Hervet, V.A., & Morrison Iii, W.R., 2021. Prospects for use of biological control of insect and mites for the food industry in North America. *Agronomy*, 11, 1969.

Jacquet, F., Jeuffroy, M.-H., Jouan, J., Le Cadre, E., Litrico, I., Malausa, T., Reboud, X., & Huyghe, C., 2022. Pesticide-free agriculture as a new paradigm for research. *Agronomy for Sustainable Development*, 42, 8.

Javaid, M.M., Mahmood, A., Nadeem, M.A., Sarwar, N., Safdar, M.E., Ahmad, M., Hasanuzzaman, M., & Ahmad, S., 2022. Weed Management Strategies in Direct Seeded Rice. *In* N. Sarwar, R. Atique Ur, S. Ahmad & M. Hasanuzzaman (eds.) *Modern Techniques of Rice Crop Production*. Singapore: Springer Singapore, 327–347.

Jehan, C., Chogne, M., Rigaud, T., & Moret, Y., 2020. Sex-specific patterns of senescence in artificial insect populations varying in sex-ratio to manipulate reproductive effort. *BMC Evolutionary Biology*, 20, 18.

Kaur, T., & Kaur, M., 2020. *Integrated Pest Management: A Paradigm for Modern Age.* IntechOpen.
Kiss, J., 2020. Implementing Biopesticides as Part of an Integrated Pest Management (IPM) Programme. *Biopesticides for Sustainable Agriculture.* Burleigh Dodds Science Publishing, 21–44.
Klassen, W., & Vreysen, M., 2021. Area-Wide Integrated Pest Management and the Sterile Insect Technique. *In Sterile Insect Technique.* CRC Press, 75–112.
Kumar, K., Gambhir, G., Dass, A., Tripathi, A.K., Singh, A., Jha, A.K., Yadava, P., Choudhary, M., & Rakshit, S., 2020. Genetically modified crops: Current status and future prospects. *Planta*, 251, 91.
Luttermoser, T.J., 2023. *Spatial and Temporal Context-Dependence in the Push-Pull System.* Cornell University.
Mahmood, A., Ullah, H., Shahzad, A., Ali, H., Ahmad, S., Zia-Ul-Haq, M., Honermeier, B., & Hasanuzzaman, M., 2013. Dry matter yield and chemical composition of sorghum cultivars with varying planting density and sowing date. *Sains Malaysiana*, 42, 1529–1538.
Mahmood, A., Wang, X., Shahzad, A.N., Fiaz, S., Ali, H., Naqve, M., Javaid, M.M., Mumtaz, S., Naseer, M., & Dong, R., 2021. Perspectives on bioenergy feedstock development in Pakistan: Challenges and opportunities. *Sustainability*, 13, 8438.
Maurya, R.P., Koranga, R., Samal, I., Chaudhary, D., Paschapur, A.U., Sreedhar, M., & Manimala, R.N., 2022. Biological control: A global perspective. *International Journal of Tropical Insect Science*, 42, 3203–3220.
Mishra, J., Dutta, V., & Arora, N.K., 2020. Biopesticides in India: Technology and sustainability linkages. *3 Biotech*, 10, 210.
Mishra, V., Singh, S., Chauhan, D., & Devi, M.S., 2022. Role of info-chemicals in sustainable pest management: A review. *Journal of Experimental Zoology India*, 25.
Moustafa, A., 2023. Smart-Insect Monitoring System Integration and Interaction via AI Cloud Deployment and GPT.
Mumtaz, T., Afzal, R., Roy, D.H., & Akhtar, S., 2023. Assessment of relative toxicity of common insecticides against adult and grubs of *Coccinella septempunctata* under laboratory settings. *Journal of Zoo Biology*, 6, 17–25.
Njeru, N.K., 2020. *Contribution of Push-Pull Cropping System to Management of Ear Rots and Mycotoxin Contamination in Maize in Western Kenya.* University of Nairobi.
O'mahony, K., 2020. Blurring boundaries: Feral rewilding, biosecurity and contested wild boar belonging in England. *Conservation and Society*, 18, 114–125.
Oberemok, V.V., Gal'chinsky, N.V., Useinov, R.Z., Novikov, I.A., Puzanova, Y.V., Filatov, R.I., Kouakou, N.J., Kouame, K.F., Kra, K.D., & Laikova, K.V., 2023. Four most pathogenic superfamilies of insect pests of suborder sternorrhyncha: Invisible superplunderers of plant vitality. *Insects*, 14, 462.
Oladosu, Y., Rafii, M.Y., Arolu, F., Chukwu, S.C., Salisu, M.A., Fagbohun, I.K., Muftaudeen, T.K., Swaray, S., & Haliru, B.S., 2022. Superabsorbent polymer hydrogels for sustainable agriculture: A review. *Horticulturae*, 8, 605.
Ostad-Ali-Askari, K., 2022. Management of risks substances and sustainable development. *Applied Water Science*, 12, 65.
Pruteanu, A., Vanghele, N., Cujbescu, D., & Nitu, M., & Gageanu, I. 2023. Review of effectiveness of visual and auditory bird scaring techniques in agriculture. *In Proceeding of the 22nd International Scientific Conference "Engineering for Rural Development",* Jelgava, Latvia.
Rasheed, A., Raza, A., Jie, H., Mahmood, A., Ma, Y., Zhao, L., Xing, H., Li, L., Hassan, M.U., & Qari, S.H., 2022. Molecular tools and their applications in developing salt-tolerant soybean (*Glycine max* L.) cultivars. *Bioengineering*, 9, 495.
Rehman, A., Saba, T., Kashif, M., Fati, S.M., Bahaj, S.A., & Chaudhry, H., 2022. A revisit of internet of things technologies for monitoring and control strategies in smart agriculture. *Agronomy*, 12, 127.
Rehman, H.U., Mahmood, A., Ishfaq, F., Javaid, M.M., Haider, A., Asif, M., Nadeem, M.A., & Nargis, J., 2023. Sustainable Crop Management for Drylands. *In* M. Hasanuzzaman (ed.) *Climate-Resilient Agriculture, Vol 1: Crop Responses and Agroecological Perspectives.* Cham: Springer International Publishing, 435–464.
Rizvi, S.a.H., George, J., Reddy, G.V.P., Zeng, X., & Guerrero, A., 2021. Latest developments in insect sex pheromone research and its application in agricultural pest management. *Insects*, 12, 484.
Sánchez-Bayo, F., 2021. Indirect effect of pesticides on insects and other arthropods. *Toxics*, 9, 177.
Sankhla, P., Sangani, S., Surve, V.H., & Raj, A., 2022. Advanced Greenhouse Horticulture: Potential Technology for Horticultural Improvement. *Agriculture & Food E-Newsletter.*
Sarkar, S., Gil, J.D.B., Keeley, J., & Jansen, K., 2021. *The Use of Pesticides in Developing Countries and their Impact on Health and the Right to Food.* European Union.

Singh, M., Pandey, N., & Sharma, O., 2023. IPM Concept and Strategies for Sustainable Agriculture. *In Integrated Pest Management in Diverse Cropping Systems.* Apple Academic Press, 31–59.

Sjah, T., & Zainuri, Z., 2020. Agricultural Supply Chain and Food Security. *In* W. Leal Filho, A.M. Azul, L. Brandli, P.G. Özuyar & T. Wall (eds.) *Zero Hunger.* Cham: Springer International Publishing, 79–88.

Sotirov, M., Pokorny, B., Kleinschmit, D., & Kanowski, P., 2020. International forest governance and policy: Institutional architecture and pathways of influence in global sustainability. *Sustainability*, 12, 7010.

Stefani, U., Schiavone, F., Laperche, B., & Burger-Helmchen, T., 2020. New tools and practices for financing novelty: A research agenda. *European Journal of Innovation Management*, 23, 314–328.

Tanveer, A., Javaid, M.M., Irfan, M., Khaliq, A., & Yaseen, M., 2015. Yield losses in chickpea with varying densities of dragon spurge (*Euphorbia dracunculoides*). *Weed Science*, 63, 522–528.

Tanveer, A., Khaliq, A., Javaid, M., Chaudhry, M., & Awan, I., 2013a. Implications of weeds of genus euphorbia for crop production: A review. *Planta Daninha*, 31, 723–731.

Tanveer, A., Mumtaz, K., Javaid, M., Chaudhry, M., Balal, R., & Khaliq, A., 2013b. Effect of ecological factors on germination of horse purslane (*Trianthema portulacastrum*). *Planta Daninha*, 31, 587–597.

Thomine, E., Mumford, J., Rusch, A., & Desneux, N., 2022. Using crop diversity to lower pesticide use: Socio-ecological approaches. *Science of the Total Environment*, 804, 150156.

Torracchi C, J.E., Morel, M.A., Tapia-Vázquez, I., Castro-Sowinski, S., Batista-García, R.A., & Yarzábal R, L.A., 2020. Fighting plant pathogens with cold-active microorganisms: Biopesticide development and agriculture intensification in cold climates. *Applied Microbiology and Biotechnology*, 104, 8243–8256.

Travičić, V., Cvanić, T., & Ćetković, G., 2023. Plant-based nano-emulsions as edible coatings in the extension of fruits and vegetables shelf life: A patent review. *Foods*, 12, 2535.

Travlos, I., Gazoulis, I., Kanatas, P., Tsekoura, A., Zannopoulos, S., & Papastylianou, P., 2020. Key factors affecting weed seeds' germination, weed emergence, and their possible role for the efficacy of false seedbed technique as weed management practice. *Frontiers in Agronomy*, 2, 1.

Tudi, M., Daniel Ruan, H., Wang, L., Lyu, J., Sadler, R., Connell, D., Chu, C., & Phung, D.T., 2021. Agriculture development, pesticide application and its impact on the environment. *International Journal of Environmental Research and Public Health*, 18, 1112.

Van Den Bosch, R., 2023. *The Pesticide Conspiracy.* University of California Press.

Vreysen, M., 2021. Monitoring sterile and wild insects in area-wide integrated pest management programmes. *Sterile Insect Technique. Principles and Practice in Area-Wide Integrated Pest Management*, 485–528.

Waheed, H., Javaid, M.M., Shahid, A., Ali, H.H., Nargis, J., & Mehmood, A., 2019. Impact of foliar-applied Hoagland's nutrient solution on growth and yield of mash bean (*Vigna mungo* L.) under different growth stages. *Journal of Plant Nutrition*, 42, 1133–1141.

Witcher, A.L., & Poudel, I., 2020. Pre-emergence herbicides and mulches for weed control in cutting propagation. *Agronomy*, 10, 1249.

Wolff, L., 2023. Regulating pests—Material politics and calculation in integrated pest management. *Environment and Planning E: Nature and Space*, 6, 455–472.

Yadav, R., 2022. Biopesticides: Current status and future prospects. Proceedings of the International Academy of Ecology and Environmental Sciences, 12, 211.

Zaller, J.G., & Zaller, J.G., 2020. What is the problem? Pesticides in our everyday life. *Daily Poison: Pesticides-an Underestimated Danger*, 1–125.

21 Bioethanol Production from Agricultural Waste

*Muhammad Saad Ullah**, *Sakina Bibi, Athar Mahmood, Muhammad Anjum Zia, Muhammad Awais, and Javaria Nargis*

21.1 INTRODUCTION

Biomass has the potential to serve as a substantial supply of fuel for transportation, such as bioethanol; its primary use is in the generation of electricity and heat via burning (Jayakumar *et al.*, 2023).

Bioethanol is an alcoholic compound that is produced by the decomposition of sugars derived from biomass sources such as sugarcane, maize, wheat, or agricultural leftovers. The significance of ethanol is increasing day by day because of several factors, such as the phenomenon of global warming and climate change, which has generated considerable attention on a global, national, and regional scale. Renewable biomass fuels, including bioethanol, biodiesel, and bio-hydrogen, generated from various sources such as sugarcane, maize, switch grass, and algae, have the potential to serve as viable alternatives to petroleum-based fuels (Kaur *et al.*, 2023). Rapid transitional development has been seen in bioethanol industry, as several nations turn their attention to renewable energy sources, particularly fuel for transportation. Ethanol is one of the viable substitutes for petrol within the transportation fuel industry. On a large scale, the primary source of fuel ethanol is sucrose, derived from sugarcane or starch which is mostly obtained from maize (Bakala *et al.*, 2023). However, both sources may not be favorable due to their suitability for food and feed purposes. The economic viability of ethanol production from grains depends upon the market demand, which serves as a source of animal feed. However, it is worth noting that the future market growth of distiller's dried grains with soluble (DDGS) may not be as substantial as that of ethanol. The cost element plays a pivotal role in the feasibility of expanding bioethanol production on a big scale.

The production of renewable fuel from lignocellulose wastes overcomes the current conflict between food and fuel resulting from grain-based bioethanol production. It has the potential to produce an estimated 442 billion liters of bioethanol per year from lignocellulose biomass, total crop residues, and wasted crops (Mahmood *et al.*, 2015; Singh *et al.*, 2022).

Ordinary ethanol and bioethanol are very similar; the key distinction lies in the raw material used and the production procedure used. Bioethanol is a very suitable alternative to fuel due to many factors. First, it has a higher octane number of 108.6 in comparison to petrol, so serving as a preventive measure against combustion-induced explosions. Furthermore, the combination of gasoline and ethanol at a proportion of 60:40 enhances the fuel economy of gasoline by 10% and leads to more environment-friendly burning, hence decreasing carbon monoxide emissions and thus mitigating air pollution (Mohammed *et al.*, 2021). Furthermore, bioethanol combustion exhibits a little lower temperature compared to other fuels; hence it's prolonging the lifespan of the vehicle engine, it exhibits exceptional efficiency, enhancing energy consumption. Furthermore, the use of bioethanol fuel has the potential to broaden the market for agricultural goods, particularly in the sugar industry, due to its reliance on essential raw materials like sugar cane, cassava, maize, and other similar resources.

* Corresponding author: muhammadsaadullah0346@gmail.com

 DOI: 10.1201/9781003498698-21

Ethanol, derived by the process of fermentation, is often regarded as an ecologically sustainable substitute for gasoline due to its superior octane number ratings and more environment-friendly combustion characteristics. The incorporation of agricultural waste as a primary source is seen as a viable strategy for mitigating waste management and production expenses, especially in nations such as Indonesia where agricultural waste is abundantly accessible. Biofuels such as bioethanol, biohydrogen are obtained from primary sources (Mahmood *et al.*, 2021; Dutta *et al.*, 2023) can be produced either by hydrolysis of starch-containing materials or obtained directly from sugar sources. Sugar conversion through the use of mineral acids is required for cellulosic sources. Thus, bioethanol holds a significant opportunity to support that transition to renewable energy sources due to the ongoing research and advancements in technology. Second-generation ethanol is the conversion of agro-industrial waste (AIW) into bioethanol. Multiple critical steps, including pretreatment, scarification, fermentation, distillation, and dehydration, are included in this process (Kumar *et al.*, 2021). Various other approaches have been investigated for bioethanol generation from AIWs (Morya *et al.*, 2023). In Mexico, many researchers have focused on improving the methodologies used to use different agro-industrial waste (a by-product of agriculture) for the synthesis of bioethanol. Comprehensive technical-economic research was conducted on the manufacture of bioethanol from sugarcane and blue agave bagasse; findings revealed that the expenses associated with ethanol production were 1.34 USD/gallon and 1.46 USD/gallon, respectively (Li *et al.*, 2023). A study conducted by Patra *et al.* (2022) evaluated three process techniques (i.e., separate hydrolysis and fermentation (SHF), simultaneous saccharification and fermentation (SSF), and pre-saccharification simultaneous saccharification and fermentation (PSSF) by using pretreated mango stem bark followed by maceration as raw material. The theoretical yields of the substrate of these strategies achieved 58.8%, 81.6%, and 84.5% to produce ethanol.

21.2 AGRICULTURAL WASTE AS A FEEDSTOCK FOR BIOETHANOL PRODUCTION

21.2.1 Harnessing Agricultural Waste

Bioenergy is an environment-friendly technology that transforms biomass into energy, using ethanol as a crucial fuel component to effectively mitigate greenhouse gas emissions by a substantial reduction of up to 18% (Barot, 2022). Bioethanol has been a viable substitute for conventional ethanol, since it is produced using distinct raw materials and production techniques. The presence of oxygen in it facilitates more environmentally friendly burning, resulting in less carbon monoxide emissions and air pollution, so providing a healthier atmosphere. The combustion of bioethanol results in lowering the temperature, hence prolonging the lifetime of vehicle engines, and exhibits notable efficiency in terms of enhanced energy production. Furthermore, this phenomenon generates novel market prospects for agricultural practitioners, particularly within the sugar industry, by harnessing primary resources such as sugar cane, cassava, and maize. The use of agricultural waste in the manufacture of bioethanol is widely acknowledged as a financially efficient approach, since the ethanol yields derived from various waste sources have been shown to achieve substantial proportions of the theoretical yields (Sofokleous *et al.*, 2022). The use of cellulosic biomass offers a favorable opportunity for the extensive production of biofuels, owing to its broad availability and the possibilities for local sourcing via established agricultural technologies and grain circulation networks. Although agricultural waste is not fully used in food processing industries, there is an increasing interest in using lignocellulose waste to produce value-added products.

Usually, a huge amount of wastes are produced from starchy crops, which are discarded without being used. Sugar crops are subjected to the phenomenon of extraction, which causes the release of sugar content. It is then undergoing through fermentation by yeast and thus results in bioethanol production. Thus these waste products contain excessive amount of organic and inorganic

TABLE 21.1
Cellulosic Composition of Different Plants

Substrate	Cellulose (%)	Hemicellulose (%)	Lignin (%)	References
Switchgrass	31–35	24–28	17–23	Wei et al. (2009), Wang et al. (2020)
Bagasse	65	–	18.4	Sarkar et al. (2012
Banana stem	63.9 (α)	65.2 (holocellulose)	18.6	Mohamad and Jai (2022)
Big bluestem	29–37	21–25	17–24	Wei et al. (2009)
Coconut	44.2 (α)	56.3 (holocellulose)	32.8	Azeta et al. (2021)
Corn straw	42.6	21.3	8.2	Panggabean et al. (2023)
Cotton seed hairs	80–85	5–20	0	Kumar et al. (2022)
Jatropha waste	56.31	17	23.91	Kolář et al. (2023)
Pineapple leaf	73.4 (α)	80.5 (holocellulose)	10.5	Huzaifah et al. (2017)
Rice straw	32–47	19–27	5–24	Zheng et al. (2012)
Wheat straw	35%–45%	20–30	8%–15%	Shenoy et al. (2011)

compounds, which is then used for biofuel production, increasing the profitability of bio refinery (Aruwajoye *et al.*, 2020).

21.2.2 Agro-Industrial Wastes' Composition

AIWs consist of complex plant lignocellulose components (Table 21.1) that are susceptible to biological breakdown by fungi, bacteria, and insects. Cellulose chains may be degraded by hydrolysis processes aided by acids, but cellulose enzymes exhibit enhanced glucose yields owing to their selectivity (Mehmood *et al.*, 2021). Moreover, pretreatment biomass exhibits promise in augmenting glucose release via the use of fungi or other organisms as a source for cellulase production. It is worth noting that filamentous fungus has a unique capacity to degrade cellulose molecules. Several microorganisms possess the ability to degrade cellulose; only a restricted subset of them generate enzymes that exhibit sufficient efficiency to break down cellulose in the absence of viable cells (Weimer, 2022). Usually, these enzymes are called cellulases. Some microbes, such as certain species of bacteria and fungi, naturally generate significant amounts of these enzymes; others may need genetic modification or optimization to augment their enzyme output for industrial purposes such as biofuel generation or waste management. The sources of sugars used in the manufacturing of bioethanol comprise a diverse range of crops, including sugar cane, sugar beet, maize, and cereals. These crops provide a substantial contribution to the overall worldwide production of bioethanol. The degradation of lignocellulose poses challenges due to its intricate composition, which consists of cellulose (approximately 45% of dry weight), hemicelluloses (30%), and lignin (25%). Cellulose resistance of degradation is associated with secondary and tertiary conformation and its association with other compounds (Ahmed *et al.*, 2022; Hassan *et al.*, 2022). In contrast, hemicellulose is composed of heterogeneous sugar residues that are more susceptible to hydrolysis. For example, lignin, which is obtained from phenolic alcohols, performs more as a structural component in the manufacture of bioethanol, showing the diverse characteristics of AIWs.

21.2.3 Raw Materials for Ethanol Production

Sugars are required to create bioethanol and can be obtained from various sources such as lignocellulose and crops. Traditional crops of bioethanol account for more than 90% of the worldwide bioethanol production. Crops that contain sucrose can be hydrolyzed to their monomeric

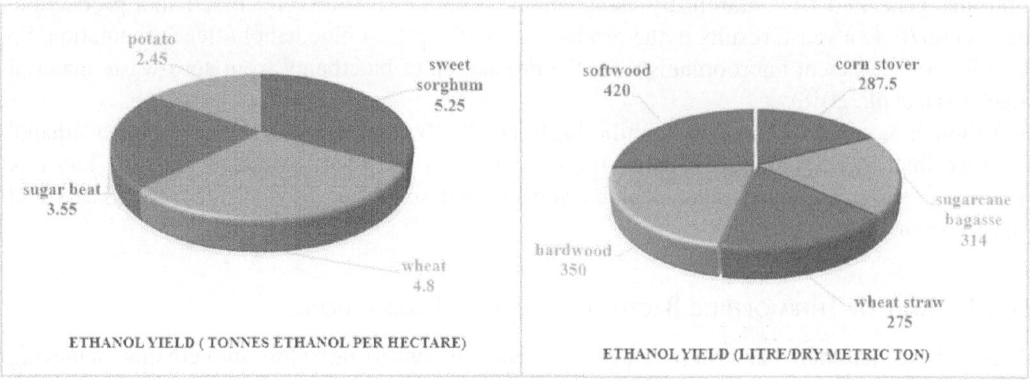

FIGURE 21.1 Bioethanol production from different agricultural wastes.

constituents. Conversely, crops that contain starch, such as maize and cereals, can be decomposed by enzymes to provide glucose. These raw materials are primarily used for ethanol production (Li *et al.*, 2022).

Cellulose is the most profuse polymer, consisting of many glucose units bonded that contain β-(1,4)-glycosidic bonds (Díez *et al.*, 2020). Cellobiose, a two-glucose-linked disaccharide, is cellulose's fundamental structural component. To form the secondary or tertiary conformation, cellulose, a hydrolysis-resistant molecule, interacts closely with lignin, hemicellulose, starch, protein, and mineral elements. It can be hydrolyzed chemically or enzymatically by using diluted or concentrated acid (Firsty *et al.*, 2023). Cellulose is more resistant to hydrolysis than hemicellulose, whereas, hemicellulose content depends on raw material origin. Lignin is one of the most widespread aromatic polymers in nature and a significant phenolic macromolecule. This substance is created by dehydrating three monomeric alcohols: trans-p-coumaryl, trans-p-coniferyl, and trans-p-sinapyl (Tirronen, 2021). These alcohols are derived from p-cinnamic acid. Different potential of agricultural products for bioethanol production have been reported in literature. Figure 21.1 represents the potential of bioethanol production from different agricultural wastes.

21.3 MICROORGANISMS FOR BIOETHANOL PRODUCTION

Microorganisms that are very efficient in producing ethanol from pure xylose in a controlled setting may not be efficient fermentors of lignocellulose hydrolysate due to the presence of inhibitors produced during hydrolysis.

Inhibitors, such as weak organic acids, for example acetate, furan derivatives, phenolic monomers derived from sucrose and lignin degradation, inhibit microbial fermentation (Kucharska *et al.*, 2018). As a result, this periodic interruption leads to an insufficient fermentation as well as a decrease in ethanol production. For lignocellulose biomass to be suitable for applications of industrial bioconversion, microorganisms should grow and produce ethanol in pretreated hemicellulose hydrolysates. However, thermophilic anaerobic ethanol generators have not been well studied on pretreated lignocellulose hydrolysate. Gut bacteria and a few yeasts, but with limited power, are the solely ones that can ferment pentose. Some yeasts, like *Pichia stipitis*, *Candida shehatae*, and *Pachysolen tannophilus*, have a limited tolerance to high ethanol concentrations, require specific environmental conditions, and are sensitive to inhibitors and changes in pH (Florentine *et al.*, 2016; Cunha *et al.*, 2019). Researchers have taken an initiative to introduce the ethanol genes into enteric bacteria.

At now, the dominant method for producing bioethanol is fermentation utilizing *Saccharomyces cerevisiae*, which is renowned for its capacity to produce significant amounts of ethanol from substrates generated from starch. By using cassava starch as a source for bioethanol production, the *S. cerevisiae,* a yeast, results in the production of 89.9 g/L of bioethanol after fermentation. So yeast is a very efficient microorganism for the production of bioethanol from agro-waste material (Kucharska *et al.*, 2018).

Although *S. cerevisiae* and mesophilic bacteria like *Zymomonas mobilis* have higher ethanol tolerance than thermophiles, the latter group of bacteria provides advantages such as less risk of contamination, improved rates of bioconversion, and simplified product recovery (Sarris and Papanikolaou, 2016).

21.3.1 Role of Thermophilic Bacteria in Ethanol Production

Thermophilic anaerobic bacteria have high efficiency in converting xylose into ethanol, achieving almost the maximum possible yield. Nevertheless, the aforementioned effectiveness has only been documented at xylose concentrations not exceeding 10 g/L. In last 20 years, multiple strains of these bacteria have been identified in diverse settings such as agricultural soils, sewage facilities, and thermal springs (Dessì *et al.*, 2018). These bacteria have been extensively researched for their potential in large-scale ethanol production. Microorganisms possess a distinct advantage due to their ability to degrade various sugars included in lignocellulose biomass. More precisely, they show potential in transforming the hemicellulose component, which comprises sugars such as xylose, arabinose, mannose, and galactose (Houfani *et al.*, 2020). Certain bacteria prefer pH values that are slightly acidic to neutral, while others flourish in environments with higher alkalinity. Furthermore, they display diverse temperature preferences, with certain individuals showing a preference for extremely high temperatures (approximately 70°C), while others tend to lean toward slightly lower temperatures (around 50°C or 60°C). In terms of ethanol tolerance, individuals display variability, although the majority is able to tolerate moderate doses. During laboratory experiments using glucose and xylose, these bacteria exhibited diverse amounts of ethanol synthesis, with some strains demonstrating superior performance compared to others (de Melo Pereira *et al.*, 2022).

21.3.2 Role of Genetically Engineered Bacteria in Ethanol Production

Several bacteria, including *Escherichia coli, Klebsiella oxytoca,* and *Z. mobilis*, have shown the capacity for industrial ethanol synthesis. *E. coli* was the first bacterium to undergo genetic alteration with the intention of facilitating ethanol production.

The ability of this to utilize diverse carbon sources, to efficiently metabolize high sugars, and the positive tolerance to ethanol were observed. In natural ways of metabolism, *E. coli* generate very little ethanol (de Melo Pereira *et al.*, 2022). The scientists have introduced *Z. mobilis* (a bacterium specialized in ethanol fermentation) genes into *E. coli.* The genetically modified KO11 strain is capable of producing ethanol from a range of substrates (including barley hulls, orange peels, and wood waste) under controlled laboratory conditions. In other words, the efficiency of producing ethanol from glucose by *Z. mobilis* is even higher compared to other organisms, mainly due to their highly active pyruvate decarboxylase (PDC)-alcohol dehydrogenase (ADH) system (Zhang *et al.*, 2019). While the wild-type strain does not ferment xylose, it has been successfully engineered to do so through the incorporation of an *E. coli* xylose-metabolizing pathway. *Z. mobilis* displays distinctive features when compared to traditional fermentative strains, including higher ethanol yield and selectivity, higher specific productivity, poor pH tolerance, and high toleration of ethanol. Although xylose from lignocellulose hydrolysates has so far been used for the production of ethanol of starch or sugar biomass, recent strains such as AX101 can also utilize xylose (although they are sensitive to inhibitors). Mutants of *K. oxytoca*, genetically engineered to incorporate the pyruvate-to-ethanol transfer (PET) operon, have redirected carbon flow into the production of ethanol. It can degrade various sugars, including hexose, pentose, cellobiose, and cello triose. This makes it an attractive candidate for the production of ethanol from substrates like corn fiber or sugar beet pulp.

21.3.3 Role of Genetically Engineered Yeast in Ethanol Production

S. cerevisiae is well known for its high ethanol production and its ability to tolerate ethanol and toxic compounds present in hydrolysate from lignocellulose biomass. However, the wild-type strain of *S. cerevisiae* has certain limitations in pentose fermentation. As a result, extensive work has been done to engineer optimized strains for this application (Favaro *et al.*, 2019). An alternative strategy is the introduction of a gene coding for xylose isomerase (XI), an enzyme that can directly convert xylose into ethanol by fermentation (Dos Santos *et al.*, 2016). However, such a strategy required the overexpression of endogenous xylulokinase (XK) for proper xylose metabolization. Co-existing transporters in the membrane interfere with xylose uptake, since xylose uptake competes with glucose uptake and vice versa, which limits the use of xylose-fermenting *S. cerevisiae*. By both high-affinity and low-affinity glucose transport systems, xylose can be consumed by *S. cerevisiae*. However, xylose absorption is much less efficient by these transporters than by glucose. To address this concern, a number of metabolic engineering efforts have been made to improve the initial xylose extrusion rate and promote the xylose metabolic pathways in *S. cerevisiae* (Kwak and Jin, 2017). Improving the transportation of xylose in *S. cerevisiae* is still a major obstacle; however, progress in metabolic engineering has shown its potential for overcoming this issue and enhancing the efficiency of xylose fermentation in different strains of *S. cerevisiae*. Figure 21.2 provides data for worldwide bioethanol production from agricultural waste.

21.4 ETHANOL PRODUCTION TECHNOLOGIES

The production of ethanol may be accomplished by chemical synthesis or fermentation. Fermentation is a process facilitated by catalysts, which are enzymes, synthesized by living cells. Ethanol production employs advanced technologies on a global scale, whereas the selection of technology is contingent upon the specific raw material undergoing fermentation. White beet and its processing products, sugar cane, citrus fruits, certain tropical plants, tree juices, and honey are examples of raw materials that contain simple sugars and are appropriate for direct fermentation. Cereals, such as rye, barley, maize, oat, wheat, and sorghum, together with vegetable bulbs, roots, seeds, fruits,

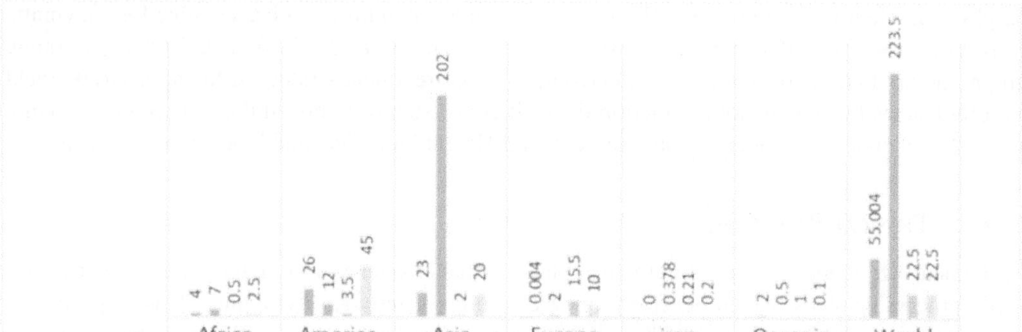

	Africa	America	Asia	Europe	iran	Oceania	World
sugar cane	4	26	23	0.004	0	2	55.004
rice straw	7	12	202	2	0.378	0.5	223.5
Barley	0.5	3.5	2	15.5	0.21	1	22.5
corn waste	2.5	45	20	10	0.2	0.1	22.5

FIGURE 21.2 Worldwide bioethanol production from agricultural waste.

lumber, grass, and moss, are among the other raw materials that include starch and polysaccharides and are used in the manufacturing of ethanol (Kumar *et al.*, 2019). The use of lignocellulose biomass, including agricultural leftovers, wasted and energy crops, fast-growing trees, wastepaper, and inappropriate cereals, for ethanol generation has significant importance yet, its widespread commercial implementation remains limited. The present global ethanol production is much outweighed by the entire potential bioethanol output derived from agricultural residues and discarded crops. The costs of manufacturing fluctuate based on the pricing of raw materials, the techniques of production, the level of refining, and the usage of bio-products and waste.

21.4.1 Pretreatment of Agricultural Waste for Bioethanol Production

Biofuel generation heavily relies on the pretreatment of biomass, especially for lignocellulose biomass consisting of cellulose, hemicellulose, and lignin. Pretreatment techniques are used with the objective of solubilizing and segregating these constituents, hence enhancing the accessibility of the residual solid biomass for further chemical or biological processing. This process degrades the cellulose and lignin matrix, which is held together by hemicellulose chains.

This leads to decreased cellulose crystallinity along with increased amorphous cellulose content, which increases its susceptibility to enzymatic hydrolysis. An effective pretreatment method aims to fulfill numerous significant goals. First, it seeks to generate sugars, directly or indirectly, through the hydrolysis of complex carbohydrates into fermentable sugars. Additionally, it acts in restricting the loss or degradation of sugar during the process and ensures maximum yield and thus efficiency in the bioethanol production process (Mittal *et al.*, 2011). Pretreatment is an essential step to fractionate the cellulosic biomass into cellulose, hemicellulose, and lignin, increasing cellulose availability through enzymatic hydrolysis. A variety of pretreatment techniques are available that are typically used individually or in multidimensional fashion to achieve favorable results. Pretreatment processes that are used for the solubilization and separation of lignocellulose biomasses try to reveal their components, including hemicellulose, lignin, and cellulose (Mohammad Rahmani *et al.*, 2022). Pretreatment modifies cellulose's structure, rendering it more susceptible to enzymatic destruction, leading to increased sugar production. So, the types of objectives could be enhancing the efficiency of enzymatic hydrolysis, limiting sugar loss, controlling by-product synthesis, and curtailing costs. Pretreatment procedures play an important role in the process of bioethanol production, encompassing physical, chemical, and biological processes that make cellulose more accessible for enzymatic hydrolysis as well as safer and more cost-effective (Alvira *et al.*, 2010). Although there are many unique factors that contribute to the effectiveness of pretreatment strategies, the appropriate yield and efficiencies for bioethanol production through conversion of lignocellulose biomass to ethanol is greatly dependent on selective combinations of different biological and biochemical aspects.

21.4.2 Physical Pretreatment

Mechanical size reduction is used in the first step of ethanol synthesis from agricultural solid wastes as pre-processing steps, which generally involve milling, grinding, or chipping. This step aims to minimize the crystallinity degree of the cellulosic material and enhance the efficacy of the subsequent processing stages. Mechanical comminution includes different processes such as wet and dry milling, vibratory ball milling, or compression milling (Cadoche and López, 1989). Power input required for mechanical comminution is determined by the initial and final particle sizes, the waste material being processed, and its moisture content. Making particle sizes smaller would optimize this further, but very small particle sizes might have some negative effect on following processing steps. In liquid-based processes, small particles can promote the aggregation of particles and more likely cause channeling which increases the specific energy consumption (Saad *et al.*, 2019). The choice of the most appropriate milling method depends on the properties of waste material, but usually, hardwood is recommended to be used with hammer mills or ball mills, whereas softwood is recommended with cutter mills.

21.4.3 Pyrolysis

Pyrolysis is a thermochemical process that can be executed without oxygen, decomposing biomass or other organic matters at temperatures over 300°C (Zhang *et al.*, 2021). It is a rapid degradation process of cellulose that produces gaseous molecules such as hydrogen (H_2) and carbon monoxide (CO), along with leftover char. The decomposition process is temperature-dependent; higher temperatures cause a quicker breakdown and generate more volatile chemicals, while lower temperatures in turn inhibit volatile chemicals production.

After the pyrolysis, the remaining char can also be further treated by leaching with water or a dilute acid solution. Another method that results in a high concentration of carbon sources in the leachate is the so-called water leaching, where char is treated with water to dissolve and remove carbonaceous molecules. This leachate is suitable for cultivating bacteria, especially those that produce bioethanol. Under certain conditions, such as using a 1 N H_2SO_4 solution and maintaining a temperature of 95°C for a period of 1 hour, the conversion rates vary between 80% and 85%. Significantly, over 50% of the final products are composed of glucose (Haldar and Purkait, 2020a). Numerous studies have shown evidence of substantial cellulose conversion into reducing sugars by moderate acid leaching. This observation underscores the efficacy of pyrolysis in conjunction with later leaching procedures for the purpose of biomass vaporization and the manufacture of bioethanol.

21.4.4 Microwave Oven and Electron Beam Irradiation Pretreatment

Microwave therapy employs a combination of thermal and non-thermal phenomena induced by microwaves in aquatic settings. It is a viable approach that capitalizes the notable thermal efficiency shown by microwaves. The thermal effect is based on sending microwave radiation to stir the polar bonds leading to generation of heat in the inner parts of the biomass itself and the external water medium. This mechanism leads to areas of high temperature development within the material as well as an explosion effect that improves the degradation of resilient structures present in lignocellulosic materials (Guo *et al.*, 2020). Moreover, heat pretreatment can create an acid environment favoring autohydrolysis by eliminating acetic acid from the lignocellulose components. Non-thermal methods, such as electron beam irradiation on the other hand, excite polar bonds due to their alignment with variable magnetic field.

During pretreatment, cellulosic biomass undergoes a series of transformations, including an increase of the specific surface area, decrease of the polymerization and crystallinity level of cellulose and hydrolytic degradation of hemicellulose as well as a partial breakdown of lignin (Haldar and Purkait, 2020b).

Microwave pretreatment is an effective way to increase the transformation of lignocellulosic biomass into valuable products like fermentable sugars. Based on its heat treatment, i.e., the pretreatment of rice straw and bagasse using microwaves and lignin removal, 43%–55% of the total reducing sugars that exist (sugars which may be utilized in the production of bioethanol or other bio refinery methods) can be extracted (Cantero *et al.*, 2019).

21.4.5 Physicochemical Pretreatment

21.4.5.1 Steam Explosion or Auto Hydrolysis

Steam explosion has been recognized as an effective pretreatment to enhance the enzymatic digestibility and consequently the catalytic reactions in hydrolysis of lignocellulose biomass. An efficient separation of biomass into fractions is provided by the pretreatment method mentioned in the study using no catalyst which helps in the generation of significant chemicals such as clavulanic acid, xylitol, and other alcohols.

The steam explosion pretreatment involves exposing biomass to high-pressure steam at temperatures of 160°C–290°C and pressures of 20–50 bar for a short period of time (Onumaegbu *et al.*,

2018). The process is later aborted with a quick decompression, bringing the ambient pressure to atmospheric level. Fast decompression occurs as the steam expands inside the lignocellulose matrix, which separates individual fibers. Physical pretreatments improve the enzymatic hydrolysis of cellulose and hemicellulose by dissolving the lignocellulose structure. A key advantage of steam explosion pretreatment is its relatively high xylose recovery, generally between 45% and 65%. Steam explosion pretreatment is attractive from the economic point of view owing to its high recovery rate, which allows the efficient use of xylose in subsequent bioconversion steps (Yu *et al.*, 2022). Overall, steam explosion pretreatment has great potential in terms of practicality as a biomass pretreatment method because it is effective, simple, and economical, facilitating the achievability of the large-scale technology of bioethanol production.

21.4.5.2 Liquid Hot Water Method

Hydrothermal processing or liquid hot water pretreatment method splits hemicellulose of lignocellulose biomass by forceful hot liquid water above saturation point. This method can release a significant amount of hemicellulosic sugars mainly as oligomers. In the preparation of liquid hot water, wood is heated between 170°C and 230°C and maintained at pressure over 5 MPa generally for 20 minutes. This process begins when water causes hemicellulose to break down, releasing oligomeric sugars (Ramos *et al.*, 2022).

Liquid hot water pretreatment may induce the formation of unwanted products, like furfural and carboxylic acids that can inhibit the development of microorganisms and interfere with ethanol production (Latif *et al.*, 2023). In spite of its disadvantages, the method presents a number of advantages as it shows considerable xylose recoveries that may vary from 88% to 98%. Its cost savings are also environmentally friendly, as no chemicals or acids are used in the pretreatment.

Researchers have explored modifications to liquid hot water pretreatment, such as employing two-stage treatments and combining it with alkaline solutions. These techniques have demonstrated potential in enhancing the efficiency of processing lignocellulose biomass.

21.4.5.3 Ammonia Fiber Explosion

Ammonia fiber explosion (AFEX) is a lignocellulose biomass pretreatment. This method consists of treating the biomass with a liquid ammonia solution and heating it to high temperatures. This mechanism changes the makeup of biomass, allowing enzymatic accessibility and breaking down complex polysaccharides into simple sugars for biofuel generation (Chundawat *et al.*, 2020). Moreover, favorable outcomes do not require meticulous biomass purification. It is handy and takes less processing time. It is especially useful for substrates with reduced lignin contents compared to sugarcane. This is when the sugars are not actually released but polysaccharides like hemicellulose and cellulose are being liberated, which can then be attacked by enzymes and converted to sugars.

The efficiency of ammonia recovery system is based on various parameters, including quantity of ammonia used, temperature, pressure, moisture content present in the biomass, and time of data collection. Studies have been carried out to maximize the efficiency of AFEX pretreatment for certain materials; under optimal conditions, cellulose and hemicellulose removal may be as high as 90% (Rodionova *et al.*, 2022). In addition, the AFEX treatment often needs a lower enzyme loading in comparison to other pretreatment procedures, which makes it a potential option for the conversion of lignocellulose biomass.

21.4.5.4 CO_2 Explosion

Explosion techniques using carbon dioxide (CO_2) exhibit similarities with steam and ammonia explosion processes, but with distinct advantages. Compared to steam explosion methods, CO_2 explosion techniques may give considerably greater conversion yields; it also considered more cost-effective as it does not yield the inhibitors that are generated by steam explosions and offers significant advantages by enhancing the efficiency and increasing the yield of sugar release from lignocellulose materials (Chadha *et al.*, 2020; Javaid *et al.*, 2022).

21.4.6 Chemical Pretreatment

There is a diverse selection of chemical pretreatment methods, such as diluted acid, alkali, ammonia, organic solvents, sulfur dioxide, carbon dioxide, and other compounds. These approaches exhibit versatility and efficacy, facilitating the efficient decomposition of lignocellulose biomass into fermentable sugars.

21.4.6.1 Acid Pretreatment

This method often utilizes concentrated or diluted acids, typically with concentrations ranging from 0.2% to 2.5% (w/w), and temperatures ranging between 130°C and 210°C. These conditions effectively hydrolyze biomass's complex polysaccharides, releasing fermentable sugars for biofuel production (Wagle *et al.*, 2022).

Sulfuric acid is often used for this particular objective, alongside hydrochloric acid, nitric acid, and phosphoric acid, among other acid kinds. The primary focus of the acid medium is on polysaccharides, namely hemicelluloses, which exhibit a higher susceptibility to hydrolysis in comparison to cellulose (Pereira *et al.*, 2017). Nevertheless, one limitation, associated with acid pretreatment, is the production of diverse inhibitors, including acetic acid, furfural, and 5-hydroxymethylfurfural, which have the potential to impede the proliferation of microbes throughout the fermentation process; as a result, hydrolysates that are meant for fermentation often need to undergo detoxification.

21.4.6.2 Alkaline Pretreatment

The process of alkaline pretreatment of lignocellulose biomass is used with the objective of breaking down the lignin matrix, hence enhancing the accessibility of cellulose and hemicellulose for enzymatic breakdown. As a result of alkali treatment, the cell wall is broken down by many processes, such as the breakdown of hemicelluloses, lignin, and silica, the breakdown of uronic and acetic esters, and the expansion of cellulose. This procedure results in a reduction in the crystallinity of cellulose as a consequence of swelling, hence promoting its enzymatic breakdown (Akhtar *et al.*, 2016; Mumtaz *et al.*, 2021). After applying alkaline pretreatment, the substrates can be fractionated into alkali-soluble lignin, hemicellulosics, and residues via alkaline pretreatment, they can thus be utilized for the generation of higher value products. The residual materials, the remaining material, which is primarily cellulose, could potentially be used in paper production or for the production of cellulose derivatives. This method utilizes a variety of hydroxides such as sodium, potassium, calcium, and ammonium. Alkaline pretreatment methods usually have less stringent temperature and pressure requirements relative to other pretreatment technologies. Through a research, the best process condition was found to be 1.5% use for 144 hours at 20°C, which resulted in 60% lignin as well as 80% hemicellulose. NaOH has proven to be a good agent that improves the digestion of hardwood significantly, reducing the dodecyl sulfated lignin from a range of 24% to 55% to a maximum of 20%, bringing on a rise of 14% to 55% (Mohsenzadeh *et al.*, 2012).

21.4.6.3 Wet Oxidation

The wet oxidation is a pretreatment process that involves the treatment of feedstock material with water and air or oxygen at temperatures exceeding 120°C (M'Arimi *et al.*, 2020). In this approach the biomass is mixed at the ratio of 1 L per 6 g of biomass. Wet oxidation promotes the transfer of hemicelluloses from the solid phase to the liquid phase without causing hydrolysis of the liberated hemicellulose molecules. Sugar oligomers are the hydrolytic degradation products of hemicellulose during the moist oxidation process. In their study, Pedersen and Meyer (2009) achieved glucose yields of 400 g/kg and xylose yields of 200 g/kg from wheat straw pretreated by wet oxidation. The yields showed here were for a 24-hour, 50°C period obtained at 36 FPU/g of Novozyme 188 1.5 L and 37 CBU/g of Novozyme 188.

21.4.6.4 Organosolv Pretreatment

This method consists of immersing the biomass in a solvent, which is an organic solvent (ethanol, methanol, or acetone) and water and a catalyst, which can be an acid or base. Then, the mixture is subjected to moderate temperatures, typically within the range of 120°C–20°C, under an applied pressure, to promote the separation and degradation of lignin (Devi *et al.*, 2021). Lignin is removed and separated from cellulose and hemicellulose fractions essentially due to interaction between the organic solvent and lignin and cellulose and hemicellulose fractions remain largely unfazed. This process has proven beneficial, especially because it can strip away lignin with high precision while maintaining cellulose and hemicellulose. The more of these components are available, the more can be used in subsequent thermochemical or biochemical conversion processes. High-purity lignin can be obtained as a valuable co-product using this approach. The integration of lignin within the course of pretreatment can contribute to increase the economic value of the lignin in addition to significantly improve the sustainability of the biomass conversion process (Yoo *et al.*, 2020; Abdullah *et al.*, 2022). As a result, including lignin within the extent of the pretreatment process adds value to its economics and dramatically improves the sustainability of biomass conversion processes.

21.4.7 Biological Pretreatment

Biological pretreatment of lignocellulose biomass is very important in ethanol production. The ligninolytic ability of the bacteria and the fungi is employed to reduce the recalcitrance of the feedstock and it facilitates the digestibility of the biomass by the hydrolytic enzymes. In biological pretreatment microorganisms are used either grown on feedstock or enzyme extract. The recent development in the field of microbiology has approved more knowledge of the current lignolytic system by an improving engineered one, identifying more enzymes for improved pretreatment and a complete use of lignocellulose biomass in bio refinering units (Hemsworth *et al.*, 2016).

The biological pretreatment methods exploit microorganisms like brown rot, white rot, and soft rot fungi to effectively degrade the lignocellulose matrix, which results in the liberation of cellulose (Hemsworth *et al.*, 2016). Compared to other decomposers, white rot fungi are known for their unparalleled ability to degrade both cellulose and lignin.

Under low-temperature conditions, application of white rot fungus 17 on bamboo culms showed promising results. Additionally, trials done on sugarcane waste using microorganisms have yielded significant reductions in cellulose and lignin levels, indicating that biological pretreatment for lignocellulosic biomass production may be efficient (Pandiyan *et al.*, 2019). Compared to acid hydrolysis, enzymatic hydrolysis requires less energy, lower temperature, lower toxicity, lower utility cost, and less wear and tear on the equipment. *Saccharification*: Saccharification is a process employed during the bioprocess of bioethanol production in which polysaccharide compounds are hydrolyzed to give sucrose monomers (Bala and Singh, 2019). Enzymatic hydrolysis occurs without the formation of inhibiting by-products. Hydrolysis through enzymes relies on the use of cellulase enzymes that have very high substrate specificity. Cellulases hydrolyze cellulose whereas hemicelluloses react with hemicellulose. A group of enzymes that work together to enzymatically hydrolyze cellulose into glucose consists of endoglucanases, exoglucanases, as well as β-glucosidases. Hemicellulolytic enzymes are involved in the degradation of hemicellulose such as xylenes, manganese, arabinofuranosidases, and glucuronidases.

Microbial organisms like *Trichoderma* and *Aspergillus*, and others, can synthesize cellulase and hemicellulase enzymes. *Trichoderma* produces many cellulases and xylenes (yet does not produce β-glucosidase). *Aspergillus* on the other hand has high efficiency in producing β-glucosidase (Imran *et al.*, 2016).

Belkacemi and Hamoudi (2003) reported a maximum scarification rate approaching up to 90% using modified maize stalk hemicellulose at 10 hours, 30°C, and pH 5. (Belkacemi and Hamoudi, 2003) observed that PEG greatly increased soft lignocellulose conversion, rising from 42% to 78% after 16 hours. This process has been best understood to occur at an ideal temperature of 50°C, and significant yields have been reported from a few separate pretreatment methods as well, such as with alkali pretreated rice straw (57%) and alkaline peroxide treated wheat straw (45%). Many factors can affect the enzymatic hydrolysis; therefore, factors such as temperature, pH, substrate concentration, enzyme loading, and surfactant inclusion must be optimized in this stage. Benefits of enzymatic hydrolysis include low toxicity and cost-effective production without inhibitory by-products. Such features give it a highly feasible alternative for bioethanol generation. Enzyme hydrolysis is crucial to bioethanol production because it can hydrolyze complex carbohydrates into simple monomers and the process consumes much less energy and also performs at moderate conditions (Chadha *et al.*, 2019; Vasić *et al.*, 2021). Cellulase is an enzyme produced by cellulolytic microorganisms and fungi and plays an important role in the hydrolysis of cellulose into glucose or galactose monomers. A significant portion of hemicellulolytic enzymes is involved in the breakdown of hemicellulose, converting it into pentoses and hexoses.

21.4.8 Enzyme Cellulase and Their Mode of Action

Microbial biomass degradation relies on the action of cellulases, which are predominant enzymes in this context. There are various complex structural types of cellulases, including endoglucanases (1,4-β-D-glucanohydrolases), eExoglucanases (which also include cellodextrinases and cellobiohydrolases), and β-glucosidases. Lynd divides cellulases into two types: the non-complex and complex. Anaerobic bacteria like *Clostridium thermocellum* produce complexed cellulases, whereas aerobic fungi and bacteria such as *Trichoderma reesei* synthesize non-complexed cellulases (Chatterjee *et al.*, 2015). To produce ethanol from cellulose, one needs to break down the cellulose fibrils that are enclosed inside plant cell wall. Cellulose is composed of both amorphous and crystalline. Cellulose crystallinity (aggregated polymer molecules) is decomposed into either crystalline or amorphous, in which crystalline form is compact, recrystallization is stable that does not degrade easily, whereas the amorphous form of the chain structure is not properly ordered but easily degrades. The degradation of both cellulosic materials requires cellulase enzymes in enzymatic hydrolysis (Furukawa and Komatsu, 2017). Eventually, this process makes it possible to extract ethanol from cellulose chains.

21.4.8.1 Endoglucanases and Exoglucanases

The enzymatic hydrolysis process is initiated by endoglucanases, which break internal bonds at non-crystalline regions of the cellulose fiber in a random fashion. This process gives rise to new reducing and non-reducing endpoints. This approach allows for the assembly of oligosaccharides of different lengths, with the incorporation of a water molecule within the 1,4-β glycosidic bond. The newly formed bonds are sensitive to the impact of cellobiohydrolases (Hosseini and Shah, 2011). Exoglucanases, also known as cellobiohydrolases, aid in the step-by-step breakdown of cellulose molecules found at both the reducing and non-reducing ends. The primary outcome of this process is the discharge of cellobiose molecules, which are then broken down by β-glucosidases by hydrolysis (Hamid *et al.*, 2015). The cellulase system consists of exoglucanases, which constitute about 40–70% of its total components. These exoglucanases have the capacity to break down crystalline cellulose by hydrolysis. Exoglycanases are enzymes that break down cellulose by hydrolyzing it at both ends, resulting in the formation of cellobiose. The enzyme β-glucosidase plays a crucial role in the hydrolysis of soluble cellobiose and other cellodextrins into glucose within the aqueous phase, so effectively removing the inhibitory effects of cellobiose (Yang *et al.*, 2011).

21.4.9 Diversity of Effective and Efficient Cellulolytic Microorganisms (Bacteria and Fungi)

Cellulolytic microorganisms have specific ability to degrade cellulose. These organisms have the capacity to degrade resistant plant cell walls by secreting a significant portion, up to 50%, of their protein during the development phase on cellulose or other biomass. Several wood-feeding insects, including silver crickets, termites, and beetles, as well as herbivorous animals such as cows, buffaloes, goats, and cockroaches, carry cellulolytic microbes in their gastrointestinal tract and bodies. The rumen of mammalian animals has an inherent cellulose-degrading system consisting of

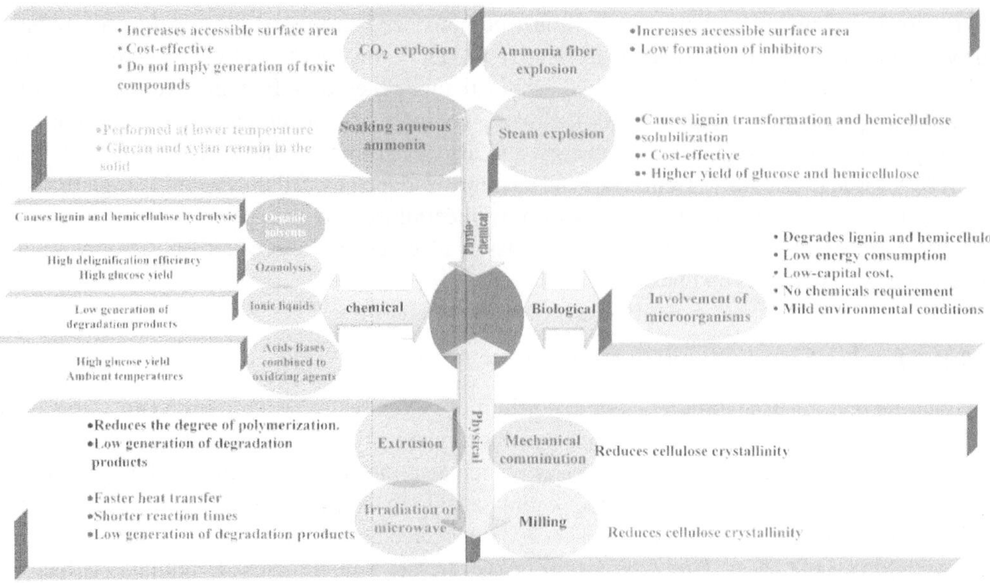

FIGURE 21.3 Various pretreatment methods of agro-waste for bioethanol production and their advantages.

diverse bacteria that possess the ability to break down cellulosic biomass derived from agricultural wastes, the organic portion of municipal solid waste, and aquatic plants (Ransom-Jones, 2014). Decomposition of cellulose-based materials is greatly aided by soil, which is home to a wide variety of germs, mostly bacteria. Bacteria, archaea, flagellates (formerly referred to as archaezoa), yeasts, and fungi have a high level of microbial activity inside their guts, particularly in the destruction of cellulose. These microbes have enzymes called cellulases that can break down cellulose. There are different methods (Figure 21.3) for pretreatment of agro-waste for bioethanol production with varying advantages.

21.5 FERMENTATION

The fermentation of saccharified biomass by different microorganisms is a crucial process in the generation of bioethanol from lignocellulosic biomass. Nevertheless, few cellulolytic microbes, such as thermophiles and mesophiles, anaerobes and aerobes, fungi, and bacteria have a surprising capacity to efficiently hydrolyze cellulose with a high crystallinity and insolubility (Abo *et al.*, 2019). Thermophilic microbes are especially important because they produce cellulase that is stable at high temperatures and retains its activity in strongly acidic or alkaline environments, even at temperatures above 90°C. The thermophilic anaerobic bacterium, Clostridium, contributes to the degradation of cellulosic plant material and has versatile fermentative properties. Thermoanaerobacterium, a specialized branch of clostridia, is one type of anaerobic microorganism able to degrade hemicellulose. They are great producers of ethanol and hydrogen from pentose carbohydrates at elevated temperatures (Mahmood *et al.*, 2013; Schuster and Chinn, 2013). Bacteria are one of the most abundant lignocellulolytic microbial taxa prevalent in microbial consortia.

These properties include broad substrate range, high ethanol yield and productivity, tolerance for high ethanol concentrations and temperature, inhibitors candidacy from hydrolysates, and cellulolytic activity.

The lignocellulosic hydrolysate undergoes two main fermentation methods: SSF and SHF. SSF improves xylose conversion into xylitol and therefore contributes to higher ethanol yield coefficient (Sudiyani *et al.*, 2019).

The approach of consolidated bioprocessing (CBP) of integrating cellulase synthesis, hydrolysis of biomass, and fermentation of ethanol in a single reactor provides a low-cost and streamlined process. In simultaneous saccharification and co-fermentation (SSCF) enzymes that are compatible with each other in terms of pH and temperature are fermented simultaneously. Many microorganisms, both naturally occurring (wild) and genetically manipulated (engineered), are used in fermentation processes, such as *S. cerevisiae*, *Z. mobilis*, *C. shehatae*, and *P. stipitis*. Many genetic engineering approaches have been adopted in the fields of fermentation to improve production, substrate consumption, and recovery rates. Thermo-tolerant bacteria, including *Kluyveromyces marxianus* and *Candida lusitaniae*, have been invented to enhance the cost-effectiveness of ethanol generation.

21.6 SCALE-UP CHALLENGES

1. Transitioning from small-scale laboratory experiments to large-scale commercial production presents numerous challenges. Processes that have proven successful in the lab may not seamlessly translate to industrial settings due to disparities in equipment, conditions, and dynamics.
2. Evaluating the energy balance and environmental footprint of bioethanol production from agricultural residues is complex. Achieving harmony between energy needs with environmental sustainability requires careful planning to minimize negative impacts.
3. Achieving profitability and financial sustainability demands careful planning, investment, and risk management. This is addressing substantial upfront costs and logistical challenges, especially in regions with limited resources. Coordinating construction and operation necessitates collaboration between the government, private investors, and local communities.
4. Creating strategies to efficiently source, handle, and process residues effectively is crucial for sustainable and reliable bioethanol production. Despite the potential of bioethanol derived from agricultural leftovers shows potential as a sustainable energy alternative, it still encounters economic challenges in competing with fossil fuels. The expenses associated with bioethanol production are contingent upon several variables, including feedstock pricing, processing technologies, energy needs, government assistance, and market demand.
5. Setting up bioethanol production facilities demands substantial capital investment in infrastructure, including processing plants, storage facilities, transportation networks, and distribution routes.
6. Moving from lab prototypes to commercial production requires further investment, validation, and optimization. Continuously investing in compliance is necessary to adjust to evolving rules and industry norms.
7. Market dynamics, influenced by factors such as government legislation, fuel costs, and consumer preferences, have an impact on strategic decisions related to production capacity, pricing, and market positioning. Addressing these challenges require to adopt a comprehensive strategy that considers technical, economic, environmental, social, and regulatory factors.

21.7 CONCLUSION

Utilizing agro-residue biomass for bioethanol production offers a financially viable and environment-friendly approach to meet global energy needs. Although sugar and starch, which are conventional sources for bioethanol production, may produce increasing amounts of ethanol, they are unable to meet the world demand. However, agricultural wastes are abundant, eco-friendly, and economical, since they do not need more land, water, or energy resources. Moreover, these chemicals have a deficiency in nutritional value, which makes them very appropriate for bioethanol production. With the decline in availability of non-renewable energy sources and the subsequent environmental issues, it is imperative to prioritize the transition to renewable and sustainable energy sources. Cellulosic biomasses possess economic viability, cost effectiveness, and ecologically advantageous attributes, rendering them a noteworthy input to the mitigation of reliance on non-renewable energy sources. Bioethanol production from cellulosic biomass exhibits promise in solving several environmental and energy crisis issues. However, the production procedure of bioethanol encounters many challenges, such as necessary pretreatment, enzymatic hydrolysis, fermentation, and distillation.

To overcome these challenges, it is essential to use efficient technologies. The pretreatment procedures used should possess the capability to efficiently liberate cellulose and hemicellulose from their intricate matrix, including lignin. In order to generate fermentable monomers, hydrolysis procedures must effectively depolymerize celluloses and hemicelluloses. The process of scarifying cellulose chains necessitates the collaborative activity of cellulase enzymes, which in turn demands the creation of effective ways for producing these enzymes. Recombinant microbial strains are required in fermentation operations when pentose and hexose sugars are fermented simultaneously. The bioethanol production business has several research gaps and challenges, such as the need for sustainable and carbon-neutral energy sources, improved breeding procedures for energy plants, enzymatic hydrolysis, specialized fermentation strains, and efficient waste treatment technologies. The complex composition of cellulose mandates expensive and time-consuming treatments for its conversion into simple sugars. Therefore, proper pretreatment techniques are required to increase the concentration of fermentable sugars. To surmount these challenges, it is essential to use cutting-edge scientific methodologies and state-of-the-art technology to generate and improve the production of bioethanol from agricultural residues. Efficient biofuel production requires the reduction of enzyme costs and the careful selection of microorganisms possessing both complexed and non-complexed types of cellulases, as well as genetically altered microbes. To achieve high production, tolerance to inhibitory chemicals, and resistance to contamination, it is important to possess resilient microorganisms that possess the ability to ferment simple sugars into ethanol. In general, the use of cellulosic biomass for the manufacture of bioethanol has the benefit of being feasible on a global scale via the utilization of pre-existing agricultural technology and grain circulation systems.

LIST OF ABBREVIATION

ADH alcohol dehydrogenase
AFEX ammonia fiber explosion
AIW agro-industrial waste
CBP consolidated bioprocessing
DDGS distiller's dried grains with soluble
PDC pyruvate decarboxylase
PET pyruvate-to-ethanol transfer
PSSF pre-scarification simultaneous scarification and fermentation
SHF separate hydrolysis and fermentation
SHCF separate hydrolysis and co-fermentation
SSF simultaneous scarification and fermentation
XR xylose reductase
XI xylose isomerase
XDH xylitol dehydrogenase

REFERENCES

Abdullah, Mahmood, A., Bibi, S., Naqve, M., Javaid, M.M., Zia, M.A., Jabbar, A., Ud-Din, W., Attia, K.A. & Khan, N., 2022. Physiological, biochemical, and yield responses of linseed (*Linum usitatissimum* L.) in α-tocopherol-mediated alleviation of salinity stress. *Frontiers in Plant Science*, 13, 867172.

Abo, B. O., Gao, M., Wang, Y., Wu, C., Ma, H., & Wang, Q., 2019. Lignocellulosic biomass for bioethanol: an overview on pretreatment, hydrolysis and fermentation processes. *Reviews on Environmental Health*, *34*(1), 57–68.

Ahmed, S.F., Mofijur, M., Chowdhury, S.N., Nahrin, M., Rafa, N., Chowdhury, A.T., Nuzhat, S. & Ong, H.C., 2022. Pathways of lignocellulosic biomass deconstruction for biofuel and value-added products production. *Fuel*, 318, 123618.

Akhtar, N., Gupta, K., Goyal, D. & Goyal, A., 2016. Recent advances in pretreatment technologies for efficient hydrolysis of lignocellulosic biomass. *Environmental Progress & Sustainable Energy*, 35, 489–511.

Althuri, A., Chintagunta, A.D., Sherpa, K.C. & Banerjee, R., 2018. Simultaneous Saccharification and Fermentation of Lignocellulosic Biomass. *In* S. Kumar & R.K. Sani (eds.) *Biorefining of Biomass to Biofuels: Opportunities and Perception*. Cham: Springer International Publishing, 265–285.

Alvira, P., Tomás-Pejó, E., Ballesteros, M. & Negro, M.J., 2010. Pretreatment technologies for an efficient bioethanol production process based on enzymatic hydrolysis: A review. *Bioresource Technology*, 101, 4851–4861.

Aruwajoye, G.S., Kassim, A., Saha, A.K. & Gueguim Kana, E.B., 2020. Prospects for the improvement of bioethanol and biohydrogen production from mixed starch-based agricultural wastes. *Energies*, 13, 6609.

Azeta, O., Ayeni, A.O., Agboola, O. & Elehinafe, F.B., 2021. A review on the sustainable energy generation from the pyrolysis of coconut biomass. *Scientific African*, 13, e00909.

Bakala, H.S., Devi, J., Ankita, L.K.S., & Kaur, S., 2023. Utilization of Wheat and Maize Waste as Biofuel Source. *In* N. Srivastava, B. Verma & P.K. Mishra (eds.) *Agroindustrial Waste for Green Fuel Application*. Singapore: Springer Nature Singapore, 27–66.

Bala, A. & Singh, B., 2019. Development of an environmental-benign process for efficient pretreatment and saccharification of *Saccharum* biomasses for bioethanol production. *Renewable Energy*, 130, 12–24.

Barot, S., 2022. Biomass and bioenergy: Resources, conversion and application. In N. Kumar & Prabhansu (eds.) *Renewable Energy for Sustainable Growth Assessment*. John Wiley & Sons, 243–262.

Belkacemi, K. & Hamoudi, S., 2003. Enzymatic hydrolysis of dissolved corn stalk hemicelluloses: Reaction kinetics and modeling. *Journal of Chemical Technology & Biotechnology*, 78, 802–808.

Cadoche, L. & López, G.D., 1989. Assessment of size reduction as a preliminary step in the production of ethanol from lignocellulosic wastes. *Biological Wastes*, 30, 153–157.

Cantero, D., Jara, R., Navarrete, A., Pelaz, L., Queiroz, J., Rodríguez-Rojo, S. & Cocero, M.J., 2019. Pretreatment processes of biomass for biorefineries: Current status and prospects. *Annual Review of Chemical and Biomolecular Engineering*, 10, 289–310.

Chadha, A., Florentine, S., Chauhan, B.S., Long, B., Jayasundera, M., Javaid, M.M. & Turville, C., 2019. Environmental factors affecting the germination and seedling emergence of two populations of an emerging agricultural weed: Wild lettuce (*Lactuca serriola*). *Crop and Pasture Science*, 70, 709–717.

Chadha, A., Florentine, S., Javaid, M., Welgama, A. & Turville, C., 2020. Influence of elements of climate change on the growth and fecundity of *Datura stramonium*. *Environmental Science and Pollution Research*, 27, 35859–35869.

Chatterjee, S., Sharma, S., Prasad, R.K., Datta, S., Dubey, D., Meghvansi, M.K., Vairale, M.G. & Veer, V., 2015. Cellulase enzyme based biodegradation of cellulosic materials: An overview. *Cellulose*, 5, 271–282.

Chundawat, S.P., Pal, R.K., Zhao, C., Campbell, T., Teymouri, F., Videto, J., Nielson, C., Wieferich, B., Sousa, L. & Dale, B.E., 2020. Ammonia fiber expansion (AFEX) pretreatment of lignocellulosic biomass. *JoVE (Journal of Visualized Experiments)*, e57488.

Cunha, J.T., Romaní, A., Costa, C.E., Sá-Correia, I. & Domingues, L., 2019. Molecular and physiological basis of *Saccharomyces cerevisiae* tolerance to adverse lignocellulose-based process conditions. *Applied Microbiology and Biotechnology*, 103, 159–175.

de Melo Pereira, G.V., Leal Maske, B., De Carvalho Neto, D.P., Da Silva Vale, A., Muynarsk, E., Binder Pagnoncelli, M.G., Karp, S.G., Bassi Pregolini, V. & Soccol, C.R., 2022. Microorganisms and Genetic Improvement for First and Second Generation Bioethanol Production. *In* C.R. Soccol, G. Amarante Guimarães Pereira, C.-G. Dussap & L. Porto De Souza Vandenberghe (eds.) *Liquid Biofuels: Bioethanol*. Cham: Springer International Publishing, 29–60.

Dessì, P., Porca, E., Waters, N.R., Lakaniemi, A.-M., Collins, G. & Lens, P.N.L., 2018. Thermophilic versus mesophilic dark fermentation in xylose-fed fluidised bed reactors: Biohydrogen production and active microbial community. *International Journal of Hydrogen Energy*, 43, 5473–5485.

Devi, A., Singh, A., Bajar, S., Pant, D. & Din, Z.U., 2021. Ethanol from lignocellulosic biomass: An in-depth analysis of pre-treatment methods, fermentation approaches and detoxification processes. *Journal of Environmental Chemical Engineering*, 9, 105798.

Díez, D., Urueña, A., Piñero, R., Barrio, A. & Tamminen, T., 2020. Determination of hemicellulose, cellulose, and lignin content in different types of biomasses by thermogravimetric analysis and pseudocomponent kinetic model (TGA-PKM method). *Processes*, 8, 1048.

Dos Santos, L.V., De Barros Grassi, M.C., Gallardo, J.C.M., Pirolla, R.A.S., Calderón, L.L., De Carvalho-Netto, O.V., Parreiras, L.S., Camargo, E.L.O., Drezza, A.L. & Missawa, S.K., 2016. Second-generation ethanol: The need is becoming a reality. *Industrial Biotechnology*, 12, 40–57.

Dutta, N., Usman, M., Ashraf, M.A., Luo, G., Gamal El-Din, M. & Zhang, S., 2023. Methods to convert lignocellulosic waste into biohydrogen, biogas, bioethanol, biodiesel and value-added chemicals: A review. *Environmental Chemistry Letters*, 21, 803–820.

Favaro, L., Jansen, T. & Van Zyl, W.H., 2019. Exploring industrial and natural *Saccharomyces cerevisiae* strains for the bio-based economy from biomass: The case of bioethanol. *Critical Reviews in Biotechnology*, 39, 800–816.

Firsty, I.M., Setyaningsih, R., Sugiwati, S., Das, A.K. & Maryana, R., 2023. *Screening of Ligninolytic Fungi and Optimization of Pretreatment Process for Bioethanol Production from Empty Palm Oil Bunches (OPEFB)*. Research Square. https://doi.org/10.21203/rs.3.rs-3782349/v1

Florentine, S.K., Weller, S., Graz, P.F., Westbrooke, M., Florentine, A., Javaid, M., Fernando, N., Chauhan, B.S. & Dowling, K., 2016. Influence of selected environmental factors on seed germination and seedling survival of the arid zone invasive species tobacco bush (*Nicotiana glauca* R. Graham). *The Rangeland Journal*, 38, 417–425.

Furukawa, S. & Komatsu, T., 2017. Intermetallic compounds: Promising inorganic materials for well-structured and electronically modified reaction environments for efficient catalysis. *ACS Catalysis*, 7, 735–765.

Guo, F., Dong, Y., Tian, B., Du, S., Liang, S., Zhou, N., Wang, Y., Chen, P. & Ruan, R., 2020. Applications of microwave energy in gas production and tar removal during biomass gasification. *Sustainable Energy & Fuels*, 4, 5927–5946.

Haldar, D. & Purkait, M.K., 2020a. Lignocellulosic conversion into value-added products: A review. *Process Biochemistry*, 89, 110–133.

Haldar, D. & Purkait, M.K., 2020b. Micro and nanocrystalline cellulose derivatives of lignocellulosic biomass: A review on synthesis, applications and advancements. *Carbohydrate Polymers*, 250, 116937.

Hamid, S.B.A., Islam, M.M. & Das, R., 2015. Cellulase biocatalysis: Key influencing factors and mode of action. *Cellulose*, 22, 2157–2182.

Hassan, M.U., Ghareeb, R.Y., Nawaz, M., Mahmood, A., Shah, A.N., Abdel-Megeed, A., Abdelsalam, N.R., Hashem, M., Alamri, S. & Thabit, M.A., 2022. Melatonin: A vital pro-tectant for crops against heat stress: Mechanisms and prospects. *Agronomy*, 12, 1116.

Hemsworth, G.R., Déjean, G., Davies, G.J. & Brumer, H., 2016. Learning from microbial strategies for polysaccharide degradation. *Biochemical Society Transactions*, 44, 94–108.

Hosseini, S.A. & Shah, N., 2011. Modelling enzymatic hydrolysis of cellulose part I: Population balance modelling of hydrolysis by endoglucanase. *Biomass and Bioenergy*, 35, 3841–3848.

Houfani, A.A., Anders, N., Spiess, A.C., Baldrian, P. & Benallaoua, S., 2020. Insights from enzymatic degradation of cellulose and hemicellulose to fermentable sugars – A review. *Biomass and Bioenergy*, 134, 105481.

Huzaifah, M., Sapuan, S., Leman, Z., Ishak, M. & Maleque, M., 2017. A review of sugar palm (*Arenga pinnata*): Application, fibre characterisation and composites. *Multidiscipline Modeling in Materials and Structures*, 13, 678–698.

Imran, M., Anwar, Z., Irshad, M., Asad, M.J. & Ashfaq, H., 2016. Cellulase production from species of fungi and bacteria from agricultural wastes and its utilization in industry: A review. *Advances in Enzyme Research*, 4, 44–55.

Javaid, M.M., Florentine, S., Mahmood, A., Wasaya, A., Javed, T., Sattar, A., Sarwar, N., Kalaji, H.M., Ahmad, H.B. & Worbel, J., 2022. Interactive effect of elevated CO_2 and drought on physiological traits of *Datura stramonium*. *Frontiers in Plant Science*, 13, 929378.

Jayakumar, M., Gindaba, G.T., Gebeyehu, K.B., Periyasamy, S., Jabesa, A., Baskar, G., John, B.I. & Pugazhendhi, A., 2023. Bioethanol production from agricultural residues as lignocellulosic biomass feedstock's waste valorization approach: A comprehensive review. *Science of the Total Environment*, 879, 163158.

Kaur, P., Singla, D., Taggar, M.S. & Sarao, L.K., 2023. Potential Technologies for Advanced Generation Biofuels from Waste Biomass. *In* N. Srivastava, B. Verma & P.K. Mishra (eds.) *Agroindustrial Waste for Green Fuel Application*. Singapore: Springer Nature Singapore, 121–168.

Kolář, V., Hrabě, P., Müller, M., Hromasová, M., Herák, D. & Sutanto, H., 2023. Influence of alkali treatment of *Jatropha curcas* L. filler on the service life of hybrid adhesive bonds under low cycle loading. *Polymers*, 15, 395.

Kucharska, K., Rybarczyk, P., Hołowacz, I., Łukajtis, R., Glinka, M. & Kamiński, M., 2018. Pretreatment of lignocellulosic materials as substrates for fermentation processes. *Molecules*, 23, 2937.

Kumar, A.S., Sankar, S.S., Godlaveeti, S.K., Kumar, D., Dheiver, S., Prasad, R., Nb, C., Nguyen, T.H.C. & Van Le, Q., 2021. Recent developments of bioethanol production. *Bioenergy Research*, 175–208.

Kumar, P., Sai Ram, C., Srivastava, J.P., Behura, A.K. & Kumar, A., 2022. Synthesis of Cotton Fiber and Its Structure. In *Natural and Synthetic Fiber Reinforced Composites: Synthesis, Properties and Applications*, John Wiley and Sons, 17–36.

Kumar, S.A., Santhosh, J., Venkatesh, C., Babu, N.R., Saravanan, N. & Parvathy, S., 2019. Microbial conversion of cellulose into bioethanol. *World Journal of Pharmaceutical Research*, 8, 733–753.

Kwak, S. & Jin, Y.-S., 2017. Production of fuels and chemicals from xylose by engineered *Saccharomyces cerevisiae*: A review and perspective. *Microbial Cell Factories*, 16, 82.

Latif, M.N., Wan Isahak, W.N.R., Samsuri, A., Hasan, S.Z., Manan, W.N. & Yaakob, Z., 2023. Recent advances in the technologies and catalytic processes of ethanol production. *Catalysts*, 13, 1093.

Li, C., Fan, M., Xie, J. & Zhang, H., 2023. Effect of NaOH-catalyzed organosolv pretreatment on the co-production of ethanol and xylose from poplar. *Industrial Crops and Products*, 200, 116774.

Li, J., Zhao, R., Xu, Y., Wu, X., Bean, S.R. & Wang, D., 2022. Fuel ethanol production from starchy grain and other crops: An overview on feedstocks, affecting factors, and technical advances. *Renewable Energy*, 188, 223–239.

M'Arimi, M.M., Mecha, C.A., Kiprop, A.K. & Ramkat, R., 2020. Recent trends in applications of advanced oxidation processes (AOPs) in bioenergy production: Review. *Renewable and Sustainable Energy Reviews*, 121, 109669.

Mahmood, A., Hussain, A., Shahzad, A.N. & Honermeier, B., 2015. Biomass and biogas yielding potential of sorghum as affected by planting density, sowing time and cultivar. *Pakistan Journal of Botany*, 47, 2401–2408.

Mahmood, A., Ullah, H., Shahzad, A., Ali, H., Ahmad, S., Zia-Ul-Haq, M., Honermeier, B. & Hasanuzzaman, M., 2013. Dry matter yield and chemical composition of sorghum cultivars with varying planting density and sowing date. *Sains Malaysiana*, 42, 1529–1538.

Mahmood, A., Wang, X., Shahzad, A.N., Fiaz, S., Ali, H., Naqve, M., Javaid, M.M., Mumtaz, S., Naseer, M. & Dong, R., 2021. Perspectives on bioenergy feedstock development in Pakistan: Challenges and opportunities. *Sustainability*, 13, 8438.

Mehmood, T., Nadeem, F., Qamar, S.A., Bilal, M. & Iqbal, H.M.N., 2021. Bioconversion of Agro-Industrial Waste into Value-Added Compounds. *In* Inamuddin & A. Khan (eds.) *Sustainable Bioconversion of Waste to Value Added Products*. Cham: Springer International Publishing, 349–368.

Mittal, A., Katahira, R., Himmel, M.E. & Johnson, D.K., 2011. Effects of alkaline or liquid-ammonia treatment on crystalline cellulose: Changes in crystalline structure and effects on enzymatic digestibility. *Biotechnology for Biofuels*, 4, 41.

Mohamad, N.N. & Jai, J., 2022. Response surface methodology for optimization of cellulose extraction from banana stem using NaOH-EDTA for pulp and papermaking. *Heliyon*, 8.

Mohammad Rahmani, A., Gahlot, P., Moustakas, K., Kazmi, A.A., Shekhar Prasad Ojha, C. & Tyagi, V.K., 2022. Pretreatment methods to enhance solubilization and anaerobic biodegradability of lignocellulosic biomass (wheat straw): Progress and challenges. *Fuel*, 319, 123726.

Mohammed, M.K., Balla, H.H., Al-Dulaimi, Z.M.H., Kareem, Z.S. & Al-Zuhairy, M.S., 2021. Effect of ethanol-gasoline blends on SI engine performance and emissions. *Case Studies in Thermal Engineering*, 25, 100891.

Mohsenzadeh, A., Jeihanipour, A., Karimi, K. & Taherzadeh, M.J., 2012. Alkali pretreatment of softwood spruce and hardwood birch by NaOH/thiourea, NaOH/urea, NaOH/urea/thiourea, and NaOH/PEG to improve ethanol and biogas production. *Journal of Chemical Technology & Biotechnology*, 87, 1209–1214.

Morya, S., Sandhu, D., Neumann, A. & James Simon, C., 2023. Sustainable Environmental Remediation by Valorization of Agro-Food Industrial Waste and By-Products. *In* P. Chowdhary & A. Raj (eds.) *Agri-Food Waste Valorisation*. Royal Society of Chemistry. 78, 70–94.

Mumtaz, S., Saleem, M.H., Hameed, M., Batool, F., Parveen, A., Amjad, S.F., Mahmood, A., Arfan, M., Ahmed, S. & Yasmin, H., 2021. Anatomical adaptations and ionic homeostasis in aquatic halophyte *Cyperus laevigatus* L. under high salinities. *Saudi Journal of Biological Sciences*, 28, 2655–2666.

Onumaegbu, C., Mooney, J., Alaswad, A. & Olabi, A.G., 2018. Pre-treatment methods for production of biofuel from microalgae biomass. *Renewable and Sustainable Energy Reviews*, 93, 16–26.

Pandiyan, K., Singh, A., Singh, S., Saxena, A.K. & Nain, L., 2019. Technological interventions for utilization of crop residues and weedy biomass for second generation bio-ethanol production. *Renewable Energy*, 132, 723–741.

Panggabean, S., Sebayang, E. & Sigalingging, R., 2023. Corn stem fiber fractionation by acid hydrolysis. In *IOP Conference Series: Earth and Environmental Science*. IOP Publishing, 012071.

Patra, D., Patra, B.R., Pattnaik, F., Hans, N. & Kushwaha, A., 2022. Chapter 5 – Recent Evolution in Green Technologies for Effective Valorization of Food and Agricultural Wastes. *In* C.M. Hussain, S. Singh & L. Goswami (eds.) *Emerging Trends to Approaching Zero Waste*. Amsterdam, Netherlands: Elsevier, 103–132.

Pedersen, M. & Meyer, A.S., 2009. Influence of substrate particle size and wet oxidation on physical surface structures and enzymatic hydrolysis of wheat straw. *Biotechnology Progress*, 25, 399–408.

Pereira, P.H.F., Waldron, K.W., Wilson, D.R., Cunha, A.P., Brito, E.S.D., Rodrigues, T.H.S., Rosa, M.F. & Azeredo, H.M.C., 2017. Wheat straw hemicelluloses added with cellulose nanocrystals and citric acid. Effect on film physical properties. *Carbohydrate Polymers*, 164, 317–324.

Ramos, A., Monteiro, E. & Rouboa, A., 2022. Biomass pre-treatment techniques for the production of biofuels using thermal conversion methods – A review. *Energy Conversion and Management*, 270, 116271.

Ransom-Jones, E., 2014. *Ecology of Fibrobacter spp. in Cellulose-Degrading Microbial Communities*. Wales, United Kingdom: Bangor University.

Rodionova, M.V., Bozieva, A.M., Zharmukhamedov, S.K., Leong, Y.K., Lan, J.C.-W., Veziroglu, A., Veziroglu, T.N., Tomo, T., Chang, J.-S. & Allakhverdiev, S.I., 2022. A comprehensive review on lignocellulosic biomass biorefinery for sustainable biofuel production. *International Journal of Hydrogen Energy*, 47, 1481–1498.

Saad, M., Kamil, M., Abdurahman, N., Yunus, R.M. & Awad, O.I., 2019. An overview of recent advances in state-of-the-art techniques in the demulsification of crude oil emulsions. *Processes*, 7, 470.

Sarkar, N., Ghosh, S.K., Bannerjee, S. & Aikat, K., 2012. Bioethanol production from agricultural wastes: An overview. *Renewable Energy*, 37, 19–27.

Sarris, D. & Papanikolaou, S., 2016. Biotechnological production of ethanol: Biochemistry, processes and technologies. *Engineering in Life Sciences*, 16, 307–329.

Schuster, B.G. & Chinn, M.S., 2013. Consolidated bioprocessing of lignocellulosic feedstocks for ethanol fuel production. *BioEnergy Research*, 6, 416–435.

Shenoy, D., Pai, A., Vikas, R.K., Neeraja, H.S., Deeksha, J.S., Nayak, C. & Rao, C.V., 2011. A study on bioethanol production from cashew apple pulp and coffee pulp waste. *Biomass and Bioenergy*, 35, 4107–4111.

Singh, S., Kumar, A., Sivakumar, N. & Verma, J.P., 2022. Deconstruction of lignocellulosic biomass for bioethanol production: Recent advances and future prospects. *Fuel*, 327, 125109.

Sofokleous, M., Christofi, A., Malamis, D., Mai, S. & Barampouti, E.M., 2022. Bioethanol and biogas production: An alternative valorisation pathway for green waste. *Chemosphere*, 296, 133970.

Sudiyani, Y., Dahnum, D., Burhani, D. & Putri, A.M.H., 2019. Chapter 10 – Evaluation and Comparison between Simultaneous Saccharification and Fermentation and Separated Hydrolysis and Fermentation Process. *In* A. Basile & F. Dalena (eds.) *Second and Third Generation of Feedstocks*. Amsterdam, Netherland: Elsevier, 273–290.

Tirronen, E., 2021. Brightness measurement of cellulosic materials. South-Eastern Finland University of Applied Sciences – Xamk. https://doi.org/10.1007/s10570-023-05553-6

Vasić, K., Knez, Ž. & Leitgeb, M., 2021. Bioethanol production by enzymatic hydrolysis from different lignocellulosic sources. *Molecules*, 26, 753.

Wagle, A., Angove, M.J., Mahara, A., Wagle, A., Mainali, B., Martins, M., Goldbeck, R. & Raj Paudel, S., 2022. Multi-stage pre-treatment of lignocellulosic biomass for multi-product biorefinery: A review. *Sustainable Energy Technologies and Assessments*, 49, 101702.

Wang, F., Shi, D., Han, J., Zhang, G., Jiang, X., Yang, M., Wu, Z., Fu, C., Li, Z., Xian, M. & Zhang, H., 2020. Comparative study on pretreatment processes for different utilization purposes of switchgrass. *ACS Omega*, 5, 21999–22007.

Wei, H., Xu, Q., Taylor, L.E., Baker, J.O., Tucker, M.P. & Ding, S.-Y., 2009. Natural paradigms of plant cell wall degradation. *Current Opinion in Biotechnology*, 20, 330–338.

Weimer, P.J., 2022. Degradation of cellulose and hemicellulose by ruminal microorganisms. *Microorganisms*, 10, 2345.

Yang, B., Dai, Z., Ding, S.-Y. & Wyman, C.E., 2011. Enzymatic hydrolysis of cellulosic biomass. *Biofuels*, 2, 421–449.

Yoo, C.G., Meng, X., Pu, Y. & Ragauskas, A.J., 2020. The critical role of lignin in lignocellulosic biomass conversion and recent pretreatment strategies: A comprehensive review. *Bioresource Technology*, 301, 122784.

Yu, Y., Wu, J., Ren, X., Lau, A., Rezaei, H., Takada, M., Bi, X. & Sokhansanj, S., 2022. Steam explosion of lignocellulosic biomass for multiple advanced bioenergy processes: A review. *Renewable and Sustainable Energy Reviews*, 154, 111871.

Zhang, C., Chao, L., Zhang, Z., Zhang, L., Li, Q., Fan, H., Zhang, S., Liu, Q., Qiao, Y., Tian, Y., Wang, Y. & Hu, X., 2021. Pyrolysis of cellulose: Evolution of functionalities and structure of bio-char versus temperature. *Renewable and Sustainable Energy Reviews*, 135, 110416.

Zhang, K., Lu, X., Li, Y., Jiang, X., Liu, L. & Wang, H., 2019. New technologies provide more metabolic engineering strategies for bioethanol production in *Zymomonas mobilis*. *Applied Microbiology and Biotechnology*, 103, 2087–2099.

Zheng, L., Hou, Y., Li, W., Yang, S., Li, Q. & Yu, Z., 2012. Biodiesel production from rice straw and restaurant waste employing black soldier fly assisted by microbes. *Energy*, 47, 225–229.

22 Nutrients Grow through Organic Soil Enrichment

Revitalizing Soil Health through Organic Nutrient Enrichment

*Asma Zafar, Ayesha Arshad, Wasim Javed, Muhammad Awais, Javaria Nargis, and Ayman El Sabagh**

22.1 INTRODUCTION

Nutrients are the basic chemical elements present in soil that are necessary for the growth, development, and structure of plants. The absence or deficiency of certain components leads to specific and recognizable symptoms. Typically, these elements are present in the soil as salts. Plants have two groups of nutrients based on their fundamental needs, i.e., micronutrients and macronutrients (van der Ent *et al.*, 2024). Soil and plant tissue investigation provides data on the presence of specific nutrients in plants, their appropriate levels in soil, and potential availability for plant uptake. Determining a desirable source for each element is beneficial for both the growth of plants and the fertility of the soil. Unfavorable consequences have been observed when the concentration of certain elements exceeds or falls below the acceptable range in soil. Under these circumstances, soil amendment approaches to supplement soil nutrients have been shown to be advantageous (Waheed *et al.*, 2019; Singh *et al.*, 2023).

Plants require a minimum of 16 elements in order to achieve normal growth and development and to successfully complete their life cycle. The elements hydrogen, carbon, and oxygen, which are required in significant quantities, are non-mineral elements obtained from the atmosphere and water. Plants exclusively absorb the other 13 elements in their mineral forms from the soil or need them to be supplemented as fertilizers or amendments. Plants need significant quantities of nitrogen, phosphorus, and potassium. These nutrients that are most often provided to plants through fertilizers are known as fundamental or primary nutrients (Guo *et al.*, 2023). The three secondary elements, namely calcium, magnesium, and sulfur, are required in fewer quantities in comparison to the primary nutrients. Liming materials often provide calcium and magnesium, whereas fertilizer materials typically provide sulfur. Rainfall also carries pollutants that add 10–20 pounds of nitrogen and sulfur per acre each year from the atmosphere, with the specific quantity depending on the local air quality (Bünemann *et al.*, 2024).

The micronutrients, that require in minute quantities, include seven important elements: boron, copper, chlorine, iron, manganese, molybdenum, and zinc. These elements are present in trace levels in both soils and plants, but their significance is just as crucial as that of the primary or secondary nutrients. The absence of one or more micronutrients may result in significant reductions in growth, productivity, and produce quality of crops. Certain soils may lack sufficient quantities of these essential elements to fulfill the plant's demands for accelerated development, good health, and optimal productivity. When faced with such situations, it is necessary to provide additional micronutrients using commercial fertilizers through soil application or foliar sprays (Babar *et al.*, 2009;

* Corresponding author: ayman.elsabagh@siirt.edu.tr

 DOI: 10.1201/9781003498698-22

Mekonen *et al.*, 2024). Moreover, there are some other elements (i.e., silicon, selenium, cobalt, or sodium) that are not considered essential elements for most plants, but these have been shown to provide several benefits, especially under stress conditions (Mittal *et al.*, 2023).

Hence, the soil provides 13 of the 15 needed to sustain higher plants. Photosynthesis can occur only when sufficient amounts of these components are available in the right ratio. If any of these key elements are absent, the health of the plant will be compromised, its production may be constrained, or it may stop growing entirely. According to the theory of limiting factors, the maximum potential of production is determined by whichever of the major factors of plant development is the most restrictive. The concept pertains to agro-ecosystems as well as natural ecosystems (Morizet-Davis *et al.*, 2023).

With extensive agriculture to grow more and more crops, multiple problems have occurred which include depleted soil fertility, imbalanced soil nutrients, soil degradation, biodiversity loss, and environmental issues. One of the significant problems in such conditions is to maintain soil fertility and plant productivity, therefore, both organic and inorganic fertilizers are utilized to restore soil essential plant nutrient deficiency. But the importance of making sure commercial fertilizer use is balanced with organic and other sustainable land management practices to build soil and improve crop yields with minimal environmental degradation cannot be overstated. Therefore, one of the efficient, cost-effective, sustainable, and eco-friendly techniques used to enhance soil fertility, improve soil structure and health, enhance water holding capacity, promote biodiversity, and lead to healthier plants and sustainable agricultural systems is the use of soil enrichers from natural and organic materials.

This chapter discusses the mechanisms and strategies that increase soil fertility and subsequently nutrition of plants containing necessary nutrients. Thus, the goal of this chapter was to summarize systematically the current knowledge regarding the dynamic interactions between organic enrichment of soil systems and nutritional development (Sharma *et al.*, 2023). It reviews multiple approaches, including organic soil enrichment approaches such as green manure, cover crops, manure, and composting. By analyzing different empirical research, case studies, practical investigations, and successful applications, we aim to explain the ways that these strategies are reinforcing soil fertility and increasing the production of probiotics in plants (Thakur *et al.*, 2023).

22.2 ORGANIC SOIL ENRICHMENT TECHNIQUES

There are currently several organic soil enhancement methods. Following are various organic soil enrichment methods that can be used to maintain soil fertility and improve soil productivity for achieving sustainable agriculture

22.2.1 Conventional Method

- Addition of organic matter to enrich soil habitat and enhance the soil food web.
- Primary consumers like bacteria fungi plant-feeding nematodes, annelids, and certain microarthropods indicate soil enrichment.
- Evaluating soil health through the total mass and quantity of soil organisms as indicators of resource quality.

22.2.2 Advanced Methods

Use of specialized extraction and evaluation methods to assess soil organism groups for more accurate resource quality analysis (Wang *et al.*, 2024).

- Fungi's ability to decompose resistant organic matter and bacteria's assimilation of soluble organic molecules for targeted enrichment.
- Nematodes as markers of substrate type and soil dynamics, with advanced methods being required to accurately measure their biomass and community composition.

- Study of nematode groups to track carbon movement in soil food web channels, such as herbivore, bacterial, and fungal pathways (Pinto *et al.*, 2024).

22.3 COMPOST

The need for food production mandates the use of intensive agricultural methods, which ultimately deteriorate soil health and give rise to several environmental concerns. Composting is regarded as a sustainable approach to waste management, since it involves the recycling of organic waste. It provides a crucial supply of organic material to improve the organic content of the soil, which is being depleted by human activities (Punia and Bharti, 2023). Composting is also recognized as a reliable way to improve various soil properties. Implementing best practices, such as prioritizing the use of organic fertilizers, may boost soil quality by improving soil attributes and providing extra advantages. The impact of applying compost to the soil is greatly affected by the inherent qualities of both the soil and the compost (the material used to make the compost), as well as the on top at which the compost is applied. It is worth noting that the effects of using this fertilizer may not be seen in the near term since the nutrients are released slowly (Gao *et al.*, 2024).

22.3.1 Optimal Composition of Nutrients in the Compost

Composting is an essential technique for improving the levels of organic matter and mineralization in soils, whether they are normal or saline. This approach is particularly important for dry and semi-dry regions, which make up the vast majority of agricultural land globally. These are characterized by low rainfall and high alkalinity. Absolutely, the most important aspect for preserving soil fertility and preventing nutrient depletion is the presence of organic matter within the soil. In addition, it plays a key role in determining several soil properties including water-holding capacity, soil fertility, biota support, soil aeration, and soil structure (Singh Yadav *et al.*, 2023). Nitrogen is known as an important fertilizer so was imparts influence on the upgrowth of plants. In typical soil conditions, the organic matter holds 95% of nitrogen and sulfur and 25% of phosphorus. Compost comprises decomposed organic waste, which naturally has relatively high nitrosine content. Thus, the application of compost increases the fertility of the soil by providing essential nutrients such as nitrogen, carbon, sulfur, and phosphorus to facilitate the growth and development of plants (Goldan *et al.*, 2023). The stability and availability of nutrients in the humus-like product are largely dependent on the specific type of raw organic materials and under what conditions the composting process has taken place. Clearly, N and C measurements are very important in the pile maturation process.

The favorable C/N ratio should be between 25 and 30, while an average bio-compost should include 33.3% organic matter (with a threshold of a minimum 20%). Compost should ideally have a water content of no more than 40%, and it is suggested that market compost has a water content of around 35%. An optimal compost pH for most agricultural purposes is about 7.5. Furthermore, the salt concentration is a crucial factor and should ideally be around 3.89 g/l of fresh substance. Compost contains around 0.62% dry matter of phosphate in the form of P_2O_5 and about 1.01% dry matter of potassium (Chaudhary *et al.*, 2023).

22.3.2 Compost Effects on Soil Enzymatic and Biological Activity

When compost boosts soil organic matter (SOM) and microbes, it is an effective fertilizer substitute. Healthy soil requires soil microbial diversity for nitrogen cycling and other biological functions. Enzymatic activities facilitate biochemical processes and reactions, resulting in a good measure of soil function. Organic fertilizers boost enzyme activity. Adding organic matter to soil considerably enhanced β-glucosidase (GLU) activity by over 200% compared to unmodified soil. Manure also boosts soil organic carbon (SOC) and enzymatic activity. GLU activity increases with total organic carbon. Microbial biomass carbon increased by 100% due to increased microbial activities, as shown by enzymatic activity studies. Applying compost to diverse soils has been proven to improve soil microbial properties, particularly biomass and respiration rate. Consequently, plant growth, development, and production improve (Du *et al.*, 2023).

Bacteria and fungi in compost may break down harmful pollutants and transform insoluble compounds into plant nutrition. This improves soil quality, sustains earthworms and other micro- and macro-fauna, and adds carbon to the environment. Compost also boosts the activity of numerous rhizosphere microorganisms, which boosts plant growth. Arbuscular mycorrhizal fungi (AMF) may be improved to absorb more phosphorus (P) and other nutrients that plants cannot easily absorb from the soil. The source of organic amendment affects microbial activity. The controlled soil and compost or manure-enriched soil exhaled different amounts of CO_2. CO_2 emission was highest at 35.8 mg CO_2/kg in 120 T/ha compost soil. Manure-derived organic material's restricted decomposition surely caused the difference. Other benefits of composting include reducing decomposition odors and eliminating disease-carrying insects and rodents (Gondal *et al.*, 2023).

Compost increases microbial biomass C, basal respiration, and several kinds of enzymatic activity in soil. The breakdown of organic materials releases nutrients gradually and consistently, explaining this phenomenon. Microorganisms may be present for longer periods. Four years following the last treatment, compost, and manure-treated soils had 20–40% more carbon in soil microbial biomass than inorganic fertilizer-treated soils. Long-term breakdown of organic materials would increase plant growth and reduce rainfall and irrigation-induced leaching (Toumi *et al.*, 2024).

22.3.3 Composting Process

Compost is a combination of soil and plant fertilizer that is used to enhance the biochemical, physical, and biological properties of the soil. It is commonly created through the decomposition of plant and food waste, recycling of organic materials, and through animal excretion. This process produces a mixture high in plant nutrients and beneficial microbes, from bacteria and protozoa to nematodes and fungi. In gardening and landscaping, horticulture, urban agriculture, and organic farming, compost increases soil fertility which in turn reduces dependence on synthetic fertilizers. Compost has many advantages, such as serving as a fertilizer to provide nutrients to crops, improving soil quality as a conditioner, raising the levels of humus or humic acid in the soil, and introducing helpful microorganisms that aid in suppressing soil pathogens and reducing soil-borne diseases (Panday *et al.*, 2024).

Composting plays a crucial role in waste management since about 20% of landfill trash consists of compostable products, such as food. These products take a longer time to break down in landfills due to the lack of oxygen. Composting is a more ecologically advantageous option compared to sending organic garbage to landfills since it decreases methane emissions caused by anaerobic conditions. Additionally, composting offers both economic and environmental advantages. Composts may serve several purposes, such as land and stream rehabilitation, wetland building, and landfill cover (Cammerino *et al.*, 2024).

22.3.4 Advancements in On-Farm Composting Technologies

Methods of composting depend on three main phases: preparation of the mixture, bio-oxidative phase, and maturation phase. Composting technology differences mostly arise from the bio-oxidative phase execution techniques, which lead to different technical and economical possibilities. Several methods are available for the bio-oxidative phase, including:

1. *Composting in Windrows or Piles Passively*: The process entails the arrangement of the raw materials into a pile or windrow, which is regularly rotated mainly to restore the porosity. Aeration is achieved by the natural flow of air through the pile. Therefore, it is necessary for the pile/windrow to be of a size that permits this natural air circulation. It is

very affordable and suitable for small farms that do not have significant area constraints (Ahamed *et al.*, 2023).

2. *Composting in Constant Aeration Static Windrow*: The main benefits of this method include quick composting time and little space needed. The system utilizes blowers to bring air into the pile, generating positive pressure to aid in oxygenation and cooling. Investigators might control it continuously or at regular intervals. This composting technique requires the use of a base layer consisting of porous materials such as wood chips or straw. This layer functions to distribute the air as it either enters or departs the aeration pipes. In addition, a protective layer made of fully decomposed compost or sawdust is used to absorb smells, deter insects, and retain moisture, ammonia, and heat (Duan *et al.*, 2023).
3. *Composting in Confined Systems*: The procedure may be carried out using wooden bins, unused storage bins, or any other appropriate receptacle, with or without a roof, such as a silo or spinning tube. The use of such technology is infrequent in the agricultural and industrial sectors due to the significant costs linked to the initial investment and continuous maintenance, making it feasible only for managing large quantities of biodegradable material (Tassinari *et al.*, 2023).

22.4 COVER CROPPING

Cover crops are used for the express objective of soil management and are recognized within many classifications. Plants known as "cover crops" are grown in fields to physically cover the soil. Originally, they were used to mitigate soil erosion during times of fallow in annual agricultural systems, when fields were not being cultivated between planting seasons (Schmaltz *et al.*, 2023). Cover crops are sometimes referred to as "green manures," "catch crops," or "living mulch." Green manure cover crops often consist of legumes that have the ability to carry out nitrogen fixation. These crops are cultivated with the explicit purpose of supplying nitrogen for the subsequent profitable crop. Catch crops refer to cover crops cultivated during times of fallow in agricultural systems to capture nutrients, particularly nitrogen, that would otherwise be lost in the absence of plants (Norberg and Aronsson, 2024).

22.4.1 Benefits of Cover Crops

Cover crops could be planted in the spaces between trees or bushes depending on the need before planting the main crops, which can better enhance the physical, chemical, and biological properties of the soil. Thus, it could result in better soil quality and greater productivity of the main crops. Utilizing cover crops as surface mulches in no-till (NT) crop production systems provides numerous advantages, such as improved nitrogen efficiency, conservation of soil moisture, prevention of soil erosion, enhancement of soil physical characteristics, increased nutrient preservation, improved soil fertility, suppression of weeds, decreased occurrence of diseases and insects, reduced potential for global warming, and higher crop yields (Farooq *et al.*, 2023). The following sections explore the many beneficial effects of cover crops on soil management and crop productivity (Figure 22.1).

22.4.1.1 Nitrogen Economy

The fundamental advantage of leguminous crops is the significant supply of nitrogen they provide. Both legume and nonlegume cover crops have an impact on the management of nitrogen fertilizer. The nitrogen fixation rate of cover crops is primarily influenced by the genetic capacity of the legume species and the quantity of nitrogen that is readily accessible to plants in the soil (Valenzuela, 2023). The process of symbiotic nitrogen fixation in legumes is carried out by two bacterial species, namely *Rhizobium* and *Bradyrhizobium*. The genus *Rhizobium* includes bacteria that grow rapidly and create acid, while *Bradyrhizobium* is slow-growing and does not make any

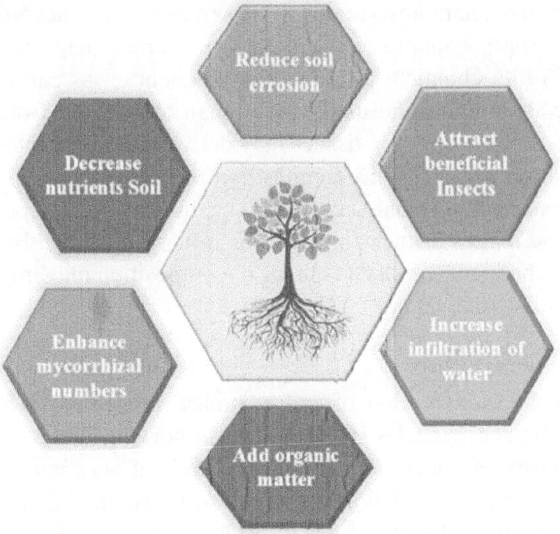

FIGURE 22.1 Examples of benefits of cover crop.

acid. Legume cover crops may sometimes give enough nitrogen (N) to provide optimum yields in succeeding non-leguminous crops (Maitra *et al.*, 2023). However, cereal crops like maize (*Zea mays* L.) that need greater levels of nitrogen usually need additional N fertilizer. It is essential to significantly reduce the amount of nitrogen fertilizer in these crops, while yet achieving the highest possible economic yields.

Cover crops prevent inorganic nitrogen from being transported away while the main crop is not growing. Cover crop waste decomposes, releasing N into the next crop. Rye (*Secale cereale* L.) may scavenge nitrogen after maize is harvested. Wong *et al.* (2023) reported that rye reduced NO_3^- leaching better than vetch (*Vicia villosa* Roth). In field cropping systems, rye controls erosion, soil compaction, and weed growth. Rye had a wider root system than vetch because it removed more nitrogen.

22.4.1.2 Increasing Organic Matter Content

SOC is one of the critical forms of terrestrial carbon pools that are important for carbon storage and exchange of carbon dioxide (CO_2) between soil and the atmosphere. SOC is oxidized during land agriculture and is subsequently released as CO_2 into the atmosphere (Murindangabo *et al.*, 2023). To accomplish the sustainable use of soil resources it is of major significance to either increase or maintain soil carbon (C), the key element of SOM. SOC stocks can be maintained or increased in a sustainable manner through appropriate land management practices, like conservation tillage with cover crops and balancing nutrient and water inputs (Nazir *et al.*, 2024).

22.4.1.3 Enhancing the Soil's Nutrient Content with Vital Minerals

Cover crops provide additional vital nutrients to future crops via the decomposition of their tissues, in addition to N. Legumes use their large root systems to investigate and extract nutrients from subsurface nutrient reservoirs. Cover crops affected soil nutrient enrichment based on dry matter production and nutrient concentration in dry tissues (Valenzuela, 2023). Cover crop biomass depends on environmental factors, crop management, and soil fertility. Moreover, the enhancement of soil fertility by the addition of nutrients mostly relies on the depth of plant roots. Therefore, the desired outcome is the transfer of nutrients from the lower soil layers to the uppermost layer of soil (Angon *et al.*, 2023).

Cover crops increase phosphate absorption by subsequent crops, according to research. Many methods increase cover crop phosphorus absorption. Cover crops may convert native and residual fertilizer phosphorus (P) into chemical forms that subsequent crops can use. Alfalfa, red clover, sweet clover, and lupine absorb phosphorus (P) better than other crops from low-phosphorus soils. The organic phosphorus in the cover crop tissues may deliver phosphorus to succeeding crops after breakdown (Kwiatkowski *et al.*, 2023). Decomposition of residue releases CO_2, which combines with soil solution water to form carbonic acid (H_2CO_3), increasing phosphorus (P) availability to following crops. This acid makes primary phosphorus minerals soluble. By inhibiting phosphorus adsorption sites, organic chemicals produced during decomposition may improve phosphorus availability in high-fixation soils (Jindo *et al.*, 2023).

22.4.1.4 Improving Soil Structure

Soil structure refers to the organization of soil particles into pores. When evaluating agricultural soils, the organization of particles is of greatest importance. The arrangement of soil particles determines the ability of water to infiltrate, the level of aeration, and the development of plant roots. Cover crop tissue degradation improves soil structure or aggregation and thereby enhances soil porosity, soil bulk density, water infiltration, and root penetration. Rashed and Van Herk (2024) report NT maize grown with crop remainders forms more soil aggregates than crop residue-free maize. Cover crops increase root activity and carbon inputs, aggregating soil and maintaining larger organic carbon reservoirs than fallow soil. Implementing tillage practices that maintain crop residues on the soil surface may prevent or limit the formation of surface crusts, improve water infiltration, decrease surface runoff and soil erosion, and increase crop productivity (Lv *et al.*, 2023).

22.4.1.5 Reducing Soil Erosion

The main factor contributing to soil deterioration and decreased agricultural productivity is the substantial depletion of topsoil resulting from erosion caused by insufficient soil management methods, including wind and water erosion. Soil erosion removes the top layer of soil, which often contains organic matter and nutrients that are not readily transported. The erosion of these topmost layers of soil ultimately reduces agricultural productivity. The effect of erosion on agricultural yields varies considerably based on the properties of the soils, crops, and management methods (Roy *et al.*, 2023).

When precipitation falls onto the surface of the soil, it forms an adhesive that decreases the capacity of the soil to allow water to pass through. The creation of a seal hinders the speed at which water enters the soil, leading to an increase in runoff and perhaps causing SOM a rise in soil erosion. Cover crops reduce soil erosion by increasing, which enhances water absorption. The mitigation of soil erosion is significant when there is an augmentation in infiltration and a decrease in runoff after each precipitation event. After cultivating soybeans, the inclusion of cover crops led to an augmentation in surface covering, the development of residue as a stabilizing agent, and a reduction in rill erosion (Mrabet, 2023).

22.4.1.6 Enhancing Soil Microbial Activity

The biological, chemical, and physical properties of the soil are interconnected and regulate soil tilth. By decomposing organic waste and storing nutrients, soil microbes help maintain soil quality and enable nutrient cycling. Cover crops may maintain soil microbes by providing moisture, heat, and carbon. Soil microbial biomass is largely bacteria and fungi. This includes soil microfauna and algae (Acharya *et al.*, 2024). Although present in modest levels (1–3% of organic carbon and 2–6% of organic nitrogen in soil), it is crucial to the transport and availability of SOM and nutrients. It stores plant nutrients during immobilization and provides nutrients during mineralization. Research shows that residual mulch increases actinomycetes, fungi, bacteria, earthworms, and nematodes compared to integrated wastes (Zhang *et al.*, 2023).

22.4.1.7 Reducing Disease and Insects

The benefits of using cover crops for disease and pest management may be attributed to the interruption of pest cycles via the cultivation of cover crops in between main crops. Research has shown that cover crops support beneficial predators and parasitoid insects, helping them manage destructive insects. Prior research has shown that the use of cover crops prior to planting cash crops may successfully control soil-borne pathogenic fungal diseases and nematodes (Sulaiman and Bello, 2024).

22.5 MULCHING

Mulching is a common method of applying materials to the field before, during, or after sowing.

The purpose of mulching is to provide support and coverage to the soil surface. Examples of materials used for mulching include plastic, agricultural waste, animal manure, sand, boulders, and cement. The primary objectives of mulching are to reduce water erosion evaporation, improve soil temperature, boost the soil's ability to retain water, and suppress weed development. Mulching enhances agricultural productivity, promotes plant development, and minimizes water use (Shaikh *et al.*, 2023).

22.5.1 Types of Mulching Materials

Mulching materials can be divided into three independent groups: organic, inorganic, and special organic mulch materials derived from agricultural waste, wood industry by-products, processed food waste, and farm animal feces. Inorganic mulching materials include polyethylene plastic sheets and synthetic polymers. Pioneered new sustainable materials such as biodegradable and photodegradable plastic films, surface coatings, and biodegradable polymer films that can deliver convenient and flexible approaches (Anwer *et al.*, 2023). Impact of various mulch types on the soil temperature in different crops is presented in Table 22.1.

22.5.1.1 Organic Mulches

Organic mulches consist of the remains of plants or animals. To maximize the benefits of organic mulch, it is advisable to apply it quickly after the crop starts to develop or during the transplanting

TABLE 22.1
Impact of Various Mulch Types on the Soil Temperature in Different Crops

Crop	Type of Mulch	Impact on Soil Temperature	Reference
Cucumber	Silver/black plastic mulch	Increase soil temperature	Mendonça *et al.* (2021)
Maize Tomatoes	Degradable film as mulch	Increase soil temperature	Yin *et al.* (2023)
Maize Potato	Transparent plastic mulch	The soil temperature boosted in plastic film mulching	Lu *et al.* (2020)
Cucumber Maize Lettuce	Black plastic mulch	The soil temperature increased more in black polyethylene mulched plots than in white-on-black polyethylene or bare ground plots	El-Beltagi *et al.* (2022)
Alfalfa Maize Wheat	Straw mulch	Reduce soil temperature fluctuations	Li *et al.* (2021)
Maize and wheat	Coupled plastic or straw mulch	Reduce soil temperature	Yin *et al.* (2020)

of vegetable seedlings, at a pace of 5 MT/ha. Organic mulches are very efficient in minimizing nitrate leaching, improving soil physical characteristics, promoting biological activities, maintaining nitrogen balance, supplying organic matter, regulating temperature and moisture retention, and preventing erosion. Employing natural elements in crop production presents challenges and requires significant workers. The extensive use of organic mulch in horticultural crop production has been restricted due to cost and logistical challenges (Salama and Geyer, 2023).

22.5.1.2 Straw

After the completion of the harvesting process, straw or agricultural residues are easily accessible. Straw mulch is a thin product that is simple to apply and use. Paddy straw is often used as a field mulch due to its ability to enhance the environment for crop production. Utilizing straw as mulch might lead to many problems. Straw mulches must be renewed annually due to their high flammability and the presence of grain seeds that might sprout and decrease soil nitrogen levels during decomposition (Biswas *et al.*, 2023).

22.5.1.3 Bark Mulches

These mulches are effective since they have a greater capacity to store moisture for an extended period, hence prolonging the availability of water to the crop. It is frequently utilized for horticulture and nurturing plants. On the other hand, its acidic properties render it unsuitable for use in vegetable crops. However, this mulch is ideal for concealing the paths that link the garden sections (Evans *et al.*, 2021).

22.5.1.4 Wood Chips

Wood chips are generated by the use of recycled wood and a wide variety of tree species. Wood chip mulches, with their elevated carbon-to-nitrogen ratio, might potentially restrict the accessibility of soil nitrogen for plant absorption throughout the breakdown process (El-Beltagi *et al.*, 2022).

22.5.1.5 Sawdust

Sawdust is a frequently used form of mulch in regions where it is readily available. It is encountered in the course of wood finishing processes. Straw has a higher nutritional value compared to it, including twice as many nutrients. The decomposition process is significantly delayed as a result of the elevated C/N ratio. The degradation of the substance will lead to a shortage of N_2 in the soil, requiring the frequent application of fertilizer. Due to its acidic attributes, it is not advisable to use it on soil with a low pH. However, it effectively retains moisture for an extended period of time (Gurmessa, 2021).

22.5.1.6 Compost

Compost is a very effective mulch and soil amendment that may be readily produced at home using various waste materials such as leaves, straw, grass, and residual plant food, among others. The use and implementation of compost in agriculture is a well-established technique. The addition of organic matter to the soil enhances its physiochemical qualities, including carbon content, hence increasing its water retention capacity and improving soil viability. Composts should not be used in vegetable fields due to their elevated nitrogen content, which increases the probability of weed development (Khan *et al.*, 2022).

22.5.1.7 Newspaper

Newspaper mulching is a cost-effective technique that reduces weed development by limiting the probability of weed seeds from the prior crop sustaining germination. The newspaper decomposes rapidly in the soil. Newspaper is a more favorable option compared to plastic since it undergoes decomposition gradually. It is more cost-effective and requires less time (Mansoor *et al.*, 2022).

22.5.2 Mulch Effects on Soil Properties

Temperature and soil moisture impact crop-soil interactions. Mulching rates affect soil moisture, organic matter, salinity, texture, porosity, and subsurface characteristics. These changes may significantly impact crop productivity. According to Bashir *et al.* (2021) organic mulches improve soil health and crop yield. Chemical properties including electrical conductivity (EC) and cation exchange capacity (CEC) improved in the soil. Compared to the control group, hardwood mulching increased SOM. Enhance SOM content in the uppermost layer (0–15 cm) of straw mulch in arid conditions. However, dissolved organic carbon rises significantly beneath black polyethylene mulch under damp conditions. Biodegradable mulches (BM) affected soil microbiological, urease, and catalase activity more than polyethylene film mulch. BM reduces soil bulk density but does not affect soil nutrients or microbial activity. Alternatively, it may improve soil quality. Climate, soil, and mulch types affect soil bulk density. Mulching improves water retention, but not without it. Various types of mulches resulted in a decrease in EC when compared to soil without any mulch. The use of black polyethylene mulch effectively limits the loss of nitrogen and organic carbon, hence enhancing soil fertility (El-Beltagi *et al.*, 2022).

22.6 BIOCHAR

Pyrolysis, a process that occurs at high temperatures without oxygen, yields three main byproducts: syngas, bio-oil, and biochar, which is often referred to as charcoal. All of these derivatives possess substantial calorific value, may be used as fuels, and facilitate the recycling and storage of organic waste for environmental advantage. Biochar was first used in agriculture over 40 years ago, but in the past decade, its unique properties and simplicity of small-scale manufacturing have increased its use. However, industrial production allows for increased utilization of all its byproducts, not only biochar but also a variety of production techniques (Wang *et al.*, 2020). Most biochar research has employed wood or municipal trash as source material for years, which has been criticized for deforestation. However, this tendency has shifted, particularly in emerging nations with increasingly diverse raw materials. Industrialized nations are also beginning to diversify biochar raw materials. Agriculture-wise, biochar has a pH between 8.5 and 10, a high C/N ratio, moderate K and P concentrations, and high porosity. Raw materials like pruning, agricultural trash, logging residues, or sewage sludge determine most of these properties (Racek *et al.*, 2020).

Biochar has limited N content, but it may add K and P to the soil. Biochar structure releases phosphorus gradually, therefore some researchers believe it may accelerate its transition to susceptible forms. Similar to other organic amendments, the structure of biochar is determined by the source material. Biochar formed from wood contains lower levels of potassium (K) and phosphorus (P) compared to biochar derived from pruning or agricultural debris. However, their application does not enhance N concentration alone. Biochar may enhance soil quality and water retention in sandy soil owing to its high organic carbon content and structure. Biochar efficacy depends on soil, climate, crop, and application. Its biggest influence will be in recovering overexploited soil. Thus, biochar's high C/N ratio may help us quickly enhance organic matter in conventional to sustainable and ecological systems. Biochar's huge specific surface area may help explain its influence on microbial populations by generating new places for bacteria. Biochar has limitations, especially due to its high pH, which immobilizes trace elements of many metals (Wang *et al.*, 2021).

Biochar has the added benefit of enhancing the properties of compost and manure, two more organic inputs used in organic farming. It adds stability and boosts the humic fulvic acid percentage when used in animal composting. Biochar enhances the plant's absorption of another nutrient, even while it is provided with another product, like compost. For optimal nutrient usage, it is ideal to employ a combination of solutions, such as charcoal and compost paired with phosphate-solubilizing bacteria, which may increase the efficiency of phosphate absorption. Additionally, it is stated that when biochar is mixed with vermicompost, green manure, or any other amendment, it has a multiplicative impact that enhances nutrient absorption, helps keep the humic fulvic acid fraction stable, and impedes the release of nutrients (Rehman *et al.*, 2023).

22.7 VERMICOMPOSTING

The earthworm's many uses as an aerator, digester, and producer of humus-rich castings make it an invaluable asset to farmers. Worm composting, also known as vermicomposting, is a simple technique that uses earthworms (the red worm *Eisenia foetida*) to quickly produce organic manure from biodegradable garbage without the need to stir the pile or deal with unpleasant odors. The worms are raised in a mixture of cow manure, dirt, and agricultural byproducts or decomposing leaves. The whole thing is turned into vermicompost or castings, which are fertilizers for all sorts of plants in gardens, lawns, and landscaping. Worms may digest over half of their body weight in waste every day due to their exceptional ability to assimilate organic waste. A simple way of composting, vermicomposting makes use of earthworms' inherent behaviors while enclosing them in containers, which simplifies feeding and harvesting their nutrient-rich byproduct for farmers (Singh and Sinha, 2022). Figure 22.2 showed the role of vermicompost in crop production.

An important benefit of using earthworms in compost production is their ability to generate a greater concentration of nutrients via their biological activity. This is in comparison to standard composting methods and enables the reuse of low-fertilization food that remains. When beginning with identical raw materials, the levels of P, K, and Ca, regularly experience a rise ranging from 30 to 120%, yielding a product that is well-suited for crops with elevated nutritional requirements. However, nitrogen does not experience this phenomenon, since it is lost by volatilization in the form of ammonium (Anas *et al.*, 2020). Earthworms have the capability to eliminate harmful organic substances that might cause issues in conventional composting systems. They are very efficient in restoring depleted agricultural soils that have reduced nutrient levels. Furthermore, the absence of pathogens in compost and the high concentration of readily accessible nutrients make it widely used in horticulture and large-scale cultivation, including grains and legumes (Panday *et al.*, 2024).

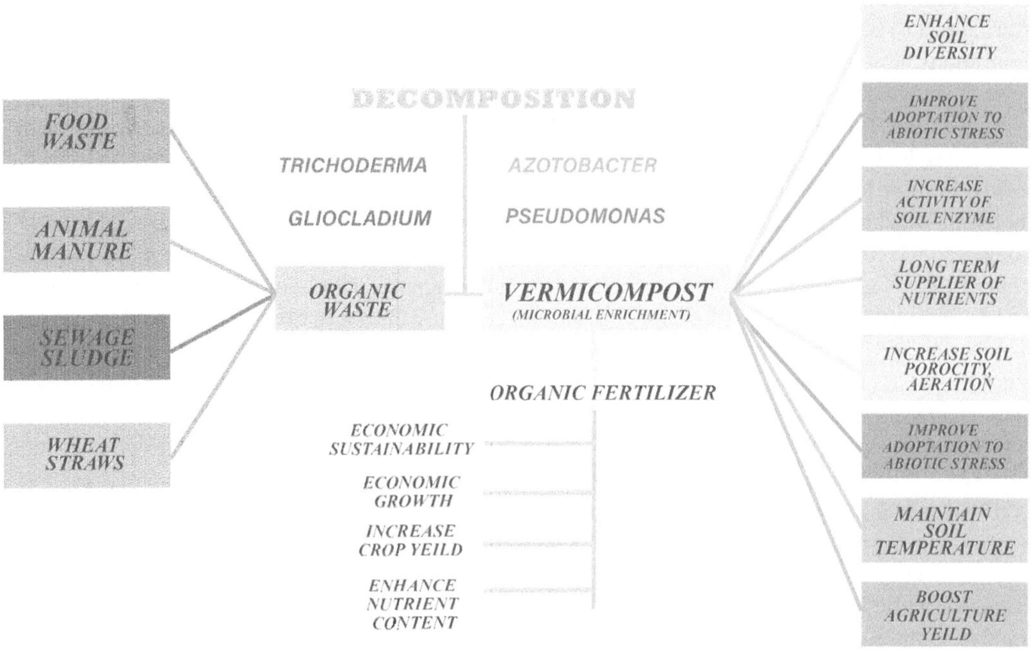

FIGURE 22.2 Vermicompost application on plants.

22.8 COMPOST TEA

The water extract of compost, called "compost tea," retains all the beneficial soluble bioactive components, making it a powerful plant stimulant and protector. However, the composting process, feedstock quality, compost tea production, utilization, management, and application dynamics affect these properties (dc Corato, 2020). The majority of compost tea nutrients are macronutrients. Secondary and micronutrient concentrations vary significantly, but they seldom fulfill crop nutritional demands. Compost tea improves crop nutrition directly and indirectly in agriculture and horticulture. Soil quality improvements were illustrated on several soils and compost teas. Compost tea-amended soils increase organic matter and microbial diversity, which has several benefits. Studies on field and container rates show substantial variance according to soil, climate, and crop type. However, applying compost tea to leaves appears to maximize nutritional advantages and plant disease resistance. Compost tea improves crop growth, soil fertility, and pest and disease control in organic agriculture (Yatoo *et al.*, 2021).

22.8.1 Process for Producing High-Quality Compost Tea

Several factors, including the compost-to-water ratio, oxygenation levels, fermentation period, and temperature, influence the formation of compost tea. Table 22.2 showed the mechanisms of action of compost tea on crop performance and nutrient status. The brewing or fermenting process involves immersing compost in water at a consistent temperature for a certain duration. Microorganisms transform insoluble nutrients into soluble nutrients during the brewing process. The variation has been linked to many elements that impact the manufacturing process

TABLE 22.2
Summary of Reported Mechanisms of Action of Compost Tea on Crop Performance and Nutrient Status

Compost Feedstock	Compost Tea Properties	Brewing Method/ Concentration	Summarized Mechanisms	References
Rice straw and animal manure	Nutrients, microbial content	Brew not specified 3% m/v	Increasing leaf nutrient absorption, directing metabolites flow	Snyder *et al.* (2020)
Tomato, escarole residue and wood chips	^{13}C NMR, nutrients, Microbial content	ACT 20%	Plant disease suppressiveness, supply of chelated nutrients and the action of humic acids and phytohormones	Savarese *et al.* (2022)
Cow manure vermicompost	Nutrients, humic acids, microbial content	Direct leachate used	Specific mention of micronutrient stimulates in tandem with humic acids and phytohormones	Oyege and Balaji Bhaskar (2023)
Fruit bunch and chicken manure compost	Nutrients, microbial content	Brew not specified 20% m/v	Beneficial microbes, presence of essential micronutrients and bioactive compounds	Kaur *et al.* (2022)
Farm refuse compost	Nutrients	ACT 10% w/v	Nutrient uptake driven by the availability of nutrients and physiological response of the plant roots	Ayub *et al.* (2024)

The factors included are as follows:

- *Compost Grades*: The organic materials or feedstocks that comprise mature compost include animal manure, landscape and agricultural plant debris, biosolids, and food waste. Each attribute has an impact on the overall quality of the mature compost. Vermin compost is a common component in several compost tea preparations. This compost has the greatest concentration of nutrients that are readily accessible. Fully decomposed compost should be chemically and biologically stable, and devoid of any disease-causing microorganisms. The emerging compost is less stable and may include pathogens (Bhunia *et al.*, 2021).
- *Brewing Time*: An ideal fermentation duration for disease control is no less than 8–16 days. Extending the brewing time enhances the extraction of a higher quantity of nutrients from the compost. This also allows for the buildup of antibiotics that trigger natural plant defense mechanisms and aid in the suppression of diseases. The perfect brewing duration of 18–24 hours coincides with the peak activity of the tea's microbial population (Haque *et al.*, 2020).
- *Nutrient Supplement*: Catalysts or microbial starters, such as kelp, fish hydrolysate, molasses, and humic acid, are included to provide essential nutrients. During the brewing process, selective enrichment of microorganisms is promoted. Fermentation nutrients have the capacity to either hinder or enhance the development rates of many organisms in both aerated compost tea (ACT) and non-aerated compost tea (NCT). Particular caution should be used while adding nutrients (Yatoo *et al.*, 2021).
- *Microbial Supplement*: Compost consists of a wide range of organisms, mostly bacteria and fungus, that play a key role in breaking down organic materials. Bacteria have the ability to survive and reproduce in surroundings that are rich in oxygen (aerobic) as well as in habitats with little or no oxygen (anaerobic). Some specific types of bacteria, called facultative anaerobes, are able to thrive in hypoxic conditions, but also flourish in aerobic surroundings (Berg *et al.*, 2022). The occurrence of anaerobic organisms in fully developed compost is likely linked to the existence of disease-suppressing characteristics. Compost proved effective in suppressing certain fungal root rot diseases when added to soil or soilless growth mediums. In the ACT, aerobic bacteria are the dominant population, but in NCT, the population consists mostly of facultative anaerobes (Gorliczay *et al.*, 2021).

22.8.2 Compost Tea Application Process

To prevent the nozzle from being restricted filtration and dilution are often required when applying tea via irrigation systems or sprayers. Filtration may eliminate suspended particles in compost tea that contain beneficial microorganisms, for both NCT and ACT. Pre-diluting the tea before applying it to the leaves may decrease the concentration of nutrients and the number of microorganisms present (Gorliczay *et al.*, 2021). When applying to the soil, use an amount of the substance that is enough to cover the region where the roots are located. Applying soil treatment assists in protecting the roots against possible invasion by root pathogens and enhances the overall health of plants. To ensure that beneficial microorganisms in the tea compete with plant diseases, it may be important to achieve maximum coverage of the leaf surface area. This requires frequent and repeated applications to maintain surface coverage (Saberi Riseh, 2024).

22.9 NATURAL ORGANIC MANURES/FERTILIZERS

Soil fertility in organic farming is sustained through cycling of organic matter. In either case, organic fertilizers that do not contain any harmful substances or metals can be used separately or in conjunction with inorganic fertilizers. Plant waste, compost, cover crops, animal manure, and other biological components form the basis of organic field productivity. Examples of organic fertilizers are blood, fish emulsion, bone meal, cottonseed meal, and wood ash (Shaji *et al.*, 2021). Organic fertilizer increases SOM and primary and secondary organic nutrients. Substances found in nature, such as those harvested from the earth, can provide soil with organic matter. Organic fertilizers can be natural or manufactured, similar to inorganic fertilizers.

Organic fertilizers differ from conventional fertilizers in the charge between nutrients due to the carbon-hydrogen bond that inhibits the release of nutrient ions. Slower nutrient release relative to inorganic fertilizers helps to increase soil physical properties, and nutrient availability, improve soil aeration, and avoid root burn and leaching. Organic fertilizers also nurture soil microbes, improving soil structure and stimulating plant growth. They are readily absorbed by plants as nitrogen, phosphate, and potash. In contrast to inorganic fertilizers being harmful to soil by polluting and degrading it, organic fertilizers protect agricultural land for years (Rashmi *et al.*, 2020).

22.9.1 Organic Manure

Organic manure mostly derives from animal waste, excluding green manure. The nutritional composition of manures varies significantly depending on the type of animal they come from. The following are the main sources of organic fertilizer in agriculture. Most farmers use organic fertilizers or manures, administering them to the field according to the soil's chemistry, crop variety, time of year, and the farmer's previous expertise and observation (Verma *et al.*, 2020).

22.9.2 Peanut Hulls Manure

Peanut shells are very ideal for mulching, a practice that entails applying a protective layer, such as straw or compost, on the ground around plants. This practice serves to mitigate excessive evaporation and erosion, while simultaneously enhancing the fertility of the soil. Additionally, peanut shells effectively hinder the development of weeds (Mechergui *et al.*, 2021). They provide a valuable supply of phosphorus (P), nitrogen (N), and potassium (K). These substances are primarily used in the production of animal feed and are also recognized for their ability to enhance crop output and perhaps improve soil fertility. A study was conducted over a period of seven months to examine the impact of compost made from peanut shells on the development of *Viola tricolor* and marigold plants. The findings revealed that the peanut shell compost had a greater influence on growth characteristics such as height, stem thickness, and dry weight of leaves, when compared to the control group (El Moussaoui and Bouqbis, 2022).

22.9.3 Poultry Manure

Poultry litter is an advantageous and valuable organic fertilizer. It enhances soil fertility and encourages the development of the root system and general plant vitality, hence lowering their susceptibility to disorders and pest infestations. Poultry dung rapidly decomposes in the soil and produces substantial heat energy (Jamaludin, 2022). Therefore, it is not recommended for usage during periods of high temperature. Implementing excellent best management practices (BMPs) may reduce the negative consequences of using chicken manure for land application. An experiment was done to determine the effectiveness of varying quantities of chicken manure (PM) on the development and production of *Citrullus lanatus*. The findings demonstrated that the use of chicken manure had a substantial positive impact on growth metrics, such as vigor and fruit yield, across both seasons (Yang *et al.*, 2023).

22.9.4 Dairy Cattle Manure

Cow dung, or cow manure, refers to the dung produced by bovines, such as domestic cattle, buffalo, bison, and yak. Cow waste is the residual material of plant matter that has undergone digestion in the animal's gastrointestinal tract. The resulting excrement has a high concentration of nutrients and is often used in urban farming (Liu *et al.*, 2022). Composted cow dung fertilizer is a very effective substrate for cultivating garden plants. They are incorporated into the soil or applied as a top layer to plants and vegetables as a fertilizer that is rich in nutrients. They are often used as first inputs due to their gradual mineralization. There is empirical data indicating that the use of animal dung may raise the pH levels of acidic soils. The soil that was treated with manure had a significantly increased pH level compared to the soil that was not fertilized (Lazcano *et al.*, 2021).

22.9.5 Application of Organic Fertilizers

Organic fertilizers may increase plant output more than inorganic fertilizers due to equal nutrients. The influence of "organic matter" on soil productivity is an enhancement that cannot be solely attributed to the supply of mineral nutrients, which is known as the direct, short-term "fertilizer effect." Evidence indicates that the use of composted manures may enhance nutrient accessibility and boost the crop-yield response curve. Moreover, the saturation value of composted manures is comparable to or even surpasses that of inorganic fertilizers. This shows a role for processes other than nitrogen delivery in plant development and a distinct response curve with a greater saturation point (Suárez *et al.*, 2022).

Soil fertility functions are maximized when organic residues and wastes are properly treated before being added to the soil. This is to prevent phytotoxicity, ensure the destruction of plant and animal pathogens, and minimize negative effects on soil properties. This means that organic fertilizers should resemble native SOM as closely as possible in terms of composition, characteristics, and behavior in soil (Gerke, 2022).

22.10 AGRONOMIC MANAGEMENT OF SOIL HEALTH AND NUTRIENT LEVELS

22.10.1 Crop Rotation

Crop rotation has been an enduring agricultural method that is well-known to every farmer. Planting the same crop in the same field every year leads to the roots of the crop depleting nutrients from the same soil depth repeatedly. This exhausts the rhizosphere and results in a decrease in crop output. To sustain attainable crop output, it is essential to use strategies to enhance soil fertility and productivity. One approach is to let the land fallow and then add nutrients using chemical fertilizers and organic manures. Allocating a substantial amount of land for an extended period of time is not sustainable owing to the rapid expansion of the global population. Crop rotation is a way for farmers to maximize their profitability while also conserving soil fertility. It entails cultivating crops in a certain order on the same plot of land (Hemathilake and Gunathilake, 2022).

Soil residual effect is one way to classify crops; for example, "exhaustive rotation" is a farming strategy that uses a wide range of crops, each of which depletes the soil's fertility due to the high quantity of plant nutrients used. Some examples of these crops are wheat, cotton, field mustard, and maize. Restorative rotation entails cultivating crops that enhance soil fertility. Leguminous and exhausting crops are included. Natural farmland management would reduce weeds in all crops (Javaid *et al.*, 2012). Crop rotation is also known to reduce weed growth. Crop rotation reduces weed populations and prevents species changes. Crop rotation changes cultivation, fertilization, herbicide application, and harvesting annually. Rotation alters the growth circumstances on an annual basis, a phenomenon that only a few numbers of weed species are capable of adapting (Ali *et al.*, 2013). Weeds face an unstable ecological environment due to the presence of clean-cultivated annual crops, densely planted grain crops, and grazed perennial sod crops (Tanveer *et al.*, 2015). Utilizing short-season weeds, such as sorghum and sudan grass, to suffocate crops may be an effective method for weed management (Islam and Sherman, 2021).

22.10.2 Benefits of Crop Rotation

Crop rotation offers a multitude of ecological and economic advantages. Specifically, it assists in the long-term maintenance of soil and crops. Currently, proceed to analyze the benefits of crop rotation and explore how farmers can benefit from them.

- *Enhanced Soil Fertility*: Due to their specific interactions with soil nutrients, various plant species produce and take in nutrients in different proportions. A well-planned rotation strategy improves soil fertility by balancing nutrient levels by using excess nutrients or restoring deficient ones. Plant waste, which includes microorganisms, increases SOM. Uncultivated land benefits from animal excrement. Pure green manure and harvested biomass improve soil fertility, and reduce synthetic fertilizer use (Launay *et al.*, 2022).
- *Improved Soil Structure*: Crop rotation greatly influences the nutrient cycling in soil, as well as the recycling of plant wastes, the creation and distribution of bio pores, and the colonization of beneficial bacteria in various soil types. These actions mitigate soil compaction, establishing optimal conditions for seeds to quickly germinate, roots to spread, water to penetrate deeply, and air to circulate freely. Hence, it is crucial to practice crop rotation to sustain soil vitality (Mehmood *et al.*, 2018; Gavrilescu, 2021).
- *Preventing Soil Erosion*: Some crops have shallow root systems, whereas others have deep ones. This depth variation increases soil porosity. Rotating high-residue plants like maize, hay, and tiny grains may minimize soil erosion by protecting topsoil with their residue. Persistent perennial grasses reduce water erosion and soil nutrient loss as well. Strategic crop rotation and perennial grass species may prevent erosion in high areas (Centeri, 2022).
- *Increase Yield and Streamlined Risk Management*: Crop rotation is recommended for stagnant land that doesn't produce enough. Through its cycles, rotation boosts plant yield and abiotic stress tolerance. Plants get plenty of nutrients from the soil, increasing crop output. Therefore, rotation variety may increase output and reduce production risks over time (Zafar *et al.*, 2023; Zaib *et al.*, 2023).
- *Nutrient Uptake Regulation and Improved Soil Nutrients*: As indicated, crop rotation renews soil nutrients, reducing fertilizer use. Because legumes contain nitrogen-fixing bacteria, planting them increases soil nitrogen fixation. By rotating grasses, grains, and oil plants, farmers may minimize fertilizer consumption and increase carbon sequestration (Muni Kumari and Durge, 2024).
- *Improved Weed Management*: Integrated weed management relies on the key practice of crop rotation. There is an abundance of weed species exhibiting diverse characteristics and life cycles, making it impractical to address it all with a single technique. Rotation and annual cover crops in an integrated system effectively reduce weed biomass and species diversity, while also decreasing the use of pesticides. By using this approach, farmers have the opportunity to shift toward NT farming, a practice that enhances soil structure and overall health (Mahmood *et al.*, 2015; Mennan *et al.*, 2020).
- *More Efficient Water Use*: Soil structure and water-holding capacity improve with crop rotation, which increases organic matter and water usage efficiency. By preserving moisture in deep soil layers, plants may utilize it during droughts. So, farmers can save irrigation water (Tanveer *et al.*, 2013; Mallareddy *et al.*, 2023).

22.11 EROSION CONTROL TECHNIQUES

Soil erosion is a natural process that affects any style of topography. Understanding the causes and consequences of soil erosion is important for the purposes of preventing it and reducing its occurrence. This chapter will discuss three processes of water, wind and tillage erosion, soil detachment, movement, and deposition. But erosion, in nature, is an incessant process. Four main factors that affect surface susceptibility to erosion: soil properties, vegetation cover, environmental conditions, and climate. Detachment transport and deposition are key processes in highland systems. The detachment of soil particles is influenced by a damaging pressure created by the impact of the raindrops and by flowing water (Dunkerley, 2020). Hydrology, topography, soil erodibility, soil transportability, soil surface cover, integrated residue, residual land use, subsurface impacts, tillage, roughness, and tillage markings are the major factors affecting upland erosion processes. Soil erosion management techniques are conceptually straightforward, but their implementation in the field can be complicated, costly, laborious, and time-consuming. Most soil erosion strategies are highly place-specific (Alizadeh *et al.*, 2022; Rasheed *et al.*, 2022).

Soil erosion is a natural process that has been negatively impacted by human activities, including agriculture. The issue is associated with notable ecological ramifications, considerable financial costs, and reduced efficiency, and therefore impacts both the availability of food and societal and economic advancement. Agriculture affects erosion rates via two mechanisms: the removal of vegetation that serves as a protective layer for the soil, and the physical alteration of topsoil through plowing. Erosion is not just caused by tillage operations; it is also impacted by the combined impacts of water and wind, with all three components mutually strengthening one another (Luo *et al.*, 2023). Erosion events on agricultural land may lead to erosion rates that exceed 100 Mg/ha/year. The solution to this challenge is found in adopting NT farming. NT farming is a technique that involves preserving a layer of vegetation on the soil surface and limiting any disruptions to the soil until it is time to plant. Therefore, it effectively mitigates many forms of soil erosion caused by the use of machinery, water, and wind, making it a significant advance in soil erosion control. The global problem of soil erosion is able to be effectively solved when farmers widely adopt NT practices (Wasaya *et al.*, 2021; Diop *et al.*, 2022).

22.11.1 Water Conservation

The water conservation concept that is addressed for dryland agriculture is as significant for irrigated agriculture. The objective of irrigation is to maximize the proportion of the water provided in order to fulfill the crop's transpiration needs. The primary factors contributing to inefficient irrigation are losses due to runoff, evaporation, and excessive deep percolation, excluding the amount necessary to maintain a safe level of rootzone salinity. Addressing these factors is key to improving irrigation water conservation. The geographical arrangement of precipitation is essentially beyond artificial influence, yet, it is essential to ensure the spatial homogeneity of irrigation practices to effectively save irrigation water (Tanveer *et al.*, 2012; Ahmed *et al.*, 2023).

The geographical and temporal distribution of irrigation is entirely determined by management and the chosen technique. Irrigation management involves the decision-making process of identifying the appropriate timing and quantity of irrigation for each stage of plant development, as well as the operation and maintenance of the irrigation system. The time of irrigation is mostly determined by the crop and soil water conditions (Bwambale *et al.*, 2022). However, the water supplier may have influence over the supply schedule, which might hinder a producer's efforts to conserve water. The optimal quantity of the application is closely linked to the timing of delivery, the requirements of the crop, and the technique used for application. Similarly, the operation and maintenance requirements are directly influenced by the manner of application and the technology used. Depending on the circumstances, the choice of application method may be determined by the type of crop being cultivated, such as the need to prevent water from coming into contact with the fruit or plant (Mahmood *et al.*, 2021; Javaid *et al.*, 2023).

22.11.2 Soil Structure Improvement

The arrangement of soil particles has significant effects on the soil conditions and the surrounding environment. The degree of stability of aggregates is frequently declared. Aggregation occurs

TABLE 22.3
Classification of Soil Types and Factors Affecting Soil Aggregation

Aggregation Factors	Soil Orders	Reference
Clay-size fractions polycationic bridges, wet-dry cycles	Vertisols	Badía-Villas *et al.* (2022)
SOM non-crystalline sesquioxide	Ultisols	Park *et al.* (2020)
Organo-metallic complexes Metallic hydroxides	Spodosols	Ferro-Vázquez *et al.* (2020)
SIC, carbonates, and weathered clay minerals	Aridisols	Stanbery *et al.* (2023)
Allophane clay, non-crystalline clay	Andisols	Arifin *et al.* (2022)
SOM	Alfisols	

when particles are reorganized, brought together, and bound together by the flocculation and cementation process. Table 22.3 presents the classification of soil types and factors affecting soil aggregation. The process is facilitated by SOC organisms, ionic bridging, clay, and carbonates (Qin *et al.*, 2020). The intricate interplay between these aggregates may either enhance or hinder the process of aggregation. Clay-sized particles are often linked to the process of aggregation via rearrangement and flocculation, while the presence of swelling clay may impair the formation of aggregates. Organo-metallic compounds and cations act as connectors between particles. Soil SOC is derived from the remains of plants, animals, and microbes, as well as the substances they release. It improves the process of aggregation by attaching together the main particles of soil. The efficacy of soil SOC in the formation of enduring aggregates is directly linked to its rate of decomposition, which is, in turn, determined by its physical and chemical safeguarding against microbial activity (Wu *et al.*, 2023).

Soil inorganic carbon (SIC) can be important for aggregation in dry and semiarid environments. SOC, as well as Ca^{2+} and Mg^{2+}, control the formation of secondary carbonates. Soil organisms release carbon dioxide (CO_2) and synthesize soil SOC that promotes the solubilization of primary carbonates. In addition, cations favor the precipitation of secondary carbonates. The precipitation of (hydr)oxides, phosphates, and carbonates increases aggregation. Cations including Si^{4+}, Fe^{3+}, Al^{3+}, and Ca^{2+} promote the synthesis of compounds that help bond primary particles. Roots and hyphae can encase particles as they replace them and exude organic compounds that bond them, which has a stabilizing effect on soil carbon sequestration. Soil structure can be affected significantly by various management practices and environmental factors. Soil aggregates and structural development are enhanced through practices that increase productivity and decrease soil disturbance (Bhattacharyya *et al.*, 2022).

22.12 ENRICHMENT OF ORGANIC SOIL FERTILIZERS/AMENDMENTS

Methods

1. *Nutrient Additions*
   Addition of nutrient-rich materials such as bone meal, blood meal, fish meal, seaweed biomass, and inorganic fertilizers.
2. *Microbial Inoculation*
   Addition of BNF, P-solubilizing, plant growth-promoting rhizobacteria (PGPR) and other beneficial bacteria, mycorrhizal fungi, algae
3. *Mineral Amendments*
   Adding mineral amendments such as rock phosphate, zeolite greensand, limestone, Epsom salt, and gypsum to provide additional nutrients

4. *Co-Composting*
   Combine use of kitchen waste, urban waste, food industry waste, sewage sludge, animal manure, agricultural straw, sawdust, etc.
5. *Mixing Amendments*
   Mixing biochar, manure, compost, crop straw, sawdust, etc.

22.13 CONCLUSION AND FUTURE PERSPECTIVES

Producing more food, feed, fuel, and fiber while enhancing environmental quality will be a huge challenge. Soil management issues, water and air quality, continue. Soil cleanses water, but farming operations can discharge nitrate nitrogen and phosphorus into nearby lakes. Ideally, better manure management systems and strategies to reduce tillage will require time for better management of soil and an increase in knowledge. Sediment loss from the fields may be transported with outside nutrients or agricultural pesticides, but this should not be concerning to everyone because these pesticides would affect soil resources over the long term. Management practices that can reduce losses from erosion may be different and it is necessary to identify management strategies that could reduce erosion losses from soils worldwide due to water and wind erosion.

Switching from a "tolerable" view of soil loss to one of improvement is critical to ensuring that future generations have sufficient land. Soil loss through erosion and tillage is increasing spatial variance across agricultural landscapes and endangering productivity. Anything that increases regional variance in soil management methods, or provides an alternative to soil erosion- or tillage-degraded regions would serve to benefit the soil and therefore humanity. These procedures must be evaluated and rooted in strong research to work across areas. The goal of this publication is to elucidate soil management practices and provide an overview of practicable options. Farming systems under consideration are changing rapidly. These worries are difficult to tackle since they branch out from a subject area problem to a regional landscape problem with numerous soils, landscapes, and microclimates. Soil water usage, nutrient extraction, and impact on soil biology must facilitate effective soil management strategies across diverse agricultural systems. Output and diversity for different applications will be difficult to boost. As part of the renewable fuel objective, switching from oil to biofuels produces concerns about the carbon balance of soils and agricultural systems.

Agriculture practices are part of the global carbon budget and help mound the balance of GMT and stores. Due to rising organic matter, the combined capacity of holding water, biological activity, and cycling of nutrients, carbon storage in soil enhances soil quality and absorption of greenhouse gas (GHG). Methane and nitrous oxide are also key GHGs along with carbon. This knowledge of how soil management regulates GHG exchange may lead to new measures to reduce GHG emissions. Cover crops should enhance carbon capture in cropping systems. Appropriate management and adaptation of cover crop systems to a wider set of soils and cropping systems can enhance carbon, water, and nutrient management, but there are few sound, cost-effective, and environmentally sustainable systems suitable for implementation.

LIST OF ABBREVIATIONS

ACT	aerated compost tea
AMF	arbuscular mycorrhizal fungi
B, Cu, Cl	boron, copper, chlorine
Fe, Mn	iron, manganese
Mo, Zn	molybdenum, zinc
BM	biodegradable mulches
BMPS	best management practices
CEC	cation exchange capacity
CO_2	carbon dioxide
EC	electrical conductivity
GHGs	greenhouse gases
GLU	glucosidase
H, C, O_2	hydrogen, carbon, oxygen
H_2CO_3	carbonic acid

NCT non-aerated compost tea
NT no till
SIC soil inorganic carbon
PGPR plant growth-promoting rhizobacteria
SOC soil organic carbon
SOM soil organic matter

REFERENCES

Acharya, P., Ghimire, R. & Acosta-Martínez, V., 2024. Cover crop-mediated soil carbon storage and soil health in semi-arid irrigated cropping systems. *Agriculture, Ecosystems & Environment*, 361, 108813.

Ahamed, M.S., Sultan, M., Monfet, D., Rahman, M.S., Zhang, Y., Zahid, A., Bilal, M., Ahsan, T.M.A. & Achour, Y., 2023. A critical review on efficient thermal environment controls in indoor vertical farming. *Journal of Cleaner Production*, 425, 138923.

Ahmed, Z., Gui, D., Murtaza, G., Yunfei, L. & Ali, S., 2023. An overview of smart irrigation management for improving water productivity under climate change in drylands. *Agronomy*, 13, 2113.

Ali, H., Tanveer, A., Nadeem, M., Javaid, M., Kashif, M. & Chadhar, A., 2013. Allelopathic effects of *Rhynchosia capitata* on germination and seedling growth of mungbean. *Planta Daninha*, 31, 501–509.

Alizadeh, M., Zabihi, H., Wolf, I.D., Langat, P.K., Pour, A.B. & Ahmad, A., 2022. Remote sensing technique and ICONA based-GIS mapping for assessing the risk of soil erosion: A case of the Rudbar Basin, Iran. *Environmental Earth Sciences*, 81, 512.

Anas, M., Liao, F., Verma, K.K., Sarwar, M.A., Mahmood, A., Chen, Z.-L., Li, Q., Zeng, X.-P., Liu, Y. & Li, Y.-R., 2020. Fate of nitrogen in agriculture and environment: Agronomic, eco-physiological and molecular approaches to improve nitrogen use efficiency. *Biological Research*, 53, 47.

Angon, P.B., Anjum, N., Akter, M.M., Kc, S., Suma, R.P. & Jannat, S., 2023. An overview of the impact of tillage and cropping systems on soil health in agricultural practices. *Advances in Agriculture*, 2023, 8861216.

Anwer, A.H., Ahtesham, A., Shoeb, M., Mashkoor, F., Ansari, M.Z., Zhu, S. & Jeong, C., 2023. State-of-the-art advances in nanocomposite and bio-nanocomposite polymeric materials: A comprehensive review. *Advances in Colloid and Interface Science*, 318, 102955.

Arifin, M., Devnita, R., Anda, M., Goenadi, D.H. & Nugraha, A., 2022. Characteristics of andisols developed from andesitic and basaltic volcanic ash in different agro-climatic zones. *Soil Systems*, 6, 78.

Ayub, A., Shabaan, M., Malik, M., Asghar, H.N., Zulfiqar, U., Ejaz, M., Alarjani, K.M. & Al Farraj, D.A., 2024. Synergistic application of *Pseudomonas strains* and compost mitigates lead (Pb) stress in sunflower (*Helianthus annuus* L.) via improved nutrient uptake, antioxidant defense and physiology. *Ecotoxicology and Environmental Safety*, 274, 116194.

Babar, B.H., Tanveer, A., Tahir, M., Aziz, A., Ahmad, A.U.H., Nadeem, M.A. & Javaid, M.M., 2009. Allelopathic potential of wild onion (*Asphodelus tenuifolius*) on the germination and seedling growth of chickpea (*Cicer arietinum*). *Weed Biology and Management*, 9, 146–151.

Badía-Villas, D., Poch, R.M., Longares, L.A., Yuste, A. & Bauluz, B., 2022. Genesis and stability of textural pedofeatures along a soil transect in the siliceous Iberian Chain (NE Spain). *CATENA*, 211, 105965.

Bashir, O., Ali, T., Baba, Z.A., Rather, G.H., Bangroo, S.A., Mukhtar, S.D., Naik, N., Mohiuddin, R., Bharati, V. & Bhat, R.A., 2021. Soil Organic Matter and Its Impact on Soil Properties and Nutrient Status. *In* G.H. Dar, R.A. Bhat, M.A. Mehmood & K.R. Hakeem (eds.) *Microbiota and Biofertilizers, Vol 2: Ecofriendly Tools for Reclamation of Degraded Soil Environs.*: Springer, Cham, Switzerland, 129–159.

Berg, J.S., Ahmerkamp, S., Pjevac, P., Hausmann, B., Milucka, J. & Kuypers, M.M.M., 2022. How low can they go? Aerobic respiration by microorganisms under apparent anoxia. *FEMS Microbiology Reviews*, 46.

Bhattacharyya, S.S., Ros, G.H., Furtak, K., Iqbal, H.M.N. & Parra-Saldívar, R., 2022. Soil carbon sequestration – An interplay between soil microbial community and soil organic matter dynamics. *Science of the Total Environment*, 815, 152928.

Bhunia, S., Bhowmik, A., Mallick, R. & Mukherjee, J., 2021. Agronomic efficiency of animal-derived organic fertilizers and their effects on biology and fertility of soil: A review. *Agronomy*, 11, 823.

Biswas, P., Mandal, S., Das, T., Dey, S., Ghorai, M., Bhattacharya, S., Ghosh, A., Nongdam, P., Kumar, V., Al-Tawaha, A.R., Bursal, E. & Dey, A., 2023. Chapter 3 – Generation of Biofuels from Rice Straw and Its Future Perspectives. *In* M.P. Shah (ed.) *Green Approach to Alternative Fuel for a Sustainable Future.* Elsevier, Amsterdam, Netherlands, 25–33.

Bünemann, E.K., Reimer, M., Smolders, E., Smith, S.R., Bigalke, M., Palmqvist, A., Brandt, K.K., Möller, K., Harder, R., Hermann, L., Speiser, B., Oudshoorn, F., Løes, A.K. & Magid, J., 2024. Do contaminants compromise the use of recycled nutrients in organic agriculture? A review and synthesis of current knowledge on contaminant concentrations, fate in the environment and risk assessment. *Science of the Total Environment*, 912, 168901.

Bwambale, E., Abagale, F.K. & Anornu, G.K., 2022. Smart irrigation monitoring and control strategies for improving water use efficiency in precision agriculture: A review. *Agricultural Water Management*, 260, 107324.

Cammerino, A.R.B., Ingaramo, M. & Monteleone, M., 2024. Complementary approaches to planning a restored coastal wetland and assessing the role of agriculture and biodiversity: An applied case study in Southern Italy. *Water*, 16, 153.

Centeri, C., 2022. Effects of grazing on water erosion, compaction and infiltration on grasslands. *Hydrology*, 9, 34.

Chaudhary, S., Sindhu, S.S., Dhanker, R. & Kumari, A., 2023. Microbes-mediated sulphur cycling in soil: Impact on soil fertility, crop production and environmental sustainability. *Microbiological Research*, 271, 127340.

de Corato, U., 2020. Agricultural waste recycling in horticultural intensive farming systems by on-farm composting and compost-based tea application improves soil quality and plant health: A review under the perspective of a circular economy. *Science of the Total Environment*, 738, 139840.

Diop, M., Chirinda, N., Beniaich, A., El Gharous, M. & El Mejahed, K., 2022. Soil and water conservation in Africa: State of play and potential role in tackling soil degradation and building soil health in agricultural lands. *Sustainability*, 14, 13425.

Du, X., Ge, Y., Zhang, Y., Hu, H., Zhang, Y., Yang, Z., Ren, X., Hu, S., Feng, H. & Song, Y., 2023. Responses of soil carbon cycling microbial functional genes to nitrogen and phosphorus addition in saline-sodic soils. *Plant and Soil*, 490, 261–277.

Duan, Y., Yang, J., Awasthi, M.K., Pandey, A. & Li, H., 2023. Chapter 3 – Innovations in Design and Operation of Aeration Devices for Composting and Vermicomposting. *In* A. Pandey, M. Awasthi & Z. Zhang (eds.) *Current Developments in Biotechnology and Bioengineering*. Elsevier, Amsterdam, Netherlands, 57–81.

Dunkerley, D., 2020. A Review of the Effects of Throughfall and Stemflow on Soil Properties and Soil Erosion. *In* I.I.J.T. Van Stan, E. Gutmann & J. Friesen (eds.) *Precipitation Partitioning by Vegetation: A Global Synthesis*. Springer Cham, Switzerland 183–214.

El-Beltagi, H.S., Basit, A., Mohamed, H.I., Ali, I., Ullah, S., Kamel, E.A.R., Shalaby, T.A., Ramadan, K.M.A., Alkhateeb, A.A. & Ghazzawy, H.S., 2022. Mulching as a sustainable water and soil saving practice in agriculture: A review. *Agronomy*, 12, 1881.

El Moussaoui, H. & Bouqbis, L., 2022. Interactive effect of biochar and bio-compost on starting growth and physiologic parameters of argan. *Sustainability*, 14, 7270.

Evans, D.L., Vis, B.N., Dunning, N.P., Graham, E. & Isendahl, C., 2021. Buried solutions: How Maya urban life substantiates soil connectivity. *Geoderma*, 387, 114925.

Farooq, M.S., Fatima, H., Rehman, O.U., Yousuf, M., Kalsoom, R., Fiaz, S., Khan, M.R., Uzair, M. & Huo, S., 2023. Major challenges in widespread adaptation of aerobic rice system and potential opportunities for future sustainability. *South African Journal of Botany*, 159, 231–251.

Ferro-Vázquez, C., Nóvoa-Muñoz, J.C., Klaminder, J., Gómez-Armesto, A. & Martínez-Cortizas, A., 2020. Comparing podzolization under different bioclimatic conditions. *Geoderma*, 377, 114581.

Gao, Y., Shao, Y., Wang, J., Hu, B., Feng, H., Qu, Z., Liu, Z., Zhang, M., Li, C. & Liu, Y., 2024. Effects of straw returning combined with blended controlled-release urea fertilizer on crop yields, greenhouse gas emissions, and net ecosystem economic benefits: A nine-year field trial. *Journal of Environmental Management*, 356, 120633.

Gavrilescu, M., 2021. Water, soil, and plants interactions in a threatened environment. *Water*, 13, 2746.

Gerke, J., 2022. The central role of soil organic matter in soil fertility and carbon storage. *Soil Systems*, 6, 33.

Goldan, E., Nedeff, V., Barsan, N., Culea, M., Panainte-Lehadus, M., Mosnegutu, E., Tomozei, C., Chitimus, D. & Irimia, O., 2023. Assessment of manure compost used as soil amendment—A review. *Processes*, 11, 1167.

Gondal, A.H., Bhat, R.A., Gómez, R.L., Areche, F.O. & Huaman, J.T., 2023. Advances in plastic pollution prevention and their fragile effects on soil, water, and air continuums. *International Journal of Environmental Science and Technology*, 20, 6897–6912.

Gorliczay, E., Boczonádi, I., Kiss, N.É., Tóth, F.A., Pabar, S.A., Biró, B., Kovács, L.R. & Tamás, J., 2021. Microbiological effectivity evaluation of new poultry farming organic waste recycling. *Agriculture*, 11, 683.

Guo, F., Guo, Y., Chen, L., Jia, W., Zhu, Y., Li, Y., Wang, H., Yao, X., Zhang, Y. & Wu, J., 2023. Multitudinous components recovery, heavy metals evolution and environmental impact of coal gasification slag: A review. *Chemosphere*, 338, 139473.

Gurmessa, B., 2021. Soil acidity challenges and the significance of liming and organic amendments in tropical agricultural lands with reference to Ethiopia. *Environment, Development and Sustainability*, 23, 77–99.

Haque, M.A., Wang, Y., Shen, Z., Li, X., Saleemi, M.K. & He, C., 2020. Mycotoxin contamination and control strategy in human, domestic animal and poultry: A review. *Microbial Pathogenesis*, 142, 104095.

Hemathilake, D.M.K.S. & Gunathilake, D.M.C.C., 2022. Chapter 31 – Agricultural Productivity and Food Supply to Meet Increased Demands. *In* R. Bhat (ed.) *Future Foods*. Academic Press, London, UK, 539–553.

Islam, R. & Sherman, B., 2021. *Cover Crops and Sustainable Agriculture*. *In* R. Islam & B. Sherman (eds), *Cover Crops and Agroecosystem Services*, CRC Press, FL, USA, 1–15.

Jamaludin, N.S., 2022. Biochar and Chicken Manure Compost. *In* G.A.M. Ali & A.S.H. Makhlouf (eds.) *Handbook of Biodegradable Materials.*: Springer Cham, Switzerland, 1–33.

Javaid, M., Haleem, A., Khan, I.H. & Suman, R., 2023. Understanding the potential applications of artificial intelligence in agriculture sector. *Advanced Agrochem*, 2, 15–30.

Javaid, M., Tanveer, A., Ahmad, R., Yaseen, M. & Khaliq, A., 2012. Optimizing activity of herbicides at reduced rate on *Emex spinosa* Campd. with adjuvants. *Planta Daninha*, 30, 425–435.

Jindo, K., Audette, Y., Olivares, F.L., Canellas, L.P., Smith, D.S. & Paul Voroney, R., 2023. Biotic and abiotic effects of soil organic matter on the phytoavailable phosphorus in soils: A review. *Chemical and Biological Technologies in Agriculture*, 10, 29.

Kaur, A., Pati, P.K., Ohri, P. & Kaur, A., 2022. Effects of vermicompost and vermicompost leachate on the biochemical and physiological response of *Withania somnifera* (L.) Dunal. *Journal of Soil Science and Plant Nutrition*, 22, 3228–3242.

Khan, B.A., Nijabat, A., Khan, M.I., Khan, I., Hashim, S., Nadeem, M.A. & Ikram, M., 2022. Implications of Mulching on Weed Management in Crops and Vegetable. *In* K. Akhtar, M. Arif, M. Riaz & H. Wang (eds.) *Mulching in Agroecosystems: Plants, Soil & Environment*. Springer Singapore, 199–213.

Kwiatkowski, C.A., Pawłowska, M., Harasim, E. & Pawłowski, L., 2023. Strategies of climate change mitigation in agriculture plant production—A critical review. *Energies*, 16, 4225.

Launay, C., Houot, S., Frédéric, S., Girault, R., Levavasseur, F., Marsac, S. & Constantin, J., 2022. Incorporating energy cover crops for biogas production into agricultural systems: Benefits and environmental impacts. A review. *Agronomy for Sustainable Development*, 42, 57.

Lazcano, C., Zhu-Barker, X. & Decock, C., 2021. Effects of organic fertilizers on the soil microorganisms responsible for N_2O emissions: A review. *Microorganisms*, 9, 983.

Li, Y., Chai, S., Chai, Y., Li, R., Lan, X., Ma, J., Cheng, H. & Chang, L., 2021. Effects of mulching on soil temperature and yield of winter wheat in the semiarid rainfed area. *Field Crops Research*, 271, 108244.

Liu, T., Klammsteiner, T., Dregulo, A.M., Kumar, V., Zhou, Y., Zhang, Z. & Awasthi, M.K., 2022. Black soldier fly larvae for organic manure recycling and its potential for a circular bioeconomy: A review. *Science of the Total Environment*, 833, 155122.

Lu, H., Xia, Z., Fu, Y., Wang, Q., Xue, J. & Chu, J., 2020. Response of soil temperature, moisture, and spring maize (*Zea mays* L.) root/shoot growth to different mulching materials in semi-arid areas of northwest China. *Agronomy*, 10, 453.

Luo, Z., Zhang, H., Pang, J., Yang, J. & Li, M., 2023. Coupling of SWAT and EPIC models to investigate the mutual feedback relationship between vegetation and soil erosion, a case study in the Huangfuchuan watershed, China. *Forests*, 14, 844.

Lv, L., Gao, Z., Liao, K., Zhu, Q. & Zhu, J., 2023. Impact of conservation tillage on the distribution of soil nutrients with depth. *Soil and Tillage Research*, 225, 105527.

Mahmood, A., Hussain, A., Shahzad, A.N. & Honermeier, B., 2015. Biomass and biogas yielding potential of sorghum as affected by planting density, sowing time and cultivar. *Pakistan Journal of Botany*, 47, 2401–2408.

Mahmood, A., Wang, X., Shahzad, A.N., Fiaz, S., Ali, H., Naqve, M., Javaid, M.M., Mumtaz, S., Naseer, M. & Dong, R., 2021. Perspectives on bioenergy feedstock development in Pakistan: Challenges and opportunities. *Sustainability*, 13, 8438.

Maitra, S., Praharaj, S., Brestic, M., Sahoo, R.K., Sagar, L., Shankar, T., Palai, J.B., Sahoo, U., Sairam, M., Pramanick, B., Nath, S., Venugopalan, V.K., Skalický, M. & Hossain, A., 2023. Rhizobium as biotechnological tools for green solutions: An environment-friendly approach for sustainable crop production in the modern era of climate change. *Current Microbiology*, 80, 219.

Mallareddy, M., Thirumalaikumar, R., Balasubramanian, P., Naseeruddin, R., Nithya, N., Mariadoss, A., Eazhilkrishna, N., Choudhary, A.K., Deiveegan, M., Subramanian, E., Padmaja, B. & Vijayakumar, S., 2023. Maximizing water use efficiency in rice farming: A comprehensive review of innovative irrigation management technologies. *Water*, 15, 1802.

Mansoor, Z., Tchuenbou-Magaia, F., Kowalczuk, M., Adamus, G., Manning, G., Parati, M., Radecka, I. & Khan, H., 2022. Polymers use as mulch films in agriculture—A review of history, problems and current trends. *Polymers*, 14, 5062.

Mechergui, T., Pardos, M., Jhariya, M.K. & Banerjee, A., 2021. Mulching and Weed Management towards Sustainability. *In* M.K. Jhariya, R.S. Meena & A. Banerjee (eds.) *Ecological Intensification of Natural Resources for Sustainable Agriculture*. Springer Singapore, 255–287.

Mehmood, A., Naeem, M., Khalid, F., Saeed, Y., Abbas, T., Jabran, K., Sarwar, M.A., Tanveer, A. & Javaid, M.M., 2018. Identification of phytotoxins in different plant parts of *Brassica napus* and their influence on mung bean. *Environmental Science and Pollution Research*, 25, 18071–18080.

Mekonen, Y., Godade, F. & Melis, B., 2024. Micronutrient limitations in crop production and strategies for improvement: A review. *Journal of Research and Education*, 2, 01–08.

Mendonça, S.R., Ávila, M.C.R., Vital, R.G., Evangelista, Z.R., Pontes, N.D.C. & Nascimento, A.D.R., 2021. The effect of different mulching on tomato development and yield. *Scientia Horticulturae*, 275, 109657.

Mennan, H., Jabran, K., Zandstra, B.H. & Pala, F., 2020. Non-chemical weed management in vegetables by using cover crops: A review. *Agronomy*, 10, 257.

Mittal, U., Kumar, V., Kukreja, S., Singh, B., Pandey, N.K. & Goutam, U., 2023. Role of beneficial elements in developing resilience to abiotic and biotic stresses in plants: Present status and future prospects. *Journal of Plant Growth Regulation*, 42, 3789–3813.

Morizet-Davis, J., Marting Vidaurre, N.A., Reinmuth, E., Rezaei-Chiyaneh, E., Schlecht, V., Schmidt, S., Singh, K., Vargas-Carpintero, R., Wagner, M. & Von Cossel, M., 2023. Ecosystem services at the farm level—Overview, synergies, trade-offs, and stakeholder analysis. *Global Challenges*, 7, 2200225.

Mrabet, R., 2023. Sustainable Agriculture for Food and Nutritional Security. *In* M. Farooq, N. Gogoi & M. Pisante (eds.) *Sustainable Agriculture and the Environment*. Academic Press, London, UK, 25–90.

Muni Kumari, A. & Durge, J.B., 2024. Improving Agricultural Carbon Sequestration Strategies by Eco Friendly Procedures for Managing Crop Residues and Weeds. *In* S. Sheraz Mahdi, R. Singh & B. Dhekale (eds.) *Adapting to Climate Change in Agriculture-Theories and Practices: Approaches for Adapting to Climate Change in Agriculture in India*.: Springer Cham, Switzerland, 361–375.

Murindangabo, Y.T., Kopecký, M., Hoang, T.N., Bernas, J., Parajuli, T., Dhakal, S., Konvalina, P., Ufitikirezi, J.D.D.M., Kaneza, G., Khanal, B.R., Dhakal, S.C. & Shrestha, A.K., 2023. Comparative analysis of soil organic matter fractions, lability, stability ratios, and carbon management index in various land use types within bharatpur catchment, Chitwan District, Nepal. *Carbon Balance and Management*, 18, 21.

Nazir, M.J., Li, G., Nazir, M.M., Zulfiqar, F., Siddique, K.H.M., Iqbal, B. & Du, D., 2024. Harnessing soil carbon sequestration to address climate change challenges in agriculture. *Soil and Tillage Research*, 237, 105959.

Norberg, L. & Aronsson, H., 2024. Effects of spring and autumn tillage, catch crops, and pig manure application on long-term nutrient leaching from a loamy sand. *European Journal of Agronomy*, 156, 127156.

Oyege, I. & Balaji Bhaskar, M.S., 2023. Effects of vermicompost on soil and plant health and promoting sustainable agriculture. *Soil Systems*, 7, 101.

Panday, D., Bhusal, N., Das, S. & Ghalehgolabbehbahani, A., 2024. Rooted in nature: The rise, challenges, and potential of organic farming and fertilizers in agroecosystems. *Sustainability*, 16, 1530.

Park, W.-P., Hyun, H.-N. & Koo, B.-J., 2020. Silicon fractionation of soluble silicon in volcanic ash soils that may affect groundwater silicon content on Jeju Island, Korea. *Water*, 12, 2686.

Pinto, A.G., Kos, T., Puškarić, J., Vrandečić, K., Benković-Lačić, T. & Brmež, M., 2024. Soil ecosystem functioning through interactions of nematodes and fungi *Trichoderma* sp. *Sustainability*, 16, 2885.

Punia, A. & Bharti, R., 2023. Loss of soil organic matter in the mining landscape and its implication to climate change. *Arabian Journal of Geosciences*, 16, 86.

Qin, W., Wang, C.-Y., Ma, Y.-X., Shen, M.-J., Li, J., Jiao, K., Tay, F.R. & Niu, L.-N., 2020. Microbe-mediated extracellular and intracellular mineralization: Environmental, industrial, and biotechnological applications. *Advanced Materials*, 32, 1907833.

Racek, J., Sevcik, J., Chorazy, T., Kucerik, J. & Hlavinek, P., 2020. Biochar – Recovery material from pyrolysis of sewage sludge: A review. *Waste and Biomass Valorization*, 11, 3677–3709.

Rashed, A. & Van Herk, W.G., 2024. Pest elaterids of North America: New insights and opportunities for management. *Annual Review of Entomology*, 69, 1–20.

Rasheed, A., Raza, A., Jie, H., Mahmood, A., Ma, Y., Zhao, L., Xing, H., Li, L., Hassan, M.U. & Qari, S.H., 2022. Molecular tools and their applications in developing salt-tolerant soybean (*Glycine max* L.) cultivars. *Bioengineering*, 9, 495.

Rashmi, I., Roy, T., Kartika, K.S., Pal, R., Coumar, V., Kala, S. & Shinoji, K.C., 2020. Organic and Inorganic Fertilizer Contaminants in Agriculture: Impact on Soil and Water Resources. *In* M. Naeem, A.A. Ansari & S.S. Gill (eds.) *Contaminants in Agriculture: Sources, Impacts and Management.*: Springer Cham, Switzerland, 3–41.

Rehman, S.U., De Castro, F., Aprile, A., Benedetti, M. & Fanizzi, F.P., 2023. Vermicompost: Enhancing plant growth and combating abiotic and biotic stress. *Agronomy*, 13, 1134.

Roy, P., Pal, S.C., Chakrabortty, R., Towfiqul Islam, A.R.M., Chowdhuri, I. & Saha, A., 2023. The role of indigenous plant species in controlling the erosion of top soil in sub-tropical environment: In-situ field observation and validation. *Journal of Hydrology*, 625, 129993.

Saberi Riseh, R., 2024. Advancing agriculture through bioresource technology: The role of cellulose-based biodegradable mulches. *International Journal of Biological Macromolecules*, 255, 128006.

Salama, K. & Geyer, M., 2023. Plastic mulch films in agriculture: Their use, environmental problems, recycling and alternatives. *Environments*, 10, 179.

Savarese, C., Cozzolino, V., Verrillo, M., Vinci, G., De Martino, A., Scopa, A. & Piccolo, A., 2022. Combination of humic biostimulants with a microbial inoculum improves lettuce productivity, nutrient uptake, and primary and secondary metabolism. *Plant and Soil*, 481, 285–314.

Schmaltz, E.M., Krammer, C., Dersch, G., Weinberger, C., Kuderna, M. & Strauss, P., 2023. The effectiveness of soil erosion measures for cropland in the Austrian Agri-environmental Programme: A national approach using local data. *Agriculture, Ecosystems & Environment*, 355, 108590.

Shaikh, T.N., Patel, J.R. & Patel, B.H., 2023. Mulching effect on water management in agro-ecological systems: A review. *AQUA – Water Infrastructure, Ecosystems and Society*, 72, 1664–1680.

Shaji, H., Chandran, V. & Mathew, L., 2021. Chapter 13 – Organic Fertilizers as a Route to Controlled Release of Nutrients. *In* F.B. Lewu, T. Volova, S. Thomas & R. K.R. (eds.) *Controlled Release Fertilizers for Sustainable Agriculture*. Academic Press, London, Uk, 231–245.

Sharma, B., Tiwari, S., Kumawat, K.C. & Cardinale, M., 2023. Nano-biofertilizers as bio-emerging strategies for sustainable agriculture development: Potentiality and their limitations. *Science of the Total Environment*, 860, 160476.

Singh, A.A., Ghosh, A., Agrawal, M. & Agrawal, S.B., 2023. Secondary metabolites responses of plants exposed to ozone: An update. *Environmental Science and Pollution Research*, 30, 88281–88312.

Singh, S. & Sinha, R.K., 2022. 7 – Vermicomposting of Organic Wastes by Earthworms: Making Wealth from Waste by Converting 'Garbage into Gold' for Farmers. *In* C. Hussain & S. Hait (eds.) *Advanced Organic Waste Management*. Elsevier, Amsterdam, Netherlands, 93–120.

Singh Yadav, S.P., Bhandari, S., Bhatta, D., Poudel, A., Bhattarai, S., Yadav, P., Ghimire, N., Paudel, P., Paudel, P., Shrestha, J. & Oli, B., 2023. Biochar application: A sustainable approach to improve soil health. *Journal of Agriculture and Food Research*, 11, 100498.

Snyder, K., Murray, C. & Wolff, B., 2020. Insulative effect of plastic mulch systems and comparison between the effects of different plant types. *Open Agriculture*, 5, 317–324.

Stanbery, C., Ghahremani, Z., Huber, D.P., Will, R., Benner, S.G., Glenn, N., Hanif, T., Spaete, L., Terhaar, D., Lohse, K.A., Seyfried, M., Freutel, W. & Pierce, J.L., 2023. Controls on the presence and storage of soil inorganic carbon in a semi-arid watershed. *CATENA*, 225, 106980.

Suárez, J.C., Anzola, J.A., Contreras, A.T., Salas, D.L., Vanegas, J.I., Urban, M.O., Beebe, S.E. & Rao, I.M., 2022. Photosynthetic and grain yield responses to intercropping of two common bean lines with maize under two types of fertilizer applications in the colombian amazon region. *Scientia Horticulturae*, 301, 111108.

Sulaiman, M.A. & Bello, S.K., 2024. Biological control of soil-borne pathogens in arid lands: A review. *Journal of Plant Diseases and Protection*, 131, 293–313.

Tanveer, A., Arshad, M.S., Ayub, M., Javaid, M.M. & Yaseen, M., 2012. Effect of temperature, light, salinity, drought stress and seeding depth on germination of *Cucumis melo* var. agrestis. *Pakistan Journal of Weed Science Research*, 18.

Tanveer, A., Javaid, M.M., Irfan, M., Khaliq, A. & Yaseen, M., 2015. Yield losses in chickpea with varying densities of dragon spurge (*Euphorbia dracunculoides*). *Weed Science*, 63, 522–528.

Tanveer, A., Mumtaz, K., Javaid, M., Chaudhry, M., Balal, R. & Khaliq, A., 2013. Effect of ecological factors on germination of horse purslane (*Trianthema portulacastrum*). *Planta Daninha*, 31, 587–597.

Tassinari, G., Bassani, A., Spigno, G., Soregaroli, C. & Drabik, D., 2023. Do biodegradable food packaging films from agro-food waste pay off? A cost-benefit analysis in the context of Europe. *Science of the Total Environment*, 856, 159101.

Thakur, N., Nigam, M., Mann, N.A., Gupta, S., Hussain, C.M., Shukla, S.K., Shah, A.A., Casini, R., Elansary, H.O. & Khan, S.A., 2023. Host-mediated gene engineering and microbiome-based technology optimization for sustainable agriculture and environment. *Functional & Integrative Genomics*, 23, 57.

Toumi, I., Ghrab, M., Zarrouk, O. & Nagaz, K., 2024. Impact of deficit irrigation strategies using saline water on soil and peach tree yield in an arid region of Tunisia. *Agriculture*, 14, 377.

Valenzuela, H., 2023. Ecological management of the nitrogen cycle in organic farms. *Nitrogen*, 4, 58–84.

van der Ent, A., Kopittke, P.M., Schat, H. & Chaney, R.L., 2024. Hydroponics in physiological studies of trace element tolerance and accumulation in plants focussing on metallophytes and hyperaccumulator plants. *Plant and Soil*. 501, 573–594.

Verma, B.C., Pramanik, P. & Bhaduri, D., 2020. Organic Fertilizers for Sustainable Soil and Environmental Management. *In* R.S. Meena (ed.) *Nutrient Dynamics for Sustainable Crop Production*. Springer, Singapore, 289–313.

Waheed, H., Javaid, M.M., Shahid, A., Ali, H.H., Nargis, J. & Mehmood, A., 2019. Impact of foliar-applied Hoagland's nutrient solution on growth and yield of mash bean (*Vigna mungo* L.) under different growth stages. *Journal of Plant Nutrition*, 42, 1133–1141.

Wang, D., Jiang, P., Zhang, H. & Yuan, W., 2020. Biochar production and applications in agro and forestry systems: A review. *Science of the Total Environment*, 723, 137775.

Wang, J., Shi, L., Zhai, L., Zhang, H., Wang, S., Zou, J., Shen, Z., Lian, C. & Chen, Y., 2021. Analysis of the long-term effectiveness of biochar immobilization remediation on heavy metal contaminated soil and the potential environmental factors weakening the remediation effect: A review. *Ecotoxicology and Environmental Safety*, 207, 111261.

Wang, L., Cheng, Y., Meftaul, I.M., Luo, F., Kabir, M.A., Doyle, R., Lin, Z. & Naidu, R., 2024. Advancing soil health: Challenges and opportunities in integrating digital imaging, spectroscopy, and machine learning for bioindicator analysis. *Analytical Chemistry*, 96, 8109–8123.

Wasaya, A., Abbas, T., Yasir, T.A., Sarwar, N., Aziz, A., Javaid, M.M. & Akram, S., 2021. Mitigating drought stress in sunflower (*Helianthus annuus* L.) through exogenous application of β-aminobutyric acid. *Journal of Soil Science and Plant Nutrition*, 21, 936–948.

Wong, K.M., Griffiths, M., Moran, A., Johnston, A., Liu, A.E., Sellers, M.A. & Topp, C.N., 2023. Cover crop cultivars and species differ in root traits potentially impacting their selection for ecosystem services. *Plant and Soil*, 500, 279–296.

Wu, S., Konhauser, K.O., Chen, B. & Huang, L., 2023. "Reactive Mineral Sink" drives soil organic matter dynamics and stabilization. *npj Materials Sustainability*, 1, 3.

Yang, H., Joo, J., Hong, E., Park, S.-J., Lee, J. & Lee, C.-G., 2023. Chicken litter-derived catalyst for persulfate activation to remove acetaminophen: An organic-waste-to-wealth strategy. *Chemical Engineering Journal*, 471, 144368.

Yatoo, A.M., Ali, M.N., Baba, Z.A. & Hassan, B., 2021. Sustainable management of diseases and pests in crops by vermicompost and vermicompost tea. A review. *Agronomy for Sustainable Development*, 41, 7.

Yin, T., Yao, Z., Yan, C., Liu, Q., Ding, X. & He, W., 2023. Maize yield reduction is more strongly related to soil moisture fluctuation than soil temperature change under biodegradable film vs plastic film mulching in a semi-arid region of northern China. *Agricultural Water Management*, 287, 108351.

Yin, W., Chai, Q., Guo, Y., Fan, Z., Hu, F., Fan, H., Zhao, C., Yu, A. & Coulter, J.A., 2020. Straw and plastic management regulate air-soil temperature amplitude and wetting-drying alternation in soil to promote intercrop productivity in arid regions. *Field Crops Research*, 249, 107758.

Zafar, S., Afzal, H., Ijaz, A., Mahmood, A., Ayub, A., Nayab, A., Hussain, S., Maqsood, U.-H., Sabir, M.A. & Zulfiqar, U., 2023. Cotton and drought stress: An updated overview for improving stress tolerance. *South African Journal of Botany*, 161, 258–268.

Zaib, M., Zeeshan, A., Aslam, S., Bano, S., Ilyas, A., Abbas, Z., Nazar, A. & Mumtaz, S., 2023. Drought stress and plants production: A review with future prospects. *International Journal of Scientific Research and Engineering Development*, 6, 1278–1293.

Zhang, P., Yuan, Y., Zhang, J., Wen, T., Wang, H., Qu, C., Tan, W., Xi, B., Hui, K. & Tang, J., 2023. Specific response of soil properties to microplastics pollution: A review. *Environmental Research*, 232, 116427.

23 Regenerative Agriculture

*Ghashia Ameen, Muaz Ameen, Adeel Shahid, Athar Mahmood, Muhammad Bilal Goheer, and Muhammad Zubair**

23.1 INTRODUCTION

Regenerative agriculture (RA) is a commonly discussed subject in conversations regarding the restructuring of the food system. Since the 1970s, Western farmers and researchers have been using this term, and it has gained popularity in recent years. Despite the fact that the term "regenerative agriculture" has been known for a while, its methods have not been given an accepted scientific meaning (Gordon *et al.*, 2023). However, a number of techniques and themes have become commonplace in the scientific community due to the rising usage and adoption of the term, and it seems that some scientific knowledge is growing. Grover *et al.* (2024) came up with the following definition after reviewing the literature in an effort to define the term: RA aims to preserve and improve soil health in order to provide agricultural areas with a variety of supporting services that go beyond the environmental to include social and economic aspects of sustainable food production. RA experts consider reducing mechanical disturbance of the soil as a crucial method for achieving soil health. This adaptation is mainly associated with soil inversion by tillage, although compaction by farm machines might also contribute (Tanveer *et al.*, 2013; Luján Soto *et al.*, 2021). Full inversion is often practiced in the mainstream agricultural context to subdue the soil and submerge present stays. Followed by one or more passes from a cultivator that creates a fine seedbed, or "tilth", for the next crop to be sown.

On the other hand, regenerative farmers either drill straight into uncultivated soil (also called "no-till") or very lightly till the top layer of the soil (also called "min-till" or reduced tillage). These techniques can enhance soil drainage and structure while lowering soil erosion. Farmers minimize soil disturbance in this way, which also lowers the rate at which soil carbon inputs are broken down by microbes (Montgomery, 2021; Abdullah *et al.*, 2022). Although there is considerable debate over the relationship between tillage and soil carbon concentration, as we cover in the Science and Uncertainties section, this may increase the organic content of the soil. Available data indicate that tillage practices alone do not significantly increase soil carbon content; instead, they merely shift the carbon in the soil to different parts of the depth profile. Organic inputs, such as manure and cover crops, play a major role in determining soil carbon gains, while tillage only temporarily lowers carbon levels in comparison to no-tillage regimes, with no significant long-term impact on soil organic content (Tanveer *et al.*, 2012; Nazir *et al.*, 2024). Weed accumulation and higher emissions of soil nitrous oxide in wet weather are additional drawbacks of decreased or no tillage.

RA is also suggested as a remedy to achieve sustainable food systems. The global food system causes 25% of human greenhouse gas (GHG) emissions, roughly one-third of terrestrial acidification, and most of the waterbody eutrophication (Schreefel *et al.*, 2020). If our food system continues to use synthetic pesticides, fertilizers, and fossil fuels and generate food debris, it seems probable that the planet's carrying capacity will be exceeded. Hence, the primary obstacle faced by mankind pertains to the production of an adequate quantity of secure and nourishing foodstuff to accommodate an expanding population while adhering to the Earth's carrying capability (Giller *et al.*, 2021).

* Corresponding author: dh24024@stu.yzu.edu.cn

DOI: 10.1201/9781003498698-23

RA has emerged as a promising alternative in producing food in literature to address these issues. RA has been suggested as an alternate method of food production that might potentially have beneficial effects on the environment and society (Newton *et al.*, 2020). The fundamental objective of RA is to boost soil health or recover severely damaged soil, therefore fostering a symbiotic relationship that improves water quality, vegetation, and food production (Tanveer *et al.*, 2013; Rhodes, 2017). Several other names were often used in close proximity to the concept of "regenerative agriculture", such as "biodynamic agriculture", "agroecological farming", "organic regenerative agriculture", "green agriculture", "alternative agriculture", "nature inclusive farming", "conservation agriculture", "carbon farming", and "sustainable agriculture" (Newton *et al.*, 2020).

The main objective of RA is to restore or regenerate soil or return soil health and ecosystem health. This model often employs methodologies that allow for the best use of natural resources, but with an emphasis on improving soil health, increasing biodiversity, and reducing carbon emissions. The method is guided by five principles elaborated in Figure 23.1, which are as follows:

1. Reduce the level of disturbance of the soil
2. Providing soil cover all year round
3. Increase how long live plants and roots remain in the soil
4. Encourage biodiversity
5. Integrate animals (Miller-Klugesherz and Sanderson, 2023).

These practices can minimize soil erosion and degradation, help to restore topsoil and fertility, supply adequate nutrients for crops with little external fertilizers, yield vigorous and productive crops with low weed and pest infestations, minimize GHG emissions, maximize farmers' economic yields, and enhance global food security (Khangura *et al.*, 2023; Ameen *et al.*, 2023a).

RA is not a technology but rather a management strategy with a focus on ecologic components to regenerate and sustain soil health, increase soil organic matter (SOM), increase the soil's carbon storage capacity, and decrease usage of synthetic fertilizers and pesticides. Examples include integrated crop management methods, such as the use of integrated pest management principles, conservation-oriented management strategies (e.g., conservation or no-till practices), and improved utilization of fungal-friendly farming methods (Omomowo and Babalola, 2019; Mukhtiar *et al.*, 2024). Appropriate crop rotations, especially rotations focusing on keeping and replacing nutrients in the soil (for example, nitrogen-fixing plant groups in crop rotations), increase climate resilience and water quality, as well as soil health. These practices enhance organic matter retention and build-up, increase water-holding capacity, promote a diversity of bioactive and efficient microbiomes to

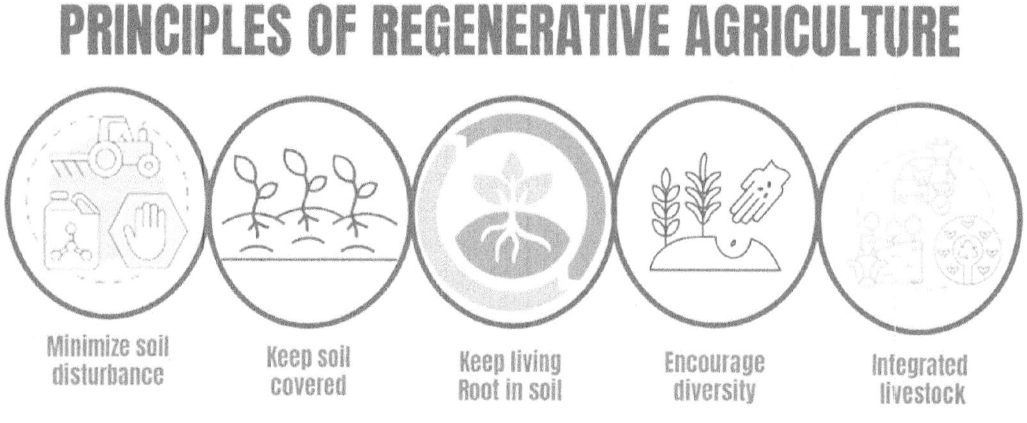

FIGURE 23.1 Principles of regenerative agriculture.

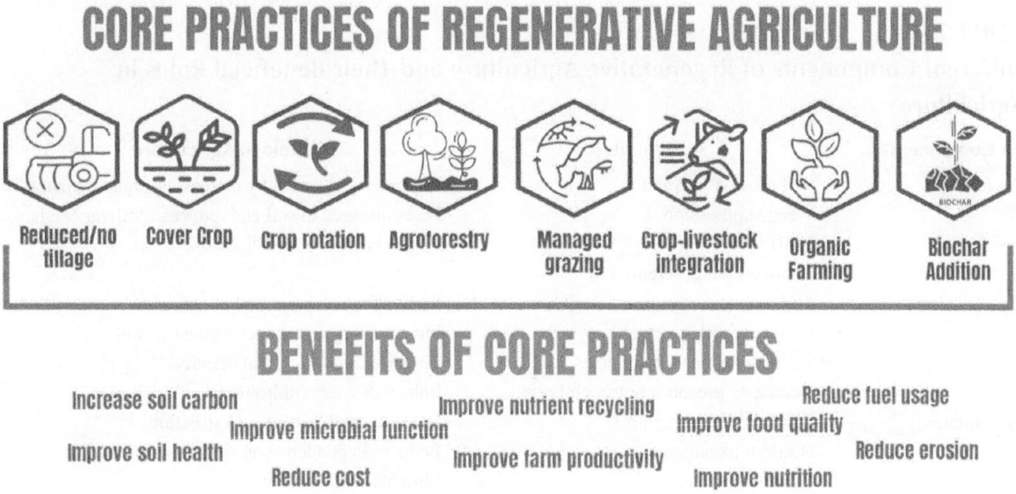

FIGURE 23.2 Core practices and their benefits.

process and immobilize nutrients, and lead to net carbon sequestration (Ahmed *et al.*, 2019; Ameen *et al.*, 2024).

Furthermore, both organic and inorganic RA systems promote integrated management practices for crops, soil, and livestock (Figure 23.2). These practices include the following:

1. Use of weed-competitive and weed-resistant varieties
2. Employing botanically diverse rotations that include legume crops that fix nitrogen
3. The use of regular applications of animal manure or composts from organic waste
4. The establishment and preservation of areas of biodiversity on farms (also referred to as "green infrastructure")
5. The use of minimum tillage
6. The reduction of the uncovered soil by vegetation (for example, through the utilization of inter-, companion, and cover crops)
7. The utilization of crop and outdoor grazing- or foraging-based livestock production systems (Table 23.1) (Rempelos *et al.*, 2021; Holka *et al.*, 2022; Ozlu *et al.*, 2022; Rempelos *et al.*, 2023).

23.2 BENEFICIAL ASPECTS CORE PRACTICE

RA and ecologically intensive farming practices advocate for the utilization of natural resources instead of artificial inputs such as fertilizers and pesticides (Muhie, 2022). The growing realization that sustainable solutions alone aren't long-term sustainable unless they also include inbuilt regeneration capabilities in agriculture (Rhodes, 2017; Mahmood *et al.*, 2021). The growing adoption of regenerative farming and ecological techniques in agriculture highlights positive aspects toward establishing sustainable food systems (Shahmohamadloo *et al.*, 2022). Benefits of different regenerative agricultural practices are discussed below.

23.2.1 Cover Crops

Cover crops refer to crops that are intentionally cultivated during times of regular crop production to improve certain ecological functions. Cover crops provide a variety of well-demonstrated advantages to both agricultural practitioners and the natural surroundings. Nevertheless, it is important

TABLE 23.1
Different Components of Regenerative Agriculture and Their Beneficial Roles in Agriculture

RA Components	Benefits	Role in Agriculture
Cover crops	Soil erosion prevention Weed suppression Nutrient cycling Improves water retention	Improves soil health by increasing organic matter Fixes nitrogen in soil and reduces fertilizer needs Suppresses pests and diseases
Conservation tillage	Soil moisture retention Preserves soil structure Reduces fuel and labor costs Decreases greenhouse gas emissions	Reduces soil erosion and compaction Increases organic matter content in soil Enhances soil microbial activity Enhances water infiltration
Crop rotation	Pest and disease control Nutrient management Weed suppression Improved soil structure and health Enhanced resilience to climate variability	Improves soil fertility and structure Reduces dependence on synthetic inputs Mitigates soil erosion Increases biodiversity on the farm
No-till farming	Reduced soil erosion Increased soil moisture retention Improved soil health Enhanced carbon sequestration Reduced greenhouse gas emissions	Preserves soil structure and organic matter Enhances water infiltration Reduces fuel and labor costs Minimizes disturbance to soil ecosystems
Agroforestry	Soil erosion prevention Carbon sequestration Improved water retention Diversification of farm income sources Nutrient cycling	Provides shade and windbreaks Enhances biodiversity on the farm Provides additional income streams (e.g., fruits, nuts) Increases wildlife habitat
Livestock integration	Nutrient cycling Soil aeration and compaction control Diversified farm income sources Improved soil structure Reduced dependency on synthetic inputs	Natural fertilization of soil Weed and pest control through grazing Utilizes marginal lands Enhances biodiversity

to note that no individual species can provide all of these advantages, which is why planting combinations are becoming more and more popular. Cover crops are widely recognized as a significant agroecological approach in the context of field crop cultivation and the maintenance of soil fertility. Cover crops are known to increase residual plant coverage on the soil surface, minimize disturbance and erosion of the soil, improve soil structure and water-stable aggregates in the soil, manage pests and weeds, promote efficient water and nutrient use in crop production systems, and enhance the overall productivity of agroecosystems and long-term C storage. This observed result can increase subsequent economic crop use efficiency (Adetunji *et al.*, 2020; Chapagain *et al.*, 2020).

23.2.2 Crops Rotation

Crop rotation involves planting many crops in a specific land in a series of growth or planting cycles (Arriaga *et al.*, 2017). Research studies have indicated that through optimizing water channels, enhancing land quality, and improving plant-microbe interactions, crop rotation techniques can significantly increase the resistance of crops against climate changes. Crop rotation with something more diverse has been shown to keep soil healthy or reduce herbivore/weed/disease pressure. Diverse crop rotation has the potential to mitigate the negative impact of climate and, in particular, heat stress and drought on maize and other crop and is likely to enhance the resilience of maize and other crops to extreme weather events such as soil moisture stress due to drought (Bowles *et al.*, 2020; Chadha *et al.*, 2020).

23.2.3 Perennials and Agroforestry

Agroforestry is an alternative form of agriculture in which farmers deliberately crop trees and bushes with their crops and livestock. This has led to a system that is ecologically advantageous, such as contributing to soil health, improving economic performance, and offering extra social benefits (Elevitch *et al.*, 2018). Agroforestry systems, defined as the purposeful inclusion of woody perennials into agricultural spaces, have been found to increase carbon sequestration while reducing GHG emissions relative to common agroecosystems (Gross *et al.*, 2022). Agroforestry has four major environmental benefits:

1. Potential for climate change mitigation through carbon sequestration
2. The preservation of biodiversity
3. The enhancement of soil health
4. The improvement of air and water quality

Due to their economic benefits and their environmental compensations, farmers around the world are keen to adopt several crops inside of agroforestry systems (Aziz *et al.*, 2018; Jose, 2019). Studies have shown that the adoption of agroforestry as an alternative to traditional agricultural methods has been demonstrated to bring about up to 40% higher soil organic carbon (SOC) stocks in the upper 30 cm of soil and 34% higher within the whole 100 cm (De Stefano and Jacobson, 2018).

23.2.4 Reduce or No-Tillage Practices

The search for sustainable crop production methods has been encouraged by inappropriate agricultural practices that deplete soil resources and reduce crop yields (Karayel and Šarauskis, 2019). The worldwide efforts among scientists involve the study of sustainable agricultural production strategies, including no-tillage agricultural systems, referred to as zero tillage or conservation agriculture (Figure 23.3). The worldwide surge in the acceptance and extension of land area under no-tillage may be attributed to countless advantages linked to this agricultural methodology. The advantages of these modern practices primarily stem from three key principles: the mitigation of soil disturbance, the enhancement of soil cover via the use of agricultural leftovers, and the promotion of species diversity via crop rotation. Consequently, enhancing the physio-chemical characteristics of soil, including SOC, total porosity, and water holding capacity, among other factors, adheres to these principles within the context of RA technology (Omara *et al.*, 2019).

23.2.5 Managed Grazing

Domestic cattle have been accused of having negative impacts on both the global environment and human well-being. Nevertheless, when properly supervised, these cattle may have a substantial impact on mitigating environmental impacts resulting from human mismanagement and negligence. Farmers are currently using alternative methods compared to the continuous grazing method shown here, where livestock are kept grazing the same places for long periods of time (Teague and Kreuter, 2020). The objective of regenerative grazing management (ReGM) is to mimic the natural dynamics of grazing to restore soil degradation and sustainable livestock production while encouraging biodiversity at the same time. Regenerative grazing, which comprises integrated planned grazing and related practices, rotates closely populated herds of livestock through a variety of paddocks in a flexible stocking method over a short period of time. This rotation allows for a consistent

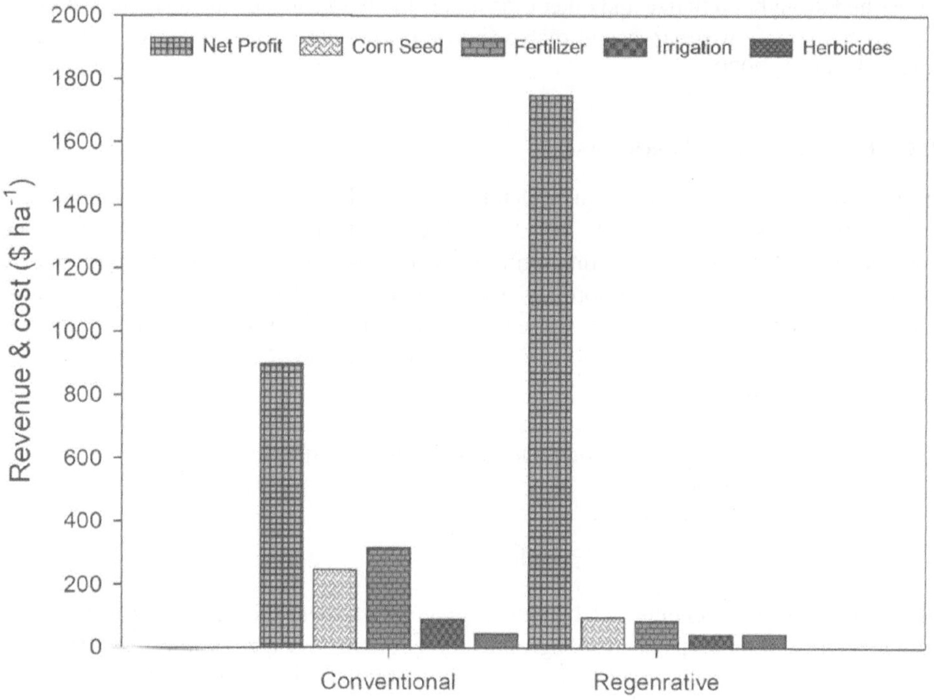

FIGURE 23.3 Regenerative agriculture natural resource conservation profitably. (Adapted and modified from LaCanne and Lundgren, 2018.)

level of uniform defoliation to the plant due to the brief grazing intervals. Long rest periods are also included to further encourage regrowth. The focused "hoof action" of herds in ReGM is considered crucial for the restoration of soils and the provision of ecological services. The presence of ReGM in soils has been shown to result in enhanced microbial bioactivity, fungal-bacterial biomass, functional diversity, and more diverse communities of microarthropods and macrofauna. The response of vegetation to ReGM exhibits variability, characterized by variations in overall plant variety, forage grass richness, and invasive species. Grass species tend to be preferred, whereas shrubs and forbs may experience depletion due to the mechanical impact of hooves. Trampling also diminishes the population of arthropods by modifying the structure of plants but also provides a good environment and sustenance for a limited number of species, such as dung beetles (Morris, 2021).

23.2.6 Crop Livestock Integration

Integrated crop-livestock systems, which are diversified systems, facilitate ecological interactions among various components of a system (such as crops, grasslands, and animals) in both space and time. These systems enable farmers to minimize input usage by implementing two strategies: (1) utilizing organic fertilization derived from livestock waste and (2) implementing diversified crop-grassland rotations for animal feed. Optimal integration, when personalized to specific local circumstances, enhances nutrient cycling by reestablishing a connection between nitrogen and carbon cycles. Additionally, it has the potential to enhance economic efficiency by reducing the production costs and mitigating the risks associated with market fluctuations. Nevertheless, the primary limitations of integrating on-farm operations are associated with the scarcity of available farm labor and a decline in the expertise and understanding needed to maximize the efficiency of both crop and animal sub-systems. In replacement of on-farm integration, it is proposed that integration might be structured at broader sizes outside the limitations of individual farms, facilitated by collaboration between specialized livestock and agricultural farms. While large organizations can achieve some benefits of on-farm integration, these benefits are significantly reduced for individual farms within the organization. The complexity of crop rotations, the skills needed, and the infrastructure required on each farm are all much less impacted compared to a single farm executing full on-farm integration (Asai *et al.*, 2018).

23.2.7 Organic Farming

Agriculture extensively uses chemical fertilizers, insecticides, and fungicides to enhance agricultural productivity. Most of the substances used are synthetic in nature, and their overutilization leads to adverse environmental consequences and human health issues. Presently, several nations are actively striving to decrease the utilization of agrochemicals. Organic agriculture is now gaining popularity as a viable and sustainable solution for conventional agricultural practices. This approach involves the use of ecologically conscious techniques, including the utilization of organic fertilizers derived from plant and animal waste, as well as insecticides formulated using plant extracts and microbiological agents (Durán-Lara *et al.*, 2020; Wasaya *et al.*, 2021; Mukhtiar *et al.*, 2023).

Crop residues refer to non-photosynthetic plant matter that remains on cultivated soils after the harvest of crops. These wastes can be used in the form of green manure, which can potentially increase soil nitrogen through biological nitrogen fixation. This process could also help improve the nitrogen supply for subsequent crops so that the dependence on the application of chemical fertilizers can be reduced (Waheed *et al.*, 2019; Zhou *et al.*, 2020). Uric acid, the main nitrogen component of bird manure (bird guano). This chemical type of fertilizer is key, cheap, and widely available. Cow dung is a type of animal dung, or another name for bovine manure, that can be found as a supplemental fertilizer used within agriculture, particularly horticulture (Das *et al.*, 2020). All essential macronutrients for fertilizer production are available in urine (Durán-Lara *et al.*, 2020).

Composting is an economically feasible, effective method for solid organic waste treatment since it is the bio-oxidative decomposition of organic matter. Composting using cattle dung has numerous agricultural benefits, such as mass and moisture reduction, disease suppression, weed seed destruction, and the production of stable and dispersible organic matter serving as a cost-effective alternative to synthetic fertilizers (Bian *et al.*, 2019; Pai *et al.*, 2019).

Over the decades, pesticide application has been so consistent and aggressive that some insects have developed resistance to them. Plant essential oils (EOs) have gained attention as alternative biopesticides instead of synthetic pesticides. Plant-based pesticides have high volatilization, short half-lives and lower toxicity to vertebrates that make them less corrosive to the environment and animals (Assadpour *et al.*, 2023).

23.2.8 Addition of Biochar

Biochar is frequently referred to as "black gold". Biochar is produced by subjecting organic waste materials, such as agricultural straw, chicken manure, and municipal sludge, to high temperatures and oxygen-deprived conditions. This process results in a remarkable improvement of the aromatic characteristics and structural stability of biochar. This process allows for a reduction of pollution and the most efficient use of each resource. Many studies indicate that biochar has large surface areas, lower bulk densities, charge densities, stable porous structures, and increased organic carbon concentrations (Nguyen *et al.*, 2019; Zafar *et al.*, 2023). The high surface area of biochar can reduce soil bulk density and increase water retention capacity in coarse-textured soil.

Biochar usage would exert a positive influence on soil chemical and biological properties, including pH and cation exchange capacity (CEC). In addition, biochar contains easily degradable carbon. Easily degradable carbon mineralizes and is leached in the soil and mineralized to be used by the microorganisms. So, biochar application can improve soil enzymes, microbial biomass, and soil biomass, which may result in increasing the availability of soil nutrients (Ullah *et al.*, 2013;

Gull *et al.*, 2023). The carbon contained within biochar can also be used to create a long-term carbon sink that helps retain carbon in the soil and reduce GHGs. It could also slow the drop in food production caused by global warming. At the same time, biochar could improve soil quality, increase crop yield in agricultural areas, and provide more ecological and economic benefits (Zhang *et al.*, 2021; Rasheed *et al.*, 2022a).

23.3 POTENTIAL FOR CARBON SEQUESTRATION IN REGENERATIVE AGRICULTURE

The ability of soils to store carbon makes them the second-highest carbon sink on Earth, holding twice as much of the gas as both the atmosphere and vegetation. This ability is influenced by land management practices, soil mineralogy, environmental factors, and soil type. Carbon sequestration occurs when the atmospheric carbon dioxide, as an output of various ecosystem services such as photosynthesis, respiration, and decomposition, enters the soil (Ramesh *et al.*, 2019). SOM is a major component of soil that aids in maintaining healthy ecosystem function and can induce resilient soil structure and productivity, and their relationships are strongly correlated with SOC levels. Generally, plant leaf litter and root biomass control SOC accumulation rate (Hong *et al.*, 2018). Carbon is also accumulated by exchanging carbon molecules between soil bacteria and plants. While decomposition and respiration are natural processes that result in carbon loss from soil ecosystems, land degradation, nutrient leaching, and erosion exacerbate these losses.

Many agricultural areas have lost their natural capacity for soil carbon sequestration as a result of the widespread use of conventional agricultural practices, such as intense tillage, prolonged monoculture, and chemical fertilizers. Around 430 million acres of land have been affected by clearing, harvesting, degradation, and abandonment. It is estimated that cultivated soils have lost between 50 and 70% of their original, pre-agricultural carbon content worldwide (Abdelrahman, 2023). Ecological succession, the process of the environment rebounding from a disturbance, has a longer time horizon for mitigating climate change and can take up to 100 years. Despite these concerning statistics, restoring agricultural soil health has the potential to be an efficient way to sequester carbon from the soil, according to growing research on climate change mitigation strategies and agricultural sustainability, specifically highlighting the possibility that grasslands and agricultural lands could contribute the most to soil carbon sequestration (Wolff *et al.*, 2018). The analysis indicates that if sustainable land management methods are applied globally, these areas could contribute 47%, or 2.3 Gt of CO_2/year, of the entire mitigation of climate change (Rasheed *et al.*, 2022b; McCauley and Barlow, 2023). The findings of Paustian *et al.* (2019) prompted organizations, businesses, and governments to understand the potential of soil and land management practices to help reduce carbon emissions. Thus, the potential for soil to mitigate climate change by soil carbon sequestration has been a focal point of interest in sustainable agriculture. Carbon sequestration in agroecosystems has the capacity to improve soil health and create more fruitful landscapes as well as help reduce atmospheric carbon dioxide concentrations (Nazir *et al.*, 2024). Land stewards are now more motivated to develop sustainable soil-focused strategies rather than focusing only on conventional land management techniques because they recognize the critical role that healthy soil plays in agriculture. Thus, the potential for soil to mitigate climate change by soil carbon sequestration has been a focal point of interest in sustainable agriculture. Carbon sequestration in agroecosystems has the capacity to improve soil health and create more fruitful landscapes as well as help reduce atmospheric carbon dioxide concentrations (M. Tahat *et al.*, 2020). As a result, RA presents a potential solution to resolve the multiple crises related to climate and biodiversity as sustainable activities are quickly marketed globally.

Land management techniques applied in industrial and agricultural sectors of production are primarily responsible for the depletion of carbon stock from earth and include using vast quantities of fossil fuel to power an assortment of heavy equipment as well as carbon extraction systems, including monocropping, extended fallow periods, and mechanized tillage. These activities lead to higher rates of microbial oxidation and degradation of existing SOM (Kenne and Kloot, 2019; Hassan *et al.*, 2022). Soil carbon sequestration addresses the problem of climate change by taking carbon from the atmosphere of atmospheric carbon dioxide and depositing it. These days, researchers are taking another look at the potential of the environment to have a constructive effect. This recognition is evident in several frameworks, such as agroecology, RA, ecosystem services, and sustainability theory (Paustian *et al.*, 2020; Ameen *et al.*, 2023b). Regenerative strategies, such as agroforestry, crop diversification, crop rotation, and native restoration initiatives, have been identified as having a favorable correlation between plant variety and carbon sequestration. Agroforestry has advantages in terms of carbon sequestration by means of robust, deep-rooted systems and the storage of carbon within biomass. The use of crop diversification and rotation strategies facilitates the return of nutrients to the soil and promotes the variety of soil microorganisms (McCauley and Barlow, 2023).

A sustainable food system can be ensured by adhering to the numerous practices and concepts delineated by the RA method. Activities that promote plant diversity have shown the ability to address these problems, given the knowledge of the detrimental effects of traditional agricultural methods and the ensuing ramifications on soil carbon storage. The results of Chen *et al.'s* (2018) study provide strong evidence that greater plant variety and soil carbon sequestration are positively correlated. In China's ecological restoration areas and degraded grasslands, it has been discovered that a greater variety of plants can sustainably replenish soil carbon stores. Two studies found that agroforestry systems that incorporate trees into agricultural landscapes have the highest rates of carbon storage. Higher soil carbon storage was obtained in crop diversification/rotation studies through the use of perennials, polycultures, and cover crops (McCauley and Barlow, 2023). Along with benefits to general soil health, biodiversity, cost-effectiveness, and yield potential, this evidence also shows that agricultural systems have many other advantages that go well beyond carbon sequestration using RA.

23.4 IMPROVED WATER QUALITY AND AVAILABILITY

Regenerative perennial agriculture involves the use of crop management techniques that effectively utilize the distinctive capabilities of perennial crops to provide significant ecological benefits, such as improved water quality and animal habitat. Perennial crops provide enhanced water quality advantages as a result of their extensive and plentiful root systems, which minimize fertilizer runoff and optimize the use of water and nutrients. Agroforestry systems have the capacity to alleviate surface runoff, soil erosion, losses of organic carbon, and the subsequent depletion of nutrients. Herbicide, pesticide, and other pollutant losses are also reduced by 49% on average as a consequence of their companion activities. The leaching of nutrients may be effectively mitigated in agricultural areas characterized by a minimal tree cover of 30% (Liu *et al.*, 2021). The use of diversified perennial systems has the potential to enhance biodiversity, enhance natural pest management, increase crop yields, and boost system resilience (Scott *et al.*, 2022). According to the estimates, nearly a 1% rise in organic matter has the potential to increase water storage capacity by up to 16,000 gallons per acre (White, 2020).

23.5 ECONOMIC VIABILITY AND FINANCIAL TOOLS FOR REGENERATIVE FARMERS

The adoption of RA systems is affected by economic factors, such as real or perceived risks (Which Risks?), to reduce the adverse impacts of traditional agricultural production systems on soil health and ecological services. While traditional agriculture may provide higher productivity and financial gains in some situations, RA systems have the potential to provide similar or even greater economic benefits and increased value per unit of land. The economic benefits of RA systems are primarily attributed to their low-input nature, particularly in terms of agrochemicals. These systems can incorporate additional organic materials, offer a high potential for soil nutrient supply, reduce the need for frequent irrigations by improving soil water-holding capacities, and minimize material loss. As a result, they contribute to the creation of balanced ecological services and improved food quality. Conversely, traditional agricultural systems are distinguished by substantial expenses for production inputs, such as machinery, personnel, chemicals, fuel, and tillage supplies. Producers often overlook these input costs, prioritizing yield exclusively, so creating a perception that a traditional system results in superior performance as compared to the RA system (Al-Kaisi and Lal, 2020).

Perennial crops provide economic advantages to growers. Presently, many perennial crops are specialty products that command premium pricing, and certain perennial systems may outperform competitive annual systems in terms of earnings. Producers have the opportunity to maximize the growing consumer demand for nutritious food items, such as tree nuts, fruits, and berries derived from shrubs. In addition, the capacity of some perennial grain crops to provide a variety of products mitigates economic risk in comparison to certain annual crops. Perennial grain crops are now less economically advantageous. This is mostly due to the short-term impact of cropping-system design and the need for repeated production from these crops. In the long run, the potential for yield enhancement via plant breeding is also a significant factor (Wolz and Delucia, 2019; Law *et al.*, 2022).

23.6 CIRCULAR ECONOMY

The circular economy (CE) approach is one that could potentially benefit sustainable, restorative, and regenerative agricultural practices, especially in the current environment of scarcity, global climate change, environmental degradation, and increasing food demands. According to studies, agricultural output must increase by 70% globally to satisfy food demand by 2050. To achieve this goal, there are two possible pathways in a typical commercial context: (1) the increase of cultivated land, occupying approximately 37% of the total available surface area, and (2) the intensification of production on the previously cultivated surface associated with a 38% overall expansion of cultivated area and a 53% increase of global water use (Aznar-Sánchez *et al.*, 2020a, 2020b).

Approximately 90% of the environmental impacts from land and water—water scarcity and biodiversity loss—can be traced back to the agricultural sector. In addition, it is also a significant contributor to human toxicity of agricultural workers to pesticides (Macarthur and Heading, 2019). In conclusion, it is important to create more research initiatives in order to identify methods for increasing the efficiency and sustainability of resource-use in agricultural production through the application of the principles of the CE. Building a circular food system is the key because the supply chain connected to food involves the vast outlay of energy, water, and GHGs. Determining the extent to which food production systems are circular is the first step toward achieving a circular food-production system (Peña *et al.*, 2021).

It places great importance on the circular food system, especially because of the absorbing GHG emissions, water, and energy within the relevant food supply chain. The evaluation of the circularity of the food production systems (De Boer and Van Ittersum, 2018).

Modern civilization has embraced precision agriculture, permaculture, RA, organic farming, and sustainable soil management have all entered into the hands of modern civilization. All things just lead back to the CE, from RA, sustainable soil management, and organic farming (Velasco-Muñoz *et al.*, 2021; Batlles-Delafuente *et al.*, 2022).

23.7 CHALLENGES AND FUTURE PERSPECTIVES

Exclusively depending on conventional farming techniques will not be enough to guarantee future worldwide food security. An all-encompassing strategy is much needed that incorporates insights from other fields such as statistics and computer science. An example is the strategic deployment of advanced technology and sensors that can predict potential issues but also gather precise geographic data. There are many different types of technologies and sensors with different characteristics that have strong potential to support the building of a resilient and sustainable food system (McCauley, 2020). Additionally, there are yield and weed mapping sensors that are capable of detecting changes to the production environment, allowing farmers to make timely assessments. Alternative sensors to monitor how a plant responds to climatic variations (temperature sensors, ground sensors) are something interesting that needs to be prioritized. Data from sensors can provide quantifiable insights and predictive analytics for the early detection of issues, such as pest or weed infestations. Remote pest activity monitoring sites that utilize passive infrared (PIR) detection in the pest traps have been adopted to make the identification of certain pests a much simpler task.

This method has the added benefit of reducing the risk of secondary poisoning when bait is used, therefore restricting potential fatalities to species that are not intended (Parsons *et al.*, 2020). Although the model and climate-smart agricultural practices offer high benefits, they might be difficult to adopt, especially in small-to-medium-scale farming systems with fewer available farm resources. With respect to making the best use of the smart agricultural services, two main criteria are set as (1) dimensions of farm area and (2) farm production cost-benefit analysis (Mclennon *et al.*, 2021). Although RA has promising opportunities, various constraints have prevented farmers and ranchers from fully adopting it. Many farmers do not have the proper skills and knowledge to apply RA practices effectively. The introduction of RA might also result in further spending on new gear, training staff, and changing production patterns to accommodate RA principles (Abson *et al.*, 2019). It can be difficult to comprehend and understand how to operate the system; it might take years (Gosnell *et al.*, 2020).

23.8 CONCLUSION

These feature the explosive growth of RA-practitioner communities and the academic field of antioxidant research. The environmental dimension of sustainability is a major concern for RA, as you have seen in this chapter. These include issues such as soil health improvement and regeneration, climate change mitigation, water quality and quantity enhancement, nutrient cycling enhancement, and resource use efficiency enhancement. These topics help improve food security through their participation in providing inputs (food, fiber, feed) as well as regulating processes like climate control, water purification, and soil erosion. They also provide important ecological processes like nutrient cycling and soil formation.

Moreover, RA has been found to have a socioeconomic dimension that improves economic growth and human health in ways linked to cultural ecosystem services. Unfortunately, there is presently no implementation framework for this socioeconomic component, and it is based on conflicting goals. To address environmental, social, and economic concerns in sustainable food production, a working definition of RA is proposed. This approach prioritizes soil conservation as a foundation for restoring and enhancing the various benefits that ecosystems provide, including food, clean water, and habitat. In order to facilitate the shift to RA, this chapter helps to define the term consistently; as a result, RA should be measured using indicators and benchmarks.

LIST OF ABBREVIATIONS

CEC cation exchange capacity
GHG greenhouse gas
RA regenerative agriculture
ReGM regenerative grazing management
SOM soil organic matter
SOC soil organic carbon

REFERENCES

Abdelrahman, M.AE., 2023. An overview of land degradation, desertification and sustainable land management using GIS and remote sensing applications. *Rendiconti Lincei. Scienze Fisiche e Naturali*, 34, 767–808.

Abdullah, Mahmood, A., Bibi, S., Naqve, M., Javaid, M.M., Zia, M.A., Jabbar, A., Ud-Din, W., Attia, K.A. & Khan, N., 2022. Physiological, biochemical, and yield responses of linseed (*Linum usitatissimum* L.) in α-tocopherol-mediated alleviation of salinity stress. *Frontiers in Plant Science*, 13, 867172.

Abson, D.J., Sherren, K. & Fischer, J., 2019. The Resilience of Australian Agricultural Landscapes Characterized by Land Sparing versus Land Sharing. *In Agricultural Resilience: Perspectives from Ecology and Economics*. Ecological Reviews (British Ecological Society), Cambridge University Press, United Kingdom, 232–252. DOI:10.1017/9781107705555.012

Adetunji, A.T., Ncube, B., Mulidzi, R. & Lewu, F.B., 2020. Management impact and benefit of cover crops on soil quality: A review. *Soil and Tillage Research*, 204, 104717.

Ahmed, A.A.Q., Odelade, K.A. & Babalola, O.O., 2019. Microbial Inoculants for Improving Carbon Sequestration in Agroecosystems to Mitigate Climate Change. In Leal Filho, W. (eds) *Handbook of Climate Change Resilience*, Springer Nature, Switzerland, 1–21.

Al-Kaisi, M.M. & Lal, R., 2020. Aligning science and policy of regenerative agriculture. *Soil Science Society of America Journal*, 84, 1808–1820.

Ameen, M., Mahmood, A., Ahmad, M., Mansoor Javaid, M., Nadeem, M.A., Asif, M., Balal, R.M. & Khan, B.A., 2023a. Impacts of Climate Change on Fruit Physiology and Quality. *In Climate-Resilient Agriculture, Vol 1: Crop Responses and Agroecological Perspectives*. Springer, 93–124.

Ameen, M., Mahmood, A., Sahkoor, A., Zia, M.A. & Ullah, M.S., 2024. The role of endophytes to combat abiotic stress in plants. *Plant Stress*, 12, 100435.

Ameen, M., Zafar, A., Javaid, M.M., Zia, M.A., Mahmood, A., Naqve, M. & Bibi, S., 2023b. Climate-Resilient Technology for Maize Production. *In* M. Hasanuzzaman (ed.) *Climate-Resilient Agriculture, Vol 2: Agro-Biotechnological Advancement for Crop Production*. Springer International Publishing, 157–188.

Arriaga, F.J., Guzman, J. & Lowery, B., 2017. Conventional Agricultural Production Systems and Soil Functions. *In Soil Health and Intensification of Agroecosytems*. Elsevier, 109–125.

Asai, M., Moraine, M., Ryschawy, J., De Wit, J., Hoshide, A.K. & Martin, G., 2018. Critical factors for crop-livestock integration beyond the farm level: A cross-analysis of worldwide case studies. *Land Use Policy*, 73, 184–194.

Assadpour, E., Can Karaça, A., Fasamanesh, M., Mahdavi, S.A., Shariat-Alavi, M., Feng, J., Kharazmi, M.S., Rehman, A. & Jafari, S.M., 2023. Application of essential oils as natural biopesticides; recent advances. *Critical Reviews in Food Science and Nutrition*, 64(19), 1–21.

Aziz, M., Ashraf, M. & Javaid, M.M., 2018. Enhancement in cotton growth and yield using novel growth promoting substances under water limited conditions. *Pakistan Journal of Botany*, 50, 1691–1701.

Aznar-Sánchez, J.A., Mendoza, J.M.F., Ingrao, C., Failla, S., Bezama, A., Nemecek, T. & Gallego-Schmid, A., 2020a. Indicators for circular economy in the agri-food sector. *Resources, Conservation and Recycling*, 163, DOI:10.1016/j.resconrec.2020.105028.

Aznar-Sánchez, J.A., Velasco-Muñoz, J.F., López-Felices, B. & Román-Sánchez, I.M., 2020b. An analysis of global research trends on greenhouse technology: Towards a sustainable agriculture. *International Journal of Environmental Research and Public Health*, 17, 664.

Batlles-Delafuente, A., Abad-Segura, E., González-Zamar, M.-D. & Cortés-García, F.J., 2022. An evolutionary approach on the framework of circular economy applied to agriculture. *Agronomy*, 12, 620.

Bian, B., Hu, X., Zhang, S., Lv, C., Yang, Z., Yang, W. & Zhang, L., 2019. Pilot-scale composting of typical multiple agricultural wastes: Parameter optimization and mechanisms. *Bioresource Technology*, 287, 121482.

Bowles, T.M., Mooshammer, M., Socolar, Y., Calderón, F., Cavigelli, M.A., Culman, S.W., Deen, W., Drury, C.F., Y Garcia, A.G. & Gaudin, A.C., 2020. Long-term evidence shows that crop-rotation diversification increases agricultural resilience to adverse growing conditions in North America. *One Earth*, 2, 284–293.

Chadha, A., Florentine, S., Javaid, M., Welgama, A. & Turville, C., 2020. Influence of elements of climate change on the growth and fecundity of *Datura stramonium*. *Environmental Science and Pollution Research*, 27, 35859–35869.

Chapagain, T., Lee, E.A. & Raizada, M.N., 2020. The potential of multi-species mixtures to diversify cover crop benefits. *Sustainability*, 12, 2058.

Chen, S., Wang, W., Xu, W., Wang, Y., Wan, H., Chen, D., Tang, Z., Tang, X., Zhou, G., Xie, Z., Zhou, D., Shangguan, Z., Huang, J., He, J.-S., Wang, Y., Sheng, J., Tang, L., Li, X., Dong, M., Wu, Y., Wang, Q., Wang, Z., Wu, J., Chapin, F.S. & Bai, Y., 2018. Plant diversity enhances productivity and soil carbon storage. *Proceedings of the National Academy of Sciences*, 115, 4027–4032.

Das, S., Chatterjee, A. & Pal, T.K., 2020. Organic farming in India: A vision towards a healthy nation. *Food Quality and Safety*, 4, 69–76.

De Boer, I.J. & Van Ittersum, M.K., 2018. *Circularity in Agricultural Production*. Wageningen University & Research. https://edepot.wur.nl/470625

De Stefano, A. & Jacobson, M.G., 2018. Soil carbon sequestration in agroforestry systems: A meta-analysis. *Agroforestry Systems*, 92, 285–299.

Durán-Lara, E.F., Valderrama, A. & Marican, A., 2020. Natural organic compounds for application in organic farming. *Agriculture*, 10, 41.

Elevitch, C.R., Mazaroli, D.N. & Ragone, D., 2018. Agroforestry standards for regenerative agriculture. *Sustainability*, 10, 3337.

Giller, K.E., Hijbeek, R., Andersson, J.A. & Sumberg, J., 2021. Regenerative agriculture: An agronomic perspective. *Outlook on Agriculture*, 50, 13–25.

Gordon, E., Davila, F. & Riedy, C., 2023. Regenerative agriculture: A potentially transformative storyline shared by nine discourses. *Sustainability Science*, 18, 1833–1849.

Gosnell, H., Charnley, S. & Stanley, P., 2020. Climate change mitigation as a co-benefit of regenerative ranching: Insights from Australia and the United States. *Interface Focus*, 10, 20200027.

Gross, C.D., Bork, E.W., Carlyle, C.N. & Chang, S.X., 2022. Agroforestry perennials reduce nitrous oxide emissions and their live and dead trees increase ecosystem carbon storage. *Global Change Biology*, 28, 5956–5972.

Grover, D., Mishra, A.K., Rani, P., Kalonia, N., Chaudhary, A. & Sharma, S., 2024. Soil Management in Sustainable Agriculture: Principles and Techniques. *In* K Pavan & Aishwarya (eds.) *Technological Approaches for Climate Smart Agriculture.* Cham: Springer International Publishing, 41–77.

Gull, T., Mahmood, A., Shaheen, C., Javaid, M.M., Zia, M.A., Naqve, M., Bibi, S., Nadeem, M.A., Ameen, M., Nargis, J. & Khan, S.R., 2023. Climate Change and Nutrient Use Efficiency of Plants. In M. Hasanuzzaman (ed.) *Climate-Resilient Agriculture, Vol 1: Crop Responses and Agroecological Perspectives.* Cham: Springer International Publishing, 291–312.

Hassan, M.U., Nawaz, M., Mahmood, A., Shah, A.A., Shah, A.N., Muhammad, F., Batool, M., Rasheed, A., Jaremko, M. & Abdelsalam, N.R., 2022. The role of zinc to mitigate heavy metals toxicity in crops. *Frontiers in Environmental Science*, 10, 990223.

Holka, M., Kowalska, J. & Jakubowska, M., 2022. Reducing carbon footprint of agriculture—Can organic farming help to mitigate climate change? *Agriculture*, 12, 1383.

Hong, Y., Chen, S., Zhang, Y., Chen, Y., Yu, L., Liu, Y., Liu, Y., Cheng, H. & Liu, Y., 2018. Rapid identification of soil organic matter level via visible and near-infrared spectroscopy: Effects of two-dimensional correlation coefficient and extreme learning machine. *Science of the Total Environment*, 644, 1232–1243.

Jose, S., 2019. Environmental Impacts and Benefits of Agroforestry. Oxford University Press, United Kingdom, *In Oxford Research Encyclopedia of Environmental Science.* https://oxfordre.com/environmentalscience/view/10.1093/acrefore/9780199389414.001.0001/acrefore-9780199389414-e-195.

Karayel, D. & Šarauskis, E., 2019. Environmental impact of no-tillage farming. *Environmental Research, Engineering and Management*, 75.

Kenne, G.J. & Kloot, R.W., 2019. The carbon sequestration potential of regenerative farming practices in South Carolina, USA. *American Journal of Climate Change*, 8, 157.

Khangura, R., Ferris, D., Wagg, C. & Bowyer, J., 2023. Regenerative agriculture—A literature review on the practices and mechanisms used to improve soil health. *Sustainability*, 15, 2338.

Lacanne, C.E. & Lundgren, J.G., 2018. Regenerative agriculture: Merging farming and natural resource conservation profitably. *PeerJ*, 6, e4428.

Law, E.P., Wayman, S., Pelzer, C.J., Culman, S.W., Gómez, M.I., Ditommaso, A. & Ryan, M.R., 2022. Multicriteria assessment of the economic and environmental sustainability characteristics of intermediate wheatgrass grown as a dual-purpose grain and forage crop. *Sustainability*, 14, 3548.

Liu, R., Thomas, B.W., Shi, X., Zhang, X., Wang, Z., & Zhang, Y., 2021. Effects of ground cover management on improving water and soil conservation in tree crop systems: A meta-analysis. *CATENA*, 199, 105085. doi:10.1016/j.catena.2020.105085.

Luján Soto, R., Martínez-Mena, M., Cuéllar Padilla, M. & De Vente, J., 2021. Restoring soil quality of woody agroecosystems in Mediterranean drylands through regenerative agriculture. *Agriculture, Ecosystems & Environment*, 306, 107191.

Macarthur, E. & Heading, H., 2019. How the circular economy tackles climate change. *Ellen MacArthur Found*, 1, 1–71.

Mahmood, A., Wang, X., Shahzad, A.N., Fiaz, S., Ali, H., Naqve, M., Javaid, M.M., Mumtaz, S., Naseer, M. & Dong, R., 2021. Perspectives on bioenergy feedstock development in Pakistan: Challenges and opportunities. *Sustainability*, 13, 8438.

McCauley, D., 2020. Cover crops, sensors, and food security: Forward-thinking ideas for the USDA's agriculture innovation agenda. *CSA News*, 65(12), 3–9.

McCauley, K. & Barlow, K., 2023. Regenerative agriculture: Increasing plant diversity and soil carbon sequestration on agricultural landscapes. *SURG Journal*, 15, 6950.

Mclennon, E., Dari, B., Jha, G., Sihi, D. & Kankarla, V., 2021. Regenerative agriculture and integrative permaculture for sustainable and technology driven global food production and security. *Agronomy Journal*, 113, 4541–4559.

Miller-Klugesherz, J.A. & Sanderson, M.R., 2023. Good for the soil, but good for the farmer? Addiction and recovery in transitions to regenerative agriculture. *Journal of Rural Studies*, 103, 103123.

Montgomery, D.R., 2021. 11 – Soil Health and the Revolutionary Potential of Conservation Agriculture. *In* A. Kassam & L. Kassam (eds.) *Rethinking Food and Agriculture*. Woodhead Publishing, 219–229.

Morris, C.D., 2021. How biodiversity-friendly is regenerative grazing? *Frontiers in Ecology and Evolution*, 9, 816374.

M. Tahat, M., M. Alananbeh, K., A. Othman, Y. & I. Leskovar, D., 2020. Soil health and sustainable agriculture. *Sustainability*, 12, 4859.

Muhie, S.H., 2022. Novel approaches and practices to sustainable agriculture. *Journal of Agriculture and Food Research*, 10, 100446.

Mukhtiar, A., Lihong, W., Mahmood, A., Ameen, M., Zia, M.A., Arshad, T., Naqve, M., Wahab, H.A., Rasheed, A. & Asghar, S., 2023. The role of endophytes and rhizobacteria to combat drought stress in wheat. *Notulae Botanicae Horti Agrobotanici Cluj-Napoca*, 51, 13453–13453.

Mukhtiar, A., Mahmood, A., Zia, M.A., Ameen, M., Dong, R., Shoujun, Y., Javaid, M.M., Khan, B.A. & Nadeem, M.A., 2024. Role of biogas slurry to reclaim soil properties providing an eco-friendly approach for crop productivity. *Bioresource Technology Reports*, 25, 101716.

Nazir, M.J., Li, G., Nazir, M.M., Zulfiqar, F., Siddique, K.H.M., Iqbal, B. & Du, D., 2024. Harnessing soil carbon sequestration to address climate change challenges in agriculture. *Soil and Tillage Research*, 237, 105959.

Newton, P., Civita, N., Frankel-Goldwater, L., Bartel, K. & Johns, C., 2020. What is regenerative agriculture? A review of scholar and practitioner definitions based on processes and outcomes. *Frontiers in Sustainable Food Systems*, 4, 577723.

Nguyen, V.-T., Nguyen, T.-B., Chen, C.-W., Hung, C.-M., Chang, J.-H. & Dong, C.-D., 2019. Influence of pyrolysis temperature on polycyclic aromatic hydrocarbons production and tetracycline adsorption behavior of biochar derived from spent coffee ground. *Bioresource Technology*, 284, 197–203.

Omara, P., Aula, L., Eickhoff, E.M., Dhillon, J.S., Lynch, T., Wehmeyer, G.B. & Raun, W., 2019. Influence of no-tillage on soil organic carbon, total soil nitrogen, and winter wheat (*Triticum aestivum* L.) grain yield. *International Journal of Agronomy*, 2019, 1–9.

Omomowo, O.I. & Babalola, O.O., 2019. Bacterial and fungal endophytes: Tiny giants with immense beneficial potential for plant growth and sustainable agricultural productivity. *Microorganisms*, 7, 481.

Ozlu, E., Arriaga, F.J., Bilen, S., Gozukara, G. & Babur, E., 2022. Carbon footprint management by agricultural practices. *Biology*, 11, 1453.

Pai, S., Ai, N. & Zheng, J., 2019. Decentralized community composting feasibility analysis for residential food waste: A Chicago case study. *Sustainable Cities and Society*, 50, 101683.

Parsons, L., Ross, R. & Robert, K., 2020. A survey on wireless sensor network technologies in pest management applications. *SN Applied Sciences*, 2, 28.

Paustian, K., Chenu, C., Conant, R., Cotrufo, F., Lal, R., Smith, P. & Soussana, J.-F., 2020. *Climate Mitigation Potential of Regenerative Agriculture Is Significant*. Regenerative Agriculture Foundation, Princeton University [accessed 2025 Jun 10]. https://searchinger. princeton. edu/sites/g/files/toruqf4701/files/tsearchi/files/paustian_et_al._response_to_wri_soil_carbon_blog_.pdf.

Paustian, K., Collier, S., Baldock, J., Burgess, R., Creque, J., Delonge, M., Dungait, J., Ellert, B., Frank, S., Goddard, T., Govaerts, B., Grundy, M., Henning, M., Izaurralde, R.C., Madaras, M., Mcconkey, B., Porzig, E., Rice, C., Searle, R., Seavy, N., Skalsky, R., Mulhern, W. & Jahn, M., 2019. Quantifying carbon for agricultural soil management: From the current status toward a global soil information system. *Carbon Management*, 10, 567–587.

Peña, C., Civit, B., Gallego-Schmid, A., Druckman, A., Pires, A.C.-, Weidema, B., Mieras, E., Wang, F., Fava, J. & Canals, L.M.I., 2021. Using life cycle assessment to achieve a circular economy. *The International Journal of Life Cycle Assessment*, 26, 215–220.

Ramesh, T., Bolan, N.S., Kirkham, M.B., Wijesekara, H., Kanchikerimath, M., Srinivasa Rao, C., Sandeep, S., Rinklebe, J., Ok, Y.S., Choudhury, B.U., Wang, H., Tang, C., Wang, X., Song, Z. & Freeman Ii, O.W., 2019. Chapter One – Soil Organic Carbon Dynamics: Impact of Land Use Changes and Management Practices: A Review. *In* D.L. Sparks (ed.) *Advances in Agronomy*. Academic Press, 1–107.

Rasheed, A., Barqawi, A.A., Mahmood, A., Nawaz, M., Shah, A.N., Bay, D.H., Alahdal, M.A., Hassan, M.U. & Qari, S.H., 2022a. CRISPR/Cas9 is a powerful tool for precise genome editing of legume crops: A review. *Molecular Biology Reports*, 49, 5595–5609.

Rasheed, A., Raza, A., Jie, H., Mahmood, A., Ma, Y., Zhao, L., Xing, H., Li, L., Hassan, M.U. & Qari, S.H., 2022b. Molecular tools and their applications in developing salt-tolerant soybean (*Glycine max* L.) cultivars. *Bioengineering*, 9, 495.

Rempelos, L., Baranski, M., Wang, J., Adams, T.N., Adebusuyi, K., Beckman, J.J., Brockbank, C.J., Douglas, B.S., Feng, T. & Greenway, J.D., 2021. Integrated soil and crop management in organic agriculture: A logical framework to ensure food quality and human health? *Agronomy*, 11, 2494.

Rempelos, L., Kabourakis, E. & Leifert, C., 2023. Innovative organic and regenerative agricultural production. *Agronomy*, 13, 1344.

Rhodes, C.J., 2017. The imperative for regenerative agriculture. *Science Progress*, 100, 80–129.

Schreefel, L., Schulte, R.P.O., De Boer, I.J.M., Schrijver, A.P. & Van Zanten, H.H.E., 2020. Regenerative agriculture – The soil is the base. *Global Food Security*, 26, 100404.

Scott, E.I., Toensmeier, E., Iutzi, F., Rosenberg, N.A., Lovell, S.T., Jordan, N.R., Peters, T.E., Akwii, E. & Broad Leib, E.M., 2022. Policy pathways for perennial agriculture. *Frontiers in Sustainable Food Systems*, 6, 983398.

Shahmohamadloo, R.S., Febria, C.M., Fraser, E.D. & Sibley, P.K., 2022. The sustainable agriculture imperative: A perspective on the need for an agrosystem approach to meet the United Nations Sustainable Development Goals by 2030. *Integrated Environmental Assessment and Management*, 18, 1199–1205.

Tanveer, A., Arshad, M.S., Ayub, M., Javaid, M.M. & Yaseen, M., 2012. Effect of temperature, light, salinity, drought stress and seeding depth on germination of *Cucumis melo* var. agrestis. *Pakistan Journal of Weed Science Research*, 18.

Tanveer, A., Mumtaz, K., Javaid, M., Chaudhry, M., Balal, R. & Khaliq, A., 2013. Effect of ecological factors on germination of horse purslane (*Trianthema portulacastrum*). *Planta Daninha*, 31, 587–597.

Teague, R. & Kreuter, U., 2020. Managing grazing to restore soil health, ecosystem function, and ecosystem services. *Frontiers in Sustainable Food Systems*, 4, 534187.

Ullah, H., Mahmood, A., Ijaz, M., Tadesse, B. & Honermeier, B., 2013. Evaluation of anise (*Pimpinella anisum* L.) accessions with regard to morphological characteristics, fruit yield, oil contents and composition. *Journal of Medicinal Plants Research*, 7, 2177–2186.

Velasco-Muñoz, J.F., Mendoza, J.M.F., Aznar-Sánchez, J.A. & Gallego-Schmid, A., 2021. Circular economy implementation in the agricultural sector: Definition, strategies and indicators. *Resources, Conservation and Recycling*, 170, 105618.

Waheed, H., Javaid, M.M., Shahid, A., Ali, H.H., Nargis, J. & Mehmood, A., 2019. Impact of foliar-applied Hoagland's nutrient solution on growth and yield of mash bean (*Vigna mungo* L.) under different growth stages. *Journal of Plant Nutrition*, 42, 1133–1141.

Wasaya, A., Abbas, T., Yasir, T.A., Sarwar, N., Aziz, A., Javaid, M.M. & Akram, S., 2021. Mitigating drought stress in sunflower (*Helianthus annuus* L.) through exogenous application of β-aminobutyric acid. *Journal of Soil Science and Plant Nutrition*, 21, 936–948.

White, C., 2020. Why regenerative agriculture? *American Journal of Economics and Sociology*, 79, 799–812.

Wolff, S., Schrammeijer, E.A., Schulp, C.J.E. & Verburg, P.H., 2018. Meeting global land restoration and protection targets: What would the world look like in 2050? *Global Environmental Change*, 52, 259–272.

Wolz, K.J. & Delucia, E.H., 2019. Black walnut alley cropping is economically competitive with row crops in the Midwest USA. *Ecological Applications*, 29, e01829.

Zafar, S., Afzal, H., Ijaz, A., Mahmood, A., Ayub, A., Nayab, A., Hussain, S., Maqsood, U.-H., Sabir, M.A. & Zulfiqar, U., 2023. Cotton and drought stress: An updated overview for improving stress tolerance. *South African Journal of Botany*, 161, 258–268.

Zhang, Y., Wang, J. & Feng, Y., 2021. The effects of biochar addition on soil physicochemical properties: A review. *CATENA*, 202, 105284.

Zhou, G., Gao, S., Lu, Y., Liao, Y., Nie, J. & Cao, W., 2020. Co-incorporation of green manure and rice straw improves rice production, soil chemical, biochemical and microbiological properties in a typical paddy field in southern China. *Soil and Tillage Research*, 197, 104499.

24 Use of Internet of Things for Modern Agriculture

Muaz Ameen, Rizwan Maqbool, Athar Mahmood, and Muhammad Bilal Goheer*

24.1 INTRODUCTION

In 1999, a British futurist named Kevin Ashton first used the term "Internet of Things" (IoT). The IoT is referred to as the "Internet of Things" because it promises a technological environment where a variety of physical objects, or "Things," such as common tools, sensors, and equipment enhanced by computing power and networking capabilities, will be able to play a role, either as widely distributed cooperative swarms of heterogeneous devices or as independently functioning units (Tran-Dang *et al.*, 2020). One industry that is anticipated to be greatly impacted by the developments in the IoT is agriculture (Figure 24.1).

According to forecasts made by the Food and Agriculture Organization (FAO) of the United Nations, there will be 8.6 billion people on the planet by 2025 and 9.6 billion by 2050. In real terms, this implies that global food production needs to rise by 70% by 2050 (Mrabet, 2023). Modernizing and intensifying agricultural operations are necessary due to the significant rise in the world's population as well as the growing demand for high-quality products. Simultaneously, it is imperative to prioritize the efficient utilization of water and other resources (Wasaya *et al.*, 2021).

The British futurist Kevin Ashton came up with the term "Internet of Things" for this technological future in 1999, where "Things"—simple machines, sensors, and devices, given computing power and networking ability—will either function as broad distributed cooperative clusters of devices, or as autonomous things in their own right (Tran-Dang *et al.*, 2020). The food sector will be one of the industries most affected by IoT innovations, according to Future Scenarios on the Internet of Things (2014) by the FAO of the United Nations. By 2025, the world's population will be 8.6 billion and could reach 9.6 billion by 2050 (Mrabet, 2023). Between now and 2050, we will have to increase global food production by 70%. We need to make agriculture more effective and modern, to cater to an exploding population with a growing demand for high-end products while using water and other natural resources more wisely.

Modern agriculture is an innovative concept that is anticipated to significantly contribute to the necessary expansion of food production in a sustainable manner. The goal of modern agriculture is to maximize productivity through process optimization and improvement. This demands rapid, precise, and distributed measurements to provide growers with a more thorough picture of the conditions in their region of cultivation, and/or to program automated machinery to optimize energy use, water use, and chemical use for plant growth and pest control (Ali *et al.*, 2013b; Monteiro *et al.*, 2021). A higher level of evaluation of scientific knowledge is possible through the organization of well-evaluated data from numerous heterogeneous systems into intelligent algorithms that perform reasoning tasks. Running locally on a low-power wireless sensor network (WSN) node is not feasible because of the complexity of the algorithms needed to handle the distributed data in real time (Mahmood *et al.*, 2021; Navarro *et al.*, 2022). However, within the framework of the IoT, all items will be linked together, enabling the effortless transfer of computing burden to the cloud or the distribution of tasks among several interconnected devices (Figure 24.2).

* Corresponding author: rizwan.maqbool@uaf.edu.pk

 DOI: 10.1201/9781003498698-24

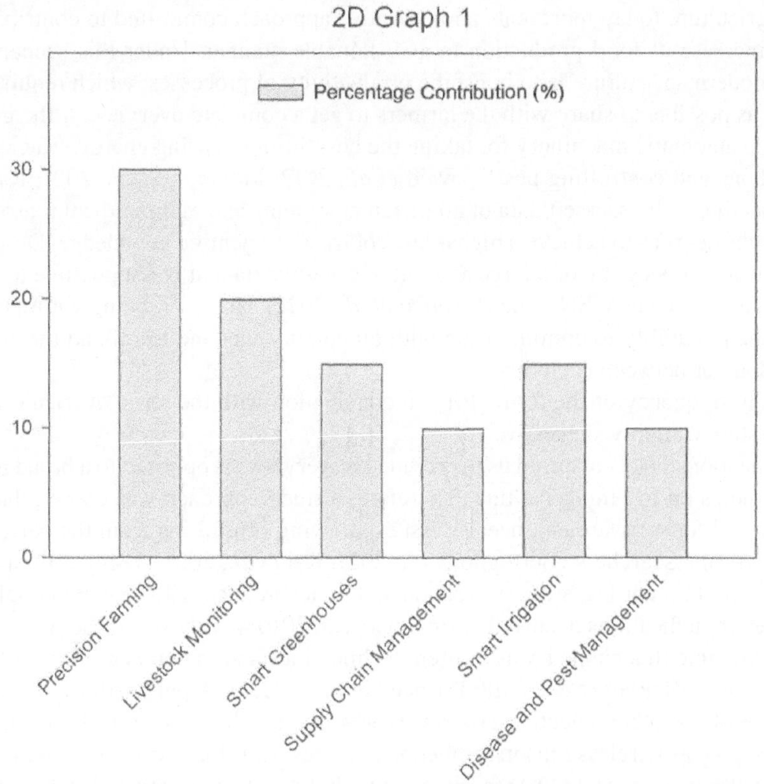

FIGURE 24.1 Percentage influence of IoT in various agricultural sectors. (From Farooq *et al.*, 2020.)

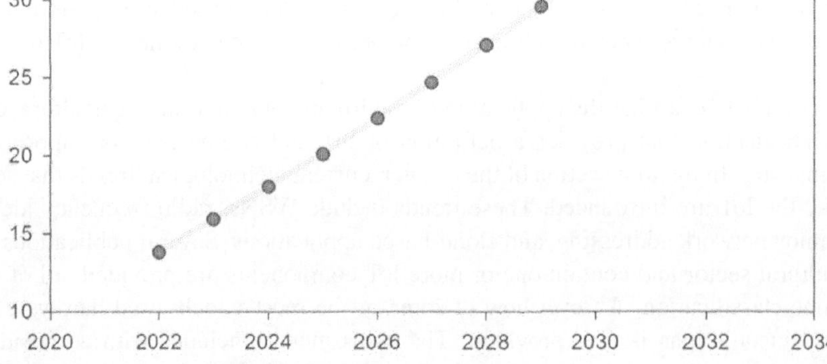

FIGURE 24.2 This data reflects the growth in the number of IoT devices worldwide from 2020 to 2023, with forecasts from 2024 to 2033. (From Statista, 2023.)

Modern agriculture today represents an innovative approach committed to contributing significantly to an increase of food production in a sustainable manner. Under this scenario, the main objective of modern agriculture is to boost the productivity of processes, which requires collecting as much data as possible to share with the farmers to get a complete overview of the growing area, and to program automatic machinery for taking the best timing in using energy, water or chemicals in growing plants and controlling pests (Javaid *et al.*, 2012; Monteiro *et al.*, 2021; Rasheed *et al.*, 2022b). Connecting well-assessed data of all different systems and making them reason together in smart algorithms helps us to achieve a higher level of trust in scientific knowledge. Despite the complexity of real-time processing of a large amount of scientific data, it is not possible to run complex algorithms on a low-power WSN node (Navarro *et al.*, 2022). However, being within the paradigm of IoT gives the possibility to communicate with all components and to offload the computational tasks to the cloud or between each other.

The growing frequency of the term "IoT" in conjunction with the term "Agriculture" in global scientific literature is highly suggestive.

The aforementioned facts inspired us to provide an overview, as opposed to a broad review, of the most recent studies on IoT in agriculture. Therefore, a study approach was chosen that was modified and enhanced for software engineering issues, drawing inspiration from the current standards utilized by medical researchers (Bertoglio *et al.*, 2021; Raj *et al.*, 2021). Using 2010 as the starting point, a selection of recent literature was conducted in accordance with this methodology. Since a sizable number of publications occurred during that year, 2010 was chosen as the starting point year. Moreover, at the time this chapter was written, technologies and methods used prior to 2010 were largely out of date (Mäntylä *et al.*, 2018; Rasheed *et al.*, 2022a). A publication's multidisciplinary nature was one of the other selection criteria. Works that combine many technologies to provide their solutions, such as wireless sensors, embedded devices, and cloud works that use multiple kinds of end devices (things) connected to the same network, were categorized as IoT-oriented (Prakash *et al.*, 2023).

The increasing frequency of the term "IoT" in association with "Agriculture" within the worldwide scientific literature is highly indicative. The aforementioned evidence has motivated us to present a synopsis, rather than an extensive evaluation, of the most recent research on IoT in agriculture. Consequently, an investigative methodology was selected that was adapted and improved for software engineering challenges, taking inspiration from the prevailing standards employed by medical researchers (Ullah *et al.*, 2013). Commencing from the year 2010, a collection of up-to-date literature has been scrutinized following this particular approach. Given the substantial volume of publications in that specific year, 2010 was designated as the initial reference year. Furthermore, as of the composition of this manuscript, technologies, and methodologies predating 2010 were predominantly obsolete (Zafar *et al.*, 2023). The interdisciplinary character of a publication was another key criterion for selection. Pieces of work that integrate various technologies to offer their solutions, such as wireless sensors, embedded devices, and cloud-based systems utilizing diverse types of end devices interconnected within the same network, were classified as IoT-focused (Shah & Razaq, 2020).

The chapter aims to examine the application of the IoT in contemporary agriculture, commencing with an introduction that provides a definition of IoT and emphasizes its importance in the agricultural industry. In the first section of the chapter, current technological trends that serve as the foundation for the IoT are introduced. These trends include WSNs, radio frequency identification (RFID), common network addressing, and cloud-based applications. Several publications that focus on the agricultural sector and contain one or more IoT components are provided, all of which fall under the same classification. An overview of some of the most widely used hardware platforms in agricultural installations is also provided. The assessment concludes with a consideration of upcoming obstacles and how they may affect the adoption of IoT. One of the objectives of this work is to provide recent progress through literature that summarizes the most recent developments in wireless communication technologies, cloud services, programming paradigms, sensor modules, and embedded devices appropriate for or optimized for use in agriculture. This aims to benefit a multidisciplinary community of researchers working on the deployment of monitoring, decision support, innovative tracking, and control systems.

The chapter seeks to explore the utilization of the IoT in modern agriculture, commencing with an introduction that furnishes a definition of IoT and underscores its significance in the agricultural sector. The initial segment of the chapter introduces current technological patterns that lay the groundwork for the IoT. These patterns encompass WSNs, RFID, standard network addressing, and cloud-based applications (Figure 24.3). Various publications focusing on the agricultural domain and encompassing one or more IoT elements are presented, all falling within the same category. Additionally, an outline of some of the most commonly utilized hardware platforms in agricultural setups is included. The evaluation culminates with a contemplation of forthcoming challenges and their potential impact on the adoption of IoT in agriculture. One of the aims of this study is to offer an up-to-date overview through literature that encapsulates the latest advancements in wireless communication technologies, cloud services, programming models, sensor components, and embedded devices tailored for or optimized for agricultural use. This endeavor seeks to be advantageous to a diverse community of researchers engaged in the implementation of monitoring, decision-making support, cutting-edge tracking, and control systems.

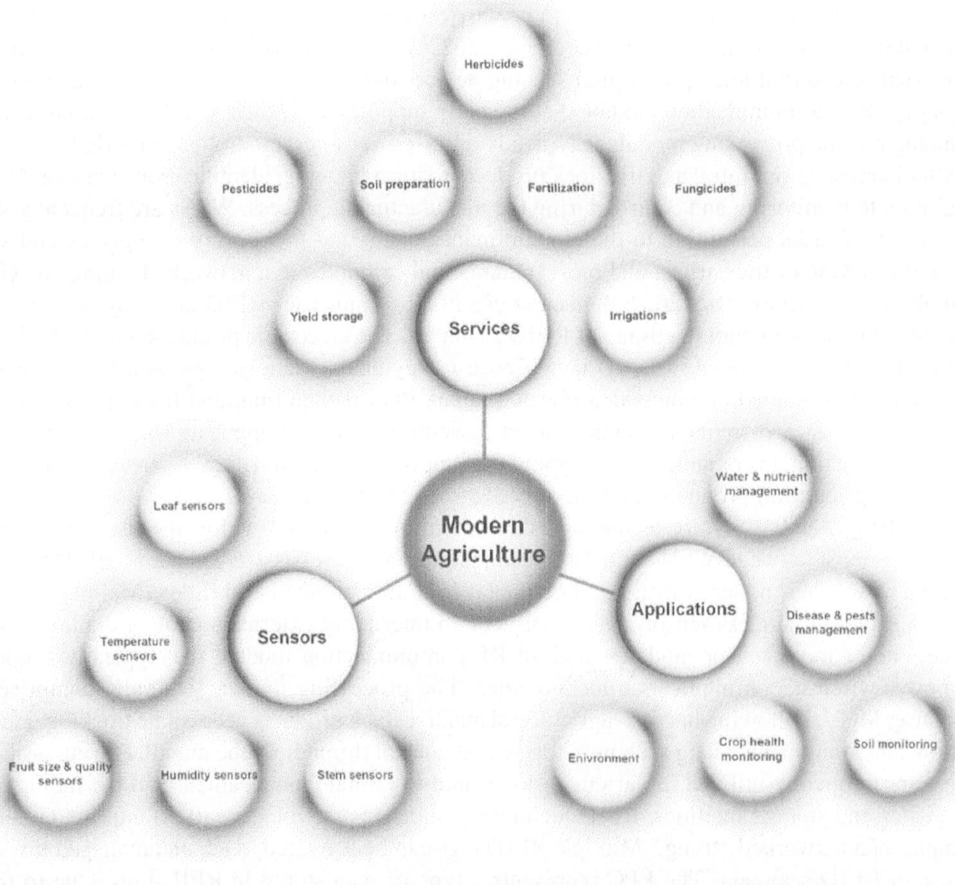

FIGURE 24.3 Graphical representation of potential applications, services, and sensors in modern agriculture.

24.2 ENABLING TECHNOLOGIES

The network layer (data transfer), the application layer (data storage and manipulation), and the perception layer (sensing) constitute the three levels that make up the IoT.

As numerous assessments demonstrate, despite significant advancements, IoT is still developing and striving to take on its final form (Anand *et al.*, 2020; Nižetić *et al.*, 2020). The IoT devices are distinguished by their innate networking potential, as denoted by the term "Internet." As the IoT is an internet of non-human objects, a lot of machine-to-machine (M2M) communication will occur, in contrast to the current internet, which is mostly an internet of end users who are humans.

The network layer, responsible for data transfer, the application layer, focused on data storage and manipulation, and the perception layer, dedicated to sensing, collectively form the triad of levels comprising the IoT. Despite notable progress, multiple evaluations establish that the IoT is still in a phase of evolution, as it continues to progress toward its ultimate state (Anand *et al.*, 2020; Nižetić *et al.*, 2020). The distinguishing feature of IoT devices lies in their inherent capability for networking, emphasized by the term "Internet." Given that the IoT predominantly involves non-human entities, a significant amount of M2M communication is anticipated to take place, unlike the current internet landscape, which primarily revolves around human end users.

24.2.1 The Perception Layer

Near field communications (NFCs), RFID, and WSN are among the technologies that we encounter at the perceptual layer. WSN and RFID technologies share some similarities because semi-passive and active RFID tags are also considered to be wireless nodes with limited computational and storage capabilities (Sun *et al.*, 2022; Rasheed *et al.*, 2022b). A typical wireless sensor node comprises an internal or external analog or digital sensing device, one or more sensor modules, and a radio frequency (RF) communication module that typically supports low-power wireless communication technologies. The processing module is typically a low-power microcontroller unit (MCU).

After harvest, agricultural and livestock products require tracking, identification, and monitoring in addition to monitoring and control during the manufacturing process. WSNs are frequently seen in a variety of tasks pertaining to the climate management and monitoring of logistics and storage facilities. One of the earliest and most fundamental examples of a networked "thing" is RFID technology (Mostaccio *et al.*, 2023; Ameen *et al.*, 2023b). Numerous RFID tags may be triggered, read, and manipulated with the usage of RFID scanners. The electronic product code (EPC) is one kind of data that is included in RFID tags. Because they enable data storage on active or passive tags (which do not need an embedded power source), item recognition, and tracking, RFID, and NFC technologies are frequently employed in agriculture. Typical applications include tracking and monitoring of items or animals throughout the supply chain, as well as assessing the quality and lifespan of agricultural products (El Matbouly *et al.*, 2022; Rasheed *et al.*, 2022c).

NFC, RFID, and WSN are technologies commonly encountered within the perceptual layer. Similarities exist between WSN and RFID technologies, as semi-passive and active RFID tags are regarded as wireless nodes with limited computational and storage capabilities (Mahmood *et al.*, 2021). A standard wireless sensor node consists of an internal or external analog or digital sensing device, one or more sensor modules, and an RF communication module that typically supports low-power wireless communication technologies. The processing module generally comprises a low-power MCU. Following harvest, agricultural and livestock products necessitate tracking, identification, and monitoring, along with monitoring and control throughout the manufacturing process. WSNs are frequently utilized in various tasks related to climate management and the monitoring of logistics and storage facilities. RFID technology stands as one of the earliest and fundamental examples of a networked "thing." Multiple RFID tags can be activated, read, and managed through the use of RFID scanners. The EPC represents a type of data stored in RFID tags. Due to their capability to store data on active or passive tags (which do not require an embedded power source), facilitate item recognition, and enable tracking, RFID and NFC technologies find common applications in agriculture. Typical uses include the tracking and monitoring of items or animals across the supply chain, as well as the evaluation of the quality and shelf life of agricultural products (Hassan *et al.*, 2022).

24.2.2 The Network Layer

Wireless sensor nodes that communicate with actual things and/or their environment make up the second layer of the IoT. By means of networks created by their communication with a gateway or with one another, these nodes establish a means of sending the data to a distant infrastructure for processing, storing, and exchanging the extracted knowledge. Sensor networks have been the subject of much study in wireless communications, including topics such as energy efficiency, networking characteristics, scalability, and resilience. Internet-enabled gateways and end-nodes may connect with one another using communication protocols created on top of wireless standards, such as 802.15.4, which allow device networking (Jain *et al.*, 2022; Saqib and Moon, 2022).

The second layer of the IoT consists of wireless sensor nodes that engage in communication with physical entities and/or their surroundings. Through the establishment of networks facilitated by interactions with a central gateway or among themselves, these nodes create a mechanism for transmitting data to a remote infrastructure for the purposes of processing, storage, and sharing of the acquired information (Figure 24.4). Extensive research has been dedicated to sensor networks within wireless communications, encompassing areas such as energy efficiency, network attributes, scalability, and robustness. Connectivity between Internet-enabled gateways and end-nodes can be achieved through communication protocols built upon wireless standards like 802.15.4, enabling device interconnection (Jain *et al.*, 2022; Saqib and Moon, 2022).

24.2.3 The Application Layer

The application layer is situated as the third stratum inside the IoT architecture. The realization of the IoT is greatly facilitated by its high relevance and the many ways in which it is significant. Solving many problems, including identifying the devices as distinct entities, is necessary for the application layer. By identifying and addressing devices, it will be possible to control and gain direct access to billions of devices worldwide through the next internet (Del Sarto *et al.*, 2022). The addressing schema is characterized by its distinctiveness in identification, dependability, durability, and capacity to handle increased demands. With its features that enable internet mobility, IPv6 may help with some of the issues related to device identification and is anticipated to be quite important in this area. The challenge is made worse by the heterogeneous nature of wireless nodes, the variety of data formats, concurrent processes, and the convergence of data from the devices (Choudhary and Meena, 2022). Supplementary to IPv6, context-aware addressing and metadata are anticipated to play a significant role in overcoming the aforementioned difficulties.

The application layer is positioned as the third level within the IoT architecture. The IoT's realization is greatly enhanced by its high relevance and the numerous ways in which it holds significance. The resolution of various issues, such as distinguishing devices as unique entities, is imperative for the application layer. Through the identification and handling of devices, there is potential to manage and establish direct connections with countless devices globally via the succeeding. The addressing scheme is recognized by its uniqueness in identification, reliability, longevity, and ability to manage growing demands. IPv6, with its functionalities supporting internet mobility, could assist in addressing certain challenges associated with device identification and is expected to be particularly crucial in this domain. The complexity is exacerbated by the diverse characteristics of wireless nodes, the range of data structures, simultaneous operations, and the amalgamation of data from the devices. In addition to IPv6, context-aware addressing and metadata are predicted to have a notable impact on surmounting the aforementioned challenges.

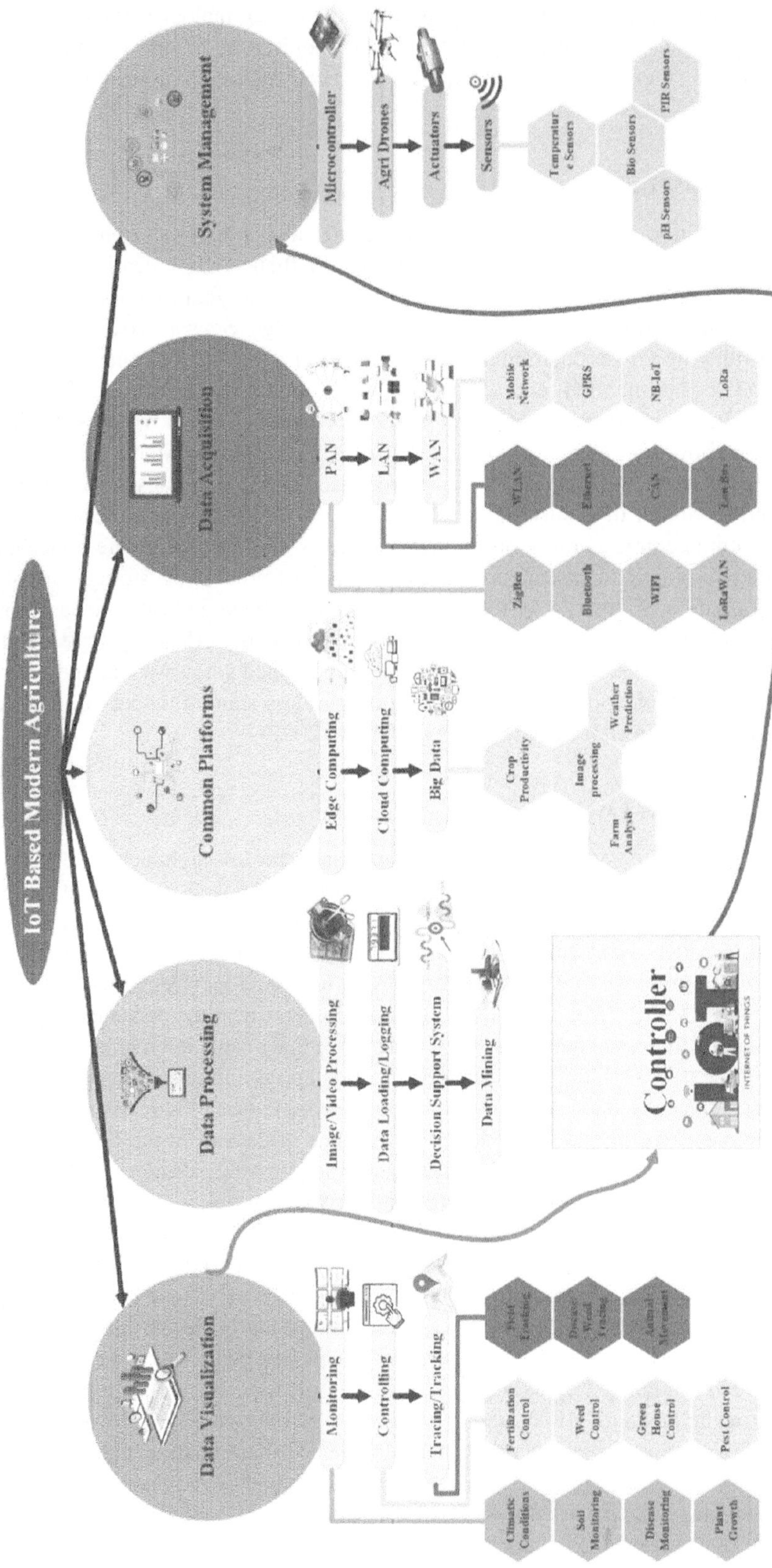

FIGURE 24.4 IoT modern agricultural framework.

24.3 APPLICATIONS IN AGRICULTURE

The large potential of these technologies in the fields of agriculture and the current trend of modern agriculture is evident from the basic review of IoT structural components described in this section. The technical transition from conventional to modern and micro-precise agriculture has been greatly aided by recent developments in sensor technology, as well as the shrinking and significant cost reduction of electronics.

Climate sensors, radiation sensors, ground sensors, and sensor-based weather stations highlight the fact that sensors and sensor data flows, which are stored and used for monitoring, knowledge mining, reasoning, and control are at the center of it all. Furthermore, there has been a rise in the market for secure and high-quality agricultural goods in recent years. Systems for logistics traceability that are reliable, dispersed, accurate, and interoperable are becoming necessary due to this development (Ali *et al.*, 2013a; Feng *et al.*, 2020). Supply chains in the agricultural and floricultural industries may be supported by supply chains that are specifically developed with the right tools, which all belong to the IoT family of technologies.

The substantial potential of these technologies within the agricultural sector and the current trajectory of modern agricultural practices are clearly evident upon examining the fundamental aspects of IoT structural components delineated in this particular section. The shift from traditional to modern and precision agriculture has been significantly facilitated by recent advancements in sensor technology, coupled with the reduction in size and cost of electronic components.

Climate sensors, radiation sensors, soil sensors, and sensor-based meteorological stations underscore the pivotal role played by sensors and the corresponding data streams, which are collected, stored, and utilized for monitoring, data analysis, inference, and management. Additionally, there has been a surge in the demand for safe and premium agricultural products in recent times. The emergence of reliable, decentralized, precise, and interoperable traceability systems for logistics is becoming imperative in response to this trend (Feng *et al.*, 2020). The agricultural and floricultural sectors' supply chains can potentially benefit from specialized IoT technologies that facilitate the development of efficient supply chain systems.

24.3.1 Monitoring and Control

Over the last several decades, sensors, both wired and wireless, have been extensively used in the field of agriculture. It is essential to accurately perceive the production environment and the plants' reactions to the climate in order to make informed choices that maximize output and improve the quality of the crops. By implementing more universal communication protocols, permitting remote internet access, and using intelligent algorithms for data meta-processing with the goal of enhancing monitoring and/or control, classic WSNs have lately transformed into IoT-friendly WSNs (Tzounis *et al.*, 2017). These days, batteries may be used to power versatile devices with high computing capacity, a small form factor, and low cost. They can run for extended periods of time with or without the help of power harvesting modules. Furthermore, the capabilities of classic WSN networking are extended by the fact that contemporary embedded devices are capable of supporting more complex networking protocols like TCP/IP and more demanding sensors like image sensors (Lata *et al.*, 2021). The monitoring and control literature can be roughly categorized as follows.

Surveillance and, in some instances, development of advance notifications, using simplified guidelines. In greenhouse horticulture, this involves the use of multi-point monitoring to detect and assimilate climate variations. Monitoring, meta-processing (server/cloud side implementations of algorithms and models), and control including user-directed control recommendations and completely automated control are all included. Monitoring with increasingly powerful end-nodes and computationally intensive sensors, including image sensors (Lata *et al.*, 2021; Ameen *et al.*, 2023a). Either the system is simply monitored using captured photos, or the images are processed on-board, at the network edge (Fog computing), or on a cloud/server-based architecture.

In recent decades, sensors, wired and wireless, have been increasingly used in agriculture; the accurate sensing of the production environment and the plant's response to climate is the best way to make the proper decisions to improve the output and quality of crops. The evolution of traditional WSNs into IoT-friendly WSNs has been facilitated by the adoption of more universal communication protocols, enabling remote internet access, and the use of intelligent algorithms for data meta-processing to improve monitoring and control capabilities (Tanveer *et al.*, 2015; Tzounis *et al.*, 2017). Modern batteries are now capable of powering versatile devices with high computational capacity, compact size, and affordability, allowing them to operate for extended durations with or without power harvesting assistance. Additionally, contemporary embedded devices can support more sophisticated networking protocols such as TCP/IP and advanced sensors like image sensors, thereby expanding the networking capabilities of classic WSNs (Lata *et al.*, 2021). The literature on monitoring and control can be broadly classified into various categories, including surveillance, early warning systems, and simplified guidelines. In greenhouse horticulture, multi-point monitoring is employed to detect and integrate climate change. The monitoring process encompasses meta-processing on the server/cloud side, control functions, user-directed recommendations, and fully automated control mechanisms. The use of powerful end-nodes and computationally intensive sensors, including image sensors, has enhanced monitoring capabilities (Lata *et al.*, 2021; Ameen *et al.*, 2023a). Image data can be either monitored through captured photos or processed on-board, at the network edge (Fog computing), or on cloud/server-based architectures.

24.3.2 Open-Field Agriculture

During open-field deployments, researchers often assess climatic conditions and place significant emphasis on monitoring soil conditions. Many authors deploy multiple sensors at various depths within the soil. Preserving water can only be achieved by optimizing irrigation to provide the precise quantity of water required by the plant. Any excess water used for irrigation is either lost in the ground or evaporates into the sky. Optical sensors have been used to gather supplementary data on crop reflectance and distant temperature monitoring, as well as to create maps of the field's conditions (Mehmood *et al.*, 2018; Hussain *et al.*, 2020). The combination of IoT with geographical information systems (GIS) has been suggested for situations where accurate mapping of the collected data is crucial. Underground WSNs (UWSNs) are a notable component of WSNs in agriculture, offering significant benefits, particularly in open-field scenarios (Mowla *et al.*, 2023).

As mentioned previously, powerful platforms are now affordable due to advancements in embedded device technology. Researchers now have the opportunity to develop more complex end devices, such as Wireless Multimedia Sensor Networks, which include sensing nodes with much greater processing capacities capable of handling very demanding peripheral devices, like image sensors. In a fog-network-like fashion, this kind of node allows more intensive local processing near the network's edge (Tzounis *et al.*, 2017). Various image processing methods are used in these works to track crop development and detect intruding animals, insects, or other plant dangers, while others employ cameras for basic security or facility monitoring purposes.

During open-field deployments, scholars frequently evaluate meteorological conditions and place considerable emphasis on monitoring soil conditions. Multiple researchers deploy several sensors at different depths within the soil. The conservation of water can solely be accomplished by optimizing irrigation to supply the precise amount of water needed by the vegetation. Any surplus water utilized for irrigation is either lost in the soil or evaporates into the atmosphere. Optical sensors have been utilized to collect additional data on crop reflectance and remote temperature monitoring, as well as to generate field condition maps (Hussain *et al.*, 2020). Current researches propose three main scenarios where IoT and GIS have been integrated to enable accurate data mapping in several industry scenarios. UWSNs are a major part of WSNs in agriculture with multiple benefits, especially in open-field agriculture (Mowla *et al.*, 2023).

As previously indicated, cost-effective robust platforms are now available due to advances in embedded device technology. Scholars now have the chance to design more sophisticated end devices, such as wireless multimedia sensor networks, which incorporate sensing nodes with significantly enhanced processing capabilities capable of managing highly demanding peripheral devices, such as image sensors. In a manner reminiscent of fog networks, this type of node enables more intensive local processing near the network's periphery (Tzounis *et al.*, 2017). Various image processing techniques are employed in these studies to monitor crop growth and identify invading animals, insects, or other plant threats, while some use cameras for basic security or facility monitoring objectives.

24.3.3 Controlled Environment Agriculture

Significant climatic fluctuations have been shown to exist in greenhouses, which can impact, if not damage, plant production. Greenhouse cultivation is characterized by a higher level of intensity, which often necessitates greater accuracy in terms of monitoring and management. Several studies have solely looked at both distant and localized monitoring (Zafari *et al.*, 2019; Costa *et al.*, 2021; Vermesan *et al.*, 2022). Most of the time, data is saved and presented graphically in different ways. Studies have been conducted that demonstrate systems utilizing meta-processing techniques alongside high-precision monitoring, with data being sent over the internet to distant infrastructures. These systems generate evaluations of the climate and/or crop condition using well-considered equations, crop, and climate models to assist growers in making better decisions or receiving early warnings.

Climatic variations of significance have been observed within greenhouse environments, potentially influencing plant productivity or causing harm. The practice of greenhouse farming is distinguished by its elevated level of intensity, often demanding heightened precision in the aspects of supervision and control. Previous research efforts have focused exclusively on remote as well as localized surveillance (Zafari *et al.*, 2019; Costa *et al.*, 2021; Vermesan *et al.*, 2022). Frequently, information is archived and visually represented in various formats. Investigations have been carried out illustrating the implementation of meta-processing methods in conjunction with highly accurate monitoring tools, transmitting data via the internet to remote facilities. These systems produce assessments of environmental conditions and/or crop health by employing sophisticated algorithms, crop models, and climatic simulations, aiding cultivators in making informed choices or receiving timely alerts.

The prevalence of agricultural-cloud IoT technologies for greenhouse monitoring and management is increasing. The end-nodes gather diverse data that is then uploaded to a cloud architecture. In this cloud environment, the data is thoroughly evaluated in a more expedient, cost-effective, reliable, and efficient manner. Due to the increasing popularity of plant factories in the context of urban controlled environment agriculture (CEA) in smart cities, numerous research studies have been conducted to explore artificial growth systems (Vatistas *et al.*, 2022). Singh (2022) introduced Optimized Irrigation Recommendation Information System (OpIRIS), a web application that utilizes established scientific information to enable accurate irrigation scheduling for greenhouses. The system communicates with remote sensors placed in the greenhouses. It included climatic sensors and high-quality equipment, specifically designed for industrial use. The collected data were then sent to the cloud infrastructure for further analysis. The device demonstrated high precision in forecasting the water requirements of crops and offered producers effective guidance on the optimal timing and dosage of fertilizer solutions for irrigation. Prior efforts to automate irrigation have relied on integrating agricultural information with a cloud and hardware that integrates IoT and RFID (Sidhu *et al.*, 2021; Gull *et al.*, 2023). This integrated system successfully achieves the dynamic allocation of resources and load balancing. Consequently, the authors see a notable increase in resource use efficiency and a substantial improvement in water quality.

The escalation in the utilization of agricultural-cloud IoT technologies for monitoring and managing greenhouses is becoming more prevalent. Diverse data is collected by the end-nodes and subsequently uploaded to a cloud framework. Within this cloud setting, the data undergoes comprehensive evaluation in a manner that is more efficient, cost-effective, and reliable. Given the rising acceptance of plant factories within the domain of urban CEA in smart cities, a multitude of research endeavors have been undertaken to investigate artificial growth systems (Vatistas *et al.*, 2022). OpIRIS, a web application introduced by Singh, leverages established scientific knowledge to facilitate precise irrigation scheduling for greenhouses. This system interacts with remote sensors positioned in the greenhouses, encompassing climatic sensors and specialized high-quality equipment tailored for industrial purposes. The amassed data is then transmitted to the cloud infrastructure for further examination. The device exhibited remarkable accuracy in predicting crop water requirements and provided producers with valuable guidance regarding the optimal timing and dosage of fertilizer solutions for irrigation. Previous attempts to automate irrigation have centered on the fusion of agricultural data with a cloud and hardware that integrates IoT and RFID (Sidhu *et al.*, 2021; Gull *et al.*, 2023). This amalgamated system effectively accomplishes the dynamic distribution of resources and equitable load distribution. As a result, the authors observe a noteworthy enhancement in resource utilization efficiency and a substantial amelioration in water quality.

Furthermore, there have been investigations conducted on the implementation of control over one or more actuator systems in greenhouses, specifically focusing on temperature and irrigation controllers, which are also referred to as wireless sensor and actuator networks (WSANs). There are two methods to accomplish the control remotely. The first method involves the farmer exerting direct control (Mohamed *et al.*, 2018). In such instances, the system administrator, guided by the recommendations provided by a Decision Support System, selects to regulate the machinery. Applications in open-field warning and control, remote monitoring, integrated pest management (IPM), and CEA fall under this area of monitoring and control (Chadha *et al.*, 2019; Shah and Razaq, 2020). Multiple research investigations are focused on achieving complete automation by directly delivering control signals, which are created via data processing, to the actuators in a closed-loop control system.

Moreover, there have been inquiries carried out regarding the management of control over one or more actuator systems within greenhouses, with a specific emphasis on temperature and irrigation regulators, commonly known as WSANs. Two approaches exist for achieving remote control. The initial approach entails the farmer directly applying control (Abdullah *et al.*, 2022). In such cases, the system administrator, under the guidance of recommendations from a decision support system, opts to oversee the equipment. This category encompasses applications in open-field alerting and regulation, distant monitoring, integrated pest control (IPC), and controlled environmental agriculture (Shah and Razaq, 2020). Numerous research studies are dedicated to attaining full automation by directly transmitting control instructions, generated through data processing, to the actuators within a closed-loop control setup.

24.3.4 Food Supply Chain Tracking

Contemporary agriculture is progressively becoming more industrialized. Food safety and quality must therefore be ensured by implementing standardized procedures at every stage of the production process, from the producer to the consumer. Systems for tracking the food supply chain (FSC) are becoming more and more popular as a result of this necessity. There are several ways in which IoT technology may significantly aid in the development, upkeep, and support of these kinds of systems. Solutions in the examined literature focus either on technology or the business aspect of the FSC (Kumar *et al.*, 2022). However, some works aim to provide solutions that satisfy both aspects. Numerous studies on supply chains have leveraged recent advancements in e-commerce. For this assessment, the emphasis was solely on FSC due to its focus on optimizing the food supply.

Within the FSC, RFID is the most prevalent IoT technology. Tracking agricultural goods is made possible with RFID tags, which function as enhanced barcodes. As per the IoT paradigm, recent research has integrated multiple sensors to enhance the status information of products when it is captured via RFID (Costa *et al.*, 2021). The dispersed nature of IoT and the asynchronous, heterogeneous information flow are prevalent problems. Therefore, for FSC tracking services, naming is essential to ensure the precise and quick retrieval of information. Because physical closeness is no longer necessary, the implementation of IoT-based infrastructure results in the virtualization of supply chains. The literature offers a number of models that examine issues within the FSC and explore how IoT technologies address them (Kumar *et al.*, 2022; Yadav *et al.*, 2022; Khan *et al.*, 2023). The advancement of technology, together with the growing strength and maturity of various technologies in the IoT, has provided researchers with the opportunity to create comprehensive systems integrating sensor modules and software infrastructures. The software of these systems is either distributed across dispersed stakeholders or hosted by cloud providers. Comprehensive systems provide intelligent plans, automated reasoning based on observed events, and automated services utilizing artificial intelligence (AI) (Aziz *et al.*, 2018). The organizational structure of a comprehensive FSC information management system and the system design strategies that optimize financial gain are covered in other publications.

Contemporary agriculture is progressively evolving toward increased industrialization. It is imperative to guarantee food safety and quality through the implementation of standardized procedures across all stages of the production process, spanning from the producer to the consumer. The rising necessity for this assurance has led to a surge in the popularity of systems designed to track the FSC. The integration of IoT technology is identified in the literature as a key enabler for advancing, maintaining, and supporting such systems. Within the realm of explored literature, solutions are predominantly categorized into those focusing on technological advancements and those addressing the business aspects of the FSC. Nevertheless, there exist endeavors aiming to offer solutions that cater to both dimensions. Recent studies in supply chain management have capitalized on the latest developments in e-commerce. The evaluation conducted in this study was exclusively centered on FSC, primarily due to its role in streamlining the food supply.

RFID stands out as the predominant IoT technology within the FSC. The utilization of RFID tags enables the tracking of agricultural products, functioning similarly to advanced barcodes. Recent scholarly investigations within the realm of IoT have incorporated a variety of sensors to enrich the status data of goods at the point of RFID data capture (Costa *et al.*, 2021). Challenges arising from the decentralized nature of IoT and the varied, unsynchronized data streams are commonly observed. Hence, in the context of FSC monitoring services, proper labeling proves crucial for ensuring accurate and prompt data retrieval. The integration of IoT-based infrastructure eliminates the necessity for physical proximity, leading to the virtualization of supply chain operations. Existing literature presents several theoretical frameworks that scrutinize FSC-related issues and assess the role of IoT technologies in addressing them (Kumar *et al.*, 2022; Yadav *et al.*, 2022; Khan *et al.*, 2023). The evolution of technology, coupled with the maturation and robustness of diverse IoT tools, has empowered researchers to develop holistic systems that amalgamate sensor components and software frameworks. The software architecture of these systems may be decentralized across multiple stakeholders or hosted on cloud platforms. These comprehensive systems furnish intelligent strategies, automated decision-making processes based on real-time data, and AI-driven services. Moreover, scholarly works delve into the organizational configuration of an exhaustive FSC information management system and the strategic approaches to optimize financial outcomes.

24.3.5 Livestock Applications

In the cattle pastures, many deployments have been accomplished. Several authors take seriously the need for an ideal habitat that can withstand harsh weather conditions that are known to reduce animal output. In certain situations, field monitoring for the best feeding techniques is included in livestock IoT, in addition to monitoring and controlling animal and climatic conditions (Akhigbe *et al.*, 2021; Monteiro *et al.*, 2021). Beehive instrumentation and analysis is another facet of livestock IoT. In addition to monitoring smells and dangerous gases, wireless sensors have been employed in animal tracking and behavioral analysis. Studies have also been conducted on the optimization of livestock deployment equipment performance in light of the particular problems that are present.

In the cattle pastures, numerous deployments have been successfully executed. Many researchers consider the necessity of an optimal environment capable of enduring severe weather conditions that have the potential to diminish animal productivity. Field monitoring for optimal feeding methods is sometimes integrated into livestock IoT applications, alongside the monitoring and regulation of animal and environmental parameters (Akhigbe *et al.*, 2021; Monteiro *et al.*, 2021). Another aspect of livestock IoT involves the instrumentation and analysis of beehives. Wireless sensors have been utilized for animal tracking and behavioral assessment, in addition to detecting odors and hazardous gases. Research has also been carried out on enhancing the performance of livestock deployment equipment to address specific challenges.

24.4 CHALLENGES IN AGRICULTURE

Various obstacles exist in the context of using IoT technology in agriculture. Foremost, the perception layer equipment has to withstand adverse climatic phenomena such as intense sunlight, high temperatures, rain or excessive humidity, strong winds, vibrations, and other hazards that might damage the electronic circuits (Chadha *et al.*, 2020; Sinha and Dhanalakshmi, 2022). The end devices' reliance on the finite power supply of batteries will determine their ability to operate continuously and dependably for extended periods of time. Hence, it is essential to have suitable programming tools and low-power capabilities to ensure the need for frequent battery replacement or station resets is avoided, especially in the context of large-scale open-field deployment (Hanif *et al.*, 2017; Saha *et al.*, 2021). A partial answer to this problem may be power harvesting; nevertheless, small power harvesting modules (solar panels, wind turbines, etc.) have a power budget that is enough to fulfill the required power consumption. In addition, a huge number of networked (internet-like) devices generate an enormous volume of data that may overwhelm small-scale server infrastructures due to resource constraints.

Numerous challenges are present when considering the implementation of IoT technology in agriculture. Primarily, the equipment at the perception layer must be capable of enduring various adverse weather conditions, including but not limited to intense sunlight, high temperatures, rain, excessive humidity, strong winds, vibrations, and other potential hazards that could potentially harm the electronic circuits (Sinha and Dhanalakshmi, 2022). The continuous and reliable operation of end devices heavily relies on the finite power supply provided by batteries, which ultimately dictates their ability to function over prolonged periods. Therefore, it is crucial to have appropriate programming tools and low-power features in place to minimize the necessity for frequent battery replacements or station resets, particularly in the scenario of large-scale deployment in open fields (Saha *et al.*, 2021). One potential solution to this issue could involve power harvesting; however, the effectiveness of small power harvesting modules (such as solar panels and wind turbines) is limited by their power budget in meeting the required power consumption levels. Furthermore, the proliferation of a vast array of networked devices, resembling traditional internet systems, results in the generation of an extensive amount of data that could strain small-scale server infrastructures due to inherent resource limitations.

24.4.1 Accessibility

It is essential that current software and hardware always remain available in any location for constructing agricultural decision support systems based on IoT technology and other devices. Resolving these issues is crucial to ensure that services are consistently available whenever and wherever they are needed. Delays and disruptions can occur if the necessary equipment is unavailable.

24.4.2 Adaptability

When building models, especially for precision farming, it's important to consider how devices will interact with each other and their environment. Some devices may struggle to adapt due to constantly changing ambient conditions or occasional hardware or communication issues.

24.4.3 Compatibility

To satisfy the requirements of fragmentation and scalability, the generated models or software should be flexible and compatible with any kind of computer.

24.4.4 Challenge

It is still a challenge for many researchers to create a strategy that maximizes profits. Scientific efforts are focused on developing cost-effective solutions by reducing the hardware and software requirements for IoT deployments. Farmers have challenges in implementing equipment and technology due to economic inequalities across countries. Thus, it is important to develop certain economic models.

24.4.5 Energy Optimization

As more devices rely on the IoT, WSNs, and other networks for communication, energy is becoming a more pressing concern. Up until now, creating and operating models have been supported by traditional energy sources. However, the growing number of gadgets has rendered the use of traditional electricity an unreliable option. While testing unconventional energy sources such as hydro-power, solar, and wind harvesting systems is important, their use in model development has not been very successful. As a result, new approaches need to be created.

24.4.6 Environmental Conditions

Agriculture makes use of a variety of terrain types. Data and services are placed in danger, and it becomes increasingly difficult to adapt to such developments. A system's accuracy is altered as a result.

24.4.7 Heterogeneity

Different devices are employed while constructing a system. The procedures and services needed by each device vary. It is crucial to establish interaction between heterogeneous modules and communication technologies in agriculture since the majority of models operate with heterogeneous equipment. The presence of heterogeneity in a network increases its complexity and might potentially result in the emergence of falsified conclusions.

24.4.8 Mobility

Most of the developed concepts or systems are static. Mobility is necessary for the framework's seamless implementation since the majority of devices and applications are portable. Ensuring consistent connection is a challenging endeavor with mobile devices.

24.4.9 Real-Time Deployment

A significant number of studies presented are not implemented or evaluated on a real-time testbed. Real-time analysis is thus necessary before system or model deployment in order to avoid post-deployment losses.

24.4.10 Reliability

Reliability is a crucial factor for IoT devices when it comes to the effective and seamless transfer of data. The devices must collect and transmit trustworthy data as choices are made in response to the data that is received and analyzed. System failures, battery problems, node failures, or other interventions continue to pose a challenge to reliability.

24.4.11 Standardization

Standardization of devices is required to fully use the technology for a wide range of applications. There is currently no validation mechanism in place for any standard formats used to describe data or operations. Errors and sporadic modifications result in inconsistent output from machine-readable programs that are prohibited from using semantic ontologies. Standardization may tackle the issues related to the interoperability of devices, applications, systems, and items.

The following key aspects must be addressed to ensure effective standardization and seamless integration of technologies:

- *Accessibility*: The perpetual availability of current software and hardware in all locations is imperative for the development of agricultural decision support systems utilizing IoT technology and other devices. Resolving these challenges is essential to guarantee uninterrupted services whenever and wherever they are required, as delays and interruptions may arise from equipment unavailability.
- *Adaptability*: When constructing models, particularly for precision agriculture, it is crucial to consider the interaction between devices and their surroundings. Some devices may encounter difficulties in adapting to dynamic ambient conditions or sporadic hardware and communication issues.
- *Compatibility*: In order to address fragmentation and scalability requirements, the produced models or software should possess flexibility and be compatible with various computing systems.
- *Challenge*: A persistent challenge for numerous researchers is devising a strategy that optimizes profits. Efforts in the scientific community are concentrated on developing cost-effective solutions by reducing hardware and software demands for IoT deployments. Economic disparities across countries present challenges for farmers in adopting equipment and technology, emphasizing the necessity to formulate specific economic models.
- *Energy Optimization*: With the increasing reliance of devices on the IoT, WSNs, and other communication networks, energy efficiency has become a pressing issue. Traditional energy sources have historically supported model creation and operation, but the surge in gadget numbers has rendered traditional electricity unreliable. While exploring alternative energy sources like hydro-power, solar, and wind harvesting systems is crucial, their integration into model development has been largely unsuccessful, necessitating the formulation of new approaches.
- *Environmental Conditions*: Agriculture operates across diverse terrains, posing risks to data and services and complicating adaptation to evolving conditions, ultimately impacting system accuracy.
- *Heterogeneity*: The construction of systems involves the utilization of diverse devices, each with distinct requirements and services. Establishing interaction between heterogeneous modules and communication technologies in agriculture is pivotal, given that most models operate with varied equipment. The presence of heterogeneity in a network elevates complexity and may lead to erroneous conclusions.
- *Mobility*: The majority of existing concepts or systems lack mobility, which is essential for seamless framework implementation due to the prevalence of portable devices and applications. Ensuring consistent connectivity with mobile devices poses a significant challenge.
- *Real-Time Deployment*: Many research studies lack real-time testing and evaluation, highlighting the necessity for real-time analysis prior to system or model deployment to mitigate post-deployment setbacks.
- *Reliability*: Ensuring the reliability of IoT devices is critical for efficient and uninterrupted data transfer. Devices must gather and transmit dependable data to support decision-making

processes based on accurate analysis. Reliability challenges persist due to system failures, battery issues, node malfunctions, and other disruptions.
- *Standardization*: Device standardization is crucial for maximizing technology utilization across various applications. The absence of a validation mechanism for standard formats describing data or operations leads to errors and irregular modifications, resulting in inconsistent output from machine-readable programs that are incompatible with semantic ontologies. Standardization efforts aim to address issues related to device, application, system, and item interoperability.

24.5 FUTURE PERSPECTIVES

The IoT is rapidly evolving, spawning new services and applications. Extensive research is underway to integrate diverse heterogeneous systems, enhance security across various levels of IoT, and analyze "Big Data" to gain deeper insights and improve business processes. National policies of global governments aim to increase the production efficiency of fresh-cut vegetables and meat while reducing costs and maintaining high-quality standards (Table 24.1). Consumers are also demanding greater transparency in production processes and are increasingly concerned about the environmental impact of products they purchase. These factors present significant opportunities for the development and widespread adoption of IoT. Estimates of IoT's potential value from 2013 to 2022 vary greatly, ranging from less than $1 trillion to over $15 trillion, not including increased revenues, cost savings across industries, and overall economic activity spurred by IoT (Tuomala, 2021).

1. The flexibility, optimization, and accuracy IoT brings to industrial and production processes contribute significantly to its added value. It is reasonable to predict that in the near future, all aspects of agricultural processes will undergo significant transformation due to IoT. The substantial economic benefits of IoT have attracted major investments from prominent firms. Recent acquisitions, such as Google's $3.2 billion purchase of Nest Labs for IoT home automation and Cisco's $1.4 billion acquisition of Jasper Technologies for an IoT cloud platform, underscore IoT's potential and appeal to large investors and technology companies (Tzounis *et al.*, 2017). However, forming partnerships is complex as organizations involved in IoT tend to invest in specific elements due to its broad scope. Therefore, collaboration, overcoming rivalry, and striving for universal standards, will eventually be necessary to advance IoT as it progresses.
2. To reduce production, maintenance, and fault tolerance costs, energy harvesting, or power optimization techniques should be developed. Farms of all sizes heavily rely on electricity and energy to operate their machinery, typically supplied by the main power grid. However, this approach is neither efficient nor sustainable. To achieve self-sufficiency on farms, it is essential to implement energy harvesting or power optimization methods throughout their infrastructure.
3. Effective sensor-based systems need to be developed for environments at high elevations. Many high-elevation regions have rugged terrain, and access to electricity remains limited in several countries. Despite these challenges, these regions have not been extensively explored in precision farming literature. The future of precision farming depends on implementing sensor-based systems tailored for these environments.
4. Creation of applications for farmers or users that are easy to use to track the health of crops and plants. The majority of farmers are from diverse languages and ethnic origins. Consequently, their understanding and methods of carrying out agricultural activities vary. In order to enhance farmer-machine interaction, frameworks that consider the local language are crucial. Addressing language and comprehension barriers will facilitate the adoption of precision agricultural techniques in places that have been hesitant to embrace them.

TABLE 24.1
Current Developments and Future Directions of Leading Technology Companies in the IoT Agriculture Sector

Organization	Initiatives and Vision
Microsoft	Microsoft has made substantial investments in the field of smart agriculture. Targeting four vital areas—climate, agriculture, biodiversity, and water—they introduced a five-year, $50 million project named "AI for Earth" (Cowls *et al.*, 2023) in 2018., targeting four crucial areas: climate, agriculture, biodiversity, and water. The primary objective is to leverage their expertise in cloud computing, Artificial Intelligence (AI), and the Internet of Things (IoT) to address agricultural challenges. Their main project, FarmBeats (Ojha *et al.*, 2021), focuses on democratizing AI services for farmers globally.
Google	Google, in collaboration with MIT Media Lab's Open Agriculture Initiative (OpenAg), envisions a future for crops and agriculture that promotes healthier food systems (Penzenstadler *et al.*, 2018). This initiative incorporates the latest cloud-based services and AI technologies within controlled environments. Noteworthy projects include Food Computer devices and the Climate Recipes plan, which integrates plants' environmental responses. The goal is to define the necessary resources for sustainable agriculture.
Watson, IBM	IBM's Watson Decision service offers an AI-based agricultural platform aimed at enhancing harvest quality, sustainability, and overall smart agriculture (Swathi Kumari and Veeramanju, 2023). By leveraging IBM's capabilities in AI, IoT, and cloud computing, this platform assists farmers in making informed decisions throughout crop stages. It represents a comprehensive ecosystem solution, from farm to fork.
Intel	Intel's Inswift (Liang *et al.*, 2024) is an IoT platform designed to enhance agricultural efficiency through high-performance architecture that delivers real-time data and analytics. Its primary purpose is to create a connected, data-driven agricultural industry, facilitating the secure collection, transmission, analysis, and utilization of key data. Inswift collaborates with growers, manufacturers, and service providers to develop IoT hardware and software solutions tailored to modern agricultural challenges.
Jasper, CISCO	Integrating automation, real-time visibility, and remote diagnostics to achieve smart agriculture, Jasper, a division of CISCO, offers a cloud-based IoT platform for the agricultural industry (Yadav *et al.*, 2022). Initiatives like as MoTech, Topcon, Observant, and Semios provide examples of how to combine sensor technology, cloud computing, and communication to increase agricultural output and efficiency.
Dell	Robots and other devices with sophisticated AI and machine learning capabilities have been introduced by Dell as part of its foray into smart agriculture. To improve IoT and data science offerings, they have teamed up with AeroFarms, a pioneer in vertical farming (Ayaz *et al.*, 2019). AeroFarms' Newark facility is 130 to 390 times more productive than traditional farms and uses 95% less water. This success is attributed to Dell's edge-to-core cloud IoT architecture.
HPE	Purdue University, in collaboration with HPE, is revolutionizing agricultural research using IoT and wireless innovations. Daily, terabytes of data are captured via sensors, cameras, and human inputs, then processed and analyzed in real time. Purdue and HPE aim to transform digital agriculture practices by blending research, innovations, and technologies such as IoT and cloud computing. An HPE supercomputer, combined with wireless sensors and edge computing, supports this initiative (Beans, 2021).
Hello Tractors	Hello Tractors, together with IBM Research, has developed an AI and blockchain-based platform tailored for African farmers (Ayaz *et al.*, 2019). The Digital Doubled Digital Wallet, a cloud-based service co-financed by IBM, is designed to support Hello Tractor's goal of equipping small-scale farmers with advanced technology and data analytics for efficient agricultural practices.
Qualcomm	AgTech has been recognized by Qualcomm Ventures (QV) as a key area for investment. Chinese agricultural data portal FarmEasy, Indian agribusiness marketplace Ninjacart, Brazilian farm management company Strider, and other Latin American (LatAm) agreements are examples of recent partnerships (Ayaz *et al.*, 2019; Khan *et al.*, 2021). QV is committed to expanding AI services in agriculture, as seen by its $15 million investment in AgTech start-up Prospera.
Farm2050	The "Farm2050" initiative addresses the challenge of increasing global food production by 70% to meet 2050 demands (Khan *et al.*, 2021). This global research platform comprises venture capitalists, start-ups, and agricultural giants, working collaboratively to accelerate the development and adoption of AgTech solutions for future agricultural needs.

5. Utilizing past and current scientific data to create decision support systems for agricultural purposes. Typically, primary data is taken into account while creating precision farming models. However, prior case studies and deployment models should be considered for research in order to create models that are both dependable and affordable. This implies that a collaborative model creation technique has to be created in order to comprehend the nature of the activity.
6. Creation of trustworthy techniques for supply chain management in modern agriculture. Using AI and other learning techniques, platforms are designed to be user-friendly from the farmers' perspective and ease of use. The term "farmer's perspective" in this context refers to the idea that conducting an audit should occur far in advance of the prototype design phase. This research provides a comprehensive understanding of the disparities in the needs of farmers and agricultural systems worldwide. Multifaceted factors such as science, experience, industry, and knowledge should be the foundation for developing concepts and techniques.
7. Creation of systems that are resilient to changing environmental and soil circumstances. Developing systems that can withstand changes in both internal and external influences is crucial since agricultural settings are harsh and constantly changing due to climatic conditions and seasonal fluctuations.
8. Introduction of cloud-based resources for trustworthy, methodical, and scientific data collection and analysis.
9. Using heterogeneous or mixed data sets to identify different plant species.
10. Design of transportable and environmentally friendly agricultural equipment control systems for both big and small farms. According to the report, first-world countries are embracing modern agriculture ideas, while most third-world and other countries continue to reject this idea because of the expense and short shelf life of goods and equipment.
11. It is anticipated that new applications or research areas will emerge as precision agricultural platforms using IoT and other technologies evolve. A list and discussion of some possible ideas and forward-thinking research paths follow an analysis of the literature.

The evolution of the IoT is occurring at a rapid pace, giving rise to novel services and applications. Ongoing research endeavors are aimed at the integration of diverse heterogeneous systems, bolstering security across various levels of IoT, and analyzing "Big Data" to extract deeper insights and enhance business processes. Global governmental policies are focused on enhancing the efficiency of producing fresh-cut vegetables and meat, cutting costs, and upholding high-quality standards. Consumers are increasingly advocating for greater transparency in production processes and expressing concerns about the environmental impact of their purchases. These factors collectively present substantial opportunities for the proliferation and adoption of IoT. The projected value of IoT from 2013 to 2022 varies significantly, with estimates ranging from less than $1 trillion to over $15 trillion, not factoring in increased revenues, cost savings, and overall economic activity driven by IoT (Tuomala, 2021).

The enhancement of industrial and production processes through the flexibility, optimization, and precision offered by IoT significantly contributes to its value proposition. It is plausible to anticipate that all facets of agricultural processes will witness substantial transformation in the foreseeable future due to IoT. The notable economic advantages of IoT have attracted substantial investments from leading corporations. Recent acquisitions, such as Google's $3.2 billion acquisition of Nest Labs for IoT home automation and Cisco's $1.4 billion acquisition of Jasper Technologies for an IoT cloud platform, underscore the potential and attractiveness of IoT to major investors and technology firms. Nevertheless, establishing partnerships in the IoT domain is intricate as organizations tend to focus on specific components due to its broad scope. Hence, fostering collaboration, transcending competition, and striving for universal standards will eventually become imperative for the advancement of IoT as it evolves. In order to mitigate production, maintenance, and fault tolerance expenses, it is imperative to develop energy harvesting or power optimization methodologies. Farms of various sizes heavily rely on electricity supplied by the main power grid to operate their machinery. However, this conventional approach is neither efficient nor sustainable. Achieving self-sufficiency on farms necessitates the implementation of energy harvesting or power optimization techniques throughout their infrastructure. Efficient sensor-based systems tailored for high-elevation environments need to be devised. Numerous high-elevation regions possess rugged terrains and face limitations in electricity access across several countries. Despite these challenges, these regions have not been extensively explored in the context of precision farming literature. The future of precision farming hinges on deploying sensor-based systems customized for these specific environments. Developing user-friendly applications for farmers or users to monitor crop and plant health is essential. The agricultural community comprises individuals from diverse linguistic and ethnic backgrounds, resulting in varying understandings and methodologies in agricultural practices. To enhance interaction between farmers and machines, frameworks that account for local languages are indispensable. Addressing language and comprehension barriers will facilitate the adoption of precision agricultural techniques in regions hesitant to embrace them. Leveraging historical and contemporary scientific data to construct decision support systems for agricultural purposes is crucial. While building precision farming models, primary data is typically considered. However, it is imperative to also incorporate past case studies and deployment models in the research process to create reliable and cost-effective models. This underscores the necessity of developing a collaborative model creation approach to comprehend the essence of the activity. Establishing trustworthy methodologies for supply chain management in modern agriculture is paramount. Platforms are designed to be user-friendly from the perspective of farmers, incorporating AI and other learning techniques. The term "farmer's perspective" in this context emphasizes the importance of conducting audits well in advance of the prototype design phase. This research provides a comprehensive understanding of the disparities in the requirements of farmers and agricultural systems worldwide. A multifaceted approach, integrating science, experience, industry insights, and knowledge, should underpin the development of concepts and methodologies. Creating systems resilient to changing environmental and soil conditions is crucial. The development of systems capable of withstanding fluctuations in both internal and external influences is essential, given the harsh and ever-changing nature of agricultural settings due to climatic variations and seasonal changes. Introducing cloud-based resources for reliable, systematic, and scientific data collection and analysis is imperative. Utilizing heterogeneous or mixed datasets for the identification of diverse plant species is a common practice in the field. The development of portable and eco-friendly control systems for agricultural equipment, catering to both large-scale and small-scale farms, is a prevalent area of research. The adoption of modern agricultural concepts is observed in first-world nations, contrasting with the hesitance displayed by many third-world and other countries due to cost implications and the limited lifespan of products and machinery. The progression of precision agricultural platforms incorporating IoT and other technological advancements is foreseen to give rise to novel applications and research domains. An exploration of potential ideas and progressive research trajectories ensues subsequent to a comprehensive examination of existing literature.

24.6 CONCLUSION

IoT is anticipated to maximize agriculture's output in a variety of ways. A micro-precision model of agricultural output is about to replace the precision model now used in farms and greenhouses. The use of distributed and ubiquitous computing, together with accurate monitoring of the facilities, would ensure the provision of ideal growing or living conditions for both crops and animals. In addition to controlling the actuators in the most effective manner, maximizing resource and utility use, autonomous systems will also be able to regulate production according to market conditions, optimizing profit and decreasing expenses in every way possible. Conversely, FSCs that are outfitted with WSN and RFID technology will have the capacity to track a product's life cycle, automatically determining actions in case of malfunction, and enhancing consumers' sense of security through an open and transparent product lifecycle information system.

The aforementioned is an optimistic approach to IoT integration in agriculture. However, there are many individual players in this notion who are set to join in. Firstly, as these technologies mature and become more fully used, local networks must be protected against network disturbance. The majority of participants in an actual IoT scenario will use various pieces of equipment with varying technological specs and/or sensor features. Clearly, there is a need to provide a certain level of interoperability, filtering, and semantic annotation of the data provided by each producer. The use of data from diverse sources is only possible through this method, in order to enhance the efficiency of collaborative decision support or expert systems. To ensure the adoption of such a system, security, anonymity, and control over information access rights are essential. Taking a broader view, much of the information regarding strategic planning in businesses or institutions is protected from disclosure or unauthorized retrieval, safeguarding the market from unconventional strategies.

The IoT is poised to enhance agricultural productivity through various means. A new micro-precision agricultural output model is on the horizon to supplant the current precision model utilized in farms and greenhouses. By leveraging distributed and ubiquitous computing alongside precise facility monitoring, optimal growth conditions for crops and animals can be ensured. Furthermore, autonomous systems will not only control actuators efficiently but will also manage resources and utilities effectively, adjusting production in alignment with market conditions to optimize profits and minimize costs. Conversely, FSCs incorporating WSNs and RFID technology will possess the capability to monitor a product's lifecycle, automatically responding to malfunctions, and boosting consumer confidence through a transparent product lifecycle information system. This optimistic perspective on integrating IoT into agriculture is supported by numerous stakeholders eager to participate. As these technologies advance and gain wider adoption, safeguarding local networks against disruptions is imperative. The diverse array of equipment utilized by participants in an IoT setting necessitates a certain degree of interoperability, filtering, and semantic labeling of data provided by each contributor. This approach facilitates the utilization of data from multiple sources to enhance the effectiveness of collaborative decision support systems or expert systems. Security, privacy, and control over data access rights are vital for the successful implementation of such a system. From a broader standpoint, much of the strategic planning information within businesses or institutions is shielded from unauthorized access, ensuring market protection against unorthodox strategies.

LIST OF ABBREVIATIONS

EPC electronic product code
FAO Food and Agriculture Organization
FSC food supply chain
GIS geographical information systems
IOT Internet of Things
IPC integrated pest control
RFID radio frequency identification
NFC near field communication
WSNs wireless sensor networks

REFERENCES

Abdullah, Mahmood, A., Bibi, S., Naqve, M., Javaid, M.M., Zia, M.A., Jabbar, A., Ud-Din, W., Attia, K.A. & Khan, N., 2022. Physiological, biochemical, and yield responses of linseed (*Linum usitatissimum* L.) in α-tocopherol-mediated alleviation of salinity stress. *Frontiers in Plant Science*, 13, 867172.

Akhigbe, B.I., Munir, K., Akinade, O., Akanbi, L. & Oyedele, L.O., 2021. IoT technologies for livestock management: A review of present status, opportunities, and future trends. *Big Data and Cognitive Computing*, 5, 10.

Ali, H., Tanveer, A., Nadeem, M., Asghar, H. & Javaid, M., 2013a. Germination ecology of *Rhynchosia capitata*: An emerging summer weed in Asia. *Planta Daninha*, 31, 249–257.

Ali, H., Tanveer, A., Nadeem, M., Javaid, M., Kashif, M. & Chadhar, A., 2013b. Allelopathic effects of *Rhynchosia capitata* on germination and seedling growth of mungbean. *Planta Daninha*, 31, 501–509.

Ameen, M., Mahmood, A., Ahmad, M., Mansoor Javaid, M., Nadeem, M.A., Asif, M., Balal, R.M. & Khan, B.A., 2023a. Impacts of Climate Change on Fruit Physiology and Quality. *In Climate-Resilient Agriculture, Vol 1: Crop Responses and Agroecological Perspectives.* Cham: Springer, 93–124.

Ameen, M., Zafar, A., Javaid, M.M., Zia, M.A., Mahmood, A., Naqve, M. & Bibi, S., 2023b. Climate-Resilient Technology for Maize Production. *In* M. Hasanuzzaman (ed.) *Climate-Resilient Agriculture, Vol 2: Agro-Biotechnological Advancement for Crop Production.* Cham: Springer International Publishing, 157–188.

Anand, P., Singh, Y., Selwal, A., Alazab, M., Tanwar, S. & Kumar, N., 2020. IoT vulnerability assessment for sustainable computing: Threats, current solutions, and open challenges. *IEEE Access*, 8, 168825–168853.

Ayaz, M., Ammad-Uddin, M., Sharif, Z., Mansour, A. & Aggoune, E.H.M., 2019. Internet-of-things (IoT)-based smart agriculture: Toward making the fields talk. *IEEE Access*, 7, 129551–129583.

Aziz, M., Ashraf, M. & Javaid, M.M., 2018. Enhancement in cotton growth and yield using novel growth promoting substances under water limited conditions. *Pakistan Journal of Botany*, 50, 1691–1701.

Beans, C., 2021. The Internet of things comes to the farm: Growers connecting to a range of data at their fingertips. *BioScience*, 71, 780–786.

Bertoglio, R., Corbo, C., Renga, F.M. & Matteucci, M., 2021. The digital agricultural revolution: A bibliometric analysis literature review. *IEEE Access*, 9, 134762–134782.

Chadha, A., Florentine, S., Chauhan, B.S., Long, B., Jayasundera, M., Javaid, M.M. & Turville, C., 2019. Environmental factors affecting the germination and seedling emergence of two populations of an emerging agricultural weed: Wild lettuce (*Lactuca serriola*). *Crop and Pasture Science*, 70, 709–717.

Chadha, A., Florentine, S., Javaid, M., Welgama, A. & Turville, C., 2020. Influence of elements of climate change on the growth and fecundity of *Datura stramonium. Environmental Science and Pollution Research*, 27, 35859–35869.

Choudhary, D.S. & Meena, G., 2022. Internet of things: Protocols, applications and security issues. *Procedia Computer Science*, 215, 274–288.

Costa, F., Genovesi, S., Borgese, M., Michel, A., Dicandia, F.A. & Manara, G., 2021. A review of RFID sensors, the new frontier of Internet of things. *Sensors*, 21, 3138.

Cowls, J., Tsamados, A., Taddeo, M. & Floridi, L., 2023. The AI gambit: Leveraging artificial intelligence to combat climate change—Opportunities, challenges, and recommendations. *AI & SOCIETY*, 38, 283–307.

Del Sarto, N., Cesaroni, F., Di Minin, A. & Piccaluga, A., 2022. One size does not fit all. Business models heterogeneity among Internet of Things architecture layers. *Technology Analysis & Strategic Management*, 34, 787–802.

El Matbouly, H., Nikbakhtnasrabadi, F. & Dahiya, R., 2022. RFID Near-Field Communication (NFC)-Based Sensing Technology in Food Quality Control. *In* P. Chandra (ed.) *Biosensing and Micro-Nano Devices: Design Aspects and Implementation in Food Industries.* Singapore: Springer Nature Singapore, 219–241.

Farooq, M.S., Riaz, S., Abid, A., Umer, T. & Zikria, Y.B., 2020. Role of IoT technology in agriculture: A systematic literature review. *Electronics*, 9, 319.

Feng, H., Wang, X., Duan, Y., Zhang, J. & Zhang, X., 2020. Applying blockchain technology to improve agri-food traceability: A review of development methods, benefits and challenges. *Journal of Cleaner Production*, 260, 121031.

Gull, T., Mahmood, A., Shaheen, C., Javaid, M.M., Zia, M.A., Naqve, M., Bibi, S., Nadeem, M.A., Ameen, M., Nargis, J. & Khan, S.R., 2023. Climate Change and Nutrient Use Efficiency of Plants. *In* M. Hasanuzzaman (ed.) *Climate-Resilient Agriculture, Vol 1: Crop Responses and Agroecological Perspectives.* Cham: Springer International Publishing, 291–312.

Hanif, Z., Naeem, M., Ali, H.H., Tanveer, A., Javaid, M.M., Peerzada, A.M. & Chauhan, B.S., 2017. Effect of environmental factors on germination of *Salsola foetida*: Potential species for rehabilitation of degraded rangelands. *Rangeland Ecology & Management*, 70, 638–643.

Hassan, M.U., Nawaz, M., Mahmood, A., Shah, A.A., Shah, A.N., Muhammad, F., Batool, M., Rasheed, A., Jaremko, M. & Abdelsalam, N.R., 2022. The role of zinc to mitigate heavy metals toxicity in crops. *Frontiers in Environmental Science*, 10, 990223.

Hussain, S., Ahmad, A., Wajid, A., Khaliq, T., Hussain, N., Mubeen, M., Farid, H.U., Imran, M., Hammad, H.M., Awais, M., Ali, A., Aslam, M., Amin, A., Akram, R., Amanet, K. & Nasim, W., 2020. Irrigation Scheduling for Cotton Cultivation. *In* S. Ahmad & M. Hasanuzzaman (eds.) *Cotton Production and Uses: Agronomy, Crop Protection, and Postharvest Technologies.* Singapore: Springer Singapore, 59–80.

Jain, V.K., Mazumdar, A.P., Faruki, P. & Govil, M.C., 2022. Congestion control in Internet of Things: Classification, challenges, and future directions. *Sustainable Computing: Informatics and Systems*, 35, 100678.

Javaid, M., Tanveer, A., Ahmad, R., Yaseen, M. & Khaliq, A., 2012. Optimizing activity of herbicides at reduced rate on *Emex spinosa* Campd. with adjuvants. *Planta Daninha*, 30, 425–435.

Khan, N., Ray, R.L., Sargani, G.R., Ihtisham, M., Khayyam, M. & Ismail, S., 2021. Current progress and future prospects of agriculture technology: Gateway to sustainable agriculture. *Sustainability*, 13, 4883.

Khan, S., Singh, R., Khan, S. & Ngah, A.H., 2023. Unearthing the barriers of Internet of things adoption in food supply chain: A developing country perspective. *Green Technologies and Sustainability*, 1, 100023.

Kumar, A., Mangla, S.K. & Kumar, P., 2022. An integrated literature review on sustainable food supply chains: Exploring research themes and future directions. *Science of the Total Environment*, 821, 153411.

Lata, S., Mehfuz, S. & Urooj, S., 2021. Secure and reliable WSN for Internet of things: Challenges and enabling technologies. *IEEE Access*, 9, 161103–161128.

Liang, H., Zhang, Z., Hu, C., Gong, Y. & Cheng, D., 2024. A survey on spatio-temporal big data analytics ecosystem: Resource management, processing platform, and applications. *IEEE Transactions on Big Data*, 10, 174–193.

Mahmood, A., Wang, X., Shahzad, A.N., Fiaz, S., Ali, H., Naqve, M., Javaid, M.M., Mumtaz, S., Naseer, M. & Dong, R., 2021. Perspectives on bioenergy feedstock development in Pakistan: Challenges and opportunities. *Sustainability*, 13, 8438.

Mäntylä, M.V., Graziotin, D. & Kuutila, M., 2018. The evolution of sentiment analysis—A review of research topics, venues, and top cited papers. *Computer Science Review*, 27, 16–32.

Mehmood, A., Naeem, M., Khalid, F., Saeed, Y., Abbas, T., Jabran, K., Sarwar, M.A., Tanveer, A. & Javaid, M.M., 2018. Identification of phytotoxins in different plant parts of *Brassica napus* and their influence on mung bean. *Environmental Science and Pollution Research*, 25, 18071–18080.

Mohamed, R.E., Saleh, A.I., Abdelrazzak, M. & Samra, A.S., 2018. Survey on wireless sensor network applications and energy efficient routing protocols. *Wireless Personal Communications*, 101, 1019–1055.

Monteiro, A., Santos, S. & Gonçalves, P., 2021. Precision agriculture for crop and livestock farming—Brief review. *Animals*, 11, 2345.

Mostaccio, A., Bianco, G.M., Marrocco, G. & Occhiuzzi, C., 2023. RFID technology for Food Industry 4.0: A review of solutions and applications. *IEEE Journal of Radio Frequency Identification*, 7, 145–157.

Mowla, M.N., Mowla, N., Shah, A.F.M.S., Rabie, K.M. & Shongwe, T., 2023. Internet of things and wireless sensor networks for smart agriculture applications: A survey. *IEEE Access*, 11, 145813–145852.

Mrabet, R., 2023. Chapter 2 – Sustainable Agriculture for Food and Nutritional Security. *In* M. Farooq, N. Gogoi & M. Pisante (eds.) *Sustainable Agriculture and the Environment*. Academic Press, 25–90.

Navarro, M., Liang, Y. & Zhong, X., 2022. Energy-efficient and balanced routing in low-power wireless sensor networks for data collection. *Ad Hoc Networks*, 127, 102766.

Nižetić, S., Šolić, P., López-De-Ipiña González-De-Artaza, D. & Patrono, L., 2020. Internet of Things (IoT): Opportunities, issues and challenges towards a smart and sustainable future. *Journal of Cleaner Production*, 274, 122877.

Ojha, T., Misra, S. & Raghuwanshi, N.S., 2021. Internet of things for agricultural applications: The state of the art. *IEEE Internet of Things Journal*, 8, 10973–10997.

Penzenstadler, B., Khakurel, J., Plojo, C.J., Sanchez, M., Marin, R. & Tran, L., 2018. Resilient smart gardens—Exploration of a blueprint. *Sustainability*, 10, 2654.

Prakash, C., Singh, L.P., Gupta, A. & Lohan, S.K., 2023. Advancements in smart farming: A comprehensive review of IoT, wireless communication, sensors, and hardware for agricultural automation. *Sensors and Actuators A: Physical*, 362, 114605.

Raj, M., Gupta, S., Chamola, V., Elhence, A., Garg, T., Atiquzzaman, M. & Niyato, D., 2021. A survey on the role of Internet of Things for adopting and promoting Agriculture 4.0. *Journal of Network and Computer Applications*, 187, 103107.

Rasheed, A., Barqawi, A.A., Mahmood, A., Nawaz, M., Shah, A.N., Bay, D.H., Alahdal, M.A., Hassan, M.U. & Qari, S.H., 2022a. CRISPR/Cas9 is a powerful tool for precise genome editing of legume crops: A review. *Molecular Biology Reports*, 49, 5595–5609.

Rasheed, A., Mahmood, A., Maqbool, R., Albaqami, M., Sher, A., Sattar, A., Bakhsh, G., Nawaz, M., Hassan, M.U. & Al-Yahyai, R., 2022b. Key insights to develop drought-resilient soybean: A review. *Journal of King Saud University-Science*, 34, 102089.

Rasheed, A., Raza, A., Jie, H., Mahmood, A., Ma, Y., Zhao, L., Xing, H., Li, L., Hassan, M.U. & Qari, S.H., 2022c. Molecular tools and their applications in developing salt-tolerant soybean (*Glycine max* L.) cultivars. *Bioengineering*, 9, 495.

Saha, H.N., Roy, R., Chakraborty, M. & Sarkar, C., 2021. IoT-Enabled Agricultural System Application, Challenges and Security Issues. *In Agricultural Informatics*. Wiley Online Library, 223–247.

Saqib, M. & Moon, A.H., 2022. A concise review on Internet of things: Architecture, enabling technologies, challenges, and applications. *International Journal of Sensors Wireless Communications and Control*, 12, 629–650.

Shah, F.M. & Razaq, M., 2020. From Agriculture to Sustainable Agriculture: Prospects for Improving Pest Management in Industrial Revolution 4.0. *In* C.M. Hussain & P. Di Sia (eds.) *Handbook of Smart Materials, Technologies, and Devices: Applications of Industry 4.0*. Cham: Springer International Publishing, 1–18.

Sidhu, R.K., Kumar, R., Rana, P.S. & Jat, M.L., 2021. Chapter Five – Automation in Drip Irrigation for Enhancing Water Use Efficiency in Cereal Systems of South Asia: Status and Prospects. *In* D.L. Sparks (ed.) *Advances in Agronomy*. Academic Press, 247–300.

Singh, A.K., 2022. Smart Farming: Applications of IoT in Agriculture. *In* C.M. Hussain & P. Di Sia (eds.) *Handbook of Smart Materials, Technologies, and Devices: Applications of Industry 4.0*. Cham: Springer International Publishing, 1655–1687.

Sinha, B.B. & Dhanalakshmi, R., 2022. Recent advancements and challenges of Internet of Things in smart agriculture: A survey. *Future Generation Computer Systems*, 126, 169–184.

Statista, 2023. *Number of Internet of Things (IoT) Connections Worldwide from 2022 to 2023, with Forecasts from 2024 to 2033*. Statista.

Sun, X., Zhao, C., Li, H., Yu, H., Zhang, J., Qiu, H., Liang, J., Wu, J., Su, M., Shi, Y. & Pan, L., 2022. Wearable near-field communication sensors for healthcare: Materials, fabrication and application. *Micromachines*, 13, 784.

Swathi Kumari, H. & Veeramanju, K.T., 2023. Revolutionizing agriculture: A case study of IBM's AI innovations. *International Journal of Applied Engineering and Management Letters (IJAEML)*, 7, 95–114.

Tanveer, A., Javaid, M.M., Irfan, M., Khaliq, A. & Yaseen, M., 2015. Yield losses in chickpea with varying densities of dragon spurge (*Euphorbia dracunculoides*). *Weed Science*, 63, 522–528.

Tran-Dang, H., Krommenacker, N., Charpentier, P. & Kim, D.S., 2020. Toward the Internet of Things for physical internet: Perspectives and challenges. *IEEE Internet of Things Journal*, 7, 4711–4736.

Tuomala, V., 2021. *IoT Productivity versus Cybersecurity: Is the Risk Worth It?* South-Eastern Finland University of Applied Sciences.

Tzounis, A., Katsoulas, N., Bartzanas, T. & Kittas, C., 2017. Internet of Things in agriculture, recent advances and future challenges. *Biosystems Engineering*, 164, 31–48.

Ullah, H., Mahmood, A., Ijaz, M., Tadesse, B. & Honermeier, B., 2013. Evaluation of anise (*Pimpinella anisum* L.) accessions with regard to morphological characteristics, fruit yield, oil contents and composition. *Journal of Medicinal Plants Research*, 7, 2177–2186.

Vatistas, C., Avgoustaki, D.D. & Bartzanas, T., 2022. A systematic literature review on controlled-environment agriculture: How vertical farms and greenhouses can influence the sustainability and footprint of urban microclimate with local food production. *Atmosphere*, 13, 1258.

Vermesan, O., Friess, P., Guillemin, P., Gusmeroli, S., Sundmaeker, H., Bassi, A., Jubert, I.S., Mazura, M., Harrison, M. & Eisenhauer, M., 2022. Internet of Things Strategic Research Roadmap. *In Internet of Things-Global Technological and Societal Trends from Smart Environments and Spaces to Green ICT*. New York: River Publishers, 9–52.

Wasaya, A., Abbas, T., Yasir, T.A., Sarwar, N., Aziz, A., Javaid, M.M. & Akram, S., 2021. Mitigating drought stress in sunflower (*Helianthus annuus* L.) through exogenous application of β-aminobutyric acid. *Journal of Soil Science and Plant Nutrition*, 21, 936–948.

Yadav, V.S., Singh, A.R., Raut, R.D., Mangla, S.K., Luthra, S. & Kumar, A., 2022. Exploring the application of Industry 4.0 technologies in the agricultural food supply chain: A systematic literature review. *Computers & Industrial Engineering*, 169, 108304.

Zafar, S., Afzal, H., Ijaz, A., Mahmood, A., Ayub, A., Nayab, A., Hussain, S., Maqsood, U.-H., Sabir, M.A. & Zulfiqar, U., 2023. Cotton and drought stress: An updated overview for improving stress tolerance. *South African Journal of Botany*, 161, 258–268.

Zafari, F., Gkelias, A. & Leung, K.K., 2019. A survey of indoor localization systems and technologies. *IEEE Communications Surveys & Tutorials*, 21, 2568–2599.

Index

Note: *Italicized* page references refer to figures and **bold** page references refer to tables.

A

Abiotic stresses, 37, 163–164, *166*
  chilling stress, 164
  drought, 163
  effect on crop plants, *253*
  high temperature, 163–164
  marginal lands, 163–164
  potential grain, 142
  resistance, **125**, 129
  salinity stress, 164
  tolerances, 128
  waterlogging stress, 164
Abscisic acid (ABA) and stress response, 191–192
Acid pre-treatment, 363
Actinobacteria, 255
Action threshold level (ATL), 344
Aerial sowing, 218
Agribot, 29
Agricultural automation and robotics, 114
Agricultural grabbing demands, 116
Agricultural waste, *see* Bioethanol production from agricultural waste
Agriculture
  better crop quality through innovation, 6
  challenges to, 1–5, *3*
  climate change and its impact, 3
  concept of sustainability in, 6–9
  current status of, 1
  declining arable land and use of marginal lands, 4
  economic and market-related challenges, 4
  economic benefits, 6
  food security and nutrition, 6
  in GDP of different countries, *2*
  market standards and consumer preferences, 5–6
  need for improved crop quality, 5–6
  nutritional demands, 5
  pest and disease resistance, 4
  population growth and food security, 5
  recent innovations in agriculture, 9
  social and labor issues, 5
  solutions and technologies, 9–15
  solutions to agriculture challenges, **10**
  water scarcity and inefficient irrigation practices, 3
Agroecology, 8, 219, 320, 407
Agroforestry, 9, 218, 220, 320–321
  fodder production systems, 220
  regenerative agriculture (RA), 401–406
  sustainable farming systems, 320–321
Agro-industrial waste (AIW), 355–356
Agronomic management of soil health and nutrient levels, 388–389
  benefits of crop rotation, 389
  crop rotation, 388
Agro-vision technologies, 24
Alfalfa aphid, 127
Alfalfa caterpillar, 127
Alkaline pretreatment, 363
Alley cropping, 218
Ammonia fiber explosion (AFEX), 362
Arable land, 4
Arbuscular mycorrhizal fungi (AMF), 377
Artificial intelligence (AI), 22, 25, 32, 71, 84, 114, 115, 116, 215, 325, 329–330, 425; *see also* Machine learning (ML)
Artificial satellites, 26–27
Ashton, Kevin, 414
Automated planting and harvesting, 12
Autonomous equipment, 27
Autonomous farming, 26
Auxins and stress response, 192
Azotobacter, 286

B

Bacillus thuringiensis (Bt), 45, 128, 131, 336, 340, *341*
Backyard forage production, 217
Baker's Rules, 105
Barren brome *(Bromus diandrus)*, 105
Basin listing, 151
Beauveria bassiana, 131
Better crop quality through innovation, 6
Better food quality, 14
Big Data, 24–25, 31, 71, 114, 115, 248–249, 429, 431
Bioaugmentation, 254–255
Biochar, 287
  nutrient management, 287
  organic soil enrichment, 383–384
Biodiversity and ecological resource management, 323–324
  crop diversification, 323–324
  habitat management for beneficial organisms, 324
  integrated pest management (IPM), 324
  pollinator conservation, 324
Biodiversity conservation, 7
Bioenergy crops, 325
Bioethanol production from agricultural waste, 354–368
  challenges, 367
  ethanol production technologies, 359–366
    acid pretreatment, 363
    alkaline pretreatment, 363
    ammonia fiber explosion (AFEX), 362
    biological pretreatment, 364–365
    chemical pretreatment, 363–364
    CO_2 Explosion, 362
    effective and efficient cellulolytic microorganisms (bacteria and fungi), 365–366
    electron beam irradiation pretreatment, 361
    endoglucanases and exoglucanases, 365
    enzyme cellulase, 365

liquid hot water method, 362
microwave therapy, 361
organosolv pretreatment, 363–364
physical pretreatment, 360
physicochemical pretreatment, 361–362
pretreatment of agricultural waste, 360
pyrolysis, 361
steam explosion/auto hydrolysis, 361–362
wet oxidation, 363
as feedstock, 355–357
agro-industrial wastes' composition, 356
harnessing agricultural waste, 355–356
raw materials for ethanol production, 356–357
fermentation, 366–367
microorganisms for, 357–359
role of genetically engineered bacteria, 358
role of genetically engineered yeast, 359
role of thermophilic bacteria, 358
overview, 354–355
Biofortification, 12
Bioleaching, 255
Biological nitrogen fixation, 323
Biological pretreatment, 364–365
Biopesticides
benefits of, 336, **337**
resistance, 131–132, **132**
Bioremediation, 246–247
Biosafety of nanomaterials, 98
Bioslurry, 287
Biosorption, 255
Biostimulation, 254–255
Biotech crop adoption and development, 46–47
Biotechnology for crop production, 11–12, 37–48
applications of, **39**
biotechnology-assisted crop improvement, 44–46
for developing countries, 47
environmental impacts of, 46–47
high yields, 45–46
irrigation, 39
overview, 37–38
recombinant DNA technology, 40
resistant to diseases, 45
resistant to insects and pests, 44–45
schematic diagram, *38*
socioeconomic implications, 46
sustainable agricultural method, 40
tillage method, 39
tools for crop improvement, 40–44
traditional cropping methods, 39
and traditional methods, 38–40
transgenic plants, 45
Biotic stresses, 37, **165**, 165–166, *166*, 339, 349
Blackgrass species *(Alopecurus myosuroides)*, 105
Blockchain technology, 26
Brome species (*Bromus* spp.), 105
Bubbler, 200

C

Carbon dioxide removal (CDR), 249
Carbon sequestration, 406–407
CAS (CRISPR-associated protein), 92
Cellulosic composition of different plants, **356**
Chatbots/digital advisors, 28
Chemical innovations in weed control, 109
Chemical pretreatment, 363–364
Chemical weed management, 109–116
data-driven technologies, 114–116
emerging technologies in weed detection/monitoring, 113–114
genetic engineering and weed resistance, 112
herbicide mixture/herbicide combinations, 110
herbigation, 109–110
nano-herbicides, 110
new generation spot sprayer technology, 110
precision weed management, 110–112
Chickpeas, 168, 171
Chilling stress, **164**
Circular economy (CE), 408
Climate change, 4
adaptation, 60
adaptation mechanisms in plants, 188–192
abscisic acid (ABA) and stress response, 191–192
auxins and stress response, 192
drought adaptation, 188
heat and cold stress adaptation, 188–189
molecular responses to climate stress, 189–191
plant hormones and climate adaptation, 191
fundamental physiologic mechanisms in crops, 184–185
photosynthesis, 184–185
respiration, 185
rubisco activase, 185
hydrological cycle, 3, 147
impact on plant physiology, 180–193
impacts of morphology, 181–184
and its impact, 3
leaf developmental responses to elevated CO_2, 183
metabolic processes, 186–188
grains quality, 187–188
oxidative stress, 186
protein denaturation, 186
mitigation, 7
overview, 180–181
reactive oxygen species, 193
reproductive developmental responses, 183–184
root developmental responses, *182*, 182–183
Climate engineering, 249–250
Climate-smart agriculture (CSA), 310
Clustered regularly interspaced short palindromic repeats (CRISPR), 271
CO_2 explosion, 362
CO_2 fertilization effect, 186
Community development, 8
Compartmental bonding, 150
Compost, 376–378
advancements in on-farm composting technologies, 377–378
composting process, 377
effects on soil enzymatic and biological activity, 376–377
optimal composition of nutrients in, 376
Composting and organic fertilization, 323
Compost tea, 385–386
Conservation agriculture (CA), 8–9, 107, 117, 320
Conservation tillage, 149–150
Consolidated bioprocessing (CBP), 367

Consultative Group on International Agricultural Research (CGIAR), 304
Contour forage production, 218
Convention for International Trade in Endangered Species (CITES), 234
Convention on Biological Diversity (CBD), 234
Cooperative Extension Service, 345–346
Cotton plant *(Gossypium hirsutum),* 96
Cover crops/cover cropping, 378–381
benefits of, 378, *379*
enhancing soil microbial activity, 380
grazing, 217
improving soil structure, 380
increasing organic matter content, 379
nitrogen economy, 378–379
reducing disease and insects, 381
reducing soil erosion, 380
soil's nutrient content with vital minerals, 379–380
CRISPR (clustered regularly interspaced short palindromic repeats), 11–12, 15, **41**, 91–92
CRISPR-associated protein 9 (CRISPR-Cas9), **41**, 42–44, 128–129, 250
CRISPR-directed nucleases, 41
Crop disease, 29
Crop diversification, 323–324
Crop livestock integration, 404–405
Crop monitoring and scouting, 12
Crop production
biotechnology, *see* Biotechnology for crop production
carbon dioxide, 181
greenhouse gases, 180
precision farming, 215
temperature, 184
Crop resilience, 96–97, 251
Crops rotation, 151
nutrient management, 289
in organic farming (OF), 58
resistance, 129–133
Crop water stress index (CWSI), 29
Cultivating resilience, 300–313
agronomic practice, 304–305
case studies and best practices, 306–310
maize production in regions vulnerable to heat stress, 308
rice cultivation in flood prone areas, 308–310
wheat cultivation in drought-prone regions, 306–307
challenges, 311–312
crop adaptation strategies, 303–304
genetic diversity and breeding for resilience, 303–304
effect of elevated CO_2 on plant physiology, 303
nutrient uptake and allocation, 303
photosynthesis and carbon assimilation, 303
water use efficiency, 303
impact of elevated CO_2 and temperature on crops, 301–303, **302**
crop yield and quality, 303
effects of temperature increase, 301
photosynthesis, 301
protein denaturation, 301–306
respiration, 301
overview, 300–301
policies and institutional support, 310–311, **312**
for climate-smart agriculture (CSA), 310
international cooperation and knowledge sharing, 310–311
technological innovations, 305–306
climate modeling, 305
crop monitoring, 305
energy efficiency, 306
food storage and preservation, 306
greenhouse management, 306
livestock management, 306
online marketplaces, 306
precision agriculture, 305
soil health management, 306
water management, 306

D

Dairy cattle manure, 387–388
Data-driven technology, 10, 20
Decision support systems (DSSs), 326–327
Deep learning (DL), 24, 25, 115
Deep tillage, 149
Deficit irrigation, 147, 172
Dichlorodiphenyltrichloroethanes (DDTs), 124
Differential GPS (DGPS), 22
Disease resistance, 124
Diseases and insect-pests, 166
Distance sensing approaches, 111
Distiller's dried grains with soluble (DDGS), 354
Diversified crop rotation (DCR), 289
Drainage management, 322
Drip irrigation, 199–201, *200*
bubbler, 200
components of the sprinkler system, 202
dripper line irrigation system, 200
enter drip irrigation, 201
farm sprinkler system, 201
field sprinkler network, 201
micro-spray irrigation system, 200
mister irrigation system, 200–201
project irrigation systems, 201–202
sprinkler system, 201
Dripper line irrigation system, 200
Drip strategy (DS), 199
Drones, 12–13, 71–85; *see also* Robots
additional benefits of, 80–81
agriculture challenges, 73
application of, *75*, 75–81
basic components of, *74*
climate variability, 73
conventional agriculture, **72–73**
crop health and irrigation monitoring, 78–79
crop monitoring, 73
diseases and weeds monitoring, 79–80
limitations of, 81
nutrient status and deficiency monitoring, 80
overview, 71–73
precision farming, 73
seed sowing from air, 76–77
soil and field analysis, 76
spraying operation in fields, 77–78
water stress monitoring, 79
Drought adaptation, 188

Drought stress, 163, **163**
Dryland farming, 141–154
climate change impacts on, 142–148
changes in crop suitability, 144
economic impacts, 145
increased frequency of extreme weather events, 144
pest and disease pressure, 144
socioeconomic effects, 145
water-use efficiency problems, 145–148
cover more crops under minimum support price schemes, 153
cover more households under crop and livestock insurance, 153
facilitate migration, 153
implications for policy, 152
implications for research priorities, 154
importance of moisture management, 148–152
increase inflow of institutional credit to dryland agriculture, 153
institute measures to cope with globalization and marginalization, 153
overview, 141–142
raise public investment in technology and infrastructure, 152
rationalize subsidies on agricultural inputs, 152

E

Eco-friendly pest management, 336–350
barriers, 345–346
basic research and evaluation, 348
benefits of biopesticides, **337**
biological control, 339
coordination and execution of federal-state relations, 347–348
cultural control, 339
defined, 336
education, 348
fundamentals, 343–344
assessing and maintaining documentation, 344
avoidance, 343
economic harm and pest activity threshold level, 344
formulating methods for management, 344
identification of pests, 343
routine monitoring and inspection mechanism, 343–344
institutional limitations, 346
integrated pest management (IPM), 341–343
IPM program into action, 344–345
lack of enough skilled workers, 346
mechanical control, 340
multifaceted impacts, 336
new developments and technology, 346–347
obstacles and prospects, 345–348
pest behavior, 337–339
behavioral dynamics of economically important pests, 337
life cycles of pests and hosts, 337–339, *338*
plant-incorporated protectants, 340–341
principal obstacles to advancement, 345
promoting use and eliminating barriers, 348
sources of user information, 345–346
suggestions for, 347
Ecological improvements in farming system (EIFS), 320
Economic diversification, 7
Economic injury level (EIL), 344
Economic threshold level (ETL), 344
Economic viability of sustainable farming practices, 327
Ecosystem services adaptability, 252
Electron beam irradiation pretreatment, 361
Electronic identification (EID) technologies, 22
Electronic product code (EPC), 418
Endoglucanases and exoglucanases, 365
Energy efficiency in farming operations, 324–325
Energy resource management, 324–325
bioenergy crops, 325
energy efficiency in farming operations, 324–325
renewable energy sources in agriculture, 325
Enter drip irrigation, 201
Environmental stress, 245–256
bioaugmentation and biostimulation, 254–255
bioleaching and biosorption, 255
climate engineering and geoengineering, 249–250
function of microbes in resolving, 254
improvement of crop resilience under, 251
manage balanced nutrition to reduce, 252
management strategies, 246–249
big data analytics, 248–249
bioremediation and phytoremediation, 246–247
ex-situ bioremediation, *247*, 247–248
innovations in technology for mitigation, 248
in-situ bioremediation, *247*, 247–248
nanotechnology, 248
remote sensing, 248–249
modern innovations in genetics and biotechnology, 250
nanoparticles (NPs) in, **246**
overview, 245–246
primary nutrients, 253–254
role of actinobacteria, 255
and secondary nutrients, 254
solution strategies adapted from nature, 251–252
stress-tolerance gene modification, 250
technologies to conserve and purify water, 250
Enzyme cellulase, 365
Equitable access to resources, 8
Equity First International principles, 12
Ethanol production technologies, 359–366
alkaline pretreatment, 363
ammonia fiber explosion (AFEX), 362
biological pretreatment, 364–365
chemical pretreatment, 363–364
CO_2 explosion, 362
effective and efficient cellulolytic microorganisms (bacteria and fungi), 365–366
electron beam irradiation pretreatment, 361
endoglucanases and exoglucanases, 365
enzyme cellulase, 365
liquid hot water method, 362
microwave therapy, 361
organosolv pretreatment, 363–364
physical pretreatment, 360
physicochemical pretreatment, 361–362
pretreatment of agricultural waste, 360
pyrolysis, 361
steam explosion/auto hydrolysis, 361–362
wet oxidation, 363
European Weed Investigation Society, 107

Evolutionary arms race, 106
Ex-situ bioremediation, *247*, 247–248
Extreme weather events, 144

F

Farmer education and extension programs, 327
Farming robots, 81; *see also* Robots/robotics
Farm integration, 217
Farm Management Information Systems (FMIS), 114
Farm management systems, 14
Farm sprinkler system, 201
Fermentation, 366–367
Fertile lands, 4
Fertilizing robots, 81; *see also* Robots/robotics
Fiber-reinforced polymer (FRP), 228
Field sprinkler network, 201
Fodder production systems, 211–222
  advantages of adopting modern agricultural technologies, 221
  aerial sowing, 218
  agroforestry, 218
  alley cropping, 218
  backyard forage production, 217
  challenges, 221–222
  changing climate patterns, 213
  contour forage production, 218
  cover crop grazing, 217
  data analytics in fodder farming, 215–216
  farm integration, 217
  fodder conservation, 216
  fodder crop breeding, 216
  genetic advancements, 219–220
    climate-adaptive traits, 220
    drought tolerance, 219
    nutritional enhancement, 220
    pest and disease resistance, 219
  greenhouses, 216
  harvesting and storage, 219
  hydroponic, 214
  importance of, 211–212
  intercropping and under-sowing, 217
  limitations and challenges, 212–213
    crop residues, 213
    land and resource constraints, 213
    land degradation, 213
    quality control, 213
    seasonal variability, 213
    weather dependency, 213
  modern agricultural technologies for, 213–216
  nutrients and grazing management, 216
  over sowing, 217
  overview, 211
  precision agriculture, 215
  role in ensuring food security and sustainability, 212
  silage and hay making, 219
  sustainable practices, 220–221
    agroforestry, 220
    conservation tillage, 220
    crop rotation, 220
    organic fodder farming, 220
    water management, 221
  technical approaches for, 216–220
  traditional foraging techniques, 212
    cut-and-carry, 212
    grazing, 212
    silage and hay making, 212
  traditional methods and challenges, 212–213
  using dual-purpose crops, 218–219
  vertical farming, 214–215
Food and Agriculture Organization (FAO), 29, 78, 90, 342, 414
Food and Drug Administration (FDA), 346
Food security and nutrition, 6
Food supply chain tracking, 424–425
French Agency for Food Safety (AFSSA), 61
Fruit-picking robots, 28

G

Gene editing, 11–12, 129
Genetically engineered bacteria, 358
Genetically engineered yeast, 359
Genetically modified (GM) crops, 54
Genetically modified organisms (GMOs), 11, 43, 129
Genetic engineering
  and biotechnology, 128–129
  and weed resistance, 112
Genetic modification (GM), 41, 92
Genome-editing techniques, 41–44, *43*, 48, 163, 271, 313
Genome editing tools, 44
Genomic selection (GS), 42
Geoengineering, 249–250
Geographic information systems (GIS), 22, 326
Global Navigation Satellite System (GNSS), 25
Global positioning systems (GPS), 11, 21–22, 112
Global Unmanned Spray System (GUSS), 26–27
Glutathione peroxidase (GPX), 252
GPS and GIS technologies, 11
Grading robots, 82
Grains quality, 187–188
Green bio-refineries (GBRs), 46
Greenhouse gases (GHGs), 248–250
Greenhouses fodder production, 216
Guidance systems, 25–26

H

Habitat management for beneficial organisms, 324
Harnessing agricultural waste, 355–356
Harvesting robots, 82
Harvest weed seed control, 106
Health and nutrition, 8
Heat and cold stress adaptation, 188–189
Heat shock proteins (HSPs), 270
Hemp fibers, 230
Herbicide mixture/herbicide combinations, 110
Herbicide reduction, 28
Herbicide resistance, 125
Herbicide-resistant (HR) weed growth, 106, 108, 112, 117
Herbigation, 109–110
Hewitt-Sperry Automatic Airplane, 74
Hierochloe spicata, 185
High temperature, 163–164
Humic substances (HS), 94, 282

Hydrogels, 151–152
Hydrological cycle, 3, 147
Hydroponic fodder production, 214

I

ICT-based agricultural gadgets, 31
Image processing procedures for weed detection, *116*
Improved crop quality, 5–6
Information and communication technology (ICT), 24, 114
Insect resistance, 124
In-situ bioremediation, *247*, 247–248
Integrated disease management (IDM), 173
Integrated farming systems (IFS), 319–320
Integrated nutrient management (INM), 322–323
Integrated pest management (IPM), 4, 7, 9, 15, 107, 126–127, 324, 341–345
  implementation strategies of, 127–128
  principles of, 127
Integrated weed management (IWM), 105–106, *107*, 107–108, 112, 118
Intercropping and under-sowing, 217
Inter-cropping/multi-cropping, 171, 289
Intergovernmental Panel on Climate Change (IPCC), 180
International crop research institute for the Semi-Arid tropics (ICRISAT), 118
International Maize and Wheat Improvement Center, 308
International Rice Research Institute (IRRI), 118, 309
International Union for Conservation of Nature and Natural Resources (IUCN), 234
Internet of Things (IoT), 14–15, 22, 24–25, 29, 30, 414–433
  applications in agriculture, 421–426
    controlled environment agriculture, 423–424
    food supply chain tracking, 424–425
    livestock applications, 425–426
    monitoring and control, 421–422
    open-field agriculture, 422–423
  based farm management systems, 14
  challenges in agriculture, 426–429
    accessibility, 426
    adaptability, 426
    challenge, 427
    compatibility, 427
    energy optimization, 427
    environmental conditions, 427
    heterogeneity, 427
    mobility, 427
    real-time deployment, 427
    reliability, 427–428
    standardization, 428–429
  enabling technologies, 418–419
    application layer, 419
    network layer, 419
    perception layer, 418–419
  modern agricultural framework, *420*
  overview, 414–417
  smart farming, 14
  supply chain traceability, 14–15
  in various agricultural sectors, *415*
"IoT Applied Greenhouse Monitoring System," 26
Irrigation management, 172, 198–206
  benefits to environment and economy, 206
  building farmer capabilities and training, 205–206
  integration of technology, 202–203
  overview, 198–199
  planning and designing, 203–205
    growing season, 205
    location, 205
    preliminary design (initial design), 203
    soil properties, 205
    thorough design (in-depth layout), 204–205
    water sources, 205
  principles of, 199–202
    drip irrigation system, 199
    primary drip irrigation source, 199–202
Irrigation systems
  for sustainable water use, 322

L

Late embryogenesis abundant (LEA), 270
Leaf developmental responses to elevated CO_2, 183
Least squares regression (LSR), 27
Lentils, 169, **169**
LiDAR, 27
Life cycles of pests and hosts, 337–339, *338*
Liquid hot water method, 362
Long-term viability, 7

M

Machine learning (ML), 21, 25, 26, 85, 113, 115, 326; *see also* Artificial intelligence (AI)
Machine vision, 111
Maize (*Zea mays* L.), 171
Maize production in regions vulnerable to heat stress, 308
Marginal lands, 4, 161–173
  abiotic stresses, 163–164
  agronomic strategies, 170
  biotic stresses, **165**, 165–166
  breeding strategies, 170
  challenges and needs, 161
  criteria for management and assessment, 162
  fallow and conservation tillage, 170–171
  limitations for pulses, 162–163
  mitigating strategies against stress, 170–173
  overview, 161
  pulses, 166–169
  use management, 161
Marker-assisted breeding (MAB), 42
Marker-assisted selection (MAS), 42
Market incentives for sustainable production, 327
Market standards and consumer preferences, 5–6
Mega nucleases, 41
Mesoporous silica nanoparticles (MSNs), 95
Metarhizium anisopliae, 131
Microbial seed coating, 271–272
Micro-spray irrigation system, 200
Microwave therapy, 361
Mister irrigation system, 200–201
Modern irrigation controller, 203
Mohyuddin, G., 115
Moisture and water conservation, 172
Moisture management, 148–152
  barriers of soil moisture conservation, 149
  basin listing, 151

compartmental bonding, 150
conservation tillage, 149–150
crop rotation, 151
deep tillage, 149
hydrogels, 151–152
mulching, 150–151
strip cropping, 151
techniques of soil moisture conservation, 149
Molecular responses to climate stress, 189–191
Mulching, 150–151, 381–383
bark mulches, 382
compost, 382
effects on soil properties, 383
newspaper, 382
organic mulches, 381–382
sawdust, 382
straw, 382
types of, 381–382
wood chips, 382
Multi-/hyper-spectral pictures, 111
Multiple-criteria decision analysis (MCDM), 328
Multi-walled carbon nanotubes (MWCNTs), 92, 269
Mung bean, 168–169
Municipal solid waste (MSW), 229
Mustard (*Brassica juncea* L.), 171
Mutational breeding, 37

N

Nanoemulsions, 95
Nano-encapsulation, 14
Nano-fertilizers, 13
Nano-herbicides, 110
Nanomaterials (NMs), 90
biosafety of, 98
manufacturing, 91
negative influence of, 96–97
for soil remediation, 93–94
Nanoparticles (NPs), 90, 97
applications of, *92*
effects on crop protection, **95**
enhancement in seed germination, *94*
in environmental stress, **246**
impact on performance of seeds and seedlings, **269**, *270*
Nano-pesticides, 13
Nano-sensors, 13, 95
Nanostructures for soil health and genetics, 93
Nanotechnology, 13–14, 90–99, 248
application of, *91*
biosafety of nanomaterials, 98
crop resilience, 96–97
evolution of, 90–91
fundamentals of, 91–95
mesoporous silica nanoparticles (MSNs), 95
nanosensors, 95
overview, 90
in plant improvement, 97
in regulating seed germination and plant growth, 92
role in crop protection, 94
seed enhancement, 268–270
silver nanoparticles, 96
for soil health and genetics, 93
for soil remediation, 93–94
National Maize Research Programs (NMRPs), 308
Natural fibers, 229, **230**
Natural organic manures/fertilizers, 386–388
application of organic fertilizers, 388
dairy cattle manure, 387–388
organic manure, 387
peanut hulls manure, 387
poultry manure, 387
Nature-based solutions (NBS), 246, 251
Near field communications (NFCs), 418
Near-infrared (NIR), 78
New generation spot sprayer technology, 110
Nitrogen economy, 378–379
Non-governmental organizations (NGOs), 62–63
Normalized Difference Vegetation Index (NDVI), 78–79, 80
Novel genetically modified technologies (nGMs), 41
Nutrient cycling in organic farming, 57
Nutrient losses, 64–65
Nutrient management, 281–293
advantages of, 291
agronomic practices, 289–290
crop rotation and diversification, 289
intercropping systems, 289
zero tillage, 289–290
eco-friendly nutrient management in soil, 284–288
Azotobacter, 286
biochar, 287
bio-compost, 284–285
bioslurry, 287
naturally occurring organic fertilizers, 284
rhizobacteria as bio-fertilizer, 285–286
seaweeds, 288
sewage sludge and biological solids, 288
vermicompost, 285
fertilizer application, 288–289
right placement, 288–289
right rate, 288
right source, 289
right timing, 288
modern strategies, 291–292
farm waste valorization and commercialization, 291
fluorescence imaging for accessing photosynthesis, 292
precision agriculture technologies, 291–292
remote sensing technologies, 292
sensor and drone technologies for real-time monitoring, 291
soil testing and plant analysis, 291
overview, 281–282
in soil using agricultural fertilizers, 283–284
inorganic agricultural fertilizers, 284
organic agricultural fertilizers, 283–284
strategies, 282
cover crop (green manure), 282
animal manure, 282
use of AI for, 292–293
Nutrient resource management, 322–323
biological nitrogen fixation, 323
composting and organic fertilization, 323
integrated nutrient management (INM), 322–323
precision agriculture techniques, 323
Nutrients and grazing management, 216
Nutritional demands, 5

O

Open-pollinated varieties (OPVs), 308
Optimal composition of nutrients, 376
Optimized Irrigation Recommendation Information System (OpIRIS), 423–424
Organic cotton, 231
Organic farming (OF), 8, 54–66, 320
  arable crops, **58**
  certification and market competition, 65–66
  challenges in, 63–66
  crop rotation in, 58
  economic benefits of, 61–62
  environmental consciousness, 59–60
  green manure in, 58
  low production rate, 63–64
  nutrient cycling in, 57
  nutrient losses, 64–65
  nutritional and health benefits, 61
  opportunities in, 59–61
  overview, 54–55
  plant disease management, 58–59
  principles and outcomes of, *56*
  principles of, 55–59
  regenerative agriculture (RA), 405
  resistance, 132–133
  social benefits of, 62–63
  soil health in, 55–57, **56**
Organic food marketing, *62*
Organic manure, 387
Organic soil enrichment
  agronomic management of soil health and nutrient levels, 388–389
    benefits of crop rotation, 389
    crop rotation, 388
  biochar, 383–384
  compost, 376–378
    advancements in on-farm composting technologies, 377–378
    composting process, 377
    effects on soil enzymatic and biological activity, 376–377
    optimal composition of nutrients in, 376
  compost tea, 385–386
    benefits of, 378
    enhancing soil microbial activity, 380
    improving soil structure, 380
    increasing organic matter content, 379
    nitrogen economy, 378–379
    reducing disease and insects, 381
    reducing soil erosion, 380
    soil's nutrient content with vital minerals, 379–380
  enrichment of organic soil fertilizers/amendments, 391–392
  erosion control techniques, 389–391
    soil structure improvement, 390–391
    water conservation, 390
  mulching, 381–383
    bark mulches, 382
    compost, 382
    effects on soil properties, 383
    newspaper, 382
    organic mulches, 381–382
    sawdust, 382
    straw, 382
    types of, 381–382
  natural organic manures/fertilizers, 386–388
    application of organic fertilizers, 388
    dairy cattle manure, 387–388
    organic manure, 387
    peanut hulls manure, 387
    poultry manure, 387
  nutrients grow through, 374–393
  overview, 374–375
  techniques, 375–376
    advanced methods, 375–376
    conventional method, 375
  vermicomposting, 384
Organosolv pretreatment, 363–364
Original equipment manufacturers (OEMs), 30
Over sowing, 217
Oxidative stress, 186

P

Peanut hulls manure, 387
Pearl millet *(Pennisetum glaucum),* 171
Permaculture, 321
Pest and disease resistance, 4
Pest management and yield monitoring, 29–30
Pest resistance, 123, 124
Phalaris minor, 107
Phaseolus vulgaris, 285
Photochemical reflectance index (PRI), 79
Photosynthesis, 184–185
Physical pretreatment of agricultural waste, 360
Physicochemical pretreatment, 361–362
Phyto-filtration, 108
Phytoremediation, 246–247
Pigeon pea (*Cajanus cajan* L.), 167, **167**
Plant-based pesticides, 131
Plant-beneficial microorganisms (PBM), 271–272, **272**
Plant growth-promoting rhizobacteria (PGPR), 284–286
Plant hormones and climate adaptation, 191
Plant-incorporated protectants, 340–341
Planting and seeding robots, 82
Plant physiology
  climate change impact on, 180–193
  effect of elevated CO_2 on, 303
    nutrient uptake and allocation, 303
    photosynthesis and carbon assimilation, 303
    water use efficiency, 303
Plant tissue culture (PTC), 47
Point of zero charge (PZC), 93
Policy support for sustainable agriculture, 327
Pollen-based selection (PBS), 42
Pollinator conservation, 324
Poppy species (*Papaver* spp.), 105
Population growth and food security, 5
Populus euramericana, 183
Poultry manure, 387
Precision agriculture (PA), 10–12, 20–32, 215, 323, 325–326
  agro-vision technologies, 24
  applications of, 27–30, *28*
  challenges, 30–31
  chatbots/digital advisors, 28

drones and robots, 73
evolution of agricultural practices, 21–23
factors affecting, **30**
fruit-picking robots, 28
future perspectives, 31–32
guidance systems, 25–26
herbicide reduction, 28
history of, *23*
integration of technology in crop management, 24–27
irrigation of crop fields, 29
machine vision (MV) technologies, 24
management platforms, 28
overview, 20–21
pest management and yield monitoring, 29–30
precision weed management, 110–112
principles of, 22, *23*
soil selection and planning, 29
unmanned aerial vehicles (UAVs), 24, 26–27
weed control robots, 28
weed detection, 28
Pretreatment of agricultural waste, 360
Preventive pest management, 339
Programmed cell death (PCD), 190
Project irrigation systems, 201–202
Protein denaturation, 186
Public Distribution System (PDS), 153
Pulses, 166–169
Pyrolysis, 361

Q

Quality protein maize (QPM) materials, 308

R

Radio frequency identification (RFID), 416–417, 418–419
Rainwater harvesting techniques, 322
Raw materials for ethanol production, 356–357
Reactive oxygen species (ROS), 97, 185, 248
Recycled fiber, 232
Regenerated cellulose fiber, 231–232
Regenerative agriculture (RA), 399–409
beneficial aspects core practice, 401–406
addition of biochar, 405–406
cover crops, 401–402
crop livestock integration, 404–405
crops rotation, 402–403
managed grazing, 403–404
organic farming, 405
perennials and agroforestry, 403
reduce/no-tillage practices, 403
carbon sequestration in, 406–407
challenges, 408–409
circular economy (CE), 408
components of, **402**
economic viability and financial tools, 407–408
improved water quality and availability, 407
overview, 399–401
principles of, 400, *400*
Regenerative grazing management (ReGM), 403–404
Remote sensing, 248–249
Remote sensing and automation systems (RASs), 26
Remote sensing and GIS applications, 326
Renewable energy sources in agriculture, 325
Reproductive developmental responses, 183–184
Resistance, 123–135
benefits in mitigating resistance, 130
biopesticides, 131–132, **132**
challenges, 134
crop rotation and diversification, 129–133
crop varieties, 128–129
development, 126
development of resistant crop varieties, 128–129
examples of successful rotation practices, 130–131
gene editing technique, 129
genetic engineering and biotechnology, 128–129
integrated pest management (IPM), 126–127
management, 126–129
organic farming methods, 132–133
overview, 123–124
types of, 124–125, **125**
Respiration, 185
Rhizobacteria as bio-fertilizer, 285–286
Rhizo-filtration, 108
Rice cultivation in flood prone areas, 308–310
Ricinus communis, 183
Robots/robotics, 81–84; *see also* Drones
advantages of, 83
application of, *82*
and autonomous systems, 26
disadvantages of, 84
farming, 81
fertilizing, 81
grading, 82
harvesting, 82
planting and seeding, 82
role in modern agriculture, 82–83
weeding, 82
Root developmental responses, *182*, 182–183
Rubisco activase, 185

S

Salinity stress, 164
SDN-based methodologies, 41
Seaweeds, 288
Seed distribution system, 171–172
Seed enhancement, 265–275
challenges of, 274–275
emerging technologies in, 268–272
biotechnology, 270–271
microbial seed coating, 271–272
nanotechnology, 268–270
overview, 265–266
sustainable practices, 272–275
eco-friendly coatings and treatments, 272–274
utilization of waste materials in seed enhancement, 274
traditional techniques, 266–268
hormonal-priming, 267
hydropriming, 267
matrix priming, 267–268
osmopriming, 267
seed coating, 268
seed priming, 266–267
Sensors for measuring soil moisture, 203
Sesame (*Sesamum indicum* L.), 171
Sewage sludge and biological solids, 288

Short-lived seeds, 105; *see also* Barren brome *(Bromus diandrus)*
Silver nanoparticles, 96
Simultaneous saccharification and co-fermentation (SSCF) enzymes, 367
Single-guide RNA (sgRNA), 42
Site-directed nucleases (SDNs), 41
Site-specific crop management, 91
Site-specific weed management (SSWM), 111
Smart farming, 14
Smart farming with Internet of Things (IoT), 326
Software ecosystem (SE) design, 31
Soil and water conservation problems, **148**
Soil erosion, 58, 144, 378, 380, 389–390, 392, 399, 409
Soil fertility, 57–58, 130–131, 133, 149
  in organic systems, 57
Soil health, 6–7, 55–57, 133
  nanotechnology for, 93
  in organic farming (OF), 55–57, **56**
  sustainability in agriculture, 6–7
  sustainable fiber production, 234
Soil moisture conservation, *145*, 149
Soil organic matter (SOM), 57
Soil tilth, 57
Solar radiation management, 249
Spraying and irrigation, 13
Sprinkler irrigation, 172, 198, 201, *202*, 205, 322
Sprinkler system (SS), 199, 201, 202
Steam explosion/auto hydrolysis, 361–362
Stress-tolerance gene modification, 250
Strigolactone biosynthesis (SB), 42
Strip cropping, 151
Sub-cellular metabolism, 96
Sub-surface irrigation system, 201
Supply chain traceability, 14–15
Surface irrigation system, 201
Sustainability in agriculture, 6–9
  economic dimensions of, 7
  environmental dimensions of, 6–7
  practices promoting, 8–9
  social dimensions of, 8
  soil health, 6–7
Sustainable crop protection (SCP), 123
Sustainable farming systems, 319–331
  agroforestry, 320–321
  biodiversity and ecological resource management, 323–324
    crop diversification, 323–324
    habitat management for beneficial organisms, 324
    integrated pest management (IPM), 324
    pollinator conservation, 324
  case studies, 328
    lessons learned and best practices, 328
    regional examples, 328
  climate change, 329
  conservation agriculture, 320
  emerging trends in, 329
  energy resource management, 324–325
    bioenergy crops, 325
    energy efficiency in farming operations, 324–325
    renewable energy sources in agriculture, 325
  integrated farming systems (IFS), 319–320
  nutrient resource management, 322–323
    biological nitrogen fixation, 323
    composting and organic fertilization, 323
    integrated nutrient management (INM), 322–323
    precision agriculture techniques, 323
  organic farming, 320
  overview, 319
  permaculture, 321
  scaling up sustainable practices, 329–330
  socio-economic aspects, 327
    economic viability of sustainable farming practices, 327
    farmer education and extension programs, 327
    market incentives for sustainable production, 327
    policy support for sustainable agriculture, 327
  technological innovations, 325–327
    decision support systems (DSSs), 326–327
    precision agriculture technologies, 325–326
    remote sensing and GIS applications, 326
    smart farming with Internet of Things (IoT), 326
  water resource management, 321–322
    drainage management, 322
    irrigation systems for sustainable water use, 322
    rainwater harvesting techniques, 322
    water use efficiency in agriculture, 321
Sustainable fiber production, 228–239
  advantages, 238
  biodiversity conservation, 234
  case studies, 237–238
  challenges in, 238–239
  chemical composition of, 232, **232**
  innovations in, 234–237
    advancements in equipment design and bio-based and non-toxic inks, 235–236
    benefits of using fiber in energy-efficient, 236–237
    energy-efficient manufacturing, 236
    low-impact dye formulation enhancements, 234–235
  organic farming, 234
  organic farming techniques, 233–234
  overview, 228–229
  role of crop rotation, 233
  soil health improvement, 234
  sustainable practices in, 233–234
  types, 229–232
    hemp fibers, 230
    kapok fiber, 230–231
    natural fibers, 229, **230**
    organic cotton, 231
    recycled fiber, 232
    regenerated cellulose fiber, 231–232
Systematic literature review (SLR), 115

T

Thermophilic bacteria, 358
Third Industrial Revolution 3.0, 22
Third-order organisms, 324
Thyavihalli Girijappa, Y.G., 237

Transcription activator-like effector nucleases (TALENs), 41
Transgenic plants, 45

U

United Nations Framework Convention on Environment and Development, 107
United States Army Signal Corps, 74
United States Department of Agriculture, 74
Unmanned aerial vehicles (UAVs), 24, 26–27, 74, 111–112
Unmanned aircraft systems (UASes), 74
Unmanned ground vehicles (UGVs), 26–27
US Environmental Protection Agency (USEPA), 229

V

Valeriana jatamansi Jones, 185
Vapor pressure deficit (VPD), 79
Variable rate application (VRA), 21–22, 24–25
Variable rate technologies (VRTs), 10, 22
Vermicompost/vermicomposting, 285
  organic soil enrichment, 384
Vertical farming, 11, 214–215
Virtual reality (VR) systems, 28
Visibility and near infrared (Vis-NIR) spectroscopy, 111

W

Water Deficit Index (WDI), 79
Water-holding capacity (WHC), 205
Waterlogging stress, 164
Water management, 7
Water resource management, 321–322
  drainage management, 322
  irrigation systems for sustainable water use, 322
  rainwater harvesting techniques, 322
  water use efficiency in agriculture, 321
Water-saving irrigation technology (WSIT), 148
Water scarcity and inefficient irrigation practices, 3
Water stress monitoring, 79
Water-use efficiency (WUE), 142, 145–148, 321
Weather monitoring and forecasting, 203
Weed control robots, 28
Weed detection/monitoring, 28, 113–114
Weed harrowing, 106
Weeding robots, 82
Weed management, 105–118, 172
  arms race, 106
  biological control, 108–109
  chemical, 109–116
  chemical innovations in, 109
  evolution of, 106–107
  future perspectives in, 116–118
  integrated, *107*, 107–108
  overview, 105–106
  strategies for controlling resistant weeds, 106
  tools/techniques for, **114**
Weeds, 165
Weed The field of science The community of America, 107
Wet oxidation, 363
Wheat cultivation in drought-prone regions, 306–307
Wireless Multimedia Sensor Networks, 422
Wireless sensor and actuator networks (WSANs), 424
Wireless sensor network (WSN), 29, 414, 416–417, 418, 421–422
Wireless sensor network (WSN)-enabled smart irrigation systems, 29
World Watch Institute, 63

Z

Zero tillage, 289–290
Zinc finger nucleases (ZFNs), 41